Deformation of Earth Materials

Much of the recent progress in the solid Earth sciences is based on the interpretation of a range of geophysical and geological observations in terms of the properties and deformation of Earth materials. One of the greatest challenges facing geoscientists in achieving this lies in finding a link between physical processes operating in minerals at the smallest length scales to geodynamic phenomena and geophysical observations across thousands of kilometers.

This graduate textbook presents a comprehensive and unified treatment of the materials science of deformation as applied to solid Earth geophysics and geology. Materials science and geophysics are integrated to help explain important recent developments, including the discovery of detailed structure in the Earth's interior by high-resolution seismic imaging, and the discovery of the unexpectedly large effects of high pressure on material properties, such as the high solubility of water in some minerals. Starting from fundamentals such as continuum mechanics and thermodynamics, the materials science of deformation of Earth materials is presented in a systematic way that covers elastic, anelastic, and viscous deformation. Although emphasis is placed on the fundamental underlying theory, advanced discussions on current debates are also included to bring readers to the cutting edge of science in this interdisciplinary area.

Deformation of Earth Materials is a textbook for graduate courses on the rheology and dynamics of the solid Earth, and will also provide a much-needed reference for geoscientists in many fields, including geology, geophysics, geochemistry, materials science, mineralogy, and ceramics. It includes review questions with solutions, which allow readers to monitor their understanding of the material presented.

Shun-ichiro Karato is a Professor in the Department of Geology and Geophysics at Yale University. His research interests include experimental and theoretical studies of the physics and chemistry of minerals, and their applications to geophysical and geological problems. Professor Karato is a Fellow of the American Geophysical Union and a recipient of the Alexander von Humboldt Prize (1995), the Japan Academy Award (1999), and the Vening Meinesz medal from the Vening Meinesz School of Geodynamics in The Netherlands (2006). He is the author of more than 160 journal articles and has written/edited seven other books.

Deformation of Earth Materials

An Introduction to the Rheology of Solid Earth

Shun-ichiro Karato
Yale University, Department of Geology & Geophysics, New Haven, CT, USA

CAMBRIDGE
UNIVERSITY PRESS

CAMBRIDGE UNIVERSITY PRESS
Cambridge, New York, Melbourne, Madrid, Cape Town,
Singapore, São Paulo, Delhi, Mexico City

Cambridge University Press
The Edinburgh Building, Cambridge CB2 8RU, UK

Published in the United States of America by Cambridge University Press, New York

www.cambridge.org
Information on this title: www.cambridge.org/9781107406056

First published 2008
First paperback edition 2011

A catalogue record for this publication is available from the British Library

ISBN 978-0-521-84404-8 Hardback
ISBN 978-1-107-40605-6 Paperback

Contents

The colour plates are between pages 118 and 119, and is also available for download in colour from www. cambridge. org/9781107406056

Preface

Understanding the microscopic physics of deformation is critical in many branches of solid Earth science. Long-term geological processes such as plate tectonics and mantle convection involve plastic deformation of Earth materials, and hence understanding the plastic properties of Earth materials is key to the study of these geological processes. Interpretation of seismological observations such as tomographic images or seismic anisotropy requires knowledge of elastic, anelastic properties of Earth materials and the processes of plastic deformation that cause anisotropic structures. Therefore there is an obvious need for understanding a range of deformation-related properties of Earth materials in solid Earth science. However, learning about deformation-related properties is challenging because deformation in various geological processes involves a variety of microscopic processes. Owing to the presence of multiple deformation mechanisms, the results obtained under some conditions may not necessarily be applicable to a geological problem that involves deformation under different conditions. Therefore in order to conduct experimental or theoretical research on deformation, one needs to have a broad knowledge of various mechanisms to define conditions under which a study is to be conducted. Similarly, when one attempts to use results of experimental or theoretical studies to understand a geological problem, one needs to evaluate the validity of applying particular results to a given geological problem. However, there was no single book available in which a broad range of the physics of deformation of materials was treated in a systematic manner that would be useful for a student (or a scientist) in solid Earth science. The motivation of writing this book was to fulfill this need.

In this book, I have attempted to provide a unified, interdisciplinary treatment of the science of deformation of Earth with an emphasis on the materials science (microscopic) approach. Fundamentals of the materials science of deformation of minerals and rocks over various time-scales are described in addition to the applications of these results to important geological and geophysical problems. Properties of materials discussed include elastic, anelastic (viscoelastic), and plastic properties. The emphasis is on an *interdisciplinary approach*, and, consequently, I have included discussions on some advanced, controversial issues where they are highly relevant to Earth science problems. They include the role of hydrogen, effects of pressure, deformation of two-phase materials, localization of deformation and the link between viscoelastic deformation and plastic flow. This book is intended to serve as a textbook for a course at a graduate level in an Earth science program, but it may also be useful for students in materials science as well as researchers in both areas. No previous knowledge of geology/geophysics or of materials science is assumed. The basics of continuum mechanics and thermodynamics are presented as far as they are relevant to the main topics of this book.

Significant progress has occurred in the study of deformation of Earth materials during the last ~30 years, mainly through experimental studies. Experimental studies on synthetic samples under well-defined chemical conditions and the theoretical interpretation of these results have played an important role in understanding the microscopic mechanisms of deformation. Important progress has also been made to expand the pressure range over which plastic deformation can be investigated, and the first low-strain anelasticity measurements have been conducted. In addition, some large-strain deformation experiments have been performed that have provided important new insights into the microstructural evolution during deformation. However, experimental data are always obtained under limited conditions and their applications to the Earth involve large extrapolation. It is critical to understand the *scaling laws* based on the physics and chemistry of deformation of materials in order to properly apply experimental data to Earth. A number of examples of such scaling laws are discussed in this book.

This book consists of three parts: Part I (Chapters 1–3) provides a general background including basic continuum mechanics, thermodynamics and phenomenological theory of deformation. Most of this part, particularly Chapters 1 and 2 contain material that can be found in many other textbooks. Therefore those who are familiar with basic continuum mechanics and thermodynamics can skip this part. Part II (Chapters 4–16) presents a detailed account of materials science of time-dependent deformation, including elastic, anelastic and plastic deformation with an emphasis on anelastic and plastic deformation. They include, not only the basics of properties of materials characterizing deformation (i.e., elasticity and viscosity (creep strength)), but also the physical principles controlling the microstructural developments (grain size and lattice-preferred orientation). Part III (Chapters 17–21) provides some applications of the materials science of deformation to important geological and geophysical problems, including the rheological structure of solid Earth and the interpretation of the pattern of material circulation in the mantle and core from geophysical observations. Specific topics covered include the lithosphere–asthenosphere structure, rheological stratification of Earth's deep mantle and a geodynamic interpretation of anomalies in seismic wave propagation. Some of the representative experimental data are summarized in tables. However, the emphasis of this book is on presenting basic theoretical concepts and consequently references to the data are not exhaustive. Many problems (with solutions) are provided to make sure a reader understands the content of this book. Some of them are advanced and these are shown by an asterisk.

The content of this book is largely based on lectures that I have given at the University of Minnesota and Yale University as well as at other institutions. I thank students and my colleagues at these institutions who have given me opportunities to improve my understanding of the subjects discussed in this book through inspiring questions. Some parts of this book have been read/reviewed by A. S. Argon, D. Bercovici, H. W. Green, S. Hier-Majumder, G. Hirth, I. Jackson, D. L. Kohlstedt, J. Korenaga, R. C. Liebermann, J.-P. Montagner, M. Nakada, C. J. Spiers, J. A. Tullis and J. A. Van Orman. However, they do not always agree with the ideas presented in this book and any mistakes are obviously my own. W. Landuyt, Z. Jiang and P. Skemer helped to prepare the figures. I should also thank the editors at Cambridge University Press for their patience. Last but not least, I thank my family, particularly my wife, Yoko, for her understanding, forbearance and support during the long gestation of this monograph. Thank you all.

Part I
General background

1 Stress and strain

The concept of stress and strain is key to the understanding of deformation. When a force is applied to a continuum medium, stress is developed inside it. Stress is the force per unit area acting on a given plane along a certain direction. For a given applied force, the stress developed in a material depends on the orientation of the plane considered. Stress can be decomposed into hydrostatic stress (pressure) and deviatoric stress. Plastic deformation (in non-porous materials) occurs due to deviatoric stress. Deformation is characterized by the deformation gradient tensor, which can be decomposed into rigid body rotation and strain. Deformation such as simple shear involves both strain and rigid body rotation and hence is referred to as rotational deformation whereas pure shear or tri-axial compression involves only strain and has no rigid body rotation and hence is referred to as irrotational deformation. In rotational deformation, the principal axes of strain rotate with respect to those of stress whereas they remain parallel in irrotational deformation. Strain can be decomposed into dilatational (volumetric) strain and shear strain. Plastic deformation (in a non-porous material) causes shear strain and not dilatational strain. Both stress and strain are second-rank tensors, and can be characterized by the orientation of the principal axes and the magnitude of the principal stress and strain and both have three invariants that do not depend on the coordinate system chosen.

Key words stress, strain, deformation gradient, vorticity, principal strain, principal stress, invariants of stress, invariants of strain, normal stress, shear stress, Mohr's circle, the Flinn diagram, foliation, lineation, coaxial deformation, non-coaxial deformation.

1.1. Stress

1.1.1. Definition of stress

This chapter provides a brief summary of the basic concept of stress and strain that is relevant to understanding plastic deformation. For a more comprehensive treatment of stress and strain, the reader may consult MALVERN (1969), MASE (1970), MEANS (1976).

In any deformed or deforming continuum material there must be a force inside it. Consider a small block of a deformed material. Forces acting on the material can be classified into two categories, i.e., a short-range force due to atomic interactions and the long-range force due to an external field such as the gravity field. Therefore the forces that act on this small block include (1) short-range forces due to the displacement of atoms within this block, (2) long-range forces such as gravity that act equally on each atom and (3) the forces that act on this block through the surface from the neighboring materials. The (small) displacements of each atom inside this region cause forces to act on surrounding atoms, but by assumption these forces are short range. Therefore one can consider them as forces between a pair of atoms A and B. However, because of Newton's law of action and counter-action, the forces acting between two atoms are anti-symmetric: $f_{AB} = -f_{BA}$ where $f_{AB\ (BA)}$

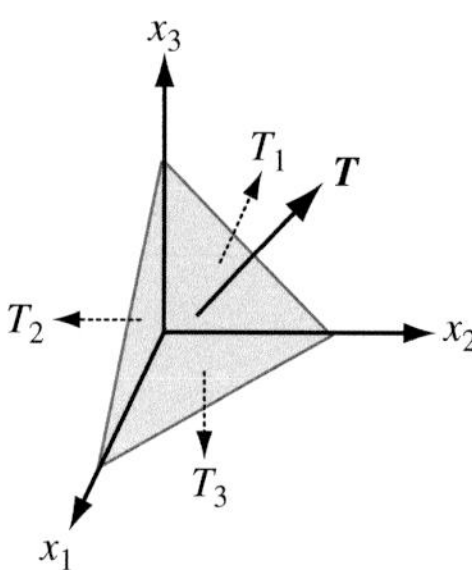

FIGURE 1.1 Forces acting on a small pyramid.

are the force exerted by atom A (B) to B (A). Consequently these forces caused by atomic displacement within a body must cancel. The long-range force is called a *body force*, but if one takes this region as small, then the magnitude of this body force will become negligible compared to the *surface force* (i.e., the third class of force above). Therefore the net force acting on the small region must be the forces across the surface of that region from the neighboring materials. To characterize this force, let us consider a small piece of block that contains a plane with the area of dS and whose normal is $\boldsymbol{n}$ ($\boldsymbol{n}$ is the unit vector). Let $\boldsymbol{T}$ be the force (per unit area) acting on the surface dS from outside this block (positive when the force is compressive) and consider the force balance (Fig. 1.1). The force balance should be attained among the force $\boldsymbol{T}$ as well as the forces $\boldsymbol{T}^{1,2,3}$ that act on the surface $dS_{1,2,3}$ respectively ($dS_{1,2,3}$ are the projected area of dS on the plane normal to the $x_{1,2,3}$ axis). Then the force balance relation for the block yields,

$$\boldsymbol{T}\,dS = \sum_{j=1}^{3} \boldsymbol{T}^j\,dS_j. \tag{1.1}$$

Now using the relation $dS_j = n_j\,dS$, one obtains,

$$T_i = \sum_{j=1}^{3} T_i^j n_j = \sum_{j=1}^{3} \sigma_{ij} n_j \tag{1.2}$$

where T_i is the ith component of the force $\boldsymbol{T}$ and σ_{ij} is the ith component of the traction $\boldsymbol{T}^j$, namely the ith component of force acting on a plane whose normal is the jth direction ($n_{ij} = T_i^j$). This is the definition of *stress*. From the balance of torque, one can also show,

$$\sigma_{ij} = \sigma_{ji}. \tag{1.3}$$

The values of stress thus defined depend on the coordinate system chosen. Let us denote quantities in a new coordinate system by a tilda, then the new coordinate and the old coordinate system are related to each other by,

$$\tilde{x}_i = \sum_{j=1}^{3} a_{ij} x_j \tag{1.4}$$

where a_{ij} is the transformation matrix that satisfies the orthonormality relation,

$$\sum_{j=1}^{3} a_{ij} a_{jm} = \delta_{im} \tag{1.5}$$

where δ_{im} is the Kronecker delta ($\delta_{im} = 1$ for $i = m$, $\delta_{im} = 0$ otherwise). Now in this new coordinate system, we may write a relation similar to equation (1.2) as,

$$\tilde{T}_i = \sum_{j=1}^{3} \tilde{\sigma}_{ij} \tilde{n}_j. \tag{1.6}$$

Noting that the traction ($\boldsymbol{T}$) transforms as a vector in the same way as the coordinate system, equation (1.4), we have,

$$\tilde{T}_i = \sum_{j=1}^{3} a_{ij} T_j. \tag{1.7}$$

Inserting equation (1.2), the relation (1.7) becomes,

$$\tilde{T}_i = \sum_{j,k=1}^{3} \sigma_{jk} a_{ij} n_k. \tag{1.8}$$

Now using the orthonormality relation (1.5), one has,

$$n_i = \sum_{j=1}^{3} a_{ji} \tilde{n}_i. \tag{1.9}$$

Inserting this relation into equation (1.8) and comparing the result with equation (1.6), one obtains,[1]

$$\tilde{\sigma}_{ij} = \sum_{k,l=1}^{3} \sigma_{kl} a_{ik} a_{jl}. \tag{1.10}$$

The quantity that follows this transformation law is referred to as *a second rank tensor*.

1.1.2. Principal stress, stress invariants

In any material, there must be a certain orientation of a plane on which the direction of traction ($\boldsymbol{T}$) is normal to it. For that direction of $\boldsymbol{n}$, one can write,

$$T_i = \sigma n_i \tag{1.11}$$

[1] In the matrix notation, $\tilde{\sigma} = A \cdot \sigma \cdot A^T$ where $A = (a_{ij})$ and $A^T = (a_{ji})$.

where σ is a scalar quantity to be determined. From equations (1.11) and (1.2),

$$\sum_{j=1}^{3}(\sigma_{ij} - \sigma\delta_{ij})n_j = 0. \tag{1.12}$$

For this equation to have a non-trivial solution other than $\boldsymbol{n} = 0$, one must have,

$$\left|\sigma_{ij} - \sigma\delta_{ij}\right| = 0 \tag{1.13}$$

where $\left|X_{ij}\right|$ is the determinant of a matrix X_{ij}. Writing equation (1.13) explicitly, one obtains,

$$\begin{vmatrix} \sigma_{11}-\sigma & \sigma_{12} & \sigma_{13} \\ \sigma_{21} & \sigma_{22}-\sigma & \sigma_{23} \\ \sigma_{31} & \sigma_{32} & \sigma_{33}-\sigma \end{vmatrix} = -\sigma^3 + I_\sigma\sigma^2 + II_\sigma\sigma + III_\sigma = 0 \tag{1.14}$$

with

$$I_\sigma = \sigma_{11} + \sigma_{22} + \sigma_{33} \tag{1.15a}$$

$$II_\sigma = -\sigma_{11}\sigma_{22} - \sigma_{11}\sigma_{33} - \sigma_{33}\sigma_{22} + \sigma_{12}^2 + \sigma_{13}^2 + \sigma_{23}^2 \tag{1.15b}$$

$$\begin{aligned} III_\sigma = {} & \sigma_{11}\sigma_{22}\sigma_{33} + 2\sigma_{12}\sigma_{23}\sigma_{31} - \sigma_{11}\sigma_{23}^2 \\ & - \sigma_{22}\sigma_{13}^2 - \sigma_{33}\sigma_{12}^2. \end{aligned} \tag{1.15c}$$

Therefore, there are three solutions to equation (1.14), $\sigma_1, \sigma_2, \sigma_3 (\sigma_1 > \sigma_2 > \sigma_3)$.These are referred to as the *principal stresses*. The corresponding $\boldsymbol{n}$ is *the orientation of principal stress*. If the stress tensor is written using the coordinate whose orientation coincides with the orientation of principal stress, then,

$$[\sigma_{ij}] = \begin{bmatrix} \sigma_1 & 0 & 0 \\ 0 & \sigma_2 & 0 \\ 0 & 0 & \sigma_3 \end{bmatrix}. \tag{1.16}$$

It is also seen that because equation (1.14) is a scalar equation, the values of I_σ, II_σ and III_σ are independent of the coordinate. These quantities are called the *invariants of stress tensor*. These quantities play important roles in the formal theory of plasticity (see Section 3.3). Equations (1.15a–c) can also be written in terms of the principal stress as,

$$I_\sigma = \sigma_1 + \sigma_2 + \sigma_3 \tag{1.17a}$$

$$II_\sigma = -\sigma_1\sigma_2 - \sigma_2\sigma_3 - \sigma_3\sigma_1 \tag{1.17b}$$

and

$$III_\sigma = \sigma_1\sigma_2\sigma_3. \tag{1.17c}$$

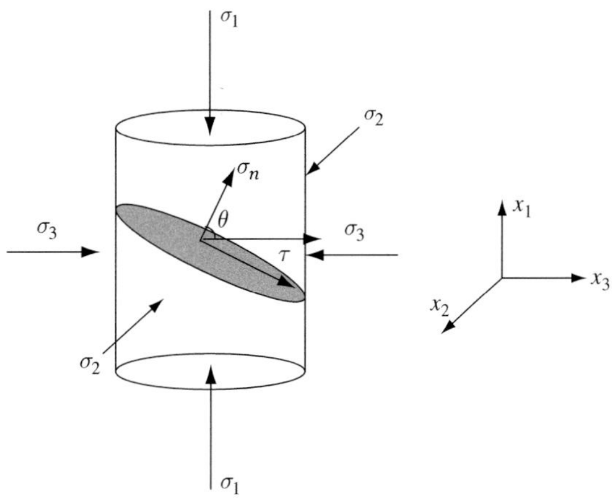

FIGURE 1.2 Geometry of normal and shear stress on a plane.

1.1.3. Normal stress, shear stress, Mohr's circle

Now let us consider the normal and shear stress on a given plane subjected to an external force (Fig. 1.2). Let x_1 be the axis parallel to the maximum compressional stress σ_1 and x_2 and x_3 be the axes perpendicular to x_1. Consider a plane whose normal is at the angle θ from x_3 (positive counterclockwise). Now, we define a new coordinate system whose x_1' axis is normal to the plane, but the x_2' axis is the same as the x_2 axis. Then the transformation matrix is,

$$[a_{ij}] = \begin{bmatrix} \cos\theta & 0 & -\sin\theta \\ 0 & 1 & 0 \\ \sin\theta & 0 & \cos\theta \end{bmatrix} \tag{1.18}$$

and hence,

$$[\tilde{\sigma}_{ij}] = \begin{bmatrix} \frac{\sigma_1+\sigma_3}{2} + \frac{\sigma_1-\sigma_3}{2}\cos 2\theta & 0 & \frac{\sigma_1-\sigma_3}{2}\sin 2\theta \\ 0 & \sigma_2 & 0 \\ \frac{\sigma_1-\sigma_3}{2}\sin 2\theta & 0 & \frac{\sigma_1+\sigma_3}{2} - \frac{\sigma_1-\sigma_3}{2}\cos 2\theta \end{bmatrix}. \tag{1.19}$$

Problem 1.1

Derive equation (1.19).

Solution

The stress tensor (1.16) can be rotated through the operation of the transformation matrix (1.18) using equation (1.10),

$$[\tilde{\sigma}_{ij}] = \begin{bmatrix} \cos\theta & 0 & -\sin\theta \\ 0 & 1 & 0 \\ \sin\theta & 0 & \cos\theta \end{bmatrix} \begin{bmatrix} \sigma_1 & 0 & 0 \\ 0 & \sigma_2 & 0 \\ 0 & 0 & \sigma_3 \end{bmatrix} \begin{bmatrix} \cos\theta & 0 & \sin\theta \\ 0 & 1 & 0 \\ -\sin\theta & 0 & \cos\theta \end{bmatrix}$$

$$= \begin{bmatrix} \frac{\sigma_1+\sigma_3}{2} + \frac{\sigma_1-\sigma_3}{2}\cos 2\theta & 0 & \frac{\sigma_1-\sigma_3}{2}\sin 2\theta \\ 0 & \sigma_2 & 0 \\ \frac{\sigma_1-\sigma_3}{2}\sin 2\theta & 0 & \frac{\sigma_1+\sigma_3}{2} - \frac{\sigma_1-\sigma_3}{2}\cos 2\theta \end{bmatrix}.$$

Therefore the shear stress τ and normal stress σ_n on this plane are

$$\tilde{\sigma}_{13} \equiv \tau = \frac{\sigma_1 - \sigma_3}{2}\sin 2\theta \tag{1.20}$$

and

$$\tilde{\sigma}_{33} \equiv \sigma_n = \frac{\sigma_1 + \sigma_3}{2} - \frac{\sigma_1 - \sigma_3}{2}\cos 2\theta \tag{1.21}$$

respectively. It follows that the maximum shear stress is on the two conjugate planes that are inclined by $\pm\pi/4$ with respect to the x_1 axis and its absolute magnitude is $(\sigma_1 - \sigma_3)/2$. Similarly, the maximum compressional stress is on a plane that is normal to the x_1 axis and its value is σ_1. It is customary to use $\sigma_1 - \sigma_3$ as (differential (or deviatoric)) stress in rock deformation literature, but the shear stress, $\tau \equiv (\sigma_1 - \sigma_3)/2$, is also often used. Eliminating θ from equations (1.20) and (1.21), one has,

$$\tau^2 + \left(\sigma_n - \frac{\sigma_1 + \sigma_3}{2}\right)^2 = \frac{1}{4}(\sigma_1 - \sigma_3)^2. \tag{1.22}$$

Thus, the normal and shear stress on planes with various orientations can be visualized on a two-dimensional plane (τ–σ_n space) as a circle whose center is located at $(0, (\sigma_1 + \sigma_3)/2)$ and the radius $(\sigma_1 - \sigma_3)/2$ (Fig. 1.3). This is called a *Mohr's circle* and plays an important role in studying the brittle fracture that is controlled by the stress state (shear–normal stress ratio; see Section 7.3).

When $\sigma_1 = \sigma_2 = \sigma_3 (= P)$, then the stress is isotropic (hydrostatic). The hydrostatic component of stress does not cause plastic flow (this is not true for porous materials, but we do not discuss porous materials here), so it is useful to define *deviatoric stress*

$$\sigma'_{ij} \equiv \sigma_{ij} - \delta_{ij}P. \tag{1.23}$$

When we discuss plastic deformation in this book, we use σ_{ij} (without prime) to mean deviatoric stress for simplicity.

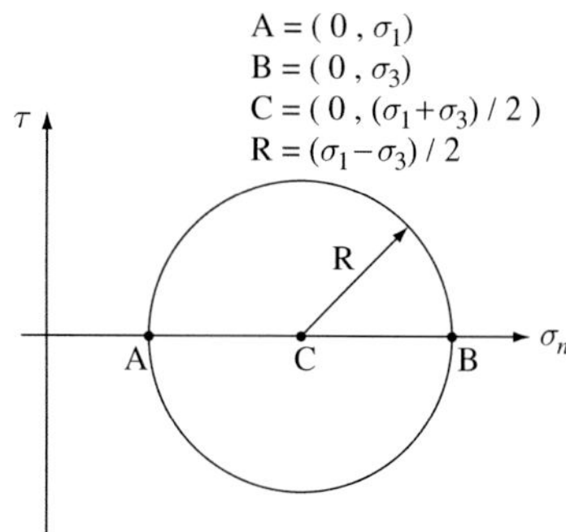

FIGURE 1.3 A Mohr circle corresponding to two-dimensional stress showing the variation of normal, σ_n, and shear stress, τ, on a plane.

Problem 1.2

Show that the second invariant of deviatoric stress can be written as $II_{\sigma'} = \frac{1}{6}\left[(\sigma_1 - \sigma_2)^2 + (\sigma_2 - \sigma_3)^2 + (\sigma_3 - \sigma_1)^2\right]$.

Solution

If one uses a coordinate system parallel to the principal axes of stress, from equation (1.15), one has $II_{\sigma'} = -\sigma'_1\sigma'_2 - \sigma'_1\sigma'_3 - \sigma'_3\sigma'_2$. Using $I_{\sigma'} = \sigma'_1 + \sigma'_2 + \sigma'_3 = 0$, one finds $I^2_{\sigma} = \sigma'^2_1 + \sigma'^2_2 + \sigma'^2_3 + 2(\sigma'_1\sigma'_2 + \sigma'_2\sigma'_3 + \sigma'_3\sigma'_1) = 0$. Therefore $II_{\sigma'} = \frac{1}{2}(\sigma'^2_1 + \sigma'^2_2 + \sigma'^2_3)$. Now, inserting $\sigma'_1 = \sigma_1 - \frac{1}{3}(\sigma_1 + \sigma_2 + \sigma_3)$ etc., one obtains $II_{\sigma'} = \frac{1}{6}\left[(\sigma_1 - \sigma_2)^2 + (\sigma_2 - \sigma_3)^2 + (\sigma_3 - \sigma_1)^2\right]$.

Problem 1.3

Show that when the stress has axial symmetry with respect to the x_1 axis (i.e., $\sigma_2 = \sigma_3$), then $\sigma_n = P + (\sigma_1 - \sigma_3)(\cos^2\theta - \frac{1}{3})$.

Solution

From (1.21), one obtains, $\sigma_n = (\sigma_1 + \sigma_3)/2 + ((\sigma_1 - \sigma_3)/2)\cos 2\theta$. Now $\cos 2\theta = 2\cos^2\theta - 1$ and $P = \frac{1}{3}(\sigma_1 + \sigma_2 + \sigma_3) = \frac{1}{3}(\sigma_1 + 2\sigma_3) = \sigma_1 - \frac{2}{3}(\sigma_1 - \sigma_3)$. Therefore $\sigma_n = P + (\sigma_1 - \sigma_3)(\cos^2\theta - \frac{1}{3})$.

Equations similar to (1.15)–(1.17) apply to the deviatoric stress.

1.2. Deformation, strain

1.2.1. Definition of strain

Deformation refers to a change in the shape of a material. Since homogeneous displacement of material points does not cause deformation, deformation must be related to *spatial variation* or *gradient* of displacement. Therefore, deformation is characterized by a displacement gradient tensor,

$$d_{ij} \equiv \frac{\partial u_i}{\partial x_j}. \tag{1.24}$$

where u_i is the displacement and x_j is the spatial coordinate (after deformation). However, this displacement gradient includes the rigid-body rotation that has nothing to do with deformation. In order to focus on deformation, let us consider two adjacent material points $P_0(\boldsymbol{X})$ and $Q_0(\boldsymbol{X}+d\boldsymbol{X})$, which will be moved to $P(\boldsymbol{x})$ and $Q(\boldsymbol{x}+d\boldsymbol{x})$ after deformation (Fig. 1.4). A small vector connecting P_0 and Q_0, $d\boldsymbol{X}$, changes to $d\boldsymbol{x}$ after deformation. Let us consider how the length of these two segments changes. The difference in the squares of the length of these small elements is given by,

$$\begin{aligned}(dx)^2-(dX)^2 &= \sum_{i=1}^{3}(dx_i)^2-\sum_{i=1}^{3}(dX_i)^2\\ &= \sum_{i,j,k=1}^{3}\left(\delta_{ij}-\frac{\partial X_k}{\partial x_i}\frac{\partial X_k}{\partial x_j}\right)dx_i\,dx_j.\end{aligned} \tag{1.25}$$

Therefore deformation is characterized by a quantity,

$$\varepsilon_{ij} \equiv \frac{1}{2}\left(\delta_{ij}-\sum_{k=1}^{3}\frac{\partial X_k}{\partial x_i}\frac{\partial X_k}{\partial x_j}\right) \tag{1.26}$$

which is the definition of *strain*, ε_{ij}. With this definition, the equation (1.25) can be written as,

$$(dx)^2-(dX)^2 \equiv 2\sum_{i,j}\varepsilon_{ij}\,dx_i\,dx_j. \tag{1.27}$$

From the definition of strain, it immediately follows that the strain is a symmetric tensor, namely,

$$\varepsilon_{ij}=\varepsilon_{ji}. \tag{1.28}$$

Now, from Fig. 1.4, one obtains,

$$du_i = dx_i - dX_i \tag{1.29}$$

hence

$$\frac{\partial u_i}{\partial x_j}=\delta_{ij}-\frac{\partial X_i}{\partial x_j}. \tag{1.30}$$

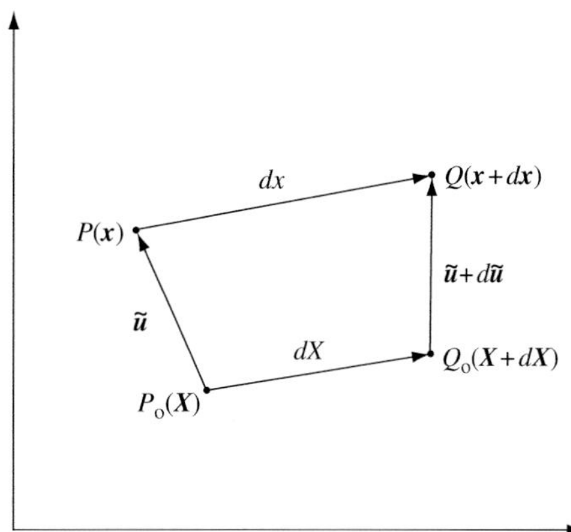

FIGURE 1.4 Deformation causes the change in relative positions of material points.

Inserting equation (1.30) into (1.26) one finds,

$$\varepsilon_{ij}=\frac{1}{2}\left(\frac{\partial u_i}{\partial x_j}+\frac{\partial u_j}{\partial x_i}-\sum_{k=1}^{3}\frac{\partial u_k}{\partial x_i}\frac{\partial u_k}{\partial x_j}\right). \tag{1.31}$$

This definition of strain uses the deformed state as a reference frame and is called the *Eulerian strain*. One can also define strain using the initial, undeformed reference state. This is referred to as the *Lagrangian strain*. For small strain, there is no difference between the Eulerian and Lagrangian strain and both are reduced to[2]

$$\varepsilon_{ij}=\frac{1}{2}\left(\frac{\partial u_i}{\partial x_j}+\frac{\partial u_j}{\partial x_i}\right). \tag{1.32}$$

1.2.2. Meaning of strain tensor

The interpretation of strain is easier in this linearized form. The displacement gradient can be decomposed into two components,

$$\frac{\partial u_i}{\partial x_j}=\frac{1}{2}\left(\frac{\partial u_i}{\partial x_j}+\frac{\partial u_j}{\partial x_i}\right)+\frac{1}{2}\left(\frac{\partial u_i}{\partial x_j}-\frac{\partial u_j}{\partial x_i}\right). \tag{1.33}$$

The first component is a symmetric part,

$$\varepsilon_{ij}=\frac{1}{2}\left(\frac{\partial u_i}{\partial x_j}+\frac{\partial u_j}{\partial x_i}\right)=\varepsilon_{ji} \tag{1.34}$$

which represents the strain (as will be shown later in this chapter).

[2] Note that in some literature, another definition of shear strain is used in which $\varepsilon_{ij}=\partial u_i/\partial x_j+\partial u_j/\partial x_i$ for $i\neq j$ and $\varepsilon_{ii}=\partial u_i/\partial x_i$; e.g., Hobbs *et al.* (1976). In such a case, the symbol γ_{ij} is often used for the non-diagonal ($i\neq j$) strain component instead of ε_{ij}.

Let us first consider the physical meaning of the second part, $\frac{1}{2}\left(\frac{\partial u_i}{\partial x_j} - \frac{\partial u_j}{\partial x_i}\right)$. The second part is an anti-symmetric tensor, namely,

$$\omega_{ij} = \frac{1}{2}\left(\frac{\partial u_i}{\partial x_j} - \frac{\partial u_j}{\partial x_i}\right) = -\omega_{ji}\,(\omega_{ii} = 0). \tag{1.35}$$

The displacement of a small vector du_j due to the operation of this matrix is given by,

$$d\tilde{u}_i^{\omega} = \sum_{j=1}^{3} \omega_{ij}\, du_j. \tag{1.36}$$

Since $\omega_{ii} = 0$, the displacement occurs only to the directions that are normal to the initial orientation. Therefore the operation of this matrix causes the rotation of material points with the axis that is normal to both ith and jth directions with the magnitude (positive clockwise),

$$\tan\theta_{ij} = -\frac{d\tilde{u}_i^{\omega}}{du_i} = -\omega_{ji} = \omega_{ij}. \tag{1.37}$$

(Again this rotation tensor is defined using the deformed state. So it is referred to as the *Eulerian rotation tensor*.) To represent this, a rotation vector is often used that is defined as,

$$\boldsymbol{\omega}(= (\omega_1, \omega_2, \omega_3)) \equiv (\omega_{23}, \omega_{31}, \omega_{12}). \tag{1.38}$$

Thus ω_i represents a rotation with respect to the ith axis. The anti-symmetric tensor, ω_{ij}, is often referred to as a *vorticity* tensor.

Now we turn to the symmetric part of displacement gradient tensor, ε_{ij}. The displacement due to the operation of ε_{ij} is,

$$d\tilde{u}_i^{\varepsilon} = \sum_{j=1}^{3} \varepsilon_{ij}\, du_j. \tag{1.39}$$

From equation (1.39), it follows that the length of a component of vector u_i^0 changes to,

$$\tilde{u}_i = (1 + \varepsilon_{ii}) u_i^0. \tag{1.40}$$

Therefore the diagonal component of strain tensor represents the change in length, so that this component of strain, ε_{ii}, is called *normal strain*. Consequently,

$$\frac{V}{V_0} = (1 + \varepsilon_{11})(1 + \varepsilon_{22})(1 + \varepsilon_{33}) \approx 1 + \varepsilon_{11} + \varepsilon_{22} + \varepsilon_{33} \tag{1.41}$$

where V_0 is initial volume and V is the final volume and the strain is assumed to be small (this assumption can be relaxed and the same argument can be applied to a finite strain, see e.g., MASE (1970)). Thus,

$$\sum_{k=1}^{3} \varepsilon_{kk} = \frac{\Delta V}{V}. \tag{1.42}$$

Obviously, normal strain can be present in deformation without a volume change. For example, $\varepsilon_{ij} = \begin{pmatrix} \varepsilon & 0 & 0 \\ 0 & -\frac{1}{2}\varepsilon & 0 \\ 0 & 0 & -\frac{1}{2}\varepsilon \end{pmatrix}$ represents an elongation along the 1-axis and contraction along the 2 and 3 axes without volume change.

Now let us consider the off-diagonal components of strain tensor. From equation (1.39), it is clear that when all the diagonal components are zero, then all the displacement vectors must be normal to the direction of the initial vector. Therefore, there is no change in length due to the off-diagonal component of strain. Note, also, that since strain is a symmetric tensor, $\varepsilon_{ij} = \varepsilon_{ji}$, the directions of rotation of two orthogonal axes are toward the opposite direction with the same magnitude (Fig. 1.5). Consequently, the angle of two orthogonal axes change from $\pi/2$ to (see Problem 1.4),

$$\frac{\pi}{2} - \tan^{-1} 2\varepsilon_{ij}. \tag{1.43}$$

Therefore, the off-diagonal components of strain tensor (i.e., ε_{ij} with $i \neq j$) represent the shape change without volume change, namely shear strain.

Problem 1.4*

Derive equation (1.43). (Assume a small strain for simplicity. The result also works for a finite strain, see MASE (1970).)

Solution

Let the small angle of rotation of the i axis to the j axis due to the operation of strain tensor be $\delta\theta_{ij}$ (positive clockwise), then (Fig. 1.5),

$$\tan\delta\theta_{ij} = \frac{d\tilde{u}_j}{du_i} \approx \delta\theta_{ij} = (\varepsilon_{ji} + \omega_{ji}) = \varepsilon_{ij} + \omega_{ij}.$$

Similarly, if the rotation of the j axis relative to the i axis is $\delta\theta_{ji}$, one obtains,

$$\tan\delta\theta_{ji} = -\frac{d\tilde{u}_i}{du_j} \approx \delta\theta_{ji} = -(\varepsilon_{ij} + \omega_{ij}) = -\varepsilon_{ij} - \omega_{ij}.$$

(Note that the rigid-body rotations of the two axes are opposite with the same magnitude.) Therefore, the net change in the angle between i and j axes is given by

$$\Delta\theta_{ij} = \delta\theta_{ij} + \delta\theta_{ji} = -2\varepsilon_{ij} \sim \tan\Delta\theta_{ij}.$$

Hence $\Delta\theta_{ij} = -\tan^{-1} 2\varepsilon_{ij}$.

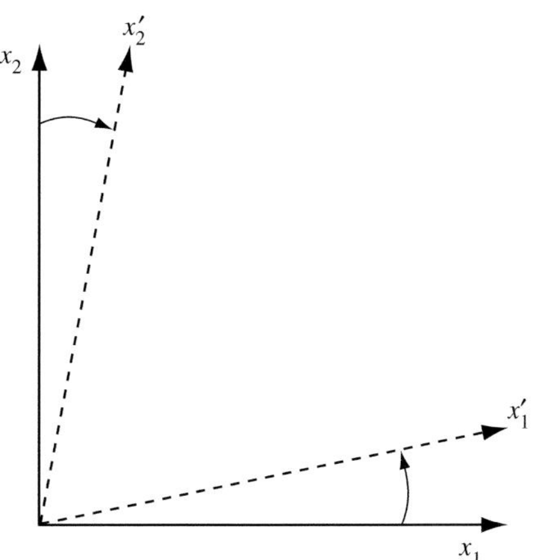

FIGURE 1.5 Geometry of shear deformation.

1.2.3. Principal strain, strain ellipsoid

We have seen two different cases for strain, one in which the displacement caused by the strain tensor is normal to the original direction of the material line and another where the displacement is normal to the original material line. In this section, we will learn that in any material and in any geometry of strain, there are three directions along which the displacement is normal to the direction of original line segment. These are referred to as the orientation of principal strain, and the magnitude of strain along these orientations are called principal strain.

One can define the principal strains ($\varepsilon_1, \varepsilon_2, \varepsilon_3$; $\varepsilon_1 > \varepsilon_2 > \varepsilon_3$) in the following way. Recall that the normal displacement along the direction i, $\delta\tilde{u}_i$, along the vector $\boldsymbol{u}$ is given by,

$$\delta\tilde{u}_i = \sum_{j=1}^{3} \varepsilon_{ij} u_j. \tag{1.44}$$

Now, let $\boldsymbol{u}$ be the direction in space along which the displacement is parallel to the direction $\boldsymbol{u}$. Then,

$$\delta\tilde{u}_i = \varepsilon u_i \tag{1.45}$$

where ε is a scalar quantity to be determined. From equations (1.44) and (1.45),

$$\sum_{j=1}^{3} (\varepsilon_{ij} - \varepsilon\delta_{ij}) u_j = 0. \tag{1.46}$$

For this equation to have a non-trivial solution other than $\boldsymbol{u} = 0$, one must have,

$$|\varepsilon_{ij} - \varepsilon\delta_{ij}| = 0 \tag{1.47}$$

where $|X_{ij}|$ is the determinant of a matrix X_{ij}. Writing equation (1.47) explicitly, one gets,

$$\begin{vmatrix} \varepsilon_{11} - \varepsilon & \varepsilon_{12} & \varepsilon_{13} \\ \varepsilon_{21} & \varepsilon_{22} - \varepsilon & \varepsilon_{32} \\ \varepsilon_{31} & \varepsilon_{32} & \varepsilon_{33} - \varepsilon \end{vmatrix} = -\varepsilon^3 + I_\varepsilon \varepsilon^2 + II_\varepsilon \varepsilon + III_\varepsilon = 0 \tag{1.48}$$

with

$$I_\varepsilon = \varepsilon_{11} + \varepsilon_{22} + \varepsilon_{33} \tag{1.49a}$$

$$II_\varepsilon = -\varepsilon_{11}\varepsilon_{22} - \varepsilon_{11}\varepsilon_{33} - \varepsilon_{33}\varepsilon_{22} + \varepsilon_{12}^2 + \varepsilon_{13}^2 + \varepsilon_{23}^2 \tag{1.49b}$$

$$III_\varepsilon = \varepsilon_{11}\varepsilon_{22}\varepsilon_{33} + 2\varepsilon_{12}\varepsilon_{23}\varepsilon_{31} - \varepsilon_{11}\varepsilon_{23}^2 - \varepsilon_{22}\varepsilon_{13}^2 - \varepsilon_{33}\varepsilon_{12}^2. \tag{1.49c}$$

Therefore, there are three solutions of equation (1.48), $\varepsilon_1, \varepsilon_2, \varepsilon_3 (\varepsilon_1 > \varepsilon_2 > \varepsilon_3)$. These are referred to as *the principal strain*. The corresponding $\boldsymbol{u}^0$ are *the orientations of principal strain*. If the strain tensor is written using the coordinate whose orientation coincides with the orientation of principal strain, then,

$$[\varepsilon_{ij}] = \begin{bmatrix} \varepsilon_1 & 0 & 0 \\ 0 & \varepsilon_2 & 0 \\ 0 & 0 & \varepsilon_3 \end{bmatrix}. \tag{1.50}$$

A strain ellipsoid is a useful way to visualize the geometry of strain. Let us consider a spherical body in a space and deform it. The shape of a sphere is described by,

$$(u_1)^2 + (u_2)^2 + (u_3)^2 = 1. \tag{1.51}$$

The shape of the sphere will change due to deformation. Let us choose a coordinate system such that the directions of 1, 2 and 3 axes coincide with the directions of principal strain. Then the length of each axis of the original sphere along each direction of the coordinate system should change to $\tilde{u}_i = (1 + \varepsilon_{ii}) u_i$, and therefore the sphere will change to an ellipsoid,

$$\frac{(\tilde{u}_1)^2}{(1+\varepsilon_1)^2} + \frac{(\tilde{u}_2)^2}{(1+\varepsilon_2)^2} + \frac{(\tilde{u}_3)^2}{(1+\varepsilon_3)^2} = 1. \tag{1.52}$$

A three-dimensional ellipsoid defined by this equation is called *a strain ellipsoid*. For example, if the shape of grains is initially spherical, then the shape of grains after deformation represents the strain ellipsoid. The strain of a rock specimen can be determined by the measurements of the shape of grains or some objects whose initial shape is inferred to be nearly spherical.

Problem 1.5*

Consider a simple shear deformation in which the displacement of material occurs only in one direction (the displacement vector is given by $\boldsymbol{u} = (\gamma y, 0, 0)$). Calculate the strain ellipsoid, and find how the principal axes of the strain ellipsoid rotate with strain. Also find the relation between the angle of tilt of the initially vertical line and the angle of the maximum elongation direction relative to the horizontal axis.

Solution

For simplicity, let us analyze the geometry in the x–y plane (normal to the shear plane) where shear occurs. Consider a circle defined by $x^2 + y^2 = 1$. By deformation, this circle changes to an ellipsoid, $(x + \gamma y)^2 + y^2 = 1$, i.e.,

$$x^2 + 2\gamma xy + (\gamma^2 + 1)y^2 = 1. \tag{1}$$

Now let us find a new coordinate system that is tilted from the original one by an angle θ (positive counter-clockwise). With this new coordinate system, $(x, y) \rightarrow (X, Y)$ with

$$\begin{pmatrix} x \\ y \end{pmatrix} = \begin{pmatrix} \cos\theta & \sin\theta \\ -\sin\theta & \cos\theta \end{pmatrix} \begin{pmatrix} X \\ Y \end{pmatrix}. \tag{2}$$

By inserting this relation into (1), one finds,

$$A_{XX}X^2 + A_{XY}XY + A_{YY}Y^2 = 1 \tag{3}$$

with

$$\begin{pmatrix} A_{XX} \\ A_{XY} \\ A_{YY} \end{pmatrix} = \begin{pmatrix} 1 + \frac{1}{2}\gamma^2 - \frac{1}{2}\gamma^2 \cos 2\theta - \gamma \sin 2\theta \\ 2\gamma(\cos 2\theta - \frac{1}{2}\gamma \sin 2\theta) \\ 1 + \frac{1}{2}\gamma^2 + \frac{1}{2}\gamma^2 \cos 2\theta + \gamma \sin 2\theta \end{pmatrix} \tag{4}$$

Now, in order to obtain the orientation in which the X–Y directions coincide with the orientations of principal strain, we set $A_{XY} = 0$, and get $\tan 2\theta = 2/\gamma$. $A_{XX} < A_{YY}$ and therefore X is the direction of maximum elongation. Because the change in the angle (φ) of the initially vertical line from the vertical direction is determined by the strain as $\tan\varphi = \gamma$, we find,

$$\begin{aligned} \tan\theta &= \frac{1}{2}(-\gamma + \sqrt{4 + \gamma^2}) \\ &= \frac{1}{2}(-\tan\varphi + \sqrt{4 + \tan^2\varphi}) \end{aligned}. \tag{5}$$

At $\gamma = 0, \theta = \pi/4$. As strain goes to infinity, $\gamma \rightarrow \infty$, i.e., $\varphi \rightarrow \pi/2$, and $\tan\theta \rightarrow 0$ hence $\theta \rightarrow 0$: the direction of maximum elongation approaches the direction of shear. $\varepsilon_1 = A_{XX}^{-1/2} - 1$ changes from 0 at $\gamma = 0$ to ∞ as $\gamma \rightarrow \infty$ and $\varepsilon_2 = A_{yy}^{-1/2} - 1$ changes from 0 at $\gamma = 0$ to –1 at $\gamma \rightarrow \infty$.

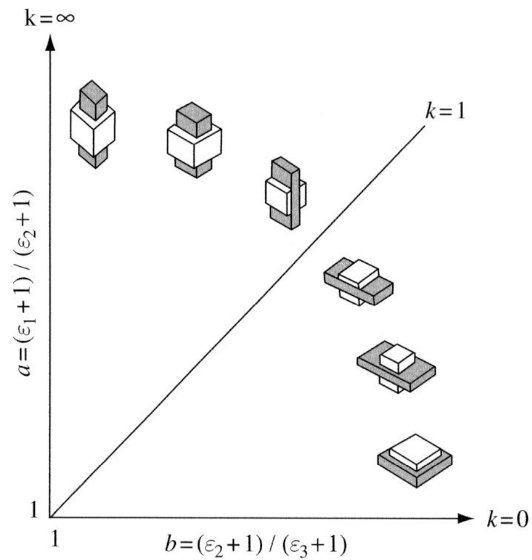

FIGURE 1.6 The Flinn diagram (after HOBBS *et al.*, 1976).

1.2.4. The Flinn diagram

The three principal strains define the geometry of the strain ellipsoid. Consequently, the shape of the strain ellipsoid is completely characterized by two ratios, $a \equiv (\varepsilon_1 + 1)/(\varepsilon_2 + 1)$ and $b \equiv (\varepsilon_2 + 1)/(\varepsilon_3 + 1)$. A diagram showing strain geometry on an a–b plane is called *the Flinn diagram* (Fig. 1.6) (FLINN, 1962). In this diagram, for points along the horizontal axis, $k \equiv (a - 1)/(b - 1) = 0$, and they correspond to the flattening strain ($\varepsilon_1 = \varepsilon_2 > \varepsilon_3 (a = 1, b > 1)$). For points along the vertical axis, $k = \infty$, and they correspond to the extensional strain ($\varepsilon_1 > \varepsilon_2 = \varepsilon_3 (b = 1, a > 1)$). For points along the central line, $k = 1$ ($a = b$, i.e., $(\varepsilon_1 + 1)/(\varepsilon_2 + 1) = (\varepsilon_2 + 1)/(\varepsilon_3 + 1)$) and deformation is plane strain (two-dimensional strain where $\varepsilon_2 = 0$), when there is no volume change during deformation (see Problem 1.6).

Problem 1.6

Show that the deformation of materials represented by the points on the line for $k = 1$ in the Flinn diagram is plane strain (two-dimensional strain) if the volume is conserved.

Solution

If the volume is conserved by deformation, then $(\varepsilon_1 + 1)(\varepsilon_2 + 1)(\varepsilon_3 + 1) = 1$ (see equation (1.41)).

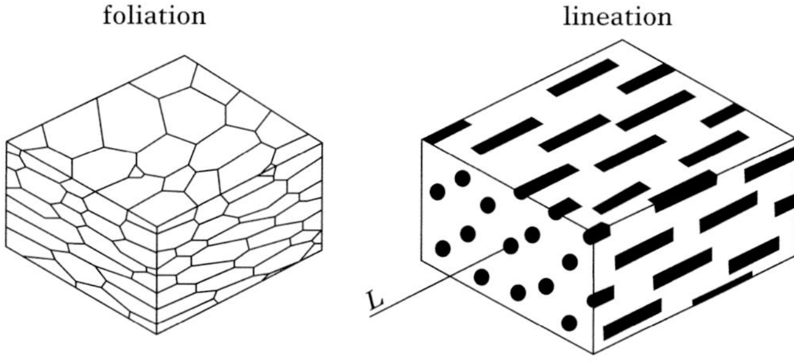

FIGURE 1.7 Typical cases of (a) foliation and (b) lineation.

Combined with the relation $(\varepsilon_1 + 1)/(\varepsilon_2 + 1) = (\varepsilon_2 + 1)/(\varepsilon_3 + 1)$, we obtain $(\varepsilon_2 + 1)^3 = 1$ and hence $\varepsilon_2 = 0$. Therefore deformation is plane strain.

1.2.5. Foliation, lineation (Fig. 1.7)

When the anisotropic microstructure of a rock is studied, it is critical to define the reference frame of the coordinate. Once one identifies a plane of reference and the reference direction on that plane, then the three orthogonal axes (parallel to lineation (X direction), normal to lineation on the foliation plane (Y direction), normal to foliation (Z direction)) define the reference frame.

Foliation is usually used to define a reference plane and *lineation* is used define a reference direction on the foliation plane. Foliation is a planar feature in a given rock, but its origin can be various (Hobbs *et al.*, 1976). The foliation plane may be defined by a plane normal to the maximum shortening strain (Fig. 1.7). Foliation can also be caused by compositional layering, grain-size variation and the orientation of platy minerals such as mica. When deformation is heterogeneous, such as the case for S-C mylonite (Lister and Snoke, 1984), one can identify two planar structures, one corresponds to the strain ellipsoid (a plane normal to maximum shortening, ε_3) and another to the shear plane.

Lineation is a linear feature that occurs repetitively in a rock. In most cases, the lineation is found on the foliation plane, although there are some exceptions. The most common is mineral lineation, which is defined by the alignment of non-spherical minerals such as clay minerals. The alignment of spinel grains in a spinel lherzolite and recrystallized orthopyroxene in a garnet lherzolite are often used to define the lineation in peridotites. One cause of lineation is strain, and in this case, the direction of lineation is parallel to the maximum elongation direction. However, there are a number of other possible causes for lineation including the preferential growth of minerals (e.g., Hobbs *et al.*, 1976).

Consequently, the interpretation of the significance of these reference frames (foliation/lineation) in natural rocks is not always unique. In particular, the question of growth origin versus deformation origin, and the strain ellipsoid versus the shear plane/shear direction can be elusive in some cases. Interpretation and identification of foliation/lineation become more difficult if the deformation geometry is not constant with time. Consequently, it is important to state clearly how one defines foliation/lineation in the structural analysis of a deformed rock. For more details on foliation and lineation, a reader is referred to a structural geology textbook such as Hobbs *et al.* (1976).

1.2.6. Various deformation geometries

The geometry of strain is completely characterized by the principal strain, and therefore a diagram such as the Flinn diagram (Fig. 1.6) can be used to define strain. However, in order to characterize the geometry of deformation completely, it is necessary to characterize the deformation gradient tensor (d_{ij} $(= \varepsilon_{ij} + \omega_{ij})$). Therefore the rotational component (vorticity tensor), ω_{ij}, must also be characterized. In this connection, it is important to distinguish between *irrotational* and *rotational deformation geometry*. Rotational deformation geometry refers to deformation in which $\omega_{ij} \neq 0$, and irrotational deformation geometry corresponds to $\omega_{ij} = 0$. The distinction between them is important at finite strain. To illustrate this point, let us consider two-dimensional deformation (Fig. 1.8). For irrotational deformation, the orientations of the principal axes of strain are always parallel to those of principal stress. Therefore such a deformation is called *coaxial deformation*. In contrast, when deformation is rotational, such as *simple shear*, the orientations of principal axes of strain rotate progressively with respect to those of the stress (see Problem 1.5). This type of

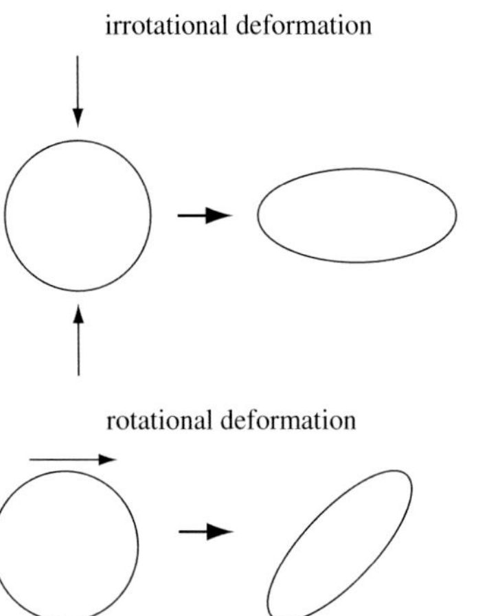

FIGURE 1.8 Irrotational and rotational deformation.

deformation is called *non-coaxial deformation*. (When deformation is infinitesimal, this distinction is not important: the principal axes of instantaneous strain are always parallel to the principal axis of stress as far as the property of the material is isotropic.)

Various methods of identifying the rotational component of deformation have been proposed (Bouchez *et al.*, 1983; Simpson and Schmid, 1983). In most of them, the nature of anisotropic microstructures, such as lattice-preferred orientation (Chapter 14), is used to infer the rotational component of deformation. However, the physical basis for inferring the rotational component is not always well established.

Some details of deformation geometries in typical experimental studies are discussed in Chapter 6.

1.2.7. Macroscopic, and microscopic stress and strain

Stress and strain in a material can be heterogeneous. Let us consider a material to which a macroscopically homogeneous stress (strain) is applied. At any point in a material, one can define a microscopic, local stress (strain). The magnitude and orientation of microscopic stress (strain) can be different from that of a macroscopic (imposed) stress (strain). This is caused by the heterogeneity of a material such as the grain-to-grain heterogeneity and the presence of defects. In particular, the grain-scale heterogeneity in stress (strain) is critical to the understanding of deformation of a polycrystalline material (see Chapters 12 and 14).

2 Thermodynamics

The nature of the deformation of materials depends on the physical and chemical state of the materials. Thermodynamics provides a rigorous way by which the physical and chemical state of materials can be characterized. A brief account is made of the concepts of thermodynamics of reversible as well as irreversible processes that are needed to understand the plastic deformation of materials and related processes. The principles governing the chemical equilibrium are outlined including the concept of chemical potential, the law of mass action, and the Clapeyron slope (i.e., the slope of a phase boundary in the pressure-temperature space). When a system is out of equilibrium, a flow of materials and/or energy occurs. The principles governing the irreversible processes are outlined. Irreversible processes often occur through thermally activated processes. The basic concepts of thermally activated processes are summarized based on the statistical physics.

Key words entropy, chemical potential, Gibbs free energy, fugacity, activity, Clapeyron slope, phase diagrams, rate theory, generalized force, the Onsager reciprocal relation.

2.1. Thermodynamics of reversible processes

Thermodynamics provides a framework by which the nature of thermochemical equilibrium is defined, and, in cases where a system is out of equilibrium, it defines the direction to which a given material will change. It gives a basis for analyzing the composition and structure of geological materials, experimental data and the way in which the experimental results should be extrapolated to Earth's interior where necessary. This chapter provides a succinct review of some of the important concepts in thermodynamics that play significant roles in understanding the deformation of materials in Earth's interior. More complete discussions on thermodynamics can be found in the textbooks such as CALLEN (1960), DE GROOT and MAZUR (1962), LANDAU and LIFSHITZ (1964) and PRIGOGINE and DEFAY (1950).

2.1.1. The first and the second principles of thermodynamics

The *first principle of thermodynamics* is the law of conservation of energy, which states that the change in the internal energy, dE, is the sum of the mechanical work done to the system, the change in the energy due to the addition of materials and the heat added to the system, namely,

$$dE = \delta W + \delta Z + \delta Q \tag{2.1}$$

where $\delta W = -P\,dV$ (the symbol δ is used to indicate a change in some quantity that depends on the path) is the mechanical work done to the system where P is the pressure, dV is the volume change, δZ is the change in internal energy due to the change in the number of atomic species, i.e.,

$$\delta Z = \sum_i \left(\frac{\partial E}{\partial n_i}\right)_{S,V,n_j} dn_i \tag{2.2}$$

where n_i is the molar amount of the ith species and δQ is the change in "heat." Thus

$$dE = -PdV + \sum_i \left(\frac{\partial E}{\partial n_i}\right)_{S,V,n_j} dn_i + \delta Q. \quad (2.3)$$

Note that "heat" is the change in energy other than the mechanical work and energy caused by the exchange of material. These two quantities (mechanical work and the energy associated with the transport of matter) are related to the *average* motion of atoms. In contrast, the third term, δQ, is related to the properties of materials that involve random motion or the *random arrangement* of atoms. The *second principle of thermodynamics* is concerned with the nature of processes related to this third term. This principle states that there exists a quantity called *entropy* that is determined by the amount of heat introduced to the system divided by temperature, namely,

$$dS = \frac{\delta Q}{T} \quad (2.4)$$

and that the entropy increases during any natural processes. When the process is *reversible* (i.e., the system is in equilibrium), the entropy will be the maximum, i.e.,

$$dS = 0 \quad (2.5)$$

whereas

$$dS > 0 \quad (2.6)$$

for irreversible processes. Equation (2.6) may be written as

$$dS = d_e S + d_i S = \frac{\delta Q}{T} + \frac{\delta Q'}{T} \quad (2.7)$$

where $d_e S = \delta Q/T$ is the entropy coming from the exterior of the system and $d_i S = \delta Q'/T$ is the entropy production inside the system. For reversible processes $\delta Q' = 0$ and for irreversible processes, $\delta Q' > 0$. From (2.3) and (2.7), one finds,

$$dE = T\,dS - P\,dV + \sum_i \left(\frac{\partial E}{\partial n_i}\right)_{S,V,n_i} dn_i - \delta Q'. \quad (2.8)$$

For equilibrium,

$$dE = T\,dS - P\,dV + \sum_i \left(\frac{\partial E}{\partial n_i}\right)_{S,V,n_i} dn_i \quad (2.9)$$

and $E = E(S, V, n_i)$.

The *enthalpy* (H), *Helmholtz free energy* (F), and the *Gibbs free energy* (G) can be defined as,

$$H = E + PV \quad (2.10a)$$

$$F = E - TS \quad (2.10b)$$

and

$$G = E - TS + PV \quad (2.10c)$$

respectively and therefore,

$$dH = T\,dS + V\,dP + \sum_i \left(\frac{\partial U}{\partial n_i}\right)_{S,V,n_j} dn_i - \delta Q' \quad (2.11a)$$

$$dF = -S\,dT - P\,dV + \sum_i \left(\frac{\partial U}{\partial n_i}\right)_{S,V,n_j} dn_i - \delta Q' \quad (2.11b)$$

and

$$dG = -S\,dT + V\,dP + \sum_i \left(\frac{\partial U}{\partial n_i}\right)_{S,V,n_j} dn_i - \delta Q'. \quad (2.11c)$$

It follows from (2.8), (2.11a)–(2.11c) that for a closed system and for constant S and V (S and P, T and V, T and P), $dE = -\delta Q'$ ($dH = -\delta Q', dF = -\delta Q', dG = -\delta Q'$) so that E (H, F, G) is minimum at equilibrium. Also from (2.8), (2.11a)–(2.11c), one obtains

$$\left(\frac{\partial E}{\partial n_i}\right)_{S,V,n_j} = \left(\frac{\partial H}{\partial n_i}\right)_{S,P,n_j} = \left(\frac{\partial F}{\partial n_i}\right)_{T,V,n_j} = \left(\frac{\partial G}{\partial n_i}\right)_{T,P,n_j} \equiv \mu_i. \quad (2.12)$$

This is the definition of the *chemical potential*. Thus at thermochemical equilibrium,

$$dE = T\,dS - P\,dV + \sum_i \mu_i\,dn_i \quad (2.13a)$$

$$dH = T\,dS + V\,dP + \sum_i \mu_i\,dn_i \quad (2.13b)$$

$$dF = -S\,dT - P\,dV + \sum_i \mu_i\,dn_i \quad (2.13c)$$

$$dG = -S\,dT + V\,dP + \sum_i \mu_i\,dn_i. \quad (2.13d)$$

From (2.13), one has

$$T = \left(\frac{\partial E}{\partial S}\right)_{V,n_i} = \left(\frac{\partial H}{\partial S}\right)_{P,n_i} \quad (2.14a)$$

$$S = -\left(\frac{\partial F}{\partial T}\right)_{V,n_i} = -\left(\frac{\partial G}{\partial T}\right)_{P,n_i} \quad (2.14b)$$

$$P = -\left(\frac{\partial E}{\partial V}\right)_{S,n_i} = -\left(\frac{\partial F}{\partial V}\right)_{T,n_i} \tag{2.14c}$$

$$V = \left(\frac{\partial H}{\partial P}\right)_{S,n_i} = \left(\frac{\partial G}{\partial P}\right)_{T,n_i}. \tag{2.14d}$$

It can be seen that the thermodynamic quantities such as T, P, S and V (and μ_i) can be derived from E, H, F and G. Therefore these quantities (E, H, F and G) are called the *thermodynamic potentials*. The thermodynamic potentials assume the minimum value at thermochemical equilibrium. Because we will mostly consider a system at constant temperature and pressure, the most frequently used thermodynamic potential is the Gibbs free energy. μ_i is the thermodynamic potential of the ith species (per unit mole). To emphasize the fact that μ_i is the thermodynamic potential of the ith species per mole, it is often called the partial molar thermodynamic potential (partial molar Gibbs free energy when the independent variables are T and P).

Using the rule of calculus, it follows from (2.13) and (2.14),

$$\left(\frac{\partial^2 E}{\partial S \partial V}\right)_{n_i} = \left(\frac{\partial^2 E}{\partial V \partial S}\right)_{n_i} \Rightarrow \left(\frac{\partial T}{\partial V}\right)_{S,n_i} = -\left(\frac{\partial P}{\partial S}\right)_{V,n_i} \tag{2.15a}$$

$$\left(\frac{\partial^2 H}{\partial S \partial P}\right)_{n_i} = \left(\frac{\partial^2 H}{\partial P \partial S}\right)_{n_i} \Rightarrow \left(\frac{\partial T}{\partial P}\right)_{S,n_i} = \left(\frac{\partial V}{\partial S}\right)_{V,n_i} \tag{2.15b}$$

$$\left(\frac{\partial^2 F}{\partial T \partial V}\right)_{n_i} = \left(\frac{\partial^2 F}{\partial V \partial T}\right)_{n_i} \Rightarrow \left(\frac{\partial S}{\partial V}\right)_{T,n_i} = \left(\frac{\partial P}{\partial T}\right)_{V,n_i} \tag{2.15c}$$

and

$$\left(\frac{\partial^2 G}{\partial T \partial P}\right)_{n_i} = \left(\frac{\partial^2 G}{\partial P \partial T}\right)_{n_i} \Rightarrow -\left(\frac{\partial S}{\partial P}\right)_{T,n_i} = \left(\frac{\partial V}{\partial T}\right)_{P,n_i}. \tag{2.15d}$$

These relations (2.15) are called *the Maxwell relations*.

Similar relations among thermodynamic variables can also be derived. Consider a quantity such as entropy that is a function of two parameters (such as temperature and pressure; this is a case for a closed system, i.e., n_i is kept constant), i.e., $Z = Z(X, Y; n_i)$, then,

$$dZ = \left(\frac{\partial Z}{\partial X}\right)_{Y,n_i} dX + \left(\frac{\partial Z}{\partial Y}\right)_{X,n_i} dY. \tag{2.16}$$

If we consider a process in which the quantity Z is kept constant, then, $dZ = 0$ and

$$\left(\frac{\partial Z}{\partial X}\right)_{Y,n_i} = -\left(\frac{\partial Y}{\partial X}\right)_{Z,n_i} \left(\frac{\partial Z}{\partial Y}\right)_{X,n_i}. \tag{2.17}$$

Examples of such a relation include

$$\left(\frac{\partial S}{\partial T}\right)_{P,n_i} = -\left(\frac{\partial P}{\partial T}\right)_{S,n_i} \left(\frac{\partial S}{\partial P}\right)_{T,n_i} \tag{2.18a}$$

$$\left(\frac{\partial T}{\partial V}\right)_{S,n_i} = -\left(\frac{\partial S}{\partial V}\right)_{T,n_i} \left(\frac{\partial T}{\partial S}\right)_{V,n_i} \tag{2.18b}$$

and

$$\left(\frac{\partial V}{\partial T}\right)_{P,n_i} = -\left(\frac{\partial V}{\partial P}\right)_{T,n_i} \left(\frac{\partial P}{\partial T}\right)_{V,n_i}. \tag{2.18c}$$

These thermodynamic identities (the Maxwell relations and the relations (2.18)) are often used in manipulating thermodynamic relationships (e.g., Chapter 4).

Now let us rewrite (2.13d) as,

$$dS = \frac{1}{T} dE + \frac{P}{T} dV + \sum_i \left(-\frac{\mu_i}{T}\right) dn_i. \tag{2.19}$$

At equilibrium, the entropy is a maximum, i.e., $dS = 0$. Consider a case where two systems (1 and 2) are in contact. In this case the condition for equilibrium can be written as

$$\begin{aligned} dS = &\frac{1}{T_1} dE_1 + \frac{1}{T_2} dE_2 + \frac{P_1}{T_1} dV_1 + \frac{P_2}{T_2} dV_2 \\ &+ \sum_i \left[\left(-\frac{\mu_1^i}{T_1}\right) dn_1^i + \left(-\frac{\mu_2^i}{T_2}\right) dn_2^i\right] = 0. \end{aligned} \tag{2.20}$$

Because we consider a system with a constant energy ($dE_1 + dE_2 = 0$), volume ($dV_1 + dV_2 = 0$) and matter ($dn_1^i + dn_2^i = 0$), (2.20) becomes,

$$\begin{aligned} dS = &\left(\frac{1}{T_1} - \frac{1}{T_2}\right) dE_1 + \left(\frac{P_1}{T_1} - \frac{P_2}{T_2}\right) dV_1 \\ &+ \sum_i \left(-\frac{\mu_1^i}{T_1} + \frac{\mu_2^i}{T_2}\right) dn_1^i = 0. \end{aligned} \tag{2.21}$$

This must occur for any arbitrary changes of internal energy (dE_1), volume (dV_1) and chemical composition (dn_1^i). Therefore when two systems (1 and 2) are in contact and in equilibrium, $1/T_1 = 1/T_2$, $P_1/T_1 = P_2/T_2$ and $\mu_1^i/T_1 = \mu_2^i/T_2$ and hence the conditions of equilibrium are

$$T_1 = T_2 \tag{2.22a}$$

$$P_1 = P_2 \tag{2.22b}$$

and

$$\mu_1^i = \mu_2^i. \tag{2.22c}$$

The variables such as temperature, pressure and the concentration of *i*th species do not depend on the size of the system. These variables are called *intensive quantities*. In contrast, quantities such as entropy, internal energy and Gibbs free energy increase linearly with the size of the system. They are called *extensive quantities*. It follows that,

$$S(\lambda E, \lambda V, \lambda n_i) = \lambda S(E, V, n_i) \tag{2.23}$$

where λ is an arbitrary parameter. Differentiating (2.23) with λ, and putting $\lambda = 1$, one obtains,

$$TS = E + PV - \sum_i \mu_i n_i. \tag{2.24}$$

Differentiating this equation, and comparing the results with equation (2.19), one finds,

$$S\,dT - V\,dP + \sum_i n_i\,d\mu_i = 0. \tag{2.25}$$

This is *the Gibbs–Duhem relation*, which shows that the intensive variables are not all independent.

The concept of *entropy* is closely related to the atomistic nature of matter, namely the fact that matter is made of a large number of atoms. A system composed of a large number of atoms may assume a large number of possible *micro-states*. All micro-states with the same macro-state (temperature, volume etc.) are equally probable. Consequently, a system most likely assumes a macro-state for which the number of corresponding micro-states is the maximum (i.e., the maximum entropy). Thus the concept of entropy must be closely related to the number of the micro-state, W, as (for the derivation of this relation see e.g., LANDAU and LIFSHITZ (1964)),

$$S = k_B \cdot \log W \tag{2.26}$$

where k_B is the Boltzmann constant.[1] The number of micro-states may be defined by the number of ways in which atoms can be distributed. When n atoms are distributed on N sites, then, $W = {}_N C_n = N!/(N-n)!n!$, and,

[1] When *log* is used in a theoretical equation in this book, the base is *e* (this is often written as *ln*). In contrast, when experimental data are plotted, the base is *10* (unless specified otherwise).

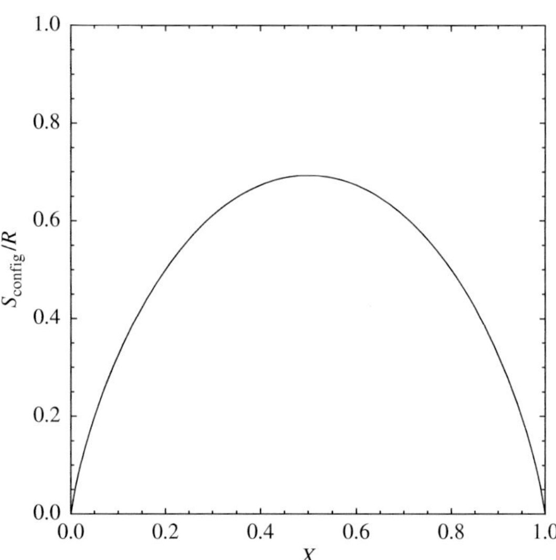

FIGURE 2.1 A plot of configurational entropy $S_{config}(x) = -R[x \log x + (1-x)\log(1-x)]$.

$$\begin{aligned} S &= k_B \cdot \log \frac{N!}{(N-n)!n!} \\ &\cong -RN_{mol}[x \log x + (1-x)\log(1-x)] \end{aligned} \tag{2.27}$$

where $x = n/N$ and $N = N_a N_{mol}$ (N_a is the Avogadro number, and N_{mol} is the molar abundance of the relevant species) where the Stirling formula, $N! \sim N \cdot \log N - N$ for $N \gg 1$ was used. The entropy corresponding to this case may be called *configurational entropy* S_{config} and is plotted as a function of concentration x in Fig. 2.1. The configurational entropy is proportional to the amount of material, and for unit mole of material, it is given by

$$S_{config} = -R[x \log x + (1-x)\log(1-x)]. \tag{2.28}$$

The micro-state of matter may also be characterized by the nature of lattice vibration; that is, matter with different frequencies of lattice vibration is considered to be in different states. The vibrational entropy defined by this is related to the frequencies of atomic vibration as (e.g., ANDERSON, 1996; BORN and HUANG, 1954, see Box 2.1),

$$S_{vib} \approx -k_B \sum_i \log\left(\frac{h\omega_i}{2\pi k_B T}\right) \tag{2.29}$$

where h is the Planck constant, k_B is the Boltzmann constant, ω_i is the (angular) frequency of lattice vibration of mode i (for a crystal that contains N atoms in the unit cell, there are $3N$ modes of lattice vibration). It can be seen that a system with a higher frequency of

Box 2.1 Lattice vibration and the vibrational entropy

Small random motion of atoms around their stable positions causes "disorder" in a material that contributes to the entropy. To calculate the contribution to entropy from lattice vibration, we note that the internal energy due to lattice vibration is given by (e.g., BORN and HUANG, 1954)

$$E = \sum_{i=1}^{3N} \frac{\sum_{n_i=0}^{\infty} (n_i h\omega_i/2\pi)\exp(n_i h\omega_i/2\pi k_B T)}{\sum_{n_i=0}^{\infty} \exp(n_i h\omega_i/2\pi k_B T)} = \sum_{i=1}^{3N} \frac{h\omega_i/2\pi}{\exp(h\omega_i/2\pi k_B T) - 1}$$

where n_i is the number of phonons of the ith mode of lattice vibration and ω_i is its (angular) frequency. Using the thermodynamic relation $F = -T\int(E/T^2)dT$, one obtains

$$F = U + k_B T \sum_i \log\left[1 - \exp\left(-\frac{h\omega_i}{2\pi k_B T}\right)\right]$$

where U is the energy of a static lattice (at $T = 0$ K). Therefore from $S = -(\partial F/\partial T)$,

$$S_{vib} = -k_B\left\{\sum_i \log\left[1 - \exp\left(-\frac{h\omega_i}{2\pi k_B T}\right)\right] - \sum_i \frac{h\omega_i}{2\pi k_B T}\frac{1}{\exp(h\omega_i/2\pi k_B T) - 1}\right\}$$
$$\approx -k_B \sum_i \log\frac{h\omega_i}{2\pi k_B T}.$$

The approximation is for high temperature, i.e., $h\omega_i/2\pi k_B T \ll 1$.

vibration has a lower entropy. When the vibrational frequency changes between two phases (A and B), then the change in entropy is given by,

$$\Delta S_{vib} \equiv S^A_{vib} - S^B_{vib} = k_B \sum_i \log\frac{\omega^B_i}{\omega^A_i} \approx R\log\frac{\omega^B_D}{\omega^A_D}. \quad (2.30)$$

where $\omega_D^{A,B}$ is a characteristic frequency of lattice vibration (the Debye frequency; see Box 4.3 in Chapter 4) of a phase A or B.

In a solid, the micro-state may be defined either by small displacements of atomic positions from their lattice sites (lattice vibration) or by large displacements that result in an exchange of atoms among various sites. Therefore the entropy may be written as,

$$S = S_{vib} + S_{config}. \quad (2.31)$$

2.1.2. Activity, fugacity

Using the equations (2.10), (2.12) and (2.28), we can write the chemical potential of a component as a function of the concentration x (for $x \ll 1$),

$$\mu(T, P, x) = \mu^0(T, P) + RT\log x \quad (2.32)$$

where μ^0 is the chemical potential for a pure phase ($x = 1$). In a system that contains several components, (2.32) can be generalized to,

$$\mu_i(T, P, x_i) = \mu_i^0(T, P) + RT\log x_i \quad (2.33)$$

where the suffix i indicates a quantity for the ith component.

Problem 2.1

Derive equation (2.32).

Solution

From (2.10) and (2.12), noting that E, V and S are the extensive variables, one obtains,

$$\mu = \left(\frac{\partial G}{\partial n_{mol}}\right)_{T,P} = \left(\frac{\partial E}{\partial n_{mol}}\right)_{T,P} + P\left(\frac{\partial V}{\partial n_{mol}}\right)_{T,P} - T\left(\frac{\partial S}{\partial n_{mol}}\right)_{T,P}$$
$$= e + Pv - Ts_{vib} - T\left(\frac{\partial S_{config}}{\partial n_{mol}}\right)$$
$$\equiv \mu^0 - T\left(\frac{\partial S_{config}}{\partial n_{mol}}\right).$$

where $\mu^0 \equiv e + Pv - Ts_{vib}$ and e, v and s_{vib} are molar internal energy, molar volume and molar (vibrational) entropy respectively.

Now, noting that $dx = dn_{mol}/N_{mol}$ (from $x = n/N = n_{mol}/N_{mol}$), it follows from (2.28), $(\partial S_{config}/\partial n_{mol})_{T,P} = -R\log(x/(1-x)) \approx -R\log x$. Therefore one obtains $\mu(T, P, x) = \mu^0(T, P) + RT\log x$.

Activity

In deriving (2.32), we made an assumption that the component under consideration has a small quantity (dilute solution) so that atoms in the component do not interact with each other or with other species. Such a material is called an *ideal solution*. In a real material where the interaction of atoms of a given component is not negligible, a modification of these relations is needed. A useful way to do this is to introduce the concept of *activity* (of the ith component), a_i, which is defined by,

$$\mu_i(T, P, a_i) = \mu_i^0(T, P) + RT \log a_i. \tag{2.34}$$

If $\mu_i^0(T, P)$ is the chemical potential of a pure phase, then by definition, for a pure system, the activity is 1 (for example, if pure Ni is present in a system, then the activity of Ni is $a_{\mathrm{Ni}} = 1$). Now we can relate (2.34) to (2.33) by introducing the activity coefficient, γ_i, defined by,

$$a_i \equiv \gamma_i x_i \tag{2.35}$$

to get

$$\mu_i(T, P, x_i) = \mu_i^0(T, P) + RT \log \gamma_i x_i. \tag{2.36}$$

The activity coefficient can be either $\gamma_i > 1$ or $\gamma_i < 1$.

Fugacity

For an ideal gas, the (molar) internal energy (e) is a function only of temperature (Joule's law), i.e., $e = e(T)$. And the enthalpy is $h = e + Pv$. Therefore using the equation of state ($Pv = RT$), one finds that enthalpy is also a function only of temperature, namely, $h = h(T)$. To get an equation for (molar) entropy, recall the relation (2.19) for a closed system,

$$ds = \frac{1}{T}\,de + \frac{P}{T}\,dv. \tag{2.37}$$

Using the definition of specific heat (c_v), this equation is translated into,

$$ds = \frac{1}{T}\frac{de}{dT}\,dT + \frac{P}{T}\,dv = \frac{c_v}{T}\,dT + \frac{P}{T}\,dv. \tag{2.38}$$

Integrating this equation, one obtains,

$$s(T, v) = s(T_0, v_0) + \int_{T_0}^{T} \frac{c_v}{T}\,dT + R \log \frac{v}{v_0}. \tag{2.39}$$

Inserting the relation $Pv = RT$, and using $\mu = h - Ts$, one has,

$$\mu^{\mathrm{id}}(P, T) = \mu^{\otimes}(P_0, T) + RT \log \frac{P}{P_0} \tag{2.40}$$

where $\mu^{\otimes}(P_0, T) = h(T) - Ts(P_0, T_0) - T\int_{T_0}^{T}(c_v/T)dT$. This equation indicates that the chemical potential (partial molar Gibbs free energy) of an ideal gas increases logarithmically with pressure. For a non-ideal gas, one can assume a similar relation, i.e.,

$$\mu(P, T) = \mu^{\otimes}(P_0, T) + RT \log \frac{f(P, T)}{P_0} \tag{2.41}$$

where $\mu^{\otimes}(T, P_0)$ is identical to the ideal gas. This is the definition of *fugacity*, f. The fugacity coefficient, ν, is often used to characterize the deviation from ideal gas,

$$f \equiv \nu P. \tag{2.42}$$

Obviously, $f \to P$ ($\nu \to 1$) as $P \to 0$.

The fugacity of a given fluid can be calculated from the equation of state. Let us integrate $\partial\mu/\partial P = v$ (v is the molar volume) to obtain

$$\mu(P, T) = \mu(P_0, T) + \int_{P_0}^{P} v(\xi, T)\,d\xi. \tag{2.43}$$

Now for an ideal gas,

$$\mu^{\mathrm{id}}(P, T) = \mu^{\mathrm{id}}(P_0, T) + \int_{P_0}^{P} v^{\mathrm{id}}(\xi, T)\,d\xi. \tag{2.44}$$

Subtracting (2.44) from (2.43), one has,

$$\mu(P, T) - \mu^{\mathrm{id}}(P, T) = \mu(P_0, T) - \mu^{\mathrm{id}}(P_0, T) + \int_{P_0}^{P} \left(v(\xi, T) - v^{\mathrm{id}}(\xi, T)\right) d\xi. \tag{2.45}$$

Noting that any fluid becomes an ideal gas at zero pressure, i.e., $\lim_{P_0 \to 0} \left(\mu(P_0, T) - \mu^{\mathrm{id}}(P_0, T)\right) = 0$ and from (2.40) and (2.41), $\mu(P, T) - \mu^{\mathrm{id}}(P, T) = RT \log(f(P, T)/P)$. Therefore one obtains

$$\log \frac{f(P, T)}{P} = \frac{1}{RT} \lim_{P_0 \to 0} \int_{P_0}^{P} \left(v(\xi, T) - v^{\mathrm{id}}(\xi, T)\right) d\xi. \tag{2.46}$$

The fugacity of a gaseous species at any T and P can be calculated from the equation of state (i.e., $v = v\,(P, T)$) using equation (2.46).

Non-ideal gas behavior occurs when the mutual distance of molecules becomes comparable to the molecular size, l_m. The mean distance of molecules in a fluid is given by $l = (v/N_{\mathrm{A}})^{1/3} = (RT/PN_{\mathrm{A}})^{1/3}$ where v is the molar volume. When $l/l_m \gg 1$, then a gas behaves like an ideal gas, whereas when $l/l_m \sim 1$, it becomes a non-ideal gas. For water, $l_m \sim 0.3$ nm and $l/l_m \sim 1$ at a pressure of $\sim$0.5 GPa (at 1673 K), whereas for hydrogen, $l_m \sim 0.1$ nm and one needs $\sim$15 GPa to see non-ideal behavior (at 1673 K) (Fig. 2.2b).

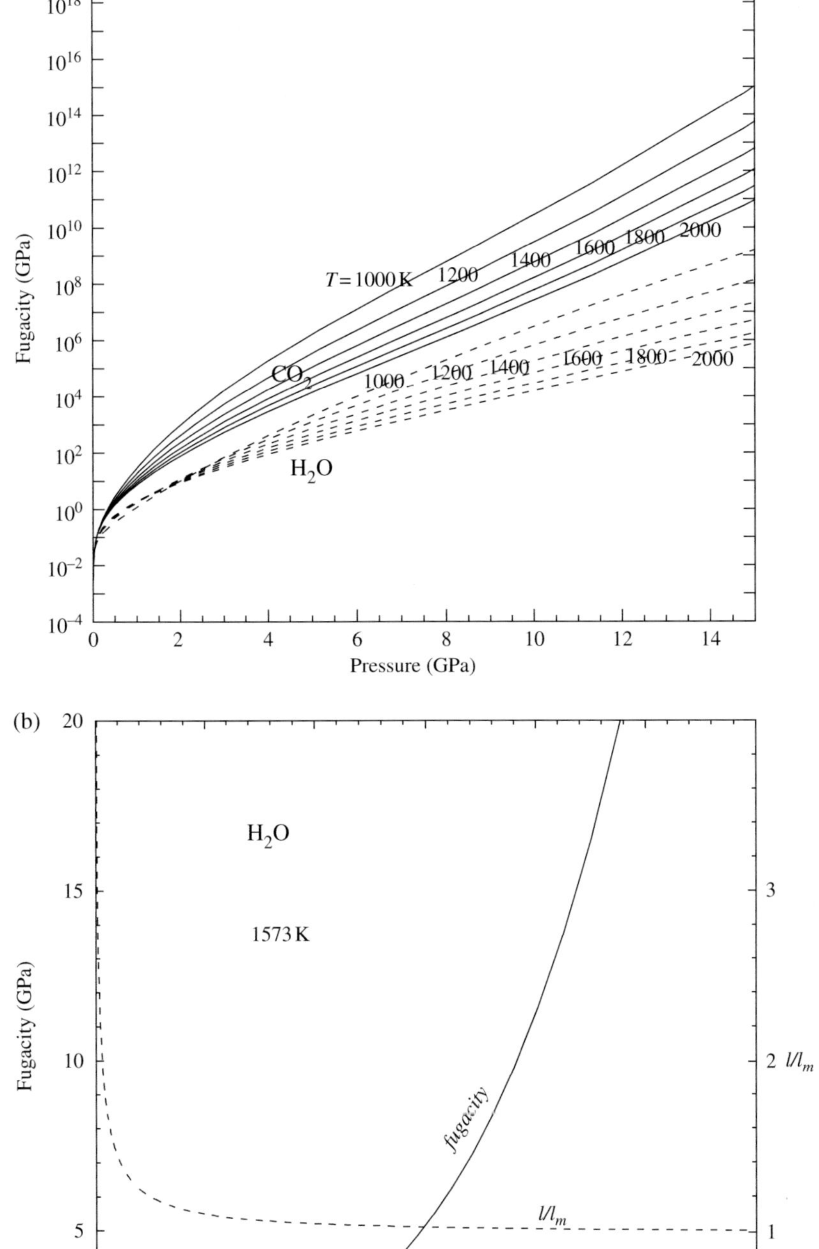

FIGURE 2.2 (a) Fugacity of water and carbon dioxide as a function of temperature (T) and pressure (P). Note the large deviation from ideal gas behavior at higher pressures (data for the equation of state are from Frost and Wood (1997a)). (b) A comparison of the fugacity of water (thin curve) with ideal gas behavior (thick curve). Significant deviation from the ideal gas behavior is seen when the mean distance of water molecules, l, is close to l_m (where l_m is the molecular size).

When a fluid behaves like an ideal gas whose equation of state is $Pv = RT$, then its fugacity defined by equation (2.41) is equal to its (partial) pressure. However, as fluids are compressed, their resistance to compression increases and the molar volume does not change with pressure as much as an equation of state of an ideal gas would imply. If the molar volume does not change with pressure, for example, then the fugacity will be an exponential function of pressure,

$$f(P,T) \approx P_0 \exp\left(\frac{v(T) \cdot P}{RT}\right). \tag{2.47}$$

Important examples are water and carbon dioxide. The fugacities of water and carbon dioxide can be calculated from the equations of state (Fig. 2.2). Water behaves like a nearly ideal gas up to ~0.3 GPa (at $T > 1000$ K), but its property starts to deviate from ideal gas behavior above ~0.5 GPa. At $P = 2$ GPa ($T = 1500$ K), for example, the fugacity of water is ~13 GPa and at $P = 3$ GPa ($T = 1500$ K), it is ~55 GPa. The large fugacity of water under high pressures means that water is chemically highly reactive under deep Earth conditions. The behavior of carbon dioxide is similar. When extrapolating laboratory data involving these fluids obtained at low pressures to higher pressures, one must take into account the non-ideal gas behavior of these fluids (see Chapter 10).

Problem 2.2

The equations of state of water and carbon dioxide are approximately given by the following formula (Frost and Wood, 1997b),

$$v(P,T) = \frac{RT}{P} - \frac{a(T)R\sqrt{T}}{(RT + b(T)P)(RT + 2b(T)P)} + b(T) + c(T)\sqrt{P} + d(T)P.$$

Where parameters (a, b, c and d) are functions of temperature, but not of pressure (see Table 2.1). Show that the fugacity of these fluids is given by

$$\log\frac{f(P,T)}{P} = \frac{a(T)}{b(T)RT\sqrt{T}}\log\frac{RT + b(T)P}{RT + 2b(T)P} + \frac{b(T)P}{RT} + \frac{2}{3}\frac{c(T)P\sqrt{P}}{RT} + \frac{d(T)P^2}{2RT}$$

and using the parameter values shown below calculate the fugacities of water and carbon dioxide for the conditions $0 < P < 20$ GPa and $1000 < T < 2000$ K.

TABLE 2.1 Equation of state parameters, $a(T)$, $b(T)$, $c(T)$ and $d(T)$ for water and carbon dioxide.

All parameters are assumed to be parabolic function of temperature: $m = m_0 + m_1 T + m_2 T^2$ (where m is a, b, c or d). Units listed in the table are for m_0. The unit for m_1 is $[m_1] = [m_0]/T$, and for m_2 is $[m_2] = [m_0]/T^2$. Units: a (m^6 Pa K$^{1/2}$ mol^{-1}), b (m^3), c (m^3 Pa$^{-1/2}$), d (m^3 Pa^{-1}).

	CO_2	H_2O
a	$a_0 = 5.373$	$a_0 = 5.395 \times 10$
	$a_1 = 5.6829 \times 10^{-3}$	$a_1 = -6.362 \times 10^{-2}$
	$a_2 = -4.045 \times 10^{-6}$	$a_2 = 2.368 \times 10^{-5}$
b	$b_0 = 4.288 \times 10^{-5}$	$b_0 = 2.7732 \times 10^{-5}$
	–	$b_1 = 2.0179 \times 10^{-8}$
	–	$b_2 = 9.2125 \times 10^{-12}$
c	$c_0 = -7.526 \times 10^{-10}$	$c_0 = -3.934 \times 10^{-10}$
	$c_1 = 1.1440 \times 10^{-13}$	$c_1 = 5.66 \times 10^{-13}$
	–	$c_2 = -2.485 \times 10^{-16}$
d	$d_0 = 3.707 \times 10^{-15}$	$d_0 = 2.186 \times 10^{-15}$
	$d_1 = 1.198 \times 10^{-20}$	$d_1 = -3.6836 \times 10^{-18}$
	$d_2 = -1.0464 \times 10^{-22}$	$d_2 = 1.6127 \times 10^{-21}$

Solution

Using equation (2.46), one obtains

$$\begin{aligned}\log\frac{f}{P} &= \frac{1}{RT}\int_0^P \left[-\frac{aR\sqrt{T}}{(RT+b\xi)(RT+2b\xi)} + b + c\sqrt{\xi} + d\xi\right] d\xi \\ &= \frac{1}{RT}\int_0^P \left\{\frac{a}{\sqrt{T}}\left[\frac{1}{(RT+b\xi)} - \frac{1}{(RT/2+b\xi)}\right] + b + c\sqrt{\xi} + d\xi\right\} d\xi\end{aligned}$$

and performing elementary integration and remembering that the parameters a, b, c and d are functions of temperature, T, one obtains

$$\log\frac{f(P,T)}{P} = \frac{a(T)}{b(T)RT\sqrt{T}}\log\frac{RT+b(T)P}{RT+2b(T)P} + \frac{b(T)P}{RT} + \frac{2}{3}\frac{c(T)P\sqrt{P}}{RT} + \frac{d(T)P^2}{2RT}.$$

Note that these gases behave like an ideal gas (i.e., $f \to P$) as $P \to 0$ as they should. At intermediate pressures ($P \sim 5$–20 GPa for water or carbon dioxide), the third term ($b(T)P/RT$) dominates and $f/P \approx \exp(b(T)P/RT)$ whereas at extreme pressures (i.e., $P \to \infty$), $f/P \approx \exp\left(d(T)P^2/2RT\right)$. The results of the fugacity calculation are shown in Fig. 2.2.

Problem 2.3

Derive equation (2.47).

Solution

Inserting the equation of state for an ideal gas, $Pv = RT$, into (2.46) and assuming $v(P, T) = v$ is constant, one has

$$\log\frac{f}{P} = \frac{1}{RT}\lim_{P_0\to 0}\left[v\cdot(P - P_0) - RT\log\frac{P}{P_0}\right]$$
$$= \frac{Pv}{RT} + \lim_{P_0\to 0}\log\frac{P_0}{P}.$$

Hence $f(P, T) \approx P_0 \exp(v(T)\cdot P/RT)$.

2.1.3. Chemical equilibrium: the law of mass action

Consider a chemical reaction,

$$\alpha_1 A_1 + \alpha_2 A_2 + \cdots = \beta_1 B_1 + \beta_2 B_2 + \cdots \qquad (2.48)$$

where A_i, B_i are chemical species and α_i, β_i are the stoichiometric coefficients (e.g., $H_2O = H_2 + \frac{1}{2}O_2$). At equilibrium for given T and P, the Gibbs free energy of the system must be a minimum with respect to the chemical reaction. When a chemical reaction described by (2.48) proceeds by a small amount, $\delta\lambda$, the concentration of each species will change as $\delta n_i = \nu_i\delta\lambda$. The condition for chemical equilibrium reads

$$\delta G = 0 = \sum_i\left(\frac{\partial G}{\partial n_i}\right)\delta n_i = \left(\sum_i \nu_i\mu_i\right)\delta\lambda \qquad (2.49)$$

and hence

$$\sum_i \nu_i\mu_i = 0 \qquad (2.50)$$

where (2.12) is used. Inserting the relation (2.32) into this equation, one finds,

$$\frac{x_{A1}^{\alpha 1}\cdot x_{A2}^{\alpha 2} - -}{x_{B1}^{\beta 1}\cdot x_{B2}^{\beta 2} - --} = \exp\left(-\frac{\alpha_1\mu_{A1}^0 + - - \beta_1\mu_{B1}^0 - -}{RT}\right). \qquad (2.51a)$$

This is called *the law of mass action* that relates the concentration of chemical species with their chemical potential. When the solution is not ideal (a case where solute atoms have a strong interaction with others), then equation (2.51a) must be modified to,

$$\frac{x_{A1}^{\alpha 1}\cdot x_{A2}^{\alpha 2}\cdots}{x_{B1}^{\beta 1}\cdot x_{B2}^{\beta 2}\cdots} = \frac{\gamma_{B1}^{\beta 1}\gamma_{B2}^{\beta 2}\cdots}{\gamma_{A1}^{\alpha 1}\gamma_{A2}^{\alpha 2}\cdots}\cdot\exp\left(-\frac{\alpha_1\mu_{A1}^0 + \cdots \beta_1\mu_{B1}^0 \cdots}{RT}\right)$$

where γ_i is the activity coefficient for the ith species defined by (2.35). These relations are frequently used in calculating the concentration of defects in minerals including point defects and trace elements.

Problem 2.4

Consider a chemical reaction $Ni + \frac{1}{2}O_2 = NiO$. The molar volumes, molar entropies and molar enthal-pies of each phase are given in Table 2.1. Calculate the oxygen fugacity for the temperature of $T = 1000 - 1600$ K and $P = 0.1$MPa $-$ 10 GPa when both Ni and NiO co-exist.

Solution

The law of mass action gives, $(f_{O_2}/P_0)^{1/2} = K(T, P)\cdot a_{NiO}/a_{Ni}$. When both Ni and NiO exist, then $a_{Ni} = a_{NiO} = 1$, so that

$$(f_{O_2}/P_0)^{1/2} = \exp\left(\left(\mu_{NiO}^0 - \mu_{Ni}^0 - \frac{1}{2}\mu_{O_2}^0\right)\Big/RT\right).$$

Now writing the temperature and pressure dependence of chemical potential explicitly and remembering that the pressure dependence of chemical potential is included in the fugacity, one has

$$\left(\frac{f_{O_2}}{P_0}\right)^{1/2} = \exp\left(\frac{e_{NiO} - e_{Ni} - \frac{1}{2}e_{O_2} + P(v_{NiO} - v_{Ni}) - T\left(s_{NiO} - s_{Ni} - \frac{1}{2}s_{O_2}\right)}{RT}\right).$$

Note that the enthalpy of the formation of NiO from the elements at zero pressure (~0.1 MPa) is $h_{NiO}^0 = e_{NiO} - e_{Ni} - \frac{1}{2}e_{O_2}$. Inserting the values of thermodynamic parameters from Table 2.2, one obtains the fugacity versus T–P relations shown in Fig. 2.3 (for simplicity, we assumed constant molar volumes for solid phases). Note that the oxygen fugacity increases with pressure.

Problem 2.5*

Discuss how the fugacity of water is controlled in a system that contains water as well as other materials such as olivine, a metal that modifies the oxygen fugacity (Hobbs, 1984; Karato *et al.*, 1986). For simplicity, assume that all the gaseous phases behave like an ideal gas.

TABLE 2.2 Thermodynamic properties of various oxides and metals relevant to the oxygen fugacity buffer.

v ($\times 10^{-6}$ m^3/mol): molar volume, h^0(kJ/mol): molar enthalpy of formation from elements, s (J/mol K): molar entropy. All quantities are at room pressure and $T = 298$ K. Molar volumes of some materials change with temperature and pressure as well as with phase transformations. However, these changes are small relative to the difference in molar volume of metals and their oxides.

	v	h^0	s
O_2	24 798	0	205.15
Fe	7.09	0	27.28
FeO	12.00	−272.04	59.80
Ni	6.59	0	29.87
NiO	10.97	−239.74	37.99
Mo	9.39	0	28.66
MoO_2	19.58	−587.85	50.02

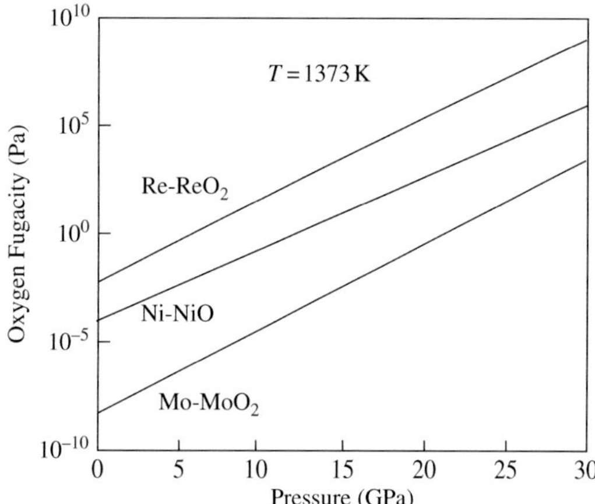

FIGURE 2.3 Oxygen fugacities corresponding to several metal-oxide buffers.

Solution

If water alone exists, then the chemical reaction that will occur is

$$H_2O = H_2 + \frac{1}{2}O_2. \qquad (1)$$

The law of mass action demands

$$f_{H_2O} P_0^{1/2} / f_{H_2} f_{O_2}^{1/2} = K_1(T, P). \qquad (2)$$

Now the total pressure of the gas must be the same as the given pressure, P, so that (assuming ideal gas behavior)

$$f_{H_2O} + f_{H_2} + f_{O_2} = P. \qquad (3)$$

In the case where only water is present, then the dissociation of one mole of water produces one mole of hydrogen and 1/2 mole of oxygen, so $f_{H_2} = 2f_{O_2}$. Inserting this into the equation for the law of mass action, and noting that one has

$$f_{H_2O} + \frac{3}{2^{2/3}} f_{H_2O}^{2/3} P_0^{1/3} K_1^{-2/3}(T, P) = P \qquad (4)$$

where for simplicity, we assume that all the gasses are ideal, so that all the fugacity coefficients are 1.This equation gives the fugacity of water when only water is present. At high pressures, exceeding ~1 GPa, the second term in this equation is small (confirm this yourself), so that $f_{H_2O} \approx P$, but when significant dissociation occurs (at lower pressures), then the water fugacity will be lower.

Now consider a case where some other species are present that also react with oxygen, hydrogen etc. For example, let us consider a case where material A (e.g., Fe) reacts with oxygen to form another mineral $A_x O_y$ (e.g., Fe_2O_3), namely,

$$xA + \frac{y}{2}O_2 = A_xO_y. \qquad (5)$$

The corresponding law of mass action gives

$$\frac{a_A^x}{a_{A_xO_y}} \frac{f_{O_2}^{y/2}}{P_0^{y/2}} = K_2(T, P). \qquad (6)$$

This means that when a substance A is present, oxygen atoms provided by the dissociation reaction of water react with A to form $A_x O_y$. Then (if enough A is present) the oxygen fugacity is controlled by reaction (5) and should be $f_{O_2}/P_0 = K_2^{2/y}(T, P)$. Inserting this into (2) and with (3) (assuming ideal gas behavior), one finds,

$$f_{H_2O} \approx \frac{P - P_0 K_2^{2/y}}{1 + K_1^{-1} K_2^{-1/y}}, \quad f_{H_2} \approx \frac{P - P_0 K_2^{2/y}}{1 + K_1 K_2^{1/y}}. \qquad (7)$$

It follows that, when the oxygen fugacity buffered by the reaction $xA + \frac{y}{2}O_2 = A_xO_y$ is low, i.e., $K_2^{1/y} K_1 \ll 1$, then $f_{H_2} \approx P \gg f_{H_2O}$, whereas when the oxygen fugacity is high ($K_2^{1/y} K_1 \gg 1$), then $f_{H_2O} \approx P - P_0 K_2^{2/y} \gg f_{H_2}$.

2.1.4. Phase transformations: the Clapeyron slope, the Ehrenfest slope

For a given chemical composition, a stable phase at a given pressure and temperature is the phase for which the Gibbs free energy is the minimum. When a material with a given chemical composition can assume several *phases*, then as the P, T conditions change, the phase with the minimum Gibbs free energy may change from one to another. In these cases, the stable phase for a material changes with these variables, and a *phase transformation* occurs. They include α to β transformation in quartz, order–disorder transformation in plagioclase, α (olivine) to β (wadsleyite) transformation in $(Mg, Fe)_2SiO_4$ and α (bcc) to ε (hcp) transformation in iron.

A phase transformation may be classified into two groups. In some cases, a phase transformation involves a change in the first derivatives of Gibbs free energy (e.g., $(\partial G/\partial T)_{P,n_i} = -S$ or $(\partial G/\partial P)_{T,n_i} = V$, where S is entropy and V is volume). This type of phase transformations is called *the first order phase transformation*. Many phase transformations in silicates and metals are of this type. In these cases, there is a change in density (molar volume) and heat is either released or absorbed upon the phase transformation (due to the change in molar entropy; recall that $T\,dS$ is the latent heat). Another is the case where there is no change in the molar volume or entropy (the first derivatives of Gibbs free energy), but changes occur only in the second derivatives. This type of phase transformations is referred to as *the second order phase transformation*. Many of the structural phase transformations belong to this class. The α to β transformation of quartz is close to this type and many structural transformations of perovskite belong to this type (e.g., GHOSE, 1985). This type of phase transformation does not involve changes in density or in entropy (hence no latent heat). Note that although there is no change in density in these types of transformation, there is a change in the elastic constants and thermal expansion (the second derivatives of Gibbs free energy), and therefore there must be a change in seismic wave velocities associated with a second-order phase transformation.

Schematic diagrams showing the change in free energy associated with a first- and a second-order transformation are shown in Fig. 2.4. In the case where a first-order transformation is considered, a material can assume two possible states. When the free energy of one phase is lower than the other, then a phase with lower free energy is more stable. Therefore if the transition from one state to the other is kinetically possible, then all the materials will transform to a phase with the lowest free energy. Note, however, that this transition involves kinetic processes over a local maximum of free energy, and therefore the transformation takes a certain time to be completed. Consequently, a *metastable phase* can exist in the case of a first-order transformation when the kinetics involved are sluggish for a given time-scale. Examples include the presence of diamond at the Earth's surface (the stable phase for carbon at the Earth's surface is graphite, so we would not have diamond if the presence of everything on Earth were controlled by thermodynamic stability), and the possible presence of metastable olivine in cold regions of subducting slabs (see Chapters 17 and 20). The situation is different for a

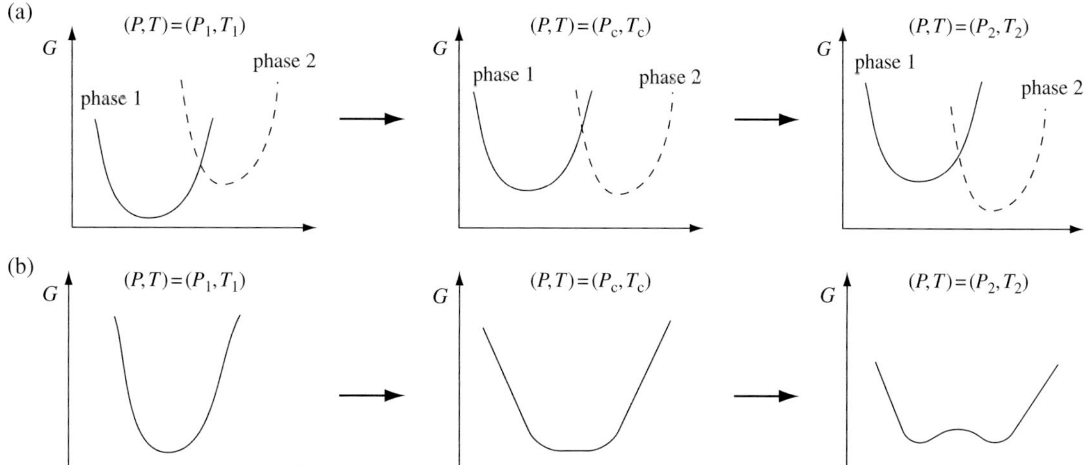

FIGURE 2.4 Free energy of a system having (a) first-order and (b) second-order phase transformations. The horizontal axis represents atomic configuration (atomic positions, crystal structure).

Box 2.2 The Gibbs phase rule

The state of a system containing c-components and p-phases can be specified by $(c-1)p+2$ variables. Two are T and P, and for each p-phase, one needs to specify the fraction of phases that requires $c-1$ variables. Now these p-phases are in chemical equilibrium, and therefore the chemical potential of each component in p-phases must be equal,

$$\mu_1^1 = \mu_2^1 = \cdots = \mu_p^1$$
$$\mu_1^2 = \mu_2^2 = \cdots = \mu_p^2$$
$$- - - - - - - - - -$$
$$\mu_1^c = \mu_2^c = \cdots = \mu_p^c$$

where μ_j^i is the chemical potential of ith component in the jth phase. This means that there are $c(p-1)$ constraints. Therefore the degree of freedom of the system, f, is

$$f=(c-1)p+2-c(p-1)=c-p+2$$

This relation is referred to as the Gibbs phase rule.

second-order transformation that occurs due to the instability of one phase. For a second order transformation, no metastable phase can exist.

A couple of points may be noted. According to the Gibbs phase rule (Box 2.2), for a material with c components, there exist $f=c-p+2$ (c, the number of components; p, the number of phases) degrees of freedom at given P and T. For example, for a single-component system ($c=1$), if three phases co-exist ($p=3$), then there are zero degrees of freedom ($f=0$). That is, there is only one set of T and P at which three phases co-exist. Similarly, when two phases co-exist in a single-component system, then there is one degree of freedom ($f=1-2+2=1$): that is, if T is changed then so is P. Therefore when two phases co-exist in a single-component system, the temperature and pressure must be related, $P=P(T)$. The slope of this curve, $(dP/dT)_{eq}$, for the first-order phase transformations is referred to as the *Clapeyron slope*.[2]

[2] In some literature, $(dT/dP)_{eq}$, is called the Clapeyron slope. It does not matter which definition one uses as far as one defines it clearly.

The situation is different for a multi-component system. Let us consider a two-component system. If two phases co-exist in such a system, there are two degrees of freedom $f=2+2-2=2$. That is T and P can be modified independently. This means that there is a space in a T–P plane where the two phases co-exist. For this reason, a multi-component system usually shows a gradual phase transition: one phase changes to another within a certain T–P range. For example, at a given pressure, melting in a multi-component system begins at a certain temperature (the solidus) and completes at another temperature (the liquidus). At a temperature between the solidus and the liquidus, solid and melt co-exist. This is why melting in Earth occurs usually as *partial melting*.

Let us derive an equation for the Clapeyron slope in terms of other thermodynamic parameters. Consider a boundary between two phases for a single-component system (univariant transformation). Along the boundary the Gibbs free energy of two phases must be identical, namely,

$$G_1 = G_2. \tag{2.52}$$

Now take the derivative *along the boundary* (the suffix n_i is omitted because we consider a single-component system) to find $dG_1 = dG_2$ along the boundary)

$$\left(\frac{\partial G_1}{\partial T}\right)_P dT + \left(\frac{\partial G_1}{\partial P}\right)_T dP = \left(\frac{\partial G_2}{\partial T}\right)_P dT + \left(\frac{\partial G_2}{\partial P}\right)_T dP. \tag{2.53}$$

Using the identities $(\partial G/\partial T)_P = -S$ and $(\partial G/\partial P)_T = V$, one obtains

$$\left(\frac{dP}{dT}\right)_{eq} = \frac{S_1-S_2}{V_1-V_2} \tag{2.54}$$

where the suffix "eq" is used to clearly indicate that the derivative is taken along the equilibrium boundary. This relation is called the *Clapeyron Clausius relation*.

Similar relations can be derived for a second-order transformation (Problem 2.6; e.g., Callen, 1960),

$$\left(\frac{dP}{dT}\right)_{eq} = \frac{\alpha_1-\alpha_2}{1/K_1-1/K_2} = \frac{C_1-C_2}{T(\alpha_1-\alpha_2)} \tag{2.55}$$

where $\alpha_{1,2}$ is the thermal expansion of 1, 2 phase, $K_{1,2}$ is the (isothermal) bulk modulus of 1, 2 phase, and $C_{1,2}$ is the specific heat (at constant pressure) of 1, 2 phase. This relation was derived by Ehrenfest (1933) and hence should be called *the Ehrenfest relation*.

Problem 2.6*

Derive equation (2.55).

Solution

For a second-order transformation, the first derivatives of the Gibbs free energy are identical for the two co-existing phases, $V_1 = V_2$, $S_1 = S_2$. Therefore, along the boundary, the following relations must be satisfied,

$$\left(\frac{\partial V_1}{\partial T}\right)_P dT + \left(\frac{\partial V_1}{\partial P}\right)_T dP = \left(\frac{\partial V_2}{\partial T}\right)_P dT + \left(\frac{\partial V_2}{\partial P}\right)_T dP \tag{1a}$$

$$\left(\frac{\partial S_1}{\partial T}\right)_P dT + \left(\frac{\partial S_1}{\partial P}\right)_T dP = \left(\frac{\partial S_2}{\partial T}\right)_P dT + \left(\frac{\partial S_2}{\partial P}\right)_T dP. \tag{1b}$$

These two equations can be combined to give,

$$\begin{bmatrix} \dfrac{\partial V_1}{\partial T} - \dfrac{\partial V_2}{\partial T} & \dfrac{\partial V_1}{\partial P} - \dfrac{\partial V_2}{\partial P} \\ \dfrac{\partial S_1}{\partial T} - \dfrac{\partial S_2}{\partial T} & \dfrac{\partial S_1}{\partial P} - \dfrac{\partial S_2}{\partial P} \end{bmatrix} \begin{bmatrix} dT \\ dP \end{bmatrix} = 0. \tag{2}$$

In order for this equation to have a non-trivial solution, the following relation must be satisfied,

$$\begin{vmatrix} \dfrac{\partial V_1}{\partial T} - \dfrac{\partial V_2}{\partial T} & \dfrac{\partial V_1}{\partial P} - \dfrac{\partial V_2}{\partial P} \\ \dfrac{\partial S_1}{\partial T} - \dfrac{\partial S_2}{\partial T} & \dfrac{\partial S_1}{\partial P} - \dfrac{\partial S_2}{\partial P} \end{vmatrix} = 0. \tag{3}$$

Therefore,

$$\left(\frac{\partial V_1}{\partial T} - \frac{\partial V_2}{\partial T}\right)^2 + \left(\frac{\partial V_1}{\partial P} - \frac{\partial V_2}{\partial P}\right)\left(\frac{\partial S_1}{\partial T} - \frac{\partial S_2}{\partial T}\right) = 0. \tag{4}$$

where the Maxwell relation (equation (2.15), $\partial S/\partial P = -\partial V/\partial T$) was used. Using the definitions of thermal expansion ($\alpha_{\text{th}} \equiv (1/V)(\partial V/\partial T)_P$), (isothermal) bulk modulus ($K \equiv -V(\partial P/\partial V)_T$) and the specific heat at constant pressure ($C \equiv T(\partial S/\partial T)_P$) and the fact that $V_1 = V_2$, one finds,

$$(\alpha_1 - \alpha_2)^2 - \frac{1}{T}\left(\frac{1}{K_1} - \frac{1}{K_2}\right)(C_1 - C_2) = 0. \tag{5}$$

Now solving equations (1a) and (1b), the slope of the phase boundary in the T–P space is given by,

$$\begin{aligned} \left(\frac{dP}{dT}\right)_{\text{eq}} &= \frac{\partial V_1/\partial T - \partial V_2/\partial T}{\partial V_1/\partial P - \partial V_2/\partial P} = \frac{\partial S_1/\partial T - \partial S_2/\partial T}{\partial S_1/\partial P - \partial S_2/\partial P} \\ &= \frac{\partial S_1/\partial T - \partial S_2/\partial T}{\partial V_1/\partial P - \partial V_2/\partial T} = \frac{\alpha_1 - \alpha_2}{1/K_1 - 1/K_2} \\ &= \frac{C_1 - C_2}{T(\alpha_1 - \alpha_2)}. \end{aligned} \tag{6}$$

2.1.5. Phase diagrams

For a given chemical composition with temperature (T) and pressure (P), there exist a certain number of phases determined by the Gibbs phase rule. A diagram showing the stable phases on a certain parameter space is referred to as a *phase diagram*. In constructing a phase diagram, one usually fixes the chemical composition, i.e., the system is assumed to be closed. For a *closed system*, the stability of each phase is solely determined by temperature and pressure. Fig. 2.5 illustrates some of the phase diagrams for binary (two-component) systems.

A phase diagram is usually constructed based on direct experimental studies. However, because the stability of each phase is determined by the chemical potential, a phase diagram can be constructed theoretically if the dependence of the chemical potential of each phase on T, P and composition (n_i) is known, i.e.,

$$\mu_A^i(T, P, n_i) = \mu_B^i(T, P, n_i) \tag{2.56}$$

where $\mu_{A,B}^i$ is the chemical potential of the ith species in phase A or B. Consider a single-component system, where a material (with a fixed composition) can assume two phases (A or B). Then the equilibrium temperature and pressure are determined by

$$e_A - Ts_A + Pv_A = e_B - Ts_B + Pv_B \tag{2.57}$$

where $e_{A,B}$, $s_{A,B}$ and $v_{A,B}$ are molar internal energy, entropy and volume of phase A and B respectively. When all the parameters ($e_{A,B}$, $s_{A,B}$ and $v_{A,B}$) are independent of temperature and pressure (this is a good approximation for liquids and solids for a small range of temperature and pressure), the phase boundary can be calculated from $e_{A,B}$, $s_{A,B}$ and $v_{A,B}$ as,

$$P = -\frac{e_A - e_B}{v_A - v_B} + \frac{s_A - s_B}{v_A - v_B} T. \tag{2.58}$$

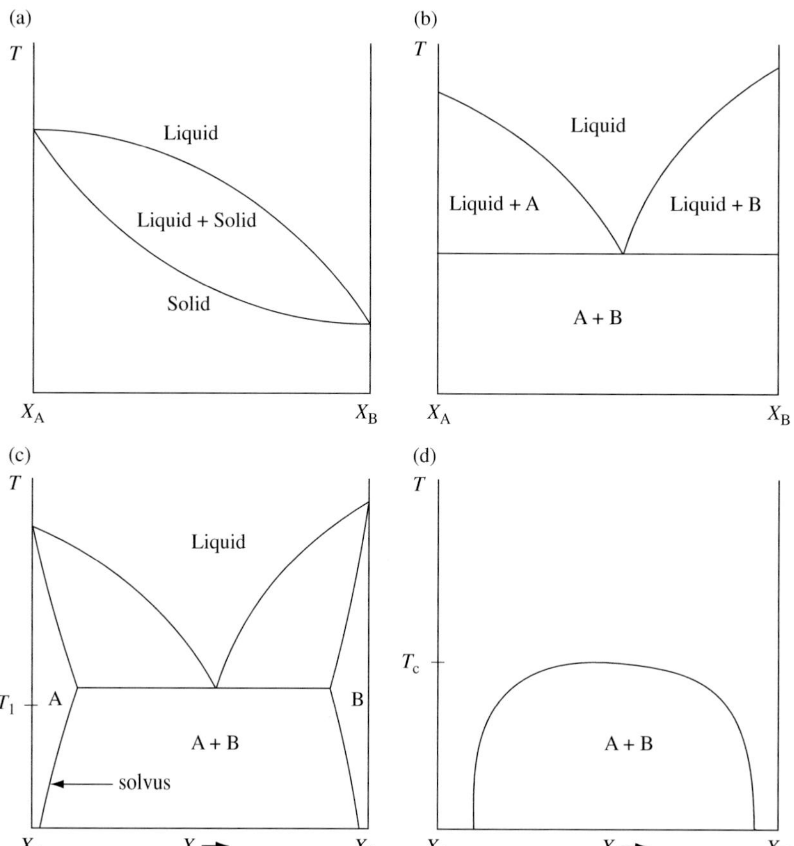

FIGURE 2.5 Typical phase diagrams for a binary system involving melting reactions. (a) A case for a solid solution, (b) a case for eutectic behavior, (c) a case for eutectic behavior with limited mutual solubility, (d) a case for two phases that have limited solubility below a critical temperature, T_c, but show complete mixing above T_c.

Problem 2.7

Calculate the phase boundaries as a function of temperature and pressure for the olivine → wadsleyite → ringwoodite (in Mg_2SiO_4) phase transformation using the values of thermodynamic parameters listed in the Table 2.3.

Solution

The phase boundaries for these two reactions can be calculated from (2.58). To calculate the thermodynamic parameters for wadsleyite → ringwoodite, use the relation $\Delta X_{\text{wad}\rightarrow\text{ring}} = \Delta X_{\text{oli}\rightarrow\text{ring}} - \Delta X_{\text{oli}\rightarrow\text{wad}}$. $P_{\text{oli}\rightarrow\text{wad}}(\text{GPa}) = 8.57 + 0.004\,27 \times T(\text{K})$ and $P_{\text{wad}\rightarrow\text{ring}}(\text{GPa}) = 12.24 + 0.006\,12 \times T(\text{K})$.

TABLE 2.3 Some thermodynamic parameters related to phase transformations (from NAVROTSKY (1994)).

Units: Δe (kJ/mol), Δs (J/mol K), Δv ($\times 10^{-6}$ m^3/mol), dP/dT (MPa/K)

	Δe	Δs	Δv	dP/dT
Mg_2SiO_4				
olivine → wadsleyite	27.1	−9.0	−3.16	2.8
olivine → ringwoodite	39.1	−15.0	−4.14	3.6
Fe_2SiO_4				
olivine → wadsleyite	9.6	−10.9	−3.20	3.4
olivine → ringwoodite	3.8	−14.0	−4.24	3.3
$MgSiO_3$				
pyroxene → garnet	35.1	−2.0	−2.83	0.71
pyroxene → ilmenite	59.4	−15.5	−4.94	3.3
ilmenite → perovskite	51.1	6.0	−1.89	−3.2
garnet → perovskite	75.0	−7.5	–	

Solid-solution, eutectic melting

When there are two or more components in the system, there are additional degrees of freedom by which the chemical potential is controlled. Consequently, the phase diagram depends on how the chemical potential of each phase varies with the composition. For simplicity let us consider a two-component system. The component $i = 1$ and 2 may assume various phases such as solid and liquid. Two cases may be distinguished. One is the case in which the two components mix well in both the solid and liquid phases. In this case, the contribution from the configurational entropy is similar for both the solid and liquid phases, and the free energy of each phase changes with composition similarly following the compositional dependence of internal energy, entropy and the molar volume. The phase diagram corresponding to this case is shown in Fig. 2.5a. In such a case, solid A and B are said to form a *solid-solution*. Another is the case where mixing occurs only in the liquid phase. In this case, the contribution from the configurational entropy is important only in the liquid phase. Consequently, the free energy of the liquid becomes low in the intermediate concentration of a given species, and therefore the solidus of the system is reduced significantly at intermediate compositions (Fig. 2.5b). Melting behavior due to this type of mixing property is called *eutectic melting*.

The solid-solution type behavior is observed when the solid phases involved have similar properties (crystal structure and chemical bonding). The examples include magnesiowüstite (MgO and FeO), olivine (fayalite Fe_2SiO_4 and forsterite Mg_2SiO_4), plagioclase feldspar (albite $NaAlSi_3O_8$ and anorthite $CaAl_2Si_2O_8$). In all of these cases, ions that have similar ionic radii are incorporated as a solid-solution in the solid phase. If the ionic radii are largely different then the solubility in the solid phase is limited and the eutectic behavior will occur. This is the case for the MgO–CaO, $MgSiO_3$–Mg_2SiO_4 systems.

Solvus

Let us now consider a two-component system in which there is a finite solubility of each phase into another in the solid state as well as in the liquid state. First, consider a system in which mixing is complete in the liquid state and a small degree of mixing also occurs in the solid state. In such a case, a phase diagram needs to be modified. A solid phase always contains, in this case, a finite amount of secondary component so that there is a modification to the phase diagram toward the end-member component representing the effects of finite solubility (Fig. 2.5c). A phase diagram for a silicate and water system at high T is an important example. Consider the equilibrium at temperature T_1 below the eutectic point. When the amount of B is small, then the only phase that exists is a phase A that contains a small amount of B. According to the Gibbs phase rule, in such a case we have $f = c - p + 2 = 3$, that is this phase, i.e., phase A with a small amount of B can exist for a range of T, P and composition. When the amount of B in the system increases, then at a certain point, the phase A can no longer dissolve all the component B and there will be two phases ($X_2 > X > X_1$). The same thing happens from another side, namely the B-phase side. Consequently the domain is divided into one-phase domains in each side of the phase diagram (A-rich or B-rich, $X < X_1, X > X_2$) and a two-phase domain ($X_2 > X > X_1$). In the latter domain, there are two phases that co-exist, and therefore the degree of freedom is $f = 2$. Consequently, if temperature and pressure are prescribed, then the chemical compositions of a material must be fixed. The boundaries between the one- and two-phase regions correspond to the solubility of each species into another.

Usually the solubility of another phase into a given phase increases with temperature, so the boundaries separating two one-phase domains will become closer as temperature rises. These boundaries are often referred to as a *solvus*. When mutual solubility is large, then at a certain temperature below the melting temperature, the two solvus curves merge. Above this critical temperature (T_c) the two phases mix completely. Above this temperature mixing occurs both in solid-state and liquid-state, and therefore the phase diagram above this temperature should look like that of a solid-solution (Fig. 2.5d). Obviously the solvus curves or any of the boundaries on a phase diagram also depend on pressure. The temperature and pressure dependence of solvus curves for various combinations of minerals is used as petrological barometers and/or thermometers (e.g., Wood and Fraser, 1976).

Effects of non-stoichiometry: a phase diagram for an open system

The phase diagram considered above assumes that the chemical composition of each phase is independent of T and P except in cases where finite solubility of one component occurs in each phase. For example, a phase diagram for (Mg, Fe)O is usually constructed assuming that this is a two-component system (MgO and FeO) assuming that the material exchange occurs only through Mg ⇔ Fe keeping the number of atoms in the (solid) system constant. This is not strictly true when the system under consideration is *open* (i.e., when the system exchanges materials with the surrounding system). In a system like XO (X = Mg etc.; O, oxygen), the ratio of the number of atoms of X and O (*stoichiometry*) can deviate from what the chemical formula would indicate. The deviation from the formal chemical formula is referred to as *non-stoichiometry*. When non-stoichiometry occurs in an ionic crystal, then charge balance must be maintained by creating another type of charged species. This is usually done by creating point defects or by incorporating another species. One example is an Fe-bearing mineral such as olivine ($(Mg, Fe)_2SiO_4$) that can have non-stoichiometry caused by a change of valence state of iron ($Fe^{2+} \Leftrightarrow Fe^{3+}$). In this case the charge balance is maintained by the change in the concentration of M-site vacancies that have a negative effective charge (see Chapter 5). Another example is a combined substitution such as $Al^{3+} + Fe^{3+} \Leftrightarrow Si^{4+} + Fe^{2+}$. In these cases, an additional variable such as oxygen fugacity or the activity of Al_2O_3 is needed to specify the degree of non-stoichiometry. The degree of non-stoichiometry in the former type of processes is usually small ($\sim 10^{-4}$ or less in olivine) but can be large in an Fe-rich compound such as FeO (in FeO the non-stoichiometry is ~8%, i.e., $Fe_{0.92}O$). Even in cases where the degree of non-stoichiometry is small, its effects on physical properties can be important. In a binary material (such as XO), the oxygen fugacity is used as an additional variable in constructing a phase diagram (Nitsan, 1974). (In a ternary system such as Mg_2SiO_4, the stoichiometry is defined by two ratios (i.e., Mg/O, Mg/Si), and hence one needs two additional parameters to completely describe the chemical state of the system. Both oxygen fugacity and the oxide activity must be specified in such a case.)

To illustrate this point, let us consider a phase diagram of Fe–O. Iron (Fe) can assume three different valence states dependent on oxygen fugacity, f_{O_2}: metallic iron Fe^0 at low oxygen fugacity, ferrous iron Fe^{2+} at intermediate oxygen fugacity and ferric iron Fe^{3+} at high oxygen fugacity, see Fig. 2.6. Each species (Fe^0, Fe^{2+} and Fe^{3+}) has a different chemical character and therefore the stable phases at different conditions will depend on the oxygen fugacity. Consequently in an Fe–O system, four compounds may be present dependent upon the oxygen fugacity, i.e., metallic iron at low oxygen fugacity, wüstite (FeO) and magnetite (Fe_3O_4)

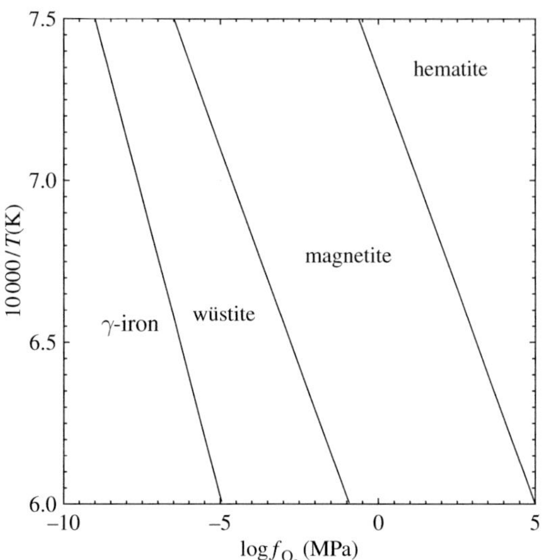

FIGURE 2.6 A phase diagram of Fe-O at 0.1 MPa.

at the intermediate oxygen fugacity and hematite (Fe_2O_3) at high oxygen fugacity. Iron in wüstite is mostly ferrous iron (Fe^{2+}), whereas in magnetite there are both ferrous iron (Fe^{2+}) and ferric iron (Fe^{3+}) and finally at high oxygen fugacity all iron changes to ferric iron (Fe^{3+}). The stability of iron-bearing olivine can be analyzed in a similar way. Olivine accepts ferrous iron but not ferric iron (ferric iron is present in olivine but only with a very small amount, ~1–10 ppm, as point defects) and therefore it is stable only within a certain range of oxygen fugacity that is determined by the stability of wüstite (FeO).

A somewhat different phase diagram applies when a given mineral favors ferric iron more than ferrous iron. In such a case, even at an oxygen fugacity in which iron would occur as FeO, iron in that mineral can be ferric iron. In some cases, the stability field of the ferric iron-bearing phase expands to a much lower oxygen fugacity, and in such a case a mineral containing ferric iron could co-exist with metallic iron. An important case is silicate perovskite that favors ferric iron, and the formation of silicate perovskite from ringwoodite leads to the formation of metallic iron (e.g., Frost *et al.*, 2004).

2.2. Some comments on the thermodynamics of a stressed system

In the usual treatment of thermodynamics, the energy change of a system due to mechanical work is treated assuming hydrostatic stress. That is $dW = -P\,dV$, i.e., the work done against pressure. An extension of such a treatment to non-hydrostatic stress conditions is

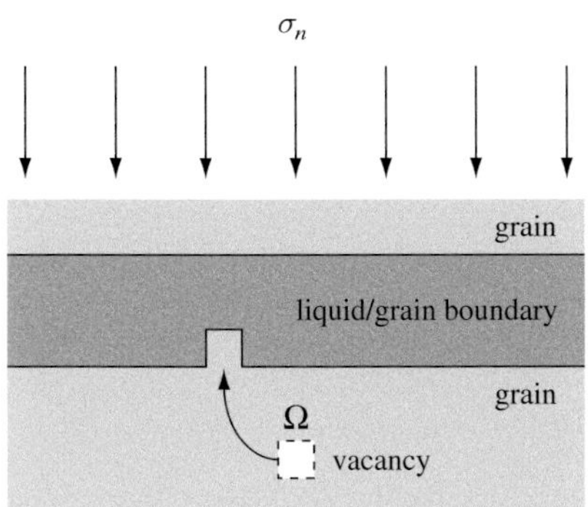

FIGURE 2.7 The process of formation of a vacancy (or dissolution of atoms) in a stressed solid at a grain boundary (or crystal–liquid interface).

needed when the deformation of materials is considered. The strain energy of stressed solid (per unit volume) is usually written as (e.g., LANDAU and LIFSHITZ, 1959; SCHMALZRIED, 1995)

$$dW = \sum_{ij} \sigma_{ij}\, d\varepsilon_{ij} \tag{2.59}$$

where σ_{ij} is stress and $d\varepsilon_{ij}$ is the strain increment. However, this expression does not capture the energy difference that plays an important role in diffusional (or pressure-solution) creep (Chapter 8). According to this equation, the energy of the system is independent of the orientation of the interface at which the reaction occurs, but the free energy change associated with a particular reaction under non-hydrostatic stress depends on the process that occurs at the crystal surface (or grain boundary). Consequently, the free energy change may depend on the orientation of the interface with respect to the applied stress at which the reaction occurs. To illustrate this point, let us consider the formation/destruction of a vacancy (dissolution/precipitation of a stressed solid) into a grain boundary (or a liquid) (Fig. 2.7). Consider the formation of a vacancy in a crystal. The formation of a vacancy is made by the removal of an atom from inside a crystal to the surface. Therefore, there is a volume expansion, and hence the formation free energy of a vacancy is often written as $G^* = E^* - TS^* + P\Omega$, where Ω is the volume of a vacancy (atom). However, to be more specific, the energy associated with vacancy formation depends on where it occurs. To see this, let us recall that vacancy creation (destruction) occurs mostly at grain boundaries. When an atom is removed from the interior of a crystal to its grain boundary, one needs to do work against *normal stress* (only normal stress comes into play because the displacement is normal to the boundary), thus

$$G^*(\boldsymbol{x}) = E^* - TS^* + \sigma_n(\boldsymbol{x}) \cdot \Omega = G_0^* + \sigma_n'(\boldsymbol{x}) \cdot \Omega \tag{2.60}$$

where E^* is the internal energy including the strain energy (e.g., (2.59)), S^* is the entropy (vibrational entropy), $\sigma_n(\mathbf{x})$ is the normal stress on the grain boundary that depends on the orientation of the boundary with respect to the applied stress, $G_0^* = E^* - TS^* + P\Omega$ and $\sigma_n'(\boldsymbol{x})$ is the deviatoric component of the normal stress ($\sigma_n'(\mathbf{x}) = \sigma_n(\mathbf{x}) - P$). Equation (2.60) defines the Gibbs free energy of a solid under stress near a boundary. Note that $G^*(\boldsymbol{x})$ is heterogeneous and there is no absolute equilibrium state under the deviatoric stress. The free energy defined by (2.60) represents only the *local equilibrium*. Also note that although the internal energy (E^*) depends on stress, its dependence is of second order (equation (2.59)) and the dominant role of stress is through the $\sigma_n(\boldsymbol{x}) \cdot \Omega$ (or $\sigma_n'(\boldsymbol{x}) \cdot \Omega$) term.

There have been extensive debates concerning the exact formula for the free energy of stressed solids (see e.g., JOHNSON and SCHMALZRIED, 1992; KAMB, 1961; PATERSON, 1973; SHIMIZU, 1992). However, except for the above well-defined case, many controversies are around the strain energy term. The contribution to this term in an actual stressed solid is usually dominated by the energy due to heterogeneous dislocation density rather than the strain energy of dislocation-free solids (e.g., POIRIER and GUILLOPÉ, 1979), and the relevance of such a controversy to real geological issues (or physical processes) is questionable.

2.3. Thermodynamics of irreversible processes

A brief account of the thermodynamics of irreversible processes is given in this section. From the microscopic point of view, *elastic deformation* occurs when atomic displacement from stable positions is limited to the small vicinity of stable positions at which the material assumes minimum free energy. When the force that causes deformation is removed, atoms move back to their original positions. Therefore elastic deformation is *reversible* (and *instantaneous*). However, statistical mechanics shows that, at finite temperatures, there is a non-zero probability that an atom can assume a position away from the equilibrium position. This random motion of atoms is referred to as *fluctuation*. Consequently, an atom could sometimes move to a new stable position over the potential barrier with a certain probability (see section 2.3). This process can be assisted by the applied stress. Once this happens, then upon the removal of the stress, the atomic configuration of the material would not revert to the original configuration instantaneously: thus the deformation becomes *irreversible* (and *time dependent* because the process now depends on the *probability* of atomic jumps).

Plastic deformation occurs due to irreversible processes. Therefore an understanding of irreversible processes is critical in the study of plastic deformation. The important issues here are the relationship between mass flux and various driving forces for mass flux and the interaction of various (*generalized*) *forces*. More detailed discussions of the thermodynamics of irreversible processes can be found in CALLEN, 1960, DE GROOT and MAZUR (1962) and GLANSDORFF and PRIGOGINE (1971).

2.3.1. Flux and the generalized forces

Irreversible motion or the flux of atoms (or energy) occurs when a system is out of equilibrium. One can surmise that some *forces* cause such a flux. Experimental observations show that in most cases the rate of motion of atoms or the flux of atoms (or energy) is linearly proportional to the force. An example of such an empirical law is *Fick's law* of diffusion (of atoms) (also see Chapter 8),

$$J = -D\nabla c. \tag{2.61}$$

where J is the flux of atoms, D is diffusion coefficient and c is the concentration of atoms (number of atoms per unit volume). Fick's law, (2.61), is given for a case where only one type of force is present (concentration gradient) and only single atomic species move. The generalization of this type of flow law is important in considering atomic transport (and plastic flow) in an ionic crystal.

In generalizing an equation such as (2.61), it is useful to examine the physical basis for atomic flow. In the previous section (section 2.1), we learned that in a system at thermochemical equilibrium, the entropy must be maximum and hence the following quantities (*intensive parameters*) must be homogeneous, $1/T$, P/T and $-\mu/T$ (equation 2.22). When the system is out of equilibrium, a flow of atoms (or energy) will occur due to the heterogeneity of these quantities. Such a flow produces entropy. Since entropy is given by the product of extensive quantities (such as energy) and intensive parameters (such as temperature), it is natural to define the gradients of these quantities as *generalized forces* that drive the motion of corresponding (extensive) quantities. Therefore we define a generalized force by

$$X \equiv \nabla \xi \tag{2.62}$$

where $\xi = 1/T,\ P/T,\ -\mu/T$ and define a linear relationship between the force and the flux, J, of an extensive quantity,

$$J = L \cdot X = L \cdot \nabla \xi \tag{2.63}$$

where L is a material constant that is often referred to as a *phenomenological coefficient*. For instance, the extensive quantity corresponding to an intensive parameter $1/T$ is the internal energy, so the flux corresponding to $\nabla(1/T)$ will be the flux of energy, i.e., heat flow. In this case, the empirical law is Fourier's law,

$$J_T = -k\nabla T \tag{2.64}$$

where J_T is the energy flux and k is thermal conductivity. Comparing Fourier's law with equations (2.62) and (2.63) ($\xi = 1/T$), one obtains

$$L = kT^2. \tag{2.65}$$

Similarly, the extensive quantity corresponding to an intensive parameter $-\mu/T$ is the number of atoms, n, so the corresponding flux is the flux of atoms. The empirical relation for this case is therefore Fick's law of atomic diffusion,

$$J = -D\nabla c \tag{2.66}$$

where D is the diffusion coefficient and c is the concentration of atoms (number of atoms per unit volume) and we have,

$$L = \frac{Dc}{R}. \tag{2.67}$$

Problem 2.8

Derive equation (2.67).

Solution

Using the relation (2.32), $\mu = \mu_0 + RT\log x = \mu_0 + RT\log(c/c_0)$, one obtains $X = -\nabla(\mu/T) = -R(\nabla c/c)$. Inserting this into (2.64), one gets $J = -D\nabla c = (Dc/R)X \equiv LX$, and therefore $L = Dc/R$.

From the above discussion, two methods of generalization of the above linear relationship are obvious. First, Fick's law describes the flux of atoms due to the concentration gradient, but more generally, atomic flux due to the gradient of chemical potential $X = -\nabla(\mu/T)$ can be considered. The chemical potential may include not only the gradient of concentration (gradient of configurational entropy) but also the gradient of free energy due to other effects such as electrostatic field. Electrostatic interaction among different diffusing species plays an important role in ionic crystals (see Chapter 8 for more details). In such a case, the electrostatic energy must be included in the chemical potential thus

$$\begin{aligned} X &= -\nabla\frac{\mu}{T} = -\nabla\frac{(\mu_0 + RT\log(c/c_0) + q\phi)}{T} \\ &= -\frac{q}{T}\nabla\phi - R\frac{\nabla c}{c} = \frac{q}{T}E - R\frac{\nabla c}{c} \end{aligned} \tag{2.68}$$

where q is the electrostatic charge of the species, ϕ is the electrostatic potential and E is the electrostatic field ($E = -\nabla\phi$).

Second, in writing equation (2.62), we considered only one force and one flux. A natural generalization of this relation is

$$J^i = \sum_j L^{ij} \cdot X^j \tag{2.69}$$

where we include various types of forces and corresponding fluxes. The phenomenological coefficients discussed above correspond to the diagonal components, L^{kk}, namely a material parameter describing the flow of intensive variables (matter (or energy)) due to the corresponding gradient of extensive parameters (generalized force). Off-diagonal components of phenomenological coefficients (L^{ij} with $i \neq j$) express the *coupling* of different fluxes. Physically these terms represent the flux of the ith variable caused by the generalized force conjugate to the jth variable. For example, the temperature gradient can cause the diffusion flux of atoms (*Soret effect*), and conversely the concentration gradient can cause the temperature gradient (*Dufour effect*). Diffusion in a multi-component system is an important example where coupling among different species can cause important effects on mass transport (e.g., Lasaga, 1997, see also Chapter 8). The entropy production rate contains various terms including the scalar (PV term), vector (heat flow, diffusional mass flow) as well as tensor (energy dissipation due to plastic flow). A coupling of flux of various variables can occur only when these variables have the same transformation with respect to the change of the coordinate system. This comes from the fact that the entropy production rate (equation 2.66) is a scalar, so that upon the transformation of the coordinate this quantity must not change.

Among the off-diagonal components that represent the coupling of different fluxes, the following symmetry relation, *the Onsager reciprocal relation*, must be satisfied,

$$L^{ij} = L^{ji} \tag{2.70}$$

if the independent fluxes are written as linear functions of the independent generalized forces. The Onsager reciprocal relation is a consequence of the symmetry of fluctuation with respect to time (for details see e.g., Callen, 1960; de Groot and Mazur, 1962; Landau and Lifshitz, 1964).

2.3.2. Some notes on the driving forces for plastic deformation

The driving forces considered in the previous section may be called *thermodynamic forces* (such as a force due to the gradient of concentration of a given species). However, under a deviatoric stress, there are forces that directly arise from the applied macroscopic (stress) field. In a solid under deviatoric stress, what forces drive large-scale atomic motion leading to plastic flow? Two cases can be distinguished. First, for an atomic species such as individual atoms or isolated point defects, there is no direct interaction between applied stress and these atomic species. In a perfect crystal, atoms are in their lattice sites without any stress or strain. Similarly for point defects, most of them are also isotropic and are not associated with deviatoric strain so that they do not interact with deviatoric stress (exceptions are defect complexes, which can have anisotropic strain field and interact with shear stress causing anelasticity; see Chapter 11). Consequently applied stress provides no *direct* driving forces for atomic species. The most important driving force for the motion of these atomic species (e.g., point defects) under deviatoric stress is the thermodynamic force, $X = -\nabla(\mu/T) = -R(\nabla c/c)$, caused by the heterogeneity of defect concentration due to the heterogeneous microstructure of a polycrystalline material (see Chapter 8). Heterogeneity of defect concentration can also occur due to the local stress heterogeneity due to the presence of a dislocation or a grain boundary (see Chapters 8 and 9). In these cases, dislocation motion or motion of atoms across a grain boundary occurs due to (*indirect*) thermodynamic force. Second is the case for dislocation motion, which is different from the case for point defects. Due to the long-range displacement field associated with it, a dislocation has a *direct* mechanical interaction with the applied stress (*Peach–Koehler force*: see Chapter 5). A dislocation moves both by the direct effect of applied stress as well as by the thermodynamic forces caused by the concentration gradient of point defects around it.

In some literature, the gradient in chemical potential that drives processes such as diffusion under stress, σ, is written as (e.g., SCHMALZRIED, 1995)

$$\nabla \frac{\mu}{T} = R\nabla \log a - \frac{\Omega}{T} \cdot \nabla \sigma \tag{2.71}$$

where a is the activity and Ω is the molar volume of the species involved. This expression is misleading and should be used with care. First, this equation would imply that the applied stress has a *direct* effect on defect motion. This is not true in many cases. For example, diffusion flux in a polycrystalline material under deviatoric stress is caused by the *indirect* thermodynamic force due to the *local* (grain scale) variation in the *normal component* of stress at the place where creation/destruction of defects occur (Chapter 8). In other words, stress in this equation should not be the macroscopic stress.[3] The use of equation (2.71) will obscure this basic physics of stress-induced kinetic processes and is not recommended. Second, in the case of diffusion flux in a polycrystalline material, the two terms in equation (2.71) are redundant if the stress dependence of concentration is included in the activity term because in this case, the driving force is a thermodynamic driving force, $\nabla(\mu/T) = R\nabla \log a$, caused by the grain-scale variation in stress, and is given by $\nabla(\mu/T) = R\nabla \log a = -(\Omega/T) \cdot (\sigma/L) \approx -(\Omega/T) \cdot \nabla\sigma$ $(a = a_0 \exp(-\sigma\Omega/RT))$.

2.3.3. Stationary (steady-state) state: the principle of minimum entropy production rate

The flow of materials (or energy) causes entropy production (energy dissipation). The rate of local entropy production is given by

$$\Phi \equiv \frac{dS}{dt} = \sum_i J^i X^i = \sum_i J^i \cdot \nabla \xi^i. \tag{2.72}$$

Inserting the relation (2.69) into (2.72), one gets

$$\Phi = \sum_{i,j} L^{ij} X^i X^j \tag{2.73}$$

where we used the Onsager reciprocal relation. Now consider a case where one force (X^k) is fixed but others are free to vary. The entropy production rate is minimum with respect to X^i $(i \neq k)$ when the following condition is met,

$$\frac{\partial \Phi}{\partial X^i} = \sum_j L^{ij} X^j = J^i = 0 \quad (i \neq k). \tag{2.74}$$

At the stationary state, all the fluxes other than the one corresponding to a fixed force (X^k) should vanish. Now from the conservation equation for an extensive parameter ξ^i, $\rho(\partial \xi^i/\partial t) = -\nabla \cdot J^i$, it is seen that the state of minimum entropy production rate (i.e., $J_i = 0$) corresponds to the stationary state $(\partial \xi^i/\partial t = 0)$. This is referred to as the *principle of minimum entropy production rate*.

2.4. Thermally activated processes

2.4.1. Absolute rate theory

Basic theory

An irreversible process involves an atomic jump from one stable position to the next. The atomic jump is often caused by a large fluctuation of atomic positions due to lattice vibration. Such a process is referred to as a thermally activated process. The energy of an atom in a crystal has a minimum value at the lattice sites. As one measures the energy of an atom from one lattice site to the neighboring site, the energy has a local maximum somewhere in the middle of the two lattice sites. The value of the energy maximum depends on the pass along which one moves an atom. The position at which the energy at the peak is minimum is called a saddle point (Fig. 2.8). Through thermal vibration, atomic positions fluctuate. Therefore, with some non-zero probability, a system can assume an atomic configuration in which an

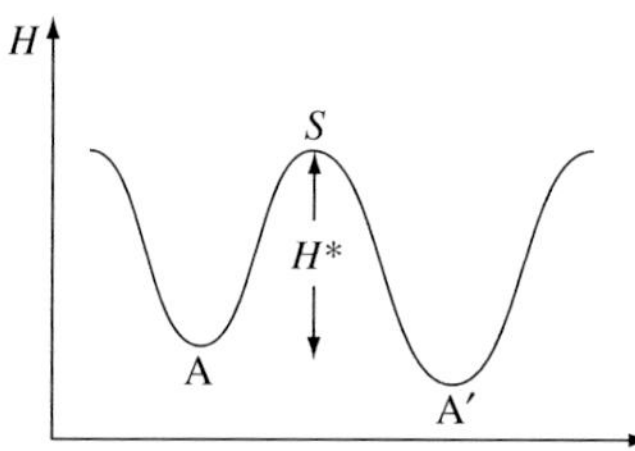

FIGURE 2.8 A plot of enthalpy as a function of atomic position associated with an atomic jump from position A to position A′. S is the saddle point. $H^* = E^* + PV^*$ is the activation enthalpy (E^* is the activation energy and V^* is the activation volume).

[3] If equation (2.71) is used, with σ being the macroscopic stress, then there would be no diffusion or deformation for a homogeneous deviatoric stress.

atom goes to a saddle point. At a saddle point, the potential energy of an atom has a maximum with respect to the spatial coordinate, and consequently, this configuration is unstable and an atom moves to the next stable position (or goes back to the original position). When the velocity of an atom at that saddle position is "positive," then the atom goes to the next stable position (a lattice site). Since this new configuration is stable, that atom does not go back to the original position instantaneously. Therefore the motion over a saddle point is irreversible.

The rate at which this occurs may be calculated by applying the principles of statistical physics and the following theory is called the *absolute rate theory*, which was developed by EYRING (1935). The application of this theory to atomic migration in a solid was made by FLYNN (1968) and VINEYARD (1957) (see also FLYNN (1972)). Noting that the activated state has different chemical bonding and hence the frequencies of lattice vibration are different, one obtains (e.g., FLYNN (1972))

$$w = \tilde{\nu} \cdot \exp\left(-\frac{H^*}{RT}\right) \tag{2.75}$$

where

$$\tilde{\nu} = \frac{\prod_{i=1}^{3N} \omega_i}{\prod_{i=2}^{3N} \omega_i'} \tag{2.76}$$

is a constant with a dimension of frequency (ω_i and ω_i' are frequencies of lattice vibration with the ith mode at the ground state (A) and at the saddle point configuration (S) respectively) and

$$H^* = H_S^* - H_A^* = (E_S - E_A) + P(V_S - V_A) \tag{2.77}$$

is the activation enthalpy ($E_{S,A}$ ($V_{S,A}$): internal energy (volume) of a system with state S and A). An interpretation of equation (2.75) is that $\tilde{\nu}$ is the "trial frequency," and $\exp(-H^*/RT)$ is the success rate of a jump. $\tilde{\nu}$ is of the order of the Debye frequency (characteristic frequency of lattice vibration, see Chapter 4) and does not change very much with pressure or temperature. The most important parameter to control the rate of process is H^*, the activation enthalpy. This term (i.e., $\exp(-H^*/RT)$) can change by orders of magnitude as temperature and/or pressure (or other chemical factors such as water fugacity) change.

The rate of atomic jump depends exponentially on temperature and pressure and hence the effects of temperature and pressure on an atomic jump are large. $E_S - E_A$ is always positive whereas $V_S - V_A$ can be positive or negative (see Chapter 10). Consequently, an atomic jump is always enhanced by higher temperature, whereas higher pressure usually suppresses an atomic jump (Chapter 10). Also note that the temperature dependence of thermally activated processes results in the time dependence of the process because these processes occur as a result of fluctuation.

Effects of stress on thermal activation

The above formula shows that the rate at which a thermally activated process occurs is strongly dependent on temperature and pressure through activation enthalpy, H^*. When an atom (or defect) has an interaction with stress, the activation enthalpy becomes dependent on the stress. This is a case for dislocations (see Chapters 5 and 9) which have an anisotropic strain field and hence have strong interactions with applied (deviatoric) stress. In such a case, the activation enthalpy becomes stress dependent, $H^*(\sigma)$. In these cases, applied stress changes the energy of a dislocation when it moves from the ground state to the saddle point configuration. The stress works on a piece of dislocation when a piece of dislocation moves to the direction of the Peach–Koehler force (Chapter 5) and hence the enthalpy will be reduced. On the other hand, when a dislocation moves in the opposite direction, the enthalpy for the barrier will increase. The net rate of motion will then be given by

$$\begin{aligned} w &= w^+ - w^- \\ &= \tilde{\nu}\left[\exp\left(-\frac{H^{*+}(\sigma)}{RT}\right) - \exp\left(-\frac{H^{*-}(\sigma)}{RT}\right)\right] \end{aligned} \tag{2.78}$$

although the backward activation is not very important at high stresses.

Because the activation enthalpy in these cases decreases with stress, there is a threshold stress, σ_c, at which activation enthalpy vanishes,

$$H^*(\sigma_c) = 0. \tag{2.79}$$

When stress reaches this level, then motion of defects is possible without the help of thermal activation. In other words, there is a finite probability of defect motion even at $T = 0$ K. Such a motion of defects is referred to as *athermal* motion. For athermal motion, fluctuation does not play an important role and hence the process becomes (nearly) time independent. A phenomenon often referred to as *yield* corresponds to this athermal motion of defects (dislocations) (the stress at which plastic deformation occurs at low temperatures is often referred to as a *yield stress*).

3 Phenomenological theory of deformation

The formal theory of deformation plays an important role in formulating energy dissipation (i.e., seismic wave attenuation) and non-linear rheological relationships. This chapter presents a brief summary of the phenomenological theory of plastic deformation. This includes the classification of deformation (elastic, viscous, plastic etc.), the mathematical formula for constitutive relations, formulation of transient creep and the mathematical formula appropriate for non-linear rheology.

Key words elasticity, visco-elasticity, anelasticity, constitutive relations, Levy–von Mises equation, mechanical equation of state, transient creep, creep response function, creep compliance function, Kramers–Kronig relation, Maxwell model, Voigt model, Zener model, Burgers model.

3.1. Classification of deformation

The response of a material to applied stress can occur in a variety of ways. Under some conditions, an *equilibrium state* can be achieved upon applying a stress, whereas time-dependent or *steady-state* deformation can also be achieved. To understand the microscopic basis for this difference, it is useful to consider the nature of atomic displacement caused by an applied stress. At static equilibrium, each atom occupies a position corresponding to the minimum potential energy (Fig. 3.1a). Upon applying a stress (a force per unit area), atoms move their positions from their stable positions. If the stress is small, or the temperature is low (or time is short), then only small, *instantaneous* displacement will occur. Consequently, when the stress is removed, atoms go back to their original positions. This type of instantaneous and recoverable deformation is called *elastic deformation*. In contrast, when a large stress is applied or stress is applied for a long time (at high temperatures), then a material responds to the stress not only instantaneously but also through delayed, time-dependent deformation and a fraction of the strain is not recoverable.

Deformation that has time-dependent and non-recoverable strain is called *non-elastic deformation*. Microscopically, this type of deformation occurs when the atomic motion is so large that atoms move, over the potential hill, to the next stable positions (Fig. 3.1b). Once atoms move over the potential hill, then even if the stress is removed, they will not go back to their original positions immediately. Note, however, that under some conditions, recoverable but time-dependent (delayed) deformation could occur. This is the case when deformation involves atomic motion over the potential hill, but deformation causes elastic strain inside a material (*back stress*). In this case, after the removal of external stress, atomic motion occurs in such a way as to reduce the internal (back) stress associated with elastic strain so that the final equilibrium state will have no permanent strain: strain is recoverable but time-dependent. This type of deformation is often referred to as *anelastic deformation*. In contrast, deformation can be time-dependent and strain is non-recoverable. This type of deformation is referred to as *viscous (or*

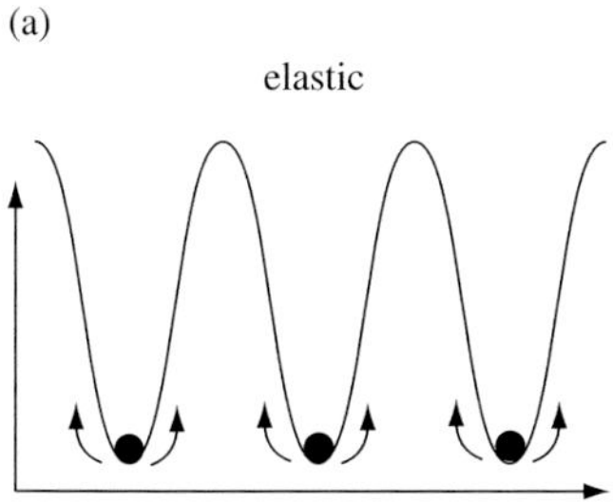

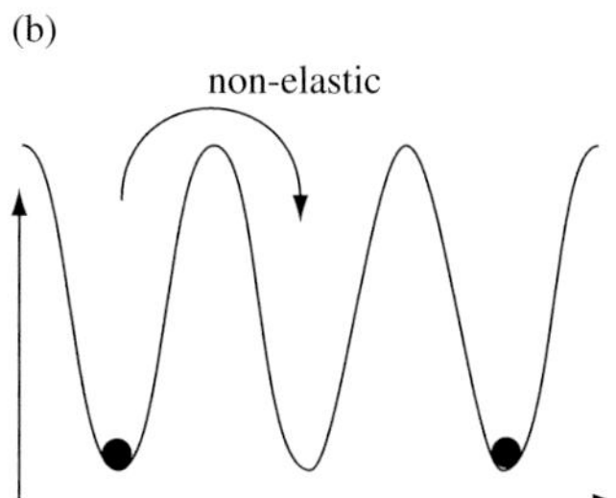

FIGURE 3.1 Atomistic view of (a) elastic (recoverable) and (b) non-elastic (non-recoverable) deformation.

plastic[1]*) deformation*. However, both anelastic and viscoelastic deformation are often collectively referred to as anelastic deformation.

Deformation can occur nearly instantaneously and be non-recoverable. Fracture or low-temperature *yielding* at high stress is nearly time-independent yet deformation is non-recoverable. This type of behavior is, however, better considered as an end-member behavior of plastic deformation (e.g., HART (1970)).

The mode of non-elastic deformation is conveniently classified into *brittle* and *plastic (ductile)* deformation. In more casual terms, brittle deformation is *fracture* and ductile deformation is *flow*. These two processes are nearly independent and therefore the non-elastic response of a material to the external force is either brittle fracture or ductile flow whichever is easier. Fracture involves the *macroscopic* breaking of chemical bonds (at the scale of *cracks*), which occurs in most cases in a localized fashion. In most cases, ductile flow involves the microscopic motion of atoms and in many cases occurs homogeneously. Consequently, microscopic, thermally activated motion of atoms (defects) controls ductile deformation, but thermally activated processes play a less important role in brittle fracture. In this book, we focus on plastic flow, but a brief summary of brittle deformation is given in Chapter 7. Extensive reviews of brittle deformation are given by PATERSON and WONG (2005) and SCHOLZ (2002).

3.2. Some general features of plastic deformation

Fig. 3.2 illustrates a typical result of mechanical tests in the plastic regime. In all cases, there is an instantaneous

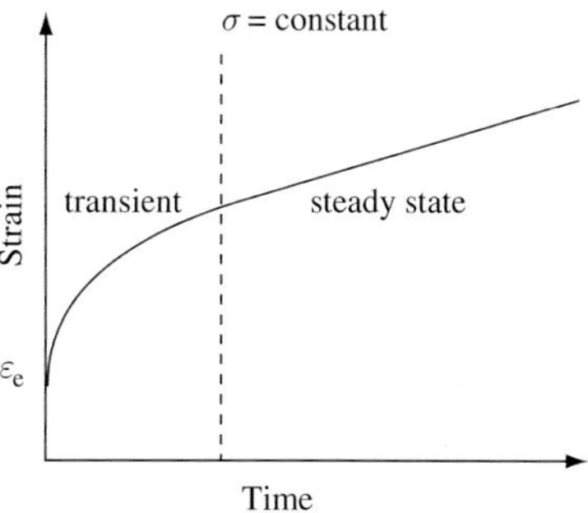

FIGURE 3.2 A schematic diagram of a strain versus time relation for a constant stress test showing various stages of deformation, where ε_e is the elastic strain. This figure corresponds to a case where work-hardening occurs in the transient stage.

elastic response yielding elastic strain, but we focus on the non-elastic component. Non-elastic deformation usually starts with a *transient period* in which the strain rate (at constant stress) or stress (at constant strain rate) changes with time. Eventually a material will usually assume *steady-state* deformation in which the strain rate (for constant stress) or stress (for constant strain rate) remains constant with time. A few points should be noted. First, the *steady state* is only approximately defined, and in fact, there are several *quasi-steady-states* of deformation for a given material. This is due to the fact that plastic deformation is sensitive to microstructures and there is a hierarchy of microstructures. As microstructures evolve, plastic properties also evolve. Consequently, when different microstructures assume quasi-steady-state at different time-scales, there will be several quasi-steady-states of deformation for a given material. One example is the deformation of a polycrystalline material in the dislocation creep regime (see Chapter 9). Dislocation microstructures at each grain level will evolve during deformation, and quasi-steady-state deformation will be reached when dislocation microstructures achieve steady state (constant dislocation density, constant

[1] In the literature, the term "plastic deformation" is often used to describe nearly time-independent yielding, and the other type of strongly time-dependent deformation is called viscous flow. However, in this book I use "plastic deformation" for any time-dependent non-recoverable deformation.

subgrain structures etc.). At the same time, orientations of individual grains will evolve with deformation to result in a lattice-preferred orientation (see Chapter 14) that will also affect plastic deformation. Consequently, another quasi-steady-state will be achieved when steady-state lattice-preferred orientation is established. Grain size may also evolve during deformation that affects the plastic flow (see Chapter 13). In most cases, these grain-scale microstructures evolve with longer time constants than dislocation microstructures. Therefore quasi-steady-states corresponding to steady-state lattice-preferred orientation and steady-state grain size will be achieved in later stages than the quasi-steady-state corresponding to steady-state dislocation microstructures. It is important to define the scale that one is talking about when one discusses the time-dependent deformation of materials. Second, under some conditions, steady-state deformation becomes unstable and localized unstable deformation occurs (Chapter 16).

In this chapter we will discuss some general (phenomenological) aspects of the mathematical description of mechanical behavior in the plastic regime. The issues that we discuss in this chapter include: (i) the formulation of non-linear flow laws (section 3.3), (ii) the formulation of transient deformation behavior (section 3.4) and (iii) the influence of ductile deformation on energy loss and on apparent elastic moduli (section 3.5).

3.3. Constitutive relationships for non-linear rheology

The plastic properties of a material are expressed in terms of a relationship between stress and strain or strain-rate. Therefore a material parameter (or parameters) that characterizes plastic deformation must express a relation between two second rank tensors (deviatoric stress σ_{ij} and deviatoric strain rate $\dot{\varepsilon}_{ij}$). In the simplest case of linear rheology at steady state, the relation should be

$$\sigma_{ij} \equiv 2\sum_{k,l} \eta_{ijkl}\dot{\varepsilon}_{kl} \tag{3.1}$$

where η_{ijkl} is the viscosity tensor. It follows that a large number of measurements along various orientations are needed to fully characterize the plastic properties of a material. However, in many cases, a material is approximately isotropic. This would be the case where one considers deformation of a polycrystal whose microstructure is nearly isotropic. In such a case, at steady state, the plastic properties of a material are characterized by a single scalar parameter, viscosity,[2] η, defined by,

$$\sigma_{ij} \equiv 2\eta\dot{\varepsilon}_{ij}. \tag{3.2}$$

Consequently, the plastic properties in such a case can be determined by measuring the one component of strain rate corresponding to one type of stress. For example, one can apply normal stress on the x_1 plane, σ_1, and observe the corresponding strain rate $\dot{\varepsilon}_1$. The viscosity can be determined by $\eta = (\sigma_1 - \sigma_3)/2(\dot{\varepsilon}_1 - \dot{\varepsilon}_3) = (\sigma_1 - \sigma_3)/3\dot{\varepsilon}_1$ (note that $\dot{\varepsilon}_3 = -\frac{1}{2}\dot{\varepsilon}_1$) where σ_3 is the lateral stress (confining pressure). Given this effective viscosity, strain rates corresponding to any other stress components can be calculated from (3.2). For a more general anisotropic rheology, one has to determine the rheological properties for a large number of orientations.

In the case of non-linear rheology, one may determine a relationship such as $\dot{\varepsilon}_1 = A(\sigma_1 - \sigma_3)^n$ $(n > 1)$. But this equation cannot be generalized to $\dot{\varepsilon}_{ij} = A\sigma_{ij}^n$, because such a relation violates the rule of transformation of tensors (if one rotates the coordinate system, the left- and the right-hand sides would change differently). If a material is isotropic, stress and strain rate must be related by a scalar quantity. Therefore, for non-linear rheology, one must have,

$$\dot{\varepsilon}_{ij} = B(\sigma_{ij}) \cdot \sigma_{ij} \tag{3.3}$$

where B is a scalar function of stress. Because B is a scalar, it must contain only the invariants of (deviatoric) stress tensor (see Chapter 1). Because the first invariant of stress tensor is hydrostatic pressure, the first invariant is usually assumed not to contribute to deformation (other than its effect on a constant term related to the rate of thermal activation, see Chapter 10). The third invariant is also assumed not to contribute to assure that energy dissipation is always positive (see equation (3.11)). In this case, the simplest form will be,

$$B = B(II_\sigma) = \bar{B} \cdot II_\sigma^{(n-1)/2}. \tag{3.4}$$

Therefore equation (3.3) becomes

$$\dot{\varepsilon}_{ij} = \bar{B} \cdot II_\sigma^{(n-1)/2}\sigma_{ij}. \tag{3.5}$$

[2] Similar to an elastic material (see Chapter 4), there are two viscosities for an isotropic material, one for shear and another for bulk deformation. However, for a perfectly dense solid, the bulk viscosity is infinite. Therefore we consider only shear viscosity here (for the bulk viscosity, see also Chapter 12).

This is referred to as the *Levy–von Mises equation.* In this case (i.e., isotropic rheology), one can translate the results from one type of deformation geometry to another deformation geometry. Note that when large internal stress exists in addition to the applied stress, the $B(II_\sigma)$ term will be largely modified by the internal stress. Consequently, the plastic flow of a material will be affected by the presence of internal stress. An application of this notion to plastic deformation during a phase transformation is discussed in Chapter 15.

Problem 3.1*

Let us consider two cases. In uni-axial (tri-axial) compression tests, one applies uni-axial stress, and measures uni-axial strain rate. Usually, the results of such tests are presented as a relationship between $\sigma_1 - \sigma_3$ and $\dot{\varepsilon}_1$ as,

$$\dot{\varepsilon}_1 = C(\sigma_1 - \sigma_3)^n \tag{1}$$

and C and n are the constants that are determined by experiments.

The second is a case of simple shear tests. In this case, the results are usually presented as a relationship between shear stress τ and shear strain rate $\dot{\gamma}$,

$$\dot{\gamma} = D\tau^n. \tag{2}$$

Show that,

$$D = 3^{(n+1)/2}C. \tag{3}$$

Solution

Under the assumption of isotropic plasticity, the rheological equation for both cases must satisfy equation (3.5). Now, the strain-rate tensor for tri-axial compression is

$$\dot{\varepsilon}_{ij} = \begin{bmatrix} \dot{\varepsilon} & 0 & 0 \\ 0 & -\frac{1}{2}\dot{\varepsilon} & 0 \\ 0 & 0 & -\frac{1}{2}\dot{\varepsilon} \end{bmatrix}$$

and the deviatoric stress tensor is

$$\sigma'_{ij} = \begin{bmatrix} \sigma_1 & 0 & 0 \\ 0 & \sigma_3 & 0 \\ 0 & 0 & \sigma_3 \end{bmatrix} - \begin{bmatrix} P & 0 & 0 \\ 0 & P & 0 \\ 0 & 0 & P \end{bmatrix}$$
$$= \begin{bmatrix} \frac{2}{3}(\sigma_1 - \sigma_3) & 0 & 0 \\ 0 & -\frac{1}{3}(\sigma_1 - \sigma_3) & 0 \\ 0 & 0 & -\frac{1}{3}(\sigma_1 - \sigma_3) \end{bmatrix}.$$

Therefore $II_\sigma = \frac{1}{3}(\sigma_1 - \sigma_3)^2$ (note that only deviatoric stress contributes to plastic flow in the ductile regime).

The flow law for the tri-axial test assumes the form,

$$\dot{\varepsilon}_1 = \dot{\varepsilon} = C(\sigma_1 - \sigma_3)^n$$

Comparing this relation with equation (3.5), $\bar{B} = (C/2)3^{(n+1)/2}$.

Now for simple shear,

$$\dot{\varepsilon}_{ij} = \begin{bmatrix} 0 & \frac{1}{2}\dot{\gamma} & 0 \\ \frac{1}{2}\dot{\gamma} & 0 & 0 \\ 0 & 0 & 0 \end{bmatrix}$$

and

$$\sigma'_{ij} = \begin{bmatrix} 0 & \tau & 0 \\ \tau & 0 & 0 \\ 0 & 0 & 0 \end{bmatrix}.$$

Therefore $II_\sigma = \tau^2$ and using (3.5), one gets $\frac{1}{2}\dot{\gamma} = \frac{C}{2}3^{(n+1)/2}\tau^n$. Hence $D = 3^{(n+1)/2}C$.

Equation (3.5) can be rewritten using the equivalent stress (a scalar)

$$\sigma_e \equiv (3II_\sigma)^{1/2} \tag{3.6}$$

to obtain

$$\dot{\varepsilon}_{ij} = 3^{-(n-1)/2} \cdot \bar{B} \cdot \sigma_e^{n-1} \cdot \sigma_{ij} \tag{3.7}$$

We can also define the equivalent strain rate (a scalar)

$$\dot{\varepsilon}_e \equiv \left(\frac{4}{3}II_{\dot{\varepsilon}}\right)^{1/2} \tag{3.8}$$

then

$$\dot{\varepsilon}_e = 2 \cdot 3^{-(n+1)/2} \cdot \bar{B} \cdot \sigma_e^n \cdot \tag{3.9}$$

These definitions, i.e., equations (3.6) and (3.8), are made to recover the relations $\sigma_e = \sigma_1 - \sigma_3$ and $\dot{\varepsilon}_e = \dot{\varepsilon}_1$ for tri-axial compression.

It is often useful to define effective viscosity, η_{eff}, by $\sigma_{ij} \equiv 2\eta_{\text{eff}}\,\dot{\varepsilon}_{ij}$, then

$$\begin{aligned}\eta_{\text{eff}} &= 2^{-1}3^{(n-1)/2}\,\bar{B}^{-1}\sigma_e^{1-n} \\ &= 2^{-1/n}3^{-(n-1)/2n}\bar{B}^{-1/n}\dot{\varepsilon}_e^{(1-n)/n}.\end{aligned} \tag{3.10}$$

The effective viscosity decreases with the increase of stress and increases with strain rate.

Plastic deformation dissipates energy. The energy dissipation per unit volume due to plastic deformation is given by

$$\Phi = \sum_{i,j} \sigma_{ij}\dot{\varepsilon}_{ij} \cdot \quad (3.11)$$

Using the equations (3.6) and (3.7), one obtains

$$\Phi = B_1\sigma_e^{n+1} = \bar{B}^{-1/n}\dot{\varepsilon}_e^{(n+1)/n}. \quad (3.12)$$

Note that the energy dissipation by plastic flow is a strong function of stress (or strain rate).

3.4. Constitutive relation for transient creep

Transient creep is a phenomenon in which the rate of deformation at a given stress (temperature and pressure) changes with time (or strain). Since all of these variables (stress, temperature, pressure) are kept constant, this temporal variation of mechanical properties must be attributed to the evolution of internal structure (or internal mechanical state such as stress distribution). Therefore a general equation for transient creep is given by

$$\sigma = F(\dot{\varepsilon}, T, P; y) \quad (3.13)$$

where y is a parameter that defines the internal structure or state of a material such as dislocation density or grain size, or the distribution of internal stress. As deformation proceeds, y evolves that leads to transient behavior. Annealing may also cause a change in y without a change in strain that leads to time-dependent (transient) behavior. The specific form of such a function depends on the microscopic mechanisms of deformation that are discussed in Chapters 8 and 9. In this section I will summarize some of the general aspects of the formulation of time-dependent deformation.

There have been some discussions as to how to formulate time-dependent deformation. The question is closely related to the issue of whether a *mechanical equation of state* exists or not (Hart, 1970; Hollomon, 1947; Zener and Hollomon, 1946). When the stress required for plastic deformation depends on the *instantaneous* values of the strain, strain rate, temperature and pressure and not on their past histories, it is said that a mechanical equation of state exists (similar to the equation of state of material, by which the density of a given material is defined only by the state variables such as temperature and pressure without any dependence on the path through which the temperature and pressure are attained). If the mechanical equation of state indeed exists, then one can predict the mechanical properties of a material including transient behavior without knowing its detailed history. Hart (1970) showed that if a mechanical equation of state exists, then y must be a unique function of strain, $y(\varepsilon)$, and should not explicitly depend on time (see also McCartney (1976)). Constitutive relations consistent with this requirement include

$$\sigma = A\varepsilon^q\dot{\varepsilon}^r + \sigma_0 \quad (3.14)$$

(e.g., Hart, 1970; Poirier, 1980) or

$$\sigma = B\left[1 - \exp\left(-\left(\frac{\varepsilon}{\varepsilon_0}\right)^s\right)\right]\dot{\varepsilon}^r + \sigma_0 \cdot \quad (3.15)$$

(e.g., Chinh *et al.*, 2004; Voce, 1948) ($q, r, s > 0$) which have been used in the literature to describe strain (work) hardening behavior. Note that equation (3.14) does not describe steady-state flow (at infinite strain, strength would become infinite which is physically meaningless), whereas (3.15) does.

However, there is no reason to believe that a mechanical equation of state exists for a range of materials or conditions, and in fact, in many cases, mechanical properties of materials indeed depend on their history and hence a mechanical equation of state does not exist. For example, when deformation occurs by grain-size sensitive creep and when grain size is controlled by grain growth, then the strength of the material depends on the entire history of that material (the initial grain size and the temperature–pressure history). Similarly, deformation superposed with pre-existing deformation depends on the pre-existing conditions such as the initial stress (this is the case for post-glacial rebound, in which deformation occurs due to stress caused by the variation of surface load in addition to the pre-existing stress due to long-term convection, see Chapter 18).

In these cases, the transient creep behavior can be described explicitly in terms of time variation of strain (at a constant stress). In many cases the experimental data can be fitted to the logarithmic constitutive equation, namely,

$$\varepsilon(t) = \varepsilon_{\mathrm{ela}} + B\log(1 + At^n) + \dot{\varepsilon}_s t \ (0 < n \le 1) \quad (3.16)$$

where $\varepsilon_{\mathrm{ela}}$ is elastic strain A and B are constants, and $\dot{\varepsilon}_s$ is the steady-state strain-rate. This equation is reduced to the following "parabolic" flow law for $At^n \ll 1$,

$$\varepsilon(t) = \varepsilon_{ela} + ABt^n + \dot{\varepsilon}_s t \cdot \quad (3.17)$$

Equation (3.16) (with $n = 1/3$) was first used by ANDRADE (1910) and is often referred to as the *Andrade creep law*. From (3.16), one has

$$\dot{\varepsilon}(t) = \frac{ABnt^{n-1}}{1 + At^n} + \dot{\varepsilon}_s \cdot \tag{3.18}$$

It is seen that in both models the strain rate decreases with time, leading to the steady-state value, $\dot{\varepsilon}_s$, at infinite time, but they yield infinite strain rate for $t \to 0$ ($\dot{\varepsilon} \to ABnt^{n-1} \to \infty$). Therefore, these models cannot be applied to short-term deformation. Some microscopic models leading to (3.16) or (3.17) are discussed in Chapter 9.

An alternative equation is an exponential formula,

$$\varepsilon(t) = \varepsilon_{\text{ela}} + C\left[1 - \exp\left(-\frac{t}{\tau}\right)\right] + \dot{\varepsilon}_s t \tag{3.19}$$

where C is a constant ($C = \tau(\dot{\varepsilon}_0 - \dot{\varepsilon}_s)$) and τ is a characteristic time. This relation is a solution of the following differential equation,

$$\frac{d\dot{\varepsilon}}{dt} = -\frac{\dot{\varepsilon} - \dot{\varepsilon}_s}{\tau}. \tag{3.20}$$

The equation of type (3.20) is a rather general equation that describes a *relaxation process* in which strain rate decreases from the initial value $\dot{\varepsilon}_0$ to the final value $\dot{\varepsilon}_s$ with a relaxation time τ. When a relaxation time has a certain distribution, and when the stress versus strain-rate relation is linear, then one can use the principle of superposition to obtain

$$\varepsilon = \varepsilon_{\text{ela}} + \dot{\varepsilon}_s t + C\int_0^{\infty}\left[1 - \exp\left(-\frac{t}{\tau}\right)\right] \cdot D(\tau)\, d\tau \tag{3.21}$$

where $D(\tau)\,d\tau$ is the fraction of relaxation time between τ and $\tau + d\tau$ ($\int_0^{\infty} D(\tau)\,d\tau = 1$). Assuming various distribution functions $D(\tau)$, one obtains a range of time dependence of strain including a parabolic flow law (equation (3.17)) (see also section 3.5 and NOWICK and BERRY (1972)). Therefore the flow laws such as (3.16) (or (3.17)) and (3.19) are not mutually exclusive.

3.5. Linear time-dependent deformation

3.5.1. General theory

Let us consider a case where deformation involves both elastic and non-elastic components. Such behavior is generally referred to as *anelasticity* or *visco-elasticity*.[3] When the amplitude of strain (or stress magnitude) is small, one can apply the principle of superposition and a systematic analysis of mechanical behavior can be performed. Such low-strain deformation involving both elastic and non-elastic components plays an important role in seismic wave propagation. This section provides a brief summary of phenomenological theory. Specific mechanisms of anelasticity are discussed in Chapter 11. Applications to seismic wave propagation are discussed in Chapter 20. A general extensive discussion on anelasticity is given by NOWICK and BERRY (1972).

Let us first consider the mechanical response in the time domain. When non-elastic deformation occurs, strain in a material includes not only instantaneous elastic strain but also delayed strain. Consider the response of a material for a small increment of stress, $d\sigma$, applied at time $t = t'$. Because the strain at time t depends on the time elapsed since the application of stress, the strain of the material will be given by

$$d\varepsilon(t) = J(t - t')\, d\sigma(t') \tag{3.22}$$

where $J(t)$ is referred to as the *creep response function*. Obviously, the response must occur after the stress is applied, hence $J(t) = 0$ for $t < 0$. The total strain corresponding to the time-dependent stress $\sigma(t)$ is given by

$$\varepsilon(t) = \int_{-\infty}^{t} J(t - t')\, d\sigma(t') = \int_{-\infty}^{t} J(t - t')\frac{d\sigma(t')}{dt'}dt'. \tag{3.23}$$

For a simple case of step-wise stress, i.e., $\sigma = \sigma_0$ for $t \geq t'$ and $\sigma = 0$ for $t < t'$, $d\sigma(t')/dt' = \sigma_0\,\delta(t')$ (where $\delta(t')$ is the Dirac delta function, $\int_{-\infty}^{\infty} A(x_0 - x)\,\delta(x)\,dx = A(x_0)$), one obtains

$$J(t) = \frac{\varepsilon(t)}{\sigma_0}. \tag{3.24}$$

So the creep response function can be obtained by solving the force balance equation for a constant stress.

Since $J(0)$ corresponds to instantaneous deformation (i.e., elastic), this is often referred to as the unrelaxed compliance, J_{U}, and $J(\infty)$ is the relaxed compliance, J_{R}, i.e.,

$$J_{\text{U}} \equiv J(0) \text{ and } J_{\text{R}} \equiv J(\infty). \tag{3.25}$$

[3] Terms such as *anelasticity* and *visco-elasticity* are used to describe deformation in which both elastic and viscous components play a role. In some literature anelasticity is referred to as a specific type of deformation in which the viscous component of strain will vanish after a long time. However, here I use the term *anelasticity* in a broad sense including both recoverable and non-recoverable deformation.

It is also useful to define the stress relaxation function $M(t)$ by

$$M(t) \equiv \frac{\sigma(t)}{\varepsilon_0}. \tag{3.26}$$

Similarly one can define the relaxed and unrelaxed modulus as

$$M_R \equiv M(\infty) = \frac{1}{J_R} \text{ and } M_U \equiv M(0) = \frac{1}{J_U}. \tag{3.27}$$

It is instructive to rewrite equation (3.23) by separating the creep response function into instantaneous and retarded components,

$$\varepsilon(t) = J_U\sigma(t) - \int_0^\infty (J(\xi) - J_U)\frac{d\sigma(t-\xi)}{d\xi}d\xi. \tag{3.28}$$

Problem 3.2

Derive equation (3.28).

Solution

By separating creep response function into two components, $J(t) = J_U + [J(t) - J_U]$, and changing the variable as $\xi \equiv t - t'$, equation (3.23) becomes

$$\varepsilon(t) = -\int_0^\infty [J_U + (J(\xi) - J_U)]\frac{d\sigma(t-\xi)}{d\xi}d\xi.$$

Therefore

$$\begin{aligned}\varepsilon(t) &= -J_U\int_\sigma^0 d\sigma(t-\xi) - \int_0^\infty (J(\xi) - J_U)\frac{d\sigma(t-\xi)}{d\xi}d\xi \\ &= J_U\sigma(t) - \int_0^\infty (J(\xi) - J_U)\frac{d\sigma(t-\xi)}{d\xi}d\xi.\end{aligned}$$

Now let us consider the mechanical response in the frequency domain. Consider a periodic stress, $\sigma = \sigma_0 \exp(i\omega t)$ (ω: frequency).[4] Then (3.23) becomes

$$\begin{aligned}\varepsilon(t) &= i\omega\, \sigma_0 \exp(i\omega t) \\ &\quad \cdot \int_{-\infty}^t J(t-t')\exp(-i\omega(t-t'))\,dt' \\ &\equiv \sigma(t)J^*(\omega)\end{aligned} \tag{3.29}$$

where we define the *creep compliance function* by

$$\begin{aligned}J^*(\omega) &\equiv i\omega\int_{-\infty}^t J(t-t')\exp[-i\omega(t-t')]\,dt' \\ &= i\omega\int_0^\infty J(t)\exp(-i\omega t)\,dt\end{aligned} \tag{3.30}$$

where we used the fact that $J(t) = 0$ for $t < 0$ (causality). Thus the creep compliance function is the Fourier transform of creep response function. Using (3.28), one can show that

$$J^*(\omega) = J_U + i\omega\int_0^\infty (J(t) - J_U)\exp(-i\omega t)\,dt. \tag{3.31}$$

Equation (3.31) means that when the response of a system has a delay (i.e., $J(t) - J_U \neq 0$), then $J^*(\omega)$ has an imaginary part and is frequency dependent (i.e., the mechanical property $\varepsilon(t)/\sigma(t)$, is frequency dependent and strain has a delay).

The creep compliance function may be written as

$$J^*(\omega) = J_1(\omega) - iJ_2(\omega) \tag{3.32}$$

with

$$J_1(\omega) = J_U + \omega\int_0^\infty (J(t) - J_U)\sin\omega t\,dt \tag{3.33a}$$

and

$$J_2(\omega) = -\omega\int_0^\infty (J(t) - J_U)\cos\omega t\,dt. \tag{3.33b}$$

Equation (3.29) can be rewritten as

$$\varepsilon(t) \equiv \varepsilon_0 \exp(i\omega t - i\delta) = J^*(\omega)\sigma_0 \exp(i\omega t) \tag{3.34}$$

where δ is the *phase lag*. From equation (3.34), one has

$$J_1(\omega) = \frac{\varepsilon_0}{\sigma_0}\cos\delta \tag{3.35a}$$

$$J_2(\omega) = \frac{\varepsilon_0}{\sigma_0}\sin\delta \tag{3.35b}$$

and

$$\tan\delta = \frac{J_2(\omega)}{J_1(\omega)}. \tag{3.35c}$$

Since the strain must arise *after* the application of stress (i.e., causality), δ is positive and $J_2(\omega)$ and $J_1(\omega)$ must be both positive.

Similarly, defining a complex modulus by

$$M^*(\omega) \equiv \frac{1}{J^*(\omega)} = M_1(\omega) + iM_2(\omega) \tag{3.36}$$

one has

[4] The use of a complex variable is for mathematical convenience. $\sigma = \sigma_0 \exp(i\omega t)$ should be understood to mean $\sigma = \mathrm{Re}[\sigma_0 \exp(i\omega t)] = \sigma_0 \cos\omega t$.

$$M_1(\omega) = \frac{\sigma_0}{\varepsilon_0}\cos\delta \tag{3.37a}$$

$$M_2(\omega) = \frac{\sigma_0}{\varepsilon_0}\sin\delta \tag{3.37b}$$

and

$$\tan\delta = \frac{M_2(\omega)}{M_1(\omega)}. \tag{3.37c}$$

The real and imaginary parts of creep response function $J^*(\omega)$ are not independent, but related to each other. This can be shown by taking the inverse Fourier transform of $J^*(\omega)$ and inserting that relation back into $J^*(\omega)$ to find (Problem 3.3)

$$J_1(\omega) - J_U = \frac{2}{\pi}\int_0^\infty \frac{J_2(x)x}{x^2-\omega^2}\,dx \tag{3.38a}$$

and

$$J_2(\omega) = -\frac{2\omega}{\pi}\int_0^\infty \frac{J_1(x)-J_U}{x^2-\omega^2}\,dx\,. \tag{3.38b}$$

The relation (3.38) is called the *Kramers–Kronig relation*. If either $J_1(\omega)$ or $J_2(\omega)$ is known for all frequency ranges, then another one (i.e., $J_2(\omega)$ or $J_1(\omega)$) can be calculated from the Kramers–Kronig relation. Similar relations can also be derived for $M_1(\omega)$ and $M_2(\omega)$.

Problem 3.3*

Derive the Kramers–Kronig relation, (3.38).

Solution

The inverse Fourier transform of equation (3.33a) yields

$$J(t) - J_U = \frac{2}{\pi}\int_0^\infty \frac{J_1(\omega)-J_U}{\omega}\sin\omega t\,d\omega.$$

Inserting this relation into equation (3.33b), one has

$$\begin{aligned}J_2(\omega) &= -\frac{2\omega}{\pi}\int_0^\infty\int_0^\infty \frac{J_1(\alpha)-J_U}{\alpha}\sin\alpha t\cdot\cos\omega t\cdot d\alpha\cdot dt\\ &= -\frac{2\omega}{\pi}\int_0^\infty \frac{J_1(\alpha)-J_U}{\alpha^2-\omega^2}d\alpha\end{aligned}$$

where the relation $\int_0^\infty \sin\alpha t\cos\omega t\,dt = \alpha/(\alpha^2-\omega^2)$ was used. Equation (3.38a) can be derived similarly.

The real and the imaginary parts of creep response function correspond to the stored energy and the energy loss during deformation. This can be shown as follows. The energy loss during one cycle of loading is given by

$$|\Delta E| = \oint \sigma\,d\varepsilon = \int_0^{2\mu/\omega}\sigma\dot{\varepsilon}\,dt = \pi J_2(\omega)\sigma_0^2 \tag{3.39}$$

and the maximum energy stored in the system is given by

$$E = \int \bar{\sigma}\,d\bar{\varepsilon} = \int_0^{\pi/2\omega}\bar{\sigma}\,\dot{\bar{\varepsilon}}\,dt = \frac{1}{2}J_1(\omega)\sigma_0^2 \tag{3.40}$$

where $\bar{\sigma}$ and $\bar{\varepsilon}$ are stress and strain for a purely elastic material.

Problem 3.4

Prove relations (3.39) and (3.40).

Solution

For a periodic stress $\sigma = \mathrm{Re}\{\sigma_0\exp(i\omega t)\} = \sigma_0\cos\omega t$. And $\dot{\varepsilon} = \mathrm{Re}\{(J_1 - iJ_2)i\omega\sigma_0(\cos\omega t + i\sin\omega t)\} = \sigma_0\omega\cdot(-J_1\sin\omega t + J_2\cos\omega t)$. Consequently, $|\Delta E| = \oint\sigma\,d\varepsilon = \int_0^{2\pi/\omega}\sigma\dot{\varepsilon}\,dt = \sigma_0^2\omega\int_0^{2\pi/\omega}(-J_1\sin\omega t\cdot\cos\omega t + J_2\cos^2\omega t)\,dt = \pi J_2\sigma_0^2$. For a periodic stress, the maximum energy is stored at 1/4 of the cycle, so that $E = \int_0^{\pi/2\omega}\bar{\sigma}\cdot\dot{\bar{\varepsilon}}\,dt = -\sigma_0^2\omega\int_0^{\pi/2\omega}J_1\sin\omega t\cdot\cos\omega t\cdot dt = \frac{1}{2}J_1\sigma_0^2$ where we used relations $\bar{\sigma} = \sigma_0\cos\omega t$ and $\dot{\bar{\varepsilon}} = -\sigma_0\omega J_1\sin\omega t$.

It is customary to use a Q-factor to measure the degree of energy loss (see also Chapter 18), which is defined by

$$Q^{-1} \equiv \frac{|\Delta E|}{2\pi E}. \tag{3.41}$$

Then using equations (3.39) and (3.40), one obtains

$$Q^{-1}(\omega) = \frac{J_2(\omega)}{J_1(\omega)} = \frac{M_2(\omega)}{M_1(\omega)}. \tag{3.42}$$

Now let us consider how attenuation affects the velocity of elastic waves. A one-dimensional equation of motion of an elastic wave in a dissipative material can be written as (e.g., Aki and Richards, 2002; Kolsky, 1956)

$$\rho\frac{\partial^2 u}{\partial t^2} = M^*(\omega)\frac{\partial^2 u}{\partial x^2} = [M_1(\omega) + iM_2(\omega)]\frac{\partial^2 u}{\partial x^2} \tag{3.43}$$

where u is the displacement. Since there is an imaginary part in the elastic constant, the energy must be dissipated and consequently the amplitude of the wave will decrease with distance. Therefore we assume the following form of displacement

$$u(x,t) = u_0 \exp\left\{-\frac{2\pi i}{\lambda}(x - V(\omega)t)\right\}\exp\{-\alpha(\omega)x\} \tag{3.44}$$

with $2\pi/\lambda = \omega/V(\omega)$ where $V(\omega)$ is the (phase) velocity of the wave and $\alpha(\omega)$ is the attenuation coefficient. Equation (3.44) can be written as

$$u(x,t) = u_0 \exp(i\omega t)\exp\{-iK(\omega)x\} \tag{3.45}$$

with

$$K(\omega) \equiv \frac{\omega}{V(\omega)} - i\alpha(\omega) \cdot \tag{3.46}$$

Inserting equations (3.45) and (3.46) into equation (3.43), one finds

$$K(\omega) = \omega\sqrt{\frac{\rho}{M^*(\omega)}} = \omega\sqrt{\frac{\rho}{M_1(\omega) + iM_2(\omega)}}. \tag{3.47}$$

From equations (3.46) and (3.47), one obtains

$$\begin{aligned} V(\omega) &= \sqrt{\frac{2\left[M_1^2(\omega) + M_2^2(\omega)\right]}{\rho\left[M_1(\omega) + \sqrt{M_1^2(\omega) + M_2^2(\omega)}\right]}} \approx \sqrt{\frac{M_1(\omega)}{\rho}} \\ &= \sqrt{\frac{J_1(\omega)}{\rho(J_1^2(\omega) + J_2^2(\omega))}} \approx \frac{1}{\sqrt{\rho J_1(\omega)}} \end{aligned} \tag{3.48}$$

and

$$\begin{aligned} \alpha(\omega) &= \sqrt{\frac{\rho}{2(M_1^2(\omega) + M_2^2(\omega))}} \frac{\omega M_2(\omega)}{M_1(\omega) + \sqrt{M_1^2(\omega) + M_2^2(\omega)}} \\ &\approx \frac{\omega}{2V(\omega)} \cdot \frac{M_2(\omega)}{M_1(\omega)} = \frac{\omega}{2V(\omega)} \cdot \frac{J_2(\omega)}{J_1(\omega)}. \end{aligned} \tag{3.49}$$

When there is energy dissipation, the elastic constants and hence elastic wave velocities become frequency dependent. This is called the *physical dispersion* (e.g., Kanamori and Anderson, 1977).

3.5.2. Some simple models

To obtain an intuitive, physical understanding of anelasticity, let us consider some simple models. Models contain both elastic and viscous elements that are combined in different ways. The manner in which they are combined is critical to the mechanical behavior of the model and the micro-physical basis for each model is discussed in Chapter 11.

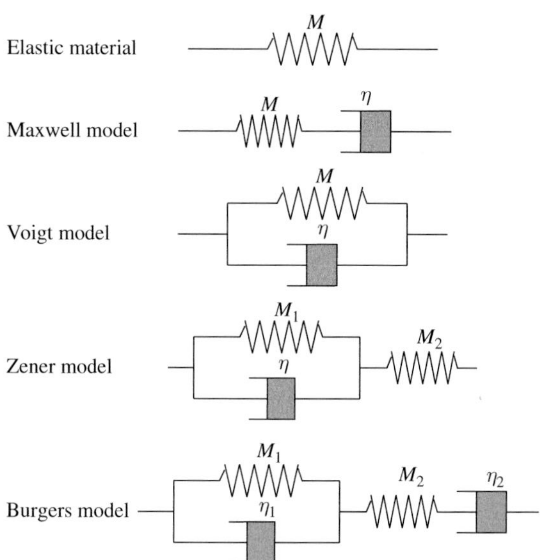

FIGURE 3.3 Various models of elastic and non-elastic deformation.

Five models are shown in Fig. 3.3 and for each, one can obtain a force balance equation by noting that the constitutive equation for an elastic element is $\sigma = M\varepsilon$ and that for a viscous element is $\sigma = \eta\dot{\varepsilon}$, where M and η are elastic constant and viscosity respectively. The results are summarized in Table 3.1 (see also Problem 3.4).

The creep response function can be obtained from the strain corresponding to a step-wise stress for each model. Similarly, the constitutive equation can be solved for periodic stress to obtain the creep compliance function. The results are summarized in Table 3.2 and the strain responses are illustrated in Fig. 3.4. Elastic wave velocity and the attenuation corresponding to each model can be calculated from the creep compliance function and the results are summarized in Table 3.3 and shown in Fig. 3.5.

For example, for the *Maxwell model*, the initial strain is elastic strain and the strain increases linearly with time due to viscous deformation, namely,

$$\varepsilon = \sigma J(t) = \frac{\sigma}{M}\left(1 + \frac{t}{\tau_M}\right) \tag{3.50}$$

where $\tau_M \equiv \eta/M$ is the Maxwell time. The viscous strain becomes comparable to the elastic strain at the Maxwell time. This is a good model for a material in which elastic and viscous deformation occur independently.

TABLE 3.1 Simple models of rheological properties and the corresponding constitutive relations.

Model	Force balance equation
Elastic material	$\sigma = M\varepsilon$
Maxwell model	$\dot{\varepsilon} = \frac{1}{\eta}\sigma + \frac{1}{M}\dot{\sigma} = \frac{1}{\eta}(\sigma + \tau_M\dot{\sigma})\ \left(\tau_M = \frac{\eta}{M}\right)$
Voigt model	$\sigma = M\varepsilon + \eta\dot{\varepsilon} = M(\varepsilon + \tau_M\dot{\varepsilon})$
Zener model	$\sigma + \tau_3\dot{\sigma} = M_\mathrm{R}(\varepsilon + \tau_\sigma\dot{\varepsilon})\ \left(\tau_3 = \frac{\eta}{M_1+M_2}, \tau_\sigma = \frac{\eta}{M_1}, M_\mathrm{R} = \frac{M_1M_2}{M_1+M_2}\right)$
Burgers model	$\frac{1}{\tau_\eta}\sigma + \left(1 + \frac{\tau_\sigma}{\tau_\varepsilon}\right)\dot{\sigma} + \tau_\varepsilon\ddot{\sigma} = M_\mathrm{R}(\dot{\varepsilon} + \tau_\sigma\ddot{\sigma})\left(\tau_\eta = \frac{\eta_2}{M_\mathrm{R}}\right)$

TABLE 3.2 Various models and corresponding creep response and creep compliance functions.

Model	Creep response function $J(t)$	Creep $J^*(\omega)$ compliance function
Maxwell model	$\frac{1}{M}\left(1 + \frac{t}{\tau_M}\right)$	$\frac{1}{M}\left(1 + \frac{1}{i\omega\tau_M}\right)$
Voigt model	$\frac{1}{M}\left[1 - \exp\left(-\frac{t}{\tau_M}\right)\right]$	$\frac{1}{M}\frac{1}{1 + i\omega\tau_M}$
Zener model	$\frac{1}{M_\mathrm{R}}\frac{\tau_\varepsilon}{\tau_\sigma} + \frac{1}{M_\mathrm{R}}\frac{\tau_\sigma - \tau_\varepsilon}{\tau_\sigma}\left[1 - \exp\left(-\frac{t}{\tau_\sigma}\right)\right]$	$\frac{1}{M_\mathrm{R}}\frac{1 + i\omega\tau_\varepsilon}{1 + i\omega\tau_\sigma}$

TABLE 3.3 Various rheological models and corresponding seismic wave velocities and attenuation.

Model	$V(\omega)$	$Q^{-1}(\omega)$
Maxwell model	$\sqrt{\frac{M}{\rho}}\sqrt{\frac{2\omega\tau_M}{\omega\tau_M+\sqrt{1+\omega^2\tau_M^2}}}$	$\frac{1}{\omega\tau_M}$
Voigt model	$\sqrt{\frac{M}{\rho}}\sqrt{\frac{2\left(1+\omega^2\tau_M^2\right)}{1+\sqrt{1+\omega^2\tau_M^2}}}$	$\omega\tau_M$
Zener model	$\sqrt{\frac{M_\mathrm{R}}{\rho}}\left(1 + \frac{\Delta}{2}\frac{\omega^2\tau^2}{1+\omega^2\tau^2}\right) \approx \sqrt{\frac{M_\mathrm{U}}{\rho}}\left(1 - \frac{\Delta}{2}\frac{1}{1 + \omega^2\tau^2}\right)$, $\left(\Delta \equiv \frac{\tau_\sigma - \tau_\varepsilon}{\tau}, \tau \equiv \sqrt{\tau_\sigma\tau_\varepsilon}\right)$	$\Delta\frac{\omega\tau}{1 + \omega^2\tau^2}$

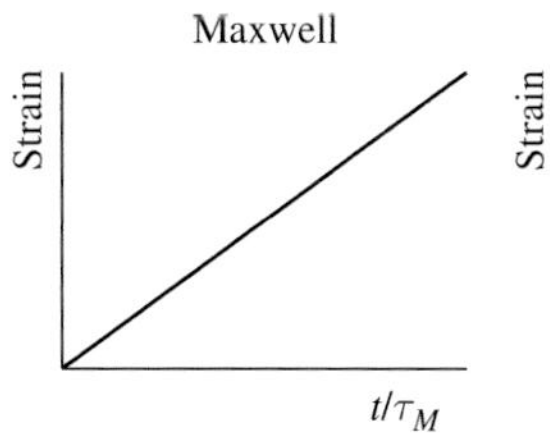

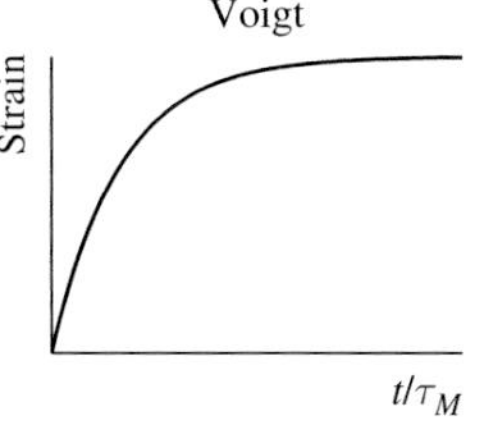

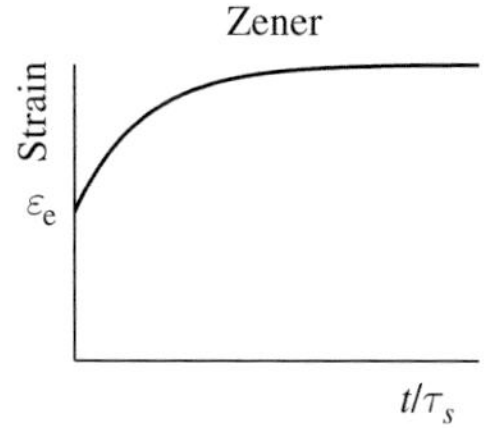

FIGURE 3.4 The strain response for a step-wise applied stress for various models.

In some cases, viscous and elastic deformation are coupled. This type of mechanical response is represented by a *Voigt model*. For the Voigt model, the extension of a viscous element causes the increase in stress in an elastic element ("back-stress") that is connected in parallel to a viscous element and therefore prevents further extension of the viscous element (examples include the viscous grain boundary sliding with elastic grains, the viscous motion of fluid in an inclusion in an elastic matrix and the viscous motion of dislocations in an elastic grain, see Chapter 11). Accordingly for the Voigt model strain will reach a

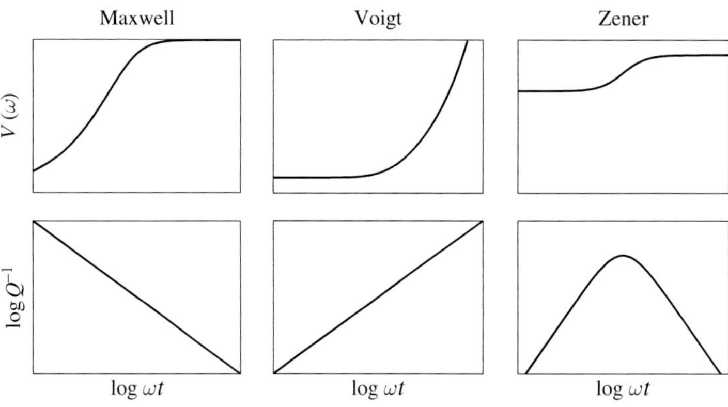

FIGURE 3.5 The frequency dependence of elastic wave velocity and attenuation.

finite value at infinite time as can be seen from the following equation,

$$\varepsilon = \sigma J(t) = \frac{\sigma}{M}\left[1 - \exp\left(-\frac{t}{\tau_M}\right)\right]. \tag{3.51}$$

The characteristic time to reach the finite ultimate strain is the same as the Maxwell time defined above ($\tau_M \equiv \eta/M$). There is no instantaneous deformation for the Voigt model because deformation of a viscous element takes some time to develop. The Voigt model is appropriate in cases where viscous deformation creates internal, back-stress that ultimately terminates viscous strain. However, there is no instantaneous response in the Voigt model, that leads to an infinite instantaneous elastic constant. This is physically unrealistic.

To rectify this one can add an elastic element in series to the Voigt model. This is the *Zener model* (also called a *standard linear solid*). The Zener model has an instantaneous deformation followed by the delayed deformation and is characterized by two elastic moduli (relaxed and unrelaxed moduli ($M_R = \frac{1}{1/M_1+1/M_2}$ and $M_U = M_2$)). The characteristic time for this delayed deformation is $\tau_\sigma \equiv \eta/M_1$. Deformation will stop at a finite strain at infinite time.

In order to have infinite viscous deformation at long term, one can add another viscous element in parallel to the Zener model. Such a model is referred to as the *Burgers model*. The Burgers model is a serial combination of the Zener and the Maxwell models and is characterized by two viscosities (short- and long-term viscosities).

When one discusses short-term rheology, the Zener model is the most important model (that is why it is called a *standard linear solid*). It has a finite elastic constant at infinite frequency (corresponding to unrelaxed modulus) and another smaller elastic modulus at zero frequency (corresponding to relaxed modulus). When anelasticity is due to the constrained motion of viscous elements in an elastic material, its behavior is described by the Zener model. However, at a long time-scale, viscous motion becomes important because the "back-stress" that causes Zener type behavior becomes ineffective (see Chapter 11 for a microscopic interpretation of this point). So the Burgers model is the most general model that describes the non-elastic behavior of materials. Such behavior has been demonstrated in some materials such as Al_2O_3 (LAKKI *et al.*, 1998).

The frequency dependence of each model can be calculated from the creep compliance function $J^*(\omega)$ (Table 3.2) and resultant frequency dependence of elastic wave velocity and attenuation can also be calculated (Table 3.3, Fig. 3.5). Note that all of these simple models show a rather strong frequency dependence for attenuation except in a narrow frequency range close to the peak frequency for the Zener model.

Problem 3.5

Derive the equations of force balance, the creep response function and the creep compliance function for the Zener model and calculate the frequency dependence of elastic wave velocity and attenuation.

Solution

A Zener model is a series combination of a Voigt model and a spring. Therefore the total strain is given by $\varepsilon = \varepsilon_1 + \varepsilon_2$ (also $\dot{\varepsilon} = \dot{\varepsilon}_1 + \dot{\varepsilon}_2$) where ε_1 is strain in the Voigt unit and ε_2 strain in the spring. Now, for a spring, $M_2\varepsilon_2 = \sigma$. In the Voigt unit, the strain in a spring and a dash-pot is the same and $M_1\varepsilon_1 = \sigma_1$

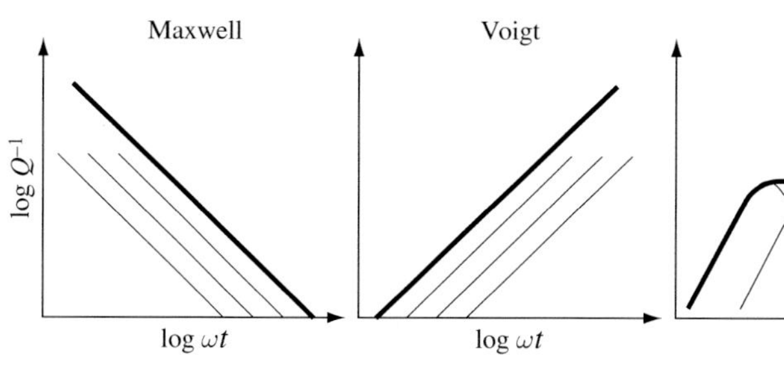

FIGURE 3.6 Effects of distributed relaxation times.

and $\eta\dot{\varepsilon}_1 = \sigma_2$, and $\sigma = \sigma_1 + \sigma_2$. Eliminating $\varepsilon_1, \dot{\varepsilon}_1, \varepsilon_2, \dot{\varepsilon}_2, \sigma_1, \dot{\sigma}_1, \sigma_2, \dot{\sigma}_2$ from these equations, one gets $\varepsilon + (\eta/M_1)\dot{\varepsilon} = (1/M_1 + 1/M_2)\sigma + (\eta/M_1M_2)\dot{\sigma}$. Now defining $\tau_\varepsilon \equiv \eta/(M_1 + M_2), \tau_\sigma \equiv \eta/M_1, M_R = M_1M_2/(M_1 + M_2)$, one obtains $\sigma + \tau_\varepsilon\dot{\sigma} = M_R(\varepsilon + \tau_\sigma\dot{\varepsilon})$. Consider a periodic stress, $\sigma = \sigma_0 \exp(i\omega t)$ and a corresponding strain, $\varepsilon = J^*(\omega)\sigma_0 \exp(i\omega t)$ to get

$$J^*(\omega) = \frac{1}{M_R}\frac{1 + i\omega\tau_\varepsilon}{1 + i\omega\tau_\sigma} = \frac{1}{M_R}\frac{1 + \omega^2\tau_\sigma\tau_\varepsilon - i\omega(\tau_\sigma - \tau_\varepsilon)}{1 + \omega^2\tau_\sigma^2}.$$

Now if one defines $\tau \equiv \sqrt{\tau_\varepsilon\tau_\sigma}$, $\Delta \equiv (\tau_\sigma - \tau_\varepsilon)/\tau$ and assuming $\Delta \ll 1$, one obtains

$$J_1(\omega) = \frac{1}{M_R}\left(1 - \Delta\frac{\omega^2\tau^2}{1 + \omega^2\tau^2}\right)$$

and

$$J_2(\omega) = \Delta\frac{1}{M_R}\frac{\omega\tau}{1 + \omega^2\tau^2}.$$

Consequently

$$Q^{-1}(\omega) = \frac{J_2(\omega)}{J_1(\omega)} \approx \Delta\frac{\omega\tau}{1 + \omega^2\tau^2}$$

and

$$V(\omega) \approx \frac{1}{\sqrt{\rho J_1(\omega)}} \approx \sqrt{\frac{M_R}{\rho}}\left(1 + \frac{\Delta}{2}\frac{\omega^2\tau^2}{1 + \omega^2\tau^2}\right) \approx \sqrt{\frac{M_U}{\rho}}\left(1 - \frac{\Delta}{2}\frac{1}{1 + \omega^2\tau^2}\right).$$

3.5.3. Effects of distributed relaxation times

Although the strong frequency dependence, as predicted by a simple model such as the Maxwell model or a Zener model, is observed in liquids, the frequency dependence of attenuation (and elastic wave velocities) in solids is in general much weaker. The distribution of relaxation times may cause a weaker frequency dependence of Q. However, this can only be the case for the Zener model. For the Maxwell or the Voigt model, distribution of relaxation times does not change the frequency dependence of Q. Therefore the observation that Q is a weak function of frequency implies that the mechanisms must involve the Zener model behavior with distributed relaxation times. The role of distributed relaxation times on frequency dependence can be easily seen in Fig. 3.6.

In order to consider the effects of distributed relaxation times, we make an assumption that the energy loss caused by each peak is additive. This is the case for small energy loss. Under this assumption, one can generalize the equation for frequency dependence of attenuation for the Zener model,

$$Q^{-1}(\omega) = \int_0^\infty \Delta\frac{\omega\tau}{1 + \omega^2\tau^2}(\tau)\, d\tau \tag{3.52}$$

where $D(\tau)\, d\tau$ is the number density of relaxation times, which is normalized as,

$$\int_0^\infty D(\tau)\, d\tau = 1\,\cdot \tag{3.53}$$

Defining $x \equiv \omega\tau$, equation (3.52) translates into

$$Q^{-1}(\omega) = \frac{\Delta}{\omega}\int_0^\infty \frac{x}{1 + x^2} D\left(\frac{x}{\omega}\right) dx. \tag{3.54}$$

It is often observed that in a certain frequency region, the frequency dependence of Q is

$$Q \sim \omega^\alpha \qquad 0 \le \alpha < 1. \tag{3.55}$$

This type of frequency dependence of Q is frequently observed particularly at high temperatures. Note, however, that this type of frequency dependence should apply only in a limited frequency range. The relation (3.54) implies $D(x/\omega) \propto \omega^{1-\alpha}$, and consequently the corresponding distribution of relaxation times have the form,

$$D(\tau) = A\tau^{\alpha-1}. \tag{3.56}$$

Obviously such a distribution function must have a limit in τ in a certain region, $\tau_2 < \tau < \tau_1$ hence,

$$\int_{\tau_2}^{\tau_1} A\tau^{\alpha-1}d\tau = \frac{A}{\alpha}(\tau_1^\alpha - \tau_2^\alpha) = 1 \quad \text{for } 0<\alpha<1 \quad (3.57a)$$

$$= A(\log\tau_1 - \log\tau_2) = 1 \text{ for } \alpha = 0 \quad (3.57b)$$

then,

$$D(\tau) = \frac{\alpha\tau^{\alpha-1}}{\tau_1^\alpha - \tau_2^\alpha} \sim \alpha\tau_1^{-\alpha}\tau^{\alpha-1} \quad \text{for } 0<\alpha<1 \quad (3.58a)$$

$$D(\tau) = \frac{1}{\tau(\log\tau_1 - \log\tau_2)} \quad \text{for } \alpha = 0. \quad (3.58b)$$

From equations (3.52) and (3.58a) or (3.58b), one gets (for $\tau_1^{-1} \ll \omega \ll \tau_2^{-1}$),

$$Q^{-1}(\omega) = \Delta(\omega\tau_1)^{-\alpha}F(\alpha) \quad \text{for } 0<\alpha<1 \quad (3.59a)$$

$$Q^{-1}(\omega) = \frac{\pi}{2}\frac{\Delta}{\log\tau_1 - \log\tau_2} \quad \text{for } \alpha = 0 \quad (3.59b)$$

with

$$F(\alpha) \equiv \alpha\int_0^\infty \frac{x^\alpha}{1+x^2}dx = \frac{\pi\alpha}{2\cos(\pi\alpha/2)}. \quad (3.60)$$

The seismic wave velocities corresponding to these relations for Q can be calculated from the complex compliance,

$$V(\omega) = V^\infty\left[1 - \frac{1}{2}\cot\frac{\alpha\pi}{2}Q^{-1}(\omega)\right] \quad \text{for } 0<\alpha<1 \quad (3.61a)$$

$$V(\omega) = V^\infty\left[1 + \frac{Q^{-1}}{\pi}\log\omega\tau_1\right] \quad \text{for } \alpha = 0 \quad (3.61b)$$

where V^∞ is the "unrelaxed" velocity (velocity when $Q^{-1} = 0$). In both cases, velocity increases with frequency (Kanamori and Anderson, 1977; Minster and Anderson, 1981).

Note that the power-law behavior of attenuation, (3.55), is related to the time dependence of transient creep. This can be understood from equation (3.31), which connects the creep response function $J(t)$ to the creep compliance function $J^*(\omega)$. The power-law behavior of attenuation, $Q^{-1} \propto \omega^{-\alpha}$, corresponds to the time dependence of transient creep, $\varepsilon = \varepsilon_{\text{ela}} + At^\alpha$ (Problem 3.6).

Problem 3.6*

Show that $\varepsilon = \varepsilon_{ela} + At^\alpha$ corresponds to $Q^{-1} \propto \omega^{-\alpha}$ (hint: use the relation $\int_0^\infty x^\alpha \cos x\, dx = -\Gamma(\alpha+1)\cdot \sin(\alpha\pi/2)$ where $\Gamma(x) \equiv \int_0^\infty y^{x-1}\exp(-y)dy$ is the gamma function).

Solution

The transient creep behavior, $\varepsilon = \varepsilon_{ela} + At^\alpha$, implies that $J(t) = J_U + Bt^\alpha$. Inserting this relation into equation (3.33), one has $J_1(\omega) = J_U + B\omega\int_0^\infty t^\alpha \sin\omega t\, dt$ and $J_2(\omega) = -B\omega\int_0^\infty t^\alpha \cos\omega t\, dt = -B\omega^{-\alpha}\int_0^\infty x^\alpha \cos x\, dx = B\omega^{-\alpha}\Gamma(\alpha+1)\sin(\alpha\pi/2)$ where $x \equiv \omega t$ and $\Gamma(x)$ is the gamma function. For weak attenuation, $J_1(\omega) \approx J_U$ hence $Q^{-1}(\omega) = J_2(\omega)/J_1(\omega) \simeq J_2(\omega)/J_U \propto \omega^{-\alpha}$.

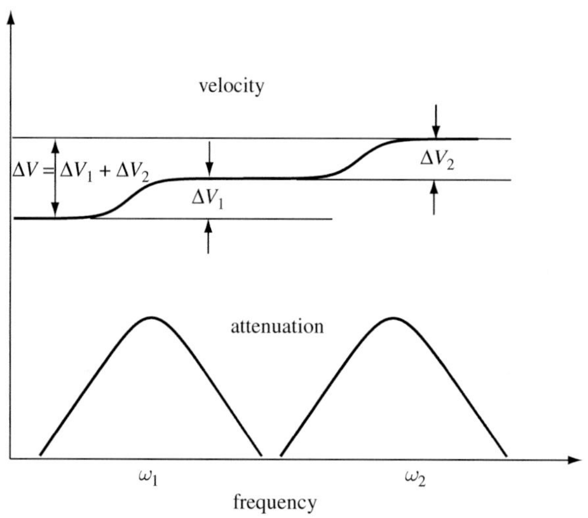

FIGURE 3.7 A schematic diagram showing the dependence of seismic wave velocity with frequency when two mechanisms of attenuation with different characteristic frequencies are present.

Relaxation times depend on thermodynamic parameters and/or microstructural parameters of a given material (for details see Chapter 11). In many cases the relaxation time depends on temperature and pressure as,

$$\tau = \tau_0 \exp\left(\frac{H^*}{RT}\right) \tag{3.62}$$

where τ_0 is the pre-exponential term, H^* is the activation enthalpy ($H^* = E^* + PV^*$; E^*, activation energy, V^*, activation volume), R is the gas constant, T is temperature, and P is pressure. Thus,

$$Q^{-1}(\omega, T, P) = \Delta(\omega\tau_{10})^{-\alpha} F(\alpha) \exp\left(-\frac{\alpha H_1^*(p)}{RT}\right) \quad \text{for } 0<\alpha<1 \tag{3.63a}$$

$$Q^{-1}(\omega, T, P) = \frac{\pi\Delta}{2} \frac{1}{\log\tau_{10} - \log\tau_{20} + (H_1^* - H_2^*)/RT} \quad \text{for } \alpha = 0 \tag{3.63b}$$

where quantities with the suffix 1 or 2 are for the upper and lower cut of characteristic times ($\tau_{1,2}$) respectively. Note that the temperature dependence of Q is directly related to the frequency dependence of Q. In particular, for frequency independent Q, temperature dependence is also small. For instance if the distribution of relaxation time is due to the pre-exponential term alone, then $H_1^* - H_2^* = 0$ and hence Q will be independent of temperature and pressure.

The above results indicate that there is one-to-one correlation between the velocity reduction $[V(\infty) - V(\omega)]/V(\infty)$ and $Q(\omega)$. This is true only when there is a single relationship on the frequency dependence of Q. Consider a simple case where there are two absorption peaks for a Zener model, ω_1 and ω_2 ($\omega_2 \gg \omega_1$). Each peak corresponds to a specific mechanism of relaxation that reduces the seismic wave velocity (see Fig. 3.7). In such a case, when the seismic wave velocities are measured near to, or at a frequency lower than ω_1, then the velocity reduction due to anelastic relaxation is caused by two mechanisms, and hence

$$\Delta V(\omega) = \Delta V_1(\omega) + \Delta V_2(\omega) \tag{3.64}$$

where $\Delta V_1(\omega)$ and $\Delta V_2(\omega)$ correspond to the velocity reduction caused by two mechanisms. In these cases, seismic wave velocities will be lower than expected from the relations such as (3.61). A potentially important situation is a case where partial melting occurs. In this case, the peak frequencies for melt-induced attenuation are likely to be higher than seismic frequencies (see Chapter 11).

Part II
Materials science of deformation

4 Elasticity

This chapter provides a brief summary of the physics of elastic constants that has important applications for the inference of the constitution of Earth's interior (Chapter 17) as well as for understanding the origin of lateral heterogeneity in seismic wave velocities (Chapter 20) and seismic anisotropy (Chapter 21). Elasticity is one of the key properties of materials that control seismic wave propagation. To the extent that elastic constants reflect the nature of chemical bonding and crystal structure, the elasticity of a given material also influences its plastic properties. Starting from the definition of elastic constants and their symmetry properties, a brief summary of the experimental techniques of measurements of elastic constants and some materials science (solid-state physics) fundamentals of elastic constants are discussed including the origin of the variation of the elastic constant with pressure and temperature (anharmonicity) and the influence of chemical composition and phase transformations on elastic constants. Birch's law on the relationship between density and elastic wave velocities and the Debye model of lattice vibration provide a unified framework to understand the relationship among various anharmonic parameters although these relations (or models) are only approximately satisfied in real materials.

Key words Hooke's law, elastic constants, bulk modulus, shear modulus, Lamé constants, Poisson's ratio, Cauchy relation, X-ray diffraction, Brillouin scattering, ultrasonic wave techniques, Birch's law, Grüneisen parameter, Anderson–Grüneisen parameter, Debye model, soft mode, polyhedral model, Pauling's rule.

4.1. Introduction

Although the main emphasis of this book is plastic deformation, an understanding of elastic deformation is important for two reasons. First, the most important means by which we explore Earth's interior is seismic wave propagation, which is controlled by the elasticity of Earth materials. Seismological observations are used to infer the constitution and dynamics of Earth's interior. In both applications, one needs to understand the way in which (isotropic average of) elastic properties change with pressure, temperature and composition (and crystal structure) as well as the origin of elastic anisotropy. Second, to the extent that elastic constants reflect the nature of chemical bonding and the crystal structure, elastic constants have important influence on plastic properties. Plastic anisotropy (i.e., the relative easiness of various slip systems) and the pressure and temperature dependence of plasticity are in part controlled by elastic properties (Chapters 8–10).

This chapter presents a brief summary of the theory and experimental observations on elasticity with the emphasis on those aspects that have important seismological applications. A more detailed account can be found in ANDERSON (1996), BORN and HUANG (1954), MARADUDIN *et al.* (1971), POIRIER (2000) and WALLACE (1972). The focus is on the elastic properties of a single crystal or the elastic properties of a homogeneous material. Specific issues of the elastic properties of polycrystalline materials including those of a partially molten material are discussed in Chapter 12. These issues become important when we consider the elastic properties of polycrystalline or polyphase materials containing highly anisotropic crystals and/or materials with highly contrasting elastic constants.

The experimental results and techniques are reviewed by BASS (1995) and LIEBERMANN (2000) and the methods and results of first-principles computational studies are summarized by KARKI *et al.* (2001).

4.2. Elastic constants

4.2.1. Hooke's law and elastic constants

When a small stress is applied to a material for a short period (or at low temperature), strain will result instantaneously and when the stress is removed there is no permanent strain. This *instantaneous* and *recoverable* deformation is referred to as *elastic deformation*. When the stress is low, then the strain, ε, is a linear function of stress, σ, namely,

$$\sigma = M\varepsilon \tag{4.1}$$

where M is the elastic constant. As is obvious from equation (4.1), the elastic constant has a dimension of stress (a commonly used unit is GPa). This relation is referred to as *Hooke's law*. In a more formal fashion, one needs to take into account the fact that both the stress and strain are (second rank) tensors (Chapter 1). For an isotropic material, there are only two modes of strain, namely volumetric and shear strain and the response of a material must be the same for all types of shear strain. Consequently one can write,

$$\sigma_{ij} = \lambda\,\delta_{ij}\sum_k \varepsilon_{kk} + 2\mu\varepsilon_{ij} \tag{4.2}$$

where λ and μ are the *Lamé constants* and δ_{ij} is Kronecker's delta ($\delta_{ij} = 1$ for $i = j$, $\delta_{ij} = 0$ for $i \neq j$). From equation (4.2), one obtains,

$$K = \lambda + \frac{2}{3}\mu \tag{4.3}$$

where K is the bulk modulus and

$$\sigma_{ij} = 2\mu\varepsilon_{ij} = 2\tau_{ij}\,(i \neq j) \tag{4.4}$$

where $\tau_{ij} = \sigma_{ij}/2$ is the shear stress (see Chapter 1) and μ is the shear modulus (rigidity).

Consider uni-axial compression along the x_1-direction ($\sigma_{11} \neq 0$ but other components of stress are 0). In this case equation (4.2) becomes

$$\sigma_{11} = \lambda\sum_k \varepsilon_{kk} + 2\mu\varepsilon_{11} \tag{4.5a}$$

and

$$\begin{aligned}\sigma_{22} = \sigma_{33} = 0 &= \lambda\sum_k \varepsilon_{kk} + 2\mu\varepsilon_{22}\\ &= \lambda\sum_k \varepsilon_{kk} + 2\mu\varepsilon_{33}.\end{aligned} \tag{4.5b}$$

Therefore if one defines *Poisson's ratio* by

$$\nu \equiv -\frac{\varepsilon_{22}}{\varepsilon_{11}} = -\frac{\varepsilon_{33}}{\varepsilon_{11}} \tag{4.6}$$

one obtains

$$\nu = \frac{\lambda}{2(\lambda+\mu)} = \frac{3K-2\mu}{2(3K+\mu)} = \frac{3(K/\mu)-2}{2(3(K/\mu)+1)}. \tag{4.7}$$

Problem 4.1

Derive equations (4.3) and (4.7) and show that for a liquid, Poisson's ratio is $\frac{1}{2}$.

Solution

From (4.2), one has $\sum_k \sigma_{kk} = (3\lambda + 2\mu)\sum_k \varepsilon_{kk} = 3P$. Now $\sum_k \varepsilon_{kk} = -\frac{\Delta V}{V}$ and $K \equiv -V\,(dP/dV) = -P(V/\Delta V)$. Therefore $K = \lambda + \frac{2}{3}\mu$.

From (4.5b), $\lambda(\varepsilon_{11} + 2\varepsilon_{22}) + 2\mu\varepsilon_{22} = 0$ and hence $\nu = -(\varepsilon_{22}/\varepsilon_{11}) = \lambda/2(\lambda+\mu)$. For a liquid, $\mu = 0$ so that $\nu = \frac{1}{2}$.

The velocities of seismic waves in an isotropic medium are given by

$$V_P = \sqrt{\frac{K+\frac{4}{3}\mu}{\rho}} \tag{4.8a}$$

and

$$V_S = \sqrt{\frac{\mu}{\rho}} \tag{4.8b}$$

where $V_{P,S}$ are *P*- and *S*-wave velocities. Therefore

$$\frac{V_P}{V_S} = \sqrt{\frac{K}{\mu} + \frac{4}{3}} = \sqrt{\frac{2(1-\nu)}{1-2\nu}} \tag{4.9}$$

or

$$\nu = \frac{(V_P/V_S)^2 - 2}{2\left[(V_P/V_S)^2 - 1\right]}. \tag{4.10}$$

For most parts of Earth's interior $\nu \approx 0.3$ (Chapter 17). For this value, $V_P/V_S = 1.87$. Some materials such as serpentine have an unusually high Poisson ratio due to the unusually low shear modulus compared to the bulk modulus (for serpentine $\nu \approx 0.35$). A high Poisson ratio is used to identify serpentine-rich regions (e.g., Kamiya and Kobayashi (2000), Omori *et al.* (2002)). The inner core has a very high Poisson ratio ($\nu \approx 0.44$) close to that of a liquid ($\nu = 0.5$), which was often attributed to the presence of partial melt (e.g., Loper and Fearn, 1983; Singh *et al.*, 2000). However, a high Poisson ratio may also be attributed to high pressure and temperature since pressure increases K more than μ and temperature decreases μ more than K. Recent first-principles calculations show that a high Poisson ratio of the inner core can be explained by a property of solid iron (Alfé *et al.*, 2002a; Steinle–Neumann *et al.*, 2001).

In most parts of this book, the approximation of elastic isotropy is good enough, but when one considers the elasticity of a crystal or seismic anisotropy, a more complete description of elastic properties is needed. In such a case, a more general form of the stress–strain relationship must be used, namely,

$$\sigma_{ij} = \sum_{kl} C_{ijkl}\varepsilon_{kl} \tag{4.11}$$

where C_{ijkl} are the elastic constants. Note that the elastic constants are the fourth-rank tensors that connect two second-rank tensors (stress and strain).

Elastic deformation does not dissipate energy and therefore by elastic deformation energy is stored in the system. The (potential) energy due to elastic deformation can be calculated as

$$U = \sum_{i,j=1}^{3} \int \sigma_{ij}\, d\varepsilon_{ij} = \frac{1}{2} \sum_{i,j,k,l=1}^{3} C_{ijkl}\varepsilon_{ij}\varepsilon_{kl}. \tag{4.12}$$

One finds that the elastic constants should not change if suffices (ij) are interchanged with (kl). Also from the symmetry of strain tensor ($\varepsilon_{ij} = \varepsilon_{ji}$, Chapter 1), one can show that the elastic constants remain unchanged when suffices (ij) are changed to (ji). Therefore

$$C_{ijkl} = C_{klij} = C_{ijlk} = C_{jikl}. \tag{4.13}$$

Because of the symmetry relationship (4.13), although the elastic constants are the fourth-rank tensors, one can write an elastic constant using two indices as C_{ij} where new indices are defined as $11 \rightarrow 1$, $22 \rightarrow 2$, $33 \rightarrow 3$, $23 \rightarrow 4$, $13 \rightarrow 5$ and $12 \rightarrow 6$. Thus $C_{1111} = C_{11}$, $C_{1213} = C_{65}$ etc. This abbreviated notation, the *Voigt notation*, is frequently used in the literature. With this notation, elastic constants can be represented by a 6×6 matrix. Note that although an elastic constant would look like a second-rank tensor (C_{ij}) with this notation, it is indeed a fourth-rank tensor and when one performs a coordinate transformation, one must go back to the full notation and follow the rule of transformation of a fourth-rank tensor (this is important when we discuss seismic anisotropy, see Chapter 21). From the symmetry relations it follows that

$$C_{ij} = C_{ji}. \tag{4.14}$$

Because of this symmetry relationship, the number of independent elastic constants is at most 21 ($= (6 \times 6 - 6)/2 + 6$). The number of independent elastic constants is further reduced by the symmetry of a material (Box 4.1, Table 4.1).

Box 4.1 Crystal structure and elastic constants

Crystal structure can be classified based on the symmetry of the unit cell and the atomic arrangement within the unit cell. The number in the parentheses is the number of independent elastic constants. Due to the different atomic arrangement within a unit cell, there are two classes of crystal symmetry for tetragonal and trigonal lattices that leads to two different numbers of independent elastic constants.

Cubic (3)	MgO, spinel (ringwoodite), garnet*
Hexagonal (5)	β-quartz, ice-I, ε-iron
Tetragonal (6,7)	Stishovite, rutile
Trigonal (6,7)	α-quartz, hematite, corundum
Orthorhombic (9)	Olivine, orthopyroxene, wadsleyite, $MgSiO_3$ perovskite**
Monoclinic (13)	Diopside, jadeite
Triclinic (21)	Albite, talc

* Garnet is usually cubic, but tetragonal garnet is reported from high-pressure synthesis (Kato and Kumazawa, 1985).

** $MgSiO_3$ perovskite is usually orthorhombic, but perovskite with a lower symmetry could exist at higher temperatures (and pressures).

TABLE 4.1 Elastic constant matrix of typical crystals.

Isotropic

$$\begin{bmatrix} C_{11} & C_{12} & C_{12} & 0 & 0 & 0 \\ C_{12} & C_{11} & C_{12} & 0 & 0 & 0 \\ C_{12} & C_{12} & C_{11} & 0 & 0 & 0 \\ 0 & 0 & 0 & \frac{1}{2}(C_{11}-C_{12}) & 0 & 0 \\ 0 & 0 & 0 & 0 & \frac{1}{2}(C_{11}-C_{12}) & 0 \\ 0 & 0 & 0 & 0 & 0 & \frac{1}{2}(C_{11}-C_{12}) \end{bmatrix}$$

Cubic

$$\begin{bmatrix} C_{11} & C_{12} & C_{12} & 0 & 0 & 0 \\ C_{12} & C_{11} & C_{12} & 0 & 0 & 0 \\ C_{12} & C_{12} & C_{11} & 0 & 0 & 0 \\ 0 & 0 & 0 & C_{44} & 0 & 0 \\ 0 & 0 & 0 & 0 & C_{44} & 0 \\ 0 & 0 & 0 & 0 & 0 & C_{44} \end{bmatrix}$$

Hexagonal

$$\begin{bmatrix} C_{11} & C_{12} & C_{13} & 0 & 0 & 0 \\ C_{12} & C_{11} & C_{13} & 0 & 0 & 0 \\ C_{13} & C_{13} & C_{33} & 0 & 0 & 0 \\ 0 & 0 & 0 & C_{44} & 0 & 0 \\ 0 & 0 & 0 & 0 & C_{44} & 0 \\ 0 & 0 & 0 & 0 & 0 & \frac{1}{2}(C_{11}-C_{12}) \end{bmatrix}$$

Orthorhombic

$$\begin{bmatrix} C_{11} & C_{12} & C_{13} & 0 & 0 & 0 \\ C_{12} & C_{22} & C_{23} & 0 & 0 & 0 \\ C_{13} & C_{23} & C_{33} & 0 & 0 & 0 \\ 0 & 0 & 0 & C_{44} & 0 & 0 \\ 0 & 0 & 0 & 0 & C_{55} & 0 \\ 0 & 0 & 0 & 0 & 0 & C_{66} \end{bmatrix}$$

4.2.2. The Cauchy relation

Under certain conditions, the elastic constant has full symmetry with respect to the indices: C_{ijkl} does not change with any permutation of indices. This leads to the following relations,

$$\begin{aligned} C_{23} &= C_{44} \\ C_{31} &= C_{55} \\ C_{12} &= C_{66} \\ C_{14} &= C_{56} \\ C_{25} &= C_{46} \\ C_{36} &= C_{45} \end{aligned} \tag{4.15}$$

which are referred to as the *Cauchy relations* (for a cubic crystal for which $C_{44} = C_{55} = C_{66}$ and $C_{23} = C_{31} = C_{12}$, these relations are reduced to $C_{12} = C_{44}$). When the Cauchy relations are met, there will be only 15 independent elastic constants. The Cauchy relation is satisfied when (1) all forces are central forces (a force that depends only on the distance between atoms but not on the angles between two (or more) chemical bonds), (2) every atom is at the center of symmetry and (3) the crystal has no internal strain (see e.g., BORN and HUANG, 1954).

Table 4.2 shows the elastic constants of some minerals at ambient conditions. The Cauchy relation is satisfied well in many alkali halides and spinel, but not in MgO and most of the silicates and many metals. Since the conditions (2) and (3) are met in these materials, this means that the inter-atomic forces in these materials contains non-central components. Contribution from covalent bonding is one cause for the non-central force.

An elastic constant represents resistance for deformation with certain geometry. Consider C_{11}. This elastic constant relates elongation strain along 1-direction (ε_{11}) ([100] direction: for a definition of directions in a crystal see Box 4.2, Fig. 4.1) with compressional stress along 1-direction (σ_{11}) (for a cubic crystal, $\sigma_{11} = C_{11}\varepsilon_{11} + C_{12}\varepsilon_{22} + C_{12}\varepsilon_{33}$). Therefore C_{11} represents the stiffness of a material for compression along the [100] direction. Similarly applying the force balance equation appropriate for σ_{12}, one gets (for a cubic crystal) $\sigma_{12} = C_{44}\,\varepsilon_{12}$. Therefore C_{44} represents the stiffness associated with the shear deformation on the (100) plane along the [010] direction (ε_{12}).

It should be emphasized that a cubic crystal is, in general, elastically anisotropic. This is different from the symmetry of properties that can be represented by a second-rank tensor (e.g., the refractive index, diffusion coefficients, thermal and electrical conductivity). For properties that can be written by a second-rank tensor, the properties are isotropic if the material has cubic symmetry. But for properties such as elastic constants and viscosity that are given by a fourth-rank tensor, the cubic symmetry does not imply isotropy. A remarkable example is MgO. MgO has highly anisotropic elasticity and its anisotropy changes with pressure, and at

TABLE 4.2 Adiabatic elastic constants of some materials at ambient conditions (unit is GPa) (from Bass (1995)).

Cubic	C_{11}	C_{12}	C_{44}
Au	191	162	42.4
Ag	122	92	45.5
Cu	169	122	75.3
α-Fe	230	135	117
NaCl	49.1	12.8	12.8
KCl	40.5	6.9	6.3
Periclase (MgO)	294	93	155
Pyrope ($Mg_3Al_2Si_3O_{12}$)	296	111	92
Spinel ($MgAl_2O_4$)	282	154	154
Ringwoodite (Mg_2SiO_4)	327	112	126

Hexagonal	C_{11}	C_{33}	C_{44}	C_{12}	C_{13}
Ice-I (H_2O)	13.5	14.9	3.09	6.5	5.9
β-SiO_2	117	110	36	16	33

Orthorhombic	C_{11}	C_{22}	C_{33}	C_{44}	C_{55}	C_{66}	C_{12}	C_{13}	C_{23}
Enstatite ($MgSiO_3$)	225	178	214	78	76	82	72	54	53
Forsterite (Mg_2SiO_4)	328	200	235	67	81	81	69	69	73
Wadsleyite (Mg_2SiO_4)	360	383	273	112	118	98	75	110	105
Perovskite ($MgSiO_3$)	515	525	435	179	202	175	117	117	139

extremely high pressure corresponding to the deep lower mantle, its anisotropy is very large: *S*-wave velocity anisotropy is $\sim 20\%$ (see section 4.7). But the thermal (and electrical) conductivity in MgO is isotropic.

4.3. Isothermal versus adiabatic elastic constants

An elastic constant represents the resistance of a material for deformation with small strain. Note that the resistance of a material for deformation (i.e., the stored energy associated with deformation) is dependent on the *process* of deformation. We can distinguish two cases: isothermal and adiabatic deformation and corresponding elastic constants. When stress is applied slowly, then as strain develops in a sample, a sample maintains thermal equilibrium with the surroundings. The elastic constant corresponding to this way of deformation is called an *isothermal elastic constant*. Alternatively, stress can be applied quickly so that no energy exchange occurs between the sample and the surroundings during deformation. The elastic constant corresponding to this type of deformation is called an *adiabatic elastic constant*. Whether deformation is isothermal or adiabatic depends on the time-scale of deformation relative to the time-scale of thermal diffusion. Let us consider a case of elastic wave propagation. The time-scale of thermal diffusion, τ_{th}, in this case is given by $\tau_{\mathrm{th}} \sim \lambda^2/\pi^2\kappa$ where λ is the wavelength of elastic waves and κ is thermal diffusivity. If $\tau_{\mathrm{th}}\omega = 2V_{P,S}\lambda/\pi\kappa \gg 1$ (where $V_{P,S}$ is elastic wave velocity, and ω is the frequency of the elastic wave or the inverse of the time-scale of deformation), deformation is adiabatic and if $\tau_{\mathrm{th}}\omega \lll 1$, deformation is isothermal. κ for most minerals is $\sim 10^{-6}$ m^2/s, and $V_{P,S} \sim 10^4$m/s , and the typical wavelength of seismic waves is $\sim 10^3$–10^4 m and the wavelength of ultrasonic waves is $\sim 10^{-3}$ m, hence $\tau_{\mathrm{th}}\omega \approx 10^{13-14}$ for seismic waves, and $\tau_{\mathrm{th}}\omega \approx 10^7$ for ultrasonic waves so for all seismic wave propagation and most laboratory studies, deformation is adiabatic. The only exception is a static measurement, such as X-ray measurement, which is isothermal.

Let us consider how isothermal elastic constants are related to adiabatic elastic constants. In general, one can write,

$$dP = \left(\frac{\partial P}{\partial V}\right)_S dV + \left(\frac{\partial P}{\partial S}\right)_V dS \tag{4.16}$$

Box 4.2 The Miller index

Crystallographic planes and orientations in a crystal are given by symbols such as (100) for plane and [010] for orientation. These are called the *Miller indices*.

A crystal is made of a periodic array of atoms, and one can define the smallest unit that preserves the symmetry of the crystal. This unit cell is defined by three basis vectors ($\boldsymbol{a}_1$, $\boldsymbol{a}_2$, $\boldsymbol{a}_3$). This set of basis vectors is used to define the coordinate system. Using this coordinate system, a plane that contains lattice points (positions of atoms) can be written as

$$hx_1 + kx_2 + lx_3 = n$$

where h, k, l and n are integers. This plane is called a (hkl) plane. For example, (100) is a plane normal to the $\boldsymbol{a}_1$ direction (the equation for this plane is $x_1 = n$). Similarly a plane $(\bar{1}10)$ corresponds to the equation $-x_1 + x_2 = n$. There are several crystallographically equivalent planes for a given structure. For example, in a cubic crystal, there are three equivalent (100)-type planes: (100), (010) and (001). A symbol {100} is used to represent such equivalent planes.

Orientations in a crystal are shown by a symbol [uvw]. u, v and w are the smallest integers that represent the direction cosines of a direction in a crystal coordinate. For example, [100] is the direction parallel to the $\boldsymbol{a}_1$ direction and [110] is the direction on the (001) plane that is 45° between the $\boldsymbol{a}_1$ and $\boldsymbol{a}_2$ directions. Similar to planes, there are three equivalent [100]-type directions ([100], [010] and [001]). A symbol ⟨100⟩ is used to denote such equivalent directions.

and

$$dT = \left(\frac{\partial T}{\partial V}\right)_S dV + \left(\frac{\partial T}{\partial S}\right)_V dS. \tag{4.17}$$

From (4.16), one gets,

$$\frac{dP}{dV} = \left(\frac{\partial P}{\partial V}\right)_S + \left(\frac{\partial P}{\partial S}\right)_V \frac{dS}{dV}. \tag{4.18}$$

For isothermal deformation ($dT = 0$) one gets from (4.17),

$$\left(\frac{\partial S}{\partial V}\right)_T = -\left(\frac{\partial T}{\partial V}\right)_S \left(\frac{\partial S}{\partial T}\right)_V. \tag{4.19}$$

Hence, from (4.18) and (4.19),

$$\begin{aligned}\left(\frac{\partial P}{\partial V}\right)_T &= \left(\frac{\partial P}{\partial V}\right)_S - \left(\frac{\partial P}{\partial S}\right)_V \left(\frac{\partial T}{\partial V}\right)_S \left(\frac{\partial S}{\partial T}\right)_V \\ &= \left(\frac{\partial P}{\partial V}\right)_S - \left(\frac{\partial P}{\partial T}\right)_V \left(\frac{\partial T}{\partial V}\right)_S.\end{aligned} \tag{4.20}$$

Using the definition, $K_{T,S} = -V(\partial P/\partial V)_{T,S}$ where $K_{T,S}$ are isothermal and adiabatic bulk modulus respectively, one has

$$K_T = K_S - \frac{\alpha_{th}^2 K_T^2 TV}{C_V} \tag{4.21}$$

where $\alpha_{th} \equiv \frac{1}{V}(\partial V/\partial T)_P$ is thermal expansion and the relations

$$\left(\frac{\partial P}{\partial T}\right)_V = \alpha_{th} K_T \tag{4.22}$$

and

$$\left(\frac{\partial T}{\partial V}\right)_S = -\frac{\alpha_{th} K_T T}{C_V} \tag{4.23}$$

have been used (Problem 4.2).

Using the definition,

$$\gamma_{th} \equiv \frac{\alpha_{th} K_T V}{C_V} \tag{4.24}$$

where γ_{th} is the (thermodynamic) *Grüneisen parameter*, and V is the molar volume, equation (4.21) can be written as (Problem 4.2),[1]

$$K_S = K_T(1 + \gamma_{th}\alpha_{th} T). \tag{4.25}$$

With the typical values of materials parameters and temperature ($\gamma_{th} \sim 1$, $\alpha_{th} \sim 3 \times 10^{-5}$ K^{-1} and $T \sim 2000$ K), K_S is larger than K_T by ~6%.

One notes that a combination of equations (4.22) and (4.24) gives $(\partial P/\partial T)_V = \alpha_{th} K_T = \gamma_{th} \frac{C_V}{V}$. Integrating this relation, one obtains

$$\begin{aligned}P_{th} = P(T_2) - P(T_1) &\approx \frac{\gamma_{th}}{V} \int_{T_1}^{T_2} C_V \, dT \\ &= \gamma_{th} \frac{E(T_2) - E(T_1)}{V}\end{aligned} \tag{4.26}$$

where $P(T_{1,2})$ is pressure at temperature of $T_{1,2}$, and $E(T_{1,2})$ is internal energy at temperature of $T_{1,2}$. P_{th} is often referred to as *thermal pressure*.

[1] For the shear modulus, there is no difference between adiabatic and isothermal modulus.

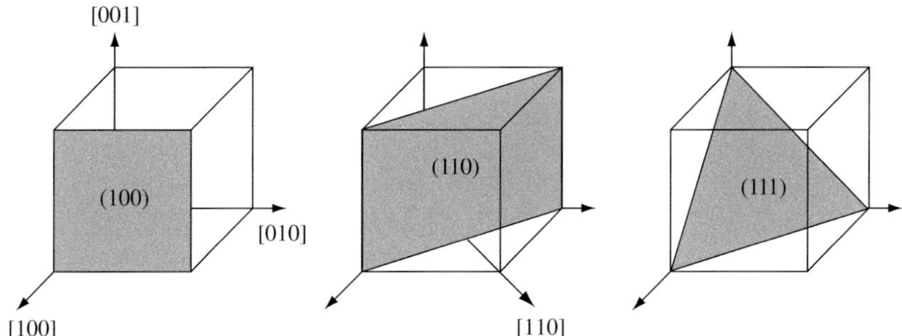

FIGURE 4.1 Definitions of crystallographic planes and orientations.

Problem 4.2*

Derive the relations (4.21) through (4.25).

Solution

If a quantity Z is a function of X and Y ($Z = Z(X, Y)$), then $dZ = (\partial Z/\partial X)_Y\, dX + (\partial Z/\partial Y)_X\, dY$. By taking $dZ=0$, one gets $0 = (\partial Z/\partial X)_Y + (\partial Z/\partial Y)_X \cdot (\partial Y/\partial X)_{dZ}$. For $Y=V$, $Z=P$ and $X=T$, we get $0=(\partial P/\partial T)_V+(\partial V/\partial T)_P(\partial P/\partial V)_T$. Using the definition of bulk modulus and thermal expansion ($K_T = -V(\partial P/\partial V)_T$ and $\alpha_{\rm th} \equiv \frac{1}{V}(\partial V/\partial T)_P$) one gets $(\partial P/\partial T)_V = \alpha_{\rm th} K_T$. Similarly for $Y=S$, $Z=T$ and $X=V$, $0=(\partial T/\partial V)_S+(\partial S/\partial V)_T(\partial T/\partial S)_V$ and using the Maxwell relation $(\partial S/\partial V)_T=(\partial P/\partial T)_V$, one gets $0=(\partial T/\partial V)_S+(\partial P/\partial T)_V(\partial T/\partial S)_V$. Now using $(\partial P/\partial T)_V = \alpha_{\rm th} K_T$ and the definition of the specific heat $C_V \equiv T(\partial S/\partial T)_V$, one obtains $(\partial T/\partial V)_S = -\alpha_{\rm th} \cdot K_T T/C_V$. Inserting these relations into equation (4.20), one obtains equation (4.21). Using the definition of the thermodynamic Grüneisen parameter, $\gamma_{\rm th} = \alpha_{\rm th} K_T V/C_V$, equation (4.21) is converted to equation (4.25), $K_S = K_T(1 + \gamma_{\rm th}\,\alpha_{\rm th} T)$.

4.4. Experimental techniques

Elastic constants in materials can be measured using a variety of techniques. Techniques may be classified into three categories: (1) static measurements, (2) dynamic measurements using wave propagation and (3) dynamic measurements using phonon–photon interactions. Understanding the basics of the experimental techniques is important because the frequency at which the elastic constants are measured is different among various techniques and elastic constants depend on frequencies (e.g., adiabatic versus isothermal elastic constants (see section 4.3), and the influence of anelasticity (see Chapters 3 and 8)). Also each technique has its own limitations and care must be exercised in using experimental data to interpret seismological observations.

4.4.1. Static measurements of elastic constants

Elastic constants of a material can be determined by X-ray diffraction. X-ray diffraction from a crystal occurs when the following condition (the Bragg condition) is met,

$$2d \sin\theta = n\lambda \tag{4.27}$$

where d is the spacing of adjacent lattice planes, θ is the angle of incident (and diffracted) X-ray with respect to the lattice plane, n is an integer and λ is the wavelength of the X-ray. Consequently, by measuring the angle θ for a given wavelength λ or by measuring λ (i.e., energy) for a fixed value of θ, one can determine the lattice spacings, d. When stress is applied, the lattice spacings will be modified. By measuring small changes in lattice spacings, one can determine the elastic constants if one knows the magnitude of stress.

When the variations in lattice spacing due to non-hydrostatic stress are measured, a full elastic constant matrix can be determined from the X-ray diffraction (Singh, 1993; Singh *et al.*, 1998; Uchida *et al.*, 1996). In this technique, one determines the spacings of many crystallographic planes in a polycrystalline sample under a given stress geometry (e.g., tri-axial compression). When a polycrystalline sample is under non-hydrostatic stress, then depending on the orientation of the plane relative to the applied macroscopic stress orientation, the values of lattice spacings will be different. Therefore by measuring the lattice spacings as a function of the orientation of the direction of a diffracted X-ray beam with respect to the applied stress, one can determine the elastic constants. One of the advantages of this technique is that when high strength X-ray beams are available (e.g., at synchrotron facilities), one can apply this technique to a small sample under high pressures. Elastic constant matrices of various materials have been determined by this technique to pressures exceeding 50 GPa (e.g., DUFFY *et al.*, 1999; MAO *et al.*, 1998). However, the uncertainties in the elastic constants determined by this technique are large because this technique relies on the subtle changes in lattice spacings and the uncertainties in the estimated stress magnitude are propagated to the uncertainties in the estimated elastic constants. For this reason, a relatively large stress is applied to a sample in this method. This causes plastic deformation that modifies the stress field around individual grains causing a systematic variation in the change in lattice spacing. This leads to a biased estimate of elastic anisotropy (WEIDNER *et al.*, 2004).

4.4.2. Ultrasonic techniques

The velocity of elastic waves can be determined by the measurement of the travel times of elastic waves similar to seismological measurements. In these measurements, the wavelength of the waves used must be significantly smaller than the sample size and consequently, the frequency of waves must be higher than $\sim V_{P,S}/L$ where L is the sample dimension. For ~ 1 mm sample with $V_{P,S} \sim 5$–10 km/s, the frequency must be higher than $\sim 10^7$ Hz. In this technique, one generates ultrasonic waves using a piezoelectric transducer and records the elastic waves that have passed through the sample using another (or the same) transducer. From the measured travel time, one can determine the elastic wave velocities from the known sample length. This technique has been used for pressures up to ~ 15 GPa and temperatures to ~ 1500 K (e.g., LIEBERMANN, 2000). For an isotropic material, the measurements of *P*- and *S*-wave velocities provide complete information on elasticity (if the density is known). For anisotropic materials such as single crystals or polycrystalline samples with anisotropic fabrics, measurements must be made along various directions and polarizations (for *S*-waves) to fully characterize the elastic properties. Uncertainties in the determined elastic constants in this technique are due to the uncertainties in travel time measurements and those in sample length estimates. High-resolution measurements of elastic wave velocities, with an accuracy of $\sim 0.1\%$, can be made using this technique when the travel times are measured using interferometric techniques (e.g., JACKSON and NIESLER, 1982; LIEBERMANN, 2000). Elastic properties determined by this technique correspond to adiabatic elastic constants. CHEN *et al.* (1997) extended this technique to GHz frequencies, which allows us to determine the elastic constants of samples with a dimension of the order of 100 μm.

4.4.3. Opto-elastic techniques

Elastic waves (phonons) interact with light (photons). One of these interactions is known as *Brillouin scattering*. At a finite temperature, phonons are excited in a crystal. These phonons interact with photons. Photons are scattered by phonons. Scattering occurs either in an elastic or an inelastic fashion. Upon inelastic scattering (i.e., scattering in which photon energy is changed), the frequency of photons is modified. The change in frequency is given by

$$\frac{\Delta\omega}{\omega} = \pm 2n\frac{V}{c}\sin\frac{\theta}{2} \tag{4.28}$$

where V is the elastic wave velocity, c is the velocity of light (photons), n is an integer and θ is the angle between the directions of incident light and scattered light (ANDERSON *et al.*, 1969). This technique can be applied to small samples (say ~ 100 μm) and a full elastic constant matrix can be determined from small samples under high pressures (WEIDNER, 1987). This technique has been applied to a number of minerals using a diamond anvil cell and elastic constants have been determined at pressures beyond 20 GPa (e.g., SINOGEIKIN and BASS, 1999; ZHA *et al.*, 1998). Elastic properties determined by this technique correspond to adiabatic elastic constants. The uncertainties of the results obtained from this technique (typically $\sim 1\%$) are larger than those of the ultrasonic measurements but are better than the static X-ray measurements. One major limitation of this technique is that it can be applied only to optically transparent materials.

Some representative results of elastic constants for typical Earth materials are summarized in Table 4.2. These data are at ambient pressure and temperature for a fixed composition. As we will learn later, pressure, temperature and composition have important effects on elastic properties.

4.4.4. Velocity estimates from the phonon density of state

Acoustic waves in a crystal are some types of phonon (see Box 4.3). The distribution of a number of phonons as a function of energy (i.e., wavenumber) is called the phonon *density of state*. Using non-elastic scattering of X-rays (or neutrons) by phonons, one can determine the phonon density of state (e.g., STURHAHN *et al.*, 1995). The low-energy part of the phonon density of state is related to the acoustic wave velocities (see Box 4.3), so that such measurements provide estimates of acoustic wave velocities. The errors in estimated velocities from this technique are high (a few %), but the advantage of this technique is that it can be applied to a small optically opaque material. This technique has been applied to estimate the sound velocities of iron by LIN *et al.* (2005).

4.5. Some general trends in elasticity: Birch's law

In general, the elastic properties of materials depend on a number of parameters, such as temperature, pressure, crystal structure and chemical composition. However, based on a large number of data, BIRCH (1961) found that, to a good approximation, the elastic properties of materials are controlled mostly by the mean atomic number and the molar volume (specific density): other factors such as the crystal structure play relatively minor roles.[2] This empirical law, *Birch's law*, captures some essence of elasticity that simplifies our understanding of the elastic properties of minerals in Earth's interior. Therefore I will first summarize Birch's law and its physical interpretation and some important applications, and then discuss the crystal structure effects and other issues that are the exceptions of Birch's law. However, I note that Birch's law is an empirical law and in some cases, this law is followed only approximately.

BIRCH (1961) established an empirical rule that the changes in elastic constants with temperature and pressure in a given material (i.e., with a given chemical composition) occur primarily through the change in density. This is referred to as *Birch's law* (Fig. 4.2). In his original paper, BIRCH (1961) summarized his results in the following formula,

$$V_{P,S}(\rho, \bar{M}) = a_{P,S}(\bar{M}) + b_{P,S} \cdot \rho \tag{4.29}$$

[2] Note, however, that the crystal structure plays an essential role for elastic *anisotropy*.

Box 4.3 The Debye model

The lattice vibration in a crystal occurs with various frequencies. The distribution of vibrational frequency is a key factor that characterizes the way in which lattice vibration influences the thermal properties of crystals. In the Debye model, the lattice vibration is approximated as a plane elastic wave, and the displacement of the atoms can be written as $\boldsymbol{u}(\boldsymbol{x}, t) = \boldsymbol{u}_0 \exp[-2\pi i(\boldsymbol{k} \cdot \boldsymbol{x} - \nu_l t)]$ where $\boldsymbol{k}$ is the wavenumber, x is distance, $\nu_l \equiv \omega_l/2\pi$ is the angular frequency of vibration with a mode l. ν_l is related to the velocity of elastic waves as $V_l = \nu_l/|\mathbf{k}|$ $(V_l \equiv V_{P,S})$, and t is time. The displacement associated with an elastic wave must be consistent with the discrete nature of a lattice. One way is to consider a wave in a finite body and impose the periodic boundary condition, i.e., $\boldsymbol{u}(x_j, t) = \boldsymbol{u}(x_j + L_j, t)$, where L_j $(j = 1, 2, 3)$ is the length of the edge of a crystal. This leads to $k_{1,2,3} = n_{1,2,3}/L_{1,2,3} = n_{1,2,3}/V^{1/3}$ where n_j is an integer and V is the volume of a crystal. Therefore, in the wave-number space ($\boldsymbol{k}$-space), the allowable wave numbers are distributed with equal distance, so that the number of wave numbers for $|\boldsymbol{k}|$ and $|\boldsymbol{k}| + d|\boldsymbol{k}|$ is $4\pi V|\boldsymbol{k}|^2 d|\boldsymbol{k}| = 4\pi V\nu^2 d\nu/V_l^3$. If one defines the distribution function of frequencies of lattice vibration in such a way that $f(\nu)\, d\nu$ is the number of modes whose frequency is between ν and $\nu + d\nu$, then $f(\nu) = 4\pi V(2/V_S^3 + 1/V_P^3)\nu^2$. The maximum frequency, ν_D, is determined by the fact that the total number of modes is $\int_0^{\nu_D} f(\nu)\, d\nu = 3N$ (N: number of atoms). Therefore $\nu_D = \left(\frac{9}{4\pi} \frac{N}{V(2/V_S^3 + 1/V_P^3)}\right)^{1/3} = k_B T_D/h$ where ν_D is the Debye frequency and T_D is the Debye temperature. Because any thermodynamic parameter is controlled by the Boltzmann factor $\exp(-h\nu/k_B T)$, and the frequency distribution is characterized by a single parameter, the Debye frequency (ν_D), any thermodynamic properties are characterized by a single non-dimensional parameter $h\nu_D/k_B T = T_D/T$. This model captures some characteristics of thermal properties of solids, and is useful in geophysics because it relates the thermal properties of solids with the elastic properties that can be inferred from seismology.

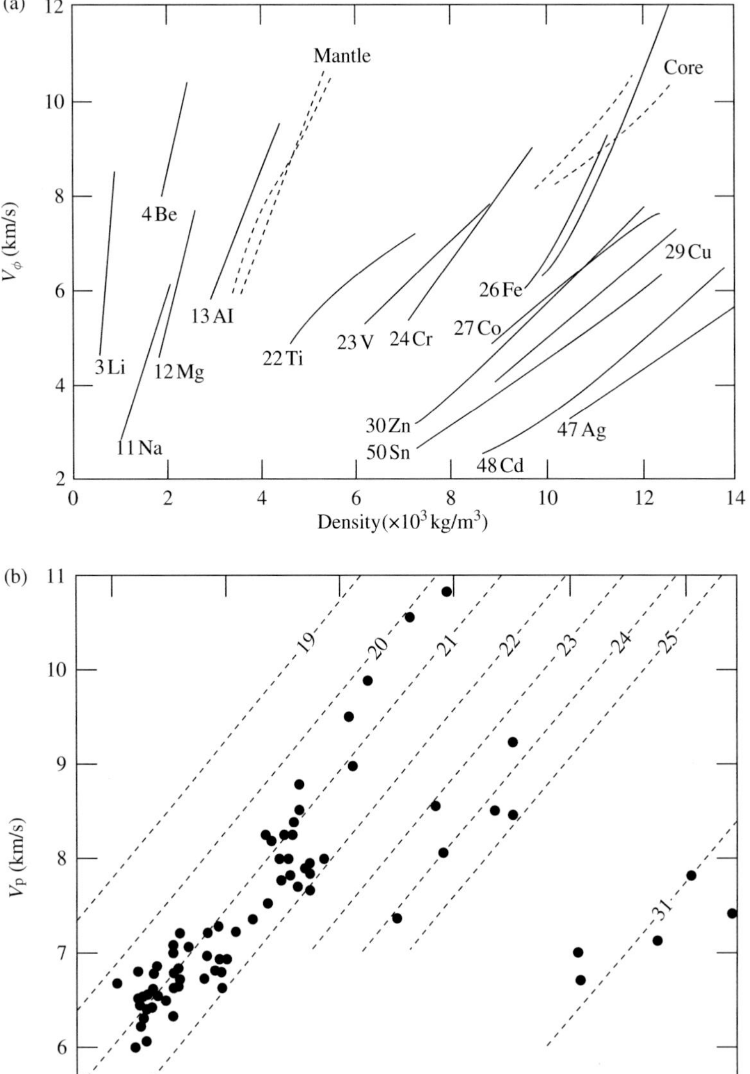

FIGURE 4.2 The Birch diagram for elastic wave velocities and density (after Birch, 1961, 1964). (a) For metals (a bulk sound velocity versus density plot) and (b) for silicates (a P-wave velocity versus density plot, numbers are the mean atomic weight). The elastic velocities are primarily determined by the density and the mean atomic number.

in which seismic wave velocities are a linear function of density and the effect of chemical composition is expressed in terms of mean atomic weight $\bar{M}$ ($\bar{M}$ for Mg_2SiO_4 is $\frac{1}{7}(24.305 \times 2 + 28.086 + 15.999 \times 4) = 20.099$; for atomic weight see Table 4.3). Although Birch's law is usually stated in terms of density and chemical composition as shown in equation (4.29), it is more instructive to write it in terms of molar volume, V, and chemical composition, X. Considering the fact that the seismic wave velocities are determined by elastic constants and density or molar volume (see equation (4.8)), one can reformulate Birch's law as,

$$C(T, P; X) = C[V(T, P); X] \tag{4.30}$$

where C is an elastic constant, X represents chemical composition such as $X_{Fe} = \mathrm{Fe}/(\mathrm{Fe} + \mathrm{Mg})$ or $\bar{M}$, and V is the molar volume. The reason for the use of the (molar) volume rather than the density is to distinguish the effects of volume change due to compression or thermal expansion from the density change due to the

TABLE 4.3 Ionic (and atomic) radii (in nm) and atomic weight of some atoms (ions) (in $\times 10^{-3}$ kg/mol).

	Ionic (or atomic) radius[1]	Atomic weight
O^{2-}	0.140 (VI)	15.999
Na^{+}	0.102 (VI)	22.990
Mg^{2+}	0.072 (VI)	24.305
Al^{3+}	0.039 (IV), 0.053 (VI)	26.982
Si^{4+}	0.026 (IV), 0.040 (VI)	28.086
K^{+}	0.138 (VI)	39.098
Ca^{2+}	0.100 (VI)	40.078
Ti^{2+}	0.086 (VI)	47.88
Cr^{3+}	0.0615 (VI)	51.996
Fe^{2+}	0.077^{HS}, 0.061^{LS} (VI)[2]	55.847
Fe^{3+}	0.0645^{HS}, 0.055^{LS} (VI)	–
Ni^{2+}	0.070 (VI)	58.693
Ni^{3+}	0.060^{HS}, 0.056^{LS} (VI)	–
Ge^{4+}	0.040 (IV)	72.6
Rb^{+}	0.149 (VI)	85.468
Sr^{2+}	0.116 (VI), 0.125 (VIII)	87.62
Nd^{3+}	0.0995 (VI), 0.112 (VIII)	144.24
Sm^{3+}	0.0964 (VI), 0.109 (VIII)	150.36
Lu^{3+}	0.0848 (VI), 0.097 (VIII)	174.967
Hf^{4+}	0.071 (VI), 0.083 (VIII)	178.49
Re^{4+}	0.063 (VI)	186.207
Os^{4+}	0.063 (VI)	190.23
Pb^{2+}	0.118 (VI), 0.129 (VIII)	207.2
Pb^{4+}	0.0775 (VI), 0.094 (VIII)	–
Th^{4+}	0.100 (VI)	232.038
U^{4+}	0.097 (VI), 0.100 (VIII)	238.029
U^{5+}	0.076 (VI)	–
U^{6+}	0.048 (IV), 0.075 (VI)	–
He	0.108 (VI)[3]	4.003
Ne	0.121 (VI)[3]	20.180
Ar	0.164 (VI)[3]	39.948
Kr	0.178 (VI)[3]	83.798
Xe	0.196 (VI)[3]	131.293

[1] Shannon–Prewitt ionic radius (from Bloss, 1971) except for rare gas elements. The roman numbers in the parentheses are the coordination numbers (i.e., the number of ions with a different electrostatic charge surrounding the particular ion).
[2] HS and LS indicate the ionic radius for high-spin state and for low-spin state respectively. Transition metal ions such as Fe^{2+} can change their spin state by compression. High-spin (low-spin) state is preferred at relatively low (high) pressures (Burns, 1970).
[3] Atomic radii for rare gas elements (from Zhang and Xu, 1995).

change in chemical composition i.e., the change in density by replacing atoms with different species (e.g., Mg ⇔ Fe).

Let us consider a case where the chemical composition (X) is fixed and consider the effects of pressure and temperature. In such a case,

$$\left(\frac{\partial \log C}{\partial T}\right)_{P,X} = \left(\frac{\partial \log C}{\partial \log V}\right)_X \left(\frac{\partial \log V}{\partial T}\right)_{P,X} = \left(\frac{\partial \log C}{\partial \log V}\right)_X \alpha_{th} \tag{4.31a}$$

and

$$\left(\frac{\partial \log C}{\partial P}\right)_{T,X} = \left(\frac{\partial \log C}{\partial \log V}\right)_X \left(\frac{\partial \log V}{\partial P}\right)_{T,X} = -\left(\frac{\partial \log C}{\partial \log V}\right)_X \frac{1}{K_T}. \tag{4.31b}$$

When Birch's law is valid, then a material (with a given composition) has the same elastic properties if the molar volume is the same, irrespective of the actual pressure and temperature. This is often referred to as *Birch's law of correspondent state*.

From relations (4.31),

$$\frac{\{\delta \log C\}_P}{\{\delta \log C\}_T} = -\frac{\delta P/K_T}{\delta T \cdot \alpha_{th}} \tag{4.32}$$

where $\{\delta \log C\}_{T,P}$ is the variation in $\log C$ due to the variation in T and P. For typical values of thermal expansion ($\sim 2 \times 10^{-5}$ K^{-1}) and of bulk modulus (~ 150 GPa), and for $\delta P = 30$ GPa and $\delta T = 1000$ K, this ratio is ~ 10. Therefore pressure has a larger effect on elastic constants (seismic wave velocities) than temperature in Earth's interior. In fact, for typical values of a pressure derivative of elastic constants ($dC/dP \sim$ 2–5), elastic constants could change by a factor of 2–5 throughout the mantle, whereas the changes in elastic constants due to temperature are of the order of 20% or less. However, temperature effects are critical when lateral variation in seismic wave velocities (i.e., seismic tomography) is considered. It should also be noted that the pressure effects are usually different for different elastic constants, C_{ij}. Consequently, the elastic anisotropy can change with pressure (Figs. 4.3b and 4.10).

Birch's law implies that the molar volume, i.e., the inter-atomic distance, plays an essential role in controlling the elastic properties. The elastic constants are related to the vibrational frequency. In the *Debye model* (Box 4.3), all lattice vibration modes are assumed to be acoustic vibration and in such a case, $\omega_i \propto V_i/a$ where V_i is the velocity of elastic

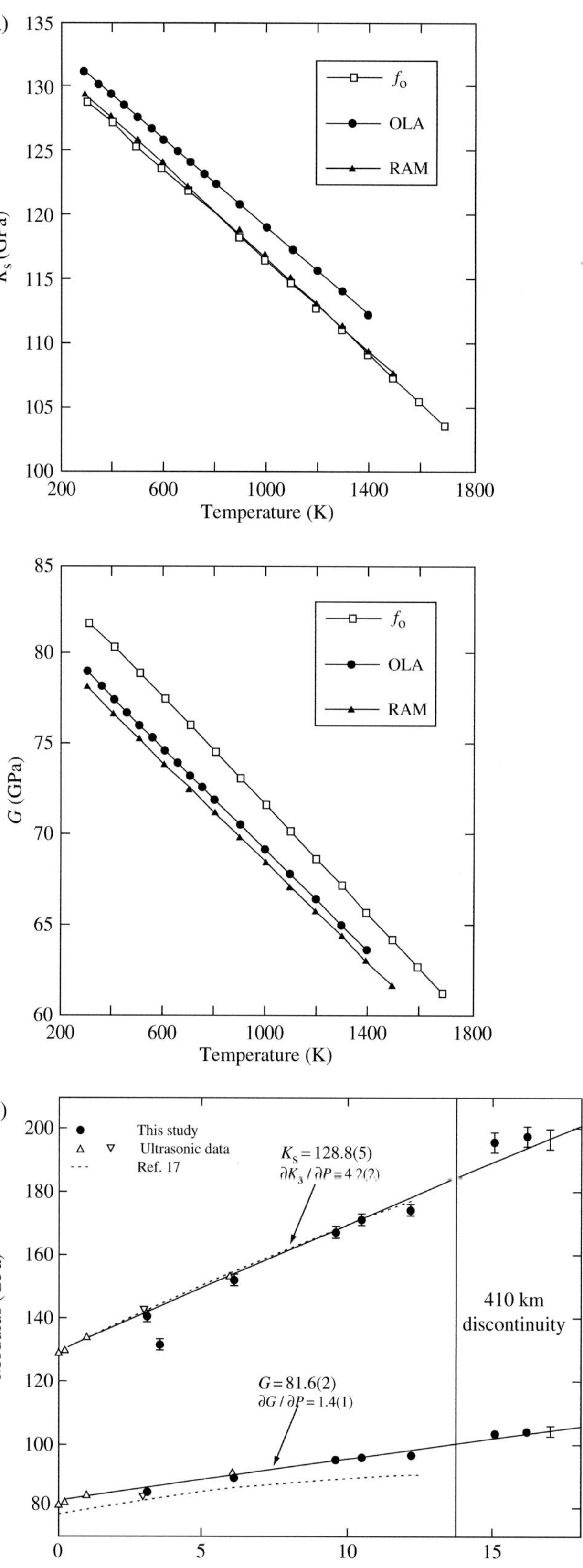

FIGURE 4.3 Elastic constants of olivine as a function of (a) temperature (Isaak, 1992) and (b) pressure (Duffy *et al.*, 1995) K_S is the adiabatic bulk modulus and G is the shear modulus (μ).

wave of ith mode and a is the inter-atomic distance. Velocity is related to elastic constants and density through $V_i = \sqrt{C_i/\rho}$ where C_i is the relevant elastic constant (for $V_S = \sqrt{\mu/\rho}$, $C_S = \mu$, and for $V_P = \sqrt{(K_S + \frac{4}{3}\mu)/\rho}$, $C_P = K_S + \frac{4}{3}\mu$, and for $V_\phi = \sqrt{K_S/\rho}$, $C_\phi = K_S$). Consequently, the Grüneisen parameter defined as the volume dependence of the frequency of lattice vibration (this is the original definition by Grüneisen (1912)),

$$\gamma \equiv -\left(\frac{\partial \log \omega}{\partial \log V}\right)_T = \left(\frac{\partial \log \omega}{\partial \log P}\right)_T \tag{4.33}$$

plays an important role in understanding the pressure and temperature dependence of elasticity. If we distinguish the Grüneisen parameter for the bulk modulus from that for the shear modulus, then

$$\begin{aligned}\gamma_K &\equiv -\left(\frac{\partial \log \omega_\phi}{\partial \log V}\right)_T = -\frac{1}{2}\left(\frac{\partial \log K_S}{\partial \log V}\right)_T - \frac{1}{6} \\ &= \frac{1}{2}\frac{K_T}{K_S}\left(\frac{\partial K_S}{\partial P}\right)_T - \frac{1}{6} \approx \frac{1}{2}\left(\frac{\partial K_S}{\partial P}\right)_T - \frac{1}{6}\end{aligned} \tag{4.34a}$$

and

$$\begin{aligned}\gamma_\mu &\equiv -\left(\frac{\partial \log \omega_S}{\partial \log V}\right)_T = -\frac{1}{2}\left(\frac{\partial \log \mu}{\partial \log V}\right)_T - \frac{1}{6} \\ &= \frac{1}{2}\frac{K_T}{\mu}\left(\frac{\partial \mu}{\partial P}\right)_T - \frac{1}{6}.\end{aligned} \tag{4.34b}$$

Or equivalently,

$$\begin{aligned}\delta_K &\equiv -\left(\frac{\partial \log K_S}{\partial \log V}\right)_T = \left(\frac{\partial \log K_S}{\partial \log \rho}\right)_T \\ &= 2\gamma_K + \frac{1}{3} \approx \left(\frac{\partial K_S}{\partial P}\right)_T\end{aligned} \tag{4.35a}$$

and

$$\begin{aligned}\delta_\mu &= -\left(\frac{\partial \log \mu}{\partial \log V}\right)_T = \left(\frac{\partial \log \mu}{\partial \log \rho}\right)_T = 2\gamma_\mu + \frac{1}{3} \\ &= \frac{K_T}{\mu}\left(\frac{\partial \mu}{\partial P}\right)_T.\end{aligned} \tag{4.35b}$$

Problem 4.3*

Show that the vibrational Grüneisen parameter defined by equation (4.33), $\gamma_{\rm vib}$ (I use subscript "vib" to clearly indicate this is a vibrational Grüneisen parameter), is identical to the thermal Grüneisen parameter defined by equation (4.24), $\gamma_{\rm th}$ (subscript "th" means thermodynamic), if the Grüneisen parameters for all the modes of lattice vibration are the same.

Solution

In order to connect the vibrational Grüneisen parameter (i.e., (4.33)) with the thermodynamic Grüneisen parameter (i.e., (4.24)), one needs to use a relation between lattice vibration and thermodynamic properties. Using the relationships $F = U + k_B T \sum_i \log[1 - \exp(-h\omega_i/2\pi k_B T)]$ (F, Helmholtz free energy; U, energy at $T = 0$ K; Box 2.1) and $S = -(\partial F/\partial T)_V$, one obtains $(\partial S/\partial V)_T = (\gamma_{\rm vib} k_B/V) \sum_i [x_i^2 \exp(x_i)/(\exp(x_i) - 1)^2]$ with $x_i \equiv h\omega_i/2\pi k_B T$, where we assumed that the Grüneisen parameters for all vibrational modes are the same, $\gamma_{\rm vib} = -\partial \log \omega_i/\partial \log V = -\partial \log \omega/\partial \log V$. Now, the internal energy is given by $E = F - T(\partial F/\partial T)_V = U + k_B T \sum_i [x_i/(\exp(x_i) - 1)]$, and hence $C_V = dE/dT = k_B \sum_i [x_i^2 \exp(x_i)/(\exp(x_i) - 1)^2]$. Therefore $(\partial S/\partial V)_T = \gamma_{vib} C_V/V$. Using the thermodynamic identities such as the Maxwell relation and the definitions of thermal expansion and bulk modulus, one obtains $(\partial P/\partial T)_V = (\partial S/\partial V)_T = -(\partial V/\partial T)_P (\partial P/\partial V)_T = \alpha_{\rm th} K_T$ (see problem 4.2). Therefore $\gamma_{\rm vib} = V\alpha_{\rm th} K_T/C_V = \gamma_{\rm th}$.

Problem 4.4

Derive equation (4.34).

Solution

Using $\omega_i \propto \frac{V_i}{a}$, one finds $\log \omega_i = \log V_i - \frac{1}{3}\log V = \frac{1}{2}\log C_i - \frac{1}{6}\log \rho$. Thus $\partial \log \omega_i/\partial \log \rho = (1/2)(\partial \log C_i/\partial \log \rho) - \frac{1}{6}$. Using $C_S = \mu$ and $C_\phi = K_S$, one obtains (4.34).

The Grüneisen parameter is often referred to as an *anharmonic parameter*, because the deviation from harmonic oscillation causes the volume dependence of elastic constants. Anharmonicity (Box 4.4) results in both temperature and pressure dependence of elastic constants. Under some assumptions, temperature and pressure dependence of elastic moduli has the same origin and hence they are related. In such a case, the variation of the bulk modulus (and other elastic moduli) occurs through the variation in displacement, i.e., volume. Both temperature and pressure change the volume, which will change the elastic constants. The values of these anharmonic parameters for some Earth materials are given in Table 4.4. From (4.35) we obtain

Box 4.4 Anharmonicity

The atomic displacement associated with elastic deformation is small, and therefore the force on each atom is approximately a linear function of displacement, namely,

$$F = -kx. \tag{1}$$

The corresponding potential energy is then given by

$$U = \tfrac{1}{2}kx^2. \tag{2}$$

The equation of motion corresponding to this type of force is given by

$$m\frac{d^2x}{dt^2} = -kx \tag{3}$$

the solution of which is

$$x = x_0 \exp(i\omega t) \tag{4}$$

with $\omega = \sqrt{k/m}$. This type of motion is referred to as *harmonic oscillation*. The frequency of harmonic oscillation is determined by k (the curvature of the potential, i.e., elastic constant) and m (mass). When the displacement becomes large, deviation from harmonic potential may occur. The nature of deviation from symmetric harmonic potential can be understood from a simple model of inter-atomic interaction (see Fig. 4.4). Such a deviation from harmonic potential is referred to as *anharmonicity*. When the potential is anharmonic, then the mean curvature of the potential that an atom feels will be dependent on the amplitude of vibration. The amplitude of vibration is determined by the thermal energy, $k_B T$, and the potential energy (see Fig. 4.4). When the potential energy is a function of the inter-atomic distance only, then, for a large volume, the amplitude of vibration at a given temperature will be large, and the mean curvature of potential will be small (i.e., small elastic constant). When a volume is reduced, then the amplitude will be small, and the curvature will be large (i.e., large elastic constant).

$$\left(\frac{\partial \log V_S}{\partial \log V}\right)_X = -\left(\frac{\partial \log V_S}{\partial \log \rho}\right)_X = -\frac{1}{2}(\delta_\mu - 1) \tag{4.36a}$$

$$\begin{aligned}\left(\frac{\partial \log V_P}{\partial \log V}\right)_X &= -\left(\frac{\partial \log V_P}{\partial \log \rho}\right)_X \\ &= -\frac{1}{2}[(1-B)(\delta_K - 1) + B(\delta_\mu - 1)]\end{aligned} \tag{4.36b}$$

with $B \equiv \frac{4}{3}\mu/(K+\frac{4}{3}\mu) = 2(1-2\nu)/3(1-\nu) = \frac{4}{3}V_S^2/V_P^2 \sim 0.4$ and $(\partial \log V_{P,S}/\partial \log V)_X$ and $(\partial \log V_{P,S}/\partial \log \rho)_X$ are the partial derivatives of P- or S-wave velocity with respect to molar volume and density for a constant chemical composition.

Note that the Grüneisen parameter defined by (4.24) (or (4.33)) is a useful parameter only to the extent that a single vibrational frequency (or two frequencies) can characterize the thermal property of a material. Such a simplification is made in the Debye model in which all modes of lattice vibration are assumed to follow a dispersion relation similar to acoustic waves. This is obviously an over simplification and one major limitation is the neglect of non-acoustic modes of lattice vibration called *optical modes*. When optical modes play an important role, the characterization of anharmonic properties using the Grüneisen parameter needs some revisions.

In the previous discussions, we assumed Birch's law. It must be emphasized that although Birch's law is a good approximation for many materials, it is an empirical law and is only approximately valid. To emphasize the distinction of real material behavior from ideal Birch's law behavior, it is instructive to write,

$$\left(\frac{\partial C}{\partial T}\right)_P = \left(\frac{\partial C}{\partial T}\right)_V + \left(\frac{\partial C}{\partial V}\right)_T \left(\frac{\partial V}{\partial T}\right)_P. \tag{4.37}$$

Here the first term, $(\partial C/\partial T)_V$, in the right-hand side of equation (4.37) represents the temperature effect at a constant volume and the second term represents the effect of thermal expansion. The first term is often referred to as the *intrinsic temperature derivative* of elastic constants (ANDERSON, 1987a), and for materials that follow Birch's law, $(\partial C/\partial T)_V = 0$. In many cases non-zero values of $(\partial C/\partial T)_V$ are found, and in these cases, the present discussions apply only approximately.

Anharmonicity causes both thermal expansion and the pressure dependence of the elastic moduli, and anharmonicity is characterized by the Grüneisen parameter. Therefore the Grüneisen parameter, thermal expansion and the pressure dependence of the elastic modulus are related. In a case where the Grüneisen parameters are identical for all modes of lattice vibration, it can be shown from equation (4.34) that,

$$\left(\frac{\partial K_S}{\partial P}\right)_T = 2\gamma_K + \frac{1}{3} = \frac{2\alpha_{th}K_T V}{C_V} + \frac{1}{3} \approx \frac{2\alpha_{th}K_T V}{3nR} + \frac{1}{3}. \tag{4.38}$$

TABLE 4.4 Anharmonic parameters of some oxides and silicates at ambient conditions (unit is $\times 10^{-5}\,K^{-1}$ for thermal expansion, α_{th}, $-(\partial \log K_S/\partial T)_P$ and $-(\partial \log \mu/\partial T)_P$, but $(\partial K_S/\partial P)_T$ and $(\partial \mu/\partial P)_T$ are non-dimensional). These values can be converted to $\delta_{K,\mu,T}$ through the relations $\delta_K \approx (\partial K_S/\partial P)_T$, $\delta_\mu = (K_T/\mu)(\partial \mu/\partial P)_P$ and $\delta_T \equiv -(1/\alpha_{th})(\partial \log K_T/\partial T)_P$. All the data are at ambient pressure and at $T \approx T_D$ (T_D Debye temperature).

	α_{th}	$\left(\frac{\partial K_S}{\partial P}\right)_T$	$\left(\frac{\partial \mu}{\partial P}\right)_T$	$-\left(\frac{\partial \log K_S}{\partial T}\right)_P$	$-\left(\frac{\partial \log \mu}{\partial T}\right)_P$	Ref.
Periclase (MgO)	3.1	4.3	2.5	15	26	(1), (2)
Pyrope ($Mg_3Al_2Si_3O_{12}$)	1.9	4.9	1.6	11	11	(1), (2)
Spinel ($MgAl_2O_4$)	2.1	5.7	0.72	13	12	(1), (2)
Ringwoodite (Mg_2SiO_4)	1.9	4.8	1.8	13	12.5	(2), (3)
Corundum (Al_2O_3)	1.6	4.3	1.8	6.0	17	(1), (2)
Enstatite ((Mg, Fe)SiO_3)	2.4	8.5	1.6	25	16	(4), (5)
Forsterite (Mg_2SiO_4)	3.1	5.0	1.8	14	17	(1), (2)
Wadsleyite ((Mg, Fe)$_2SiO_4$)	1.9	4.3	1.4	11	15	(2), (6)
Perovskite ($MgSiO_3$)	3.0	3.9	1.6	13	14	(7), (8)

(1): Anderson and Isaak (1995)
(2): Bass (1995)
(3): Jackson *et al.* (2000b)
(4): Frisillo and Barsch (1972)
(5): Flesh *et al.* (1998)
(6): Mayama *et al.* (2004)
(7): Aizawa *et al.* (2004)
(8): Jackson (1998)

Problem 4.5

Derive equation (4.38).

Solution

From (4.34), one has $(\partial \log K_S/\partial \log V)_T = -2\gamma_K - \frac{1}{3} = -(\partial K_S/\partial P)_T$. Now if we assume that the Grüneisen parameter for the bulk modulus is the same as that for all other modes of lattice vibration, then we can use the relation (4.24) to get $(\partial K_S/\partial P)_T = 2\gamma + \frac{1}{3} = 2\alpha_{th} K_T V/C_V + \frac{1}{3}$. At high temperatures, $C_V \approx 3nR$ where n is the number of atoms per unit formula ($n = 7$ for olivine Mg_2SiO_4), therefore $(\partial K_S/\partial P)_T \approx 2\alpha_{th} K_T V/3nR + \frac{1}{3}$.

If Birch's law is satisfied, i.e., if the intrinsic temperature derivative is negligible, then using the definitions of the Grüneisen parameter and the relations (4.35) and (4.36), the temperature and pressure derivatives of seismic wave velocities can be written in terms of the bulk modulus, thermal expansion and the Grüneisen parameters as

$$\left(\frac{\partial \log V_S}{\partial P}\right)_T = \frac{\delta_\mu - 1}{2K_T} \tag{4.39a}$$

$$\left(\frac{\partial \log V_P}{\partial P}\right)_T = \frac{(1-B)(\delta_K - 1) + B(\delta_\mu - 1)}{2K_T} \tag{4.39b}$$

and

$$\left(\frac{\partial \log V_S}{\partial T}\right)_P = -\frac{(\delta_\mu - 1)\alpha_{th}}{2} \tag{4.40a}$$

$$\left(\frac{\partial \log V_P}{\partial T}\right)_P = -\frac{\left[(1-B)(\delta_K - 1) + B(\delta_\mu - 1)\right]\alpha_{th}}{2}. \tag{4.40b}$$

A few points may be noted here. As noted before, the influence of pressure is large in comparison to that of temperature. Furthermore, as seen from equation (4.39), the influence of pressure on seismic wave velocities (and elastic constants) is non-linear (i.e., $(\partial \log V_{P,S}/\partial P)_T$ changes with pressure) because K_T changes with pressure: pressure derivatives decrease

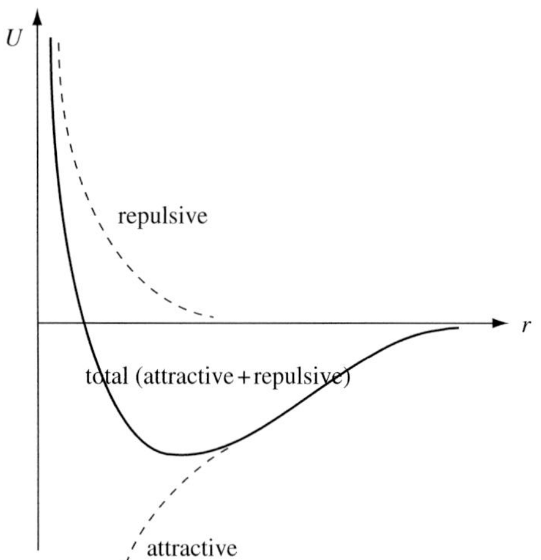

FIGURE 4.4 A schematic drawing of the inter-atomic potential. The inter-atomic potential is composed of an attractive potential due to Coulombic interaction and a short-range repulsive potential caused by quantum mechanical effects (the Pauli exclusion principle). The Coulombic potential changes with atomic distance as $\propto 1/r$, whereas the repulsive potential changes more rapidly with atomic distance. Consequently, the inter-atomic potential has an asymmetric shape around the equilibrium atomic distance. This asymmetry causes the pressure and temperature dependence of elastic moduli and is responsible for thermal expansion.

significantly with pressure. In contrast, the influence of temperature is linear (i.e., $\left(\partial \log V_{P,S}/\partial T\right)_P$ is constant) at above $\sim \frac{1}{2}T_D(T_D$, Debye temperature) where the Grüneisen parameters ($\delta_{\mu,K}$) and thermal expansion are nearly independent of temperature. However, thermal expansion vanishes at $T = 0\,\mathrm{K}$, and increases strongly with temperature below $\sim \frac{1}{2}T_D(T_D$, Debye temperature). Consequently, the temperature derivatives are 0 at $T = 0\,\mathrm{K}$ and reach asymptotic values above $\sim \frac{1}{2}T_D$ where thermal expansion becomes insensitive to temperature.

Now let us consider how these parameters change with pressure. To a first approximation, elastic constants such as the bulk modulus increase linearly with pressure (see Fig. 4.3). In this case $(\partial K_S/\partial P)_T$ is constant (typically $(\partial K_S/\partial P)_T \sim 4$, see Table 4.4) and the Grüneisen parameter is independent of pressure. But from (4.24), it can be seen that as pressure increases, the bulk modulus increases significantly and therefore thermal expansion will decrease. The pressure dependence of thermal expansion can be calculated as follows. Taking the derivative of thermal expansion with respect to pressure, one obtains,

$$\left(\frac{\partial \alpha_{th}}{\partial P}\right)_T = -\frac{\alpha_{th}}{K_T}\delta_T \tag{4.41}$$

where

$$\delta_T \equiv -\frac{1}{\alpha_{th}}\left(\frac{\partial \log K_T}{\partial T}\right)_P = \left(\frac{\partial \log K_T}{\partial \log \rho}\right)_P. \tag{4.42}$$

δ_T is referred to as the *Anderson–Grüneisen parameter* which takes a value of $\sim$4–6 for most materials and is nearly independent of temperature or pressure (see ANDERSON, 1996). If δ_T is independent of density, then equation (4.41) can be integrated to get,

$$\frac{\alpha_{th}}{(\alpha_{th})_0} = \left(\frac{\rho_0}{\rho}\right)^{\delta_T} \tag{4.43a}$$

where $(\alpha_{th})_0$ and ρ_0 are thermal expansion and density at a reference state. Alternatively, if the Anderson–Grüneisen parameter changes weakly with density as $\delta_T \cdot \rho$ is constant as is the case for the Grüneisen parameter (see the next section), then

$$\frac{\alpha_{th}}{(\alpha_{th})_0} = \exp\left[-\delta_T^0\left(1 - \frac{\rho_0}{\rho}\right)\right] \tag{4.43b}$$

where δ_T^0 is the Anderson–Grüneisen parameter at the reference density. These relations indicate that thermal expansion decreases significantly with the increase in density.

Problem 4.6

Derive equation (4.43), and calculate the difference in thermal expansion between the shallow region (720 km depth) and the very bottom (2890 km) of the lower mantle using the PREM density model (see Table 17.1, Chapter 17) using a value of δ_T (or δ_T^0), = 5.

Solution

By differentiating thermal expansion ($\alpha_{th} \equiv \partial \log V/\partial T$) with pressure, one obtains

$$(\partial \alpha_{th}/\partial P)_T = \frac{\partial(\partial \log V/\partial P)_T}{\partial T} = -\frac{\partial}{\partial T}\frac{1}{K_T} = \frac{1}{K_T^2}\frac{\partial K_T}{\partial T}.$$

Therefore using $\delta_T \equiv -(1/\alpha)(\partial \log K_T/\partial T) = (\partial \log K_T/\partial \log \rho)_P$, $(\partial \log \alpha_{th}/\partial P)_T = -(1/K_T)\delta_T$. Noting $dP = K_T d\log \rho$, one gets $(\partial \log \alpha_{th}/\partial \log \rho)_T = -\delta_T$ and therefore if δ_T is independent of density, then $\alpha_{th}/(\alpha_{th})_0 = (\rho_0/\rho)^{\delta_T}$. If δ_T changes with density as $\delta_T \cdot \rho$ is constant, then $(\partial \log \alpha_{th}/\partial \log \rho)_T = -\delta_T = -\delta_T^0(\rho_0/\rho)$. Integrating this relation, one obtains $\alpha_{th}/(\alpha_{th})_0 = \exp\left[-\delta_T^0(1 - \rho_0/\rho)\right]$. In PREM, the density at 720 km is 4410 kg/m^3 whereas the density at

2890 km is 5570 kg/m^3. Therefore $\alpha_{th}(2890\text{km})/\alpha_{th}(720\text{km}) = 0.31$ from (4.43a) or $= 0.35$ from (4.43b).

Let us consider how the Grüneisen parameter changes with pressure. By differentiating equation (4.24) ($\gamma = \alpha_{th} K_T V / C_V$) by density (for a constant chemical composition, X) assuming C_V is constant (i.e., at high temperature behavior), one has

$$\left(\frac{\partial \log \gamma}{\partial \log \rho}\right)_T = \left(\frac{\partial \log \alpha_{th}}{\partial \log \rho}\right)_T + \left(\frac{\partial \log K_T}{\partial \log \rho}\right)_T - 1. \quad (4.44)$$

Using the relations $(\partial \log \alpha_{th}/\partial \log \rho)_T = -\delta_T$, and $(\partial \log K_T/\partial \log \rho)_T = \partial K_T/\partial P$, one gets $(\partial \log \alpha_{th}/\partial \log \rho)_T + (\partial \log K_T/\partial \log \rho)_T - 1 = -\delta_T + (\partial K_T/\partial P)_T - 1 \equiv -q$.

If $q \equiv \delta_T - (\partial K_T/\partial P)_T + 1$ is independent of density, one can integrate (4.44) to get

$$\gamma/\gamma_0 = \left(\frac{\rho_0}{\rho}\right)^q. \quad (4.45)$$

Now if $(\partial K_T/\partial T)_V = 0$, $(\partial K_T/\partial T)_P = (\partial K_T/\partial T)_V + (\partial K_T/\partial V)_P(\partial V/\partial T)_P = (\partial K_T/\partial V)_P(\partial V/\partial T)_P$. Using the definitions $\alpha_{th} \equiv (1/V)(\partial V/\partial T)$ and $\delta_T \equiv -(1/\alpha_{th}) \cdot (\partial \log K_T/\partial T)_P$ and the relation $(\partial K_T/\partial V)_P = -(\partial K_T/\partial P)_T(K_T/V)$, one gets $\delta_T = (\partial K_T/\partial P)_T$. So $q = \delta_T - (\partial K_T/\partial P)_T + 1 = 1$. In such a case, the Grüneisen parameter decreases with depth in Earth. There are not many experimental studies on the pressure (density) dependence of Grüneisen parameters, but results of theoretical calculations support the notion that Grüneisen parameters decrease with pressure (density) (Fig. 4.5). In general, the pressure dependence of the Grüneisen parameter can be different for the Grüneisen parameters for different modes. This causes the pressure dependence of $(\delta \log V_S/\delta \log V_P)$ which has an important implication for the interpretation of seismic tomography (see Chapter 20).

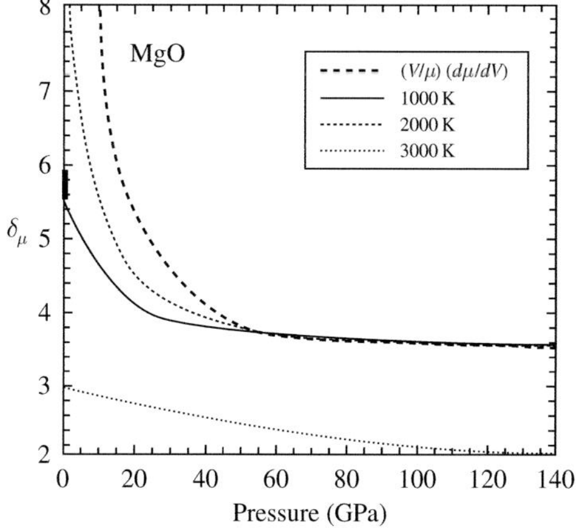

FIGURE 4.5 Results of theoretical calculations on the pressure dependence of the Grüneisen parameter, δ_μ, in MgO (KARATO and KARKI, 2001). δ_μ decreases with pressure.

We have seen that all the parameters that control the temperature and pressure dependence of seismic wave velocities change with density (pressure): thermal expansion decreases significantly with density, the bulk modulus increases significantly with density, while the Grüneisen parameter decreases slightly with density. Consequently, the absolute values of the temperature derivatives of seismic wave velocities decrease significantly with density, i.e., with depth. This has important implications for the interpretation of the lateral variation of seismic wave velocities (KARATO and KARKI, 2001, see Chapter 20).

I should emphasize, however, that the assumption of Birch's law is not always justifiable. In these cases, the above discussions are not valid in a strict sense, and one needs to use the results of direct measurements of elastic wave velocities at various pressures and temperatures.

4.6. Effects of chemical composition

In section 4.5, we focused on one aspect of Birch's law, namely the effects of molar volume (density) on elastic constants for the same chemical composition. Now let us focus on the effects of chemical composition at a fixed temperature and pressure. Birch's law shows that the elastic constants or seismic wave velocities of a material at a given temperature and pressure are controlled by chemical composition mainly through the mean atomic weight, $\bar{M}$. Such a relation holds very nicely for bulk sound velocity defined as $V_\phi \equiv \sqrt{V_P^2 - \frac{4}{3}V_S^2} = \sqrt{K_S/\rho}$. The bulk sound velocity for materials with a given crystal structure changes with mean atomic weight as (e.g., CHUNG (1972), Fig. 4.6),

$$V_\phi \propto \frac{1}{\sqrt{\bar{M}}}. \quad (4.46)$$

A theoretical justification for this relation can be found if one uses the Debye model (SHANKLAND, 1977). In the Debye model the frequency of lattice vibration is related to elastic constants as,

$$\omega \propto \sqrt{\frac{K_{P,S}}{m}} \propto V_{P,S} = \sqrt{\frac{C_{P,S}}{\rho}} \quad (4.47)$$

where m is the atomic mass, which is proportional to $\bar{M}$, $\rho = n\bar{M}/V$ where n is the number of atoms for

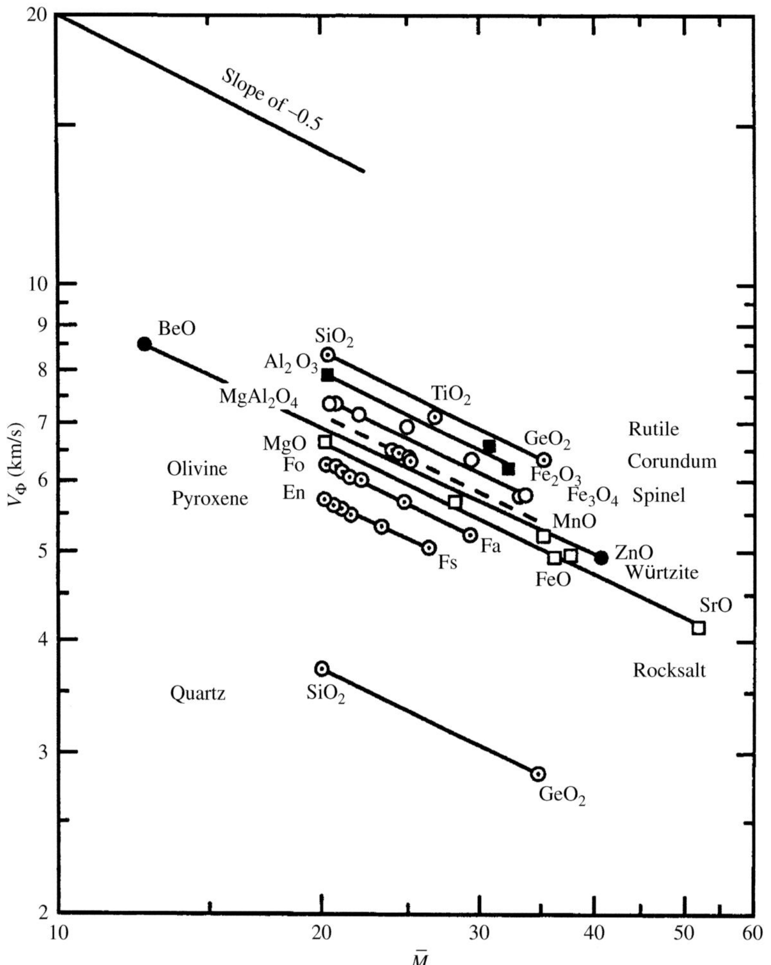

FIGURE 4.6 The relation between bulk sound velocity and mean atomic weight for materials with various crystal structures. The relation (4.46) is well satisfied (after CHUNG, 1972).

one formula (7 for olivine Mg_2SiO_4), and V is the molar volume. Therefore relation (4.46) implies that the bulk modulus is independent of chemical composition. Let us consider how such a relation is explained on an atomic basis. Let us assume the Born–Meyer type inter-atomic potential for an ionic solid, namely,

$$U(r) = -\frac{AZ_cZ_ae^2}{r} + B\exp\left(-\frac{r}{b}\right) \tag{4.48}$$

where r is the inter-atomic distance, $-AZ_cZ_ae^2/r$ is the potential energy associated with the Coulombic interaction among cations and anions (A is the Madelung constant that depends only on the crystal structure), $Z_{c,a}e$ is the electrostatic charge on cation and anion respectively, and $B\exp(-r/b)$ is the repulsive potential (B and b are parameters representing the strength of repulsive potential and the characteristic length at which repulsive potential becomes strong). Using the definition of the bulk modulus, $K = -V(\partial P/\partial V)$, we find (see Problem 4.7)

$$V_\phi = \sqrt{\frac{K}{\rho}} = \sqrt{\frac{AZ_cZ_ae^2(d/b-2)}{3dn\bar{M}}} = \frac{1}{\sqrt{\bar{M}}}\sqrt{\frac{AZ_cZ_ae^2(d/b-2)}{3dn}}. \tag{4.49}$$

where d is the inter-ionic distance. For many materials, d/b is practically constant ($d/b \sim 6$–10; see e.g., TOSI, 1964). The mean atomic distance, d, changes only slightly with the variation of atomic species. Consequently, $\sqrt{AZ_cZ_ae^2(d/b-2)/3dn}$ is nearly independent of mean atomic weight. This reflects the fact that the strength of inter-atomic potential is essentially controlled by the nature of outer electrons and therefore insensitive to the atomic weight that determines the nature of core electrons. The situation for the

TABLE 4.5 Dependence of elastic constants on chemical composition (after SPEZIALE *et al.* 2005).

	X^3	$\frac{\partial \log V_P}{\partial X}$	$\frac{\partial \log V_S}{\partial X}$	$\frac{\partial \log V_P}{\partial \log \rho}$	$\frac{\partial \log V_S}{\partial \log \rho}$
Olivine	Fe/(Fe + Mg)	−0.24	−0.37	−0.67	−1.1
Garnet[1]	Fe/(Fe + Mg)	−0.09	−0.08	−0.52	−0.48
Garnet[2]	Ca/(Ca + Mg)	0.03	0.08	−1.6	−3.7
Orthopyroxene	Fe/(Fe + Mg)	−0.21	−0.30	−0.92	−1.3
Wadsleyite	Fe/(Fe + Mg)	−0.37	−0.52	−1.2	−1.6
Ringwoodite	Fe/(Fe + Mg)	−0.18	−0.28	−0.48	−0.74
Majorite–pyrope	Al/(Fe + Mg + Si)[4]	0.03	0.04	1.2	1.5

[1] Fixed Ca/Mg
[2] Fixed Fe/Mg
[3] Molar fraction
[4] Mg at VI coordinated site, Si at VI coordinated site

shear modulus is not as simple as the bulk modulus because it depends on the angle of the chemical bonds as well as the distance between atoms, but to the extent that the nature of atomic bonding is largely controlled by outer electrons, the shear modulus is also not very sensitive to the atomic weight.[3]

Problem 4.7

Derive equation (4.49) from the Born–Meyer type inter-atomic potential, i.e.,

$$U(r) = -\frac{AZ_cZ_ae^2}{r} + B\exp\left(-\frac{r}{b}\right) \tag{1}$$

where r is the inter-atomic distance, $-AZ_cZ_ae^2/r$ is the potential energy associated with the Coulombic interaction among cations and anions (A is the Madelung constant that depends only on the crystal structure), $Z_{c,a}e$ is the electrostatic charge on cation and anion respectively, and $B\exp(-r/b)$ is the repulsive potential (B and b are parameters representing the strength of repulsive potential and the characteristic length at which repulsive potential becomes strong).

Solution

From the requirement of mechanical equilibrium, we have (at $P=0$), $P = 0 = -\partial U/\partial V$. Therefore $B = (AbZ_cZ_ae^2/d^2)\exp(d/b)$ where d is the equilibrium inter-atomic distance. From the definition of the bulk modulus, we get

$$K = -\frac{1}{9\beta d^2}\left[\frac{4AZ_cZ_ae^2}{d^2} - \frac{B}{b}\left(2+\frac{d}{b}\right)\exp\left(-\frac{d}{b}\right)\right]$$

where $V = \beta d^3$. Using the relation $B = (AbZ_cZ_ae^2/d^2)\exp(d/b)$, one gets

$$K = \frac{AZ_cZ_ae^2}{9\beta d^4}(d/b - 2). \tag{2}$$

Now inserting this relation and $\rho = n\bar{M}/V$ into $V_\phi = \sqrt{K/\rho}$, one gets $V_\phi = (1/\sqrt{\bar{M}})\sqrt{AZ_cZ_ae^2(d/b-2)/3dn}$.

When relation (4.46) is satisfied and if the variation of elastic wave velocities is due to the variation in chemical composition through mean atomic weight, we have

$$\frac{\partial \log V_\phi}{\partial \log \rho} \approx \left(\frac{\partial \log V_\phi}{\partial \log \rho}\right)_T = -\frac{1}{2}. \tag{4.50}$$

Similar relations also apply to $V_{P,S}$. In all cases, as far as the variation in chemical composition affects mainly the density but not the elastic constants, then we have (for variation in chemical composition)

$$\frac{\delta \log V_{P,S,\phi}}{\delta \log \rho} \approx \left(\frac{\partial \log V_{P,S,\phi}}{\partial \log \rho}\right)_T < 0. \tag{4.51}$$

This is in marked contrast to the effects of temperature and pressure for which $(\partial \log V_{p,s,\phi}/\partial \log \rho)_X > 0$. The geodynamic significance of this difference is discussed in Chapter 20. Table 4.5 summarizes some of the experimental data on the composition dependence of elastic constants (elastic wave velocities).

Some subtleties may be worth commenting on. Our discussion on the independence of bulk modulus on chemical composition is valid when the changing

[3] However, because the shear modulus is sensitive to some subtle feature of inter-atomic bonding, the shear modulus is more sensitive to chemical composition than the bulk modulus.

chemical composition does not change the inter-atomic distance d too much and the electrostatic charge remains the same (see equation (4.49)). This is true when a change in chemical composition occurs through the replacement of one ion with another with a similar radius and an identical charge (for ionic radii of typical ions see Table 4.3). An example is the change in $x_{Fe} = Fe/(Fe + Mg)$. The ionic radii of Mg^{2+} and Fe^{2+} ions are not very different (0.072 nm for Mg^{2+} versus 0.077 nm for Fe^{2+}), but they have a large difference in atomic weight (24.31 g/mol versus 55.85 g/mol). In such a case the change in density has a much larger effect than the change in the bulk modulus. However, when the chemical composition changes through the variation of $x_{Ca} = Ca/(Ca + Mg + Fe)$, the situation is different. In this case both the ionic radii and the atomic weight are significantly different (the ionic radius and the atomic weight of Ca^{2+} and Mg^{2+} are 0.100 nm and 40.08 g/mol, and 0.072 nm and 24.31 g/mol respectively). Consequently, in this case, the change in interatomic distance plays an important role as well as the change in density. The Ca^{2+} ion has another unique property. Because its size is so large compared to Mg^{2+} and Fe^{2+}, the addition of Ca^{2+} modifies the structure of a crystal and a compound containing a large amount of the Ca^{2+} ion tends to have a crystal structure that is different from those with only Mg^{2+} and Fe^{2+}. An example is that $CaSiO_3$ assumes a cubic perovskite structure whereas $(Mg, Fe)SiO_3$ assumes an orthorhombic perovskite structure under most of the lower mantle conditions. Another well-known example is pyroxene. Ca-poor pyroxene assumes the orthorhombic structure (orthopyroxene) whereas Ca-rich pyroxene assumes the monoclinic structure (clinopyroxene) at low pressures. Since the shear modulus is sensitive to the subtle difference in crystal structure, changing $x_{Ca} = Ca/(Ca + Mg + Fe)$ can cause a large difference in the shear modulus relative to the bulk modulus (e.g., KARATO and KARKI, 2001).

4.7. Elastic constants in several crystal structures

4.7.1. A polyhedral model of ionic crystals and elastic anisotropy

The elasticity of silicate minerals can be examined based on a polyhedral model. The basis of this model is the concept that the crystal structure of silicates can be understood in terms of a mixture of coordination

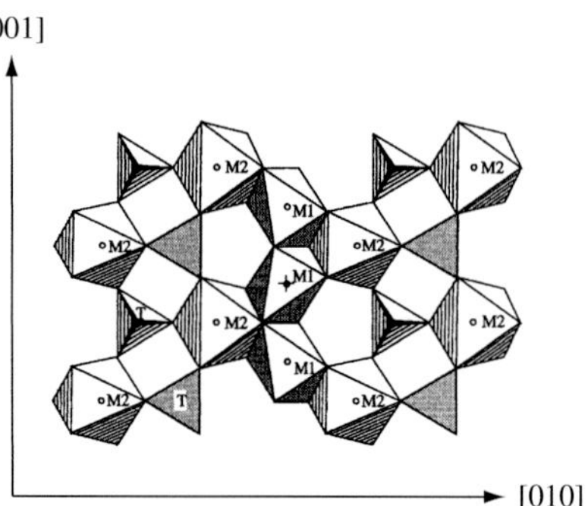

FIGURE 4.7 The structure of a silicate such as olivine can be viewed as a mixture of coordination polyhedra. Tetrahedra (T) are made of Si and O, and octahedra (M1 and M2) are made of Mg and O.

polyhedra, which are composed of a cation in their centers surrounded by anions, in most cases oxygen ions (see Fig. 4.7, Box 4.5). This "asymmetry" of cations and anions is due to the fact that anions (oxygen ions) are much larger than cations. The chemical bonding in a coordination polyhedron is stronger than the chemical bonding between them. For instance, a mineral olivine, Mg_2SiO_4, can be viewed as being made of an arrangement of SiO_4 and MgO_6 polyhedra. In brief, in this model, one considers a material as being composed of a collection of relatively strong polyhedra that are connected by relatively weak forces. In this model, the elastic constants of a material are controlled by the stiffness of individual polyhedra and the change in the geometry of polyhedra. The fundamental reason for this is the tendency in an ionic crystal for charge balance to be maintained locally between adjacent ions with different charges. This is one of the *Pauling rules* controlling the structure of ionic crystals (Box 4.5; PAULING (1960)).

To understand the basis for this model, let us first look at the *bond strength* of various chemical bonds as measured by the incompressibility (bulk modulus) of each bond (Fig. 4.8; HAZEN and FINGER, 1979). The bulk modulus for each bond is a measure of bond strength. HAZEN and FINGER (1979) compiled the data on bond strength based on the results of X-ray diffraction studies and found that the bond strength of ionic bonds has a clear correlation with the electrostatic charges of ions and the bond length (the distance between ions) but they also found that correction must be made for "effective ionic charges" on each ion as[4]

[4] HAZEN and FINGER (1979) originally proposed a relation $K_{poly} \propto S^2 Z_c Z_a e^2/d^3$, but this expression is dimensionally incorrect and equation (4.52) is more appropriate.

Box 4.5 Pauling's rules for the structure and chemical bonding in an ionic crystal (Pauling, 1960)

Most minerals are ionic crystals. In ionic crystals, electrostatic interaction largely controls their structure. Pauling formulated the principles of crystal structures of ionic solids as follows:

(1) A *coordination polyhedron* of atoms is formed about each cation, the bond length is the sum of ionic radii, and the coordination number of the cation is determined by the radius ratio.
(2) The total strength of the bonds that reach an anion from all neighboring cations is equal to the charges of the cation.
(3) Shared edges and, especially, shared faces of coordination polyhedra destabilize a structure. This effect is strongest for cations of high charge and small coordination number, especially if near the lower limit of the radius ratio for that coordination.
(4) Cations of high valency and small coordination number tend not to share polyhedral elements.
(5) The number of essentially different kinds of constituent in a crystal structure tends to be small.

These are empirical rules that give a basis for the concepts of *ionic radii, bond length*, and a broad area of materials science, studying crystal structures and properties of crystals, based on these concepts is called *crystal chemistry*. Although these concepts are empirical and hence the validity is limited, they provide a useful guideline for understanding a range of important physical and chemical properties of ionic crystals (for details see Bloss, 1971 and Navrotsky, 1994).

$$K_{poly} \propto \frac{S^2 Z_c Z_a e^2}{d^4} \tag{4.52}$$

where d is the *bond length* (distance) between cations and anions and the parameter S is introduced to take account of *ionicity* ($S = 0.75$ for halides, $S = 0.5$ for oxides and $S = 0.2$ for carbide: a small value of S means that the effective charge is small, i.e., the chemical bonding is more covalent). Consequently, the Si–O is the strongest chemical bond, followed by the Al–O, Mg–O and K–O. Relation (4.52) is essentially identical to equation (2) in Problem 4.7.

Elastic anisotropy of silicates can be understood through a polyhedral model (e.g., Bass *et al.*, 1984; Webb and Jackson, 1990). For instance, the olivine (Mg_2SiO_4) structure can be considered as an array of MgO_6 and SiO_4 polyhedra (Fig. 4.7). Anisotropy in elastic constants can be analyzed using three principles: (1) the SiO_4 polyhedron is stronger than the MgO_6 polyhedron, (2) when strong polyhedra are linked, then a linkage that has a degree of freedom of bending (tilting) is weaker than a linkage without a degree of freedom of bending and (3) when one deforms a series of connected elements the overall stiffness is an appropriate average of stiffness of elements (see Chapter 12). From Fig. 4.7 it can be seen that the compression along the [001] (as well as [100]) direction requires the compression of a network of strong SiO_4 polyhedra as well as of stronger MgO_6 octahedra (M1 octahedra: in olivine there are two types of MgO_6 octahedra M1 and M2 octahedra, M1 is stronger than M2 octahedra (Levien and Prewitt, 1981)), whereas the compression along the [010] direction can be accommodated by the deformation of weak MgO_6 polyhedra. This provides a simple explanation for $C_{11} > C_{33} > C_{22}$.

The effect of changes in the coordination number (number of other ions surrounding an ion) can also be incorporated in the polyhedral model. When the coordination number is increased, the bond length will also increase. Consequently, the bulk modulus of a polyhedron will decrease. The reason for this is that when a larger number of ions are involved in making the chemical bonds around an ion, then the charge of that ion is "shared" with a larger number of surrounding ions and the effective charge that can be used for an individual bond decreases. This is Pauling's rule (2) (see Box 4.5).

4.7.2. Elastic anisotropy caused by intrinsic anisotropy of bonding

For materials with a simpler structure, a polyhedral model does not apply. This includes some metals and simple ionic solids such as alkali halides or oxides. Let us consider two cases.

Elasticity of hcp metals

For obvious reasons, the concept of coordination polyhedra does not apply to metals. In metals elastic constants are determined by the nature of free electrons as well as the interaction of core electrons at each atom (electrons bound to each nucleus). Although the polyhedral model does not apply to metals, to a large degree, the stiffness ("bond strength") can be

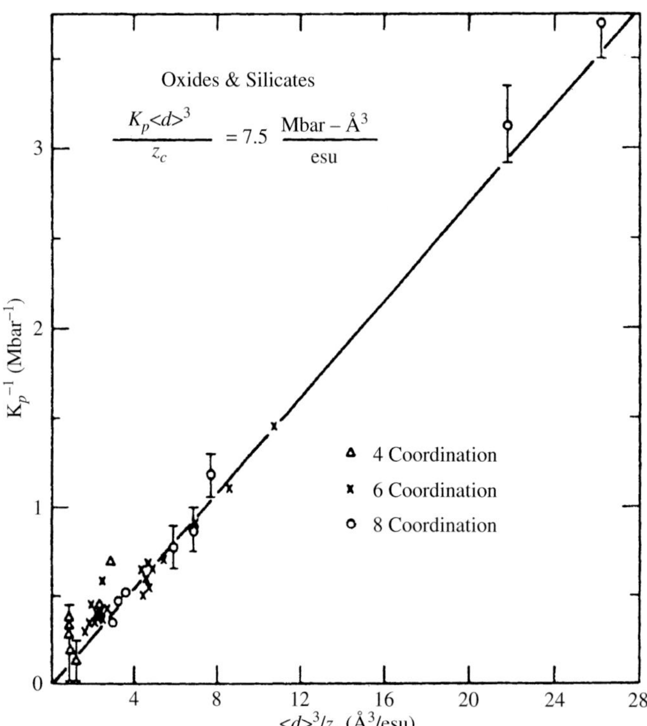

FIGURE 4.8 The relation between the bond strength (K_p) and the bond length (d is the inter-atomic distance) (after HAZEN and FINGER, 1979).

interpreted in terms of atomic distance ("bond-length") similarly to the case for ionic crystals. The shorter the atomic distance, the larger the stiffness. One example is the elastic anisotropy in hcp (hexagonal close packed) metals. The hcp structure is characterized by two lattice constants, a and c. For the ideal hcp structure $c/a = \sqrt{8/3} = 1.633$. Fig. 4.9 shows a correlation between C_{11}/C_{33} and c/a. There is a positive correlation supporting the above notion. STEINLE–NEUMANN *et al.* (2001) calculated the elastic constants of hcp Fe (a high-pressure phase of Fe and a likely constituent of the inner core) and showed that c/a increases significantly with temperature and C_{11}/C_{33} also increases, leading to a change in elastic anisotropy from $C_{11} < C_{33}$ at low temperatures to $C_{11} > C_{33}$ at high temperatures. However, the recent study by GANNARELLI *et al.* (2005) showed a much more modest influence of temperature on the c/a ratio and elastic anisotropy.

Elasticity of materials with B1 and B2 structures

Simple ionic crystals such as MgO cannot be described by coordination polyhedra. All cation–anion bonds are identical in these structures, and there is no way to define a coordination polyhedron. Even in these simple ionic crystals, a marked elastic anisotropy is often observed. For materials with B1 structure (NaCl structure), both C_{11} and C_{12} increase more or less linearly with pressure, whereas C_{44} does not increase with pressure very much, and in some cases even decreases with pressure. Recall that in a cubic crystal like B1 structure, there are two shear moduli, C_{44} and $(C_{11} - C_{12})/2$. The anomalous behavior of C_{44} of (Mg, Fe)O leads to large depth variation (pressure dependence) of elastic anisotropy of this material (KARATO, 1998e; KARKI *et al.*, 1997) (see Fig. 4.10). A similar pressure-induced softening is observed in materials with the spinel structure (see ANDERSON (1996), ANDERSON and LIEBERMANN (1970)). However, such an unusual behavior of C_{44} is not observed in materials with B2 structure (CsCl structure). ANDERSON (1968) showed that such anomalous behavior of the shear elastic constant tends to occur for ionic crystals with a low coordination number.

4.8. Effects of phase transformations

A phase transformation changes elastic properties. The nature of changes in elastic properties should be examined for the first- and second-order transformations separately. When a first-order phase transformation occurs, both phases are dynamically stable but a phase transformation occurs due to the difference in

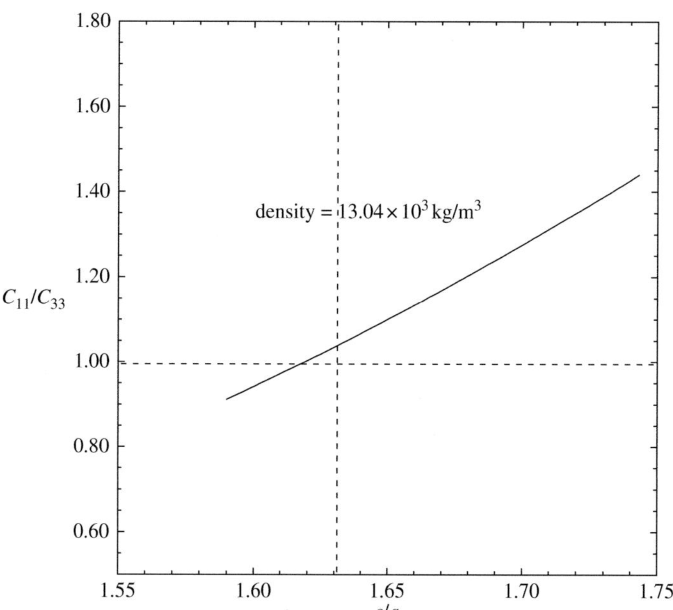

FIGURE 4.9 Elastic anisotropy (C_{11}/C_{33}) and c/a relationship in some hcp metals for a constant density (after STEINLE–NEUMANN *et al.*, 2001).

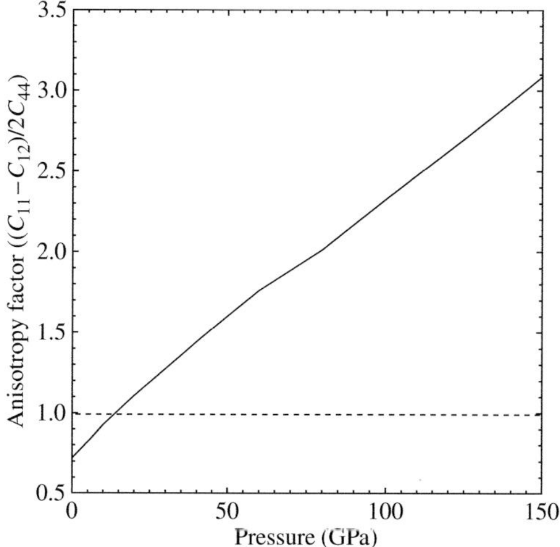

FIGURE 4.10 Elastic anisotropy in B1 structure (e.g., MgO) as a function of pressure (calculated from KARKI *et al.*, 1997).

free energy of these two phases. Therefore, a change in elasticity in this case is simply the result of the difference in crystal structures. Consequently, to the extent that Birch's law describes the elasticity, there is positive correlation between density and elastic wave velocities, and a denser phase will have higher elastic wave velocities (elastic constants). This is approximately true as seen in Fig. 4.11 (LIEBERMANN, 1982; LIEBERMANN and RINGWOOD, 1973). However, such a trend is modified when a phase transformation involves a change in coordination (DAVIES, 1974; LIEBERMANN and RINGWOOD, 1973). Examples include the B1 (NaCl structure) to B2 (CsCl structure) transformation in which the coordination of cations changes from 6 to 8 and the transformation from coesite to stishovite in which the coordination of silicon with respect to oxygen changes from 4 to 6. In these cases, the effects of weakened bond strength reduces the degree to which the elastic constants (elastic wave velocities) increase associated with density increase. However, in almost all cases, transformation to a denser structure results in the increase in elastic constants (or elastic wave velocities). This is in marked contrast to the case of plasticity. In plasticity, density plays only a small role and consequently, denser phases can often have smaller plastic strength (see KARATO (1989c) and Chapter 15).

The situation is different in the case of a *second-order phase transformation*. A first-order phase transformation occurs when a free energy of one phase (one crystal structure) becomes higher than another phase (crystal structure). In such a case, infinitesimal atomic displacement in one phase keeps a phase stable, whereas a finite atomic displacement (that occurs by large stress or thermal activation) will bring a material to another phase that is more stable than the other. In such a case, an infinitesimal displacement that is controlled by elastic constants is stable and nothing unusual happens. In contrast to this first-order phase

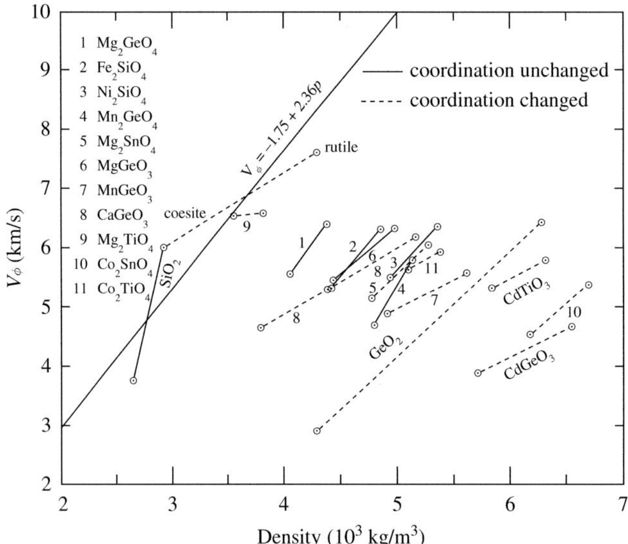

FIGURE 4.11 Elastic wave velocity versus density relationship for materials with various polymorphs (LIEBERMANN, 1982). A transformation to a denser structure invariably results in an increase in elastic wave velocities. However the degree to which the elastic wave velocities increase with density (associated with a phase transformation) is significantly smaller for transformations that involve a large increase in the coordination number.

transformation, there is another type of phase transformation. A material can become unstable for a small displacement of atoms along a certain direction, leading to a new phase (structure). This type of phase transformation is referred to as the second-order phase transformation (Chapter 2). When this instability occurs there is no resistance force against atomic displacement along that direction and therefore the frequency of lattice vibration corresponding to the displacement of atoms along this particular direction goes to zero. Consequently, one of the elastic constants vanishes. This softened mode of lattice vibration is referred to as a *soft mode*. The first theoretical treatment of such lattice instability was made by BORN (1940) (see also BORN and HUANG, 1954) and a more general theory was developed by Landau (LANDAU and LIFSHITZ, 1964; see also CARPENTER, 2006; GHOSE, 1985). For instance, the vibrational frequency of the Γ_{25} mode (lattice vibration with the Γ_{25} mode corresponds to the rotation of TiO_6 octahedra) goes to 0 at 110 K in $SrTiO_3$ associated with a transformation from a cubic structure (at higher T) to a tetragonal structure. This phase transformation is caused by the rotation of TiO_6 octahedra and it occurs because the crystal becomes unstable against the rotation of TiO_6 octahedra. Similarly, stishovite (a high-pressure polymorph of quartz) undergoes a second-order phase transformation at high pressure ($\sim$ 50 GPa) that is associated with a soft mode (ANDRAULT *et al.*, 1998; SHIEH *et al.*, 2002). The α- to β-quartz transformation is close to the second-order transformation (the volume change is very small, and this transformation is associated with the rotation of SiO_4 tetrahedra). The frequencies of some phonons in quartz become low near the transformation temperature implying some degree of softening of these modes (BOYSEN *et al.*, 1980).

It must also be noted that the softening of atomic displacement is often associated with the increase in the dielectric constant (in the extreme case, the dielectric constant becomes infinite: e.g., KITTEL (1986)). A dielectric constant has an important effect on the concentration of point defects (see Chapters 5 and 10). Both the reduction of elastic constants and the increase of the dielectric constant are likely to enhance defect-related processes such as diffusion and dislocation motion. However, not many experimental or theoretical studies have been made on this issue (see Chapter 15 and CHAKLADER, 1963; SCHMIDT *et al.*, 2003; WHITE and KNIPE, 1978).

5 Crystalline defects

Crystalline defects such as point defects (vacancies, interstitial atoms, hydrogen-related defects etc.), dislocations and grain boundaries play essential roles in plastic deformation. This chapter gives a brief review of the fundamentals of crystalline defects. The relation between the concentration of point defects and temperature, pressure and chemical environment is discussed based on thermodynamic principles. Some basic properties of dislocations including the slip systems, dislocation energy and the important properties of grain boundaries and subgrain boundaries are summarized.

Key words point defects, dislocations, grain boundaries, subgrain boundaries, vacancies, interstitials, stoichiometry, ferric iron, slip systems, Burgers vector, Peach–Koehler force, Schmid factor.

5.1. Defects and plastic deformation: general introduction

A crystalline material is made of a nearly perfect periodic array of atoms. When a force is applied, individual atoms move away from their stable positions. This will cause a restoring force. When the statistical fluctuation caused by atomic vibration is neglected and when deformation is homogeneous, then the (homogeneous) displacement of atoms due to a reasonable level of stress is so small that upon the removal of force, the atoms will return to their original positions, hence deformation is *elastic*. Plastic deformation occurs when atoms move to the next stable position, so that after the removal of external force they do not go back to their original positions. The force necessary to move atoms to the next stable positions is determined by the inter-atomic potential energy $U(x)$ by,

$$F_{\text{plastic}} = -\left(\frac{\partial U}{\partial x}\right)_{\max}. \tag{5.1}$$

If we assume a periodic atomic potential $U = U_0 \cdot \cos(2\pi x / a)$ (a, lattice spacing), then,

$$F_{\text{plastic}} = \frac{2\pi U_0}{a}. \tag{5.2}$$

Now, consider a small elastic deformation. For a small displacement, the potential energy must be related to the elastic constants. Thus,

$$\sigma = \frac{F_{\text{elastic}}}{a^2} = \mu\varepsilon \tag{5.3}$$

where $\varepsilon = x / a$ is the strain and μ is the shear modulus. Now,

$$\begin{aligned} F_{\text{elastic}} &= -\lim_{x \to 0}\left(\frac{\partial U}{\partial x}\right) = \left(\frac{2\pi}{a}\right) U_0 \sin\left(\frac{2\pi x}{a}\right) \\ &\approx \left(\frac{2\pi}{a}\right)^2 U_0 x. \end{aligned} \tag{5.4}$$

Hence

$$U_0 = \frac{\mu a^3}{(2\pi)^2}. \tag{5.5}$$

Therefore, the stress needed to deform a material by a *homogeneous, coherent motion* of atoms without the help of thermal agitation would be given by,

$$\sigma_{\text{plastic}} = \frac{F_{\text{plastic}}}{a^2} \approx \frac{\mu}{2\pi} \tag{5.6}$$

which is ≈10–50 GPa for typical metals and silicates. This is obviously much higher than what is empirically found (<1 GPa for most materials). Thus the coherent motion of atoms without the help of thermal activation cannot be responsible for plastic deformation of solids. *Defects* or *incoherent motion* of atoms are needed to explain the strength of real materials. Incoherent motion of atoms is in most cases associated with defects. So in short, defects play a central role in plastic deformation. Thermally activated motion of atoms is one of the important types of incoherent motion that is assisted by defects (Section 2.3). Therefore the rate of plastic deformation is controlled largely by the thermally activated *stochastic processes* and hence is time and temperature dependent.

Defects within a grain include point defects, dislocations and stacking faults. Grain boundaries can also be considered as defects. Atomic bonding around these defects is largely disturbed and atoms are much more mobile in and around these defects. Therefore the motion of these defects can cause non-recoverable deformation, namely permanent plastic deformation.

Now let us consider the motion of a defect. A defect could move recoverably. However, because the atomic bonding around a defect is weak, a large non-recoverable motion of defects can also occur. This is particularly the case at high temperatures. This is due to thermally activated processes (section 2.4).

At a very general level, one can obtain the following rate equation by considering that the rate of deformation must be proportional to the amount of strain caused by a unit motion of a defect, a number of defects and their mobility. The strain rate caused by the motion of defects can be written as,

$$\dot{\varepsilon} = \Delta\varepsilon \cdot K(T, P, C, \sigma, \varepsilon, t) \cdot N(T, P, C, \varepsilon, \sigma, t) \quad (5.7)$$

where K is the rate (velocity) of motion of defects, $\Delta\varepsilon$ is the plastic strain associated with a unit motion of defect, N is the number density of defects, T is temperature, P is pressure, C is a parameter that characterizes the chemical environment and t is time.

The creep constitutive relation (equation (5.7)) depends on how K and N depend on various parameters. The manner in which K and N vary with T, P, C, σ etc. depends on the specific defects involved. In general, the relation tends to be highly non-linear when larger defects are involved (cracks, dislocations) and/or under higher stresses.

5.2. Point defects

5.2.1. Generalities

Definition

Defects at a scale of individual atoms are called *point defects*. They include vacant atomic sites (vacancies), atoms that occur at sites that are normally not occupied by atoms (interstitial atoms), ions with electrostatic charges that are different from normal charges (aliovalent ions (e.g., Fe^{3+} at Fe^{2+} site)) and impurity atoms (Fig. 5.1). Point defects may be classified into two types based on the thermodynamic origin: one is the *thermal* defect (or defects of thermal origin), which is formed by thermal activation in a closed system, and the other is the *chemical* defect (defects of chemical origin), which is formed as a result of interaction with the environment that is created only in an open system. The terms *intrinsic* defects or *extrinsic* defects are often used but it is recommended to use *thermal* and *chemical* to describe their origin more clearly. The former type of defect includes vacancies and interstitial atoms and the latter type of defect includes Fe^{3+} at Fe^{2+} site and impurity atoms. Water (hydrogen)-related defects also belong to the second category.

Equilibrium concentration of point defects

In both cases, the concentration of point defects is usually controlled by the thermochemical equilibrium. In other words, at a certain finite temperature (and pressure) and with a certain chemical environment (such as oxygen fugacity), the concentration of point

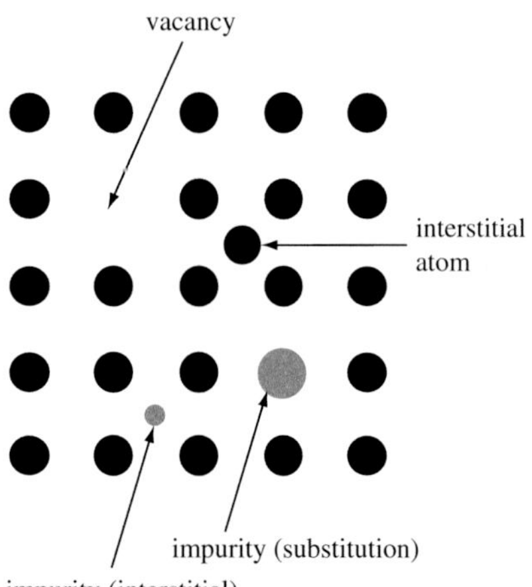

FIGURE 5.1 Point defects in a crystal (a vacancy, an interstitial atom, and an impurity atom are shown).

defects is determined by the thermochemical parameters. One should also note, however, that the concentration of point defects is determined by thermochemical parameters only when the system is in chemical equilibrium. The kinetics to achieve chemical equilibrium takes a finite time, and hence non-equilibrium concentration can be realized under certain conditions.

The basic principle that controls the concentration of point defects can be understood as follows. Consider a case where one creates a certain number, n, of point defects in a crystal that contains N lattice sites. By replacing n atoms with point defects, the free energy of the crystal is increased by $n \cdot g_f$ where g_f is the change in free energy due to the replacement of one atom with a point defect.[1] g_f is therefore referred to as the free energy of formation of a defect. However, by introducing a certain number (n) of point defects in a crystal, the (configurational) entropy (see Chapter 2) of the crystal increases. Thus the net change in the Gibbs free energy of the crystal is,

$$\Delta G(n) = n \cdot g_f - T \cdot \Delta S_{\mathrm{conf}}(n). \tag{5.8}$$

The configurational entropy, $\Delta S_{\mathrm{conf}}(n)$, is given by (see Chapter 2),

$$\Delta S_{\mathrm{conf}}(n) = k_{\mathrm{B}} \log \frac{N!}{(N-n)!n!} \tag{5.9}$$

where k_{B} is the Boltzmann constant. The equilibrium state of the system is the one that minimizes the Gibbs free energy so that, the equilibrium number of defects, n, is determined by the equation,

$$\frac{\partial \Delta G(n)}{\partial n} = 0. \tag{5.10}$$

Inserting the relation (5.9) and using the Stirling formula, $\log \cdot n! \approx n \cdot \log \cdot n - n$, for large n, one obtains,

$$\frac{n}{N} = \frac{\exp\left(-g_f/k_{\mathrm{B}}T\right)}{1+\exp\left(-g_f/k_{\mathrm{B}}T\right)} \approx \exp\left(-g_f/k_{\mathrm{B}}T\right). \tag{5.11}$$

[1] Note that here I use free energy per defect, rather than free energy per one mole of defect to clearly illustrate the statistical nature of defect formation. It should also be noted that g_f does not include the configurational entropy.

Problem 5.1

Derive equation (5.11).

Solution

From equations (5.8) and (5.9), one has, $\Delta G(n) = ng_f - T\Delta S_{\mathrm{conf}}(n)$. Using the Stirling formula, this equation becomes,

$$\Delta G(n) \approx n \cdot g_f - k_{\mathrm{B}}T[N \log N - (N-n)\log(N-n) - n \log n].$$

Taking the derivative of this equation with respect to n,

$$\frac{\partial \Delta G(n)}{\partial n} = g_f - k_{\mathrm{B}}T[\log(N-n) - \log n] = 0.$$

Therefore

$$\frac{n}{N-n} = \exp\left(-\frac{g_f}{k_{\mathrm{B}}T}\right).$$

The free energy for the formation of a defect in the above treatment is defined for one defect. In the literature, the free energy of the formation of a defect is often defined for one mole ($N_{\mathrm{A}} = 6.02 \times 10^{23}$ species, the Avogadro number). In this case, equation (5.11) can be re-written as,

$$\frac{n}{N} = \exp\left(-\frac{G_f}{RT}\right) \tag{5.12}$$

where $R \equiv N_{\mathrm{A}} \cdot k_{\mathrm{B}}$ is the gas constant and $G_f = N_{\mathrm{A}} \cdot g_f$ (G_f is the free energy of one mole of the defect).

The free energy of formation of (one mole of) a point defect can be written as,

$$G_f = E_f + PV_f - TS_f \tag{5.13}$$

where E_f is the internal energy for formation of one mole of the defect, V_f is the volume of formation of one mole of the defect $\left(V_f = \left(\partial G_f/\partial P\right)_T\right)$ and S_f is the entropy for formation of one mole of the defect $\left(S_f = -\left(\partial G_f/\partial T\right)_P\right)$. Physically, E_f corresponds to the change in energy of a crystal caused by the change in chemical bonding associated with the formation of one mole of the defect. The formation volume, V_f, corresponds to the pressure dependence of formation free energy, G_f, which includes an explicit change in crystal volume as well as the pressure dependence of some physical constants such as elastic constants (for details see Chapter 10). The formation entropy, S_f, represents the change in the vibrational entropy of a crystal caused by the formation of a defect. Combining (5.12) and (5.13), we obtain,

$$\frac{n}{N} = \exp\left(\frac{S_f}{R}\right) \cdot \exp\left(-\frac{E_f + PV_f}{RT}\right) \equiv \left(\frac{n}{N}\right)_0 \exp\left(-\frac{E_f + PV_f}{RT}\right). \tag{5.14}$$

The formation entropy of a defect, S_f, is due to the change in the frequencies of lattice vibration and is $\approx R$, so the first term, $(n/N)_0$, is on the order of unity (Shewmon, 1989). The formation energy varies from one type of defect to another and is ~50–300 kJ/mol in many oxides (e.g., Sammis *et al.*, 1981; Shewmon, 1989). The formation volume also varies from one type of defect to another and will be discussed in Chapter 10. However, in most cases it is on the order of molar volume (~(5–50) × 10^{-6} m^3/mol). The formation volume is usually positive but in some materials (such as ice I), it takes a negative value. The concentration of a point defect is shown as a function of temperature and pressure for a range of E_f and V_f in Fig. 5.2. The important messages to learn are:

(a) the concentration of point defects increases exponentially with temperature;
(b) the concentration of point defects changes significantly with pressure; when the activation volume for formation is positive (negative), the concentration decreases (increases) with pressure.

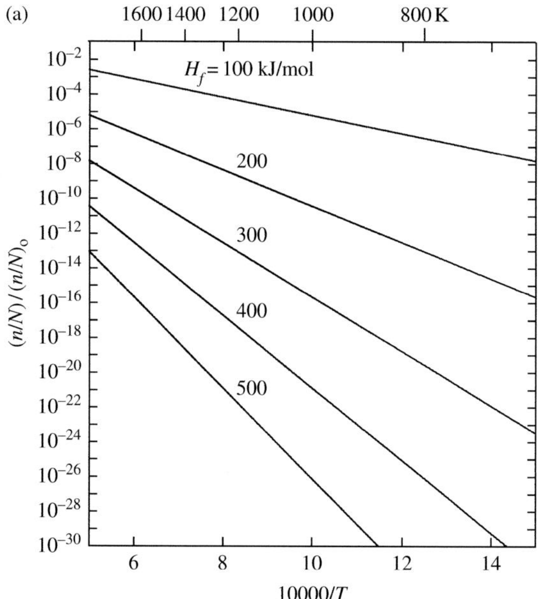

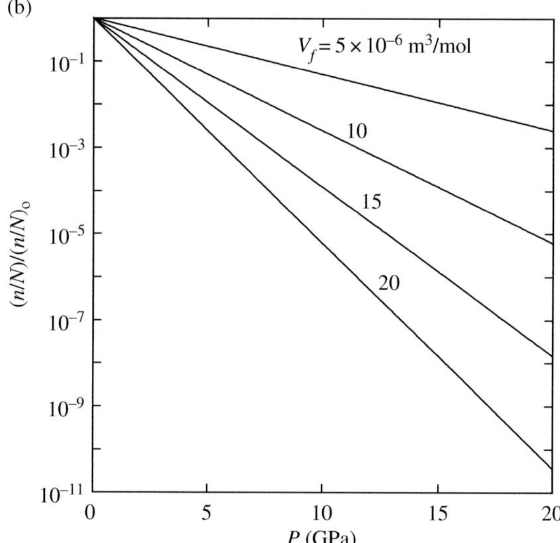

FIGURE 5.2 Influence of (a) temperature and (b) pressure on the concentration of point defects, n/N (for (b) $T = 1500$ K).

5.2.2. Point defects in ionic solids

The above treatment is quite general and can be applied to any material. However, some special points may be noted as to the point defects in ionic crystals. Most minerals are classified as ionic solids (with some contribution from covalent bonding especially for Si–O bonds) in terms of chemical bonding. In these materials, the majority of bonding energy is due to the electrostatic interaction and therefore the requirement for charge balance has a strong control on the concentration of defects.

A Schottky pair

Consider, for simplicity, an ionic solid with a chemical formula AX (A, cation; X, anion; MgO is an example). One can have vacancies at the A (cation) site or vacancies at the X (anion) site. The ease of forming a vacancy at the A site is usually different from the ease of forming a vacancy at the X site. This would cause different concentrations of vacancies at the A and X sites. However, if the concentrations of A- and X-site vacancies are different then there will be a net charge in a crystal, which will increase its energy enormously. Therefore the concentration of positively charged defects is very nearly equal to that of negatively charged defects, and the *charge neutrality condition* is

Box 5.1 The Kröger–Vink notation of point defects

In considering a chemical reaction involving point defects in an ionic crystal, it is important to consider the conservation of lattice sites and of electrostatic charge. The Kröger–Vink notation of point defects is a useful tool to clearly define a point defect and indicate its effective charge *relative to the perfect lattice*. A Kröger–Vink notation contains three types of information: X_Y^Z. X indicates the atomic species that occupies the lattice site Y, and Z represents the effective charge (• is effective + charge, $''$ is effective 2− charge and × is neutral). Thus for example, $Fe_M^\bullet$, $Fe_M^\times$ in olivine $(Mg, Fe)_2SiO_4$ indicate a ferric iron (Fe^{3+}) at M-site, a ferrous iron (Fe^{2+}) respectively, and V_M'' indicates a vacancy at M-site with an effective charge of 2− (relative to the perfect lattice).

well satisfied in most cases.[2] Therefore the unit process of the formation of vacancies in an ionic crystal is a formation of a pair of defects (*a Schottky pair*),

$$\text{null} \Leftrightarrow V_A'' + V_x^{\bullet\bullet} \tag{5.15}$$

where we used the Kröger–Vink notation of point defects (Box 5.1) and "null" indicates a perfect lattice. Because of the requirement of charge balance,

$$[V_A''] = [V_x^{\bullet\bullet}]. \tag{5.16}$$

In this case, one can consider a formation energy for a mole of a pair of defects V_A'' and $V_x^{\bullet\bullet}$, a mole of Schottky pair, G_S. Applying the law of mass action, one gets,

$$[V_A''] \cdot [V_x^{\bullet\bullet}] = \exp\left(-\frac{G_s}{RT}\right) = K_{17} \tag{5.17}$$

where K_{17} is the equilibrium constant for equation (5.17). From (5.16) and (5.17),

$$[V_A''] = [V_x^{\bullet\bullet}] = \exp\left(-\frac{G_s}{2RT}\right) = K_{17}^{1/2}. \tag{5.18}$$

In a more general case, such as Mg_2SiO_4 olivine, the above formulation must be generalized. For a compound $A_\alpha B_\beta C_\gamma - - - -$, the reaction to make a Schottky defect is

$$\text{null} \Leftrightarrow \alpha V_A + \beta V_B + \gamma V_C \cdots \tag{5.19}$$

The condition of preservation of stoichiometry requires,

$$\frac{[V_A]}{\alpha} = \frac{[V_B]}{\beta} = \frac{[V_C]}{\gamma} = \cdots = [V_S] \tag{5.20}$$

where $[V_S]$ is the concentration of a Schottky defect of a compound. The law of mass action (see Chapter 2) yields,

$$[V_A]^\alpha . [V_A]^\beta . [V_A]^\gamma \cdots = \exp\left(-\frac{G_S}{RT}\right) \tag{5.21}$$

hence,

$$[V_S] = (\alpha^\alpha \beta^\beta \gamma^\gamma \cdots)^{1/(\alpha+\beta+\gamma+-)} . \exp\left(-\frac{G_S}{(\alpha+\beta+\gamma+\cdots)RT}\right). \tag{5.22}$$

Therefore,

$$E_f^* = \frac{E_S}{(\alpha+\beta+\gamma+\cdots)} \tag{5.23}$$

and

$$V_f^* = \frac{V_S^*}{(\alpha+\beta+\gamma+\cdots)} \tag{5.24}$$

with

$$V_f^* = \psi \Omega_M \tag{5.25}$$

where Ω_M is the molar volume and ψ is a factor on the order of unity (for details see Chapter 10).

A Frenkel pair

Another example of a pair of charged defects is *a Frenkel pair*, a pair of a vacancy and an interstitial atom. For example, a Frenkel pair at A-sublattice in a compound AX is formed through a reaction,

$$\text{null} \Leftrightarrow V_A'' + A_I^{\bullet\bullet}. \tag{5.26}$$

Thus,

$$[V_A''][A_I^{\bullet\bullet}] = \exp\left(-\frac{G_F}{RT}\right) = K_{26} \tag{5.27}$$

and if the Frenkel pair is the dominant defect, then the charge balance must be maintained by the vacancy and interstitial atom, $[V_A''] = [A_I^{\bullet\bullet}]$, thus

$$[V_A''] = [A_I^{\bullet\bullet}] = \exp\left(-\frac{G_F}{2RT}\right). \tag{5.28}$$

Note that the volume change associated with a Frenkel pair is smaller than that of a Schottky pair.

[2] This is true only for the bulk of a crystal. The charge neutrality condition can be violated near defects. Dislocation lines and grain boundaries can be charged in an ionic crystal (see Box 5.2).

Consequently, interstitial atoms tend to be dominant defects under high-pressure conditions.

Defect formation in an open system (the Kröger–Vink diagram)

In the above formulation, we assumed that there is no exchange of atoms between the sample and the surroundings, and therefore the charge balance must be maintained internally. Let us now consider a case where a sample can react with the surroundings. A system is referred to an open system when an exchange of materials (atoms) occurs between the system and the surroundings. An example is a crystal that is in contact with some fluids (say water or air). At high temperatures, some components of the fluids can be dissolved into the crystal and conversely some components of the crystal go into the fluid. Such an exchange of materials often results in the change in stoichiometry of the crystal and hence the formation of point defects. In these cases, the change in free energy includes the change associated with material exchange.

We consider two cases in which the influence of oxygen fugacity and oxide activity (the oxide component of a compound: for Mg_2SiO_4, one can think of this compound as made from MgO and SiO_2 (or $MgSiO_3$)) changes the stoichiometry and modifies the concentration of point defects. Another important case is the defect formation in a system that contains some impurities. An important case is hydrogen (water) as impurity. This is an important subject and is discussed in Chapter 10 in detail.

- Case 1. Point defect chemistry of an Fe-bearing system: the effect of oxygen fugacity.

In a compound containing Fe, an exchange of oxygen with the surrounding atmosphere occurs easily because of the ease for Fe to change its valence state. The change in the valence state of Fe is associated with the change in the stoichiometry of the given compound and hence other point defects. Consequently, many physical properties in these materials are dependent on oxygen fugacity, f_{O_2}. Let us consider, for the sake of simplicity, a system such as (Mg, Fe)O (a similar discussion applies to a system such as $(Mg, Fe)_2SiO_4$ as far as the $(Mg, Fe)O/SiO_2$ ratio is fixed). We can think of a reaction of oxygen with a crystal in two steps. First is the ionization of an oxygen molecule, which creates an oxygen ion, O^{2-}, plus electron holes, $h^\bullet$ (an electron hole is an electronic defect that is created in an insulator (or semiconductor) when an electron is transferred from the valence band to the conduction band or to an impurity level. Therefore it has a positive effective charge. See a textbook of solid state physics, e.g., Kittel (1986)),

$$\frac{1}{2}O_2 \Leftrightarrow O^{2-} + 2h^\bullet. \qquad (5.29)$$

Second, let us carry an oxygen ion and two electron holes to a crystal. Note that the addition of an oxygen ion (to the surface) creates an M-site vacancy (V_M''). Therefore the reaction of the oxygen ion with an oxide crystal can be written as

$$\frac{1}{2}O_2 \Leftrightarrow 2h^\bullet + V_M'' + O_O^\times \qquad (5.30)$$

where the Kröger–Vink notation of point defects is used (see Box 5.1). These electron holes react with ferrous iron ($Fe_M^\times$) to form ferric iron ($Fe_M^\bullet$),

$$Fe_M^\times + h^\bullet = Fe_M^\bullet. \qquad (5.31)$$

Combining equations (5.30) and (5.31),

$$\frac{1}{2}O_2 + 2Fe_M^\times \Leftrightarrow O_O^\times + V_M'' + 2Fe_M^\bullet. \qquad (5.32)$$

Applying the law of mass action (Chapter 2),

$$f_{O_2}^{1/2} = K_{32}[V_M''][Fe_M^\bullet]^2 \qquad (5.33)$$

where K_{32} is the relevant equilibrium constant where we used the fact that $[O_O^\times] \sim 1$.

Such a reaction changes the concentration of other charged defects. Consequently, the formation of these defects modifies the charge balance. Because the electrostatic interaction energy is large, the charge neutrality condition must always be satisfied, namely,

$$[Fe_M^\bullet] + [h^\bullet] + 2[V_O^{\bullet\bullet}] + \cdots = [e'] + 2[V_M''] + 4[V_{Si}''''] + \cdots. \qquad (5.34)$$

In many cases, one type of defect among positively (or negatively) charged defects has much higher concentration than others and the equation (5.34) can be simplified. For example, under some conditions in Fe-bearing minerals, the dominant defects are ferric iron and the vacancy at the M (Mg or Fe)-site. Then the condition for charge neutrality (equation 5.34) is simplified to,

$$[Fe_M^\bullet] = 2[V_M''] \qquad (5.35)$$

thus,

$$[Fe_M^\bullet] = 2[V_M''] \propto f_{O_2}^{1/6}. \qquad (5.36)$$

Given the above formula, the concentration of other defects can also be determined through the law of mass action. For example, the concentration of M-site interstitial atoms can be calculated by considering a reaction (the formation of a Frenkel pair),

$$\text{null} \Leftrightarrow V_M'' + M_\text{I}^{\bullet\bullet} \tag{5.37}$$

and an associated relation from the law of mass action,

$$[V_M''][M_\text{I}^{\bullet\bullet}] = K_{37} \tag{5.38}$$

thus,

$$[M_\text{I}^{\bullet\bullet}] \propto f_{\text{O}_2}^{-1/6}. \tag{5.39}$$

The concentration of electrons can be calculated from the law of mass action for the formation of an electron-hole pair,

$$\text{null} \Leftrightarrow e' + h^\bullet \tag{5.40}$$

$$[e'][h^\bullet] = K_{40} \tag{5.41}$$

hence

$$[e'] \propto f_{\text{O}_2}^{-1/6}. \tag{5.42}$$

Because of their positive dependence on f_{O_2}, the concentration of M-site vacancies, ferric iron or electron holes will increase when f_{O_2} increases. In contrast, the concentration of some defects such as M-site interstitial atoms or electrons will decrease. Then the dominant types of defect will change, and the charge balance will be controlled by those new defects. A diagram that illustrates the overall nature of point defect populations as a function of f_{O_2} etc. is called the Kröger–Vink diagram (Fig. 5.3).

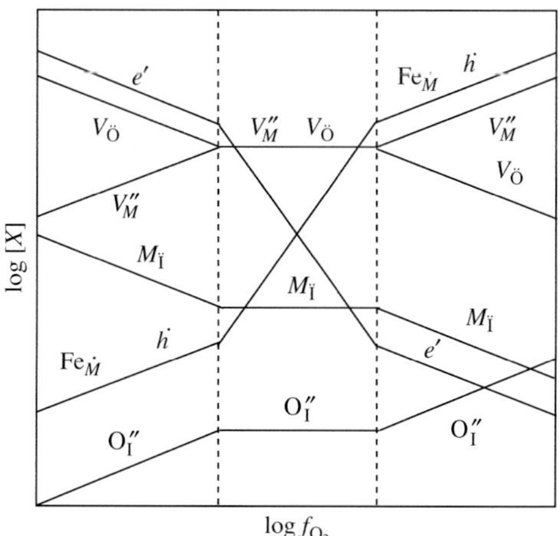

FIGURE 5.3 A Kröger–Vink diagram for (Mg, Fe)O. The M-site may be occupied by Mg or Fe.

Problem 5.2

Construct a Kröger–Vink diagram (concentration versus f_{O_2} diagram) for (Mg, Fe)O. Consider the following types of point defect: $e', h^\bullet, V_M'', V_\text{O}^{\bullet\bullet}, M_\text{I}^{\bullet\bullet}, \text{O}_\text{I}'', \text{Fe}_M^\bullet$. Assume that the concentration of ferric iron ($\text{Fe}_M^\bullet$) is higher than that of electron holes ($h^\bullet$). Consider first a regime in which the dominant defects are V_M'' and $\text{Fe}_M^\bullet$. Calculate the relationship between the defect concentration and f_{O_2}. The calculation will predict that the assumption for the dominant defects is violated at lower f_{O_2}. What are the likely dominant defects at lower f_{O_2}?

Solution

See Fig. 5.3.

It is also noted that because the concentration of ferric iron increases with f_{O_2}, at a certain f_{O_2}, the concentration of ferric iron becomes so large that a given crystal become unstable relative to another structure in which ferric iron can occupy the crystallographic sites with low free energy. Similarly, at lower f_{O_2} a phase containing metallic iron becomes more stable. Consequently, a mineral that contains transition metals such as Fe^{2+} is stable only under a certain range of f_{O_2} (see Chapter 2).

- Case 2: point defects in a ternary system: effects of oxide activity.

In a ternary system such as Mg_2SiO_4, some properties depend on the oxide activity as well as the oxygen fugacity. The oxide activity is controlled by the presence of secondary phases. For example, when olivine (Mg_2SiO_4) co-exists with pyroxene (MgSiO_3) then the activity of MgO is low but the activity of SiO_2 is high. To show the change in the Mg/Si ratio explicitly, we may rewrite equation (5.30),

$$\frac{1}{2}\text{O}_2 + M_M^\times \Leftrightarrow V_M'' + 2h^\bullet + M\text{O (surface)}. \tag{5.43}$$

This equation is essentially the same as equation (5.30) but the lattice site conservation is explicitly expressed. This equation means that an oxygen molecule reacts with Mg in the crystal to form MgO on the surface. By doing so, an Mg atom is removed from the M-site leaving a vacancy (V_M'').

Applying the law of mass action to equation (5.43), one gets,

$$f_{O_2}^{1/2} = K_{43}[V_M''][h^\bullet]^2 a_{MO} \tag{5.44}$$

where a_{MO} is the activity of MO (MgO). When the charge neutrality condition is given by $[Fe_M^\bullet] = 2[V_M'']$, one has,

$$[Fe_M^\bullet] \propto [h^\bullet] \propto [V_M''] \propto f_{O_2}^{1/6} a_{MO}^{1/3}. \tag{5.45}$$

Problem 5.3

Derive equation (5.45).

Solution

When the charge balance is maintained by ferric iron and M-site vacancies, then equation (5.35) must be satisfied. The application of the law of mass action to (5.31) yields, $[Fe_M^\bullet] = 2[V_M''] = K_{30}^{-1}[Fe_M^\times][h^\bullet]$. Inserting this into (5.44), and solving for $[Fe_M^\bullet]$, one gets, $[Fe_M^\bullet] = f_{O_2}^{1/6} a_{MO}^{1/3} \left(\frac{1}{2} K_{43} K_{30}^2\right)^{-1/3} [Fe_M^\times]^{2/3}$. The concentrations of other defects can be calculated by a similar manner as in the case of (Mg, Fe)O (Problem 5.2).

An important point to note here is that the concentration of point defects in a ternary system such as $(Mg, Fe)_2SiO_4$ depends not only on the oxygen fugacity but also on the oxide activity, a_{MO}. The oxide activity is, in practice, set by the bulk chemistry of the system. Consider a chemical reaction,

$$MgO + MgSiO_3 = Mg_2SiO_4. \tag{5.46}$$

The law of mass action gives,

$$a_{MgO} a_{MgSiO_3} = K_{46}(T, P).a_{Mg_2SiO_4} \tag{5.47}$$

where $K_{46}(T, P)$ is the equilibrium constant for the reaction (5.46). Because olivine is present, $a_{Mg_2SiO_4} = 1$. When excess MgO is present in the system, then $a_{MgO} = 1$ and $a_{MgSiO_3} = K_{46}(T, P)$. When excess orthopyroxene ($MgSiO_3$) is present, then $a_{MgSiO_3} = 1$ and $a_{MgO} = K_{46}(T, P)$. When there is excess MgO ($MgSiO_3$), then the concentration of vacancies of Mg and O-sites (Si-site) will be reduced.

In addition to the oxygen fugacity and the oxide activities discussed above, the activity of water (hydrogen oxide) can play an important role in controlling the density (and mobility) of defects in silicate minerals. Because of its importance, the issues of hydrogen-related defects are treated separately in Chapter 10.

5.3. Dislocations

The concept of crystal dislocations plays an essential role in the plastic flow of solids. This chapter summarizes the basic concepts of dislocations including its definition, geometrical properties and the stress–strain associated with them. The dynamics of dislocation motion is reviewed in Chapter 9. For a more comprehensive description of the theory of crystal dislocations, a reader can refer to a number of excellent textbooks including *Dislocations and Plastic Flow in Crystals* (Cottrell, 1953), *Theory of Crystal Dislocations* (Nabarro, 1967b), and *Theory of Dislocations* (Hirth and Lothe, 1982).

5.3.1. Generalities: definition of crystal dislocations, slip systems

Experimental observations on single crystals often show that plastic deformation occurs heterogeneously by crystalline *slip*. Slip occurs on a certain plane along a certain direction, and therefore a slip is characterized by the *slip plane* and the *slip direction*, together defining a *slip system*.

If slip were to occur on a certain plane by the homogeneous relative motion of materials on each side, then a large number of chemical bonds must be cut simultaneously. The stress necessary for such a deformation was calculated and is on the order of ~10% of the elastic modulus (see equation (5.6)) that far exceeds the actual strength of materials (as shown at the beginning of this chapter). Recognizing this discrepancy, Polanyi (1934), Orowan (1934) and Taylor (1934) independently proposed that the actual plastic deformation occurs by the propagation of a slip front. The slip front is a line (not necessarily a straight line) that separates, on a crystallographic plane, the region that has already slipped from regions that have not yet slipped (Fig. 5.4). This line is called a *dislocation*. Since the unit step of slip in this manner involves cutting a limited number of chemical bonds, it will occur more easily than a homogeneous slip.

A portion of dislocation where the dislocation line direction is perpendicular to the slip direction is called an *edge dislocation*. If it is parallel to the slip direction,

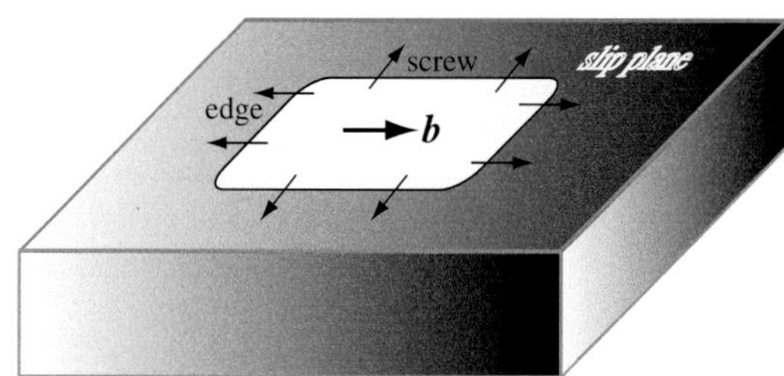

FIGURE 5.4 Slip and dislocation. Dislocation lines are the boundaries on a slip plane between the slipped (white region) and unslipped portions. A thick arrow shows the direction of the relative motion of the bottom and the top parts of a crystal between which the slip occurs. Small arrows show the direction of propagation of the slip. The velocity of propagation depends on the dislocation mobility that can be anisotropic. The dislocation line has a certain angle with respect to the slip direction. A dislocation line that is parallel to the slip direction is called a screw dislocation, and a dislocation line that is normal to the slip direction is called an edge dislocation (a dislocation line that is neither parallel nor normal to the slip direction is called a mixed dislocation).

it is called a *screw dislocation*. A dislocation with a more general orientation is called a *mixed dislocation*. The atomic distortion around each dislocation is shown in Fig. 5.4.

The incremental slip in a crystal associated with the propagation of a slip front is characterized by a vector $\boldsymbol{b}$ called the *Burgers vector*. The Burgers vector is one of the vectors of the relative movement of atoms by which the periodicity of the lattice is preserved. (Occasionally one observes slip vectors that are smaller than these. In this case, one disturbs crystal periodicity by the propagation of this dislocation. These dislocations are called *partial dislocations* and they are associated with *stacking faults*, a misalignment of atomic packing on a plane.)

The easiness of slip for various slip systems of a given crystal is determined by the crystal structure and the nature of chemical bonding (Table 5.1). In simple metals, easy slip planes are usually the planes in which the atomic packing is dense. For example, in the fcc (face-centered cubic) structure, the {111} planes are the dense-pack plane and are the dominant glide planes. Similarly in the hcp (hexagonal-close-pack) structure, the basal plane (the (0001) plane) is the easy slip plane. The choice of slip systems in silicates and ionic crystals is more complicated. For example, electrostatic interaction is an important factor that controls the easy slip plane in ionic crystals. In the NaCl type crystal, the densest packing plane is {100}, but the easy glide plane is {110} in most cases. This is considered to be due to the effects of electrostatic interaction: the glide on the {110} plane brings ions with the same charge so close that it is not favored. Thus, a favorable glide plane in an NaCl type crystal is dependent on the nature of chemical bonding. In crystals with strong ionic bonding (e.g., NaCl) the {110} planes are the favorable glide planes. Whereas in crystals with less ionic bonding (e.g., PbS), the {100} planes are the favorable glide planes (e.g., NABARRO, 1967b). In silicate minerals in which Si is surrounded by four oxygen ions (low-pressure form of silicates), the preferred glide planes are those for which the relative motion of atoms along them does not cut strong Si–O bonds.

In olivine, [100](010) is the favored slip system at low-stress, high-temperature conditions. However, slip systems with $\boldsymbol{b} = [001]$ are favored under high-stress conditions (CARTER and AVÉ LALLEMANT, 1970). It is also found that slip systems with $\boldsymbol{b} = [001]$ are also favored under high water fugacity conditions (JUNG and KARATO, 2001b). An important point to emphasize here is that *the dominant slip system(s) in a given material can change when the physical/chemical conditions of deformation change*. Changes in dominant slip system(s) result in major changes in the nature of deformation fabrics (lattice-preferred orientation, LPO; Chapter 14).

Recently, there have been some attempts to calculate the dominant slip systems in minerals from "first-principles" (e.g., DURINCK *et al.*, 2005a, 2005b; OGANOV *et al.*, 2005). In these studies, the authors calculated the energy associated with the *homogeneous shear* in a perfect crystal along a certain direction on a certain plane and considered that the strength calculated from such a procedure corresponds to the strength of each slip system. The strength corresponding to homogeneous shear is different from the strength controlled by dislocation motion, and therefore this approach is not appropriate in predicting the dominant slip systems.

Because of the plastic anisotropy due to slip, only some components of applied force cause plastic deformation. This concept can be quantified by the quantity called the *Schmid factor*, *S*, which is defined by (Fig. 5.5),

$$\sigma_{\mathrm{RSS}} \equiv S\sigma_a \qquad (5.48)$$

where σ_a is the applied stress (F/A) and σ_{RSS} is the resolved shear stress, i.e., the component of stress that actually acts on a given slip system. The Schmid factor is related to the orientation of slip plane and slip direction relative to the direction of applied force,

$$S = \cos\psi \cdot \cos\lambda \qquad (5.49)$$

TABLE 5.1 Slip systems of typical materials.

Crystal structure	Burgers vector (glide direction)	Glide plane
fcc metal	$\frac{1}{2}\langle 110\rangle$	$\{111\}, \{110\}, \{100\}$
bcc metal	$\frac{1}{2}\langle 111\rangle, \langle 100\rangle$	$\{110\}, \{112\}, \{123\}$
hcp metal	$\frac{1}{3}\langle 11\bar{2}0\rangle, \langle 0001\rangle, \frac{1}{3}\langle 11\bar{2}3\rangle$	$(0001), \{1\bar{1}00\}$
B1 (NaCl-type)	$\langle 110\rangle$	$\{110\}, \{100\}, \{111\}$
Quartz	$\frac{1}{3}\langle 11\bar{2}0\rangle, \langle 0001\rangle, \frac{1}{3}\langle 11\bar{2}3\rangle$	$(0001), \{10\bar{1}0\}, \{10\bar{1}1\}$
Spinel	$\frac{1}{2}\langle 110\rangle$	$\{110\}, \{111\}, \{100\}$
Garnet	$\frac{1}{2}\langle 111\rangle, \langle 100\rangle$	$\{110\}$
Olivine	[100], [001]	(010), (100), (001), $\{0kl\}$
Orthopyroxene	[001]	(100), (010)
Clinopyroxene	[001]	(100)
Wadsleyite	[100], [001], $\frac{1}{2}\langle 111\rangle$	$\{0kl\}, \{1\bar{1}0\}$
Perovskite (cubic)	$\langle 100\rangle, \langle 110\rangle$	$\{100\}, \{110\}$
Perovskite ($CaTiO_3$)[a]	[100]	(010)
Ilmenite	$\frac{1}{3}\langle 11\bar{2}0\rangle$	(0001)

[a] Dominant slip systems in the orthorhombic phase inferred from lattice-preferred orientation (Karato *et al.*, 1995b).

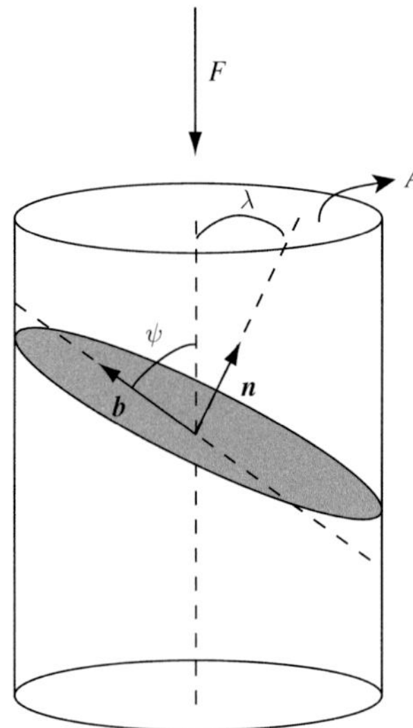

FIGURE 5.5 Definition of the Schmid factor.
The shaded plane is the glide plane (with normal ***n***) and the glide direction is ***b***.

where $\cos\psi$ is the direction cosine between the direction of the applied force and the slip direction and $\cos\lambda$ is the direction cosine between the direction of the applied force and the slip plane normal. The $\cos\lambda$ term comes into equation (5.49) because only the component of applied force along the slip direction is effective in producing slip. The $\cos\lambda$ term comes into equation (5.49) because applied force acting on a glide plane acts on the area of $A/\cos\psi$.

Problem 5.4

Derive the equation (5.49) and show that the Schmid factor is less than or equal to 1/2.

Solution

The component of the force along the slip direction is $F\cos\lambda$. This component of force causes the slip. This force applies to the plane with an area of $A/\cos\psi$. Therefore the resolved shear stress is $F\cos\lambda/(A/\cos\psi) = (F/A)\cos\lambda\cos\psi$. Now without loss of generality, we can choose the coordinate system such that $\boldsymbol{n} = (0,0,1)$. We define the y-axis such that the force $\boldsymbol{F}$ is in the y–z plane. Thus the unit vector parallel to the applied force is $\boldsymbol{F} = (0, \sin\psi, \cos\psi)$. The slip vector $\boldsymbol{b}$ must be on the x–y plane, thus $\boldsymbol{b} = (\cos\theta, \sin\theta, 0)$. Therefore $S = (\boldsymbol{F}\cdot\boldsymbol{n})(\boldsymbol{F}\cdot\boldsymbol{b}) = \cos\psi\sin\psi\sin\theta = \frac{1}{2}\sin 2\psi\sin\theta \le \frac{1}{2}$.

By its definition, it is clear that the motion of a dislocation results in strain. The nature of strain caused by dislocation motion is determined by the geometry of the slip system. The deformation is localized along the dislocation line, but by taking a statistical average, one can think of a homogeneous deformation field due to the motion of homogeneously distributed dislocations. Under this assumption, the displacement due to the glide motion of homogeneously distributed dislocations is given by,

$$\boldsymbol{u} = \gamma(\boldsymbol{r} \cdot \boldsymbol{n})\boldsymbol{l} \tag{5.50}$$

i.e., $u_i = \gamma(\sum_j x_j n_j) l_i$ where γ is strain, $\boldsymbol{l}$ is the unit vector parallel to the Burgers vector ($\boldsymbol{l} \equiv \boldsymbol{b}/b$) and $\boldsymbol{n}$ is the slip plane normal (unit vector). Here we use the fact that the displacement is parallel to the Burgers vector and the amount of displacement increases linearly with the distance from a glide plane. Therefore,

$$\varepsilon_{ij} = \frac{1}{2}\left(\frac{\partial u_i}{\partial x_j} + \frac{\partial u_j}{\partial x_i}\right) = \frac{1}{2}\gamma(n_j l_i + n_i l_j) \tag{5.51}$$

and

$$\omega_{ij} = \frac{1}{2}\left(\frac{\partial u_i}{\partial x_j} - \frac{\partial u_j}{\partial x_i}\right) = \frac{1}{2}\gamma(n_j l_i - n_i l_j) \tag{5.52}$$

Note that deformation due to slip is simple shear, which is associated with *rigid body rotation*. This rigid body rotation is essential for the rotation of crystallographic axes during the deformation of a polycrystalline material that causes lattice-preferred orientation (LPO) (Chapter 14).

Deformation by dislocation motion can also occur by its motion being perpendicular to its glide plane. Consider the motion of a large number of edge dislocations normal to their glide planes (*climb*). In this case, extra half-planes are added or removed. The displacement of material is parallel to the glide direction and the deformation geometry is *pure shear*,

$$\boldsymbol{u} = \gamma(\boldsymbol{r} \cdot \boldsymbol{l})\boldsymbol{l} \tag{5.53}$$

i.e., $u_i = \gamma\left(\sum_j x_j l_j\right) l_i$ and hence,

$$\varepsilon_{ij} = \gamma l_i l_j \tag{5.54}$$

and

$$\omega_{ij} = 0. \tag{5.55}$$

There is no rigid body rotation associated with dislocation climb.

von Mises condition and the independent slip systems

Plastic deformation by dislocation motion occurs only with a limited geometry. Therefore if only one slip system operates, then only one type of simple shear deformation can occur for a given crystal. Homogeneous deformation of a polycrystalline material cannot occur by dislocation glide in such a case. For a homogeneous deformation of a polycrystalline material to occur by dislocation motion, a certain number of independent slip systems must operate. Consider a polycrystalline material made of crystals with random orientation. If homogeneous deformation were to occur in such a material, a crystal must be able to change its shape in an arbitrary fashion. Consequently, the deformation by dislocation glide must be able to create any arbitrary strain. The total strain due to different slip systems can be written as,

$$\varepsilon_{ij}^T = \sum_k \varepsilon_{ij}^k = \sum_k \frac{1}{2}\gamma^k(n_j^k l_i^k + n_i^k l_j^k) \tag{5.56}$$

where ε_{ij}^k is strain caused by the kth slip system. Since the volume is conserved by plastic deformation,

$$\varepsilon_{11}^T + \varepsilon_{22}^T + \varepsilon_{33}^T = 0. \tag{5.57}$$

Equations (5.56) and (5.57) give a set of five equations to determine five unknowns, γ^k. Consequently, five independent slip systems, i.e., five independent sets of ($\boldsymbol{n}$, $\boldsymbol{l}$) are needed to achieve homogeneous deformation. This condition is referred to as the *von Mises condition*. When this condition is not met, a polycrystalline material may fracture (if pressure is low) or deform inhomogeneously. The independent slip systems are those slip systems that the strain tensors from them cannot be produced by a combination of others. Let us consider the case of an NaCl-type crystal (e.g., (Mg, Fe)O). Consider deformation by $\langle 1\bar{1}0\rangle\{110\}$ slip systems. Let us consider first the slip due to $[\bar{1}01](101)$, and define the coordinate system parallel to $\langle 100\rangle$ axes. In this case, using equation (5.51), one can show that the strain due to this slip system is $\begin{bmatrix} -1 & 0 & 0 \\ 0 & 0 & 0 \\ 0 & 0 & 1 \end{bmatrix}$. Similarly, the strain due to the $[0\bar{1}1]$ (011) and $[\bar{1}10]$ (110) is $\begin{bmatrix} 0 & 0 & 0 \\ 0 & -1 & 0 \\ 0 & 0 & 1 \end{bmatrix}$ and $\begin{bmatrix} -1 & 0 & 0 \\ 0 & 1 & 0 \\ 0 & 0 & 0 \end{bmatrix}$. It is clear that only two of them are independent. As is obvious from the above analysis, the off-diagonal components of strain are all zero, which means that the deformation changing the angles of two different crystallographic axes cannot occur by this set of slip systems. Similarly it can be shown that the $\langle 1\bar{1}0\rangle\{100\}$ system has three independent slip systems. Thus a combination of the $\langle 1\bar{1}0\rangle\{110\}$ and the $\langle 1\bar{1}0\rangle\{100\}$ slip systems provides five independent strain components needed for the von Mises condition (Problem 5.5).

Note, however, that this condition can be relaxed in several cases. First, if heterogeneous deformation is allowed, the von Mises condition is relaxed and only four independent slip systems are needed (HUTCHINSON, 1976). Second, contribution from other modes of deformation can also relax this condition. One important example is the contribution from diffusional creep, which is important under most Earth conditions as well as experimental conditions with small grain size. The other is small-strain deformation such as the deformation associated with the post-glacial rebound in which elastic strain can make an important contribution (KARATO, 1998c). The contribution from dislocation climb may also be important (DURHAM *et al.*, 1977; TAKESHITA *et al.*, 1990). In these cases, deformation can occur using a smaller number of slip systems. Under these circumstances, the rate of deformation will be significantly higher than that for large-strain deformation.

Because of the von Mises condition, (large-strain) steady-state creep of a polycrystalline material with several slip systems is usually rate-controlled by the rate of the hardest slip system (HUTCHINSON, 1976, 1977). Therefore in a study of the plastic deformation of a single crystal, one needs to investigate the deformation of all the possible slip systems including hard ones to obtain useful information for the strength of a polycrystalline aggregate.

Problem 5.5*

Show that the strain caused by the $\langle 1\bar{1}0\rangle\{110\}$ slip systems in the NaCl structure has only two independent strain components and the $\langle 1\bar{1}0\rangle\{100\}$ system in the NaCl structure has three independent slip systems so that the combination of the two sets make a homogenous deformation of a polycrystal possible.

Solution

For the $\langle\bar{1}10\rangle(101)$ slip system, $n=\left(1/\sqrt{2},0,1/\sqrt{2}\right)$, $l=\left(-1/\sqrt{2},0,1/\sqrt{2}\right)$. Now using equation (5.51), we find $\varepsilon_{11}=-\gamma, \varepsilon_{12}=\varepsilon_{13}=\varepsilon_{22}=0$ and $\varepsilon_{33}=\gamma$. Therefore the strain is $\varepsilon_{ij}^1=\begin{bmatrix}-1&0&0\\0&0&0\\0&0&1\end{bmatrix}$. Similarly the strain due to the $[0\bar{1}1](011)$ slip system is $\varepsilon_{ij}^2=\begin{bmatrix}0&0&0\\0&-1&0\\0&0&1\end{bmatrix}$ and the strain due to the $[\bar{1}10](110)$ slip system is $\varepsilon_{ij}^3=\begin{bmatrix}-1&0&0\\0&1&0\\0&0&0\end{bmatrix}$. Therefore $\varepsilon_{ij}^1-\varepsilon_{ij}^2-\varepsilon_{ij}^3=0$ and hence one of these strain components can be made by a linear combination of the others, so that only two components are independent.

For the $\langle\bar{1}10\rangle\{100\}$ slip systems, there are six physically different systems (note that the slip direction and the slip plane normal must be perpendicular), i.e., $[\bar{1}10](100)$, $[110](100)$, $[\bar{1}01](010)$, $[101](010)$, $[\bar{1}10](001)$, $[110](001)$. Among them, $[011](100)$ produces $\varepsilon_{ij}^4=\begin{bmatrix}0&1&1\\1&0&0\\1&0&0\end{bmatrix}$, $[101](010)$ $\varepsilon_{ij}^5=\begin{bmatrix}0&1&0\\1&0&1\\0&1&0\end{bmatrix}$ and $[110](001)$ $\varepsilon_{ij}^6=\begin{bmatrix}0&0&1\\0&0&1\\1&1&0\end{bmatrix}$. These three are independent, and they cause the change in angles of crystallographic axes (but not the elongation along the crystallographic axes because $\varepsilon_{11}=\varepsilon_{22}=\varepsilon_{33}=0$). Thus a combination of $\langle 1\bar{1}0\rangle\{110\}$ and $\langle 1\bar{1}0\rangle\{100\}$ slip systems provides the five independent slip systems.

Stress–strain field, energy of a dislocation, force on a dislocation

A dislocation is associated with a stress–strain field. The stress–strain field associated with a dislocation is dependent on the nature of dislocation, either edge or screw (or mixed). The stress–strain field around a dislocation is characterized by the stress and strain tensor and therefore dependent on the direction with respect to dislocation line and the Burgers vector. One can find a full treatment of the stress–strain field of a dislocation in the textbooks listed previously. If we use a simpler scalar formula,

$$\sigma=\frac{\mu b}{2\pi K r} \tag{5.58}$$

where μ is the shear modulus, b is the length of the Burgers vector, $K=1-\nu$ (ν, Poisson's ratio) for edge dislocation and $K=1$ for screw dislocation, r is the distance from the dislocation line.[3] An important point is that a dislocation has a long-range stress (strain) field, and consequently, dislocation–dislocation interaction plays an important role in plastic deformation. In contrast, a strain field caused by a point defect varies with distance as $\sigma \propto (1/r^2)$ (e.g., FLYNN, 1972).

There are three consequences of a stress (and strain) field of a dislocation. First, because of the stress–strain field around a dislocation, a dislocation has energy that is proportional to its length. Therefore a dislocation is associated with line tension. Knowing the stress (and strain) field associated with a dislocation, the energy of a dislocation (per unit length) can be calculated as (COTTRELL, 1953),[4]

$$E = \frac{\mu b^2}{4\pi K}\log\left(\frac{R}{b_0}\right) \sim \mu b^2 \tag{5.59}$$

where b_0 is the radius of the dislocation core and R is the upper limit for integration. Note that in this calculation, the energy associated with the central regions of the dislocation line (dislocation core, i.e., a region $r < b_0$) is not included. In this region, elastic theory does not apply and an atomistic calculation is needed. A detailed calculation shows that the contribution to dislocation energy from the dislocation core is usually ~10–20 % (e.g., COTTRELL, 1953). So equation (5.59) (i.e., $E \approx \mu b^2$) is a good approximation for most purposes. In deriving equation (5.59), we made an assumption that the crystal is isotropic. This assumption is not valid in many geological materials, and the elastic anisotropy affects the relative values of dislocation energy for different slip systems which can influence the relative ease of different slip systems.[5]

[3] The stress is obviously a tensor, and therefore it depends on orientation. If z is the axis parallel to the dislocation, and x is the direction of the Burgers vector, then the stress field around a screw dislocation is

$$\sigma_{xz} = -\frac{\mu b}{2\pi}\frac{y}{x^2+y^2}, \sigma_{yz} = \frac{\mu b}{2\pi}\frac{x}{x^2+y^2}$$

and for an edge dislocation

$$\sigma_{xx} = -\frac{\mu b}{2\pi(1-\nu)}\frac{y(3x^2+y^2)}{(x^2+y^2)^2}, \sigma_{yy} = \frac{\mu b}{2\pi(1-\nu)}\frac{y(x^2-y^2)}{(x^2+y^2)^2},$$

$$\sigma_{zz} = \nu(\sigma_{zz}+\sigma_{yy}), \quad \sigma_{xy} = \frac{\mu b}{2\pi(1-\nu)}\frac{x(x^2-y^2)}{(x^2+y^2)^2}$$

other components are 0.

[4] The equation (5.59) would imply that energy of a dislocation in an infinite medium would be infinite (because $\log R \to \infty$ as $R \to \infty$). However, in practice, there must be other dislocations in a given material, therefore R would be about the mean distance of dislocations. The mean distance of dislocations in typical Earth materials is 0.1–10 μm and hence $\log(R/b_0) = 7$–12. Therefore for a good approximation, $E \approx \mu b^2$.

[5] Note that the relative ease of different slip systems should also depend on the slip plane.

Problem 5.6

Derive equation (5.59).

Solution

The strain energy per unit volume is given by $\int \sigma\, d\varepsilon = \mu\varepsilon^2/2K = K\sigma^2/2\mu = (b^2/8\pi^2)(\mu/Kr^2)$. Now we integrate this along a cylinder (parallel to the dislocation) with length l to get

$$E' = \int dE' = \int_{b_0}^{R} \frac{b^2}{8\pi^2}\frac{\mu}{Kr^2} 2\pi r l\, dr = \frac{l\mu b^2}{4\pi K}\log\frac{R}{b_0}.$$

Therefore the energy per unit length is $E = \frac{E'}{l} = \left(\frac{\mu b^2}{4\pi K}\right)\log\left(\frac{R}{b_0}\right) \sim \mu b^2$.

Problem 5.7

Using equation (5.59), calculate the change in energy per unit volume due to dislocations for dislocation densities ranging from 10^{10}–10^{14} m^{-2} (assume $b = 0.5$ nm, $\mu = 150$ GPa). Discuss how the strain energy due to dislocations affects the condition for a phase transformation from olivine to wadsleyite (a high-pressure polymorph of olivine) assuming that dislocations are generated only in olivine. Use the following parameters: $U_1 - U_2 = -27.1$ kJ/mol, $S_1 - S_2 = 9.0$ J/mol K, $V_1 - V_2 = 3.16 \times 10^{-6}$ m^3/mol, $V_1 = 43.67 \times 10^{-6}$ m^3/mol and $V_2 = 40.54 \times 10^{-6}$ m^3/mol (where 1 refers to olivine and 2 refers to wadsleyite).

Solution

The Gibbs free energy of each phase is given by,

$$G_{1,2} = U_{1,2} - TS_{1,2} + (P + E_{1,2})V_{1,2}$$

where $E_{1,2}$ are the strain energies of 1, 2 phases. The condition for a phase transformation is therefore,

$$P = P_0 + \Delta P$$

with

$$P = -\frac{U_1 - U_2 - T(S_1 - S_2)}{V_1 - V_2} = 13.7\,\text{GPa}$$

and

$$\Delta P = -\frac{E_1 V_2 - E_2 V_2}{V_1 - V_2} = -(5.2 - 520)\,\text{MPa}$$

where we used an assumption that dislocations are present only in olivine, so $E_1 = \rho\mu b^2$ and $E_2 = 0$.

Therefore the phase boundary will be shifted to a shallower depth, the magnitude of which is on the order of $\sim$0.15–15 km depending on the dislocation density.

Second, because a dislocation has a long-range stress–strain field, dislocations strongly interact each other. This strong interaction is a cause of various mechanical behaviors of materials such as *work hardening* (see Chapter 9).

Third, a dislocation is associated with the relative displacement, $\boldsymbol{b}$, and consequently, some work will be done on a material when a dislocation moves under applied stress. From this the force acting on a dislocation line can be calculated as follows. Consider a small motion of a dislocation segment $\boldsymbol{dl}$ by $\boldsymbol{dx}$. This motion will sweep an area segment of $d\boldsymbol{A} = \boldsymbol{dl} \times \boldsymbol{dx}$. Now on this plane, the applied stress σ will exert a force $d\boldsymbol{F} = \sigma d\boldsymbol{A}$ $(dF_i = \sum \sigma_{ij} dA_j)$. By the motion of a dislocation, material across the glide plane will move by $\boldsymbol{b}$. Therefore the work done by this force is

$$dW = \boldsymbol{b} \cdot d\boldsymbol{F} = \boldsymbol{b} \cdot (\sigma d\boldsymbol{A}) = (\boldsymbol{b}\sigma) \cdot d\boldsymbol{A} \\ = (\boldsymbol{b}\sigma) \cdot (d\boldsymbol{l} \times d\boldsymbol{x}) = [(\boldsymbol{b}\sigma) \times d\boldsymbol{l}] \cdot d\boldsymbol{x} \quad (5.60)$$

where we used the fact that the stress is a symmetric tensor $(\boldsymbol{b} \cdot (\sigma\, d\boldsymbol{A}) = \sum_{i,j} b_j \sigma_{ij}\, dA_j = \sum_{i,j} b_j \sigma_{ij} dA_j = (\boldsymbol{b}\sigma) \cdot d\boldsymbol{A})$, and the vector identity $\boldsymbol{A} \cdot (\boldsymbol{B} \times \boldsymbol{C}) = (\boldsymbol{A} \times \boldsymbol{B}) \cdot \boldsymbol{C}$. Therefore the force acting per unit length of a dislocation is

$$\boldsymbol{f} = (\boldsymbol{b}\sigma) \times \frac{d\boldsymbol{l}}{dl} = (\boldsymbol{b}\sigma) \times \boldsymbol{e}_l \quad (5.61)$$

where $\boldsymbol{e}_l \equiv d\boldsymbol{l}/dl$ is the unit vector along the dislocation line. This is called the *Peach–Koehler force*. Note that the direction of the force is normal to the direction of a dislocation line, and the magnitude of the force is the same for all the directions on the glide plane (i.e., the force tends to expand a dislocation loop; Problem 5.8). In a simple scalar form,

$$f = \sigma b. \quad (5.62)$$

Problem 5.8*

Show that the force on the dislocation is normal to the dislocation line and the magnitude is the same for all portions, and derive relation (5.49) (definition of the Schmid factor) from relation (5.61).

Solution

Let the x–z plane be the glide plane of dislocation, and let the z-axis be parallel to the dislocation line. The Burgers vector (slip direction) $\boldsymbol{b}$ is on the x–z plane, and let φ be the angle between $\boldsymbol{b}$ and the dislocation line. Then $\boldsymbol{b}/b = (\sin\varphi, 0, \cos\varphi)$. Then applying equation (5.61), the force is $\boldsymbol{f} = b(\sin\varphi \cdot \sigma_{xy} + \cos\varphi \cdot \sigma_{yz}, \sin\varphi \cdot \sigma_{xx} + \cos\varphi \cdot \sigma_{xz}, 0)$. Therefore there is a force on the x–z plane normal to the dislocation line, f_x, and there is also a force normal to the glide plane, f_y (this force does not cause glide, but could cause dislocation climb). Now let the angle between the force and the normal of the glide plane be ψ and the angle between the projection of the force on the x–z plane and the z-axis be θ. Then, $\sigma_{xy} = (F/A) \sin\psi \cos\psi \sin\theta$ and $\sigma_{yz} = (F/A) \sin\psi \cdot \cos\psi \cos\theta$, and the direction of force is $\boldsymbol{F}/F = (\sin\psi \sin\theta, \cos\psi, \sin\psi \cos\theta)$. So the direction cosine between the slip direction and the force direction is $(\boldsymbol{F}/F) \cdot (\boldsymbol{b}/b) \equiv \cos\lambda = \sin\psi(\sin\theta \sin\varphi + \cos\theta \cos\varphi) = \sin\psi \cos(\theta - \varphi)$. Therefore $f_x = b(\sin\varphi \cdot \sigma_{xy} + \cos\varphi \cdot \sigma_{yz}) = (bF/A)\cos(\theta - \varphi)\sin\psi \cdot \cos\psi \equiv b\sigma_{RSS}$, which does not depend on the position of a dislocation and $\sigma_{RSS} = (F/A)\cos\lambda \cos\psi \equiv (F/A)S$. Hence the Schmid factor, S, is given by $S = \cos\lambda \cos\psi$ which agrees with (5.49) as it should.

Partial dislocations, dissociation of a dislocation

Dissociation of a dislocation into partial dislocations is commonly observed in oxides and silicates (as well as in some metals). Dissociation occurs when a perfect dislocation has a large energy (long Burgers vector) and when there is a mechanism to decompose the Burgers vector into shorter ones. For instance, when there are three possible Burgers vectors in a given crystal among which the following relation holds,

$$\boldsymbol{b}_1 = \boldsymbol{b}_2 + \boldsymbol{b}_3 \quad (5.63)$$

then a dislocation with the Burgers vector $\boldsymbol{b}_1$ can be split into two dislocations with Burgers vectors $\boldsymbol{b}_2$ and $\boldsymbol{b}_3$. This dislocation splitting occurs when the splitting reduces the dislocation energy. If the dislocation energy relation (5.59) for an isotropic material is applied, then the condition for splitting (Frank energy criterion) is given by

$$b_1^2 > b_2^2 + b_3^2 \quad (5.64)$$

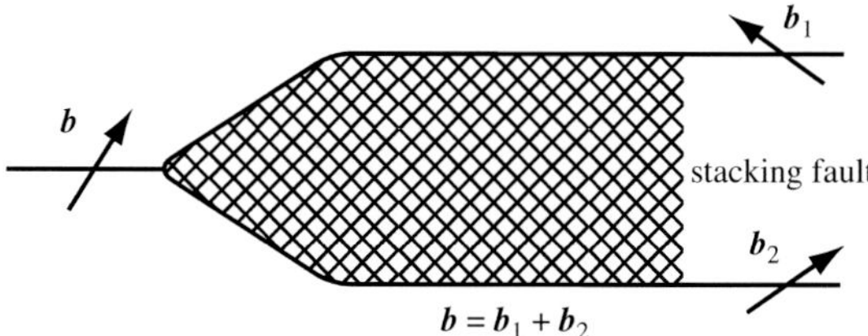

FIGURE 5.6 Partial dislocations and a stacking fault.

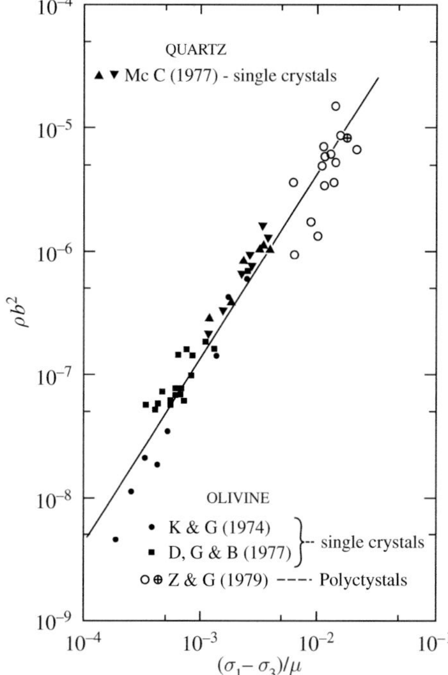

FIGURE 5.7 The dislocation density versus stress relationship (after Kohlstedt and Weathers, 1980).

Examples of dislocation splitting include $\frac{1}{2}[111] = \frac{1}{4}[111] + \frac{1}{4}[111]$ in garnet (Allen *et al.*, 1987) and $[010] = \frac{1}{4}[01\bar{1}] + \frac{1}{4}[01\bar{1}] + \frac{1}{4}[011] + \frac{1}{4}[011]$ in olivine (Fujino *et al.*, 1992). Fujino *et al.* (1992) discussed the influence of elastic anisotropy in estimating the energetics of dislocation splitting.

Note that the Burgers vectors of split dislocations are not that of the displacement vectors corresponding to a unit cell. These dislocations are called *partial dislocations* and the reaction such as (5.63) is called the *dissociation of a dislocation* (Fig. 5.6). In such a case, the atomic arrangement in the plane between two partial dislocations is different from that of a perfect crystal. This planar defect where the atomic arrangement deviates from that of a perfect lattice is called a *stacking fault*. The partial dislocations repel each other and the width of the stacking fault is determined by the balance due to the repulsive force and the attractive force that tends to reduce the width of a stacking fault.

Dissociation of a dislocation occurs both on the glide plane (glide dissociation) or on a plane perpendicular to it (climb dissociation).

Dislocation density versus stress relationship

Unlike the density of point defects, dislocation density is not controlled by thermodynamic equilibrium. This is due to its small configurational entropy because of its large dimension as compared to point defects. Instead, the dislocation density is primarily controlled by the applied stress (other sources for dislocations include crystal growth). In an ideal steady-state deformation, the dislocation density is determined by the balance between the applied stress and the stress around a dislocation, namely,

$$\sigma = \frac{\mu b}{r} \sim S\sigma_a \tag{5.65}$$

hence

$$\rho = \left(\frac{1}{r}\right)^2 = \left(\frac{S}{b}\right)^2 \left(\frac{\sigma_a}{\mu}\right)^2. \tag{5.66}$$

Thus the dislocation density, at a steady state, is proportional to the square of applied stress. Such a relation is approximately consistent with observations (Fig. 5.7) although the observed stress exponent is often smaller than 2. Note that the dislocation density is only weakly dependent on pressure and temperature through[6] $\rho(P, T)/\rho(P_0, T_0) = \left(\mu(P_0, T_0)/\mu(P, T)\right)^2$.

Problem 5.9*

Dislocation density is often measured on a section by counting the number of dislocations that intersect the plane, N. Show that if the dislocation orientation is random, then the dislocation density is given by $\rho = 2N/A$ where A is the area (e.g., Underwood, 1969).

Solution

Let us consider a cube (whose edge length is l) containing the total length, L, of dislocations.

[6] Pressure and temperature will modify the strain partitioning among different slip systems, which have some effects on the dislocation density through the change in the Schmid factor, S.

Consider a small segment of a dislocation line whose length is ΔL. Now let θ be the angle between the direction of dislocation and the normal to the plane on which the measurement is done. Then the length of this segment of dislocation projected to the line normal to the plane is $\Delta L \cdot \cos\theta$. Now let us consider cutting these lines by planes whose distance is Δl. Then the number of points at which the dislocation lines are cut by these planes is $\frac{\Delta L \cdot \cos\theta}{\Delta l}$. Consequently, the number of points per unit area is $N/A = (\sum \Delta L)\langle\cos\theta\rangle/\Delta l \cdot l^2 = L\langle\cos\theta\rangle/l^2\Delta l = (L/V)$ $\langle\cos\theta\rangle$. Now $\langle\cos\theta\rangle = \int_0^{\pi/2}\cos\theta\, f(\theta)\, d\theta = \int_0^{\pi/2}\cos\theta$ $\sin\theta\, d\theta = \frac{1}{2}$ where $f(\theta)\, d\theta = \sin\theta\, d\theta$ is the probability that a line has an angle $\theta \sim \theta + d\theta$. Therefore $\rho \equiv L/V = 2N/A$. Note that the assumption of random orientation distribution is usually not valid because of the crystallographic control of dislocation lines that can cause large errors in dislocation density measurements. This error can be reduced if the total length of the dislocation lines for a given volume is measured (Karato and Jung, 2003).

Geometrically necessary dislocations

When a strain gradient is present in a material, then the density of dislocations can be controlled by strain rather than stress. For example in a deforming polycrystalline aggregate, the *strain gradient* is present near grain boundaries due to the contrast in the plastic properties of the grains in contact. At high enough temperatures, this strain gradient can be accommodated by the accumulation of dislocations. Cottrell (1964) termed them as *geometrically necessary dislocations*, and the implications of these dislocations on mechanical properties were examined by Ashby (1970). The density of geometrically necessary dislocations can be calculated by considering the geometry of deformation. Consider a case where bending occurs in a crystal (a similar analysis applies to other deformation geometries). Let us define the z-axis as the axis around which the bending occurs, and let the x- and y-axes be defined as shown in Fig. 5.8. Let us consider a small cross section on the z-plane whose edge lengths are dx and dy. The bending causes the relative displacement along the y-direction of

$$du = \gamma(y + dy) \cdot dx - \gamma(y) \cdot dx = \frac{\partial\gamma}{\partial y} \cdot dx \cdot dy \qquad (5.67)$$

where γ is the plastic strain. Now the same displacement can also be caused by the presence of

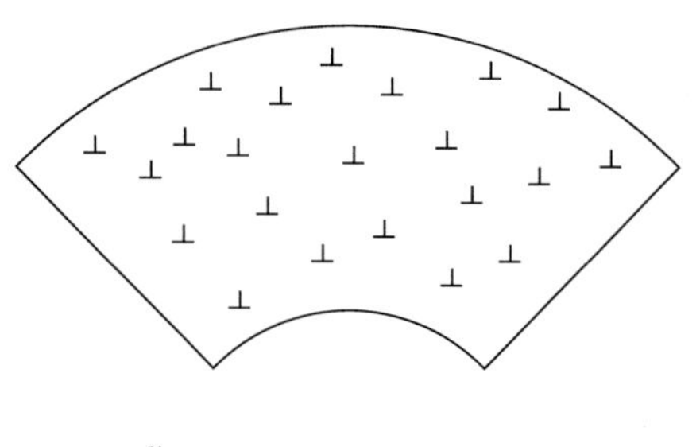

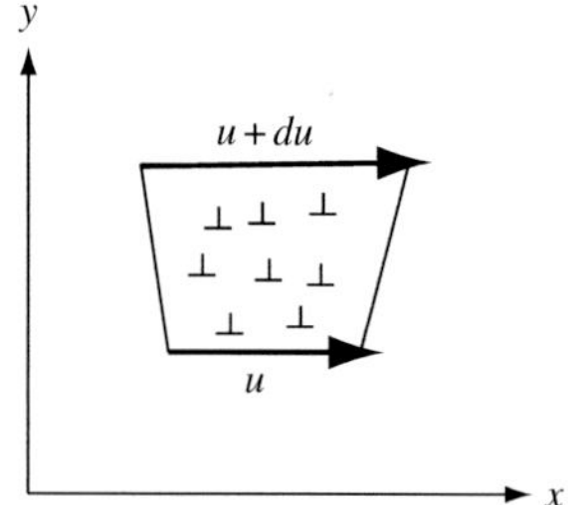

FIGURE 5.8 Geometrically necessary dislocations.

dislocations. Recall that each dislocation causes a displacement of a crystal of magnitude b. If dislocations of the same Burgers vector are present in an area (of a cross section) $dx \cdot dy$, then there are a total of $\rho_G\, dx \cdot dy$ dislocations (ρ_G is the density of geometrically necessary dislocations) in this area so the relative displacement of materials between lines separated by dy parallel to the Burgers vector (x-direction) is

$$du = \rho_G b \cdot dx \cdot dy. \qquad (5.68)$$

From equations (5.67) and (5.68), we get

$$\rho_G = \frac{1}{b}\frac{\partial\gamma}{\partial y}. \qquad (5.69)$$

The density of geometrically necessary dislocations can be high (for $\partial\gamma/\partial y = 0.01/10\ \mu\text{m} = 10^3\text{m}^{-1}$, and $b = 10^{-9}\text{m}$, $\rho_G \sim 10^{12}\text{m}^{-2}$). Geometrically necessary dislocations may play an important role in strain accommodation in a polycrystal as well as in the development of dynamic recrystallization (Chapter 13). Enhanced deformation associated with a phase transformation may also involve geometrically necessary dislocations (see Chapter 15).

Kinks and jogs

Dislocation motion may be classified into two categories: motion on a glide plane and motion away from the glide plane. In both cases, dislocation motion occurs through the motion of small steps. A small step on a dislocation line on a slip plane is called a *kink*. Similarly, a step on a dislocation line out of its slip plane is called a *jog*. Because of their small size, the concept of statistical equilibrium by the competition of energy increase with configurational entropy works in such a case.

Note that unlike dislocation density, kink or jog density is in most cases controlled by thermal equilibrium (because of their small dimension). Thus,

$$c_{k,j} = \frac{1}{b}\exp\left(-\frac{G^*_{k,j}}{RT}\right) \tag{5.70}$$

where $G^*_{k,j}$ is the formation free energy of a kink or a jog. Essentially, the formation energy of a kink (or a jog) is the excess energy associated with the increase in the dislocation line. Since the excess length is $\sim b$, the formation energy of a kink or a jog is given roughly by (for a material with a large Peierls stress, the formation energy of a kink or a jog depends also on the Peierls stress, see HIRTH and LOTHE (1982), and Chapter 9)

$$G^*_{k,j} \approx \mu b^3. \tag{5.71}$$

Kink density described by equation (5.70) corresponds to the density of thermal kinks. Kinks may also be present on a dislocation due to geometrical reasons. For example, if a dislocation line is tilted from the minimum of the Peierls potential, then there must be kinks with density given by

$$c_k = \frac{\theta}{a} \tag{5.72}$$

where a is the spacing of the minimum of the Peierls potential. These are called *geometrical kinks*. Short-term, low strain deformation could occur by the motion of geometrical kinks (KARATO, 1998a). In these cases, only kink migration is needed for deformation. When deformation occurs in a steady-state fashion, one needs to continuously create kinks by thermal activation. Therefore dislocation motion in such a case involves both the formation and migration of kinks.

Charges on dislocations, dislocation-point defect interaction

The kink or jog density may also depend upon the chemical environment such as oxygen partial pressure and/or water fugacity. There is evidence that dislocations in ionic solids have an electrostatic charge. HOBBS (1983) and JAOUL (1990) proposed that dislocations in some silicate minerals (such as olivine) may have electrostatic charges. The direct evidence for electrostatic charges on dislocations in KCl was reported by COLOMBO *et al.* (1982) and KATAOKA *et al.* (1983, 1984a, 1984b) who showed that the strength of KCl doped with $CaCl_2$ changed as an electric field was applied. In these cases, the density of kinks (also jogs) depends on the concentration of charged point defects and the velocity of dislocations becomes sensitive to the chemical environment.

Electric charges on dislocations may develop due to the imbalance of formation energies of cation and anion defects. In ionic crystals, point defects can be created for cation as well as anion sites. Because of the charge neutrality requirement, the concentrations of positively and negatively charged defects must be the same in a crystal. However, locally near the source/sink of defects, the concentrations of positively charged defects can be different from those of negatively charged defects. This arises due to the fact that the formation energy for positively charged defects can be different from that of negatively charged defects. In these cases, a dislocation will be associated with unequal amounts of electrostatic charges, a global charge balance being maintained by the accumulation of oppositely charged defects around a dislocation (ESHELBY *et al.*, 1958; BROWN, 1961) (see Box 5.2). Kinks (or jogs) on dislocations in ionic crystals can also contain electrostatic charges for geometrical reasons (e.g., HIRTH and LOTHE, 1982).

Charges on dislocations may also be introduced through the interaction of a dislocation with electrons in semiconductors (LOUCHET and GEORGE, 1983). Dislocations in semiconductors act as donors or acceptors for electrons. In both cases, charges on dislocations are through kinks (or jogs). In other words, any processes that change the electronic state of these materials has a strong influence on the density of kinks (or jogs) and therefore affects the mobility of dislocations.

Observation of dislocations

A dislocation is associated with a long-range, strong strain field. Consequently, a number of techniques can be used to observe a dislocation. They include:

(1) Decoration technique
Chemical reactions occur faster along the dislocation line than other dislocation-free portions of a crystal. Consequently, certain chemical species can be diffused along dislocation lines to "decorate" them. If these species have certain optical characteristics, then one can see dislocation lines under an optical microscope. An example is the oxidation decoration of $(Mg, Fe)_2SiO_4$ olivine (KOHLSTEDT

Box 5.2 Electric charge on grain boundaries and dislocations

Extended defects such as dislocations and grain boundaries in ionic solids can contain an electric charge. An extended defect is the site at which point defects are formed. But the formation energy of cation defect is usually different from that of anion defect. Therefore near the extended defect, there is an imbalance of concentration in charged defects. Let us consider a case of (Mg, Fe)O where the formation of cation vacancy, V_M'', is easier than anion vacancies. Formation of a cation (anion) vacancy near a grain boundary creates an excess cation (anion) in the grain boundary, null $\Leftrightarrow M_M^{\bullet\bullet}$(boundary) $+ V_M''$(crystal) (or null $\Leftrightarrow O_O''$(boundary) $+ V_O^{\bullet\bullet}$(crystal)). Since a larger amount of cation vacancies are formed at the extended defects, there will be positive charge in the boundary and negative charge near the defect. The redistribution of charged species under these circumstances can be analyzed by noting that the concentration of any charged defect is determined by its intrinsic concentration and the electrostatic interaction, namely, $[X] = [X]_0 \exp(-Ze\phi/RT)$ where $[X]$ is the concentration of defect X, $[X]_0$ is the concentration of defect X for zero potential, Ze is the electrostatic charge of the defect, and ϕ is the electrostatic potential. Far from the boundary (dislocation), the charge neutrality condition must be met, namely,

$$
\begin{aligned}
[V_M'']_\infty &= [V_M'']_0 \exp\left(\frac{2e\phi_\infty}{RT}\right) \\
&= [V_O^{\bullet\bullet}]_\infty = [V_O^{\bullet\bullet}]_0 \exp\left[-\frac{2e\phi_\infty}{RT}\right]
\end{aligned} \tag{1}
$$

where $[V_M'']_0 = \exp(G_{V_M''}/RT), [V_O^{\bullet\bullet}]_0 = \exp(G_{V_O^{\bullet\bullet}}/RT)$ and quantities with ∞ mean those far from the extended defect. Thus, $\phi_\infty = (G_{V_M''} - G_{V_O^{\bullet\bullet}})/4e < 0$. Since the electric potential is caused by the space charge in the crystal, if the potential is negative in the crystal, the potential on the boundary must be positive (this corresponds to an excess $M_M^{\bullet\bullet}$ (the net charge of a material must be neutral)). The distribution of charge is determined by the distribution of electric potential that must satisfy the Poisson equation, namely,

$$
\begin{aligned}
\nabla^2 \phi &= -\frac{4\pi \sum q}{\varepsilon_0} \\
&= -\frac{8\pi e([V_O^{\bullet\bullet}] - [V_M''])}{\varepsilon_0}
\end{aligned} \tag{2}
$$

where ε_o is the static dielectric constant and q is the electric charge. Inserting $[V_M''] = [V_M'']_\infty \exp(2e\cdot(\phi - \phi_\infty)/RT), [V_O^{\bullet\bullet}] = [V_O^{\bullet\bullet}]_\infty \exp(-2e(\phi - \phi_\infty)/RT)$ into equation (2), and assuming $2e(\phi - \phi_\infty)/RT \ll 1$, equation (2) is transformed to

$$
\nabla^2(\phi - \phi_\infty) = k^2(\phi - \phi_\infty) \tag{3}
$$

with $k^2 \equiv 16\pi e^2 [V_M'']_\infty / \varepsilon_0 RT$. For a grain boundary, this equation can be written as $(d^2/dx^2)(\phi - \phi_\infty) = K^2(\phi - \phi_\infty)$ where x is normal to the boundary and has a solution of the form $\phi - \phi_\infty \propto \exp(-\kappa x)$. A similar analysis can be made for a dislocation, in which case, we use a cylindrical coordinate, namely, $(d^2/dr^2 + (1/r)(d/dr))(\phi - \phi_{\infty}) = \kappa^2(\phi - \phi_\infty)$ and the solution has the form $\phi - \phi_\infty \propto K_0(\kappa r)$ where $K_0(\kappa r)$ is the Bessel function. In both cases the charge distribution has a characteristic length λ given by

$$
\begin{aligned}
\lambda = \kappa^{-1} &= \sqrt{\frac{\varepsilon_0 RT}{16\pi e^2 [V_M'']_\infty}} \\
&= \sqrt{\frac{\varepsilon_0 RT \exp((G_{V_M''} + G_{V_O^{\bullet\bullet}})/2RT)}{16\pi e^2}}.
\end{aligned} \tag{4}
$$

Let us now consider a case where charge balance is controlled by impurity atoms such as $Fe^{\bullet}$ and cation vacancies (and therefore we ignore anion vacancies). Far from the boundary (or a dislocation), the spatial distribution of charge due to V_M'' and $Fe^{\bullet}$ in the crystal must satisfy the charge neutrality condition,

$$2[V_M'']_\infty = 2[V_M'']_0 \exp\left(\frac{2e\phi_\infty}{RT}\right) = [\mathrm{Fe}_M^\bullet]_\infty = [\mathrm{Fe}_M^\bullet]_0 \exp\left[-\frac{e\phi_\infty}{RT}\right]. \qquad (5)$$

Using $[V_M'']_0 = \exp\left(-G_{V_M''}/RT\right)$, we get from equation (5)

$$3e\phi_\infty = G_{V_M''} + RT\log\frac{[\mathrm{Fe}_M^\bullet]_0}{2}. \qquad (6)$$

Note that the sign of electrostatic potential, ϕ_∞, away from the extended defect changes with temperature and the impurity content (the potential is neutral when $2[V_M'']_0 = [\mathrm{Fe}_M^\bullet]_0$ or at a temperature $T_0 = -G_{V_M''}/R\log\frac{[\mathrm{Fe}_M^\bullet]_0}{2}$ called the *iso-electric temperature*). The potential is positive for a large value of $[\mathrm{Fe}_M^\bullet]$ but negative for a low $[\mathrm{Fe}_M^\bullet]$. Positive potential means that the boundary (dislocation) is negatively charged. This is due to the fact that $[\mathrm{Fe}_M^\bullet]$ in the crystal is higher than $2[V_M'']_0$ so $[V_O^{\bullet\bullet}]_0$ must be reduced and hence excess oxygen ions will be present on the boundary (dislocation) leading to a negative boundary potential. The spatial distribution of charged defects and potential can be obtained by solving the Poisson equation and one can obtain

$$\lambda = \kappa^{-1} = \sqrt{\frac{\varepsilon_0 RT}{12\pi e^2\left[\mathrm{Fe}_M^\bullet\right]_\infty}} = \sqrt{\frac{\varepsilon_0 RT\exp\left(e\phi_\infty/RT\right)}{12\pi e^2\left[\mathrm{Fe}_M^\bullet\right]_0}}. \qquad (7)$$

As can be seen from (4) and (7), the effective thickness of a grain boundary (or dislocation core) is sensitive to temperature (and impurity content).

et al., 1976). This is a handy technique that can be used with a minimum facility (all one needs are a simple furnace to heat a sample to ~1200 K and an optical microscope). A three-dimensional image of the dislocations can be observed by this technique. Decorated dislocations can also be observed by a scanning electron microscope, which increases the resolution of dislocation observations (Karato, 1987a) (Fig. 5.9a).

(2) Etch pit technique
A chemical reaction at the surface of a crystal often occurs selectively at a point where a dislocation line intersects the surface. This often results in the formation of etch pits. Etch pits can be observed either by an optical microscope or by a scanning electron microscope. Some acid (HCl, H_2NO_3, HF etc.) is used for the etching purpose depending on the material.

(3) Transmission electron microscopy (TEM)
The distortion of a crystal lattice by a dislocation line results in the distortion of the way in which electron beam is diffracted. Using a well-aligned electron beam, one can investigate the distortion of small regions of lattice by dislocations. In this technique, one can determine the nature of lattice distortion and therefore the details of the dislocation structure (such as the Burgers vector). The spatial resolution of TEM is high. With a conventional imaging technique, images with a spatial resolution of better than 0.1 μm are routine (Fig. 5.9b). With more sophisticated high-resolution imaging techniques, direct imaging of the atomistic arrangement is possible (Fig. 5.9c). For details of transmission electron microscopy see e.g., Williams and Carter (1996).

Dislocation density measurement

Density of dislocations is defined by,

$$\rho \equiv \frac{\sum l}{V} \qquad (5.73)$$

where $\sum l$ is the total length of dislocations, and V is the volume of a crystal. The dislocation density can be determined using various techniques. In most cases, observation is made on a two-dimensional plane, so that the correction for orientation (when one counts the number of dislocations that cross the given plane) (Problem 5.9). However, the uncertainties due to this correction factor are large. A more accurate

(a)

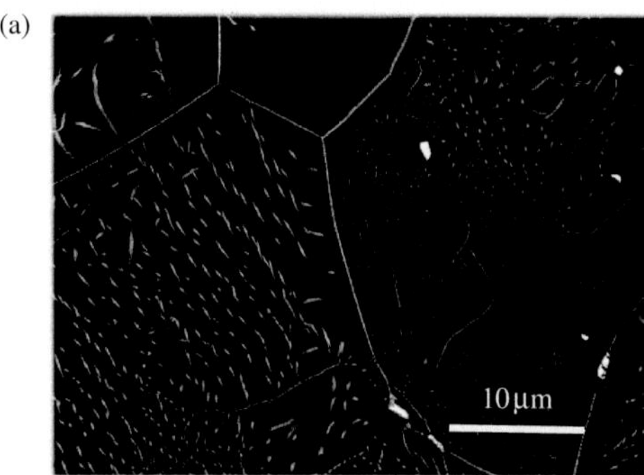

(b)

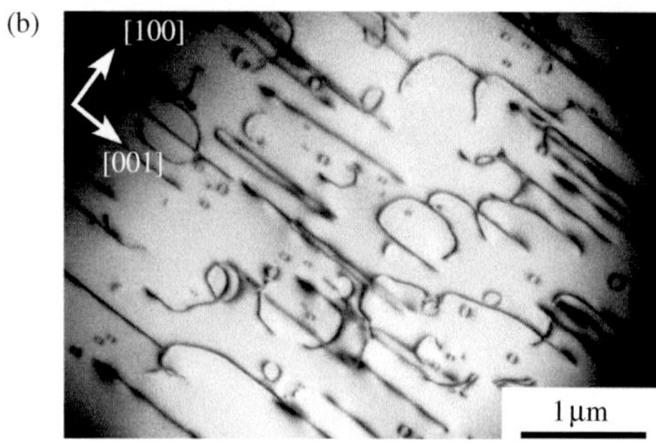

(c)

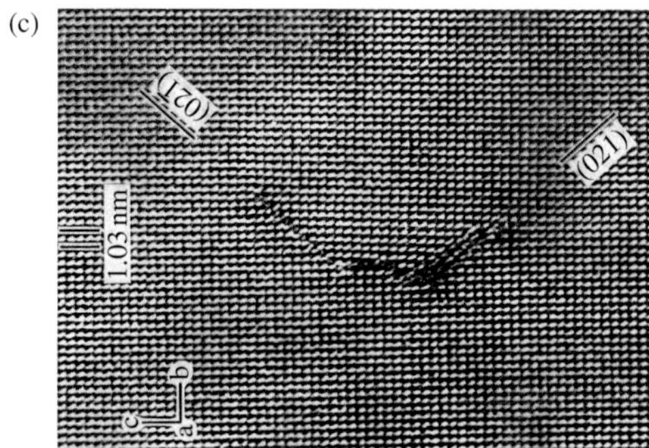

FIGURE 5.9 Various techniques to observe dislocations (all in olivine). (a) Scanning electron microscopy on decorated dislocations, (b) transmission electron microscopy (TEM, bright field image; courtesy of Jun-ichi Ando), (c) high-resolution TEM (courtesy of Kiyoshi Fujino).

measurement is a three-dimensional measurement of a total dislocation length in a given volume of a crystal. This is possible even for a two-dimensional image if the effective thickness of a layer from which an image is obtained is known (e.g., KARATO and LEE, 1999).

5.4. Grain boundaries

5.4.1. Generalities: geometry and boundary energy

A grain boundary is a planar defect that separates two portions of crystals with different orientations. Grain boundaries may conveniently be classified into two categories, namely low-angle boundaries and high-angle boundaries. Low-angle boundaries (or subgrain boundaries) are those that can be modeled as arrays of dislocations. They may further be classified into two categories depending on the types of dislocations that a grain boundary is made of. A (low-angle) grain boundary made of an array of edge dislocations is called a *tilt boundary*. A (low-angle) boundary made of screw dislocations is referred to as a *twist boundary*. In both cases, the misfit angle,[7] θ, is related to the spacing of the dislocations, d, as (Fig. 5.10a),

$$\theta \sim \frac{b}{d}. \tag{5.74}$$

The energy of a low-angle boundary depends on the misfit angle. This can be easily calculated by combining the equation for dislocation energy (5.59) with relation (5.74) to get,

$$\gamma = \frac{\mu b}{4\pi K}\theta\left(\log\frac{b}{b_0} - \log\theta\right). \tag{5.75}$$

The energy versus angle relationship is shown in Fig. 5.10.

Thus a low-angle grain boundary can be easily specified based on the orientation of the crystal and the slip system(s) of dislocations. The characterization of more general, high-angle boundaries is more complicated. In order to define a grain boundary, one needs five parameters: two to define the orientation of a plane in one crystal, and three to define the orientation of another crystal. However, if one focuses on one type of boundary such as a tilt boundary, then one can use one parameter, the misfit angle between the two grains, to fully characterize the boundary. In Fig. 5.11, we use the misorientation associated with a tilt boundary to plot the energy as a function of the misfit angle. Note that up to $\sim 20^\circ$, equation (5.75) reproduces the experimental data well. Above this angle, the mean distance of the dislocations will be smaller than the atomic spacing so that the concept of dislocation will lose its meaning. Also note that there are several minima in energy as a function of misfit angle that correspond to

[7] For any pair of grains, one can define the misorientation by noting that for any pair of grains, there is an orientation, the rotation around which makes the orientations of two grains identical.

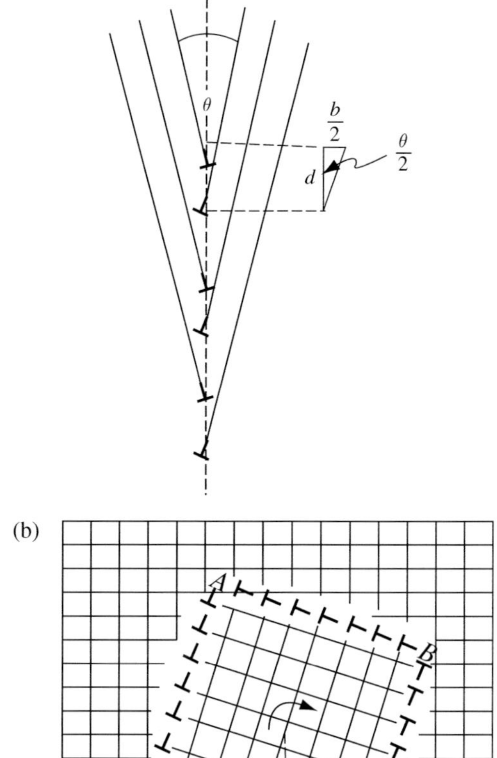

FIGURE 5.10 Dislocations and (a) a tilt boundary and (b) a twist boundary.

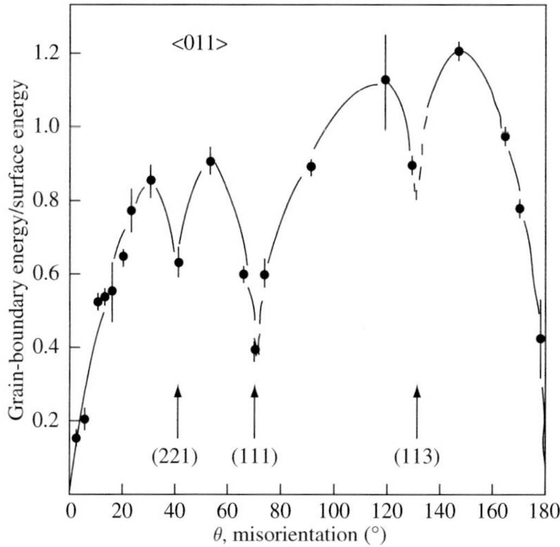

FIGURE 5.11 Energy of grain boundaries (< 011 > tilt boundaries) as a function of the misfit angle (a case for NiO; after DHALENNE *et al.*, 1982).

some specific misorientation at which the atomic arrangement in neighboring grains has small differences. These two lattices are often referred to as the *coincident-site-lattice* (CSL).

Problem 5.10

Derive equation (5.75).

Solution

The energy of a (low-angle) grain boundary is given by (total length of dislocations per unit area) $\times$ (dislocation energy per unit length). Now if the misfit angle of a grain boundary is θ, then using equation (5.74), the total length of dislocations in unit area of the boundary is $1/d$. Therefore the energy per unit area of a grain boundary is given by $\gamma = (\mu b^2/4\pi dK)\log(R/b_0)$. Now, the limit of integration R in this case is $\sim d$. Therefore $\gamma = (\mu b^2/4\pi dK)\log(d/b_0) = (\mu b/4\pi K)\theta(\log(b/b_0) - \log\theta)$.

Problem 5.11

Discuss how to infer the slip systems from the observations on low-angle boundaries.

Solution

A tilt boundary is made of periodic array of dislocations (Fig. 5.10a). The normal to the tilt boundary, $\boldsymbol{n}_t$, is parallel to the Burgers vector, so $\boldsymbol{b} \parallel \boldsymbol{n}_t$. The slip plane normal ($\boldsymbol{n}$) is normal to both the dislocation line ($\boldsymbol{l}$) and the tilt plane normal, $\boldsymbol{n} \parallel (\boldsymbol{l} \times \boldsymbol{n}_t)$. A twist boundary is made of two sets of screw dislocations. Therefore from the directions of line of dislocations, one can infer two Burgers vectors, $\boldsymbol{b} \parallel \boldsymbol{l}$, but one cannot determine the glide plane.

As can be seen from equation (5.74), the mean spacing between dislocations becomes smaller as the misfit angle increases. At a certain point, it becomes meaningless to use the concept of dislocations when the distance between dislocations becomes comparable to the atomic spacing. This occurs when the misfit angle is $\sim$10–20° ($b/d \approx \frac{1}{3}$–$\frac{1}{5}$). A grain boundary whose misfit angle exceeds this may be classified as a high-angle boundary.[8] From (5.75), it is easy to show that the typical energy of a high-angle grain boundary is $\sim 0.01 \times \mu b \sim 1\ \text{J/m}^2$. Regions separated by low-angle boundaries are often referred to as *subgrains*. However, the distinction between subgrains and grains may not always be clear. A convenient way to distinguish them may be to use the energy–misfit angle curve. Due to the overlap of dislocation cores at high angles, the actual grain boundary energy versus misfit angle relation starts to deviate from (5.75) at a certain angle. Boundaries exceeding this angle could be called grain boundaries and those with misfit angles less than this angle may be defined as subboundaries. A grain boundary is a planar defect that separates two portions of crystals with different orientations. Grain boundaries have higher energy than a perfect lattice. This means that *a grain boundary is associated with surface (boundary) tension*. One of the consequences of this fact is that each grain with a different radius has extra energy due to surface tension. That is, a grain with a given diameter has an excess pressure. This excess pressure (energy per unit volume) is given by,

$$\Delta P = \frac{2\gamma}{r} \tag{5.76}$$

where γ is the grain boundary energy and r is the radius of the grain.

Problem 5.12

Derive equation (5.76) and calculate the excess pressure (energy per unit volume) for $r = 1\ \mu\text{m}$–$10\ \text{mm}$ (this is a plausible range of grain size in the crust and the mantle; see Chapter 15) and $\gamma = 1\ \text{Jm}^{-2}$. Compare this with the energy due to dislocations.

Solution

Consider the excess energy associated with the volume expansion dv when a pressure difference of ΔP is present. The work done against excess pressure must be equal to the work done against the increase in surface energy, so that,

$$\Delta P\, dv = \gamma\, dA$$

where dA is the change in the area. Now $dv = 4\pi r^2 dr$ and $dA = 8\pi r dr$ and therefore,

$$\Delta P = \frac{2\gamma}{r}.$$

For the range of the radius of grains ($r = 1\ \mu\text{m}$–$10\ \text{mm}$), this gives, $\Delta P = 2 \times 10^2 - 2 \times 10^6$ Pa. These values are comparable to those due to dislocations ($4 \times 10^2 - 10^6$ Pa; see Problem 5.8) but the relative magnitude of these two contributions depends on the dislocation density and grain size (see also Chapter 13).

Because each grain boundary is associated with the surface (interface) tension, when several grain boundaries meet, the geometry of a grain boundary must be such that the surface tension is balanced. This equilibrium geometry is achieved when the material transport is effective (i.e., at high temperatures or when mass transport through a liquid phase occurs). Let us consider a three-grain junction (Fig. 5.12). Three grain boundaries meet at a line. On this line, three forces act from the surface tension due to three grain boundaries. The force balance equation reads,

$$\mathbf{F}_1 + \mathbf{F}_2 + \mathbf{F}_3 = \mathbf{0} \tag{5.77}$$

where F_i is the force due to the surface (interface) tension on each boundary, i.e., $F_i \propto \gamma\gamma_i$ where γ_i is the grain boundary energy of the ith boundary. Therefore these forces must be on a single plane, and their magnitudes must satisfy,

$$\frac{\gamma_1}{\sin\theta_1} = \frac{\gamma_2}{\sin\theta_2} = \frac{\gamma_3}{\sin\theta_3}. \tag{5.78}$$

In particular, when the grain boundary energy is isotropic, then $\theta_1 = \theta_2 = \theta_3 = 2\pi/3$.

Note that the relation (5.78) provides a means of estimating the grain boundary energy. In most cases, this equation gives a *relative* energy from the measured angles of grain boundary intersections. However, when one of the boundaries is a subboundary, then knowing

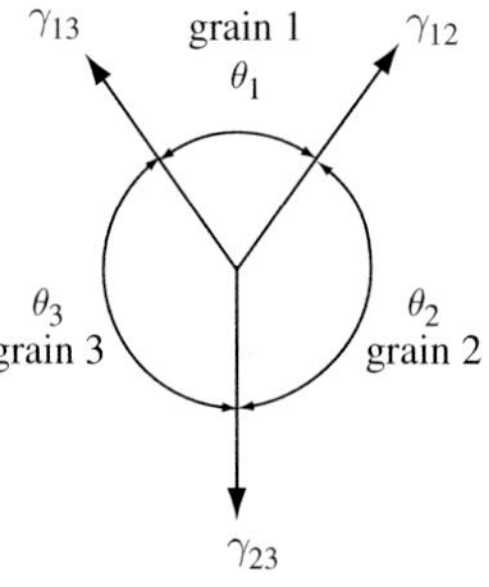

FIGURE 5.12 Force balance at grain boundaries.

[8] A low-angle boundary is also referred to as *subboundary*, and a high-angle boundary is simply referred to as grain boundary.

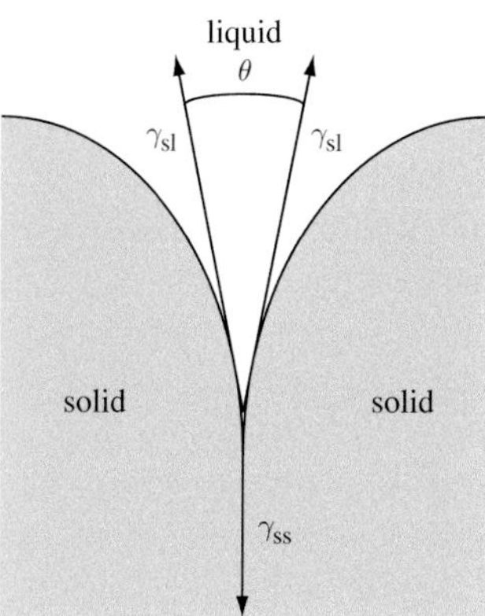

FIGURE 5.13 The definition of a dihedral angle.

the spacing of dislocations, one can calculate the absolute value of energy of this subboundary. Consequently from (5.78), one can calculate the *absolute values* of grain boundary energy. This technique has been used to estimate the grain boundary energy in olivine by DUYSTER and STÖCKHERT (2001).

The force balance argument can also be applied to a liquid–solid contact (Fig. 5.13). Assuming isotropic boundary energy, one gets,

$$\frac{\gamma_{ss}}{2\gamma_{sl}} = \cos\frac{\theta}{2}. \tag{5.79}$$

The angle between liquid and solid contact, θ, is referred to as the *dihedral angle*. When $\gamma_{ss}/\gamma_{sl} \geq 2$, then $\theta = 0$. That is the liquid completely wets the grain boundary. When $2 > \gamma_{ss}/\gamma_{sl} \geq \sqrt{3}$, then $\pi/3 \geq \theta > 0$. Then the liquid has a negative curvature and assumes a continuous tube at the three-grain junctions. When $\sqrt{3} > \gamma_{ss}/\gamma_{sl} \geq 1$, then $2\pi/3 \geq \theta > \pi/3$ and the liquid will occur at four-grain junctions with negative curvature. In this case, the liquid phase can be connected only at a large liquid fraction. Finally when $1 > \gamma_{ss}/\gamma_{sl}$, then $\theta > 2\pi/3$ and the liquid occurs at isolated pockets at four-grain junctions with positive curvature (see Fig. 5.13). The morphology of liquid in a liquid–solid mixture has an important influence on the physical properties (see Chapters 12 and 20).

5.4.2. Thickness of grain boundaries

Grain boundaries are the high-diffusivity paths. The degree to which diffusion is enhanced by grain boundaries depends partly on the "width" of the grain boundaries. The most direct technique for determining the width of grain boundaries is the high-resolution transmission electron microscopy (TEM) (Fig. 5.14). In most cases, the thickness of grain boundaries determined by high-resolution TEM is a few times the unit cell dimension (e.g., RICOULT and KOHLSTEDT, 1983). However, the effective thickness can be larger than the structural thickness when grain boundaries have an electrostatic charge (see below). MISTLER and COBLE (1974) proposed a method for estimating grain boundary width from the experimental data on grain boundary migration and grain boundary diffusion.

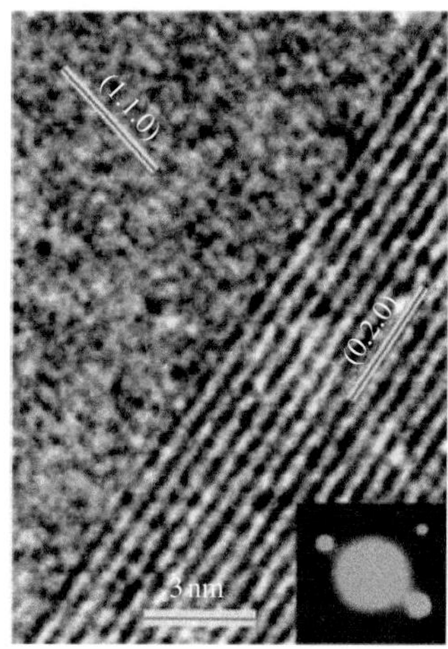

FIGURE 5.14 A high-resolution TEM image of a grain boundary (from HIRAGA *et al.*, 2002).

5.4.3. Grain boundary ledges (steps)

Grain boundaries are not always smooth. Under some conditions grain boundaries contain a large number of irregular points. These are called *ledges* or *steps*. Chemical reactions involving atomic diffusion may occur preferentially at these sites. Consequently, grain boundaries with smooth morphology will have lower mobility compared to other boundaries. Under some conditions the number of ledges (steps) is controlled by thermochemical equilibrium, and grain boundaries tend to have more ledges (steps) at high temperatures.

5.4.4. Grain boundary charges

Similar to dislocations, grain boundaries in ionic solids can have static charges (KINGERY, 1974a, 1974b). The basic physics is common, that is the imbalance in formation energies of cation and anion defects results in local excess charge that is compensated for by the Debye–Hückel atmosphere of space charge (a cloud

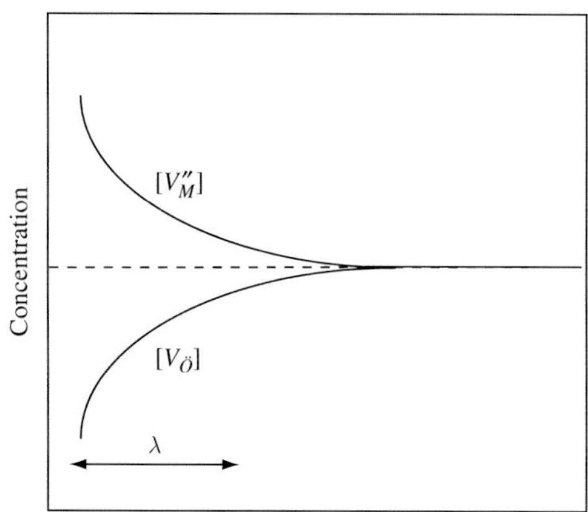

FIGURE 5.15 A schematic diagram showing spatial distribution of electric charge near an extended defect (grain boundary or dislocation).

of charged species around a charged defect that neutralizes the net electrostatic charge). KLIEWER and KOEHLER (1965) developed a basic theory for space charge associated with an extended defect such as grain boundaries (see also theories by ESHELBY *et al.* (1958) and BROWN (1961) on charged dislocations), (see Box 5.2). KINGERY (1974a, 1974b) showed that grain boundaries in a ceramic material (Al_2O_3) have electrostatic charges by demonstrating the migration of grain boundaries by an applied electric field.

The electrostatic charge on a grain boundary has an influence on the properties of a grain boundary. The electrostatic charge controls the concentration of impurities on the grain boundary (or vice versa). The effective width of a grain boundary in which transport properties are enhanced will be the Debye–Hückel length $\lambda = \sqrt{\varepsilon RT/4(Z+Z')\pi N'q^2}$ (ε, dielectric constant; N', number density of impurity atoms; q, electrostatic charge of charged defects; Z, Z', effective valence of two charged defects) that can be much thicker than the structural width seen by high-resolution TEM (HRTEM) images (Fig. 5.15).

5.4.5. Impurities on grain boundaries

Grain boundaries have more-or-less disordered structures and therefore can accommodate relatively large amounts of impurities. Segregation of impurities on grain boundaries is well known (e.g., CHIANG and TAKAGI, 1990; HIRAGA *et al.*, 2004). YAN *et al.* (1983) investigated the influence of electrostatic charge (space charge) and elastic misfit on the segregation of impurities on grain boundaries. Impurities on grain boundaries have important control on grain boundary migration (see Chapter 13). YOSHIDA *et al.* (2002) investigated the role of grain boundary impurities on diffusional creep in Al_2O_3. The concept of grain boundary segregation of impurities can also be applied to the distribution of trace elements. In geochemistry, the distribution of trace elements such as K, U, Th, Rb, Sr, Nd, Sm is used to infer the evolution of Earth. A key physical parameter in such an argument is the distribution coefficients, namely the partitioning of an element between melt and solid. If grain boundaries contain a significant amount of these elements, then the distribution coefficients are no longer the material parameter but they become dependent on the grain size (HIRAGA *et al.*, 2004). Hydrogen is also likely to be present preferentially on grain boundaries.

6 Experimental techniques for study of plastic deformation

An understanding of the experimental techniques of study of plastic deformation is important not only for those who wish to conduct deformation experiments but also for those who would like to apply experimental data on plastic deformation for geological, geophysical and geochemical studies. There is a range of techniques that have been used and each of them has some advantages and limitations. This chapter provides a brief summary of the basic techniques used in experimental studies of plastic deformation. They include mechanical tests under high pressure and temperature using a standard tri-axial testing apparatus as well as large-strain torsion apparatus and some of the recently developed ultrahigh-pressure (>10 GPa) deformation apparatus. Key issues in sample preparation and characterization are also reviewed including the control of the chemical environment.

Key words gas-medium deformation apparatus, solid-medium deformation apparatus, load cell, X-ray *in-situ* stress/strain measurement, constant stress (strain-rate) test, hardness test, stress-relaxation test, stress-dip test, compression (tension) test, torsion test.

6.1. Introduction

Experimental techniques of deformation studies in Earth sciences involve (1) the generation and characterization of Earth-like thermodynamic conditions (pressure, temperature, chemical environment), (2) the controlled generation of differential stress (or strain) and (3) the measurements of stress and strain. Therefore an experimental set-up usually consists of an *environmental chamber* (*sample assembly*) in which a desired thermodynamic environment (e.g., pressure, temperature, fugacity of water, fugacity of oxygen etc.) is established, and an *actuator* to generate differential stress, and some *sensors* to measure strain and/or stress, temperature and pressure (Fig. 6.1). A detailed account of a moderately high-pressure (<3 GPa) deformation apparatus is given by Tullis and Tullis (1986). Paterson (1990) describes some details of a gas-medium deformation apparatus (<0.5 GPa) and Weidner *et al.* (1998) discusses the stress measurements under high pressure using X-ray diffraction techniques. Durham *et al.* (2002), Wang *et al.* (2003) and Xu *et al.* (2005) provide a detailed account of the new development of high-pressure deformation apparatus beyond ~5 GPa and Paterson and Olgaard (2000) give a detailed account of an apparatus for torsion tests under low-pressure (<0.5 GPa) conditions.

6.2. Sample preparation and characterization

6.2.1. Sample preparation

Sample preparation is not a trivial part of the experimental study of plastic deformation. Samples may be classified into single crystals and polycrystals. Studies on single crystals will be desirable when the microscopic processes of deformation involving crystal

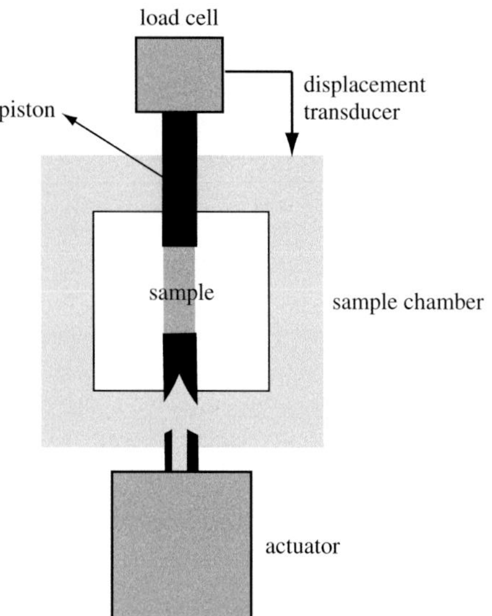

FIGURE 6.1 A typical set-up for a deformation experiment. A deformation apparatus consists of: (1) an environmental chamber (a high-pressure vessel with a furnace), (2) an actuator and (3) various sensors. When a load cell is located outside a sample chamber, friction at the piston and chamber wall causes large errors in the stress estimate. Sensors such as a load cell may be located inside a sample chamber to reduce the errors. Also when a high-intensity X-ray is used, then both stress and strain can be measured *in situ* from X-ray diffraction and X-ray absorption imaging.

dislocations and/or twinning are to be investigated. The crystallographic orientation of a sample can be determined using X-ray or electron-beam diffraction techniques and the plastic deformation of a sample should be investigated for several orientations. Usually, the plastic properties of a given crystal are anisotropic (Chapters 5 and 9) and therefore the plastic properties of a crystal must be investigated for several orientations including those in which a hard slip system(s) is activated. This point must be emphasized because it is not easy to activate the hard slip system(s) and one tends to focus on the flow behavior of a soft slip system(s) only. In such a case, the results may not be applicable to the deformation of a polycrystal because the strength of the hard slip system(s) usually controls the strength of a polycrystal (Chapter 9).

Deformation experiments on single crystals do not provide information on grain boundary processes. Experiments on polycrystalline samples must be carried out to investigate the role of grain boundary processes. In these cases, samples with controlled grain size must be prepared. This can be done by crushing grains followed by sorting the grain size (e.g., Karato *et al.*, 1986). Fine-grained specimens can also be made from chemical reactions in which the grain size is controlled by the competition of nucleation and growth (see Chapter 13). Samples should not have large porosity: otherwise, the collapse of the pore space will affect the mechanical data and the results are hard to interpret. Also when a powder sample is used in a high-pressure experiment, pressurization creates a high density of defects (dislocations, crushed powders) that complicates the interpretation of mechanical data. A dense polycrystalline sample can be prepared by pressing powders in a capsule by high pressure, temperature apparatus (hot-isostatic-pressing (HIP)). A high confining pressure provides a large driving force for compaction, and hence a dense material can be made without much grain growth. If a sample is annealed after densification at a high enough temperature (for a long time), most of the defects created during pressurization can be removed. It is important to check the degree of annealing by the microstructural observation of annealed (and undeformed) samples. A small amount of water enhances compaction and grain growth, but an excessive amount of water prevents both processes (Karato, 1989b). Care must be taken not to add too much water.

In uni-axial (or tri-axial) compression tests, a specimen is cut to a rectangular or cylindrical shape. The length to diameter ratio is usually chosen to be ~2–2.5. If this ratio is significantly smaller than 2, then the friction between the sample and the piston at the interface significantly affects the results. If this ratio is larger than 2.5, a sample tends to buckle. When a polycrystalline material is tested, the grain size must be small enough so that a large number of grains are contained in a sample to allow a statistically meaningful average be determined. As a rule of thumb, the diameter of the sample in a tri-axial test must be larger than ~10 times the grain diameter.

In shear deformation experiments using a sandwich method, the sample thickness must be small enough to allow nearly simple shear geometry of deformation: a thick sample tends to show a large amount of shortening. At the same time, a layer of sample should contain a large enough number of grains in a cross section, say more than ~10 grains across the section. Usually, a sample is cut from a cylindrical sample at 45° thus a sample is a thin oblate shape.

A sample for a torsion test has a cylindrical shape with or without a hole at the center. A cylindrical sample with a hole may be used in a torsion test that minimizes the gradient in strain (rate) and stress across a sample (Paterson and Olgaard, 2000; Xu *et al.*, 2005).

In a deformation experiment, force is transmitted to a sample from the actuator through a piston. The material for a piston must be significantly stronger than the sample. In both a sandwich method and a torsion test, the coupling between the sample and the piston must be strong: otherwise the torque would not be transmitted to the sample and a slip would occur at the interface. This slip is a serious problem when the sample strength is a significant fraction of, or exceeds, the confining pressure. In these cases, grooves may be made on the piston surfaces to minimize the slip (KAMB, 1972; ZHANG and KARATO, 1995; ZHANG *et al.*, 2000). To make sure that the slip does not occur, it is recommended to insert a strain marker (a soft thin layer of material which can be a foil of metal or can be a coated material) in the sample. The rotation of the strain marker provides a measure of the magnitude of the shear strain of a sample. If the amount of rotation agrees with the value calculated by the displacement of a piston or the rotation angle of the piston, then one can conclude that a significant slip did not occur. Also, the sample thickness must be measured both before and after an experiment. The change in sample thickness gives the magnitude of shortening strain.

When a fluid (a gas or a liquid) is used as a pressure medium, a polycrystalline sample needs to be placed in a jacket made of fluid-impermeable material to prevent the penetration of the fluid into grain boundaries. In most cases in a high-temperature deformation experiment, a metal jacket such as an iron jacket is used. A metal jacket supports some load that needs to be corrected. Also a metal jacket will control the chemical environment such as oxygen fugacity (see section 6.3.3). When one wants to control the oxygen fugacity for a large range, then one can use a jacket made of two layers, one inside to control the oxygen fugacity and another one outside to seal the sample from the fluid.

6.2.2. Microstructural characterization

Microstructures of a sample must be characterized as much as possible. For a polycrystalline specimen, the porosity of a sample must be significantly lower than the total strain. Grain size must be measured including its distribution. Grain size can change during deformation due either to grain growth or to dynamic recrystallization. Therefore grain size should be measured both before and after each experiment.

Usually grain size is measured on a two-dimensional section. A correction must be made for the sectioning effects. The average grain size can be determined from the average length of intercepts. In this technique, the average length of intercepts of a line with grain boundaries is measured on two-dimensional sections. The average length of intercepts is proportional to the average grain size as

$$\bar{L} = \beta \bar{l} \tag{6.1}$$

where $\bar{L}$ is the average grain size, $\bar{l}$ is the intercept length and β is a constant that depends on the grain shape. For nearly spherical grains, $\beta \sim 1.5$ (MENDELSON, 1969). Several hundred or more grains need to be measured to obtain statistically meaningful data on grain size. Determining the grain-size distribution from the observations on a two-dimensional plane is not trivial. The distribution of grain sizes on a two-dimensional section is not the same as the distribution of grain size in three dimensions. One method developed by CAHN and FULLMAN (1956) is to calculate the grain-size distribution from the observation on a two-dimensional section in which some assumption must be made concerning the grain shape. For spherical grains, the grain-size distribution is given by

$$N(L) = \frac{2}{\pi} \frac{n(x)}{x^2} - \frac{2}{\pi x} \frac{dn(x)}{dx} \tag{6.2}$$

where $N(L)\,dL$ is the number of spheres with diameters between L and $L + dL$, $n(x)\,dx$ is the total number of intercepts per unit length with the length between x and $x + dx$. Another method is the Schwartz–Saltykov method in which one uses the distribution of diameters of grains on a two-dimensional section to infer the three-dimensional distribution of grain size. A detailed account of the measurements of three-dimensional microstructures on a two-dimensional section of a specimen is given in UNDERWOOD (1969). Various software packages such as the NIH software are available to make digital measurements of microstructures.

Problem 6.1*

Derive equation (6.2).

Solution

Let $N(L)\,dL$ be the number of spheres per unit volume with size between L and $L + dL$. Given this distribution, what is the number of intercepts per unit length of a chord whose length is between x and $x + dx$? Let us first note that the intercept length between x and $x + dx$ will be produced when a chord intersects a sphere at a distance between X and $X + dX$ where

$X^2 + x^2/4 = L^2/4$. The cross sectional area corresponding to this intersection is $2\pi\ XN(L)\,dX\,dL$. Therefore the expected number of intersections with the length between x and $x + dx$ is $2\pi\ XN(L)\,dX\,dL$. Translating this into the total number of intersects between x and $x + dx$, one obtains $2\pi XN(L)\ dX\ dL = 2\pi XN(L)(\partial X/\partial x)_L\ dx\ dL$. Now from $X^2 + x^2/4 = L^2/4$, one has $(\partial X/\partial x)_L = -x/4X$. Therefore $2\pi X \cdot N(L) \cdot (\partial X/\partial x)_L\ dx \cdot dL = -\frac{\pi}{2} xN(L)\ dx \cdot dL$. The intercept with x and $x + dx$ can come from any sphere whose diameter is larger than x. Therefore the total number per unit length of intersections with length between x and $x + dx$ is given by $n(x)\ dx = -(\pi/2)x \cdot dx \int_x^\infty N(L)\ dL$. Differentiating this equation by x, one has $dn/dx = -(\pi/2)\int_x^\infty N(L)\ dL - (\pi/2)xN(x) = n/x - (\pi/2)xN(x)$. Rearranging this for $N(L)$, one obtains equation (6.2).

Deformation microstructures must be carefully characterized using a range of equipment (optical microscope, scanning and transmission electron microscopes (SEM, TEM) etc.). Grain-boundary morphology including subgrain structures and dislocation microstructures (Burgers vectors, glide planes, morphology) are among the important microstructures that need to be characterized. For details see the appropriate chapters (Chapters 5, 8, 9, 13 and 14).

6.3. Control of thermochemical environment and its characterization

6.3.1. Pressure generation and its measurements

Pressure is an important thermodynamic variable in the study of deformation of materials (see Chapter 10). When rheological experiments are made on a polycrystalline sample, grain boundary cracking could occur if a differential stress is applied at a low confining pressure. Therefore to suppress cracking, one inserts a sample in a jacket and a jacketed specimen is placed in a high-pressure environment (a rule of thumb is that cracking is not very important if the confining pressure significantly exceeds the differential stress). Also, a confining pressure is important to minimize "cavitation" during high-temperature creep that affects the creep behavior (e.g., Chen and Argon, 1981). In addition, Earth's interior is at high pressure and temperature so there is an obvious motivation to investigate the rheological properties under high pressures (and temperatures). It should be noted that various thermodynamic variables change with pressure (and temperature). One notable example is the fugacity of water (and carbon dioxide) that increases significantly with pressure (Pitzer and Sterner, 1994, Frost and Wood, 1997a, 1997b) (see Chapter 2). Also, important phenomena including some of the phase transformations in silicate minerals (e.g., olivine to wadsleyite transformation) occur only under high pressures (Chapters 2, 15 and 17). Consequently, studies on the effects of water and on phase transformations (in silicate minerals) on rheology must be conducted under high pressures.

Pressures to ~0.3 GPa are routinely achieved by pumping a gas (such as argon) into a vessel. The vessel must be sealed to keep the pressure. An "O" ring is used to seal the high-pressure gas, particularly for the moving portions. However, sealing by an "O" ring becomes difficult above ~0.5 GPa (because most polymers used for "O" rings become brittle at high pressures). Pressure exceeding ~1 GPa is usually obtained by compressing a sample surrounded by a soft solid-medium such as pyrophyllite or MgO. A sample assembly is squeezed by a hard material such as tungsten carbide through a hydraulic press.

A *pore–fluid pressure* can be controlled with an additional device. A sample may be connected through a porous end-piece to an additional hydraulic system and the fluid pressure there can be controlled by a pump independent of the pressure of the solid portion.

Pressure is measured by a pressure transducer at a low-pressure level at which the pressure can be measured in the fluids (<1 GPa), but at higher pressures (in a solid-medium apparatus), pressure is usually estimated based on the calibration using phase diagrams or equations of state of some standard materials. When a synchrotron X-ray facility is used, one can measure the density of some standard material by X-ray diffraction (this could be the sample itself) and estimate the pressure from the equation of state of that material.

6.3.2. Generation of high temperature and its measurements

High temperature (high temperature relative to the melting temperature of a sample) is needed to conduct deformation experiments in the plastic regime. In most cases, a resistance heater is used to increase the temperature. The design of a furnace constitutes an important part of deformation studies. Generating a high temperature ($T > 1300$ K) in a gas-medium apparatus is not trivial. One needs to minimize the convection to obtain a high enough efficiency of a furnace. This is accomplished by putting porous ceramic materials in the space surrounding a furnace winding (a porous ceramic must be packed tightly to minimize the convection). By controlling the power of two or more different portions of a furnace independently, one can minimize the temperature gradient. A temperature exceeding ~1600 K is hard to obtain in a gas-medium apparatus. In contrast, a temperature exceeding ~2000 K can be obtained in a solid-medium apparatus although the temperature gradient tends to be higher in a solid-medium apparatus because of its smaller size and lower flexibility in the control of furnace power (as compared to a gas-medium apparatus). Temperature is usually measured by a thermocouple. A thermocouple is made of a junction of two metals, and when this junction is placed at a certain temperature, then an electrostatic voltage is generated. By measuring the difference in the voltage between two junctions (one is called a "cold junction" usually at room temperature), one can measure the temperature. A variety of thermocouples are available, and one must choose an appropriate one for a planned range of temperature.

Three issues need to be considered in terms of temperature. First, an experiment must be carried out after the temperature distribution is nearly at a steady state. Otherwise, a long-term variation in temperature causes distortion of an apparatus by thermal expansion leading to large errors in measured strain. Second, the temperature distribution in a sample must be as homogeneous as possible. Plastic properties are highly sensitive to temperature and therefore a large temperature gradient in a sample can cause large errors. Third, the voltage output from a thermocouple depends not only on temperature but also on pressure. Getting and Kennedy (1970) investigated the effects of pressure on two types of thermocouples. The pressure effects on the thermocouple output depend on the type of thermocouple. For the pressure effects on thermocouple readings, see also Li *et al.* (2003a).

6.3.3. Control and characterization of the chemical environment

In Chapter 5, we learned that the concentration of defects in minerals depends on the chemical environment. Consequently, the nature of plastic deformation is sensitive to the chemical environment. Among many aspects, three parameters are important for plastic deformation: oxygen fugacity, oxide activity and water fugacity. In an atmospheric pressure experiment, oxygen fugacity is usually controlled by a mixture of gas such as CO/CO_2. The chemical reaction,

$$CO + \frac{1}{2}O_2 = CO_2. \tag{6.3}$$

The chemical equilibrium demands

$$f_{O_2} = K_3^2(T)\frac{f_{CO_2}^2}{f_{CO}^2} \tag{6.4}$$

where f_{O_2} is oxygen fugacity, f_{CO_2} is the fugacity of carbon dioxide, f_{CO} is the fugacity of carbon monoxide, and K_3 is the equilibrium constant for reaction (6.3). For a given temperature, the oxygen fugacity can be controlled by changing the ratio of $[CO_2]/[CO]$ because $f_{CO_2} \propto [CO_2]$ and $f_{CO} \propto [CO]$ at low pressures. In high-pressure experiments, a gas-mixture cannot be used, and instead, the oxygen fugacity is controlled by the solid–solid reaction such as,

$$xM + \frac{y}{2}O_2 = M_xO_y. \tag{6.5}$$

The law of mass action corresponding to this reaction yields

$$f_{O_2}^{y/2} = K_5(T, P)\frac{a_{M_xO_y}}{a_M^x} \tag{6.6}$$

where $K_5(T, P)$ is the equilibrium constant for the reaction (6.5). When both M and M_xO_y phases are present, the activity of each phase is 1, so $f_{O_2} = K_5^{2/y}(T, P)$ (one needs to demonstrate that both of these two phases exist after a run to confirm that this reaction buffered the oxygen fugacity). Various pairs of metals and metal oxides can be used to control a range of f_{O_2}. A limitation of this technique is that only discrete values of f_{O_2} can be realized. However, some trick can be used to extend the range of f_{O_2}. For instance, a solid solution rather than end-member materials can be used to achieve the values of f_{O_2} between end-member values. Note that the oxygen fugacity corresponding to a given oxide–metal buffer changes with temperature and pressure.

Similarly the oxide activity can be controlled by placing an oxide in a sample assembly. For instance, if an oxide M_xO_y is present, then $a_{M_xO_y} = 1$. In a system that contains more than two oxide components (e.g., Mg_2SiO_4), oxide activity must also be controlled. Oxide activity can be controlled by placing an oxide next to a sample. For example, the oxide activity of a system involving olivine, Mg_2SiO_4, is determined by the following reaction,

$$Mg_2SiO_4 = MgSiO_3 + MgO. \quad (6.7)$$

Applying the law of mass action, one has

$$\frac{a_{Mg_2SiO_4}}{a_{MgSiO_3} \cdot a_{MgO}} = K_7(T, P). \quad (6.8)$$

Therefore if MgO is placed next to olivine, then at chemical equilibrium, the activity of oxide is such that $a_{MgO} = 1$ and from the thermodynamic data one obtains $a_{MgSiO_3} \approx 0.1$ (at $P = 0.1$ MPa and $T = 1500$ K).

Water fugacity is usually controlled by adding excess water at various pressures (and temperatures). Excess water may be added by the decomposition of some hydrous minerals or simply adding free water. In these experiments, it is important to make sure that water (hydrogen) be kept in the sample space and melting did not occur. Hydrogen is easy to diffuse through the capsules and the use of capsule material with low hydrogen diffusivity (and solubility) is recommended. Pt, or Au–Pd (or Ag–Pd) is known to be a good material for a capsule by which much of the hydrogen is effectively kept in the sample space. Melting should be avoided because melting often leads to a reduced water content in the sample due to high water (hydrogen) solubility in the melt if the system is under saturated with water. When an experiment is to be made on a sample containing water (hydrogen), water (hydrogen) content must be measured both before and after an experiment. Hydrogen loss during an experiment often occurs and in such a case the interpretation of the result is complicated. Conversely, hydrogen may penetrate into a sample from the pressure medium. Water fugacity may be controlled below the maximum level by using a buffering material. For example, if a hydrous mineral co-exists with the anhydrous counter-part, hydrogen fugacity must be buffered to a value controlled by chemical equilibrium. An example is the reaction between brucite ($Mg(OH)_2$) and MgO, $MgO + H_2O = Mg(OH)_2$ that could buffer the water fugacity. However, the ability of this type of reaction to control the water fugacity depends on the degree to which one can retain hydrogen in the system and this method has not been explored yet.

The measurement of these thermochemical parameters is not trivial. At one atmospheric pressure, the oxygen fugacity can be measured by an electrochemical method using a zirconia sensor (e.g., SATO, 1971). At high pressures, a chemical reaction such as the reaction between Fe-bearing minerals and Pt can be used to infer f_{O_2}(e.g., RUBIE *et al.*, 1993). Water content can be determined by infrared absorption spectroscopy or by using SIMS (secondary-ion-mass-spectrometer) (see Chapter 10). It should be noted that the water content (water fugacity) of a sample is not easy to control. Water can diffuse into or out from a sample easily. It is strongly recommended to determine the water content of a sample both before and after each experiment.

6.4. Generation and measurements of stress and strain

6.4.1. Generation of deviatoric stress and strain

Except for deformation associated with compaction in a porous material, plastic deformation occurs due to deviatoric stress. The simplest way of generating a deviatoric stress is *dead weight loading* in which load is applied by a mass being placed on top of a sample (e.g., KOHLSTEDT and GOETZE, 1974). The stress, $\sigma_1 = mg/A$ (m, mass; g, acceleration due to gravity; A, cross section area of loading), is generated by a load of mass, m. By this method a very accurate stress can be applied to a sample, but the magnitude of stress is limited. If a loading piston must be sealed, then the friction will limit the accuracy of stress measurement for this method. When a large force is needed to generate a deviatoric stress (e.g., at high confining pressures), one needs to use some type of *actuator*. An actuator is made of either an electric motor (and gears) or a hydraulic pump to move a piston (either advance/retract or rotate a piston) in a controlled manner. The most common mode of operation is to provide a constant displacement or rotation rate, yielding an approximately constant strain rate (after the correction for the change of sample geometry and apparatus distortion if necessary). If one wants to operate a deformation apparatus with an actuator with a constant load mode (approximately constant stress mode) then one needs a feedback mechanism such as a servo-control system. The load output must be fed to an actuator driver circuit to maintain a constant load.

A deformation apparatus at high pressures exceeding ~3 GPa is difficult to design. The main reason is that in a conventional design in which a deformation experiment is made by moving a piston through a pressure medium, a mobile piston must support both confining the pressure and load needed for deformation. Because this mobile piston must move in an area in which it is not well supported from the surrounding material, the piston tends to fail if the stress exerted on the piston exceeds the fracture strength of the piston, which depends on the material and geometry of the piston. For tungsten carbide the fracture strength is several GPa, and consequently, deformation experiments above a few GPa are difficult. A deformation-DIA (D-DIA) is designed to conduct deformation experiments above a few GPa. In this apparatus, in addition to four pistons that provide $\sigma_2(=\sigma_3)$, two deformation pistons can be moved independently from these four to provide σ_1. The maximum pressure of operation is increased by using short tapered pistons (Wang *et al.*, 2003). However mobile pistons in this apparatus must move in a poorly supported area and the maximum pressure of operation with tungsten carbide pistons is limited to ~10 GPa. An alternative type of apparatus was designed by Yamazaki and Karato (2001a) and Xu *et al.* (2005). In this apparatus a rotational actuator is attached to one of the anvils of a Drickamer apparatus (rotational Drickamer apparatus, RDA). In a Drickamer apparatus, a high pressure can be generated using tapered anvils that are supported by a gasket. Deformation experiments are made by rotating an anvil, and therefore the deformation piston (anvil) is supported (almost) exactly the same way as in static experiments. Consequently, deformation experiments can be performed to much higher pressures than by other types of apparatus (at the time of writing (2007) the maximum pressure of operation with RDA is ~18 GPa (at temperature of ~2000 K)).

Deviatoric stress can also be generated by squeezing a sample in an anisotropic fashion. For example, in a multi-anvil apparatus, deviatoric stress can be generated if the strength of materials in a sample assembly is different along different directions (Bussod *et al.*, 1993; Karato and Rubie, 1997). Similarly, deviatoric stress is generated in a diamond anvil cell due to axial compression (Kinsland and Bassett, 1977; Sung *et al.*, 1977; Mao *et al.*, 1998). In these cases, the stress is not generated in a well-controlled fashion, and both stress and strain rate tend to be high and not well defined.

Table 6.1 summarizes some characteristics of various deformation apparatus.

6.4.2. Measurements of stress and strain

The strain of a sample is usually measured by the displacement of a piston relative to a fixed portion of the apparatus. Various displacement sensors can be used, but the most common is the linear-variable-displacement-transducer (LVDT), by which displacement can be converted to a voltage output through an inductance-based transducer. With this type of measurement, one must be careful about the long-term variation in the length of various portions of the apparatus due to the variation of temperature (and pressure). Actual measurements must be conducted at conditions where the drift in displacement is negligible in comparison to the actual strain rates that one expects. It is also important to make corrections for the deformation of the apparatus itself. An apparatus will be deformed when a stress is applied to the sample. The deformation of apparatus is determined by its stiffness that depends on the elastic constants of materials and the geometry of the frame. The stiffness of the apparatus must be measured and corrected for the mechanical tests with a constant strain rate. When an X-ray image of a sample is taken (using a high intensity X-ray facility such as synchrotron radiation facility), then the sample strain can be measured directly from the X-ray images (Chen *et al.*, 2004).

The stress measurements are usually conducted using various types of force transducer (called a *load cell*). The most common one is the strain-gauge-based force transducer in which the applied force causes elastic strain of gauges that changes their resistance. The change in resistance, in turn, can be monitored as a voltage output from a bridge circuit. When a force transducer is placed outside a high-pressure apparatus, the force that one measures includes the resistance for the motion of a piston at the pressure seal or other portions of the sample assembly. In most cases, this resistance is large and accurate measurements of stress from outside a pressure vessel are difficult. In some cases, the correction for resistance forces is made through "touch-point" measurements (Fig. 6.2) (Borch and Green, 1989; Tingle *et al.*, 1993). In such a measurement, after the piston contacts the sample, the piston is backed up (removed from the sample by a small amount). When a sample is surrounded by a liquid, the liquid will penetrate into the gap between the piston and the sample. Now one moves the piston down to squeeze the sample. Before the piston touches the sample, the load measured by the load cell reflects mostly the friction. When the piston touches the sample, there is a sharp increase in load. The load now includes the sample strength as well as friction. Assuming that friction does not change with displacement, one can determine the strength of a sample as a function of strain. A key to the success of this stress measurement is the sharpness of the touch point and the validity of the assumption that the friction is insensitive to displacement. A significant improvement to the stress measurements has been achieved by this technique.

TABLE 6.1 Some characteristics of various deformation apparatus.

	P (GPa)	T (K)	Stress measurement	Note	Ref.
dead-weight creep apparatus	10^{-4}	< 2000	from weight of a load	low f_{H_2O}	(1)
gas-medium apparatus*	<0.5	<1600	internal load cell	limited f_{H_2O}	(2)
Griggs-type apparatus	<3	<1600	external load cell	limited strain	(3)
deformation-DIA	<10**	<1600	X-ray	limited strain	(4)
rotational Drickamer apparatus	<18**	<2000	X-ray	unlimited strain	(5)
multianvil stress-relaxation	<23	<2000	estimate (or X-ray)	non-steady state	(6)
diamond anvil	<200	<1000	X-ray	non-steady state, very high stress	(7)

(1): Carter *et al.* (1980) and Kohlstedt and Goetze (1974).
(2): Paterson (1970) and Paterson (1990).
(3): Tullis and Tullis (1986).
(4): Wang *et al.* (2003).
(5): Xu *et al.* (2005) and Yamazaki and Karato (2001a).
(6): Bussod *et al.* (1993), Fujimura *et al.* (1981), and Karato and Rubie (1997).
(7): Kinsland and Bassett (1977), Mao *et al.* (1998), and Sung *et al.* (1977).
*A rotational actuator can be attached to a gas apparatus that allows an unlimited strain deformation experiment. The major limitation is the low maximum pressure. The effects of pressure and water, both of which are very important in Earth science applications of deformation experiments, cannot be determined precisely enough if a gas-medium apparatus is used (for details, see Chapter 10).
** This limit is for tungsten carbide anvils. The pressure range can be expanded by using anvils with a harder material (e.g., diamond or cubic boron nitride).

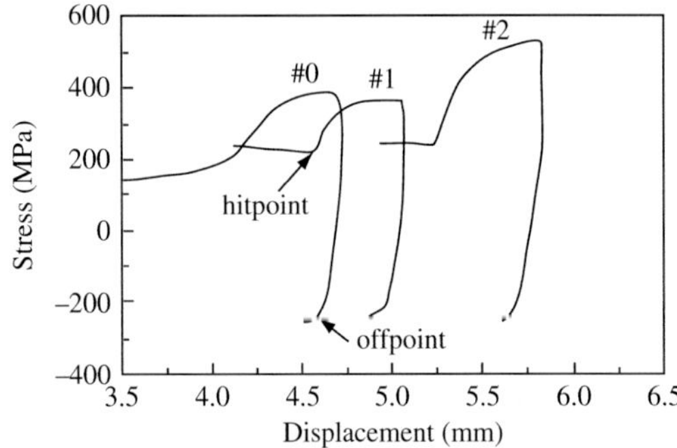

FIGURE 6.2 Some examples showing the variation of the external load cell reading during deformation experiments (modified from Borch and Green, 1989).

Although this liquid-cell sample assembly is an important improvement to the stress measurement in a solid-medium apparatus, uncertainties remain as to the nature of friction that is not always reproducible. Paterson (1970) developed an *internal load cell* for a gas-medium high-pressure deformation apparatus. In this case, the load cell is placed inside a high-pressure vessel and there is no influence from friction. Consequently, the resolution of stress measurements by this internal load cell is nearly the same as that at room pressure and far better than that using an external load cell. A major limitation of this technique, however, is that it can only be applied to a gas-medium apparatus and the practical limit of pressure is ~0.5 GPa (above this pressure, sealing at an interface between the moving piston and the pressure vessel becomes very difficult).

Totally different methods of stress measurements have been developed by Weidner (1998) using X-ray diffraction obtained at synchrotron radiation facilities. These techniques use the effects of stress to modify the spacings of atomic planes (lattice spacing). Since high-intensity X-rays generated by synchrotron radiation can penetrate the sample through pressure media, these techniques can be used under very high pressures.

Two techniques have been developed. The first is stress measurement by the broadening of X-ray peaks. If stress develops in a polycrystalline aggregate, then individual grains will be subject to different stress states. Consequently, the values of lattice spacing become distributed. This causes broadening of X-ray peak positions. In most earlier studies, this technique was used for granular materials where large contact stress develops at grain-to-grain contacts by high pressure. Weidner and his colleagues applied this technique to characterize the rheology of materials at pressures to ~20 GPa (Chen *et al.*, 1998, 2002b; Xu *et al.*, 2003). However the nature of deformation at grain contacts is highly complicated and the relevance of such data to geological problems is questionable.

The second method uses the dependence of lattice spacing on the orientation of crystals with respect to the orientation of applied macroscopic stress. When a sample is under hydrostatic equilibrium, the lattice spacing corresponding to a specific crystallographic orientation has a single value, d_{hkl}. When a crystal is under deviatoric stress, however, the lattice spacing depends on the local stress state. It will be smaller (larger) for grains where the crystal plane is normal to the maximum (minimum) compression stress. By conducting X-ray diffraction measurements at various angles, one can determined the lattice spacing as a function of orientation with respect to the macroscopic stress (Fig. 6.3). The amplitude of the deviation of lattice spacing from a standard value corresponding to a given temperature and pressure is proportional to the magnitude and geometry of deviatoric stress and the orientation of the lattice plane from which the X-ray is diffracted. Thus, in general, the degree to which the lattice spacing is modified by stress is a function of stress as well as the crystallographic index of the lattice plane and the stiffness of the lattice plane, namely,

$$\frac{d(hkl) - d_0(hkl)}{d_0(hkl)} = F\big(\sigma_{ij}, C(hkl), \alpha\big) \tag{6.9}$$

where $d(hkl)$ is the lattice spacing corresponding to the crystallographic orientation of (hkl), σ_{ij} is the stress tensor, $C(hkl)$ is the elastic constant corresponding to the compression of the (hkl) plane, and α is a parameter characterizing the stress–strain distribution.

The lattice strain $[d(hkl) - d_0(hkl)]/d_0(hkl)$ can be separated into compressional and shear components. The compressional component provides an estimate of hydrostatic pressure (from the known equation of state). The shear component gives an estimate of deviatoric stress. In order to convert the shear component of

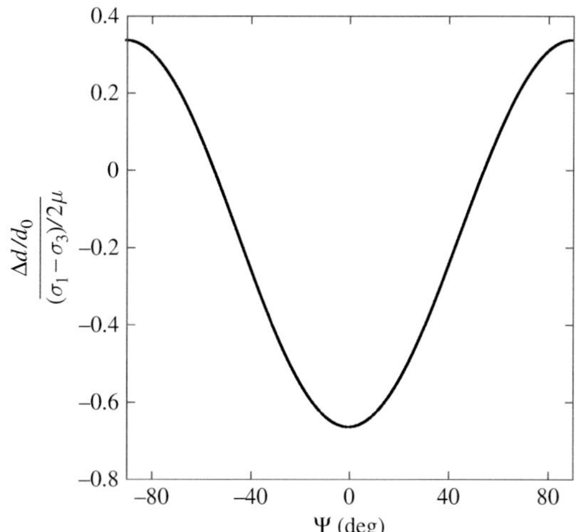

FIGURE 6.3 Variation of the lattice strain, $\Delta d/d_0$, as a function of the direction of the diffracting plane relative to the maximum compression direction, Ψ (where $\Psi = 0$ corresponds to the maximum compression direction), for uni-axial compression.

lattice strain, one needs to have a model that relates the grain-scale strain, $[d(hkl) - d_0(hkl)]/d_0(hkl)$, to the macroscopic stress. Such a relation has been established for the elastic deformation of a polycrystalline aggregate (Singh, 1993; Funamori *et al.*, 1994; Uchida *et al.*, 1996). Singh (1993) and Funamori *et al.* (1994) showed that when deformation is elastic, then the nature of the stress–strain distribution (i.e., the parameter α) can be determined from the X-ray diffraction data for multiple crystallographic planes to determine the magnitude of deviatoric stress (for axially symmetric stress). When the microscopic stress is identical to the macroscopic stress (Reuss or Sachs state, see Chapter 12), then the equation is simplified and for uni-axial (tri-axial) compression, it becomes (Problem 6.2)

$$\frac{d(hkl) - d_0(hkl)}{d_0(hkl)} = -\frac{\sigma_1 - \sigma_3}{2\mu_{hkl}}\left(\cos^2\Psi - \frac{1}{3}\right) \tag{6.10}$$

where μ_{hkl} is the appropriate shear modulus (the relevant shear modulus) and Ψ is the angle of normal of the diffraction plane with respect to the direction of maximum compression. A complete expression for μ_{hkl} corresponding to elastic deformation for various crystal symmetries is given by Uchida *et al.* (1996).

However, Li *et al.* (2004a) and Weidner *et al.* (2004) showed that the assumption of elastic accommodation of strain is often violated under high-pressure (and high-temperature) experiments where plastic deformation plays an important role. However, a model equivalent to that of Singh (1993) and Funamori *et al.* (1994)

has not been formulated for plastic deformation. The role of plastic deformation on microscopic stress becomes important when the plastic anisotropy is high. This results in some uncertainties in the stress measurements from lattice strain determined by X-ray diffraction.

Despite the issue of grain-scale stress–strain distribution, this technique of stress measurement has a much firmer basis than the stress measurements from X-ray peak broadening. Combined with the strain measurements by X-ray imaging, this synchrotron-based technique provides a promising new method for the quantitative study of deformation of materials under deep Earth conditions (e.g., CHEN *et al.*, 2004).

An alternative method for estimating stress is to use some microstructural "paleo-piezometers" (see Chapters 5 and 13). BUSSOD *et al.* (1993) used a dynamically recrystallized grain size versus stress relationship (see Chapter 13) to estimate stress and KARATO and JUNG (2003) used the dislocation density versus stress relationship. The former relationship is sensitive to thermochemical variables (Chapter 13) whereas the latter is insensitive to these variables (Chapter 5). Consequently, the dislocation density gives a reasonable estimate of stress in deforming materials, although a precise measurement of the dislocation density is not always straightforward. For olivine, an SEM-based technique of dislocation density measurements provides a highly accurate determination of dislocation density and hence stress (KARATO and JUNG, 2003).[1] In contrast, the use of recrystallized grain size is not recommended because this paleo-piezometer is sensitive to various thermochemical parameters (see Chapter 13) and does not provide a reliable estimate of stress.

Problem 6.2*

Derive equation (6.10).

Solution

If the coordinate is chosen such that the x_1 direction is parallel to the maximum compression direction, then the stress tensor is given by

$$\begin{pmatrix} P & 0 & 0 \\ 0 & P & 0 \\ 0 & 0 & P \end{pmatrix} + \begin{pmatrix} \frac{2}{3}(\sigma_1 - \sigma_3) & 0 & 0 \\ 0 & -\frac{1}{3}(\sigma_1 - \sigma_3) & 0 \\ 0 & 0 & -\frac{1}{3}(\sigma_1 - \sigma_3) \end{pmatrix}$$

The first component gives rise to the linear compressional strain, $\varepsilon_P = -P/3K_T$, and the second term the shear strain. In order to calculate the lattice strain due to the second term (i.e., shear stress), let us recall that the lattice strain due to the normal component of a shear stress, σ_n, on a plane is the sum of the normal strain caused by the normal stress and the strain due to the traction on planes normal to that plane. Consequently, $\varepsilon_{hkl} = -(1+\nu_{hkl})\sigma_n/E_{hkl} = -\sigma_n/2\mu_{hkl}$ (where E_{hkl}, Young's modulus of deformation normal to the (*hkl*) plane; ν_{hkl}, Poisson's ratio of deformation normal to the (*hkl*) plane). Now recall that the normal stress to the plane where the normal is oriented Ψ from the x_1 direction is $\sigma_n = (\sigma_1 - \sigma_3)(\cos^2\Psi - \frac{1}{3})$ (Chapter 1, Problem 1.3). Therefore $\varepsilon_{hkl} = -[(\sigma_1 - \sigma_3)/2\mu_{hkl}](\cos^2\Psi - \frac{1}{3})$.

6.5. Methods of mechanical tests

Plastic deformation is a mode of deformation where strain is irrecoverable, and time dependent. In other words, the mechanical properties in this mode are dependent on time and/or strain. In general, we seek to find a relationship between strain, ε_{ij}, and stress, σ_{ij}, and their time derivatives, namely,

$$F(\varepsilon_{ij}, \dot{\varepsilon}_{ij}, \cdots, \sigma_{ij}, \dot{\sigma}_{ij}, \cdots; t) = 0 \tag{6.11}$$

where $\dot{\varepsilon}_{ij} = d\varepsilon_{ij}/dt$ and $\dot{\sigma}_{ij} = d\sigma_{ij}/dt$. In an experimental study, some variables (e.g., strain rate, $\dot{\varepsilon}_{ij}$) are imposed and one measures other dependent variables (e.g., stress).

6.5.1. Constant stress, constant strain-rate tests

When the strain rate is constant,

$$F_{\dot{\varepsilon}}(\sigma, \dot{\sigma}, \cdots, t; \dot{\varepsilon}) = 0. \tag{6.12}$$

In this case one measures the time variation of stress, $\sigma(t)$, for a given strain rate to determine the mechanical property of a given material. Similarly, when stress is given, then one determines the variation of strain with time, i.e.,

$$F_{\sigma}(\varepsilon, \dot{\varepsilon}, \cdots, t; \sigma) = 0. \tag{6.13}$$

Typical results from these two types of experiment are illustrated in Fig. 6.4. For a constant strain-rate test, the initial t–$\sigma(t)$ curve represents an elastic

[1] In this technique, the total *length* of dislocation lines per unit volume is measured as opposed to a conventional two-dimensional measurement of a *number* of dislocations on a section. Consequently, errors caused by the uncertainties in dislocation orientation are largely eliminated resulting in a much higher resolution in dislocation density measurements.

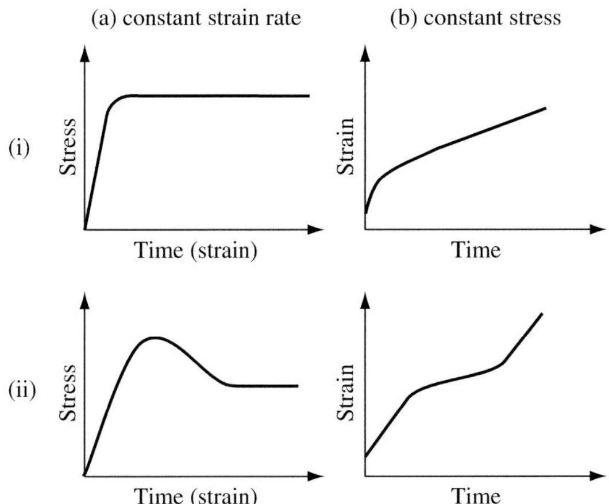

FIGURE 6.4 Typical results of constant (a) strain rate and (b) constant stress tests.

response. This is followed by a transient behavior, and finally steady-state behavior is usually obtained at which the stress remains constant. For a constant load (stress) test, there is an instantaneous elastic response followed by a transient response, and finally the displacement–time curve becomes linear (steady-state deformation).

Either of these experiments provides the data to characterize the mechanical properties of materials. As far as the rheological data at steady states are concerned, the choice of the mode of deformation experiments (either constant stress or constant strain rate) does not matter. These two modes provide the same results. However, there are several differences between the two modes of deformation experiments. First, the determination of activation energy and volume is more straightforward for a constant stress test than for a constant strain-rate test. Consider a power-law rheology,

$$\dot{\varepsilon} = A\sigma^n \exp\left(-\frac{E^* + PV^*}{RT}\right). \tag{6.14}$$

When stress is kept constant, then the activation enthalpy ($E^* + PV^*$) can be determined by the temperature variation of strain rate, namely,

$$\left(\frac{\partial \log \dot{\varepsilon}}{\partial T}\right)_{\sigma,P} = \frac{E^* + PV^*}{RT^2}. \tag{6.15}$$

Similarly, the activation volume can be determined by measuring the pressure dependence of strain rate,

$$\left(\frac{\partial \log \dot{\varepsilon}}{\partial P}\right)_{\sigma,T} = -\frac{V^*}{RT}. \tag{6.16}$$

However, such a measurement for constant strain-rate experiments can be made only after one knows the stress exponent, n. In constant strain-rate experiments, the activation enthalpy $E^* + PV^*$ is related to the temperature dependence of strength as

$$\left(\frac{\partial \log \sigma}{\partial T}\right)_{\dot{\varepsilon},P} = \frac{E^* + PV^*}{nRT^2}. \tag{6.17}$$

Similarly, the activation volume V^* is related to the pressure dependence of strength as

$$\left(\frac{\partial \log \sigma}{\partial P}\right)_{\dot{\varepsilon},T} = -\frac{V^*}{nRT}. \tag{6.18}$$

Consequently, the error in the stress exponent propagates to the error in E^* and V^* measurements.

Second, the practical identification of "steady-state" deformation is more difficult in a constant stress test than in a constant strain-rate test. This is particularly true when an experiment is made for small strains. In a constant stress test, it is important to calculate strain rates as a function of strain to assure that a steady state is indeed achieved.

Third, when performing a constant strain-rate test, one needs to know the deformation of apparatus due to the change in stress to obtain a true deformation of the sample. This can be made by measuring the apparatus distortion by applying a load on a sample with known strength (such as an elastic spring). This correction is not needed for a constant stress test, because the apparatus distortion is constant.

Fourth, technically speaking, a constant strain-rate test is easier to perform than a constant stress test. For a constant stress test, one needs to have a good control of stress that can be made only when a high-resolution servo-controlled actuator is available and the load acting on the sample is measured accurately. In contrast, a constant strain-rate test is readily performed with a constant speed of actuator motion that is easier to achieve.

Finally, one should note that strictly speaking, both so-called constant stress and constant strain-rate tests are usually performed not truly at a constant stress nor a constant strain rate. These tests are usually made at a constant load or a constant displacement rate. Since sample geometry (i.e., cross section area) changes with deformation, some corrections are needed to convert them to truly constant stress or strain-rate conditions.

In many cases, results at "steady state" are of particular interest. To confirm that the experimental results correspond to steady state, one must conduct deformation experiments to relatively large strains (at least ~10%) to determine the mechanical response of a specimen ranging from transient to steady-state behavior (see Chapter 3). For a constant strain-rate test, the steady state is relatively easily identified by the (approximately) constant load on the stress–strain curve (obviously the change in sample geometry must be corrected for). Identification of steady-state deformation is not trivial for a constant stress (load) test. In this type of test, the raw data are the displacement versus time curves. When one looks at only a small portion of such curves, any such a curve would look like a straight line even though deformation may still be in the transient stage. To identify the steady state, one needs to calculate the strain rates as a function of time (strain) and confirm that the strain rate is indeed (nearly) independent of strain (time). A steady state is not attained at low strains. This will result in a systematic error in the estimated parameters. A typical case is work-hardening behavior (stress needed to deform a material at constant strain rate increases with strain). In such a case, if the strain rate or creep strength of material is determined at roughly the same strain, then the stress dependence of strain rate (i.e., stress exponent) tends to be underestimated. Similarly, when the transient period (strain) corresponding to work-hardening increases with pressure, then the pressure dependence of deformation (such as the activation volume, see Chapter 10) determined from low-strain, transient creep data will be systematically lower than the true value corresponding to the steady state.

Determining a complete mechanical property such as a stress–strain curve (or strain–time curve) for a large span of strain is time consuming but a critical first step in any mechanical test of a given material. Such a result will tell us at which strain "steady-state" deformation is achieved for a given material under a certain range of conditions. Only after this type of test, does one know how much strain one needs to determine "steady-state" mechanical properties. Unfortunately this important step is often skipped. In such a case, the results from such a study do not necessarily correspond to "steady-state" flow law and can result in serious systematic errors (see also BLUM *et al.*, 2002).

Problem 6.3

Explain how the results of constant stress experiments correspond to the results of constant strain-rate tests for a given same material as shown in Fig. 6.4.

Solution

A stress–strain curve shown in Fig. 6.4a.i is characterized by the initial increase in strength followed by a steady-state strength. For a constant stress experiment, the initial part will correspond to the decreasing strain rate with time (work hardening) followed by a linear relation of strain with time (Fig. 6.4b.i). The stress–strain curve shown in Fig. 6.4a.ii is characterized by a peak in stress followed by a steady-state stress. If the initial part corresponds to elastic deformation, then this corresponds to a displacement–time curve that has an initial softening period (work softening) followed by a linear displacement–time relation shown in Fig. 6.4b.ii.

6.5.2. Stress-relaxation tests

The constant stress or constant strain-rate tests are most common. However, in some cases, other more complicated modes of test are performed. One of which is *the stress-relaxation test*. In this case, a sample is loaded at a given condition and then the motion of a piston is terminated. The stress accumulated in the sample is relaxed by plastic flow. The manner in which stress relaxation occurs provides some information about the rheological properties of a material. One of the advantages of this technique is that it allows us to investigate plastic properties at very slow strain rates. Note, however, that neither stress nor strain rate is constant during deformation in such a test, and therefore the mechanical data obtained may not represent the steady-state rheology of a given material.

In the stress-relaxation test, the motion of a motor (actuator) is terminated. Therefore the total strain is the sum of elastic strain of the machine, ε_M, elastic and plastic strain of a sample, ε_E and ε_P respectively,

$$\varepsilon = \varepsilon_M + \varepsilon_E + \varepsilon_P \tag{6.19}$$

where $\varepsilon_M = \sigma/k$ and $\varepsilon_E = \sigma/M$. Now taking the time derivative of equation (6.19), and considering the fact that the total strain is kept constant, one has,

$$\dot{\varepsilon}_P = -\left(\frac{1}{k} + \frac{1}{M}\right)\dot{\sigma}. \tag{6.20}$$

Thus, during a stress-relaxation test (see Fig. 6.5), the elastic strain in the sample and the machine changes to the plastic strain of the sample. Therefore by measuring the time dependence of stress, $\dot{\sigma}$, one can determine

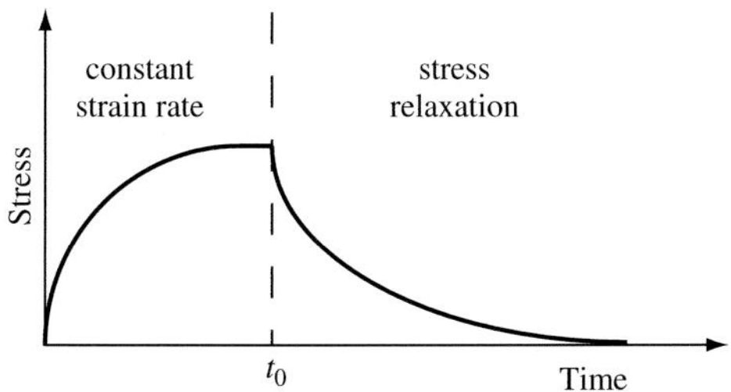

FIGURE 6.5 A typical result of a stress-relaxation test. At $t = t_0$, the advancement of the piston is stopped and stress relaxation occurs.

the rheological properties of a sample. A few points need to be noted.

(1) In order for this technique to be useful in obtaining the rheological properties of a sample, the stiffness of the machine must be significantly higher than the elastic constant of the sample ($k \gg M$). Otherwise, the signal will be very small and it is difficult to obtain useful data on the rheological properties of the sample.
(2) Since a small elastic strain is converted to plastic strain, the strain rates associated with stress-relaxation tests are usually much smaller than those in typical mechanical tests particularly at the later stage of a test. Consequently, this type of test allows us to obtain some insights into the rheology at small strain rates.
(3) However, both stress and strain change with time in this type of test. Therefore the rheological flow law may not represent the steady-state flow law.

6.5.3. Stress-dip tests

Another method that has been used to investigate the transient response of a material is so-called *stress-dip tests*. In this technique, the stress is suddenly decreased and the response of a material to this stress dip is measured. If the stress dip is small and the response of the material is measured immediately after the stress dip, then one may be able to assume that the transient response occurs without changing the microstructure (such as the dislocation density). Therefore, this technique provides some clues as to the effects of microstructures (such as dislocation interaction) on rheological properties. In particular, this test has been used to infer the influence of (microstructure-induced) *internal stress* in dislocation creep (see Chapter 9).

6.5.4. Indentation hardness tests

Some insights into the "strength" of materials can be obtained from the indentation tests. In this test, one indents a sample surface by a hard material (typically diamond) and the area of indentation for a given load is determined. The load divided by the area of indentation (which has a dimension of stress) gives a measure of "hardness." For example, when a pyramid-shaped indenter (called the Vickers indenter) is used, the hardness is defined by,

$$H = \frac{L}{A} \tag{6.21}$$

where L is load (N) and A is area (m^2) (A is the actual contact area between the indenter and a sample, so for the Vickers indenter (a pyramid shape indenter), $A = d^2/2\sin(68°) = d^2/1.854$ where d is the diameter of indentation mark). So the unit of hardness is N/m^2.

Hardness provides some measure of the strength of materials. However, the relation between hardness and other well-defined plastic properties is not well known. Gilman (1985) gives an excellent review of the physics of hardness tests. Briefly, hardness is controlled by the resistance of materials against plastic flow. When a nearly isotropic indenter such as the Vickers indenter is used, the hardness is controlled by the activation of multiple slip systems and hence is approximately proportional to the strength of polycrystalline materials (Evans and Goetze, 1979). In contrast, when a highly anisotropic indenter such as the Knoop indenter is used, then the hardness becomes highly dependent on the orientation of indenter with respect to the crystallographic orientation of a sample. This provides a means to constrain active slip systems (Brookes *et al.*, 1971).

One of the advantages of this technique is that it can be applied to a small sample. For a small load (say 0.1–0.2 N), the diameter of indentation for a typical silicate is ~10 μm, so that a sample as small as ~50 μm can be used (e.g., Karato *et al.*, 1990). The variation of strength across a grain boundary was determined by this technique and Gilman (1985) called this technique the *mechanical microprobe*. Another benefit of this technique is that one can activate dislocation motion even at relatively low temperatures because cracking is minimized due to high local confining pressure caused by indentation. Accordingly this technique provides a unique method for estimating the low-temperature plasticity such as the Peierls stress (e.g., Evans and Goetze, 1979).

6.6. Various deformation geometries

6.6.1. Uni-axial (tri-axial) compression/tension

This is the most popular geometry of mechanical testing. In ceramic or Earth sciences, typical samples are brittle, and so the tri-axial testing machine is usually used to apply pressure to prevent brittle fracture. One determines the relationship between the shortening strain $\varepsilon_1(t)$ and the compressional stress $\sigma_1(t)$ (x_1 being the direction of applied force). In this mode of deformation, the direction of the principal axis of strain is parallel to the direction of principal stress (*co-axial deformation*). In most cases, tri-axial compression tests are conducted on a jacketed specimen. Tension tests can also be made, one of the advantages of this mode of test is that one can reach larger strains (Rutter, 1998). In this mode, $\sigma_1(=\sigma_2)$ is confining pressure, and one determines the relationship between $\varepsilon_3(t)$ and $\sigma_3(t)$.

6.6.2. Simple shear deformation

Saw-cut sample assembly (a sandwich method)

Nearly simple shear tests can be performed using a tri-axial testing apparatus by sandwiching a thin slice of a sample between saw-cut pistons (Fig. 6.6b). Upon uni-axial (tri-axial) compression, a sample sandwiched between two pistons will be sheared. A sample will also be shortened, but if the sample thickness is sufficiently small, extrusion is difficult and deformation is approximately simple shear. Two points must be carefully examined. First, deformation geometry is not strictly simple shear. Some component of uni-axial compression is present due to extrusion. Therefore, the deformation geometry must be examined by measuring the change in sample thickness and the amount of shear through the measurements of rotation of a strain marker.

The conditions of deformation are nearly homogeneous in this test except at sample edges. Toward the end of a sample, a sample (and a piston) extrudes toward a supporting material. Because the two pistons move laterally to each other, the effective area to support a load changes with strain. Because of these complications, the maximum strain to be reached by this technique is limited to ~2–3 (depending on the thickness of a sample), although much higher strains (~8) can be obtained by repeating shear experiments (e.g., Yamazaki and Karato, 2002).

In these tests, one measures the uni-axial load and uni-axial shortening, and converts them to shear stress ($\tau \equiv (\sigma_1 - \sigma_3)/2$) and shear strain ($\gamma \equiv 2(\varepsilon_1 - \varepsilon_3)$) (see Chapter 1).

Torsion tests

When a sample is placed between two pistons and the one piston is rotated with respect to another, then the sample will be deformed by simple shear. Given a sample thickness, h, and the rate of rotation, ω, the strain rate of a sample is given by,

$$\dot{\varepsilon}(r) = \frac{\omega r}{h} \tag{6.22}$$

That is, the strain rate is heterogeneous. Consequently, the stress is also heterogeneous,

$$\sigma(r) = \left(\frac{\omega}{Ah}\right)^{1/n} r^{1/n} \tag{6.23}$$

where $\dot{\varepsilon} = A\sigma^n$. The torque needed to rotate the sample at the angular velocity of ω is given by,

$$T = \int_0^R \int_0^{2\pi} r^2 \sigma(r)\, dr\, d\theta. \tag{6.24}$$

If A does not change with r, then, by inserting the relation (6.23) into (6.24), one has

$$T = \frac{2\pi n}{3n+1}\left(\frac{\omega}{Ah}\right)^{1/n} R^{(3n+1)/n}. \tag{6.25}$$

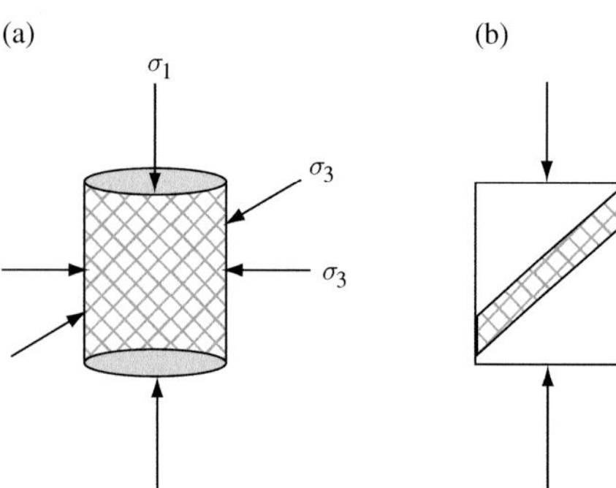

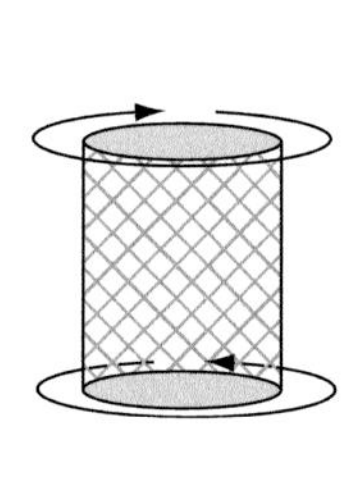

FIGURE 6.6 Typical deformation geometries. (a) Tri-axial compression (or tension), (b) saw-cut deformation for a (nearly) simple shear deformation and (c) a torsion test.

Paterson and Olgaard (2000) discussed how to retrieve rheological data from torque measurements.

When the flow laws determined by one method are compared with the flow laws determined by another method, an appropriate conversion must be made. Such a conversion is possible only when some assumptions concerning the symmetry of flow properties are made. In a general case of anisotropic rheology, a large number of independent experiments with different geometries are needed and they are independent. Under the assumption of isotropic rheology, one way to make a conversion is to cast rheological flow laws into an isotropic formula, namely the Levy–von Mises flow law (equation (3.5), Chapter 3).

7 Brittle deformation, brittle–plastic and brittle–ductile transition

Non-elastic deformation is conveniently classified into brittle and plastic (ductile) deformation based on the scale of discontinuous deformation. Although the main focus of this book is plastic (ductile) deformation, some knowledge of brittle deformation is critical to the understanding of rheological stratification of the lithosphere. Also an understanding of brittle to plastic transition is essential for the understanding of the origin of deep and intermediate earthquakes. This chapter provides a brief summary of brittle deformation and the major differences between brittle and plastic deformation. The nature of brittle–plastic transition is discussed based on the different dependence of strength in two regimes on temperature and pressure and other parameters.

Key words micro-cracks, faulting, Coulomb–Navier's law, Byerlee's law, pore fluid pressure, brittle fracture, plastic flow, brittle–ductile transition, semi-brittle deformation, brittle–plastic transition.

7.1. Brittle fracture and plastic flow: a general introduction

When a stress is applied to a material, a material will deform *elastically* when the stress is small or when stress is applied at low temperatures (or for a short time). Deformation of a material in this case is due to a small displacement of atoms from their stable positions. Upon applying stress, atoms move to new equilibrium positions instantaneously, and upon the removal of this stress, they move back to their original positions. The nature of elastic deformation and the elastic properties of materials are discussed in Chapter 4.

Non-elastic deformation involves atomic movements to the next stable positions. Once atoms move to the next stable positions, then strain is not recoverable immediately. Such a style of deformation occurs when the stress is high and/or the stress is applied at high temperatures (for a long time). Non-elastic deformation involves "defects" around which atomic bonding is weak. Non-elastic deformation can be classified into two categories based on the scales of defects (Table 7.1): (i) deformation involving cracks or pores (cataclastic deformation) and (ii) deformation involving point defects, dislocations and grain boundaries (crystal plasticity). A *crack* (or pore) is a defect around which atomic bonding is completely broken. As a result, there is a large volume expansion associated with formation of these defects, and there is large stress concentration and a resultant long-range stress field surrounding cracks or pores. Consequently, interaction among cracks (or pores) is strong and this often leads to localized deformation (*brittle fracture*). Other defects such as point defects, dislocations and grain boundaries ("crystalline defects") are defects around which atomic bonding is only partially broken. These defects are associated with a relatively short-range stress–strain field. Consequently, the interaction is weak among these defects, and in most cases, deformation occurs homogeneously when these small-scale defects are involved.

Since relatively large-volume expansion is involved in cataclastic deformation, hydrostatic pressure

TABLE 7.1 Classification of style (regime) of deformation.

	Homogeneous	Localized
Point defects Dislocations Grain boundaries (crystal–plastic)	A_1 (plastic flow)	A_2 (ductile faulting)
Cracks Pores (cataclastic)	B_1 (cataclastic flow)	B_2 (brittle fracture)

suppresses these processes much more than deformation involving crystalline defects. Consequently, there is a transition from cataclastic deformation in the shallow part of Earth to crystalline plasticity in the deep part of Earth. Traditionally, this transition was loosely called a *brittle–ductile transition* implying transition from localized cataclastic deformation to homogeneous deformation caused by crystalline defects (B_2 to A_1 in Table 7.1). However, cataclastic deformation (deformation involving cracks and pores) can occur homogeneously in a stable fashion when crack growth or pore coalescence is limited (B_1 in Table 7.1). In other words, deformation in regime B_1 corresponds to a case where cracks (or pores) nucleate but do not grow to cause instability. In contrast, deformation involving crystalline defects occurs in most cases in a homogeneous stable fashion. However, under some limited conditions, deformation involving these small-scale defects can become unstable and localized deformation associated with a fault can occur (A_2 in Table 7.1; Chapter 16).

So in short, each mechanism of deformation involving different types of defects can further be classified into four categories from the point of view of homogeneity of deformation (Table 7.1). RUTTER (1986) and EVANS *et al.* (1990) provide some discussion on the nomenclature for deformation regimes noting these complications (see also KOHLSTEDT *et al.* (1995)). I recommend that A_1 be called "plastic flow," A_2 "ductile faulting," B_1 "(distributed) cataclastic flow" and B_2 "brittle fracture." Note the term "flow" is used here to describe homogeneous deformation.

The distinction between the different regimes summarized in this table is not always very sharp. There is a broad range of conditions under which deformation involves both crystalline defects and cracks. This regime is often referred to as a "semi-brittle" regime in which deformation often occurs in a localized fashion (a regime between A_2 and B_2).

In this chapter, I will provide a brief summary of cataclastic deformation with an emphasis on brittle fracture (mode B_2) and the physical processes of transition from brittle fracture (localized deformation involving cracks and pores) to plastic flow (homogeneous deformation involving crystalline defects; mode A_1) including a discussion of some of the roles of intermediate regimes such as cataclastic flow (mode B_1) and ductile faulting (A_2). No justification can be made, in this short space, for an extensive discussion of this important topic. For a more complete discussion of cataclastic deformation and the transition between brittle fracture to plastic flow, the reader is referred to excellent textbooks such as PATERSON and WONG (2005; for materials science bases) and SCHOLZ (2002; for geological/geophysical applications).

7.2. Brittle fracture

7.2.1. Micro-cracks and faulting

When a deviatoric stress is applied to a heterogeneous, porous material, stress is concentrated at certain points (e.g., grain boundaries or cracks or pores). A high-stress concentration would lead to the breaking of chemical bonds, creating new surfaces. New cracks are formed and/or pre-existing cracks grow. Cracks have a long-range stress–strain field. Consequently, their interaction is strong. Brittle fracture starts with the formation of micro-cracks. Micro-cracks interact to form a *fault*, and finally large-scale displacement occurs along a fault. Consequently, the process of brittle fracture may be classified into the *nucleation* of a fault through micro-crack interaction, and the *growth* of a fault through sliding. When one deforms an intact rock without pre-existing faults, then in order to cause fracture, one needs to nucleate micro-cracks and let them interact. In other words, one needs both the nucleation and growth of faults to fracture an intact rock. On the other hand, if a piece of rock has pre-existing faults, then the failure of this rock involves only the propagation (growth) of these faults. Consequently, the fracture strength of an intact rock is higher than that of a rock that contains pre-existing faults (e.g., PATERSON and WONG, 2005). In most of Earth's lithosphere, it is likely that rocks contain pre-existing faults, and in such a case, the strength of rocks in the brittle regime is controlled by the resistance against fault motion (i.e., growth of cracks). For this reason, the brittle strength of the lithosphere is usually calculated based on the resistance against fault motion.

7.2.2. Coulomb–Navier's law of fracture strength

Coulomb–Navier's law of friction

The resistance to fault motion is described by an empirical relation, Coulomb–Navier's law (e.g., PATERSON and WONG, 2005; SCHOLZ, 2002), which states that the motion along a fault occurs when the magnitude of shear stress on the fault exceeds a certain critical value that increases linearly with the normal stress on the fault, namely

$$\tau_{\text{fric}} = \tau_c + \mu_f \sigma_n \tag{7.1}$$

where τ_c is "the cohesive strength" and μ_f is the friction coefficient and σ_n is the normal stress acting on the fault plane. τ_c represents the resistance to fault motion without any normal stress. Therefore τ_c is the force per unit area of the fault plane due to the chemical bonding, hence the name cohesive strength. The second term represents the "true" frictional resistance that is proportional to the normal stress. The normal stress increases with pressure. For instance, in the very deep interior of Earth where the deviatoric stress is much less than the confining pressure, then $\sigma_n \approx P$ (see Problem 1.2). Consequently, brittle fracture becomes difficult in the deep interior of Earth.

When a large amount of fluid is present, then the effective normal stress will be reduced and hence,

$$\tau_{\text{fric}} = \tau_c + \mu_f(\sigma_n - P_f) \approx \tau_c + \mu_f(P - P_f) \tag{7.2}$$

where P_f is the pressure of pore fluid (pore pressure) and the approximation is valid for the case where deviatoric stress is significantly smaller than the confining pressure. Thus a significant reduction in strength occurs when a pore fluid is present.

Byerlee's law, the role of pore fluids and a simple model of brittle fracture strength

Based on a large number of experiments, BYERLEE (1978) found that the two parameters τ_c and μ_f are nearly independent of the materials ($\tau_c \approx 50$ MPa, $\mu_f \approx 0.6$–0.8), the temperature and the rate of deformation. (Strictly speaking, the friction coefficient is weakly dependent on the rate of deformation. This rate dependence of the friction coefficient controls the stability of fault motion (e.g., SCHOLZ, 2002). However, the variation of the friction coefficient with the rate of deformation is usually small (~a few % or less). This empirical law of the constancy of the friction coefficient is referred to as *Byerlee's law* (BYERLEE, 1978).

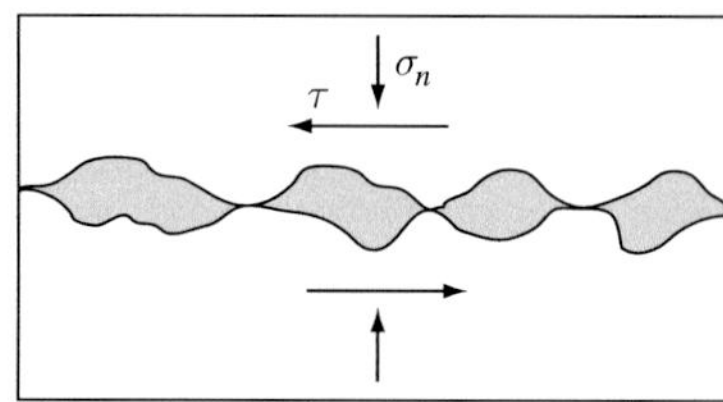

FIGURE 7.1 A schematic drawing of the structure of a fault surface. The contact between two sides of a material occurs at small irregularities.

Some essence of brittle fracture strength can be understood using a simple model. Consider the surface of a fault plane that is made of irregular topography (Fig. 7.1). The two sides of irregular surface are in contact with each other only at certain contact areas. Let A be a total surface area, and A_c a contact surface area. When a force acts on this fault area, the force across the interface must be supported by these contact points. Therefore the force balance normal to the surface yields,

$$\sigma_\perp A_c = \sigma_n A \tag{7.3}$$

where $\sigma_\perp$ is the yield stress of material against normal stress and σ_n is the normal stress on the fault plane (the definition of *yield stress* is somewhat ambiguous, but at the low-temperature, high strain-rate conditions considered here, materials can support a certain deviatoric stress that is nearly independent of temperature and strain rate (for more detail, see Chapter 9). Similarly, the force balance parallel to the surface yields,

$$\sigma_\parallel A_c = \tau A \tag{7.4}$$

where $\sigma_\parallel$ is the yield stress for shear stress and τ_c is the cohesive strength. Dividing equation (7.4) by (7.3), one gets,

$$\tau_{\text{fric}} = \frac{\sigma_\parallel}{\sigma_\perp} \sigma_n. \tag{7.5}$$

If we add a strength at zero normal stress ("cohesive strength," τ_c), then this formula becomes identical to Coulomb–Navier's law, (7.1), with $\mu_f \equiv \sigma_\parallel / \sigma_\perp$. This model provides a simple explanation for a number of fundamental properties of rock friction. In this model, the friction coefficient is given by the *ratio* of strength for two different geometries. Therefore if the material on the fault is not plastically anisotropic, then the friction coefficient should be close to 1 and hence independent of rock type, temperature and strain rate as observed. This model also predicts that if a material on the fault (fault gouge) is highly anisotropic (e.g., clay minerals), then the friction coefficient will be small (e.g., SHIMAMOTO and LOGAN, 1981).

It may be mentioned here that although it is remarkable that the friction coefficient is nearly independent of materials and also insensitive to temperature and the rate of slip, there is a subtle dependence of friction coefficients on these parameters (e.g., DIETERICH, 1978; RUINA, 1983; MARONE, 1998; PATERSON and WONG, 2005). A particularly important issue is the sensitivity of the friction coefficient to the rate of the slip and the slip distance sometimes called *rate and state dependent friction*. The microscopic basis for the rate and state dependent friction has not been studied in detail although it likely involves some thermally activated processes such as plastic deformation and diffusion-controlled crack healing (see e.g., NAKATANI, 2001). The dependence of friction coefficients on the rate of slip and other parameters is critical to the stability of slip. Briefly, if the friction coefficient decreases with slip rate or displacement, then slip will be unstable (and vice versa) (e.g., DIETERICH, 1978; RUINA, 1983; MARONE, 1998; PATERSON and WONG, 2005). However, the amplitude of variation of the friction coefficients with these variables is rather small (~a few %). Therefore for most purposes of estimating the "strength," one can assume a constant friction coefficient.

Now let us consider a case where a fraction f of the interface is occupied with a fluid with pressure P_{pore}. Under these conditions, the brittle fracture strength is significantly reduced. To understand the origin of this effect, let us consider the model shown in Fig. 7.1. The force balance equation (7.3) becomes

$$\sigma_n = P_{pore} f + (1-f)\sigma_s \tag{7.6}$$

where σ_s is the normal stress supported by the solid portion. Similarly, equation (7.4) should be modified to

$$\tau = (1-f)\tau_s \tag{7.7}$$

where τ_s is the stress in the solid part. Now we assume that for the stress in the solid part, Coulomb–Navier's law (7.1) works, i.e., $\tau_s = \tau^0_{fric} = \tau_c + \mu_f \sigma_s$. Then we get,

$$\tau_{fric} = (1-f)\tau_c + \mu_f(\sigma_n - fP_{pore}). \tag{7.8}$$

A few comments may be made on relation (7.8). First, the role of the pore fluid pressure on the strength depends on f, the fraction of the interface filled with the fluid. Consequently, when the volume fraction of the fluid is small, say 0.1–1% (this would be the case for most partially molten materials; see Chapter 12), then the mechanical effects of fluids will be minor. However, when the volume fraction of fluids increases, f becomes large. Note that the value of f also depends on the geometry of fluid (dihedral angle; see Chapter 12). When the dihedral angle of a fluid with solid is zero, $f \approx 1$. The dihedral angles between aqueous fluids and rocks are typically non-zero (e.g., HOLNESS, 1993) except at very high pressures (above a few GPa, see Chapter 12). This means that a pore pressure high enough to cause faulting is either due to a large amount of fluid (a fluid phase tends to be sucked to a high porosity region, and therefore the amount of fluid on the fault plane can be much higher than the average amount of fluid) or due to the non-equilibrium geometry of fluids.

The strength of a material in the brittle regime depends on the stress state, namely either compression or tension. To see this point, let us consider a two-dimensional case. We consider a plane whose normal is at an angle θ from the x-axis (direction of the maximum compressive stress) and calculate the normal (σ_n) and shear stress (τ) acting on it. We take the x-axis along the direction of σ_1 (the maximum compressive stress) and the y-axis along the direction of σ_3 (the minimum compressive stress), and let θ be the angle from the x-axis. Inserting the relation $\tau = \frac{1}{2}(\sigma_1 - \sigma_3)\sin 2\theta$ and $\sigma_n = \frac{1}{2}(\sigma_1 + \sigma_3) - \frac{1}{2}(\sigma_1 - \sigma_3)\cos 2\theta$ (see Chapter 1, equations (1.20) and (1.21)) into equation (7.1), we find that the faulting occurs when the following condition is met,

$$\tau_c + \frac{1}{2}\mu_f(\sigma_1 + \sigma_3) = \frac{1}{2}(\sigma_1 - \sigma_3)(\sin 2\theta + \mu_f \cos 2\theta). \tag{7.9}$$

As one increases differential stress $\sigma_1 - \sigma_3$ at a constant confining pressure $\frac{1}{2}(\sigma_1 + \sigma_3)$, this condition is first met at the angle where $\sin 2\theta + \mu_f \cos 2\theta$ is the maximum. Now $\sin 2\theta + \mu_f \cos 2\theta$ has the maximum at

$$\tan 2\theta = \frac{1}{\mu_f}. \tag{7.10}$$

Thus the orientation of the fault plane depends on the value of the friction coefficient. With a typical value of $\mu_f = 0.6$–0.8, 25°–30° from the compression axis ($\theta = 45°$ ($\pi/4$) for $\mu_f = 0$).

Problem 7.1*

Show that the above discussion is equivalent to the Mohr-circle construction discussed in Chapter 1, in which one draws a circle representing $(\sigma_n(\theta), \tau(\theta))$ and compares the values of $(\sigma_n(\theta), \tau(\theta))$ with the Coulomb–Navier criterion. The fracture occurs at the condition where the Mohr-circle and the line for the Coulomb–Navier criterion first intersect.

Solution

The Mohr circle is defined by $\tau^2 + \left(\sigma_n - \frac{1}{2}(\sigma_1 + \sigma_3)\right)^2 = \frac{1}{4}(\sigma_1 - \sigma_3)^2$ (equation (1.23)). The Coulomb–Navier condition for faulting is $\tau = \tau_c + \mu_f \sigma_n$. These two intersect when $\tau = \tau_c + \mu_f \sigma_n$ is the tangent of a circle $\tau^2 + \left(\sigma_n - \frac{1}{2}(\sigma_1 + \sigma_3)\right)^2 = \frac{1}{4}(\sigma_1 - \sigma_3)^2$. The tangent of this circle is given by

$$\tau = \tau_0 + \frac{\sigma_{n_0}\left(\sigma_{n_0} - \frac{1}{2}(\sigma_1 + \sigma_3)\right)}{\tau_0} - \frac{\sigma_{n_0} - \frac{1}{2}(\sigma_1 + \sigma_3)}{\tau_0}\sigma_n.$$

Consequently, $\mu_f = -\left(\sigma_{n_0} - \frac{1}{2}(\sigma_1 + \sigma_3)\right)/\tau_0$. Now using the relations (1.21) and (1.22) (Chapter 1), $\left(\sigma_{n_0} - \frac{1}{2}(\sigma_1 + \sigma_3)\right)/\tau_0 = -1/\tan 2\theta$. Therefore $1/\mu_f = \tan 2\theta$ and the two discussions are equivalent.

For this value of θ, one has,

$$\tau_c = -\frac{1}{2}\mu_f(\sigma_1 + \sigma_3) + \frac{1}{2}(\sigma_1 - \sigma_3)\sqrt{\mu_f^2 + 1}. \tag{7.11}$$

For tension, $\sigma_1 = 0$, $\sigma_3 = -T_0$, thus,

$$T_0\left(\mu_f + \sqrt{\mu_f^2 + 1}\right) = 2\tau_c. \tag{7.12}$$

For compression, $\sigma_1 = C_0$, $\sigma_3 = 0$, thus,

$$C_0\left(\sqrt{\mu_f^2 + 1} - \mu_f\right) = 2\tau_c. \tag{7.13}$$

Therefore,

$$\frac{C_0}{T_0} = \frac{\sqrt{\mu_f^2 + 1} + \mu_f}{\sqrt{\mu_f^2 + 1} - \mu_f}. \tag{7.14}$$

We conclude that *the brittle (fracture) strength of a material is stronger for compression than for tension*

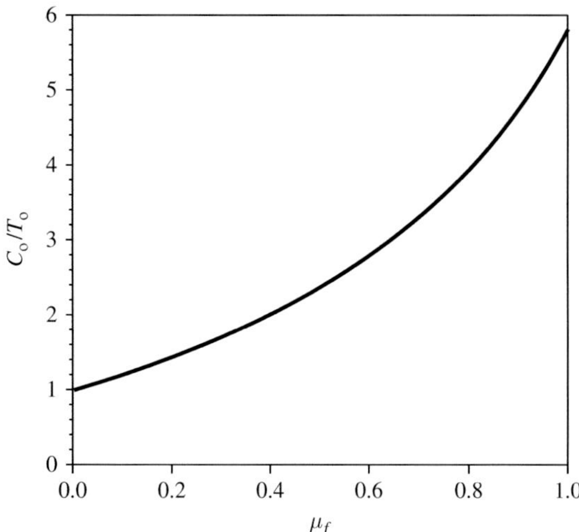

FIGURE 7.2 The compressional strength/tensional strength ratio in the brittle regime.

(Fig. 7.2) (for this reason, any architecture, such as a bridge, is designed such that most of the components are under compressional stress).

7.3. Transitions between different regimes of deformation

7.3.1. Generalities

The strength of a material in the brittle fracture regime (B_2) is insensitive to strain rate, temperature and materials, but increases nearly linearly with pressure (see equation (7.1) or (7.2)). In contrast, in another mode of non-elastic deformation, namely the plastic flow (A_1), the strength of materials is highly sensitive to temperature and weakly sensitive to pressure (under low-pressure conditions).[1] Therefore when the physical conditions of deformation are changed, then the dominant mechanism of non-elastic deformation will change. To a good approximation, cataclastic and crystal–plastic deformation occur independently (this is not strictly true as will be discussed later in this section). Consequently, the easier one out of the two mechanisms (the one that gives a lower strength) will be the dominant mechanism of deformation.

[1] In much of the literature, it is said that the strength in the plastic flow regime is insensitive to pressure. This is misleading because pressure effect on plastic flow can be enormous under high-pressure conditions (Chapter 10). Plastic flow strength is insensitive to pressure only at low pressures.

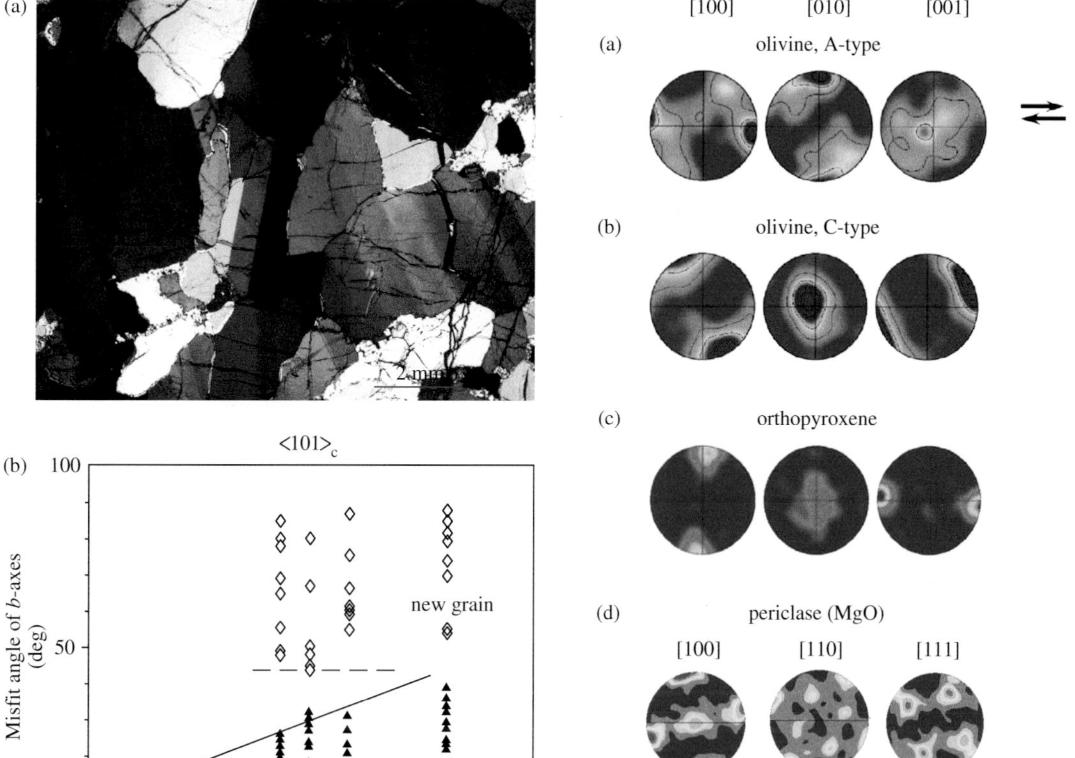

PLATE 1 Recrystallization due to subgrain rotation. (a) Microstructure of a deformed peridotite (from Avachinsky Volcano, Kamchatka) showing the development of subgrains in olivine (seen as a slightly different color), (b) the relation between the misfit angles between subgrains and the strain (Toriumi and Karato, 1985).

PLATE 3 Some examples of pole figures showing the distribution of crystallographic orientations in the coordinate system defined with respect to the deformation geometry. The color code indicates the density of data (red, high; blue, low). (a) Olivine experimentally deformed in simple shear (water-poor, low-stress, high-temperature conditions), (b) olivine experimentally deformed in simple shear (water-rich, low-stress) (from Jung *et al.*, 2006), (c) naturally deformed orthopyroxene showing a girdle type LPO (from Skemer *et al.*, 2006), (d) MgO experimentally deformed in simple shear (from Long *et al.*, 2006).

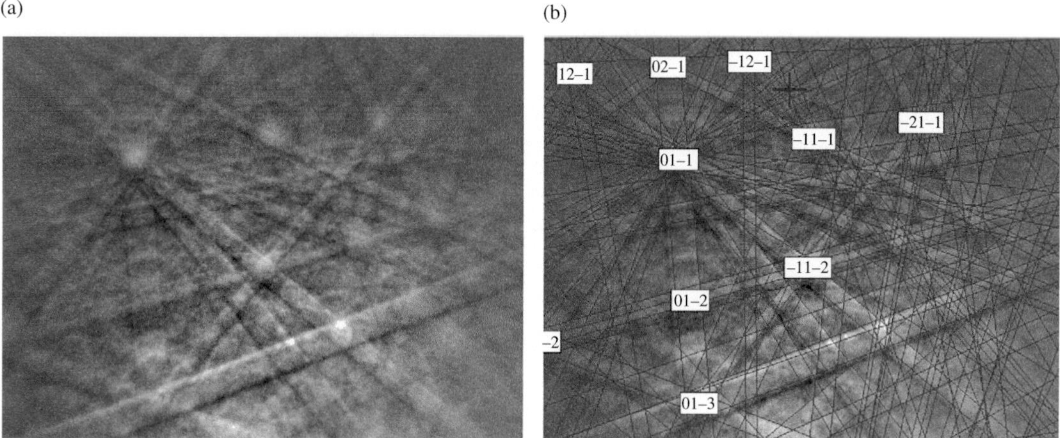

PLATE 2 (a) A Kikuchi pattern of deformed olivine and (b) a comparison of a calculated Kikuchi pattern with the observed one.

This plate section is also available for download in colour from www.cambridge.org/9781107406056

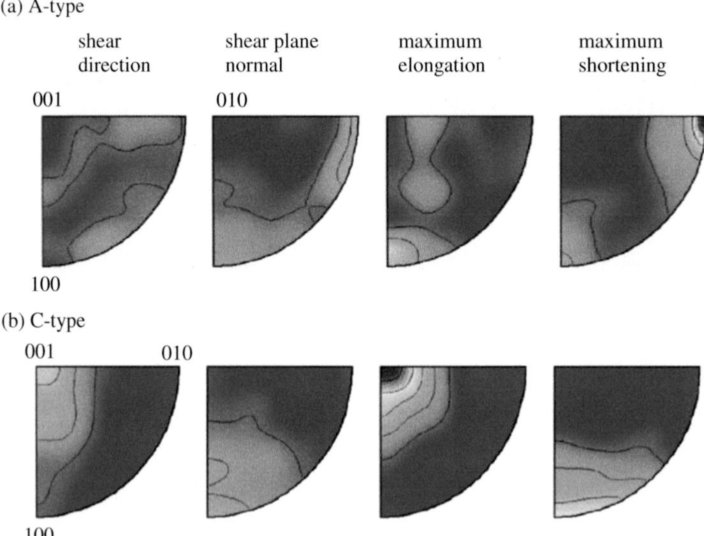

PLATE 4 Examples of inverse pole figures showing the distribution of orientations in space relative to the crystallographic coordinate. (a) Olivine deformed in the A-type regime (the same sample as shown in Fig. 14.2a) and (b) olivine deformed in the C-type regime (the same sample as in Fig. 14.2b). The color code indicates the density of data (red, high; blue, low).

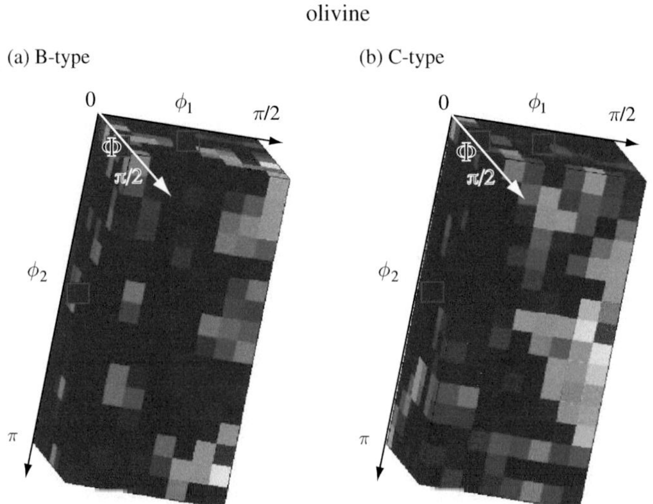

PLATE 5 Some examples of the orientation distribution function (ODF). (a) Olivine deformed in the B-type regime and (b) olivine deformed in the C-type regime. The distribution of crystallographic orientation of individual grains is shown in the Euler angle space. The color code indicates the density of data (red, high; blue, low).

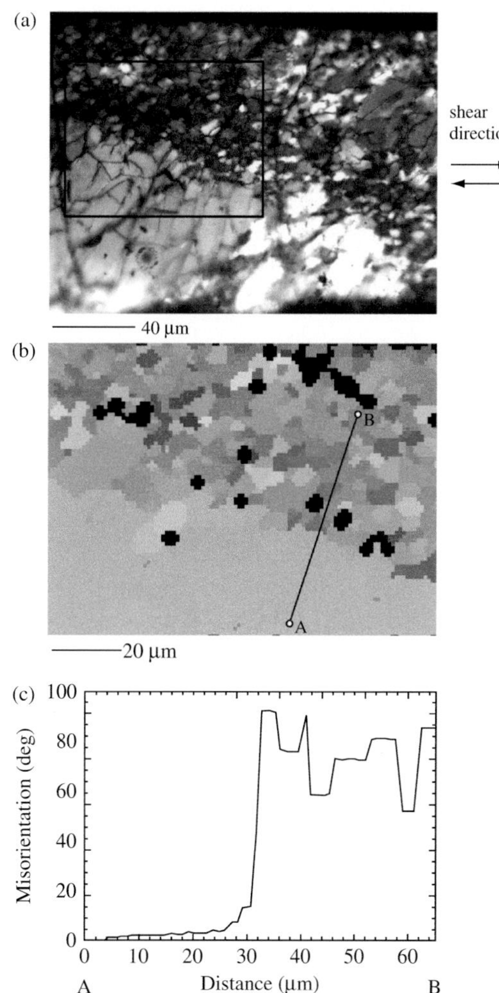

PLATE 6 Orientation mapping. Orientations of each grain can be color-coded and the spatial distribution of orientation is visualized (from JUNG *et al.*, 2006).

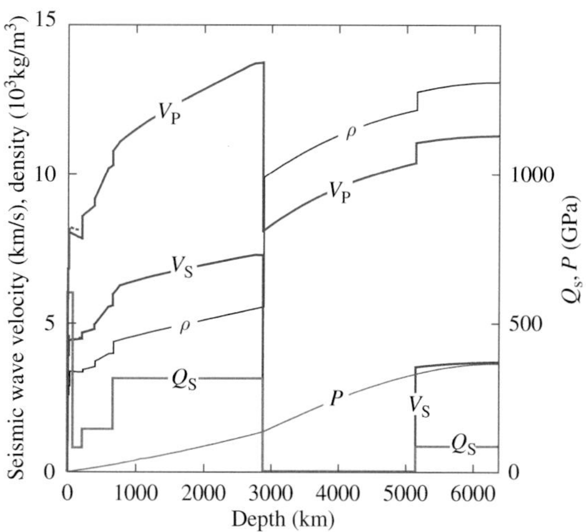

PLATE 7 The PREM model for gross Earth structure (after DZIEWONSKI and ANDERSON, 1981). V_P, P-wave velocity; V_S, S-wave velocity; ρ, density; P, pressure; Q_S, Q for the S-wave.

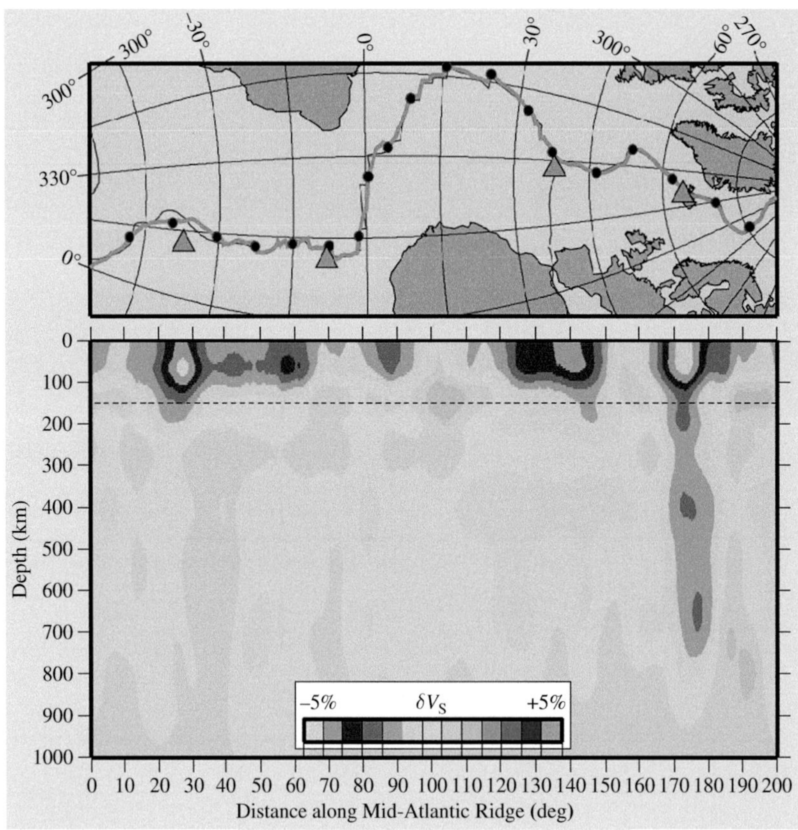

PLATE 8 A result of high-resolution velocity tomography (S-wave velocity anomalies) in the upper mantle along the mid-Atlantic ridge (MONTAGNER and RITSEMA, 2001). Low-velocity anomalies beneath ridges are in most cases shallow (< 150 km), whereas in regions near hotspots, low-velocity anomalies occur in the deeper portions.

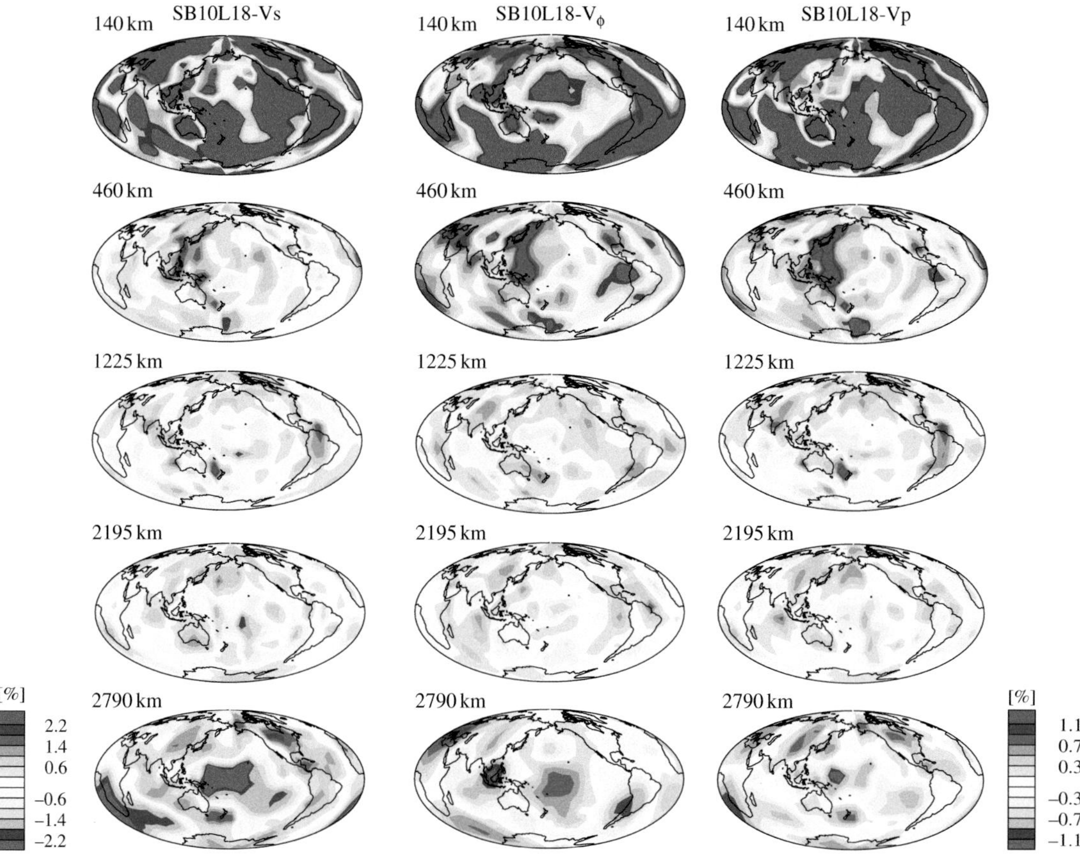

PLATE 9 Results of global seismic tomography of P- (V_P), S-wave (V_S) and bulk sound velocities (V_ϕ) as a function of depth (after MASTERS *et al.*, 2000).

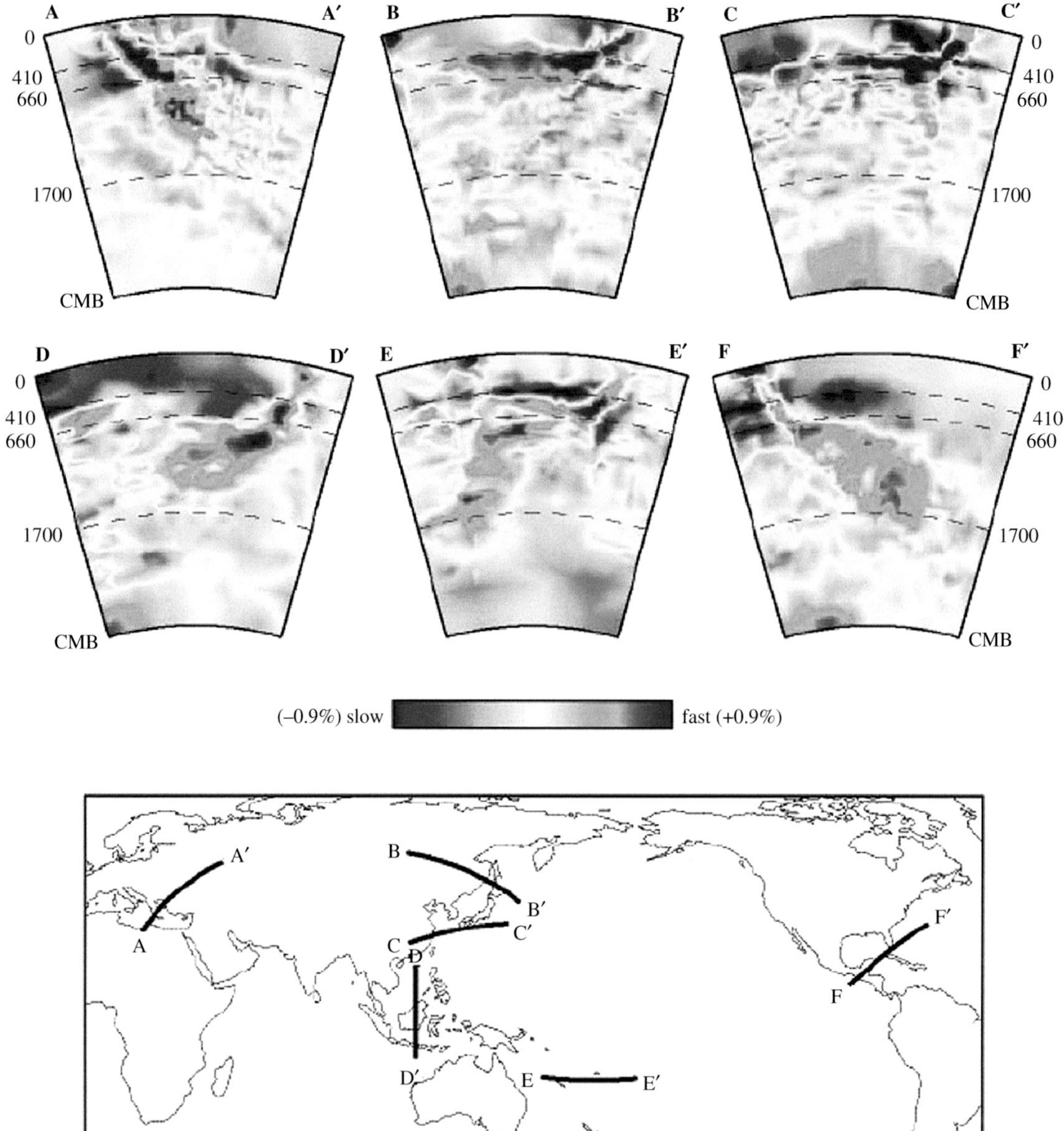

PLATE 10 Examples of high-resolution velocity tomography for subduction zones. The pattern of fast velocity anomalies is modified in the transition zone in a variety of ways depending on regions (Romanowicz, 2003).

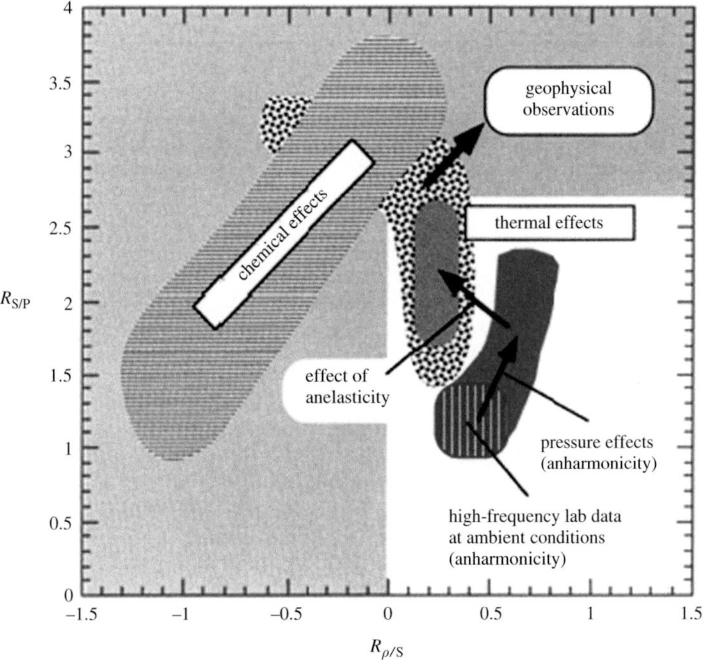

PLATE 11 A plot of $R_{S/P}$ against $R_{\rho/S}$ predicted from the thermal and chemical origin of heterogeneity as compared with actual observations (after KARATO and KARKI, 2001).

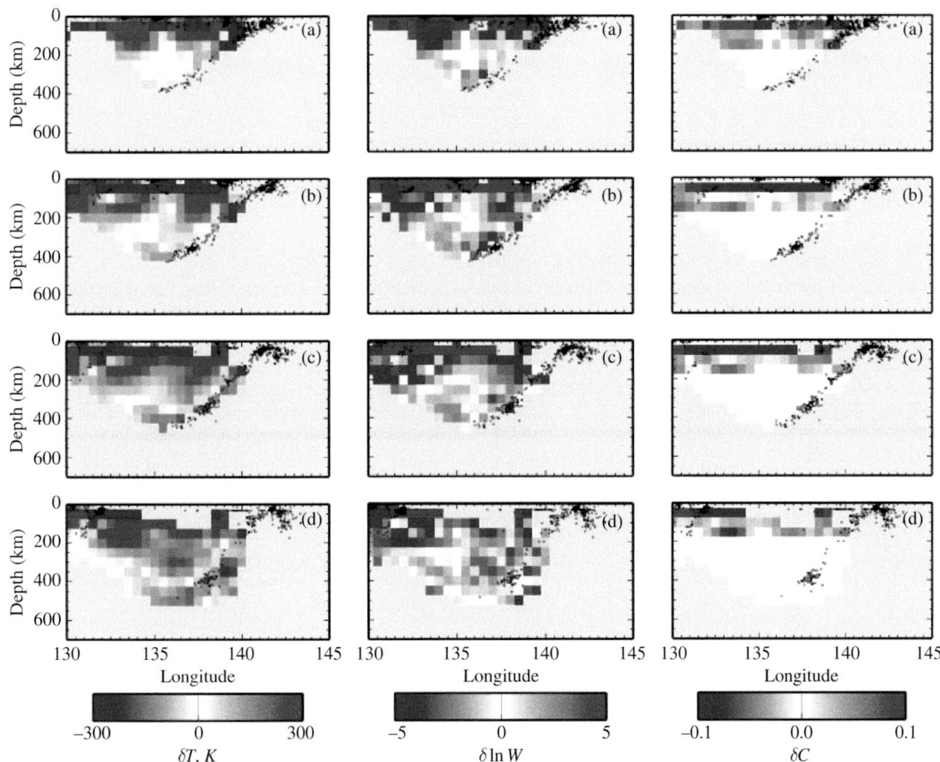

PLATE 12 Inversion of velocity and attenuation tomography in terms of temperature (T), major element composition (C; Mg#) and water content (W) anomalies in the upper mantle beneath the Philippine Sea (SHITO *et al.*, 2006). High-resolution tomography is available both for seismic wave velocities and attenuation from this region. A region of unusually high water content is identified in the deep upper mantle where attenuation is high but velocities are not so low.

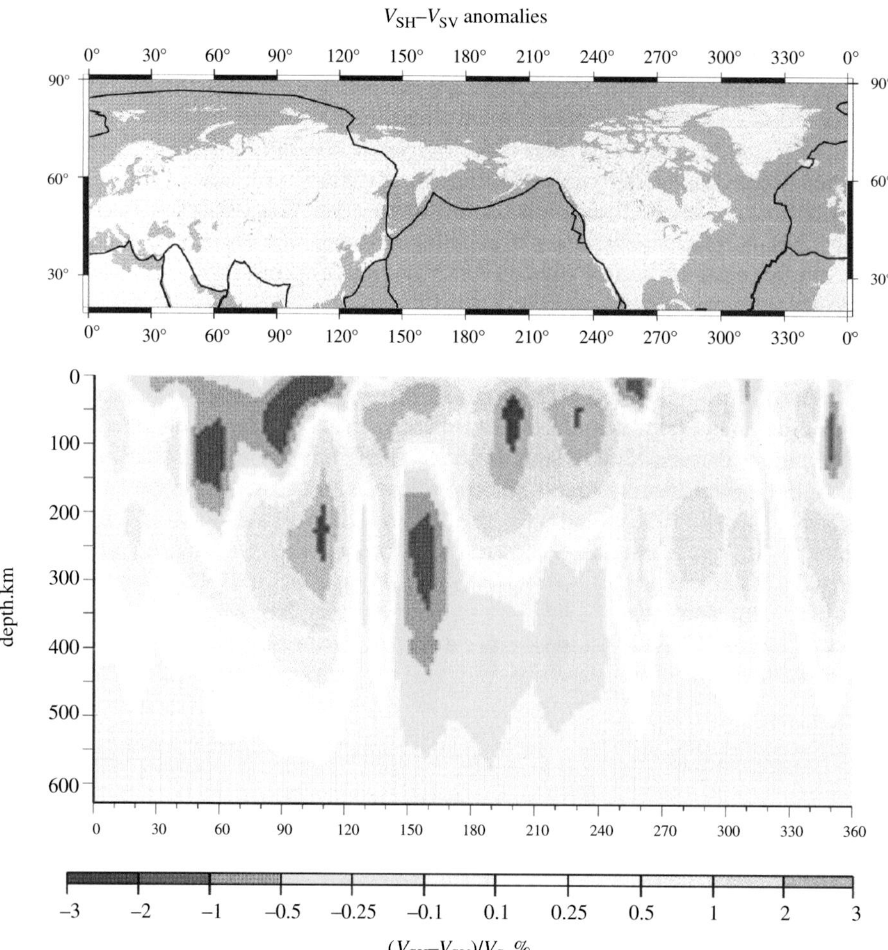

PLATE 13 The V_{SH}/V_{SV} polarization anisotropy of the upper mantle (after Montagner and Guillot, 2000).

The deviation of V_{SH}/V_{SV} polarization anisotropy in the upper mantle from the PREM model is shown along a cross section shown in the top map (the unit is %).

TABLE 7.2 Summary of characteristics of brittle and plastic deformation.

Sensitivity to	Brittle fracture (regime B_2)[2]	Plastic flow (regime A_1)[2]
Pressure	~Linear	~Exponential
Temperature	Weak	Strong (exponential)
Materials	Weak	Strong
Rate of deformation	Weak	Strong
Water content	Strong	Strong
Stress state[1]	Yes	Weak[3]

[1] Stress state means the geometry of the stress tensor such as compression, tension or shear.
[2] For the definition of the regimes, see Table 7.1.
[3] Strength in the plastic regime depends on the stress state only when plasticity is anisotropic. In many materials, the degree of plastic anisotropy is weak, and the strength depends only weakly on the stress state in the plastic regime.

The strength of a material in the plastic flow regime, τ_{plas}, is highly sensitive to temperature (see Chapters 8–10). In many cases, the following power-law relationship holds between the stress and strain rate for plastic flow, namely,

$$\tau_{plas} \propto \dot{\varepsilon}^{1/n} \exp\left(\frac{E^* + PV^*}{nRT}\right) \approx \dot{\varepsilon}^{1/n} \exp\left(\frac{E^*}{nRT}\right) \cdot \left(1 + \frac{PV^*}{nRT}\right) \quad \left(\text{for} \frac{PV^*}{nRT} \ll 1\right) \tag{7.15}$$

where $\dot{\varepsilon}$ is strain rate, n is stress exponent ($n = 1-5$), E^* is activation energy, V^* is activation volume, T is temperature, P is pressure and R is the gas constant. From equation (7.15), it follows that the strength in the plastic flow regime decreases significantly with depth mainly due to the rapid increase in temperature with depth. In contrast, the strength of materials in the brittle fracture regime (B_2) is almost independent of temperature, but increases nearly linearly with pressure (see equation (7.1)). (Basic differences between brittle fracture (regime B_2) and plastic deformation (regime A_1) are summarized in Table 7.2.) It should be emphasized that relation (7.15) is applicable only to a specific case, i.e., a power-law rheology. At relatively low temperatures and/or high stress, the strength τ_{plas} becomes much less sensitive to temperature and pressure. The physical mechanisms responsible for this type of behavior include deformation by dislocation glide over the Peierls potential and mechanical twinning. For details see Chapter 9.

In a simplified view for the transition between deformation by crystalline defects and by cracks (or pores), one assumes that (i) these two mechanisms of deformation are independent (i.e., the one with smaller strength dominates), and (ii) the strength corresponding to deformation by cracks (or pores) is given by Coulomb–Navier's law (equation (7.1) i.e., regime B_2). Consequently, the plastic flow dominates at high temperatures and high pressures, i.e., in the deeper portions of Earth (Fig. 7.3). Because both temperature and pressure increase with depth, the dominant mode of non-elastic deformation changes from brittle fracture to plastic flow as one goes from the shallow to deep portions of Earth (see Chapter 19). Such a diagram was first proposed by Goetze and Evans (1979). In the simple model described above, the transition conditions for plastic flow to brittle fracture (between B_2 and A_1) is given by

$$\tau_{plas} = \tau_{fric} = \frac{\sigma_1 - \sigma_3}{2} \approx \mu_f \sigma_n \approx \mu_f P. \tag{7.16}$$

Note that this criterion refers to the growth of a fault but does not refer to the nucleation of faults (or the nucleation of cracks). Consequently, this condition corresponds to a condition at which the stress needed for plastic flow is the same as the stress needed for frictional sliding. In other words, this condition is for the transition from plastic flow (A_1) to brittle fracture (B_2) (point **A** in Fig. 7.3).

However, as discussed in section 7.1, there are complications in the way in which deformation mechanisms or the mode of deformation change with physical conditions. A key point is the notion that homogeneous ("ductile") deformation can occur even in the cataclastic deformation where deformation occurs by cracks and pores, and that localized

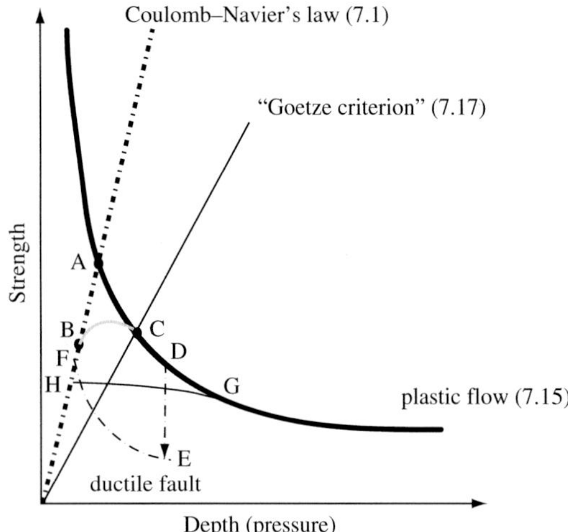

FIGURE 7.3 A schematic diagram showing the conditions for a brittle–plastic and brittle–ductile transition.

The thick solid curve shows the depth variation of strength corresponding to plastic flow and the thick broken line shows the pressure dependence of strength corresponding to Coulomb–Navier's law. Shown together are the line corresponding to the "Goetze criterion" and the strength–depth relation for the ductile fault. The thick gray curve is the estimated strength for the intermediate regime where cracking occurs but deformation is homogeneous (regime (B_1)). Point A corresponds to $A_1 \to B_2$ transition, B to $B_1 \to B_2$ and C to the $A_1 \to B_1$ transition respectively. The trend from D–E–F shows a strength evolution corresponding to ductile faulting (point D corresponds to the $A_1 \to A_2$ transition). The curve G–H represents a trend for "yielding." The position of this curve depends on the mechanism of "yielding." Yielding due to the Peierls mechanism provides a modest reduction of strength, whereas when twinning at a low stress is possible (e.g., in orthopyroxene), then the reduction of strength is substantial.

deformation can occur even in a case where deformation is by crystalline defects.

Consequently, one can define four different regimes (not two), and therefore one can imagine various combinations to describe the transition from homogeneous crystalline plastic deformation (mode A_1) to localized deformation by cracks (mode B_2). The details of this gradual transition are not fully understood and consequently, there is no widely accepted formula to describe this gradual transition.

A concept that captures some aspects of gradual transition is the so-called "Goetze criterion" (KOHLSTEDT *et al.*, 1995), namely,

$$\sigma_1 - \sigma_3 = 2\tau_{plas} = P. \tag{7.17}$$

One may interpret that relation (7.17) defines the conditions at which the stress associated with plastic flow becomes high enough for crack nucleation, which is suppressed by confining pressure (due to volume expansion; see e.g., KOHLSTEDT *et al.*, 1995).[2] An experimental basis for this criterion is given by EDMOND and PATERSON (1972). In other words, equation (7.17) represents the conditions corresponding to a transition between regimes A_1 and B_1. For a typical value of $\mu_f = 0.7$, $\sigma_1 - \sigma_3 \approx 1.4P$ for equation (7.16) and $\sigma_1 - \sigma_3 = P$ for equation (7.17). Consequently, the region $P < \sigma_1 - \sigma_3 < 1.4P$ may be interpreted as non-localized or semi-brittle deformation involving homogeneously distributed micro-cracks (e.g., KOHLSTEDT *et al.*, 1995).

Note, however, that the physical basis for the Goetze criterion is not well defined and that the difference between Coulomb–Navier's law and the Goetze criterion is small and therefore the introduction of the Goetze criterion helps reduce the strength only slightly. In addition to these criteria, one may also introduce a criterion corresponding to the onset of shear localization in the plastic flow regime (A_1 to A_2 transition; see Chapter 16). When shear localization occurs during deformation by crystalline defects, deformation will occur along a fault that is made of fine-grained materials without a large pore space. Strength along such a ductile fault will be temperature sensitive and pressure insensitive but is significantly lower than that corresponding to homogeneous shear for coarse-grained regions. The curve D–E–F in Fig. 7.3 represents the strength corresponding to deformation along a ductile fault, but the conditions for ductile faulting and the strength corresponding to ductile faulting are very poorly constrained (Chapter 16).

7.3.2. Semi-brittle regime

When one constructs a strength–depth diagram assuming either brittle fracture or plastic flow occurs (equation (7.16)), then the dominant mechanism of deformation would change abruptly from brittle fracture to plastic flow and the peak strength would be very

[2] KOHLSTEDT *et al.* (1995) recommend that the transition defined by equation (7.17) be called the "brittle–plastic transition (BPT)," and the transition defined by equation (7.16) "brittle–ductile transition (BDT)." However, these terminologies are confusing because the conditions shown by equation (7.17) define the conditions at which micro-cracking starts, but at this condition micro-cracks are distributed uniformly. Uniform deformation by micro-cracking is often referred to as "ductile" deformation (e.g., RUTTER, 1986). Likewise, the transition defined by equation (7.16) corresponds to the transition between A_1 and B_2, but according to RUTTER (1986)'s terminology, brittle–ductile transition could correspond to the transition between B_1 and B_2.

high, ~500–1000 MPa (KOHLSTEDT *et al.*, 1995, see Fig. 19.4). Such a sharp transition is an artifact of a simplifying assumption that relations such as (7.1) (or (7.2)) and (7.15) would apply to the whole range of deformation behavior. At conditions near the peak (modest temperature and high stress), the microscopic processes assumed for equations (7.1) and (7.15) are probably no longer fully active.

For example, at modest temperatures, the crack tip will be blunted by plastic flow and the stress concentration will be reduced. This will reduce the stress–strain field around the cracks hence reduce their interactions, leading to a more stable deformation. Cracks and dislocations co-exist in most of the materials deformed in the semi-brittle regime (e.g., EVANS *et al.*, 1990; JIN *et al.*, 1998). On the other hand, dislocation motion in grains may cause stress concentration due to dislocation pile up (ZENER, 1948b; STROH, 1954, 1955). When a large number of dislocations are piled up at grain boundaries, then a high-stress concentration occurs that may cause crack nucleation if pressure is low enough. Also, in the plastic flow regime, a simple power-law creep formula such as (7.15) is not a good approximation at high stresses (see Chapter 9 for more details). Under high-stress conditions, strain rate will depend exponentially on stress and hence the temperature dependence of strength is weaker. Furthermore, deformation may occur in a localized fashion even under the conditions where cracking is inhibited (by high confining pressure) when temperature in not high (see Chapter 16 for more details on shear localization). These additional processes that occur in the modest temperatures reduce the strength in both regimes, and hence make the strength substantially lower than would be expected from a simple model.

However, because of the complications of the interactions of various processes, the constitutive relationship appropriate for this regime has not been well established. CHESTER (1988) proposed the following equation that interpolates the behavior of brittle fracture and plastic flow regime, namely,

$$\tau_{sb} = \phi\tau_{plas} + (1 - \phi)\tau_{fric} \tag{7.18}$$

where τ_{sb} is the strength in the semi-brittle regime, $\tau_{fric} = \tau_c + \mu_f\sigma_n$ and $\tau_{plas} \propto \dot{\varepsilon}^{1/n} \exp\left((E_P^* + PV^*)/nRT\right)$ and with $\phi = \tanh(\alpha\sigma_n)$ and α is an empirical constant (CHESTER (1988) chose $\alpha = 0.0014\,\mathrm{MPa}^{-1}$ for halite). As normal stress (pressure) increases, $\phi \to 1$ and $\tau_{sb} \to \tau_{plas}$, and as normal stress decreases, $\phi \to 0$ and $\tau_{sb} \to \tau_{fric}$. Although this formula conveniently represents a transition from brittle fracture to plastic flow when normal stress (pressure) increases, this is entirely an empirical formula and there is no strong theoretical basis for this formula. For example, a general behavior that plastic flow is favored at high temperatures is not represented by this equation. Furthermore equation (7.18) is not consistent with a view that plastic flow and brittle failure are parallel (independent) processes.

An alternative way to obtain a constitutive relation for a semi-brittle regime is to assume that the strain rate in this regime includes a contribution from two processes,

$$\dot{\varepsilon}_{sb} = \dot{\varepsilon}_{plas} + \dot{\varepsilon}_{catac} \tag{7.19}$$

where $\dot{\varepsilon}_{catac}$ is the strain rate due to cataclastic deformation and $\dot{\varepsilon}_{plas}$ is the strain rate due to plastic deformation. In this equation, the influence of temperature (and pressure) on the brittle fracture to plastic flow transition is captured because $\dot{\varepsilon}_{plas}$ increases with temperature more rapidly than $\dot{\varepsilon}_{catac}$ and $\dot{\varepsilon}_{catac}$ decreases with pressure more rapidly than $\dot{\varepsilon}_{plas}$. A difficulty in this approach is that the strain rate due to cataclastic deformation is not always well defined particularly in the brittle fracture regime. In the brittle fracture regime, $\dot{\varepsilon}_{catac} = 0$ for $\tau < \tau_{fric}$, whereas for $\tau = \tau_{fric}$ any value of $\dot{\varepsilon}_{catac}$ is possible. Therefore, in this model, the strength for the semi-brittle regime is simply either plastic flow strength for $\tau < \tau_{plas}$ or brittle strength at $\tau = \tau_{fric}$.

What is missing is a quantitative model for the *interaction* between fracture and flow and the physical description of the flow law corresponding to the localized deformation in the plastic flow regime. From a microscopic point of view, this interaction involves the interaction between cracks and dislocations near the crack tip, where high-stress concentration occurs. In most silicates, high-dislocation density around crack tips will enhance ductility either by enhancing plastic deformation and/or reducing the tendency for crack coalescence that leads to faulting. Some theoretical studies have been conducted to understand the interaction of dislocations (or plasticity) with cracks (e.g., KELLY *et al.*, 1967; YOKOBORI, 1968). However, no generally accepted quantitative models have been formulated to describe the strength in the semi-brittle regime as a function of thermomechanical variables.

Shear localization in the crystal–plastic deformation regime will also reduce the strength. However, the conditions under which shear localization occurs in crystal–plastic deformation are not well understood and the physical description of deformation mechanisms along a ductile fault is highly incomplete (see Chapter 16).

7.3.3. Geological observations on the brittle–plastic transition

The cataclastic to crystal–plastic deformation transition can be investigated by the structural observations on exposed sections of the crust (and the upper mantle). Sibson (1975, 1977) provides a summary of structural observations on such a transition (see also Scholz, 2002). Evidence of plastic flow in minerals increases with depth, but in more detail, the degree to which plastic deformation occurs is mineral specific. In the (upper continental) crust, quartz first shows evidence of plastic flow (at ~570 K) followed by plagioclase feldspar (at ~720 K). In the brittle fracture and semi-brittle regime, deformation is localized, and localized deformation in the semi-brittle regime often leads to intensive shear heating causing formation of a totally molten and quenched rock (pseudotachylyte). It is important to note that any structural observations on naturally deformed rocks that are related to stress magnitude in the plastic flow regime or semi-brittle regime (recrystallized grain size, dislocation density, twinning etc.) usually show the maximum stress of ~100–200 MPa or so. This is significantly smaller than the maximum stress inferred from a simple model such as Kohlstedt *et al.* (1995; i.e., ~500–1000 MPa). The large discrepancy between a simple model and the geological observations suggests that a simplified model such as the one proposed by Kohlstedt *et al.* (1995) does not capture some essence of deformation at relatively low temperature and high stress. Three points may be noted. First, as discussed in section 7.3.2, deformation in the semi-brittle regime is not well characterized and there remain large uncertainties in quantifying the strength in this regime. Second, deformation in the plastic flow regime can be localized under some conditions (Chapter 16). If localization occurs then the steady-state flow law (equation (7.15)) does no longer apply and the actual strength can be significantly lower than that predicted by equation (7.15). Third, the flow law represented by equation (7.15) will not apply at high stress and/or low temperature conditions (see curve G–H in Fig. 7.3). "Yielding" caused by deformation due to the Peierls mechanism or twinning will limit the deviatoric stress supported by any materials under these conditions (see Chapter 9) such as semi-brittle deformation.

8 Diffusion and diffusional creep

Diffusional creep is an important mechanism of plastic deformation in a polycrystalline material at relatively low stress and small grain size. There is evidence that diffusional creep plays an important role in some regions of Earth. At high temperatures, atoms move from their stable positions with some probability due to thermally activated processes. This is referred to as diffusion. The driving force for diffusion is the gradient in chemical potential including the concentration gradient caused by the contact of materials with different chemical compositions or by the stress gradient at grain boundaries created by the applied stress. Consequently, the rate of deformation due to diffusive mass transport is sensitive to diffusion coefficient as well as grain size: the rate of deformation is faster for a smaller grain size. Similar to other processes, diffusional mass transport involves a number of parallel (independent) and sequential (dependent) processes. As a result, the interplay of various diffusing species can be complicated and this also results in a complicated variation in grain-size sensitivity with grain size. Deformation of a polycrystalline material is associated with grain boundary sliding. Large-strain plastic flow involving grain-boundary sliding is sometimes referred to as superplastic flow. Materials science models of superplastic flow are reviewed and some geological significance is discussed. Finally, transient diffusional creep caused by the stress redistribution and its possible roles in small-strain deformation in Earth are discussed.

Key words point defects, diffusion, Fick's law, high-diffusivity path, chemical reaction, Nabarro–Herring creep, Coble creep, pressure-solution creep, grain-boundary sliding, superplasticity.

8.1. Fick's law

At finite temperatures, atoms move around their stable positions due to thermal vibration. As a result, they jump into the next stable positions with a finite probability. Therefore, initially concentrated atomic species will "diffuse out" as time goes on. This process is called *diffusion*. Consequently, atoms tend to diffuse from a highly concentrated region to a less concentrated region. An empirical law to describe this phenomenon is called *Fick's first law of diffusion* which states that the atomic flux is linearly proportional to the concentration gradient, namely,

$$J = -D\frac{\partial c}{\partial x} \tag{8.1}$$

where J is the flux of atoms, c is the number of atoms per unit volume and D is the diffusion coefficient (this is the definition of diffusion coefficient). Therefore the diffusion coefficient, D, has a dimension of m^2s^{-1} in SI units. Diffusion occurs not only under concentration gradient, but also under some other generalized forces including electrostatic force (Chapter 2). Equation (8.1) can then be rewritten in terms of generalized force, X,

$$J = L \cdot X \tag{8.2}$$

with

$$X \equiv -\nabla\left(\frac{\mu}{T}\right) \tag{8.3}$$

where μ is the chemical potential that is related to the concentration of the species,[1] c, as (see Chapter 2),

$$\mu = \bar{\mu}^0 + RT\log c \tag{8.4}$$

Consequently, equation (8.2) may be rewritten as,

$$J = -L\frac{\partial}{\partial x}\left(\frac{\mu}{T}\right) \tag{8.5}$$

with

$$L = \frac{Dc}{R} \tag{8.6}$$

where R is the gas constant.

This is a useful expression when the diffusion equation is to be generalized. For example, the free energy of an atomic species may be modified by the free energies of other species if there is some interaction between them. In this case, the diffusion flux of one atomic species will depend on the concentration (gradient) of other species as well. Also, the free energy of an atomic species may include interaction with the electric field. Note that, strictly speaking, the diffusion coefficient, D, is a (second-rank) tensor that connects two vector quantities, "force" and "flux." Therefore it should be written as D_{ij} where the suffix ij indicates the directions in space. That is, D is, in general, anisotropic. Similarly, the coefficient L may also represent the flux of different species (J^i) corresponding to different forces (X^j). In such cases, we will use the notation, $J^i = \sum_j L^{ij}X^j$ where the symbol L^{ij} is a transport coefficient that represents the coupling of the force X^j and the flux J^i.

Combining Fick's first law of diffusion with the equation of mass conservation, $\partial c/\partial t = -\sum_i (\partial J_i/\partial x_i)$, we obtain *Fick's second law of diffusion*,

$$\frac{\partial c}{\partial t} = \sum_{ij}\frac{\partial}{\partial x_i}D_{ij}\frac{\partial c}{\partial x_j}. \tag{8.7}$$

If diffusion coefficient is spatially homogeneous,

$$\frac{\partial c}{\partial t} = \sum_{ij} D_{ij}\frac{\partial^2 c}{\partial x_i \partial x_j}, \tag{8.8}$$

[1] In Chapter 2, the concentration x (i.e., the fraction of possible atomic sites) is used to define μ^0. Since $x = c/(N/V), \bar{\mu}^0 = \mu^0 - RT\log(N/V)$ (N: number of atomic sites, V: volume).

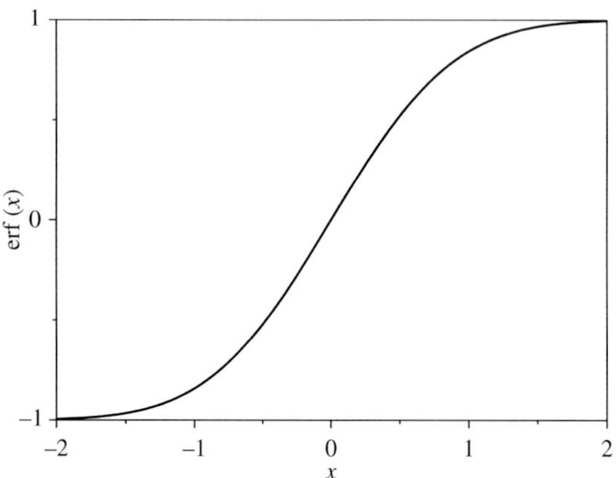

FIGURE 8.1 A graph showing an error function.

and if diffusion coefficient is isotropic,

$$\frac{\partial c}{\partial t} = D\sum_i \frac{\partial^2 c}{\partial x_i^2}. \tag{8.9}$$

Equation (8.7), or (8.8) or (8.9) together with the boundary and initial conditions determines the concentration profile $c(x_i, t)$. Let us consider a case where two materials are in contact at a surface ($x = 0$) and the initial concentration of a given species is $c = 0$ for $x < 0$ and $c = c_0$ for $x > 0$. The solution of equation (8.9) for a one-dimensional problem is,

$$c(x,t) = \frac{c_0}{2}\left[1 + \mathrm{erf}\left(\frac{x}{2\sqrt{Dt}}\right)\right] \tag{8.10}$$

where erf (y) is the error function defined by

$$\mathrm{erf}\,(y) = \frac{2}{\sqrt{\pi}}\int_0^y \exp(-z^2)\,dz. \tag{8.11}$$

Equation (8.10) indicates that the concentration profile $c(x,t)$ controlled by diffusion is characterized by a non-dimensional parameter $\xi \equiv x/2\sqrt{Dt}$. The function $c(x,t)$ given by equation (8.10) is plotted in Fig. 8.1 showing that at $\xi = \pm 1$ ($x = \pm 2\sqrt{Dt}$), the concentration of a given species becomes ∼92% of the value at infinite x.

Problem 8.1

Show that equation (8.10) is the solution of the diffusion equation subject to the initial conditions defined above. (Note erf $(\infty) = 1 = -$erf $(-\infty)$.)

Solution

From (8.11), one gets, $d\mathrm{erf}\,(y)/dy = (2/\sqrt{\pi})\exp(-y^2)$ and $d^2\mathrm{erf}\,(y)/dy^2 = -(4y/\sqrt{\pi})\exp(-y^2)$. Applying the chain rule of differentiation one gets $\partial c/\partial t = \frac{1}{2}c_0(d\text{erf } \xi/d\xi)(\partial\xi/\partial t) = -(c_0 x/4\sqrt{\pi D})\exp(-\xi^2)t^{-3/2}$. Similarly, $\partial c/\partial x = \frac{1}{2}c_0\,(d\text{erf } \xi/d\xi)(\partial\xi/\partial x) = c_0/\sqrt{\pi Dt}\cdot \exp(-\xi^2)$ and $\partial^2 c/\partial x^2 = -(c_0 x/4D\sqrt{\pi D})t^{-3/2}\exp(-\xi^2)$. Therefore, $\partial c/\partial t = D(\partial^2 c/\partial x^2)$ and remembering that $\text{erf}(\infty) = 1$ and $\text{erf}(-\infty) = -\text{erf}(\infty) = -1$, one can show that $c(x,0) = \frac{1}{2}c_0[1+\text{erf}(\infty)] = c_0$ for $x > 0$, and $c(x,0) = \frac{1}{2}c_0[1+\text{erf}(-\infty)] = 0$ for $x < 0$. Therefore the initial conditions are met.

8.2. Diffusion and point defects

Statistical mechanics of diffusion (random walk) shows that (e.g., SHEWMON, 1989),

$$D = \frac{1}{6}a^2\Gamma \tag{8.12}$$

where Γ is the frequency of jump and a is the distance of jump (a numerical factor 1/6 comes for the case of 6-coordinated site). An atomic jump in a crystal to the next site occurs at an appreciable rate only when the neighboring site is vacant or only when the jump of an interstitial atom is considered. Thus the jump frequency (probability) of an atom in a crystal is proportional to the probability of finding a defect multiplied by the probability of atomic jump when a defect is present, namely,

$$\Gamma = \Gamma_f \Gamma_m \tag{8.13}$$

where $\Gamma_f = c_d/c = n_d$ is the fraction of sites occupied by the point defect (c_d, defect concentration; c, atomic concentration) and Γ_m is the probability of atomic jump when the neighboring site is vacant. Atomic jumps occur as a consequence of statistical fluctuation of atomic positions. Thermal vibration causes fluctuation of atomic positions, and consequently, any atoms (or the group of atoms) can jump to the next stable position over a barrier with a finite probability. Such a process is referred to as a thermally activated process and the rate of atomic jumps increases significantly with temperature (see Chapter 2). The basic concept is that an atomic jump occurs over a potential barrier (G_m) and the probability at which a jump occurs is related to the probability that an atom occupies a saddle point through which a jump occurs (G_m is the free energy difference between saddle point and the initial position). Thus,

$$D = \frac{1}{6}a^2\frac{c_d}{c}\Gamma_m = \frac{1}{6}a^2\nu\cdot\exp\left(-\frac{G_f+G_m}{RT}\right) \tag{8.14}$$

where G_f is the formation free energy of defect and G_m is the migration free energy and $\Gamma_m = \nu\exp(-G_m/RT)$ (ν is the frequency of lattice vibration).

Equation (8.14) can be written as,

$$Dc = D^d c^d \tag{8.15}$$

where D^d is the diffusion coefficient of defects, and c^d is the concentration of defects (the fraction of lattice sites occupied by defects),

$$D^d = \frac{1}{6}a^2\nu\cdot\exp\left(-\frac{G_m}{RT}\right). \tag{8.16}$$

Note that the diffusion coefficient of defects (D^d) is much larger than that of atoms because a defect can always jump to the next atomic site but an atom can do this only when it has a defect in the next site (from (8.15), the difference is a factor of c/c^d). In many cases, such kinetics as the kinetics of equilibration with respect to f_{O_2} are controlled by the diffusion of defects (such as vacancies) rather than that of oxygen ions and are hence very fast (KARATO and SATO, 1982) (see Problem 8.3).

The diffusion of defects and that of atoms are related. Let us write the flux of atoms J^a and the flux of defects (of that atom) J^d as

$$J^a = -L^{aa}\frac{1}{T}\frac{\partial\mu^a}{\partial x} - L^{ad}\frac{1}{T}\frac{\partial\mu^d}{\partial x} \tag{8.17a}$$

$$J^d = -L^{da}\frac{1}{T}\frac{\partial\mu^a}{\partial x} - L^{dd}\frac{1}{T}\frac{\partial\mu^d}{\partial x} \tag{8.17b}$$

respectively, where J^{ij} are the coefficients related to the diffusion coefficient by (8.6). Now since the total number of atomic species is conserved,

$$J^a + J^d = 0 \tag{8.18}$$

In order for this relation to be satisfied for arbitrary chemical potential gradients, one must have,

$$L^{aa} = -L^{ad} = -L^{da} = L^{dd} \tag{8.19}$$

where we used the Onsager reciprocal relation, $L^{ad} = L^{da}$ (Chapter 2).[2] Inserting equation (8.19) into equations (8.17a,b), one obtains,

[2] Because $L = Dc/R$, $L^{aa} = L^{dd}$ is equivalent to equation (8.15).

$$J^a = -L^{aa} \frac{1}{T} \frac{\partial(\mu^a - \mu^d)}{\partial x} = -J^d. \quad (8.20)$$

8.3. High-diffusivity paths

In a real material, diffusion of atoms can occur through the bulk of crystal (bulk diffusion (or volume diffusion)), or through some regions that have largely different diffusion coefficients due to the difference in structures. They include grain boundaries and dislocations. Diffusion through these regions is usually faster than diffusion through the bulk and they are referred to as high-diffusivity paths. The enhancement of diffusion through these regions is, however, selective to particular species. In simple ionic solids such as alkali halides or oxides, diffusion of anions is highly enhanced by high-diffusivity paths, but diffusion of cations is not much enhanced (e.g., GORDON, 1973; CANNON and COBLE, 1975).

The degree to which diffusion is enhanced is proportional to the volume fraction of these regions. When one estimates the effective diffusion coefficient involving a large number of high-diffusivity paths, one can use the following relationship,

$$D^{\text{eff}} = D^V + \frac{\pi\delta}{L} D^B \text{ for grain-boundary diffusion} \quad (8.21)$$

where δ is the thickness of grain boundary, and L is grain size, and

$$D^{\text{eff}} = D^V + \pi\rho b^2 D^{\text{pipe}} \text{ for pipe diffusion (diffusion along dislocations)} \quad (8.22)$$

where ρ is the dislocation density and b is the length of the Burgers vector. In these equations, it is assumed that the area of the high-diffusion coefficient near grain boundaries is limited to thickness δ and the area of the high-diffusion coefficient near dislocations is limited to the regions within the Burgers vector. Obviously this assumption is a rough approximation and the proportional constants can be off from these equations by a factor of 3–5.

Usually, the activation energy for diffusion along high-diffusivity paths is smaller than that for bulk diffusion. As a result, diffusion along high-diffusivity paths tends to be important at relatively low temperatures.

When one considers diffusion of a specific ion (say Mg in MgO or in olivine), there are several possible defects that may contribute to diffusion. The effects of the chemical environment on diffusion come mainly from the effects of the chemical environment on point defect concentrations. Some of the diffusion coefficients in olivine are summarized in Fig. 8.2. Note that diffusion coefficients of various species in a given material can be different. In MgO, for example, diffusion of Mg is faster than that of O through the bulk (e.g., GORDON, 1973). However, diffusion of O is significantly enhanced by grain boundaries whereas that of Mg is not. Consequently, when the grain size is small enough, diffusion of O becomes faster than that of Mg. A similar

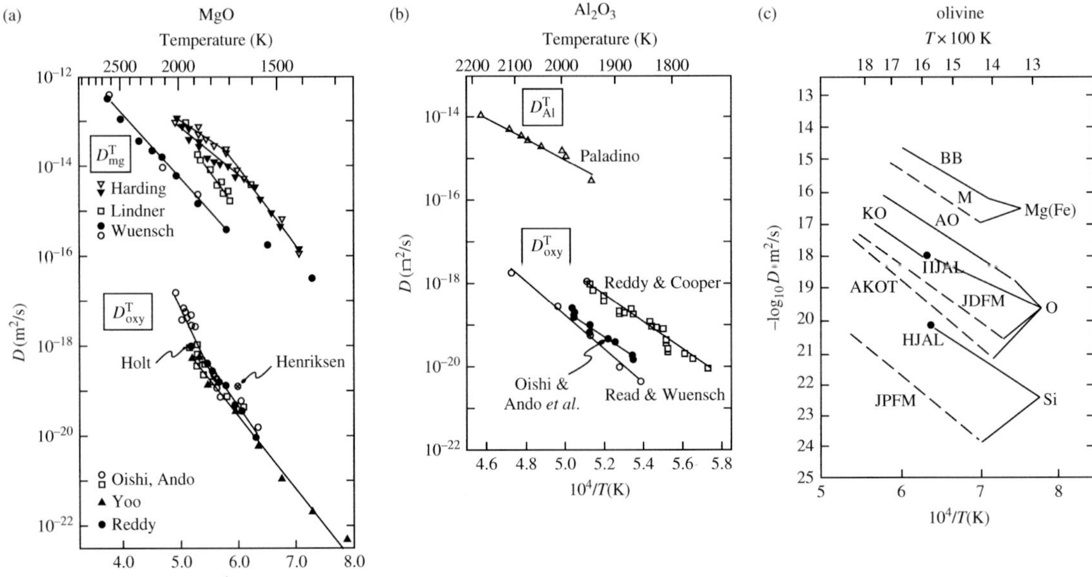

FIGURE 8.2 Self-diffusion coefficients of several ionic species in single crystals of MgO, Al_2O_3 and olivine: (a) MgO (ANDO, 1989), (b) Al_2O_3 (ANDO, 1989), (c) olivine (KARATO, 1989a).

case is found in Al_2O_3, that is diffusion of Al is faster than that of O through the bulk. However, diffusion of O is significantly enhanced by the presence of grain boundaries whereas that of Al is not. Consequently, when the grain size is small, diffusion of O will become faster than that of Al (e.g., GORDON, 1973; CANNON and COBLE, 1975). In Mg_2SiO_4 olivine, diffusion of Mg is the fastest, and the diffusion of Si is the slowest through the bulk of grains. Again the presence of grain boundaries may offset the relative ease of diffusion of various species in olivine. Note also that diffusion in most minerals is sensitive to the chemical environment. Consequently, the results such as shown in Fig. 8.2 need to be modified when the diffusion coefficients are compared in different chemical environments. The most notable factor related to the chemical environment is the fugacity of water (hydrogen). In many cases, the presence of water significantly enhances diffusion, but the details, such as the relative degree of enhancement for different species, are not well known. For the effects of water see also Chapter 10.

8.4. Self-diffusion, chemical diffusion

Above, we considered the diffusion of only one species in isolation. Such an assumption would be applicable to the case where an isotope of one species diffuses in the matrix. Because the isotopes of a given species (e.g., ^{17}O and ^{16}O, isotopes of oxygen with atomic mass of 17 and 16 respectively) have exactly the same (outer) electron distribution, their chemical characteristics are identical. Consequently, when a gradient in the concentration of one isotope is present, the motion of the isotope through the matrix does not cause any changes in energy, and there is no interaction between isotopes. Therefore, the diffusion of an isotope can be treated as diffusion caused solely by the concentration gradient without being affected by any chemical or electrostatic interactions and is therefore identical to the diffusion of that atomic species without any interactions. Consequently the diffusion coefficient of an isotope of a given atomic species is often referred to as the *self-diffusion coefficient* (or *tracer diffusion coefficient*) of the atom. Another important diffusion coefficient is the *chemical diffusion coefficient* that controls the diffusion of two different chemical species (e.g., Mg and Fe). In this latter case, the motion of one species affects the motion of another species because of chemical interactions. The self-diffusion coefficient is an intrinsic property of a given species in a given material. The chemical diffusion coefficient can be calculated from the self-diffusion coefficients of the atomic species involved and the nature of their mutual interactions. The rate of atomic diffusion involved in the plastic deformation can also be calculated from the self-diffusion coefficients by knowing the nature of interaction of all diffusing species.

Now let us consider a case where two atomic species (A and B) occupy the same lattice site. An example is $A = \text{Mg}$ and $B = \text{Fe}$. In these cases, diffusion of one species must be accompanied by that of another, and in general, there could be some chemical interactions between the two diffusing species. This type of diffusion, i.e., diffusion of two (or more) atomic species is referred to as *chemical diffusion*. The case such as the diffusion of Mg and Fe in (Mg, Fe)O or in $(Mg, Fe)_2SiO_4$ is a simple example. In these cases, diffusion would occur when the concentration of Mg (Fe) in a given material is not in chemical equilibrium with the surroundings. For example, when MgO is located next to FeO at high temperature, then Fe diffuses into MgO from FeO and Mg diffuses into FeO from MgO. Since both Mg and Fe have 2+ charge, the diffusion of these species does not change the electric field of the sample. To analyze such a problem, let us define the *chemical diffusion coefficient*, $\tilde{D}$, by

$$J = -\tilde{D}\frac{\partial c}{\partial x} \tag{8.23}$$

where c is the concentration of one species and J is the flux of that species. In calculating the flux, we must take into account the fact that the boundary between the two crystals moves as diffusion occurs. If we denote v as the velocity of motion of the boundary, and if there is no direct interaction between A and B atoms (i.e., $L^{AB} = 0$) then using (8.20),

$$J^A = -\frac{L^{AA}}{T}\frac{\partial(\mu^A - \mu^d)}{\partial x} + vc^A \tag{8.24a}$$

$$J^B = -\frac{L^{BB}}{T}\frac{\partial(\mu^B - \mu^d)}{\partial x} + vc^B. \tag{8.24b}$$

But because the concentration of defects is not affected by the diffusion of A and B atoms,[3] $(\partial\mu^d/\partial x) = 0$ and the

[3] This is valid when the diffusion of A and B atoms does not affect the concentration of atoms in other atomic sites. In an ionic solid such as (Mg, Fe)O, chemical diffusion of Mg and Fe will create imbalance of concentration of oxygen as a result of difference in the rate of diffusion of Mg and Fe. Consequently, if the oxygen diffusion rate is much slower than that of Mg and Fe (this is true in single crystals but not in most polycrystals), then the diffusion rate is adjusted to make Mg and Fe diffusion the same, i.e., the effective diffusion coefficient becomes the same as the one given by (8.36) (see CHEN and PETERSON, 1973).

total concentration of A and B atoms does not change with time ($\partial c/\partial t = 0$),

$$\frac{\partial c^A}{\partial t} + \frac{\partial c^B}{\partial t} = \frac{\partial c}{\partial t} = 0 = -\frac{\partial}{\partial x}(J^A + J^B) = \frac{\partial}{\partial x}\left(D^A \frac{\partial c^A}{\partial x} + D^B \frac{\partial c^B}{\partial x} - cv\right) \quad (8.25)$$

where we used the relation $L^{AA} = D^A c^A/R, L^{BB} = D^B c^B/R$ (equation (8.6)) and $\mu^{A,B} = \bar{\mu}^{A0,B0} + RT \log c^{A,B}$ (equation (8.4)). Therefore, $D^A(\partial c^A/\partial x) + D^B(\partial c^B/\partial x) - cv = C$ where C is a constant that is independent of x. The value of C depends on the choice of coordinate system, which can be set to $C = 0$ by choosing a coordinate that is fixed to a marker (i.e., $J^A + J^B = 0$), thus

$$v = \frac{1}{c}\left(D^A \frac{\partial c^A}{\partial x} + D^B \frac{\partial c^B}{\partial x}\right). \quad (8.26)$$

Inserting this into (8.24a) and noting $\partial c^A/\partial x = -\partial c^B/\partial x$, one obtains

$$\frac{\partial c^A}{\partial t} = \frac{\partial}{\partial x}\tilde{D}\frac{\partial c^A}{\partial x} \quad (8.27)$$

with

$$\tilde{D} = x^B D^A + x^A D^B \quad (8.28)$$

where $x^{A,B}$ are the fraction of sites occupied by an A (B) ion ($x^A + x^B = 1$; $x^A = c^A/c, x^B = c^B/c$), $D^{A,B}$ is the self-diffusion coefficient of an A(B) ion, and it is assumed that species A and B behave like an ideal solution. Equation (8.28) means that the effective diffusion coefficient for the chemical diffusion is simply the arithmetic mean of two diffusion coefficients (e.g., Chen and Peterson, 1973).

Problem 8.2

Derive equation (8.24). (Assume there is no direct interaction between A and B atoms (i.e., $L^{AB} = 0$).)

Solution

The flux of A, B atoms and the defects at that site (d) are given by

$$J^A = -\frac{L^{AA}}{T}\frac{\partial \mu^A}{\partial x} - \frac{L^{AB}}{T}\frac{\partial \mu^B}{\partial x} - \frac{L^{Ad}}{T}\frac{\partial \mu^d}{\partial x} + vc^A$$
$$J^B = -\frac{L^{BA}}{T}\frac{\partial \mu^A}{\partial x} - \frac{L^{BB}}{T}\frac{\partial \mu^B}{\partial x} - \frac{L^{Bd}}{T}\frac{\partial \mu^d}{\partial x} + vc^B$$
$$J^d = -\frac{L^{dA}}{T}\frac{\partial \mu^A}{\partial x} - \frac{L^{dB}}{T}\frac{\partial \mu^B}{\partial x} - \frac{L^{dd}}{T}\frac{\partial \mu^d}{\partial x} + vc^d.$$

Now, the total flux of the lattice site must be conserved, $J^A + J^B + J^d = v(c^A + c^B + c^d)$. Therefore, $L^{AA} + L^{AB} + L^{Ad} = 0$, $L^{BA} + L^{BB} + L^{Bd} = 0$ and $L^{dA} + L^{dB} + L^{dd} = 0$. Using the Onsager reciprocal relation ($L^{AB} = L^{BA}, L^{Ad} = L^{dA}, L^{Bd} = L^{dB}$),

$$J^A = -\frac{L^{AA}}{T}\frac{\partial(\mu^A - \mu^d)}{\partial x} - \frac{L^{AB}}{T}\frac{\partial(\mu^B - \mu^d)}{\partial x} + vc^A$$
$$J^B = -\frac{L^{BA}}{T}\frac{\partial(\mu^A - \mu^d)}{\partial x} - \frac{L^{BB}}{T}\frac{\partial(\mu^B - \mu^d)}{\partial x} + vc^B.$$

If one makes an assumption that the direct coupling between A and B atoms is negligible, then $L^{AB} = L^{BA} = 0$ and we get equation (8.24).

Now let us consider a case in which the diffusing atomic species have different electrostatic charges from the matrix material. An example is the diffusion of protons, $H^\bullet$, into a silicate or an oxide. The incorporation of a charged species must be accompanied with the diffusion of another species that has a charge with different sign. The chemical reaction involved may be written as,

$$2Fe_M^\bullet + H_2O = 2H^\bullet + \frac{1}{2}O_2 + 2Fe_M^\times. \quad (8.29)$$

That is, upon the reaction of water with a mineral, hydrogen atoms (protons) are dissolved and ferric iron ($Fe_M^\bullet$) is reduced to ferrous iron ($Fe_M^\times$). Defect species that have excess charges are $Fe_M^\bullet$ and $H^\bullet$. If the charge balance is maintained by these defects, the concentrations of these defects must satisfy $\delta[H^\bullet] = \delta[Fe_M^\bullet]$ where $\delta[X]$ indicates the change in concentration of a defect X. Such a situation can be generalized and we can consider a general chemical reaction such as (8.29) involving two charged species A and B with concentrations c^A and c^B with stoichiometric coefficients α and β (A is $H^\bullet$, B is $Fe_M^\bullet$, and $\alpha = 1, \beta = 1$ in the above example) respectively. The charge neutrality condition for such a reaction reads,

$$\frac{c^A}{c^B} = \frac{\alpha}{\beta}. \quad (8.30)$$

The flux of each charged species is given by,

$$J^A = -\frac{D^A c^A}{RT}\left(\frac{\partial \mu^A}{\partial x} + q^A \frac{\partial \phi}{\partial x}\right) + vc^A \quad (8.31a)$$

$$J^B = -\frac{D^B c^B}{RT}\left(\frac{\partial \mu^B}{\partial x} + q^B \frac{\partial \phi}{\partial x}\right) + vc^B \quad (8.31b)$$

where ϕ is the electrostatic potential caused by the local heterogeneity in charge distribution, υ is the velocity of the boundary and q^A and q^B are electric charges. Note that in contrast to the previous case, we include the interaction between A and B through the electrostatic potential. The charge neutrality condition demands,

$$\alpha q^A + \beta q^B = 0. \tag{8.32}$$

When a macroscopic electric field is not applied, then the total electric current must vanish, i.e.,

$$q^A J^A + q^B J^B = 0. \tag{8.33}$$

Using (8.30), (8.31) and (8.33), one obtains

$$\begin{aligned}\frac{\partial \varphi}{\partial x} &= -\frac{D^A - D^B}{D^A + (\alpha/\beta)D^B}\frac{1}{q^A}\frac{\partial \mu^A}{\partial x} \\ &= \frac{D^A - D^B}{D^A + (\alpha/\beta)D^B}\frac{1}{q^B}\frac{\alpha}{\beta}\frac{\partial \mu^B}{\partial x}\end{aligned} \tag{8.34}$$

where we used the relation $\partial\mu^A/\partial x = \partial\mu^B/\partial x$ that can be obtained from $\mu^k = \bar{\mu}_0^k + RT\log c^k$ (equation (8.4)) and (8.30).[4] The internal electric potential (called the *Nernst field*) is caused by the difference in diffusion coefficients of different charged species. Inserting this into (8.31a), and defining the effective diffusion coefficient, $\tilde{D}^A$, by

$$J^A = -\frac{\tilde{D}^A c^A}{R}\frac{1}{T}\frac{\partial \mu^A}{\partial x} \tag{8.35}$$

one gets

$$\tilde{D}^A = \frac{(\alpha + \beta)}{\alpha/D^A + \beta/D^B}. \tag{8.36}$$

The same equation can be derived for $\tilde{D}^B$ defined by $J^B \equiv -(\tilde{D}^B c^B/R)(1/T)(\partial\mu^B/\partial x)$ (i.e., $\tilde{D}^B = \tilde{D}^A$).

Physically, this equation means that because of the requirement of charge neutrality, reaction such as (8.29) requires diffusion of *both* positively and negatively charged species. Consequently, the slower of the two diffusing species controls the overall rate of reaction. For instance, the incorporation of hydrogen (proton) in Mg_2SiO_4 requires the counter diffusion of negatively charged species such as M-site vacancies and/or electrons.

Problem 8.3*

Consider the oxidation kinetics of (Mg, Fe)O. Write a chemical reaction equation for oxidation and find the effective diffusion coefficient for oxidation in terms of diffusion coefficients of atomic species involved.

Solution

As we learned in Chapter 5, the oxidation kinetics involves the attachment (or removal) of an oxygen molecule on (from) the surface and the resultant creation (destruction) of electron holes (by the change of oxygen molecule to an oxygen ion), $h^\bullet$, and M (metal)-site vacancies, V_M'', namely,

$$\frac{1}{2}O_2 = O_O^\times + V_M'' + 2h^\bullet.$$

Thus, at the surface, we will have excess V_M'' and $h^\bullet$ both of which have excess charges. Therefore the diffusion of these species must occur satisfying the charge balance so we use equation (8.32) ($A = V_M'', B = h^\bullet, \alpha = 1, \beta = 2$), to get

$$\tilde{D} = \frac{3}{2/D_{h^\bullet} + 1/D_{V_M''}} = \frac{3D_{h^\bullet}D_{V_M''}}{2D_{V_M''} + D_{h^\bullet}}.$$

Consequently, the kinetics of oxidation in this case are controlled by the diffusion of protons or M-site vacancies, the slower of these species.

Equation (8.36) can be extended to a reaction that involves three species that have different electrostatic charges (e.g., the reaction between MgO and SiO_2 to form Mg_2SiO_4). In such a case, one can choose the lattice positions of the slowest moving species in the resultant compound as a reference frame and obtain a result similar to (8.36) in which the two species are those with the fastest and the intermediate values of diffusion coefficients. Consequently, the rate of reaction involving three differently charged species is determined largely by the diffusion of a species that has the *intermediate* value of the diffusion coefficient (e.g., Kingery *et al.*, 1976).

8.5. Grain-size sensitive creep (diffusional creep, superplasticity)

8.5.1. Experimental observations and historical notes

When a deviatoric stress is applied to a polycrystalline material at low stresses and high temperatures, plastic deformation occurs, with a rate that is inversely

[4] The velocity, υ, in equation (8.31) disappears when equation (8.33) is used together with equations (8.30) and (8.32). In other words, the choice of the reference frame does not matter for the chemical diffusion as far as charge neutrality is maintained.

proportional to grain size and only weakly dependent on applied stress, namely,

$$\dot{\varepsilon} = A\frac{\sigma^n}{L^m}\exp\left(-\frac{E^* + PV^*}{RT}\right) \tag{8.37}$$

where A is a constant, σ is deviatoric stress, L is grain size, E^* and V^* are the activation energy and volume respectively and the parameters n and m take values of $n = 1{-}2$, $m = 1{-}3$. Both the stress and grain-size dependence of strain rate are distinct from another important mechanism of deformation, dislocation creep (Chapter 9), in which stress sensitivity is higher and strain rate does not depend on grain size. And as we will see, these two classes of deformation mechanisms are independent in most cases. Consequently, the deformation mechanisms described by equation (8.37) dominate in materials with small grain sizes and/or at low stresses (see section 9.9).

As is obvious from the sensitivity to grain size, grain-boundary processes play an important role in this type of deformation. This section reviews theoretical models and experimental observations of grain-size sensitive creep including (1) diffusional creep, (2) grain-boundary sliding and (3) "superplasticity."

Plastic deformation caused by stress-induced diffusional mass transport was first formulated by Nabarro (1948) and then more formally by Herring (1950). The basic physics of this process was well established in these papers. Interestingly, Nabarro (1948) suggested the possible importance of diffusional creep in Earth's interior in his first paper on diffusional creep.[5] Coble (1963) modified these theories by including mass transport along grain boundaries. The importance of grain-boundary sliding was first analyzed by Lifshitz (1963) in detail. Raj and Ashby (1971) made a detailed and elegant analysis of the interplay between diffusional mass transport and grain-boundary sliding. The role of the change in shape of grains due to grain-boundary energy-driven processes was discussed by Ashby and Verrall (1973), Ashby *et al.* (1978) and Spingarn *et al.* (1979), which provides a model for *superplasticity* in which large deformation occurs without significant change in grain shape. An alternative model for superplasticity was proposed by Mukherjee (1971) in which the role of dislocations created at grain boundaries is emphasized. The consequence of charge balance in mass transport in ionic compounds was discussed by Ruoff (1965). Jaoul (1990) proposed a new theory for diffusional mass transport in an ionic compound that predicts a behavior different from previous models. As I will discuss later, this new theory contains a fundamental error and the classic treatment by Ruoff (1965) (see also Stocker and Ashby, 1973) is still valid. Ashby (1969) and Arzt *et al.* (1983) discussed the importance of interface reaction in diffusional creep.

From a geological or geophysical point of view, Gordon (1965) was among the first to point out the importance of diffusional creep based on semi-quantitative estimates of viscosity. Elliott (1973) discussed the possible importance of diffusional creep based mainly on geological observations. The role of a pressure-solution creep, one type of diffusional creep involving a fluid-phase transport has been discussed including (Elliott, 1973; Rutter, 1976; Rutter, 1983; Green, 1984; Spiers *et al.*, 1990; Shimizu, 1994). Stocker and Ashby (1973) examined the relative importance of several mechanisms of deformation in Earth's upper mantle and concluded that the contribution from diffusional creep is in most cases not important. However, their conclusion was not based on direct experimental study. The first systematic study on diffusional creep in Earth materials was made by Karato *et al.* (1986). Based on the experimental results, Karato *et al.* (1986) and Karato and Wu (1993) suggested that diffusional creep may play an important role in deep portions of Earth's upper mantle. The importance of diffusional creep in the lower mantle was discussed by Karato *et al.* (1995b) and Yamazaki and Karato (2001b).

8.5.2. Basic theory of diffusional creep

Nabarro–Herring creep, Coble creep

When a differential stress is applied to a polycrystalline material, a heterogeneous stress state is created due to heterogeneous mechanical properties. Grain boundaries are weaker than the grains themselves, and hence mutual sliding of grains along grain boundaries occurs upon the application of stress. This grain-boundary sliding causes high stress concentration at grain boundaries such that stress states near grain boundaries will depend on the orientation of the grain boundaries relative to the stress. Because grain boundaries are a

[5] If the "corresponding state" argument (namely homologous temperature scaling, SK) could be extended to include crystals held at their melting points under external pressures comparable with their internal pressures, this estimate could be used as a lower (upper?, SK) limit to the viscosity of the lithosphere (the asthenosphere, SK) of the earth which is believed to be a few degrees below its melting point at all depths. This extension would require further discussion (Nabarro, 1948).

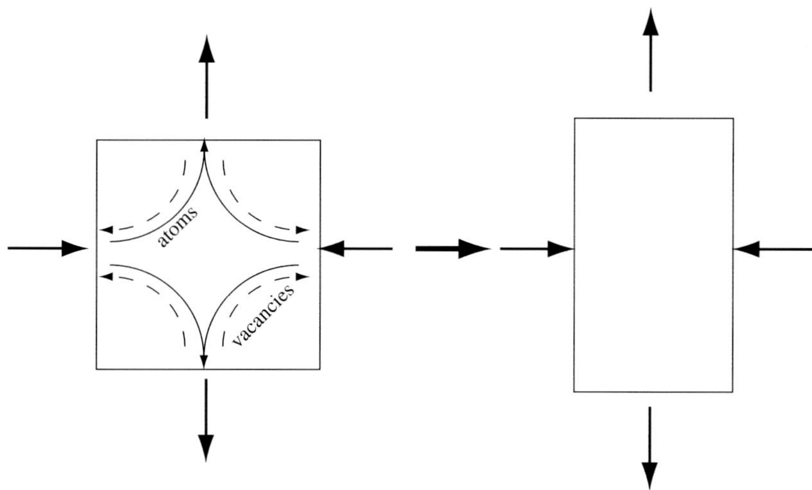

FIGURE 8.3 A conceptual drawing showing the process of diffusional creep.

good source/sink for defects,[6] the concentration of point defects near grain boundaries will be dependent upon the stress state at grain boundaries. The free energy for formation of defects at grain boundaries is different from that in the bulk. There is an extra work associated with the stress, $\sigma_n\Omega$. Thus,

$$c^d(\sigma) = c^d(0) \cdot \exp\left(\frac{\sigma_n\Omega}{RT}\right) \tag{8.38}$$

with $c^d(0) = c \cdot \exp(-G_f/RT)$.

Because the stress states are dependent upon the orientation of grain boundaries, this causes spatial distribution of vacancy concentration at the scale of grain size (L). Therefore the flux of defects will occur, causing macroscopic strain (Fig. 8.3). This strain will relax the stress at grain boundaries, leading to the steady-state distribution of stress (Raj and Ashby, 1971). This mass transport involves diffusion of both defects and atoms, so we must use the formula that we developed before, that is the flux of atoms in these cases is given by equation (8.20), i.e.,[7] $J^a = -(L^{aa}/T)(\partial(\mu^a - \mu^d)/\partial x) = -J^d$. Now using the relation between the chemical potential and the concentration of a given species, i.e., $\mu = \bar{\mu}^0 + RT\log c$ (equation (2.17)), and the relation between atomic diffusion coefficient and diffusion coefficient of defects, $Dc = D^d c^d$ (equation (8.15)), one gets,

$$\begin{aligned} J^a &= -\frac{L^{aa}}{T}\frac{\partial(\mu^a - \mu^d)}{\partial x} = D\frac{(c - c^d)}{c^d}\frac{\partial c^d}{\partial x} \\ &\approx D^d\frac{\partial c^d}{\partial x} = -J^d. \end{aligned} \tag{8.39}$$

By noting $\partial c^d/\partial x \approx [c^d(\sigma_n) - c^d(-\sigma_n)]/L$, where L is grain size, equation (8.39) becomes,

$$J^a \approx D^d c^d(0)\frac{\exp(\sigma_n\Omega/RT) - \exp(-\sigma_n\Omega/RT)}{L} \tag{8.40}$$

where we used relation (8.38) and the fact that the difference in concentration of defects occurs at a scale of grain size, L. Now, for a typical case, $\sigma_n\Omega/RT \ll 1$, hence,

$$J^a \approx \frac{2D^d c^d(0)}{L}\frac{\sigma_n\Omega}{RT} = \frac{2Dc}{L}\frac{\sigma_n\Omega}{RT} \tag{8.41}$$

The rate of change in the length of crystal is given by,

$$\delta\dot{L} = \frac{\delta\dot{V}}{L^2} = \frac{\Omega J^a}{N_A} \tag{8.42}$$

where $\delta\dot{V}$ is the volumetric rate of addition of atoms to a grain boundary ($\delta\dot{V} = J^a L^2(\Omega/N_A)$). Therefore,

$$\dot{\varepsilon} = \frac{\delta\dot{L}}{L} = \frac{\Omega J_a}{LN_A} = \frac{2Dc\Omega}{L^2 N_A}\frac{\sigma_n\Omega}{RT} = \frac{2D}{L^2}\frac{\sigma_n\Omega}{RT} \tag{8.43}$$

where we used a relation $c\Omega = N_A$. A more detailed analysis gives,

[6] This assumption is valid only under some conditions. Cases where this assumption is not valid will be discussed later in this chapter.

[7] Strictly speaking the chemical potential of a defect or atomic species in non-hydrostatically stressed solids can be defined only at the vicinity of grain boundaries or other extended defects where the concentrations of species can be modified. Consequently, the use of chemical potential above must be understood as an approximation $\partial\mu/\partial x \approx (\mu^+ - \mu^-)/L$ where μ^{+-} are the chemical potentials at the grain boundary with different orientations and L is grain size.

$$\dot{\varepsilon} = \alpha \frac{D}{L^2} \frac{\sigma \Omega}{RT} \tag{8.44}$$

with $\alpha \approx 14$ where $\sigma \equiv \sigma_1 - \sigma_3$. Creep due to diffusional mass transport through the bulk of grains is referred to as *Nabarro–Herring creep*. Note that the stress versus strain-rate relationship is linear and that the strain rate is inversely proportional to (the second power of) grain size.

In the original Nabarro–Herring creep model, diffusional mass transport in the bulk of crystals is considered. The extension of this model to the cases where high-diffusivity paths make an important contribution is straightforward. For instance, when diffusion along grain boundaries is also important, then we replace the diffusion coefficient with an effective diffusion coefficient given by (8.21) to get,

$$\dot{\varepsilon} = \alpha \frac{1}{L^2} \left(D^V + \frac{\pi \delta}{L} D^B \right) \frac{\sigma \Omega}{RT}. \tag{8.45}$$

If $(\pi\delta/L)D^B \gg D^V$, i.e., at small grain size and/or at low T, then,

$$\dot{\varepsilon} = \alpha \pi \frac{\delta D^B}{L^3} \frac{\sigma \Omega}{RT}. \tag{8.46}$$

Thus in these cases, the grain-size dependence of strain rate changes from $\dot{\varepsilon} \propto 1/L^3$ at small grain size to $\dot{\varepsilon} \propto 1/L^2$ at coarser grain size. Note that strain rate in this case is a linear function of stress, but the strain rate is more sensitive to grain size than that in the Nabarro–Herring creep. This type of creep mechanism is referred to as the *Coble creep* (COBLE, 1963). As we will learn later, in diffusional creep in a compound, the variation in grain-size dependence of strain rate with grain size can be different from this simple trend. A similar case is also found where the interface reaction controls the rate of deformation. Typical results of diffusional creep are shown in Fig. 8.4.

The geometry of deformation by diffusional mass transport is irrotational: the directions of principal strain are parallel to those of applied stress. This is in contrast to deformation due to a dislocation motion that is rotational (simple shear) constrained by the crystallographic nature of slip (see Chapter 5). Consequently deformation due to diffusional creep does not usually result in the lattice-preferred orientation (see Chapter 14, see also BONS and DEN BROK (2000) for an exceptional case).[8]

[8] However, when diffusion is highly anisotropic, then a certain degree of lattice-preferred orientation can develop. BONS and DEN BROK (2000) presented a case in which lattice-preferred orientation develops during pressure-solution creep when the rate of dissolution/precipitation is strongly anisotropic.

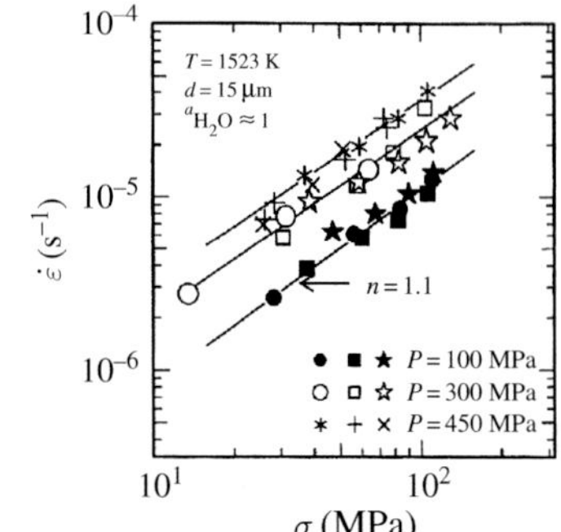

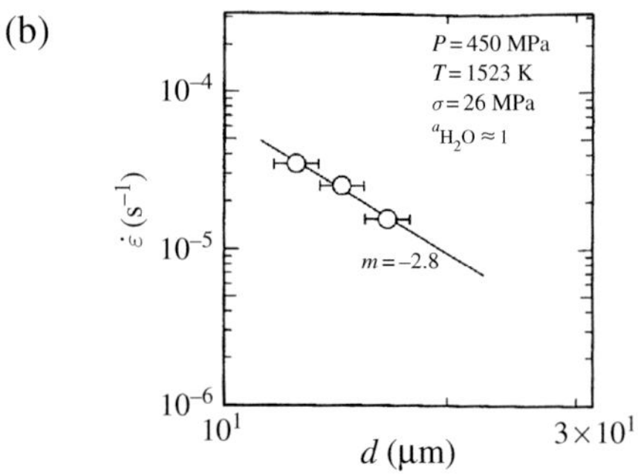

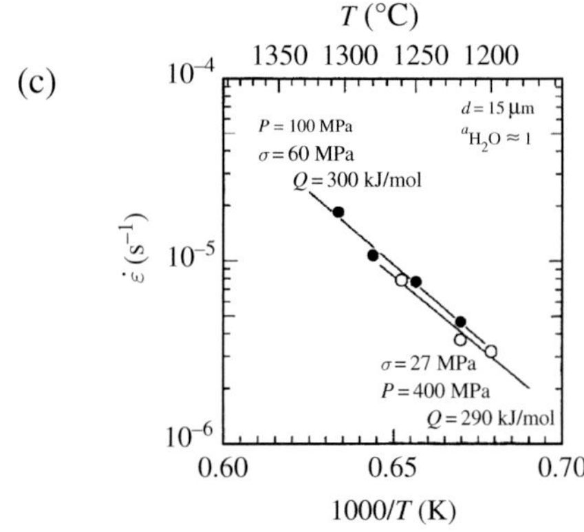

FIGURE 8.4 Experimental results showing evidence for diffusional creep in olivine under water-saturated conditions. (a) The stress dependence of the strain-rate relationship, (b) The grain-size dependence relationship, and (c) The temperature dependence (where d is the grain size, Q is the activation enthalpy, data from MEI and KOHLSTEDT, 2000a).

Diffusional creep in a compound

In ionic solids, many species must diffuse to achieve mass transport. For instance, in MgO, we must consider diffusion of Mg as well as O. Diffusion of which species controls the rate of deformation? Let us consider a compound $A_{\alpha_1} A_{\alpha_2} A_{\alpha_3} \cdots$. If one species diffuses faster than others, then this species will assume excess concentration at grain boundaries that gives rise to a "counter effect" to retard diffusion of that species by for example the effects of electrostatic field.

In a multi-component system, the relation (8.39) should be extended to include the interaction of different diffusing species. To simplify the analysis, let us assume that this interaction is solely through electrostatic field.[9] Then one has,

$$J_i = -\frac{D_i c_i}{RT}\left[\frac{\partial(\mu_i - \mu_i^d)}{\partial x} + q_i\frac{\partial \phi}{\partial x}\right] \tag{8.47}$$

where J_i is the flux of ith species, c_i is the number of atoms (of the ith species) per unit volume, D_i is the diffusion coefficient, μ_i^d is the chemical potential of the vacancies at the lattice site, μ_i is the chemical potential, q_i is the electrostatic charge of the ith species and ϕ is the electrostatic potential caused by the diffusion of ionic species.

The charge balance requirement reads

$$\sum_i \alpha_i q_i = 0. \tag{8.48}$$

At a steady state, the stoichiometry must be preserved,[10] and hence

$$c_i = \alpha_i c \tag{8.49}$$

and

$$J_i = \alpha_i J. \tag{8.50}$$

Note that equations (8.48) and (8.50) guarantee that there is no net electric current,

$$\sum_i q_i J_i = 0. \tag{8.51}$$

From (8.47) and (8.50),

$$-\frac{1}{RT}\left[\frac{\partial(\mu_i - \mu_i^d)}{\partial x} + q_i\frac{\partial \phi}{\partial x}\right] = \frac{J}{D_i c}. \tag{8.52}$$

Using the relation (8.4), the chemical potential term can be written as $(1/RT)\partial(\mu_i - \mu_i^d)/\partial x \approx -(1/c_i^d)\cdot(\partial c_i^d/\partial x)$ (see (8.39)). The spatial variation of defect concentration comes from the spatial variation of normal stress at grain boundaries (equation 8.38). Since the stoichiometry is preserved by assumption, the generation or destruction of defects at grain boundaries must occur in such a way that the change in defect concentration satisfies the relation similar to (8.39), i.e.,

$$c_i^d = \alpha_i c^d. \tag{8.53}$$

Therefore

$$\frac{1}{RT}\frac{\partial(\mu_i - \mu_i^d)}{\partial x} = \frac{\partial \log(c_i/c_i^d)}{\partial x} = \frac{\partial \log(c/c^d)}{\partial x} \equiv \frac{1}{RT}\frac{\partial(\tilde{\mu} - \tilde{\mu}^d)}{\partial x} \tag{8.54}$$

where $\tilde{\mu}$ and $\tilde{\mu}^d$ are the chemical potential of a compound and of a defect group of a compound (e.g., Schottky defect). Equation (8.54) means that the driving force for diffusion caused by the concentration gradient is the same for all species.[11] Operating $\sum_i \alpha_i$ to equation (8.52) and using (8.48), one can eliminate the electrostatic potential to get

$$-\frac{1}{RT}\frac{\partial(\tilde{\mu} - \tilde{\mu}^d)}{\partial x}\sum_i \alpha_i = \frac{J}{c}\sum_i \frac{\alpha_i}{D_i}. \tag{8.55}$$

Now we define the effective molecular diffusion coefficient for diffusional creep in a compound, $\tilde{D}$, by

$$J \equiv \frac{\tilde{D} c}{RT}\frac{\partial(\tilde{\mu} - \tilde{\mu}^d)}{\partial x} \tag{8.56}$$

hence,

$$\tilde{D} = \frac{\sum_i \alpha_i}{\sum_i (\alpha_i/D_i)}. \tag{8.57}$$

Thus, the creep constitutive relation for a compound is given by the same equation as (8.44), the atomic volume being replaced with a molar volume and the diffusion coefficient being replaced with the one given by equation (8.57). For example, for MgO, the effective diffusion coefficient is

[9] The following derivation is similar to the derivation of chemical diffusion coefficient for the case of diffusion of different species with different electrostatic charges. The only difference is that for deformation the driving force for diffusion is the chemical potential gradient of vacancies (defects), $(1/RT)\ \partial(\mu_i - \mu_i^d)/\partial x \approx -(1/c_i^d)\ (\partial c_i^d)/\partial x$.

[10] This assumption can be challenged when the mechanical work done by stress becomes comparable to the work needed to cause chemical decomposition (see the later part of this chapter).

[11] Jaoul (1990) obtained a different relation,

$$\frac{1}{RT}\frac{\partial \tilde{\mu}}{\partial x} \approx -\frac{\alpha^j}{c}\frac{\partial c_j^d(\sigma)}{\partial x} \propto c_j^d(0)$$

where c_j^d is the concentration of the *most abundant defect* and α^j is the stoichiometric coefficient of that atomic species. Therefore, in his theory, the most abundant defect (M-site vacancy in olivine) also plays a special role in creep. However, his treatment is incorrect because he made an approximation to ignore all other defects in the chemical potential *before* taking the derivative of chemical potential with respect to the space coordinate: although one type of defect may dominate in the chemical potential, it may not dominate in the *gradient* of chemical potential, and in fact it does not because of the requirement of preservation of stoichiometry (equation (8.54)).

$$\tilde{D} = \frac{2}{1/D_{Mg} + 1/D_O} = \frac{2D_{Mg}D_O}{D_{Mg} + D_O}$$

and for olivine Mg_2SiO_4,

$$\tilde{D} = \frac{7}{\frac{2}{D_{Mg}} + \frac{1}{D_{Si}} + \frac{4}{D_O}} = \frac{7D_{Mg}D_{Si}D_O}{2D_{Si}D_O + D_{Mg}D_O + 4D_{Mg}D_{Si}}.$$

The physical meaning of equation (8.57) is that because of the assumption that the stoichiometry must be preserved during creep, the slowest diffusing species controls the overall rate of creep.

How valid is the assumption of conservation of stoichiometry? This assumption must be very accurately true for a binary ionic solid such as MgO and Al_2O_3, because changing the ratio of number of cations and anions requires a huge energy ($\Delta h = 608.5$ kJ/mol for MgO; $\Delta h = 1693.4$ kJ/mol for Al_2O_3 (Navrotsky, 1994)) compared to the work done by the stress, $\sigma\Omega$ (0.05–50 kJ/mol for $\sigma = 1$–1000 MPa, $\Omega \sim (20$–$40) \times 10^{-6}\,m^3$/mol). In contrast, the energy associated with changing the stoichiometric ratio in a ternary system such as Mg_2SiO_4 is much less. For instance, the decomposition of Mg_2SiO_4 (olivine) into $2MgO + SiO_2$ is associated with only $\Delta h = 60.7$ kJ/mol and $Mg_3Al_2Si_3O_{12}$ (pyrope garnet) into $3MgO + Al_2O_3 + 3SiO_2$ is $\Delta h = 74.2$ kJ/mol (Navrotsky, 1994). The large difference between the energies associated with two types of decomposition is due to the fact that the decomposition of a binary compound into ionic species includes electrostatic energy whereas the decomposition of a ternary compound into several binary oxides does not involve electrostatic energy. Consequently, if a high enough stress is applied, then stress-induced decomposition of material can occur in a ternary compound. According to Dimos *et al.* (1988), this critical stress is dependent on the difference in diffusion coefficients between different species and is given by

$$\sigma_c \approx \frac{|\Delta h|}{\Omega}\left|\frac{\tilde{D}}{\Delta D}\right| \tag{8.58}$$

where ΔD is the difference in diffusion coefficients. $\sigma_c \approx 100$–1000 MPa for most geological materials. Ozawa (1989) reported evidence for stress-induced chemical segregation in the spinel grains in a naturally deformed spinel lherzolite.

When multiple diffusion paths are present, diffusion coefficient in equation (8.57) must be replaced with that of effective diffusion coefficient for individual ion. Thus, the diffusion coefficient of each species in equation (8.57) must be replaced with

$$D^k_{eff} = D^k_V + \sum_i A^k_i D^k_i \tag{8.59}$$

where D^k_V is the volume diffusion coefficient of the kth species, A^k_i is the non-dimensional geometrical factor for the ith high-diffusivity path ($A^k_i = \pi\delta/L$ for grain-boundary diffusion), and D^k_i is the diffusion coefficient of the kth species along the ith high-diffusivity path. The point here is that the diffusion of a given species along different paths is independent (parallel processes), whereas the diffusion of different species are dependent (sequential processes).

Problem 8.4*

Consider the effects of grain-boundary diffusion on the constitutive law for a polycrystalline MgO. Experimental studies show that the diffusion of oxygen is much slower than that of magnesium through the bulk, whereas the diffusion of oxygen is significantly enhanced by the presence of grain boundaries, but the effects of grain boundaries are negligible for the diffusion of magnesium. Discuss how the grain-size sensitivity of strain rate changes with grain size.

Solution

The effective diffusion coefficients of Mg and O are given by $D^{Mg}_{eff} = D^{Mg}_V$ and $D^O_{eff} = D^O_V + (\pi\delta/L)D^O_B$. The effective diffusion coefficients of Mg and O are plotted as a function of grain size, L (Fig. 8.5). One can recognize three regimes. In the largest grain-size regime, i.e.,

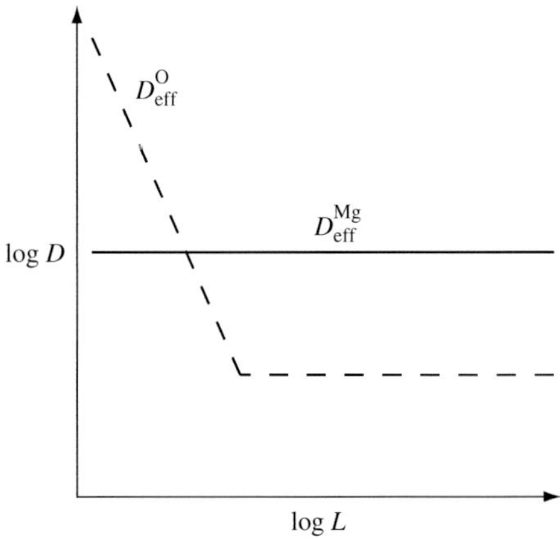

FIGURE 8.5 A diagram showing the variation of effective diffusion coefficients of the cation (Mg) and the anion (O) in oxides with grain size.

$$\frac{\pi\delta}{L}D_B^{\rm O} \ll D_V^{\rm O} \ll D_V^{\rm Mg},$$

$$\tilde{D} = \frac{2D_V^{\rm Mg}(D_V^{\rm O} + (\pi\delta/L)D_B^{\rm O})}{D_V^{\rm Mg} + D_V^{\rm O} + (\pi\delta/L)D_B^{\rm O}} \approx 2D_V^{\rm O}.$$

In the intermediate grain-size regime,

$$D_V^{\rm O} \ll \frac{\pi\delta}{L}D_B^{\rm O} \ll D_V^{\rm Mg},$$

$$\tilde{D} = \frac{2D_V^{\rm Mg}(D_V^{\rm O} + (\pi\delta/L)D_B^{\rm O})}{D_V^{\rm Mg} + D_V^{\rm O} + (\pi\delta/L)D_B^{\rm O}} \approx 2\frac{\pi\delta}{L}D_B^{\rm O}.$$

In the smallest grain-size regime,

$$\frac{\pi\delta}{L}D_B^{\rm O} \gg D_V^{\rm Mg} \gg D_V^{\rm O},$$

$$\tilde{D} = \frac{2D_V^{\rm Mg}(D_V^{\rm O} + (\pi\delta/L)D_B^{\rm O})}{D_V^{\rm Mg} + D_V^{\rm O} + (\pi\delta/L)D_B^{\rm O}} \approx 2D_V^{\rm Mg}.$$

Consequently, the grain-size dependence of strain rate changes with grain size from $\propto 1/L^2$ for the largest grain size, to $\propto 1/L^3$ for the intermediate grain-size regime and finally, $\propto 1/L^2$ for the smallest grain-size regime. Note that this trend is different in a simple case, where the grain-size dependence of strain rate changes from $\propto 1/L^2$ for large grain size to $\propto 1/L^3$ for small grain size.

Interface reaction-controlled creep

Diffusional creep involves two processes, namely the development of concentration gradient of defects at the grain-scale as a result of heterogeneity of stress, and the resultant diffusion of atoms (defects) from one region of grains to others. Consequently, to complete the process of deformation by diffusional creep, one needs both the interfacial reaction to establish concentration gradient and resultant diffusional flow. In the previous analysis of diffusional creep, we have made an assumption that the grain boundaries act as perfect source or sink for defects and the interface reaction is faster than diffusion. This assumption may not be valid under some conditions.

The equivalent strain rate corresponding to surface reaction (i.e., the strain rate corresponding to an infinite diffusion coefficient) is given by ASHBY (1969) and ARZT *et al.* (1983),

$$\dot{\varepsilon}_{\rm reac} = A\frac{(\sigma - \sigma_{\rm th})^n}{L} \tag{8.60}$$

where A is a constant, $n = 1$–2 and $\sigma_{\rm th}$ is the threshold stress that depends on the structure of the grain boundaries. The threshold stress exists in the interface reaction, because the creation or destruction of defects involves the change in the length of boundary dislocations. In the models of ARZT *et al.* (1983) and ASHBY (1969), it is determined by the balance between the dislocation line tension (see Chapter 5) and the external stress, namely,

$$\sigma_{\rm th} \approx \frac{\mu b_b}{2L} \tag{8.61}$$

where μ is the shear modulus (please do not confuse this with the chemical potential), $b_b \approx b/3$ is the length of the Burgers vector of grain-boundary dislocations ($\sigma_{\rm th} \sim 10^3$–$10^5$ Pa for typical minerals; $\mu = 10^{11}$ Pa, $b_b = 0.2$ nm, $L = 0.1$–10 mm). Because the interface reactions to create or destroy defects and the diffusional mass transport must occur sequentially, the overall strain rate is given by,

$$\dot{\varepsilon} = \frac{1}{1/\dot{\varepsilon}_{\rm diff} + 1/\dot{\varepsilon}_{\rm reac}} \tag{8.62}$$

where $\dot{\varepsilon}_{\rm diff}$ is strain rate due to diffusional creep (equation (8.45)). Interface control can become important in cases where diffusional mass transport is easy. One example is diffusional creep through fluid-phase mass transport, i.e., pressure-solution creep. SPIERS *et al.* (1990) reported evidence for interface reaction-control in pressure-solution creep in NaCl.

Pressure-solution creep

When a fluid phase is present, diffusional mass transport through a fluid phase makes an important contribution. At grain boundaries under compression, solid materials have higher free energy than those at the tensional side, and consequently materials are dissolved at the compressional side and precipitated at the tensional side of grain boundaries. This mechanism of creep is similar to that of Coble creep and is often referred to as *pressure-solution creep* or *dissolution–precipitation creep*. Evidence of this type of deformation is found in sedimentary and metamorphic rocks (see e.g., ELLIOTT, 1973; DICK and SINTON, 1979; RUTTER, 1983).

Although the mechanisms of pressure-solution creep are similar to that of Coble creep and hence its constitutive relation is similar, there are a few points that are worth mentioning. First, the role of the fluid phase on deformation depends on the geometry of the fluid phase in an aggregate. In order for a fluid phase to have a large effect on deformation, it must wet the boundary efficiently. Experimental studies on the wetting behavior of aqueous fluids showed that in most cases the dihedral angles are larger than 60° and aqueous fluids do not wet the grain boundaries of silicate minerals such as quartz or olivine at static equilibrium (e.g., WATSON and BRENAN, 1987). Consequently, the effective wetting, which is required for the enhanced deformation by fluid-phase mass transport, will occur only under special conditions where rocks are highly fractured, or in rocks that are undergoing *diagenesis* (compaction at the presence of fluids). In other words, *pressure-solution creep occurs effectively only in rocks that do not have equilibrium microstructure*. URAI *et al.* (1986b) emphasized the role of stress to wet the grain boundaries that are otherwise devoid of fluids.

Second, given an assumption that most grain boundaries are wetted, then a conceptual question is how can one maintain deviatoric stress at grain boundaries when a fluid phase is present on grain boundaries? If all the grain boundaries contain a fluid phase, it is not straightforward to maintain deviatoric stress. Several models have been proposed to solve this paradox. (1) A fluid film may support a finite deviatoric stress when the film is very thin. There are some reports that show that a very thin fluid film (say on the order of nm thick) can have a much higher viscosity than a bulk of the fluid (HORN *et al.*, 1989) (see also NAKASHIMA, 1995; NAKASHIMA *et al.*, 2004), and in that case a finite stress can be supported (RUTTER, 1976). (2) Grain boundaries may assume an *island structure*, that is grain boundaries with a fluid phase may be composed of patches that connect two grains and between the patches a fluid phase is present (RAJ and CHUNG, 1981). (3) The deviatoric stress is supported by solid–solid contacts, and the dissolution actually occurs not at the contacts but at the regions near the contacts due to the increased free energy caused by higher strain energy due to higher density of dislocations (GREEN, 1984; TADA and SIEVER, 1986; TADA and SIEVER, 1987). The first model is at best speculative, and if indeed a very thin film with high viscosity is required for pressure-solution creep to occur, then the effectiveness of pressure-solution creep becomes questionable. The presence of a thin film of fluid along grain boundaries is sometimes proposed (e.g., DRURY and FITZ GERALD, 1996), but evidence against such fluid phases is also reported (HIRAGA *et al.*, 2002). The second model is also speculative and neither a theoretical basis nor experimental support for such a structure is firmly confirmed. The third model is a viable one that has some experimental support (TADA and SIEVER, 1987).

Third, because fluid-phase mass transport is fast, the kinetics of deformation by pressure-solution creep is often controlled by the interface reaction. This is particularly the case where the interface reaction is sluggish. Spiers and his colleagues have conducted extensive experimental and theoretical studies to identify the mechanisms of pressure-solution creep, and emphasized the importance of interface reaction-control. For a review see SPIERS *et al.* (2004).

Fourth, a mixture of a fluid and a solid is not usually stable. When such a mixture is in a gravitational field, compaction occurs due to the difference in density. Compaction itself may occur by pressure-solution creep. Consequently, deformation by pressure-solution creep is not generally steady state.

Fifth, which is related to the fourth point above, deformation by pressure-solution creep may occur in a system in which a fluid phase is not confined in a system (an open system). In this case deformation will occur through the removal of material by a fluid without being associated with precipitation.

8.5.3. Small-strain and large-strain phenomena

The analysis presented above is for steady-state creep. Models of steady-state diffusional creep assume (1) the local stress at grain scale is a steady state, and (2) the grain morphology is a steady state. These assumptions can be violated under some important geological/geophysical conditions. This section discusses two cases, first diffusional creep at small strains (strains of the order of elastic strain), and second diffusional creep at very large strains (superplasticity). The first case is important in the deformation associated with post-glacial rebound (see also Chapter 18), and the second case is important in many geological cases.

Grain-boundary sliding and "superplasticity"

Diffusive mass transport changes the shape of individual grains. To maintain the coherency at grain boundaries, grains must slide past each other. This can be shown as follows (see Fig. 8.6).

The velocity of mass transport at a point in grain 1 is given by $v_1 = v_1^0 + D\nabla c_1$ where v_1^0 is the velocity of the center of gravity of grain one and the second term represents the displacement due to diffusive flow. If we consider the tangential component of velocity, then $\nabla = \nabla_t$. At grain boundaries, $c_1 = c_2$. Thus,

$$v_2 - v_1 = v_2^0 - v_1^0 + D(\nabla c_2 - \nabla c_1) = v_2^0 - v_1^0. \quad (8.63)$$

Therefore, the self-consistent change of shape of grains and their displacement in the process of diffusional creep are necessarily accompanied by slip along grain boundaries (LIFSHITZ, 1963).

The interplay between grain-boundary sliding and diffusional creep was analyzed by RAJ and ASHBY

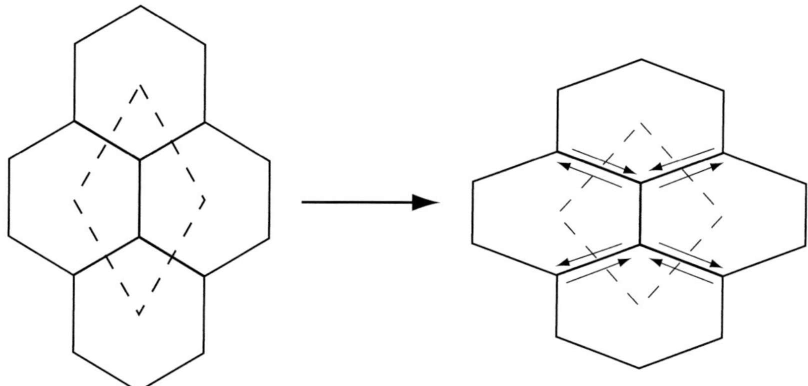

FIGURE 8.6 Grain-boundary sliding is necessarily associated with diffusional creep (Lifshitz sliding).

(1971). Grain-boundary sliding and diffusional mass transport are both needed for continuing deformation. If plastic flow (either by diffusion or dislocation creep) does not occur, then grain-boundary sliding will increase the stress in the surrounding grains due to elastic strain and sliding eventually stops. The continuing deformation to large strain is only possible with the help of plastic flow. Consequently, if one has a constitutive relation for diffusional mass transport ($\dot{\varepsilon}_{\text{diff}}$) or dislocation creep ($\dot{\varepsilon}_{\text{disl}}$) corresponding to a case where there is no resistance to grain-boundary sliding and deformation due to grain-boundary sliding ($\dot{\varepsilon}_{\text{gbs}}$) with no resistance to plastic deformation, then the overall rate of deformation ($\dot{\varepsilon}$) is given by,

$$\dot{\varepsilon}^{-1} = \left(\dot{\varepsilon}_{\text{diff}} + \dot{\varepsilon}_{\text{disl}}\right)^{-1} + \dot{\varepsilon}_{\text{gbs}}^{-1}. \tag{8.64}$$

This is because diffusion and dislocation creep are independent, but these processes and grain-boundary sliding are dependent. Results supporting this type of formulation have been reported by Goldsby and Kohlstedt (2001).

Models of superplasticity

Superplasticity refers to a phenomenon in which large strain deformation is achieved without strain localization (or failure), usually in tension. As such it is defined by phenomenological criteria without specifying microscopic mechanisms (e.g., Nieh *et al.*, 1997). The interest in superplasticity is mainly from an engineering point of view: one can form materials to various shapes under superplastic conditions.

There have been some interests on superplasticity in the geological community. However, the significance of superplasticity in the geological sciences is different from that in engineering. I will first provide a brief historical note on engineering literature on superplasticity and will discuss geological significance.

In short, superplasticity refers to deformation in which large strains can be achieved without necking instability. As such the term superplasticity is relevant only for tensional deformation. Deformation in geology involves not only tensional deformation but also compressional or shear deformation. Therefore, superplasticity as defined above is not very relevant to geology (one can extend the discussion on stability to other deformation geometries, but instability caused by geometrical reasons will not occur for simple shear).

In essence, the characteristics of superplastic deformation can be summarized as follows:

(1) deformation is stable to large strains;
(2) grains remain nearly equiaxial after large strains; and
(3) the stress exponent is small ($n = 1$–2).

The stability of deformation in tension is analyzed in Chapter 16 following the analysis of Hart (1967), which indicates that the stable deformation in tension is possible when n is small (strictly speaking $n \leq 1$). As such the stability of deformation to large strains is nothing but a result of a small stress exponent, and any mechanisms that have a small stress exponent can result in "superplastic" behavior.[12] As a consequence of stability of deformation, deformation to extremely large strains is often observed, and in these cases, a marked evidence of *grain-boundary sliding* is usually found. Therefore the most important issue regarding

[12] Deformation with a small stress exponent ($n = 1$) can occur during a phase transformation or deformation during thermal cycling. Large-strain, stable deformation caused by a phase transformation is sometimes referred to as *transformation plasticity* (for details see Chapter 15), and we will focus on superplasticity due to small grain size in this section (this is often referred to as *structural superplasticity*).

superplasticity is to explain the contribution of grain-boundary sliding in plastic deformation.

Ashby–Verrall model (Ashby and Verrall, 1973)

(a heterogeneous grain-switching model for large strain diffusional creep)

At large strains, the shape of grains will change, which results in unstable grain-boundary morphology. Then grain-boundary energy-driven processes to modify grain-boundary morphology will operate. In essence, "superplasticity" occurs, in this model, because of the interplay of deformation and grain-boundary energy-driven processes. Grain-boundary energy-driven processes tend to keep grain shape equant. If this rearrangement does not occur, then grain elongation will retard deformation (Green, 1970).

A grain switching event will occur at *large strains* when, due to *grain-boundary sliding*, two triple junctions meet (center of Fig. 8.7). At this point, a four-grain corner will be formed but it is unstable and will transform into a new configuration (right-hand side of Fig. 8.7). The essence of this grain-switching event is therefore the interface (grain-boundary) energy-driven processes. In other words, plastic deformation leads to an energetically unstable grain-boundary morphology (four-grain junction), which must be modified if grain-boundary migration occurs. Ashby and Verrall (1973) analyzed this process and obtained a constitutive relation that is similar to diffusional creep, but the magnitude of strain rate is larger than that of diffusional creep.

The rate of deformation involving grain neighbor-switching events could be controlled by various processes. Ashby and Verrall (1973) considered that grain-boundary sliding is relatively easy and therefore either the diffusional mass transport or the grain-boundary reaction can control the rate of deformation. The constitutive relation they derived is identical to that for diffusional creep except for a numerical factor.

Ashby *et al.* (1978) and Spingarn and Nix (1978) extended the above model to homogeneous deformation (Fig. 8.7b).

Mukherjee model (Mukherjee, 1971)

Grain-boundary sliding accommodated by dislocation motion may control the rate of deformation. These models (see e.g., Mukherjee, 1971) usually lead to a higher stress exponent, typically $n = 2$. In Mukherjee's

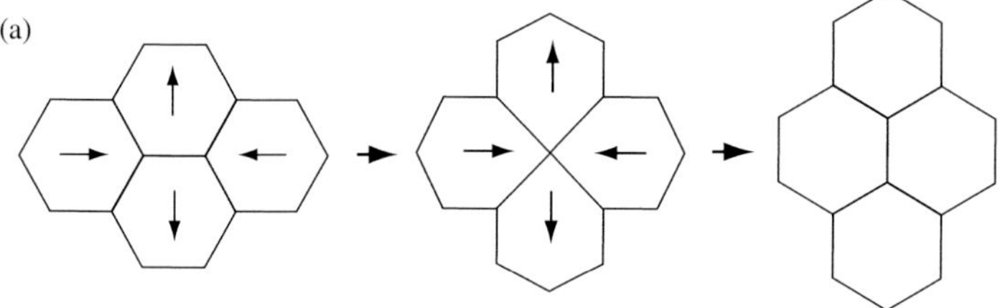

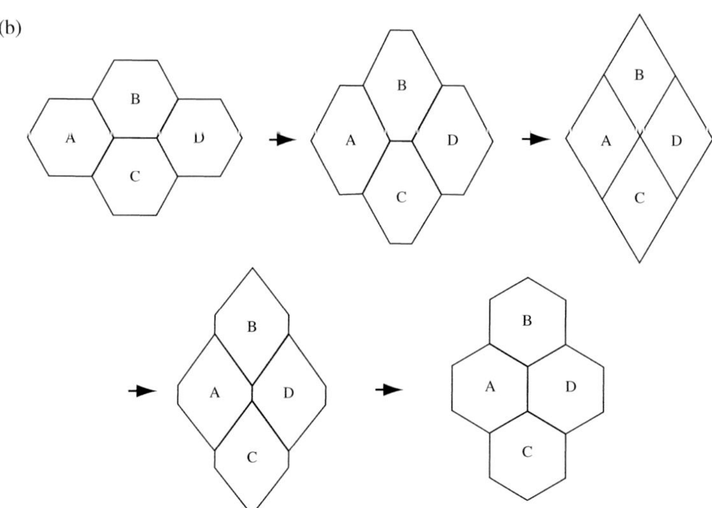

FIGURE 8.7 (a) A grain-switching event leading to superplasticity (Ashby and Verrall, 1973), (b) a homogeneous grain-switching model for superplasticity (after Ashby *et al.*, 1978).

model, deformation occurs by grain-boundary sliding and sliding is controlled by dislocation motion within the grains. Resistance to sliding is concentrated at ledges. This high stress generates dislocations. They pile up at opposite grain boundaries. The back-stress eventually prevents the ledge acting as a source of dislocations. Thus continuous deformation is possible only when these dislocations are removed by recovery. The recovery rate is proportional to stress acting on piled-up dislocations, which leads to $n = 2$ dependence of strain rate,

$$\dot{\varepsilon} \propto \frac{\sigma^2}{L^2}. \tag{8.65}$$

The creep constitutive relation is similar to (8.65), i.e., $\dot{\varepsilon} \propto \sigma^n/L^m$ with $n > 1$ and $m = 1$–3 is often reported (e.g., Goldsby and Kohlstedt, 2001; Nieh *et al.*, 1997; Hirth and Kohlstedt, 2003).

Note, however, that non-linear grain-size sensitive creep constitutive law can be obtained when deformation occurs under the conditions close to the dislocation–diffusional creep boundary. The total strain rate is the sum of strain rate due to diffusional creep and dislocation creep, $\dot{\varepsilon} = A_1(\sigma/L^m) + A_2\sigma^n$. If the deformation conditions are close to the boundary between dislocation and diffusional creep regimes, then the experimental results could be fitted to a single flow law of a form $\dot{\varepsilon} = A(\sigma^p/L^q)$ for a narrow range of stress with $1 < p < n$ and $0 < q < m$. In these cases, the stress exponent and grain-size exponent that one obtains have little physical significance.

Problem 8.5

Show that if a material deforms at the condition where contributions from diffusional creep and dislocation creep are nearly equal, then if the data are fitted to a flow law $\dot{\varepsilon} = A\sigma^p/L^q$, then one obtains $p \approx (1+n)/2$ and $q \approx m/2$ where n is the stress exponent ($\dot{\varepsilon} = A_2\sigma^n$) in the dislocation creep regime and m is the grain-size exponent ($\dot{\varepsilon} = A_1(\sigma/L^m)$) in the diffusional creep regime.

Solution

The flow law in this case can be written as $\dot{\varepsilon} = A_1(\sigma/L^m) + A_2\sigma^n$ where the first term is the contribution from diffusional creep and the second term from dislocation creep. Let us fit the data using a power-law equation $\dot{\varepsilon} = A(\sigma^p/L^q)$. Then $A_1(\sigma/L^m) + A_2\sigma^n = A(\sigma^p/L^q)$ and therefore the apparent stress exponent p is given by

$$p = \frac{\partial \log(A_1(\sigma/L^m) + A_2\sigma^n)}{\partial \log \sigma} = \frac{A_1(\sigma/L^m) + A_2 n\sigma^n}{A_1(\sigma/L^m) + A_2\sigma^n}.$$

Now at conditions near the boundary between dislocation and diffusional creep regimes, $A_1(\sigma/L^m) \approx A_2\sigma^n$. Inserting this relation into the above relation, one obtains $p \approx \frac{1}{2}(1+n)$. Similarly,

$$q = -\frac{\partial \log(A_1(\sigma/L^m) + A_2\sigma^n)}{\partial \log L} = \frac{mA_1(\sigma/L^m)}{A_1(\sigma/L^m) + A_2\sigma^n} \approx \frac{m}{2}.$$

Superplasticity (geological significance)

Superplasticity is characterized by a small stress exponent and significant grain-boundary sliding. These two points have the following geological/geophysical significance. First, a small stress exponent means that flow is more stable and time dependence of convection is weak in comparison to highly non-linear rheology (e.g., Christensen, 1989). Second, significant grain-boundary sliding has two important consequences. (i) When deformation involves significant grain-boundary sliding, the rate of deformation is sensitive to grain size. In this case, grain-size reduction can cause rheological weakening and shear localization. (ii) Significant grain-boundary sliding destroys the pre-existing fabric (lattice-preferred orientation), leading to very weak seismic anisotropy. Note, however, there is no satisfactory theory for how the fabric is destroyed by superplastic deformation.

Note that the grain-size sensitivity of deformation tends to result in *localization (and hence instability) of deformation* (as opposed to stable deformation that is the essence of superplasticity in the engineering context). The main reason for this apparent discrepancy lies in the fact that in realistic geological contexts, grain size is neither constant nor homogeneous, but it changes with time and space due to deformation and/or reactions (including phase transformations). As a result, a portion of rocks that has smaller grain size deform more and more grain-size reduction would occur leading to localization (for more details on the processes controlling grain size see Chapter 13).

The terms *superplasticity* and *diffusional creep* are used in a lot of the literature with some distinctions but they are used to describe basically the same phenomenon. Poirier (1985) stated that "it (diffusional creep) obviously does not (lead to superplasticity), for the only reason that the creep rate is very low (page 205)." This statement is misleading and not appropriate. It is not the rate of deformation in some

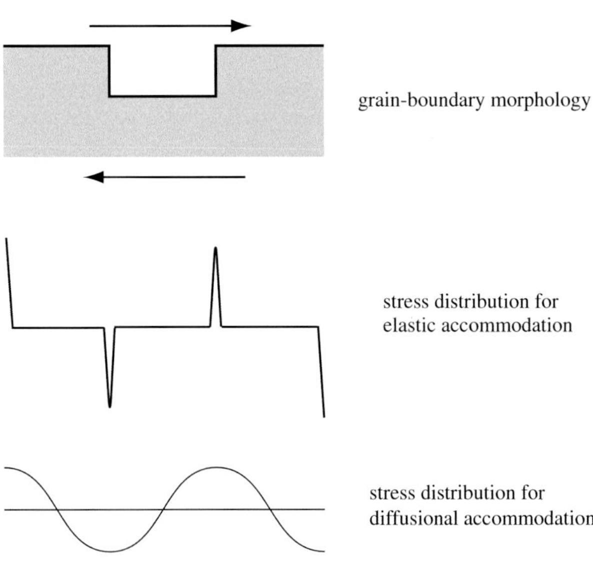

FIGURE 8.8 The stress distribution near grain boundaries corresponding to elastic or diffusional accommodation. Arrows indicate the shear direction. The stress distribution immediately after the application of stress is determined by elastic strain and has a large stress concentration at grain boundaries. This high stress concentration is relaxed after a certain time and at steady state the stress distribution is controlled by diffusive mass transport. As a result the initial strain rate is much faster than the steady-state strain rate and the steady state is attained after a relaxation time of $\sim \varepsilon_e/\dot{\varepsilon}_s$.

specific mechanisms that allows superplastic deformation. It is (i) the low-stress exponent and (ii) the grain-boundary sliding (or more appropriately grain switching) that allow stable large strain deformation (superplasticity). Therefore, in the case of diffusional creep, the classic model neglecting grain-switching events does not explain superplasticity. However, new models of (large-strain) diffusional creep such as RAJ and ASHBY (1971), ASHBY and VERRALL (1973), ASHBY *et al.* (1978) or SPINGARN and NIX (1978) in which diffusional creep *associated with grain-boundary sliding* or *grain switching* are modeled can naturally explain "superplasticity." In fact, ASHBY *et al.* (1978) and SPINGARN and NIX (1978) showed that a grain-switching event *necessarily occurs* at large strains during diffusional creep. Therefore it is appropriate to consider that diffusional creep at large strains that is associated with grain-switching events is one of the possible mechanisms of superplasticity.

Transient phenomena in diffusional creep (Lifshitz–Shikin theory)

The driving force for diffusional creep is the gradient in vacancy concentration at grain boundaries caused by the variation in local stress at grain boundaries. The local stress at grain boundaries is, in turn, determined by the nature of strain accommodation. Upon loading, stress distribution will initially be controlled by elastic deformation. Therefore initial creep rate will be controlled by the stress distribution corresponding to elastic deformation. However, diffusional mass transport will change the stress distribution. Thus transient behavior will occur. The change in stress distribution is schematically shown in Fig. 8.8 (after RAJ and ASHBY, 1971). Elastic accommodation results in a sharp stress concentration near corners, which is relaxed by diffusional mass transport.

Thus, the creep behavior can be characterized by initial high strain rate $\dot{\varepsilon}_0$ that gradually decreases to a steady-state value, $\dot{\varepsilon}_\infty$. Therefore the constitutive equation is given by,

$$\dot{\varepsilon} = \dot{\varepsilon}_0 + (\dot{\varepsilon}_\infty - \dot{\varepsilon}_0)\left[1 - \exp\left(-\frac{t}{\tau}\right)\right] \tag{8.66}$$

where τ is the relaxation time of stress at grain boundaries, which is in turn determined by creep rate. Thus we need to know two parameters: $\dot{\varepsilon}_\infty/\dot{\varepsilon}_0$ and relaxation time, τ. The relaxation time is the time to relax elastic strain by diffusive flow. Therefore it is given by LIFSHITZ and SHIKIN (1965)

$$\tau \sim \frac{\sigma}{E\dot{\varepsilon}} \sim \frac{L^2 RT}{D\Omega c E} \sim \frac{\eta}{E} \tag{8.67}$$

where E is the elastic modulus and η is the viscosity of the material. The experimental study by GORDON and TERWILLINGER (1972) showed $\dot{\varepsilon}_0 \sim 10\dot{\varepsilon}_\infty$ and the initial creep rate is linearly dependent upon stress.

The relaxation time is $\sim 10^{21}$Pa s$/10^{11}$Pa $\sim 10^{10}$ s $\sim 10^3$y for a typical mantle. For a high-viscosity region ($\eta \sim 10^{22}$ Pa s), it is $\sim 10^4$ y. The time-scale of post-glacial rebound is 10^3–10^4 y. Therefore the conclusion is that transient behavior is only marginally important for typical mantle or for soft regions, but it is significant for hard regions such as the deep lower mantle.

8.5.4. Several issues on diffusional creep

How to identify diffusional creep or superplasticity
In laboratory studies, the identification of diffusional creep can most clearly be made by determining the stress exponent ($n=1$) and the grain-size exponent ($m=2$–3). $n=1$ and $m=2$–3 will be the strong evidence for diffusional creep (diffusional creep accommodated by grain-boundary sliding) ($n=2$, $m=2$–3 will be observed with dislocation creep accommodated by grain-boundary sliding).

When there is strong evidence for large strain but grain shape remains nearly equant, this suggests an importance of grain-boundary sliding. However, other processes can also result in the absence of grain elongation. They include energy-driven grain-boundary migration and dynamic recrystallization (associated with dislocation creep). Therefore the absence of grain elongation cannot provide unequivocal evidence for superplasticity.

Inference of diffusional creep or superplasticity from naturally deformed rocks is more difficult. The strongest evidence for diffusional creep or superplasticity is the absence of strong lattice-preferred orientation despite large strain (Boullier and Gueguen, 1975; Behrmann and Mainprice, 1987). If there is strong evidence for large strain, yet lattice-preferred orientation is weak, then one may conclude that grain-boundary sliding (superplasticity) has made a significant contribution to strain. However, the correlation between the presence/absence of lattice-preferred orientation and superplasticity (i.e., grain-boundary sliding) is not entirely clear. There are strong cases where no lattice-preferred orientation develops or even the pre-existing lattice-preferred orientation is destroyed by superplastic flow (e.g., Edington *et al.*, 1976; Karato *et al.*, 1995b) but there are some reports showing strong lattice-preferred orientation during grain-size sensitive creep (e.g., Pieri *et al.*, 2001). It is likely that the strong lattice-preferred orientation observed in the latter study is caused by the processes of dynamic recrystallization (dislocation glide and/or grain-boundary migration) and the dominant deformation mechanism is not grain-boundary sliding (see Chapters 13 and 14 for more details). An exception to this is a case where there is strong anisotropy in diffusion and/or the rate of dissolution/precipitation. In these cases, a certain degree of lattice-preferred orientation may develop by deformation due to diffusional creep (Bons and Den Brok, 2000).

In some cases, evidence for diffusional mass transport can be identified from a "denuded zone" from which the concentration of certain elements are reduced (Poirier, 1985, p. 197). This occurs in certain alloys (or solid-solutions) in which the diffusion of certain atoms is much slower than others. In such a case, stress-induced diffusion results in an enhanced concentration of easier-to-diffuse elements at boundaries with a particular orientation relative to stress. Ozawa (1989) observed stress-induced chemical zoning in spinel in a deformed peridotite, similar to stress-induced (kinetic) demixing (Dimos *et al.*, 1988) providing evidence for diffusive mass transport. A similar observation includes "pressure shadow" in which the deposition of minerals at one side of the grains is observed in certain rocks that were deformed at the presence of a fluid phase (e.g., Elliott, 1973). However, there has been some controversy as to the implications of denuded zones for diffusional creep (e.g., Ruano *et al.*, 1993; Wolfenstein *et al.*, 1993; Bilde-Sorenson and Smith, 1994; Burton and Reynolds, 1994; Greenwood, 1994). Some authors pointed out discrepancies between experimental observations and simple models of diffusional creep (Ruano *et al.*, 1993; Wolfenstein *et al.*, 1993). These discrepancies are, however, probably the consequence of some complications in diffusional creep such as the imperfect action of boundaries as source/sink for defects, or some contributions from other processes such as grain-boundary sliding.

Four-grain junctions (Fig. 8.6) are evidence for grain-switching events that are unique to superplasticity (Ashby and Verrall, 1973; Karato *et al.*, 1998; Goldsby and Kohlstedt, 2001). Thus a frequent observation of four-grain junctions can be taken as evidence for superplasticity. However, the absence of four-grain junctions in naturally deformed rocks should *not* be considered as evidence *against* superplasticity. In naturally deformed rocks, post-deformational annealing is likely to have modified these small-scale microstructures.

Some issues on the extrapolation of experimental data on diffusional creep
Experimental observations on diffusional creep are limited to samples with a small grain size (in most cases ~10 μm or less). The typical grain size in Earth's

mantle is ~100 μm to a few mm (see Chapter 13), and therefore significant extrapolation is needed to apply experimental data to evaluate deformation in Earth's mantle (and crust). In many studies, experimental data on diffusional creep show Coble creep behavior, i.e., $m=3$ (e.g., HIRTH and KOHLSTEDT, 1995a; RYBACKI and DRESEN, 2000). However, the extrapolation of these data to coarser grain sizes in Earth assuming $m=3$ is not necessarily valid. As shown by GORDON (1973) and CANNON and COBLE (1975) (see also Problem 8.4), the grain-size exponent, m, for diffusional creep in ionic compounds probably changes as grain size changes. An extrapolation to a coarser grain size assuming $m=3$ leads to an underestimation of strain rates due to diffusional creep. Note also that if the grain-size exponent m changes as in the case of MgO and Al_2O_3, then it will also imply that the rate-controlling diffusing species changes with grain size. Since the sensitivity of the diffusion coefficient on the chemical environment such as water fugacity is different among different diffusing species (and also different diffusion mechanisms, see Chapter 10), it is possible that the sensitivity of diffusion creep on chemical environment (such as water fugacity) determined by laboratory experiments is different from that for natural conditions.

9 Dislocation creep

Plastic deformation can occur by a collective motion of atoms as crystal dislocations. Evidence for dislocation creep in Earth is abundant although other mechanisms such as diffusional creep (see Chapter 8) dominate under some conditions. The rate of deformation due to dislocation motion is proportional to dislocation density and velocity (the Orowan equation). In most cases dislocation density increases with applied stress and dislocation velocity also increases with stress leading to a non-linear relationship between stress and strain rate. However, steady-state dislocation density is achieved only after a certain time or strain and therefore a significant period of transient creep is often observed in dislocation creep. The dislocation velocity for its glide motion is controlled by a variety of resistance forces including the intrinsic resistance caused by the crystal lattice (the Peierls stress, reorganization of dissociated (partial) dislocations), the interaction with impurity (solute) atoms, and mutual interaction. Dislocation motion out of its glide plane (climb) is controlled by the diffusion of atoms. In both glide and climb, the motion of dislocation often occurs in a step-wise manner through the motion of kinks and jogs respectively whose density is in most cases controlled by the thermochemical equilibrium. Consequently, the velocity of dislocations is often sensitive to thermochemical environment such as oxygen and water fugacity. Creep due to dislocation motion involves a number of processes many of which must occur sequentially. As a result, the slowest of these processes usually controls the overall rate of deformation. In the final section, a brief summary of the concept of a deformation mechanism map is presented. A deformation mechanism gives a guide to infer dominant mechanisms of deformation under a broad range of conditions.

Key words dislocation glide, dislocation climb, recovery, work hardening, Peierls stress, the Orowan equation, Weertman model, Cottrell atmosphere, dissociation, partial dislocations, stacking fault, Harper–Dorn creep, internal stress, deformation mechanism map, twinning.

9.1. General experimental observations on dislocation creep

Historically, the importance of dislocations in plastic deformation was recognized (in 1934 by Taylor, Polanyi and Orowan; see Orowan, 1934; Polanyi, 1934; Taylor, 1934) more than 10 years before the importance of atomic diffusion in plastic deformation was appreciated (Nabarro, 1948). Significant progress was made on the fundamentals of dislocation motions during the 1960s and important models of high-temperature creep were proposed (e.g., Weertman, 1968). About the same time, David Griggs at UCLA started extensive studies on plastic deformation of minerals and rocks (see the Griggs volume (Heard *et al.*, 1972)).

Experimental studies on rock deformation on natural coarse-grained rocks (or single crystals) in these early days showed that, in most cases, plastic deformation occurs by dislocation motion and results in characteristic microstructures. The deformation microstructures of most naturally deformed rocks are similar to those found in experimentally deformed rocks in the dislocation creep regime, which provides a strong evidence for dislocation creep in Earth's interior. Based partly on this notion, WEERTMAN (1970) suggested the importance of dislocation creep in Earth's mantle and pointed out the importance of stress- or strain-rate-dependent effective viscosity in modeling the depth variation of viscosity. Similarly, STOCKER and ASHBY (1973) concluded that dislocation creep plays an important role in Earth's upper mantle. Using the seismological observations of the distribution of anisotropic microstructures, KARATO (1998d) and KARATO and WU (1993) suggested that dislocation creep plays an important role in the boundary layers (i.e., the top and the bottom layers) in the mantle.

The important characteristics of dislocation creep include:

(1) the relation between (steady-state) strain-rate and stress is in most cases non-linear (Fig. 9.1);[1]
(2) deformation is usually highly anisotropic at the level of grains;
(3) a significant transient period exists due to the evolution of dislocation structures (density, distribution);
(4) grain-scale microstructures develop during dislocation creep including the formation of new grain boundaries, strain-induced grain-boundary migration (Chapter 13) and lattice- (crystallographic) preferred orientation (Chapter 14);
(5) the activation energy of creep agrees with that of self-diffusion (of the slowest species) in many simple materials (metals and alkali halides), although such a correlation is not clear in oxides and silicates; and
(6) in almost all the silicate minerals so far studied, plastic deformation by dislocation creep is significantly enhanced by the presence of a trace amount of water (hydrogen) (Chapter 10).

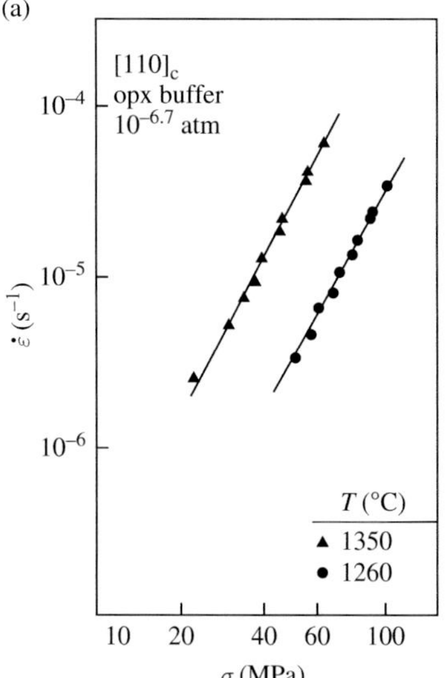

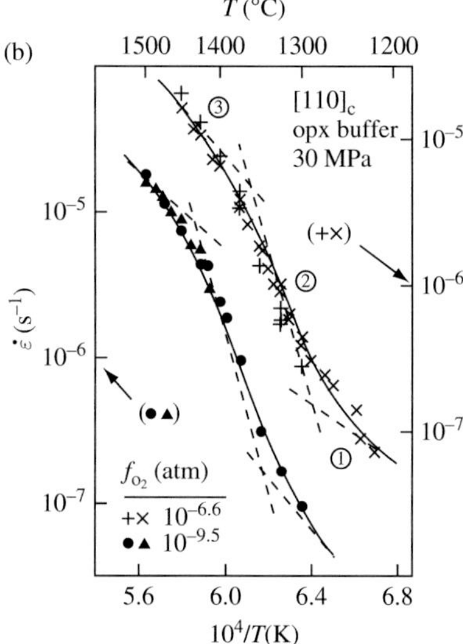

FIGURE 9.1 (a) Stress versus strain rate and (b) inverse temperature versus strain rate relationships in a single crystal of olivine oriented for the [100](010) slip system (room pressure, buffered by orthopyroxene) (BAI et al., 1991).

The microscopic origin of non-linear rheology is that the strain rate in dislocation creep is proportional to both dislocation density and velocity, both of which are stress dependent. As a consequence of the non-linear relationship between stress and strain rate, the effective viscosity corresponding to dislocation creep is stress dependent (strain-rate dependent (see Chapter 3)). Consequently, when estimating the variation in effective viscosity for dislocation creep, one needs to specify either the stress or strain rate (see Chapter 19). However, the dislocation density has a definitive value corresponding to applied stress only at a steady state. The establishment of steady-state dislocation density requires certain reorganization of dislocation microstructures, and therefore there is a significant transient period (strain) in dislocation creep.

[1] There is one important exception to this rule. In certain cases where dislocation density is independent of applied stress, then strain rate is a linear function of stress (*Harper–Dorn creep*). This mechanism is discussed in section 9.5.

In this chapter we will learn the basic physics of dislocation creep, starting from the fundamental relationship among dislocation density, velocity and strain rate, i.e., *the Orowan equation*. The Orowan equation is the starting point for most dislocation creep models. This equation indicates that it is the processes controlling dislocation density and the velocity of dislocation motion that control the strain rate. Therefore in this chapter we will focus on these two topics, the dynamics of dislocation motion and the processes controlling the dislocation density (dislocation multiplication and annihilation), which are essential to the understanding of dislocation creep.

9.2. The Orowan equation

Let us consider the deformation of a crystal by the motion of dislocations. For simplicity, let us assume that dislocations are distributed uniformly. In such a case, the deformation of a crystal can be calculated by considering the deformation of a small region of a crystal (of size L) that contains a single dislocation (Fig. 9.2). When a dislocation has moved through a crystal with a dimension L, then an average strain of $\sim b/L$ is created. Thus, when a dislocation moves a small distance ΔL, then an increment of strain will be

$$\Delta\varepsilon \sim \frac{b}{L}\frac{\Delta L}{L} = \rho b \Delta L \tag{9.1}$$

where I used a relation $\rho = 1/L^2$ (ρ, dislocation density).

If this strain increment is created during a time Δt,

$$\dot{\varepsilon} = \frac{\Delta\varepsilon}{\Delta t} = b\rho\frac{\Delta L}{\Delta t} = b\rho v. \tag{9.2}$$

This is the *Orowan equation* (in the tensor formulation, the Orowan equation can be written as $\dot{\varepsilon}_{ij} = \frac{1}{2}\rho(b_i v_j + b_j v_i)$, which comes from $\varepsilon_{ij} = \frac{1}{2}\gamma(n_j l_i + n_i l_j)$, see Chapter 5).

The Orowan equation is the basis for most of the models of plastic deformation by dislocation motion.

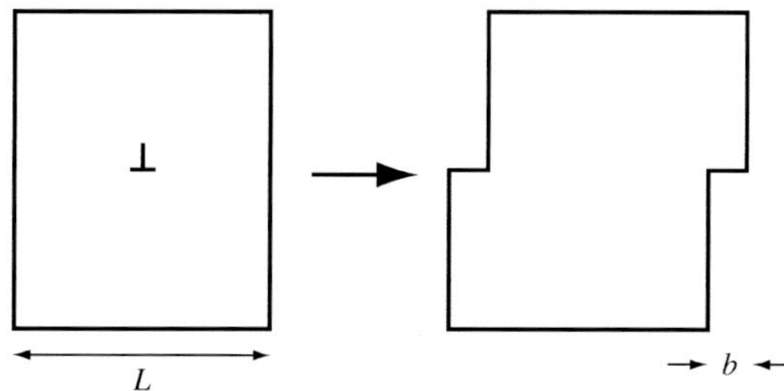

FIGURE 9.2 Motion of a dislocation across a crystal causes the relative displacement of two parts of a crystal of *b*.

The physical meaning of the Orowan equation is obvious. The rate of deformation is proportional to the amount of unit displacement caused by dislocation, b, dislocation density, ρ, and the (average) velocity of dislocation motion, v. Among these parameters, the length of the Burgers vector is nearly independent of either the physical or chemical conditions (throughout the conditions in Earth, it will change less than ~10%). The dislocation density at a steady state is determined by the magnitude of stress (equation (5.66))[2] and is largely independent of materials, temperature and pressure (and other thermochemical conditions). In contrast, the dislocation velocity is a complicated function of stress, thermochemical conditions (T, P, fugacity of water etc.). *Complications of dislocation creep come mostly from the variation of dislocation velocity.* Dependence of dislocation velocity on dislocation density gives rise to *work hardening* (or *work softening*) and *recovery*. Temperature and pressure dependence of creep are also through the velocity term. Similarly the effects of the chemical environment are mainly through its effect on dislocation velocity.

9.3. Dynamics of dislocation motion

When a deviatoric stress is applied, a force acts on a dislocation (see equation (5.61)); consequently a dislocation moves. The velocity of a dislocation is determined by the balance of a force due to stress and the resistance force(s). The force acting on a dislocation includes the forces caused by external stress as well as stress due to other dislocations (internal stress). Therefore the resistance to dislocation motion includes: (1) the intrinsic resistance (the Peierls stress), (2) the extrinsic resistance (resistance due to impurity or solid-solution atoms) and (3) the mutual interactions. Dislocation motions controlled by the Peierls stress or impurities are insensitive to strain magnitude, whereas dislocation motion controlled by mutual interactions is sensitive to the strain because mutual interaction depends on dislocation density and configuration, both of which evolve with strain. A dislocation must overcome all of these resistance forces and therefore the largest resistance force controls the rate of deformation. In fcc (face-centered-cubic) metals (such as Au, Cu), where chemical bonding is weak and the unit cell dimension is small, the Peierls stress is small

[2] There are some cases in which dislocation density is independent of applied stress. In such a case strain rate is linearly proportional to applied stress (Harper–Dorn creep).

and the impurities or the mutual interactions are the most important resistance forces. In such a case, significant transient creep is often observed, which is caused by the increase in dislocation interactions with strain. Also strain hardening (*work hardening*) is often observed in such a material, caused by the increase in resistance forces due to the increase in dislocation density. In materials where chemical bonding is strong (covalent bonding) (Si, Ge, diamond, SiO_2) or the unit cell is large (garnet), the Peierls stress is high and has important effects on the rate of dislocation motion. In the extreme case where the Peierls stress is very large, dislocation velocity is nearly independent of strain, but dislocation density can evolve with strain, causing the transient creep characterized by *strain (work) softening* (see later). In most silicate minerals, the nature of chemical bonding is between ionic and covalent and unit cell dimension is large. Consequently, the Peierls stress is relatively large and the dislocation glide is in general difficult.

9.3.1. Intrinsic resistance

The Peierls stress

When a dislocation is present in a crystalline lattice there must be extra energy associated with it (see Chapter 5). This extra energy varies with distance with a period of the unit cell of the crystalline lattice. This potential energy of dislocation is referred to as the *Peierls potential* (self-energy of a dislocation per unit length), $\phi_P(x)$. Consequently, when a dislocation moves through a crystal, there is a force on it that is given by the spatial derivative of potential energy, $f = -(\partial \phi_P/\partial x)(= \sigma b)$. The maximum of this force gives the resistance for motion of dislocation over the Peierls barrier to the next stable position,

$$\sigma_P = \frac{1}{b}\left|\frac{\partial \phi_P}{\partial x}\right|_{max} \tag{9.3}$$

and is called the *Peierls stress*. The Peierls stress is the stress needed to move a dislocation in a crystal without the help of thermal activation. It is determined by the crystal structure and chemical bonding and hence is intrinsic to a given material. For a simple sinusoidal potential, $\phi_P \propto \mu \cdot \cos(2\pi x/b)$ (μ, shear modulus; b, the length of the Burgers vector), the Peierls potential is related to the crystal structure and the elastic constant, which is determined by the strength of chemical bonds (for the derivation of this formula see a standard textbook on crystal dislocations: e.g., COTTRELL, 1953; NABARRO, 1967b; HIRTH and LOTHE, 1982),

$$\sigma_P = \frac{2\mu}{1-\nu}\exp\left(-\frac{2\pi}{1-\nu}\frac{h}{b}\right) \tag{9.4}$$

where h is the distance between two planes between which dislocation glide occurs and ν is the Poisson's ratio. This relation indicates that among the possible slip systems, the one that has the largest h/b will have the smallest Peierls stress.

Such a sinusoidal potential is appropriate for metallic or ionic bonding. For covalent bonding, the potential has a sharp minimum near the lattice sites, hence the Peierls stress will be higher than given by equation (9.4). The effects of form of inter-atomic potential (namely the effects of chemical bonding) on the Peierls stress were investigated by FOREMAN *et al.* (1951). TAKEUCHI and SUZUKI (1988) reviewed experimental data on the Peierls stress in a large number of materials and showed that the variation in the Peierls stress with materials can be explained mostly by the geometrical factor through (9.4) despite this oversimplification (Fig. 9.3).

The motion of a dislocation over the Peierls potential does not occur homogeneously. Rather it occurs step-wise through the formation of a "step" and its motion (Fig. 9.4) through the nucleation and migration of *kinks*. Therefore glide velocity of an edge dislocation v_g is proportional to the number of kinks and their velocity. If c_k is the density of a kink pair (number of kink pairs per unit length), then the mean distance between the pairs of kinks is $l_k \approx 1/c_k$. If kinks spread by this distance, then a dislocation line will move by a distance, b. The time needed for this to occur is $\tau \approx l_k/v_k = 1/c_k v_k$. Therefore the velocity of the glide motion of a dislocation by the nucleation and migration of kinks is given by

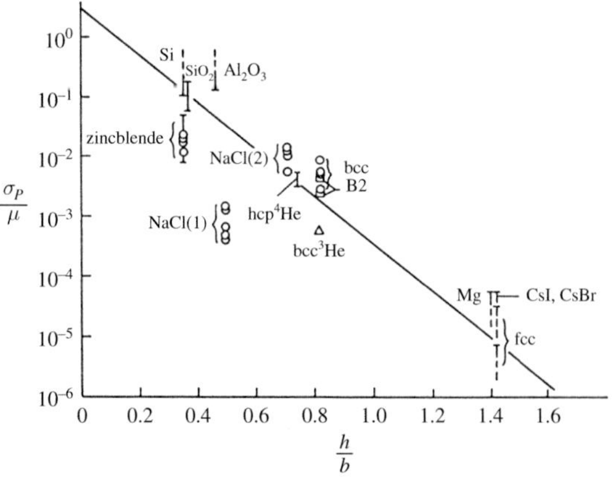

FIGURE 9.3 The Peierls stress versus h/b relation for various materials (after TAKEUCHI and SUZUKI, 1988).

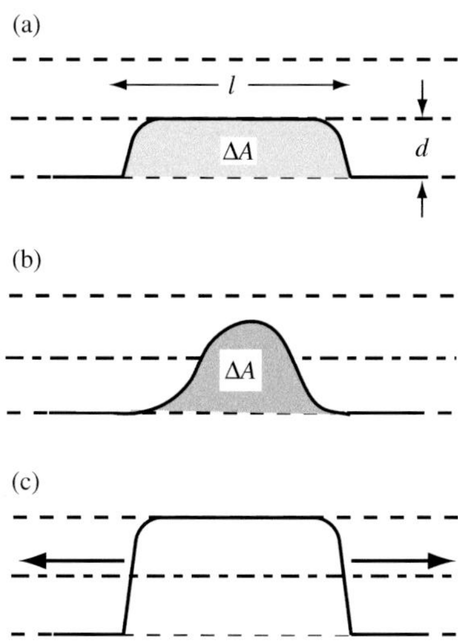

FIGURE 9.4 Dislocation glide through the nucleation and migration of a pair of kinks.

(a) A saddle-point configuration for low stress. A pair of kinks separated by a distance l is formed. ΔA is the area swept by a dislocation segment that has moved from the ground state (in the Peierls valley) to the saddle point. $\sigma b\, \Delta A$ is the work done by the applied stress to cause kink nucleation. (Kink migration also occurs over a Peierls potential hill that is normal to the Peierls potential hill for nucleation. This Peierls potential is often referred to as the Peierls potential of the second kind.)

(b) A saddle-point configuration for high stress. At a saddle point two kinks are not well separated leading to a different stress dependence for ΔA compared with the case for low stress.

(c) After nucleation of a kink (pair), segments of the kink will migrate to achieve dislocation glide.

$$v_g = \frac{b}{\tau} = c_k b v_k \tag{9.5}$$

where c_k is the kink density (number of kinks per unit length) and v_k is the kink velocity. Kinks may be formed by thermal activation or they may exist by deformation for geometrical reasons (*geometrical kinks*; see Chapter 5). In a case where kinks are formed by thermal activation, the kink density is sensitive to the temperature and chemical environment (see Chapter 5), $c_k = c_k(T, P; C)$ where C denotes parameters related to the chemical environment such as oxygen and water fugacity. The kink velocity is also a function of thermochemical variables as well as stress, $v_k = v_k(T, P; C)$.

At a high stress, effects of stress to change the potential barrier for kink formation (and migration) become important and hence the following equation applies,[3]

$$v_g \propto \exp\left(-\frac{H^*(\sigma; C)}{RT}\right) \tag{9.6}$$

where $H^*(\sigma;\ C)$ is the activation energy for a double-kink nucleation (and migration). The reason for the stress dependence of the activation enthalpy is the fact that the geometry of dislocation at the saddle point depends on the stress and that the activation enthalpy includes the mechanical energy done by the applied stress (see Fig. 9.4), namely,

$$\begin{aligned} H^*(\sigma; C) &= H_0^*(C) - \sigma b \cdot \Delta A(\sigma; C) \\ &= H_0^*(C)[1 - B(\sigma; C) \cdot \sigma] \end{aligned} \tag{9.7}$$

where ΔA is the area swept by a segment of dislocation when it has moved from the ground state to the activated state (called an *activation area*) and $B(\sigma; C) = b \cdot \Delta A(\sigma; C)/H_0^*$.

To understand how $H^*(\sigma;\ C)$ depends on stress, let us consider a few simple cases. In a case where $B(\sigma; C)$ is independent of stress, then the activation enthalpy is a linear function of stress, namely,

$$H^*(\sigma; C) = H_0^*(C)[1 - B(C) \cdot \sigma]. \tag{9.8}$$

However, when a dislocation moves over a Peierls potential, the activation area is in most cases stress dependent because the shape of a dislocation moving over the Peierls potential depends on the stress. Imagine a straight dislocation line. Dislocation glide occurs through the formation (nucleation) of kinks and their migration. When applied stress is low, relative to the Peierls stress, kink nucleation occurs through the formation of a pair of discrete kinks. Once they are formed, then two kinks will migrate to opposite directions. Let us first calculate the enthalpy of formation of a pair of kinks. The formation enthalpy of a pair of kinks is the enthalpy associated with the saddle point configuration as shown in Fig. 9.4a. The enthalpy difference between this configuration and the ground state configuration (a dislocation lies on the Peierls potential minimum) is

$$H^*(\sigma, l) = 2H_k^* + H_{\text{int}}^*(l) - \sigma b l d \tag{9.9}$$

where l is the distance between two kinks, H_k^* is the formation enthalpy of a kink, $H_{\text{int}}^*(l)$ is the interaction enthalpy of kinks, σ is the applied stress and d is the

[3] At low stresses, the backward motion of dislocation must also be considered, so that $v_g \propto \exp(-\mathrm{H}^{*+}(\sigma)/\mathrm{RT}) - \exp(-H^{*-}(\sigma)/RT)$ where $H^{*+,-}(\sigma)$ is activation enthalpy for forward and backward motion. If the activation enthalpy depends on stress as $H^{*+,-}(\sigma) = H_0^*(1 \mp \sigma/\sigma_0)$ then $v_g \propto \exp(-H_0^*/RT) \sinh((H_0^*/RT)(\sigma/\sigma_0)) \approx 2\exp(-H_0^*/RT)(H_0^*/RT) \cdot \sigma/\sigma_0$ as $\sigma/\sigma_0 \longrightarrow 0$.

distance of neighboring Peierls potential minima ($d \sim b$). H_k^* consists of the excess enthalpy associated with the formation of a kink due to the increase in the line energy and the Peierls energy, and the line energy in turn depends on the Peierls energy because the geometry of a kink is controlled by the balance of line tension and the force caused by the Peierls potential. Thus

$$H_k^* = \sqrt{\frac{K}{2\pi}\mu b^3}\sqrt{\frac{\sigma_P}{\mu}} \tag{9.10}$$

where K is a constant of order unity, and the kink-kink interaction enthalpy is given by HIRTH and LOTHE (1982),

$$H_{\text{int}}^* = -\frac{K}{8\pi}\frac{\mu b^2 d^2}{l}. \tag{9.11}$$

Taking the derivative of equation (9.9) with respect to l, one finds that the enthalpy $H^*(\sigma, l)$ has a maximum of

$$H^*(\sigma, l_m) = H_0^*\left(1 - \sqrt{\frac{\sigma}{\sigma_P}}\right) \tag{9.12}$$

at $l_m = \frac{1}{2}\sqrt{K/2\pi}\sqrt{\mu b d/\sigma}$ where $H_0^* = 2H_k^* = \sqrt{2K/\pi}\cdot\mu b^3\sqrt{\sigma_P/\mu}$. This maximum enthalpy is the enthalpy for thermal activation needed to nucleate a pair of kinks.

When a high stress is applied, then l_m becomes smaller and the saddle point configuration occurs at the middle of two adjacent Peierls potential minima (Fig. 9.4b), and the stress dependence of activation enthalpy becomes stronger. In this case, one needs to find the geometry of dislocation at the saddle point configuration considering the balance of line tension and the force caused by the applied stress, σb. A detailed analysis yields (see e.g., KOCKS *et al.*, 1975),

$$H^*(\sigma) = H_0^*\left(1 - \frac{\sigma}{\sigma_P}\right)^2. \tag{9.13}$$

So, we have three cases for the stress dependence of the activation enthalpy, equations (9.8), (9.12) and (9.13). Other functional relationships can be found for other models, but in all models, the activation enthalpy is a function of σ/σ_C, and $H^*(\sigma/\sigma_C) = 0$ for $\sigma = \sigma_C$ (for the cases of (9.12) or (9.13), $\sigma_C = \sigma_P$). A generic form that satisfies these relations is

$$H^*\left(\frac{\sigma}{\sigma_C}\right) = H_0^*\left(1 - \left(\frac{\sigma}{\sigma_C}\right)^q\right)^s \tag{9.14}$$

where q and s are non-dimensional parameters with $0 < q \leq 1$ and $1 \leq s \leq 2$ (e.g., KOCKS *et al.*, 1975; FROST and ASHBY, 1982). Note that the critical stress, σ_C (σ_P), is sensitive to the chemical environment such as the presence of water and H_0^* depends on σ_P, therefore both H_0^* and $(1 - (\sigma/\sigma_C)^q)^s$ terms depend on the chemical environment. Experimental data at relatively low temperature and high stress can be fitted to the flow law formula involving the stress-dependent activation enthalpy such as (9.14). In fitting experimental data, a range of parameters (q and s) may be used but the choice of these parameters does not have large effects (Fig. 9.5).

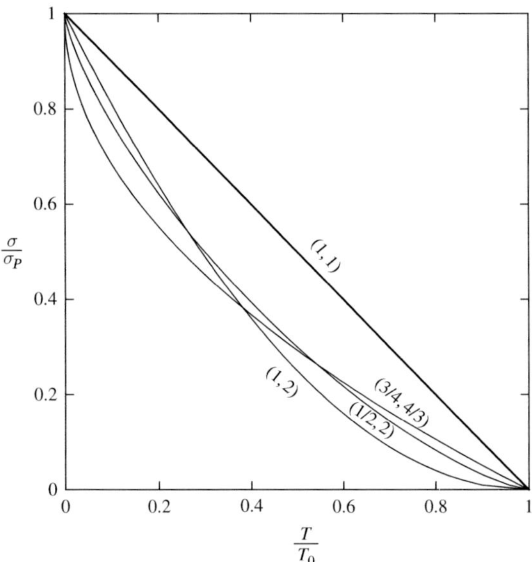

FIGURE 9.5 The stress needed to move a dislocation with a given velocity as a function of temperature in the Peierls mechanism. Numbers in parentheses indicate (q, s).

When the applied stress reaches σ_C, then dislocation motion is possible even at $T = 0$ K. In this sense, σ_C corresponds to the "yields stress." When this critical stress is due to the motion of a dislocation over the Peierls potential, this stress corresponds to *the Peierls stress* (σ_P). Note that although equation (9.14) formally allows stress higher than σ_C, physically such a stress is meaningless because a material cannot support stress higher than this value. Consequently the stress σ_C can be understood to be the absolute maximum strength that a given material can support (with a finite number of dislocations).

The activation enthalpy for dislocation glide at zero stress (i.e., H_0^*) is highly sensitive to the length of the Burgers vector as well as the Peierls stress. Both of which are sensitive to the crystallographic nature of each slip system. This causes strong anisotropy in dislocation glide. The chemical environment such as water fugacity can change H_0^* through the change in the

Peierls stress (Chapter 10). For example, there is evidence that water enhances the deformation of silicates with particular slip system(s) more than others (e.g., Blacic, 1975; Mackwell *et al.*, 1985) and this could partly be due to the anisotropic enhancement of dislocation glide.

Heggie and Jones (1986) developed a theoretical model and showed that hydrogen reduces the Peierls stress in quartz (see Chapter 10). Similarly Haasen (1979), Hirsch (1979) and Jones (1980) proposed models for the influence of impurities on dislocation mobility in semiconductors. In these cases, impurities affect dislocation motion either through the change in kink density or kink mobility (or both). Hobbs (1981) suggested that similar mechanisms may apply to silicate minerals. When a dislocation is split into partial dislocations, then the length of the Burgers vector is reduced that reduces H_0^*. If the excess work needed to re-organize a stacking fault is not large, then dislocation splitting can result in softening. Consequently, in a material where the Peierls stress plays an important role, splitting of dislocations can result in weakening. In contrast, in a material where the Peierls stress is low, splitting usually leads to hardening.

In most silicate minerals, the Peierls stress is large because of largely covalent (plus some ionic component of) bonding and of large unit cell dimensions. In these cases, dislocation glide is difficult and may control the overall rate of plastic deformation. In a material with a high Peierls stress, deformation with a small strain may occur through the migration of pre-existing "geometrical kinks" (see Chapter 5) and therefore rheology at small strains will be different from steady-state rheology. This transient behavior is a possible mechanism of seismic wave attenuation (Karato, 1998a) (see also Chapter 11).

When the Peierls stress is large, the direct mechanical effects of impurity atoms (which are important in materials with low Peierls stress) are not important. In these materials, the effects of impurities would likely be through indirect effects including the effects to modify the concentration of point defects that enhance diffusion and the effects to reduce the Peierls stress by reducing the chemical bonds.

Thermal and athermal motion of defects

When the activation enthalpy is stress dependent and decreases with stress, then there is a critical stress at which the activation enthalpy for defect motion becomes zero,

$$H^{*+,-}(\sigma_c) = 0. \tag{9.15}$$

If the relation between stress and activation enthalpy is linear, this stress is given by

$$\sigma_c = \frac{H^*}{b\Delta A}. \tag{9.16}$$

When stress reaches this level, then motion of defects is possible without the help of thermal activation. In other words, there is a finite probability of defect motion even at $T = 0$ K. Such a motion of defects is referred to as *athermal* motion.

Effects of dissociation

The dissociation of a dislocation into two (or more) *partial dislocations* is often seen in silicate minerals (as well as metals and oxides) (see Chapter 5). Dissociation usually results in extra resistance for dislocation motion and is often invoked to explain work hardening (e.g., Cottrell, 1953). For instance, if dissociation occurs on a screw dislocation, then the cross-slip motion of screw dislocation occurs only when the stacking fault shrinks (Escaig, 1968). Thus an extra energy is needed for a cross-slip of a dissociated screw dislocation. Poirier and Vergobbi (1978) suggested that creep in olivine might be controlled by the motion of screw dislocations involving the shrinkage of a stacking fault.

However, the correlation between dissociation and creep is not straightforward. In some cases, dissociation could lead to weakening. For example, in a crystal with a large Peierls stress due to a large Burgers vector, dissociation can result in a reduction in the Peierls stress, leading to weakening (e.g., Cordier and Doukhan, 1995). The dynamics of motion of dissociated dislocations with a high Peierls stress is discussed by Möller (1978).

9.3.2. Extrinsic resistance

Interaction with impurity atoms

Dislocations are associated with strain field. Therefore dislocations have strong interaction with impurity atoms that have strain field. In addition, if a dislocation has an electrostatic charge then it also interacts with charged defects.

In these cases, a dislocation will be associated with either a higher or lower concentration of certain atomic species including vacancies. Consequently, a motion of a dislocation must be associated with diffusion of these atomic species. Let us consider a case for impurity atoms with different atomic (ionic) radius from that of

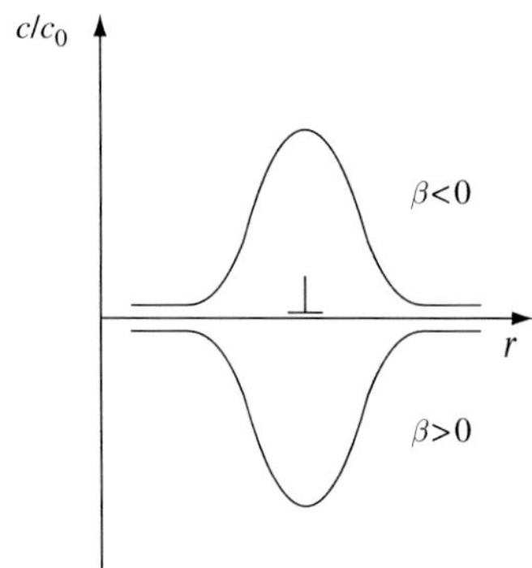

FIGURE 9.6 The distribution of impurities around a dislocation.

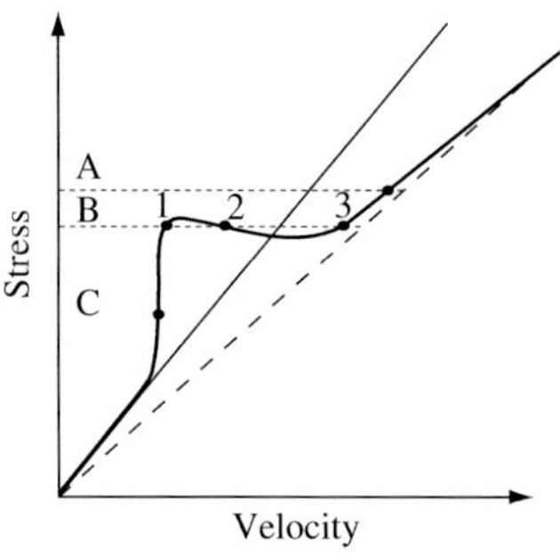

FIGURE 9.7 A stress versus velocity diagram for a dislocation interacting with impurity atoms.

the host atoms. In this case, elastic interaction occurs between a dislocation and impurity atoms. The strength of this interaction can be characterized by the interaction energy that is characterized by a parameter,

$$\beta = \frac{\mu b \Delta V}{3\pi}\frac{1+\nu}{1-\nu} \tag{9.17}$$

where ν is the Poisson ratio and

$$\Delta V = V_s - V_a \tag{9.18}$$

where V_s is the molar volume of solute (impurity atoms) and V_a is the molar volume of atoms in the host crystal. Consequently, the distribution of impurity atoms around a dislocation is modified to (Fig. 9.6),

$$c = c_0 \exp\left(-\frac{\beta \sin\theta}{rRT}\right). \tag{9.19}$$

When a dislocation is accompanied with impurity atoms, then the rate of dislocation motion is controlled by the dislocation–impurity atom interaction and the diffusion of impurity atoms. First, when a modest stress is applied at high temperatures, a dislocation moves together with impurity atoms. In such a case, the velocity (mobility) of dislocation is controlled by the rate of diffusion of impurity atoms. In this regime, the dislocation velocity is proportional to the diffusion coefficient of impurities, and inversely proportional to the concentration of impurity atoms and the strength of dislocation–impurity interaction. To obtain the relation between dislocation velocity and diffusion coefficient, we start from the general relation between flux and force (see Chapter 2),

$$J = \frac{D(c-c_0)}{RT} f = (c-c_0)v \tag{9.20}$$

where f is the force on the particle. Therefore if there are N impurity atoms around a dislocation (per unit length), then the total force, F, needed to move these atoms will be

$$F = \frac{RTN}{D} v = \sigma b. \tag{9.21}$$

The number of impurity atoms around a dislocation (per unit length) must now be calculated. The distance from a dislocation where the concentration of impurity atoms is different from the rest of the crystal is $\sim\beta/RT$ (see (9.19)). Most of the impurity atoms are removed (collected) from this region by a moving dislocation if the potential is repulsive (attractive), so $N \approx c_0(\beta/RT)^2$ (for details see HIRTH and LOTHE, 1982), and hence

$$v = \frac{\alpha DRT\sigma b}{c_0\beta^2} \tag{9.22}$$

where α is a constant of order unity. Second, if a very large stress is applied, then impurity atoms are broken away from a dislocation, then a dislocation becomes free from impurity atoms and moves fast. The critical stress for breakaway depends on the concentration of impurity atoms and the strength of binding of impurities to a dislocation. Now recall that the extra concentration of impurity atoms around a dislocation per unit length is given by $|c-c_0|b^2 \approx c_0[1-\exp(-\beta/rRT)]b^2 \approx c_0(\beta b/RT)$. The force needed to remove one impurity from the potential well is given by $\approx |\partial W/\partial r| \approx \beta/b^2$ (W: interaction potential). Consequently, the force (per unit length) needed to break away impurity atoms is $\approx |c-c_0|b^2|\partial W/\partial r| \approx c_0\beta^2/bRT = \sigma_c b$. So that

$$\sigma_c \approx \frac{c_0\beta^2}{b^2RT}. \tag{9.23}$$

Above this stress, the dislocation velocity is controlled by the intrinsic resistance such as the Peierls stress (or phonon drag). Thus the relation between dislocation velocity and stress would look like Fig. 9.7. At high temperatures where atoms are mobile, there are two branches in the force versus dislocation velocity diagram. One slow branch corresponds to the *Cottrell drag* case where dislocation drags impurity atmosphere. The second branch corresponds to *intrinsic resistance* such as the Peierls mechanism. A transition from slow, Cottrell drag branch to fast intrinsic resistance branch occurs at a certain stress level, leading to a catastrophic change in dislocation velocity. This catastrophic transition sometimes leads to an unstable deformation called the *Portevin–Le Chatelier effect* (fluctuation of stress during a constant strain-rate deformation; Chapter 16). A similar behavior is also observed in grain-boundary migration (see Chapter 13).

Problem 9.1

Explain the cause for unstable dislocation motion using Fig. 9.7 (see also the discussion on Fig. 13.4 on page 234–235).

Solution

A solution of equation (9.20) is stable (unstable) if a small increase in dislocation velocity results in increase (decrease) in stress (force). Therefore in a stress–velocity diagram, if the slope is positive (i.e., $\partial F/\partial v > 0$), then dislocation motion is stable. The stress (force) versus dislocation velocity curve shown in Fig. 9.7 has three regions. In a region between point 1 and 3, $\partial F/\partial v < 0$ and dislocation motion is unstable. Physically, this corresponds to the fact that when dislocation velocity becomes faster, more impurity atoms are broken away from the dislocation core so that stress (force) becomes smaller. In essence, the unstable motion of dislocation is caused by breakaway of impurities from the dislocation line.

Suzuki effect

SUZUKI (1962) proposed that the motion of a dislocation in a solid-solution alloy may be controlled by the interaction of stacking faults with alloying elements. In many minerals a certain degree of solid-solution and stacking faults (partial dislocations) are commonly observed. Therefore a similar effect may play some role in minerals (e.g., SMITH, 1985). The stacking fault (associated with partial dislocations) has different atomic packing from the perfect lattice and hence has an extra energy. Therefore the concentration of an alloying element is likely different in the stacking fault from that in the matrix. Consequently, when a partial dislocation moves, it must exchange alloying elements with the matrix and this diffusion-controlled exchange may contribute to the resistance to dislocation motion.

SUZUKI (1962) showed that the resistance (stress) due to this effect is given by

$$\sigma_S = \frac{\gamma(c_0) - \gamma(c_1)}{b} \approx \frac{\partial \gamma}{\partial c}\frac{\Delta c}{b} \tag{9.24}$$

where $\gamma(c)$ is the energy of stacking fault with the concentration of solid-solution c. A typical value of $\partial\gamma/\partial c$ is $\sim 0.1\ \mathrm{Jm^{-2}}$, and therefore one gets $\sigma_S \sim 10$ MPa for $\Delta c = 0.1$, $b = 0.5$ nm.

9.3.3. Resistance due to mutual interaction

Dislocation motion is caused by the stress on a dislocation line. This stress includes applied stress but also the stress caused by other dislocations. The stress caused by other dislocations play an important role in dislocation motion during creep because dislocations have a long-range elastic strain field. Dislocation–dislocation interaction can influence dislocation motion by other mechanisms. They include formation of jogs formed by the crossing of two screw dislocations, and the formation of dislocation junctions.

Problem 9.2

Explain how a jog can be formed when two screw dislocations cut each other and explain why a jog on a screw dislocation results in resistance for dislocation motion.

Solution

Consider a screw dislocation that cuts another screw dislocation whose orientation is normal to another one. The atomic displacement along a screw dislocation is parallel to the dislocation line, and therefore there will be a step, a jog, in each dislocation normal to the dislocation line. At a jog on a screw dislocation, the dislocation is an edge dislocation. Consequently, this portion can move only along a particular plane and the motion away from its glide plane requires diffusion.

The exact microscopic mechanism in which dislocation–dislocation interaction modifies the dislocation velocity is therefore complicated and depends on a particular mechanism. However, dislocation–dislocation interaction usually reduces the dislocation velocity and the degree to which dislocation velocity is reduced increases with dislocation density. Consequently, the effect of dislocation–dislocation interaction is often parameterized by replacing the stress with the effective stress, $\sigma_{\text{eff}} \equiv \sigma - \sigma_i$,

$$v = v(\sigma_{\text{eff}}) \equiv v(\sigma - \sigma_i) \tag{9.25}$$

where

$$\sigma_i \approx \mu b \sqrt{\rho} \tag{9.26}$$

is the *internal stress*. This relation can be understood by remembering that the mean interaction stress between dislocations is given by $\sigma_i = \langle \sigma \rangle \approx \mu b / \langle r \rangle = \mu b \sqrt{\rho}$. Implicit in this formulation is that the basic microscopic physics of dislocation motion is unaffected by the dislocation–dislocation interaction and therefore the stress versus dislocation velocity relation is the same as that for an isolated dislocation. Consequently, strictly speaking this formulation is appropriate only for the case of mechanical interactions rather than the case for jog formation and/or node formation.

When the internal stress is the most important resistance force, then no dislocation motion would occur when the increase in dislocation density due to deformation results in $\sigma \approx \sigma_i = \alpha \mu b \sqrt{\rho}$. In such a case, continuing deformation is possible only when dislocation density is reduced by *recovery*. Theories of creep corresponding to such a case will be discussed later in this chapter.

9.3.4. Dislocation climb, cross-slip

Climb of an edge dislocation

Motion of an edge dislocation out of its glide plane is referred to as *climb*. Climb of an edge dislocation involves removal or addition of atoms (Fig. 9.8), and needs diffusion of atoms away from or to the dislocation. Therefore this mode of dislocation motion occurs effectively only at relatively high temperatures (more than half the melting temperature). The climb motion of a dislocation occurs step by step, similar to the glide through the migration of a *jog* (a jog is a step on a dislocation line that is normal to the glide plane, Fig. 9.8). The velocity of dislocation climb is therefore proportional to the density of jogs, c_j (a number of jogs per unit length of a dislocation), and its velocity, v_j,

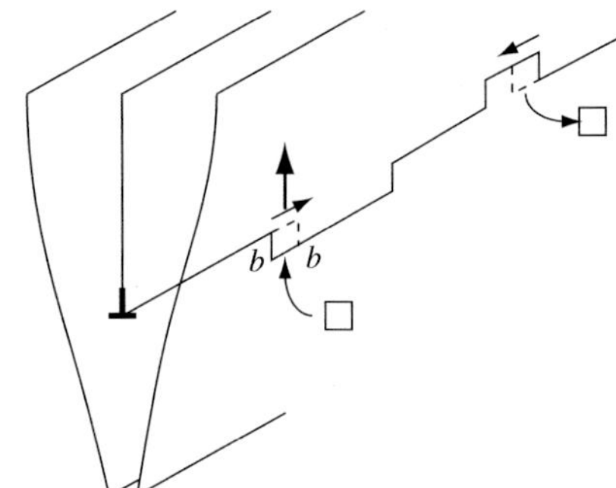

FIGURE 9.8 The atomic process of edge dislocation climb. Climbing of an edge dislocation requires atomic diffusion (after Poirier, 1985).

$$v_c = c_j b v_j. \tag{9.27}$$

In most cases the jog density is determined by thermodynamic equilibrium and is dependent on thermochemical parameters (temperature, water fugacity etc., see Chapter 5).

Let us calculate the velocity of jog migration, v_j. The rate at which a jog moves is controlled by the rate at which vacancies are emitted or attached to a dislocation line and therefore the velocity of jog migration is controlled by diffusion of vacancies (and hence atoms). When an external stress is applied, then the energy of formation of a vacancy at the dislocation core becomes stress dependent. Consequently, the concentration of vacancies along a dislocation line is modified to

$$c_v = c_v^0 \exp\left(\frac{\sigma\Omega}{RT}\right). \tag{9.28}$$

Consequently, a gradient in vacancy concentration is created by the stress that leads to diffusion and hence results in the migration of jogs. To determine how fast a jog migrates, we will first solve the diffusion equation for vacancies with the boundary condition that the vacancy concentration at the dislocation core is $c_v = c_v^0 \exp(\sigma\Omega/RT)$, and the concentration of vacancies at a large distance L (L is approximately the mean distance of dislocations) is c_v^0. At steady state, the vacancy concentration must satisfy the Laplace equation,

$$\nabla^2 c_v = 0 \tag{9.29}$$

for isotropic diffusion coefficient. Solving this equation with appropriate boundary conditions, one gets,

$$\begin{aligned} c_v(r) &= c_v^0 \left[1 + \frac{(\exp(\sigma\Omega/RT) - 1)\log(r/L)}{\log(b/L)}\right] \\ &\approx c_v^0 \left[1 + \frac{(\sigma\Omega/RT)\log(r/L)}{\log(b/L)}\right] \end{aligned} \tag{9.30}$$

where we made an approximation $\sigma\Omega/RT \ll 1$.

Problem 9.3

Derive equation (9.30).

Solution

The Laplace equation in the cylindrical coordinate is $\left((1/r)(\partial/\partial r)r(\partial/\partial r) + (1/r^2)(\partial^2/\partial\theta^2)\right)c_v = 0$. With the axial symmetry, this equation is reduced to $(1/r)(d/dr)r(d/dr)c_v = 0$. The solution of this equation has the following form, $c_v(r) = C_1 \log r + C_2$ where C_1 and C_2 are the integration constants that must be determined by the boundary conditions. The boundary conditions for this case are $c_v(b) = c_v^0 \exp(\sigma\Omega/RT)$ and $c_v(L) = c_v^0$ which give $C_1 = \frac{c_v^0[\exp(\sigma\Omega/RT)-1]}{\log(b/L)}$ and $C_2 = c_v^0\left[1 - \frac{\exp(\sigma\Omega/RT)-1}{\log(b/L)}\log L\right]$. Thus we get (9.30).

Consequently, there is a vacancy current per unit length of a dislocation,

$$I = -2\pi r D_v \frac{\partial c_v}{\partial r} = \frac{2\pi\sigma\Omega D_v c_v^0}{RT\log(L/b)} = \frac{2\pi D c\sigma\Omega}{RT\log(L/b)} = \frac{2\pi D\sigma N_A}{RT\log(L/b)} \quad (9.31)$$

where D_v is the diffusion coefficient of vacancies and we used the relations $D_v c_v = Dc$ and $c\Omega = N_A$. Because this is a flux of vacancies per unit length, the flux for one jog (whose width is $\sim b$) is Ib. This flux makes a displacement of a jog by b. Therefore, the jog velocity is related to the vacancy flux as,

$$v_j = Ib^2 \quad (9.32)$$

and hence

$$v_c = c_j b v_j = \frac{2\pi\sigma\Omega D c_j}{RT\log(L/b)} \propto D c_j \quad (9.33)$$

where we used a relation $b^3 N_A = \Omega$. The dependence of climb velocity on L is annoying, but for a reasonable range of dislocation density, $\log(L/b)$ is only weakly dependent on dislocation density and $\log(L/b) \sim 4-5$.

The velocity of dislocation climb is proportional to the diffusion coefficient and the jog concentration. Therefore the dependence of dislocation climb velocity on temperature and chemical environment is through their influence on diffusion coefficient and jog concentration. If the activation enthalpy for jog formation is not small then $c_j \propto \exp\left(-H_{jf}^*/RT\right)$ (H_{jf}^*: enthalpy for jog formation), and $H_{\text{climb}}^* = H_{jf}^* + H_{\text{diff}}^*$. If the formation energy of a jog is small, then the mean distance of jogs will be smaller than the characteristic length of diffusion of atoms along a dislocation line, then all sites along a dislocation line can operate as a source/sink of atoms. In such a case, a dislocation line is said to be "saturated" with jogs and $v_c \propto D$, and hence $H_{\text{climb}}^* = H_{\text{diff}}^*$.

Note that the diffusion coefficient in equation (9.33) is an average diffusion coefficient normal to the dislocation line, $D_\perp$. Anisotropy in $D_\perp$ and in jog formation energy causes anisotropy in the climb velocity of a dislocation.

Problem 9.4*

Consider a pair of edge dislocations with a distance l whose extra half-plane is common but extend to different directions (Fig. 9.9). Such a pair is called a *dislocation dipole*. Since these two dislocations with opposite signs attract each other (at a long distance), they will climb and finally annihilate. During this climb, atoms (vacancies) must diffuse out (in) from this region. Assuming that a dislocation dipole is in a cylinder with a radius $\sim L'(L' \gg l)$, show the velocity of dislocation climb is given by $v_c = \frac{Dc_j\mu b\Omega}{(1-\nu)lRT\log\left(L'/\sqrt{bl}\right)}$.

Solution

The interaction of two dislocations is, in general, complicated because the interaction depends on the orientations and relative positions of two dislocations. However, when the two edge dislocations are separated more than $\sim b$ on parallel planes with a distance l, then stress between two dislocations is attractive and is given by $\sigma = \mu b/2\pi(1-\nu)l$ (equation (5.65)). Therefore the

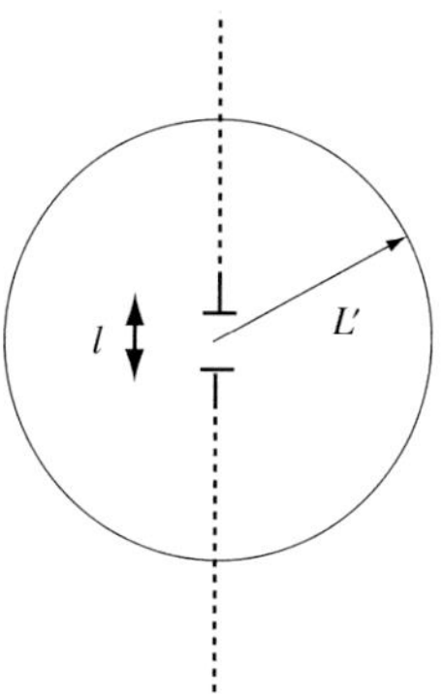

FIGURE 9.9 Climb motion of a pair of edge dislocations.

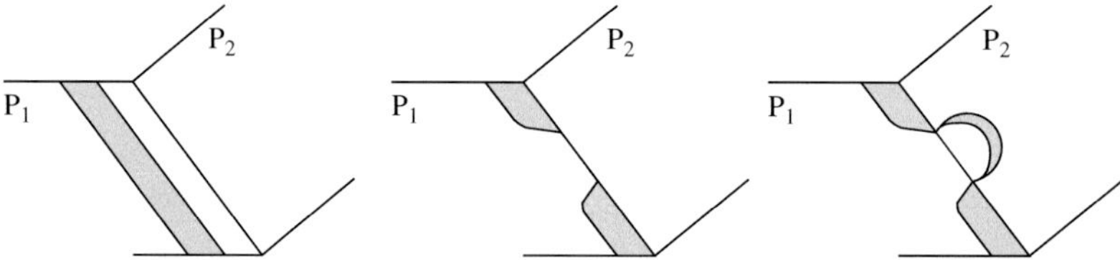

FIGURE 9.10 Cross-slip of a dissociated screw dislocation (after POIRIER, 1985). P1, P2 denote different slip planes. Hatched regions show stacking faults. In order for a screw dislocation to change its slip plane by cross-slip, the stacking fault must constrict locally.

vacancy concentration near the dislocation core is modified by this stress to $c_v = c_v^0 \exp(\mu b\Omega/2\pi(1-\nu)l) \approx c_v^0(1+\mu b\Omega/2\pi(1-\nu)l)$. As a result, vacancy concentration at the dislocations is modified (to the same degree), and consequently, there must be a flux of vacancies out of this region containing a dislocation dipole. To determine the steady-state flux, we first consider vacancy concentration profile corresponding to a single dislocation subjected to the stress caused by the mutual interaction. Since the diffusion equation is linear, the vacancy concentration can be determined by the superposition of two solutions, each of which corresponds to the vacancy distribution by one dislocation. Let us assume the following form $c_v(r) = C_1 \log r + C_2$ which is valid for $r \gg l$. An obvious boundary condition is $c_v(L') = c_v^0$ so that $c_v(r) = C_1 \log(r/L') + c_v^0$. Another boundary condition is that at the cores of each dislocation (at distance $\sim b$ from each dislocation), the concentration of vacancy must be $c_v \approx c_v^0(1+\mu b\Omega/2\pi(1-\nu)l)$. Using the superposition of solutions for each dislocation, one finds $c_v(r) = (c_v^0 \mu b\Omega/4\pi(1-\nu)l)\,(\log(r/L')/\log(\sqrt{bl}/L')) + c_v^0$. The vacancy flux I can be calculated as $I = -2\pi r D_v(\partial c_v/\partial r) = D_v c_v^0 \mu b\Omega/2(1-\nu)lRT \log(L'/\sqrt{bl})$. Using the relations $v_j = Ib^2$ and $v_c = c_j b v_j$, one obtains $v_c = Dc_j \mu b\Omega/(1-\nu)lRT \log(L'/\sqrt{bl})$. This is similar to equation (9.33) with minor modifications

Cross-slip of a screw dislocation

The glide plane is defined by the unit vector normal to the plane $\boldsymbol{n} = \boldsymbol{\xi} \times \boldsymbol{b}$ where $\boldsymbol{\xi}$ is the vector parallel to the dislocation line and $\boldsymbol{b}$ is the Burgers vector. For a screw dislocation, $\boldsymbol{\xi} \| \boldsymbol{b}$ and therefore $\boldsymbol{n} = 0$ and there is no unique glide plane. Consequently, a glide motion of a screw component of dislocation can occur on any crystallographic planes. The glide out of a glide plane is called *cross-slip* and is more difficult than the glide on a glide lane if a dislocation is dissociated.

When a dislocation is dissociated, then the motion of a (screw) dislocation involves the motion of two partial dislocations. Except for a special case in which the Burgers vectors of two dislocations (partial dislocations) are parallel to the original undissociated screw dislocation, cross-slip involves climb component of partial dislocations. Consequently, cross-slip may occur when the dissociated portion becomes undissociated by the shrink of a stacking fault. This is an energetically unfavorable process therefore it occurs only through thermal activation (Fig. 9.10; ESCAIG, 1968). POIRIER (1976a) emphasized the role of cross-slip as a process of dislocation recovery similar to dislocation climb. Both cross-slip of a screw dislocation and climb of an edge dislocation are needed for a dislocation glide loop to move its glide plane.

9.4. Dislocation multiplication, annihilation

We have so far considered the factors that control dislocation velocity. Let us now consider processes by which dislocation density is controlled.

9.4.1. Dislocation multiplication

Dislocations can be introduced into a crystal either through growth process (from a melt or from another phase due to a phase transformation) or by deformation (applied stress). A dislocation can move by an applied stress, but as far as its motion is homogeneous, there is no change in the length of a dislocation. In order to increase the length of a dislocation, dislocation motion must be heterogeneous (this is analogous to the generation of a magnetic field by dynamo (e.g., MERRILL *et al.*, 1998)). Let us consider a case where a dislocation is pinned at two points (Fig. 9.11a). Such pinning may occur for various reasons including the presence of a jog on a screw dislocation, a dislocation node, or an immobile impurity atom. In such a case, an applied stress will move a dislocation line between the pinning points until they form a loop. This process can continue to cause the

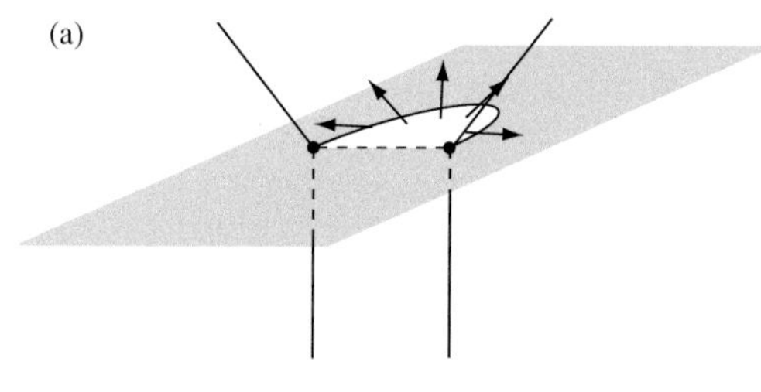

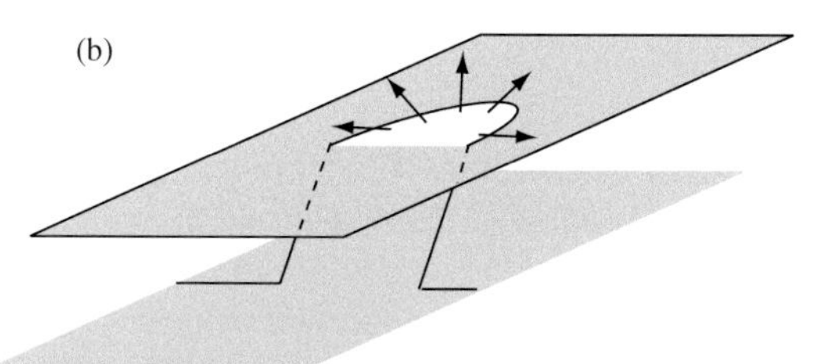

FIGURE 9.11 Processes of dislocation multiplication: (a) Frank–Read source, (b) double cross-slip.

increase in the length of dislocation (i.e., dislocation density). Such a source of dislocations is called the *Frank–Read* source. In order for this mechanism to work, one needs mechanisms for pinning. Also because of the *line tension* of a dislocation, as discussed in Chapter 5, the magnitude of applied stress must exceed

$$\sigma_{FR} \approx \frac{\mu b}{l} \tag{9.34}$$

where l is the distance of pinning points. A similar process is a double cross-slip (Fig. 9.11b). In both cases, the rate of dislocation multiplication is proportional to dislocation velocity and the number of dislocation sources, that is proportional to dislocation density, namely,

$$\dot{\rho}^{+} = AN\upsilon_g \tag{9.35a}$$

where N is the number density of dislocation source. When dislocations are multiplied by the Frank–Read source or by double cross-slip, $N \propto \rho$ and therefore

$$\dot{\rho}^{+} = A'\rho\upsilon_g. \tag{9.35b}$$

9.4.2. Recovery, annealing of dislocations

Dislocation can be annihilated. Two mechanisms are important. First, dislocations can be annihilated at *grain boundaries* (or subgrain boundaries). This happens when a dislocation reaches a grain boundary or when a moving boundary sweeps materials that contain dislocations. The latter process is particularly important in dynamic recrystallization (see Chapter 13 for details). Second, when dislocations with different signs collide, they disappear through dislocation climb (see Problem 9.4). Third, once dislocations are organized as a regular array, then they become immobile, and they no longer play an active role in deformation (note that the dislocation density in the Orowan equation is that of *mobile* dislocations). Collectively these processes are referred to as *recovery processes*. The rate at which dislocation density decreases in the first mechanism is proportional to the glide velocity and dislocation density, whereas for the second mechanism it is proportional to the climb velocity and the number of dipoles and hence to the second power of dislocation density, namely,

$$\dot{\rho}^{-} = B_1\rho\upsilon_g \text{ for grain-boundary annihilation} \tag{9.36a}$$

and

$$\dot{\rho}^{-} = B_2\rho^2 \text{ for dislocation–dislocation interaction} \tag{9.36b}$$

where B_2 is proportional to the diffusion coefficient.

9.4.3. Evolution of dislocation density and transient creep

The dislocation density changes with time following

$$\dot{\rho} = \dot{\rho}^{+} - \dot{\rho}^{-}. \tag{9.37}$$

Given specific mechanisms of dislocation multiplication and annihilation, and the knowledge of dependence of dislocation velocity on dislocation density, one can integrate equation (9.37) and calculate the strain versus time (or stress versus strain) relation.

To solve equation (9.37), one needs to know the relation between dislocation velocity and density. In the initial stage of work hardening (or at low temperatures), one may assume that the dislocation density does not change so much and hence dislocation velocity is independent of dislocation density (e.g., Johnston and Gilman, 1959; Johnston, 1962) see also Li (1963), Reppich *et al.* (1964), Webster (1966a, 1966b), Sumino (1974), and Karato (1977). Under these conditions equation (9.37) becomes

$$\dot{\rho} = A'\upsilon_g\rho - B_2\rho^2 = k_1\rho - k_2\rho^2 \tag{9.38}$$

where $k_1 = A'\upsilon$ and $k_2 = B_2$. Here I assumed that the dislocation multiplication is by a Frank–Read source and the annihilation is by mutual interaction (k_1 is proportional to the glide mobility and k_2 is proportional to the climb mobility, i.e., diffusion coefficient). If $k_{1,2}$ are independent of time, then equation (9.38) can be integrated to obtain $\rho(t)$. The dislocation density,

$\rho(t)$, can be inserted into the Orowan equation, $\dot{\varepsilon}(t) = \rho(t)bv$, to obtain (Problem 9.5)

$$\varepsilon = \varepsilon_0 + \dot{\varepsilon}_s t + \frac{\dot{\varepsilon}_s}{k_1}\log\left[1 + \frac{\dot{\varepsilon}_0 - \dot{\varepsilon}_s}{\dot{\varepsilon}_s}(1 - \exp(-k_1 t))\right] \tag{9.39}$$

where $\dot{\varepsilon}_0$ is the initial strain rate ($\dot{\varepsilon}_0 = \rho_0 bv$) and $\dot{\varepsilon}_s$ is the steady-state strain rate ($\dot{\varepsilon}_s = \rho_s bv$). For a short time (i.e., $k_1 t \ll 1$),

$$\varepsilon \approx \varepsilon_0 + \dot{\varepsilon}_s t + \frac{\dot{\varepsilon}_s}{k_1}\log\left(1 + \frac{\dot{\varepsilon}_0 - \dot{\varepsilon}_s}{\dot{\varepsilon}_s}k_1 t\right). \tag{9.40}$$

The assumption that $k_{1,2}$ is independent of time is valid only at low temperatures or the initial stage of deformation. In fact this logarithmic transient creep is often found at low temperatures (e.g., GAROFALO, 1965).

Problem 9.5

Derive equation (9.39).

Solution

Rearranging equation (9.38), one has $(1/\rho - 1/(\rho - k_1/k_2))\, d\rho = k_1\, dt$. If $k_{1,2}$ are independent of time, then this equation can be integrated to obtain $\rho = (k_1/k_2)\frac{1}{1-(1-(k_1/k_2)(1/\rho_0))\exp(-k_1 t)}$, where ρ_0 is the initial dislocation density. Inserting this relation into the Orowan equation, one gets $\dot{\varepsilon} = \frac{\dot{\varepsilon}_s}{1-[(\dot{\varepsilon}_0-\dot{\varepsilon}_s)/\dot{\varepsilon}_0]\exp(-k_1 t)}$ ($\dot{\varepsilon}_0 = \rho_0 bv$ (initial strain rate) and $\dot{\varepsilon}_s = \rho_s bv = (k_1/k_2)\cdot bv$ (steady-state strain rate)). When $k_{1,2}$ are independent of time, dislocation velocity is independent of time, and therefore integrating this equation once more (by changing the variable, $x \equiv A\exp(-k_1 t)$), one obtains, $\varepsilon = \varepsilon_0 + \dot{\varepsilon}_s t + (\dot{\varepsilon}_s/k_1)\log[1 + [(\dot{\varepsilon}_0 - \dot{\varepsilon}_s)/\dot{\varepsilon}_s]\cdot(1 - \exp(-k_1 t))]$.

At the later stage of transient creep (i.e., $(\dot{\varepsilon}_0 - \dot{\varepsilon}_s)/\dot{\varepsilon}_s \ll 1$), equation (9.39) becomes

$$\varepsilon = \varepsilon_0 + \dot{\varepsilon}_s t + [(\dot{\varepsilon}_0 - \dot{\varepsilon}_s)/k_1][1 - \exp(-k_1 t)]. \tag{9.41}$$

This form of transient creep (exponential transient creep) is frequently observed at high temperatures (e.g., AMIN *et al.*, 1970). However the derivation of (9.41) from (9.39) is not valid because the assumption that dislocation velocity is independent of dislocation density used to derive (9.39) is not appropriate in the later stage of transient creep. Furthermore, equation (9.41) implies that the relaxation time will be controlled by the glide velocity of dislocations, whereas the experimental data at high temperatures show that the relaxation time is controlled by the rate-controlling processes of dislocation recovery (in most cases the climb velocity; see AMIN *et al.*, 1970).

AKULOV (1964) presented a physical model to derive an exponential transient creep law from dislocation dynamics. He noted that equation (9.38) can be modified to

$$\frac{d\rho}{d\varepsilon} = \frac{d\rho}{dt}\frac{dt}{d\varepsilon} = \frac{k_1\rho - k_2\rho^2}{\rho bv} = \alpha_0 - \beta_1\rho = -\beta_1(\rho - \rho_s) \tag{9.42}$$

where $\alpha_0 = A'v_g/bv \propto v_g/v$, $\beta_1 = B_2/v \propto v_c/v$ (v_g, velocity of dislocation for multiplication (glide velocity); v_c, velocity of dislocation climb) and $\rho_s = \alpha_0/\beta_1$. At high temperatures, dislocation velocity is climb-controlled so β_1 is independent of strain and dislocation density. Under these conditions, equation (9.42) can be integrated to yield,

$$\dot{\varepsilon} = \dot{\varepsilon}_s + (\dot{\varepsilon}_0 - \dot{\varepsilon}_s)\exp(-\beta_1\varepsilon) \tag{9.43}$$

where $\dot{\varepsilon}_0 = \rho_0 bv$ and $\dot{\varepsilon}_s = \rho_s bv$. Changing the variable using $y \equiv [(\dot{\varepsilon}_0 - \dot{\varepsilon}_s)/\dot{\varepsilon}_s]\exp(-\beta_1\varepsilon)$, this can be modified to

$$\dot{\varepsilon} = \frac{\dot{\varepsilon}_s}{1 - [(\dot{\varepsilon}_0 - \dot{\varepsilon}_s)/\dot{\varepsilon}_0]\exp(-\beta_1\dot{\varepsilon}_s t)} \tag{9.44}$$

and hence

$$\varepsilon = \varepsilon_0 + \dot{\varepsilon}_s t + \frac{\dot{\varepsilon}_0 - \dot{\varepsilon}_s}{\beta_1\dot{\varepsilon}_s}[1 - \exp(-\beta_1\dot{\varepsilon}_s t)]. \tag{9.45}$$

This formula represents the transient creep behavior controlled by a relaxation process with the relaxation time of $\tau = 1/\beta_1\dot{\varepsilon}_s$, and is shown to fit a large number of data on high-temperature transient creep in metals and minerals (AMIN *et al.*, 1970; SMITH and CARPENTER, 1987). AMIN *et al.* (1970) derived equation (9.45) assuming an empirical relation (Problem 9.6)

$$\frac{d\dot{\varepsilon}}{dt} = -\frac{\dot{\varepsilon} - \dot{\varepsilon}_s}{\tau} \tag{9.46}$$

with $\tau \propto 1/\dot{\varepsilon}_s$.

Problem 9.6

Derive equation (9.45) from equation (9.46).

Solution

Integrating equation (9.46), one has $\dot{\varepsilon} = (\dot{\varepsilon}_0 - \dot{\varepsilon}_s)\cdot\exp(-t/\tau) + \dot{\varepsilon}_s$. Integrating this once more $\varepsilon = \varepsilon_0 + \dot{\varepsilon}_s t +$

$\tau(\dot{\varepsilon}_0 - \dot{\varepsilon}_s)[1 - \exp(-t/\tau)]$. If one inserts $\tau \propto 1/\dot{\varepsilon}_s$, then equation (9.45) can be obtained.

The above analysis for high-temperature transient creep formula can further be expanded to explain *Andrade creep*, i.e., $\varepsilon \propto t^{\beta}$ with $\beta < 1$ which is observed at modest to high temperatures (see Chapter 3). One way is to generalize equation (9.42) to

$$\frac{d\rho}{d\varepsilon} = \sum_m (\alpha_m - \beta_m)\rho^m \tag{9.47}$$

where α_m, β_m are relevant constants (Akulov, 1964). In a special case, where only one term dominates, i.e., $d\rho/d\varepsilon = -\beta_m \rho^m$and for large strain, one obtains

$$\varepsilon \propto t^{(m-1)/m}. \tag{9.48}$$

With $m = \frac{3}{2}$, one has $\varepsilon \propto t^{1/3}$, a transient creep behavior frequently observed in metals (Garofalo, 1965). The same relation can also be derived from (9.45) if one assumes a distribution of relaxation times (Minster and Anderson, 1981). In either case, note that the Andrade creep formula can be derived only for a certain range of time (or strain). As is obvious from the above derivation, the Andrade creep formula is not a good approximation for short time (small strain). In fact for $t \to 0$, the Andrade creep formula would predict an infinite strain rate that is not physical.

Let us example an implication of dislocation dynamics on the steady-state dislocation density. At high temperatures or at conditions close to steady-state creep, the assumption of a constant dislocation velocity is no longer valid. Under these conditions, the mutual interaction of dislocations to control dislocation velocity becomes important, and hence equation (9.38) needs to be modified. A commonly used relation is[4]

$$v_g = C(\sigma - \sigma_i) = v_0 - k\sqrt{\rho} \tag{9.49}$$

with $v_0 = C\sigma$ and $k = C\mu b$. Thus with the same assumption for the dislocation multiplication and annihilation as (9.38), one has

$$\dot{\rho} = A'\rho\left(v_0 - k\sqrt{\rho}\right) - B_2\rho^2. \tag{9.50}$$

At steady state, $A'\left(v_0 - k\sqrt{\rho_{ss}}\right) - B_2\rho_{ss} = 0$ and hence

$$\rho_{ss} = b^{-2}\left(\frac{\sigma}{\mu}\right)^2 \text{ for } \sigma \ll \frac{A'C\mu^2 b^2}{4B_2}. \tag{9.51}$$

This provides a theoretical basis for the dislocation density versus stress relationship,[5] and physically this means that under the condition of slow dislocation recovery (i.e., $\sigma \ll A'C\mu^2b^2/4B_2$), the dislocation density reaches a point where applied stress nearly balances with the internal stress. The characteristic time for equilibrium of dislocation density is given by $\tau \approx 1/B_2\rho$ for $\sigma \ll A'C\mu^2b^2/4B_2$. Essentially, this corresponds to a case where $\dot{\rho}^+ \gg \dot{\rho}^-$ so that at steady state, $\dot{\rho}^+ \propto \sigma - \sigma_i \approx 0$.

9.5. Models for steady-state dislocation creep

9.5.1. Generalities

The previous sections provide a general theoretical framework for modeling plastic deformation by dislocation motion. Given the dynamics of dislocation multiplication and recovery, such a theory provides us with a useful means to understand the variety of behavior in plastic deformation in various materials.

For example, in metals with a low Peierls stress (e.g., fcc metals, such as Cu and Au), dislocation glide in pure materials is easy, and these materials are soft. However, with further deformation these materials become harder due to the increase in resistance for dislocation motion caused by the increase in dislocation density (*work hardening*). Steady-state deformation is possible only at high temperatures, where dislocation density can be reduced to maintain steady-state deformation by dislocation motion. In short, in these materials, intrinsic dislocation glide is easy, but mutual interaction among dislocations makes their motion difficult, and steady-state deformation is possible only when dislocation density is maintained at a steady-state value due to dislocation annihilation via dislocation *climb* (or cross-slip).

In a material where the intrinsic resistance for dislocation glide (the *Peierls stress*) is large, dislocation multiplication by glide is difficult. In these cases, the rate of dislocation multiplication by glide may control the rate of deformation. The deformation of (dry) quartz in which the Peierls stress is high due to strongly covalent bonding belongs to this class. Glide-controlled steady-state deformation is also observed in some metallic alloys.

[4] Webster (1966b) used a relation $v_g = v_0 - k\rho$ in his analyses.

[5] Equation (9.38) would give $\rho_{ss} = k_1/k_2$. In this case the dislocation density would be sensitive to temperature, which is not consistent with the observation.

The variation in deformation behavior described above can be investigated through the formulation discussed in the previous section by varying the various parameters. In this section, I will review various models of steady-state creep based on some specific models for dislocation motion and structure.

9.5.2. Some models

The analysis of interaction between dislocation multiplication by glide and annihilation by climb (or by cross-slip) can be simplified in the following way if only the *steady-state* deformation is concerned. The whole process of dislocation multiplication and annihilation may be considered to be made of two steps, i.e., glide of dislocation by a distance L and annihilation by climb that eliminates the internal stress and allows the next glide. In this scheme the average velocity of dislocation can be written as,

$$v = \frac{L}{t_g + t_c} \tag{9.52}$$

where L: distance of unit motion of dislocation before dislocations stop at certain obstacles. If $t_g \ll t_c$ (glide is easy, climb is difficult) then you have climb-controlled creep. This is the case for metals and simple ionic solids. If $t_c \ll t_g$ (glide is difficult) then you have glide-controlled creep. This is the case for covalent materials and/or materials with complicated crystal structure. Detailed discussions on the models of high-temperature creep can be found in Poirier (1976b, 1985).

Climb control model

When dislocation climb is more difficult, then

$$v = \frac{L}{t_g + t_c} \sim \frac{L}{t_c} = \frac{L}{d} v_c \tag{9.53}$$

where d is the distance between glide planes. Using the Orowan equation, we get,

$$\dot{\varepsilon} = \rho b v = \rho b \frac{L}{d} v_c. \tag{9.54}$$

Subgrain boundary recovery model

In a material that has been deformed at steady-state creep, dislocations often show an organized structure (Fig. 9.12) in which many of them occur as *subgrain boundaries*. Several models have been proposed in which it is assumed that the rate of steady-state deformation is controlled by dislocation recovery and the dislocation recovery occurs in subgrain boundaries.

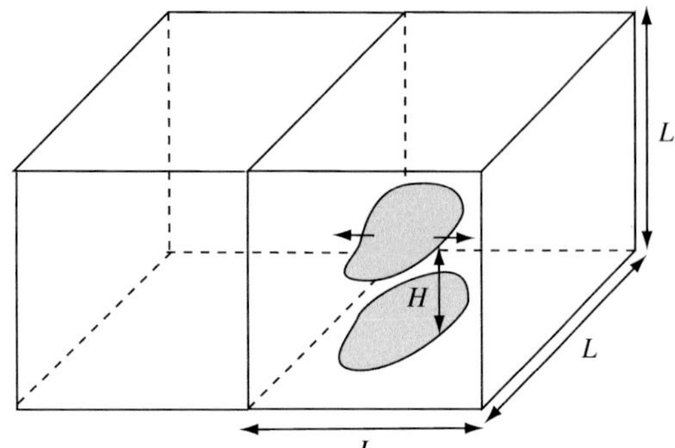

FIGURE 9.12 A subboundary recovery model of creep. Dislocation loops whose mutual distance is H expand and when they reach the subboundary, they annihilate through dislocation climb. Formation of loops is due to the operation of some dislocation sources (see Fig. 9.11). The rate of annihilation at subboundaries is assumed to be the rate-controlling step.

Therefore the starting point of this type of model is the relation (9.42) and one calculates the steady-state density of mobile dislocations and the rate of recovery.

Let us consider a model in which dislocations are distributed as loops in a subgrain with the size L (in this model there is no mention of the source of dislocations). This block is surrounded by subboundaries. Dislocations move within subgrains to create strain, but their density is kept constant by the recovery at subgrain boundaries. If the mean distance of loops is H, then the density of mobile dislocations (dislocations in a subgrain) is given by

$$\rho = NL = \frac{L}{H}\frac{1}{L^3}L = \frac{1}{HL} \tag{9.55}$$

where N is the number of loops in the cell. Now let us assume that the internal stress caused by mutual interaction of dislocations ($\sigma_i \approx \mu b/H$) is balanced with the applied stress. Thus,

$$H = \frac{\mu b}{\sigma}. \tag{9.56}$$

The distance that dislocations must climb to cause annihilation is $\approx \frac{1}{2}H$. Therefore using (9.54), one gets,

$$\dot{\varepsilon} = b\frac{1}{HL}\frac{2L}{H}v_c = 2\left(\frac{\sigma}{\mu b}\right)^2 b v_c = \frac{\alpha D c_j \mu \Omega}{b^2 RT}\left(\frac{\sigma}{\mu}\right)^3. \tag{9.57}$$

In this model, the size of subgrains, L, disappears from the final equation. If one uses an empirical relation $L \approx b\mu/\sigma$, then equations (9.55) and (9.56) yield $\rho \approx b^{-2}(\sigma/\mu)^2$.

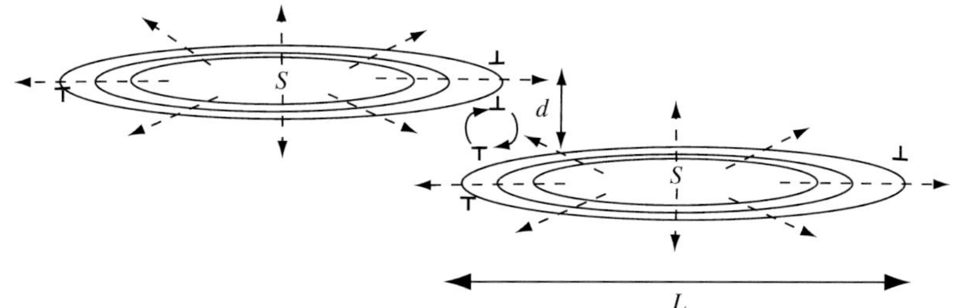

FIGURE 9.13 Weertman's model of steady-state, recovery-controlled dislocation creep. Dislocation loops are generated by the source (S), and they expand. When two loops meet at a distance d, their motion stops until a dislocation is annihilated by climb (see Fig. 9.9). When a pair of dislocations is annihilated (through climb motion), then the propagation of the next loop becomes possible to maintain "steady-state" creep.

Weertman model

WEERTMAN (1968) proposed a model of steady-state creep in which some details of dislocation multiplication and annihilation processes are considered. In this model, both dislocation multiplication and annihilation are considered to occur homogeneously in a crystal. He proposed that dislocations are generated from some sources and propagate on a given glide plane (Fig. 9.13). Creep strain is due to dislocation glide. Dislocations on neighboring planes interact with each other to make pairs of dislocations (dislocation dipoles). Once a dislocation dipole is made, dislocations do not glide any more. Further dislocation glide is possible only after dislocations are annihilated through climb. Thus the whole process is rate-controlled by climb of dislocations and hence by diffusion.

The starting point is the modified Orowan equation (9.54). Let us first consider a number (or the density) of dislocations. There is one source per volume of $\pi L^2 d$ and thus, if M is the density of dislocation source, then

$$M = \frac{1}{\pi L^2 d} \tag{9.58}$$

The density of mobile dislocation is given by (a number of sources per unit volume) × (the length of dislocations per one dipole) × (number of dipoles for one source). The number of dislocation dipoles is here because a dislocation becomes *mobile* only when annihilation occurs at a dipole, i.e., the rate of dislocation motion. For one source, the length of dislocation is πL. The number of dipoles per source is given by $\sim L/d$ (at a dipole, there must be a balance of stress due to neighboring dislocation on the same plane ($\sim 1/L$) and the stress between two dislocations on separate planes ($\sim 1/d$)). Therefore,

$$\rho = M \cdot \pi L \cdot \frac{L}{d} = \frac{\pi M L^2}{d} \tag{9.59}$$

From (9.58) and (9.59), one gets $\rho = 1/d^2$. Thus,

$$\dot{\varepsilon} \sim b\frac{L}{d^3} v_c \sim \frac{b}{d^{\frac{7}{3}} M^{\frac{1}{2}}} v_c \tag{9.60}$$

Now from Problem 9.4, $v_c \sim Dc_j \sigma$ and $d \sim b(\mu/\sigma)$ (from $\rho \approx b^{-2}(\sigma/\mu)^2$). Therefore,

$$\dot{\varepsilon} \sim \left(\frac{\sigma}{\mu}\right)^{9/2} \frac{Dc_j}{M^{1/2}} \tag{9.61}$$

$n = 4.5$ if M (the density of dislocation source) is independent of stress (n is larger than 3 (a standard value), because of the factor L/d). However, if M is dependent on stress, then other stress dependence will result. For example, if one assumes $M \sim \rho \sim \sigma^2$, then $n = 3.5$. If one assumes $M \sim \rho/d \sim \sigma^3$, then $n = 3$.

Nabarro model

NABARRO (1967a) proposed a model of dislocation creep in which strain is caused by dislocation climb. Contribution from dislocation climb was demonstrated by the analysis of shape change in single crystals of olivine by DURHAM *et al.* (1977). Similarly, DUCLOS *et al.* (1978) inferred important contribution from dislocation climb in oxide spinel deformed at high homologous temperature based on transmission electron microscopy study of dislocation structures. The rate of deformation in this model is proportional to the rate of dislocation climb and hence to the diffusion coefficient, and is to the third power of stress ($n = 3$). Deformation in this case does not have a rotational component, and hence no lattice-preferred orientation will develop (see Chapters 5 and 14).

Influence of pipe diffusion

Some modification can be made for the recovery-controlled model by introducing the effects of dislocation core diffusion. The velocity of dislocation climb is controlled by diffusion. Diffusion can occur, not only through the bulk of a crystal, but also along the dislocation lines (see Chapter 8). If the influence of diffusion along a dislocation line is included, equation (9.32) can be replaced with

$$v_c = \frac{2\pi(D + D^{\text{pipe}}\pi b^2 \rho)c_j\sigma\Omega}{bRT\log(L/b)} = \frac{2\pi\left(D + D^{\text{pipe}}\pi(\sigma/\mu)^2\right)c_j\sigma\Omega}{bRT\log(L/b)} \tag{9.62}$$

which predicts $\dot{\varepsilon} \propto \sigma^n$ at low stress and $\dot{\varepsilon} \propto \sigma^{n+2}$ at high stress where n is the stress exponent in the model where bulk diffusion controls the rate of dislocation climb.

The major features of climb-controlled creep can be summarized as follows. (1) The stress and strain-rate are related by a non-linear relation. If a power-law formula is used then, $n = 3$–5. (2) The activation energy of creep is the same as that of activation energy of diffusion if c_j is constant. In crystals with a high Peierls stress, c_j is dependent upon temperature through $c_j \sim \exp(-H_j^*/RT)$ where H_j^* is enthalpy for jog formation. In this case, $H^*_{\text{creep}} = H^*_{\text{diff}} + H_j^*$ where H^*_{diff} is activation energy for diffusion. (3) Processes that enhance diffusion will also enhance creep. (4) Processes that increase jog density will enhance creep in materials with a high Peierls stress. These last two points are important in relation to the enhancement of creep rate by water (see Chapter 10).

Glide control model

When dislocation glide is difficult, then glide itself will control the rate of deformation. Crystals with a high Peierls stress or solid-solution with a large misfit (class I alloys) (because of large interaction of dislocations with solute atoms that resists glide) belong to this category.

When dislocation glide is more difficult, then

$$v = \frac{L}{t_g + t_c} \sim \frac{L}{t_g} = v_g. \tag{9.63}$$

Using the Orowan equation, we find,

$$\dot{\varepsilon} = \rho b v = \rho b v_g. \tag{9.64}$$

The creep rate is directly related to the dislocation density and dislocation velocity. There is no term related to dislocation geometry (such as L/d in equation (9.54)). Consequently the stress dependence of strain rate comes from that of a dislocation density, $\rho \approx b^{-2}(\sigma/\mu)^2$, and that of dislocation glide velocity. The activation enthalpy for creep is that of dislocation glide. WEERTMAN (1957) proposed a model for glide-controlled high-temperature creep. He considered both viscous dislocation motion (e.g., motion of a dislocation controlled by solute drag) and dislocation motion controlled by the Peierls stress. In both cases, the dislocation density is considered to be controlled by the balance between internal stress and applied stress and hence, $\rho \approx b^{-2}(\sigma/\mu)^2$. At low stress, the velocity is nearly linearly dependent and hence the stress exponent is $n \sim 3$. A glide-controlled creep model involving solute atoms was proposed by TAKEUCHI and ARGON (1976).

In geological materials, glide-controlled creep could occur due to the high Peierls stress. In this case, the rate of deformation depends on stress in an exponential fashion at high stresses. This occurs due to the stress dependence of activation free energy (see section 9.3.1).

Harper–Dorn creep

There are some reports suggesting that plastic deformation of a single crystal shows linear relationship between stress and strain rate under some conditions (HARPER and DORN, 1957). In these cases, the dislocation density is independent of applied stress and such a mechanism of creep is called the *Harper–Dorn creep*. In other words, the dislocation density is not determined by the applied stress but it assumes a *frozen-in* value. If this mechanism works at low stresses, then the extrapolation of high-stress data (on power-law creep) assuming that the only competing mechanism is linear diffusional creep would be misleading. There have been some suggestions that this mechanism might operate in Earth (e.g., LANGDON *et al.*, 1982; WANG, 1994; WANG *et al.*, 1994; VAN ORMAN, 2004).

The most unusual feature of Harper–Dorn creep is that the dislocation density is independent of applied stress. There are several proposals to explain this. For instance, WEERTMAN and BLACIC (1984) proposed that the fluctuation of temperature gives rise to a chemical stress at a dislocation that might exceed the applied stress. In other words, they suggested that the observed linear rheological behavior is an artifact of experimental conditions and is not a material property. There is no experimental support for this notion, however.

LANGDON and YAVARI (1982) reviewed several mechanisms by which the dislocation density is independent of applied stress. They start from the Orowan equation and discuss that if the stress dependence of $\dot{\rho}^+$ is equal to that of $\dot{\rho}^-$, then steady-state dislocation density will be independent of stress and is determined by the number of dislocation sources. However, when dislocation density is controlled by processes other than the balance of internal stress with applied stress, dislocation density is sensitive to kinetic parameters such as diffusion coefficient and hence to temperature (and other chemical environment). For example, in a model preferred by LANGDON and YAVARI (1982), dislocation density will be proportional to diffusion coefficient, which is inconsistent with the observations.

ARDELL (1997) applied a dislocation network model for creep to explain Harper–Dorn creep (see also PRZYSTUPA and ARDELL, 2002). In the dislocation network model of creep, creep involves motion of "free" dislocations in subgrains and their interaction with dislocations on subgrains that form networks. "Free" dislocations glide into subboundaries where free dislocations and dislocations in the network interact. Assuming the balance between the stress around a free dislocation and stress around a network dislocation, the density of free dislocations is given by $\rho \propto 1/l^2$ where l is the mean spacing of network dislocations. Now under the conditions where network dislocations are easy to change their mean spacing, force balance demands $\sigma \propto 1/l$ leading to $\rho \propto \sigma^2$. A dislocation network consists of a number of nodes where the relation $\boldsymbol{b} = \sum \boldsymbol{b}_i$ is satisfied (see Chapter 5). Coarsening of a network involves destruction of some nodes. ARDELL (1997) argued that network coarsening is no longer possible when possible dislocation reactions that allow network coarsening are exhausted. If this happens then l and hence the dislocation density becomes independent of applied stress, and would explain the linear relation between stress and strain rate. Although this model explains observations in some materials with high symmetry where a number of reactions exist, a dislocation network is not observed in some minerals with low symmetry. In order to have a network, one needs to have a reaction among various dislocations satisfying $\boldsymbol{b} = \sum \boldsymbol{b}_i$ but such a reaction is not observed in olivine for example.

NABARRO (1989) proposed that the dislocation density is controlled by the Peierls stress if the applied stress is below the Peierls stress, i.e., $\rho \approx b^{-2}(\sigma_P/\mu)^2$ (i.e., it is independent of applied stress). J. N. Wang presented some discussions to support the Nabarro model (e.g., WANG and NIEH, 1995). However, no explanation is provided as to why dislocation density is independent of applied stress below the Peierls stress (many materials show non-linear dislocation creep with stress-dependent dislocation density at a stress below the Peierls stress). In addition, the choice of experimental data is unclear and the observational basis for Harper–Dorn creep in this series of work is rather weak. For example, WANG (1994) and WANG *et al.* (1994) cited some experimental data on olivine and quartz. However, their interpretation of the experimental data is not convincing. WANG (1994) interpreted the low stress portions of data by KOHLSTEDT and GOETZE (1974) and KARATO *et al.* (1986) in terms of Harper–Dorn creep (linear dislocation creep), but the later detailed follow-up studies clearly showed that the linear rheological behavior observed by KARATO *et al.* (1986) is indeed due to diffusional creep (e.g., HIRTH and KOHLSTEDT, 1995a; MEI and KOHLSTEDT, 2000a) and the stress exponent for dislocation creep is $n \sim 3.5$ down to the lowest stress so far investigated (e.g., BAI *et al.*, 1991).

In summary, there is no experimental support for Harper–Dorn creep in geological materials and a well accepted model to explain stress-independent dislocation density has not been proposed. I conclude that physical processes of Harper–Dorn creep remain elusive and the possibility of this mechanism in Earth and planetary interior is at best speculative at this stage. For a critical review of Harper–Dorn creep see also BLUM *et al.* (2002).

9.6. Low-temperature plasticity (power-law breakdown)

In the above models, the rate of deformation is considered to be controlled by the slower of two successive processes, namely dislocation glide and climb (equation (9.58)). The constitutive relation for these models is characterized by the power-law relationship between stress and strain rate, $\dot{\varepsilon} \propto \sigma^n$. Deviation from such a relation is often observed at high stress (or at low temperatures) in which strain rate increases with stress more rapidly than a power-law relation predicts. In these cases, the constitutive relation is best characterized by assuming that the activation enthalpy is stress dependent, i.e., $\dot{\varepsilon} \propto \exp[-H^*(\sigma)/RT]$ (or $\dot{\varepsilon} \propto \sigma^2 \cdot \exp[-H^*(\sigma)/RT]$). In all cases, the activation enthalpy in this regime decreases with stress and there is a threshold stress at which the activation enthalpy becomes zero. When stress reaches this level, deformation is possible without thermal activation (athermal motion of dislocations). Under these conditions, the strength is only weakly dependent on temperature and strain rate, and the strength in this regime is often referred to as *yield strength* (or *yield stress*). The yield strength is therefore nearly identical to the Peierls stress in materials with relatively strong chemical bonds.

This transition can be interpreted in two different ways. First is the case where creep is always controlled by glide because glide is more difficult than climb (recovery). In this case, the creep rate is always controlled by the following type of formula, $\dot{\varepsilon} \propto \sigma^2 \exp[-H^*(\sigma)/RT]$, and the transition to exponential creep law occurs when the stress dependence of activation enthalpy is strong (i.e., at high stress). This is a likely case for crystals with a high Peierls stress. However, if this explanation is valid, then at low temperatures where climb is very difficult, a switch to climb-control would occur. In contrast, the exponential flow law applies to very low temperatures as shown experimentally by EVANS and GOETZE (1979). Indeed, the exponential flow law appropriate for the Peierls mechanism fits to their experimental data very well down to room temperature and is used to determine the Peierls stress. In such a case, an alternative model proposed by FROST and ASHBY (1982) may work better in which it is presumed that the competition of glide and climb will no longer occur at low-temperature and high-stress regimes. A microscopic basis for this model has not been well examined but likely involves the stress-assisted *unpinning* of dislocations from an obstacle such as a dislocation dipole (this is analogous to a model of unpinning proposed by KARATO and SPETZLER (1990) to explain transition from anelasticity to visco-elasticity). In this model, the low-temperature, high-stress exponential creep and high-temperature power-law creep are alternative mechanisms, whichever is easier will control the strength. This model is used to construct a deformation mechanism map (ASHBY, 1972; FROST and ASHBY, 1982) (for more details of "power-law breakdown," see also TSENN and CARTER (1987)).

Deformation by *twinning* is a similar case. When two crystals share a lattice point on a common plane but other atomic positions of two crystals (or two portions of a crystal) have mirror symmetry with respect to this common plane, they are referred to as twins (Fig. 9.14). In certain crystals (calcite, quartz, orthopyroxene,[6] orthorhombic perovskite) twinning occurs by the shear stress. In most cases, the rate of deformation by mechanical twinning is controlled by thermally activated nucleation and the appropriate constitutive equation is

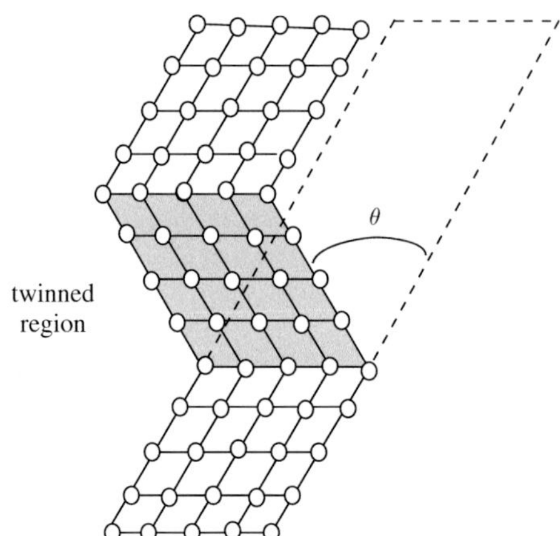

FIGURE 9.14 Deformation by twinning (compare with Fig. 9.2). Twinning results in shear strain, $\gamma = \tan\theta$, that is determined by the crystallography of twinning.

$$\dot{\varepsilon} = A \cdot \exp\left[-\frac{H_0^*(1-\sigma/\sigma_T)}{RT}\right] \tag{9.65}$$

where A is a constant, H_0^* is the activation enthalpy at zero stress, and σ_T is the critical stress for nucleation of twins at $T = 0$ K (e.g., FROST and ASHBY, 1982). The magnitude of σ_T tends to be smaller than that of the Peierls stress in many crystals for easy twinning. In these cases, stress supported by these minerals is limited to σ_T.

9.7. Deformation of a polycrystalline aggregate by dislocation creep

A single crystal has a number of different slip systems with different rheological properties. Therefore, at a single crystal level, rheological properties are anisotropic: rheological properties depend on the orientation of a crystal relative to the stress. Therefore when a polycrystalline aggregate is subjected to an external stress, individual grains will be deformed differently that will also modify the stress distribution at the level of grains. Both stress and strain will be heterogeneous in a deforming polycrystal and the actual stress and strain distribution is highly complicated (see Chapter 12 for some details). However, a simple end-member case can be considered. VON MISES (1928) showed that when strain is homogeneous, then in order for a polycrystalline aggregate to deform by slip, there must be five independent slip systems (see also Chapter 5). In this case, stress at individual grains will be heterogeneous, and higher stress will be supported in grains in which deformation occurs

[6] In case of pyroxene, finite strain is created associated with a transformation from orthoenstatite to clinoenstatite under the shear stress. This is not twinning, but the kinetics and geometry of deformation by this process are similar to those involved in deformation by twinning.

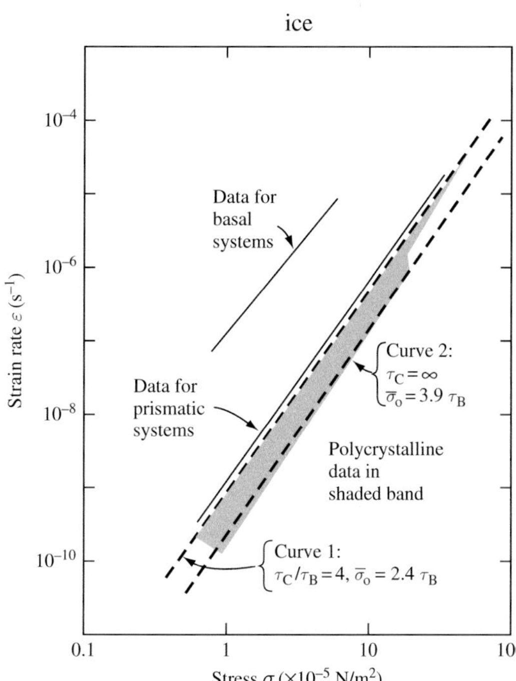

FIGURE 9.15 A comparison of single crystal stress versus strain-rate relations with that of a polycrystal (for ice) (after HUTCHINSON, 1977).

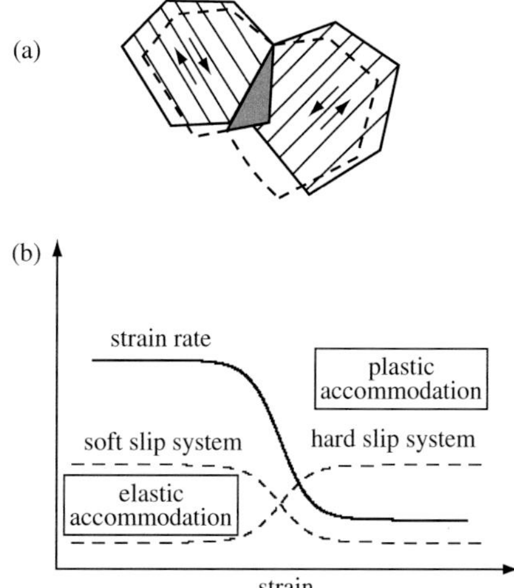

FIGURE 9.16 Strain partitioning in a deforming polycrystal. (a) At small strain, strain at grain boundaries can occur by elastic accommodation, whereas it involves the operation of hard slip systems at large strain. (b) The strength of a polycrystal is controlled by that of a soft slip system for small strains, whereas for large strains, the strength is controlled by that of the hard slip system(s).

mostly by hard slip system(s). When deformation proceeds to a large strain, then strain accommodation just mentioned becomes important and consequently, the strength of a polycrystalline aggregate at large strains is largely controlled by that of hard slip system(s) (HUTCHINSON, 1976, 1977) (Fig. 9.15).

However, the processes of strain accommodation among various grains can be different at small strains. DUVAL *et al.* (1978, 1983) showed that the creep strength of a polycrystalline ice at large strains is controlled by that of hard slip systems (prismatic and/or pyramidal slip systems), whereas the creep strength at small strains is controlled largely by a weak slip system (basal slip system). This can be understood if one recalls that strain accommodation at small strains can be made by elastic strains. At small strains, plastic deformation occurs first in grains which are orientated favorably for a soft slip system(s). These grains deform causing strain mismatch with surrounding grains. This strain mismatch can be accommodated by elastic strain if the total strain is small enough. At this stage, therefore, the strength of a polycrystalline aggregate is largely controlled by that of a weak slip system. Eventually, stress associated with elastic strain becomes so large that other harder slip systems start to operate. Eventually most of the deformation and strain accommodation will be accomplished by plastic deformation and at this stage, the overall strength is controlled by that of the hard slip systems. Therefore there will be a gradual transition from soft slip system-control to hard slip system-control as deformation proceeds (Fig. 9.16). The critical strain at which this transition occurs is approximately an elastic strain. This model predicts that at small strains, there must be recoverable strain (anelastic deformation) because a part of strain is elastic. This point was demonstrated by DUVAL *et al.* (1978). KARATO (1998c) formulated this model and argued that since the strain magnitude associated with the post-glacial rebound is not far from elastic strain, transient creep may play an important role in post-glacial rebound (see Chapter 19). A similar process was discussed by BUSSOD and CHRISTIE (1991) and by HIRTH and KOHLSTEDT (1995b) in relation to dislocation creep in fine-grained peridotite and dunite respectively.

Important messages to be learned are (1) in order to understand the creep strength of a polycrystal from the experimental studies on single crystals, one needs to investigate the rheological properties of hard slip systems as well as those of soft ones. This point must be emphasized, because in many experimental studies on deformation of single crystals only the plastic flow for soft slip systems were determined (see Chapter 6). (2) There is a unique mechanism for transient creep in deformation of a polycrystal. Deformation of a polycrystalline aggregate at strains near or less than the elastic strain occurs only through the operation of soft slip systems.

9.8. How to identify the microscopic mechanisms of creep

From a purely pragmatic point of view of a geologist or geophysicist, finding a constitutive relation under a given physical and chemical state is enough and there is no actual need for identifying the microscopic mechanisms of deformation (such as climb versus glide-control). However, understanding the microscopic mechanisms is critical when one wants to understand the significance of limited experimental data for a broad range of geophysical questions. For example, if dislocation glide is slower than climb, mechanical data from a laboratory study often show strain softening (controlled by dislocation multiplication). In contrast if dislocation climb is more difficult than glide, then work hardening will occur. However, in Earth transient behavior will not be important as far as deformation occurs to large strain in a quasi-steady-state fashion. A functional form to represent the dependence of creep rate on water content is different between these two cases (see Chapter 10).

9.8.1. From steady-state flow law

Non-linear dependence of strain rate on stress is a strong indication for dislocation creep. When the activation enthalpy is stress dependent, it suggests an important contribution of dislocation glide. However the detailed values of stress exponent (say $n=3$ versus $n=4$) do not provide useful information as to microscopic mechanisms of deformation, although the difference between $n=1$ and $n=3$ is important in understanding the mechanisms of deformation. Similarly, the values of activation enthalpy are not very useful in constraining the microscopic mechanisms of dislocation creep. In contrast, the dependence of strain rate on some thermochemical parameters such as oxygen fugacity, oxide activity, water fugacity provide useful constraints on the microscopic mechanisms of deformation (e.g., Bai *et al.*, 1991; Karato and Jung, 2003).

9.8.2. From dislocation microstructures

Observations of dislocation microstructures are often the key to the identification of microscopic mechanisms of deformation. (1) Slip systems (glide directions, i.e., the Burgers vectors, and glide planes) can be determined by the observations by transmission electron microscopy (operating slip systems can also be inferred from SEM or optical observations of subboundaries). (2) Among the edge versus screw dislocations, those dominantly observed are the ones that move slower, and hence are rate-controlling dislocations. (3) The morphology of dislocation lines provides some clue as to the importance of intrinsic resistance for dislocation motion (by the Peierls stress): if the nucleation of kinks is much more difficult than the migration of kinks, dislocations will show linear morphology along certain crystallographic directions. This is an indication of a high Peierls stress. (4) When dislocation lines are not confined on a glide plane, it implies that dislocation climb and/or cross-slip are active.

9.8.3. From a stress-dip test

A stress-dip test is a useful technique in distinguishing glide-control from a recovery (climb)-control model. The basic idea behind this test is the concept of "internal stress." When the creep rate is controlled by recovery, then during steady-state deformation, the applied stress is nearly balanced with the internal stress. Consequently, if one reduces the stress (a stress dip) after steady-state creep, then the applied stress would be less than the internal stress and therefore deformation will *completely* stop until internal stress is reduced. In contrast, if internal stress is substantially lower than the applied stress, then strain rate after a reduction of stress (stress dip) will be finite. This will be the case when creep rate is controlled by dislocation glide. Stress-dip tests were applied to metals (Kurishita *et al.*, 1989) as well as to oxides (Wang *et al.*, 1993, 1996).

9.9. Summary of dislocation creep models and a deformation mechanism map

Summary of high-temperature creep models

Except for poorly understood Harper–Dorn creep, most of the high-temperature creep models summarized above show a power-law relationship between (deviatoric) stress and strain rate, $\dot{\varepsilon} \propto \sigma^n$. The value of stress exponent n is typically 3–5. Among these values of n, $n=3$ is the most natural value coming from $\dot{\varepsilon} = \rho b v$ with $\rho \propto \sigma^2$ and $v \propto \sigma$. Consequently, a power law $\dot{\varepsilon} \propto \sigma^3$ is considered to be "canonical" high-temperature creep behavior (e.g., Weertman and Weertman, 1975). In other words, power-law creep with $n > 3$ involves some specific processes of dislocation multiplication and/or recovery.

Cannon and Sherby (1973) and Cannon and Langdon (1988) suggested that the stress exponent, n, may depend on properties of materials. For example, Cannon and Sherby (1973) argued that creep with $n = 3$ is observed in ceramics with $r_{anion}/r_{cation} > 2$ ($r_{anion,cation}$ is ionic radius of anion (cation)) while in ceramics with $r_{anion}/r_{cation} > 2$ creep with $n = 5$ is commonly observed. Similarly, Cannon and Langdon (1988) argued that creep with $n = 5$ occurs in materials that satisfy the von Mises condition (more than five independent slip systems), whereas creep with $n = 5$ occurs in materials that have less than five independent slip systems. However, the correlation of stress exponent with properties of materials is not clear. For example, for MgO for which extensive experimental studies have been conducted, the stress exponent is $n = 3$–4 (e.g., Frost and Ashby, 1982; Yamazaki and Karato, 2002), although it has $r_{anion}/r_{cation} = 1.80 < 2$ and the von Mises condition is satisfied. I conclude that there is no clear trend in the stress exponent for power-law creep. Various possible mechanisms by which the stress exponent takes a value other than 3 have already been discussed including the stress dependence of dislocation sources, change in diffusion path (bulk versus dislocation core diffusion).

The role of jogs in recovery processes or the degree to which dislocation glide controls the rate of deformation of minerals is unclear. Recently, Kohlstedt (2006) proposed a simple diffusion-controlled recovery model for olivine, but the experimental and theoretical bases for such a model are weak. For example, the observed large anisotropy suggests an important role of jogs or the important contribution from dislocation glide. In fact, the interpretation of fabric transitions in olivine (Chapter 14) suggests an important role of glide-controlled creep under high-stress and/or low-temperature conditions.

Deformation mechanism maps

We have learned that there are a number of mechanisms by which a given material can be deformed plastically. A deformation mechanism map is a useful tool to identify a dominant mechanism of deformation for a given material under a variety of physical (and chemical) conditions. A deformation map is constructed based on a number of experimental data on plastic deformation for a variety of mechanisms, and using the relationships between different mechanisms, one can identify the dominant mechanism of deformation as a function of some key parameters.

A deformation mechanism map (for steady-state deformation) for a given material can be constructed as follows (a detailed account of deformation mechanism maps is provided by Frost and Ashby (1982)):

1. Write down the steady-state constitutive equation for possible mechanisms under consideration using available experimental data.
2. Consider which mechanisms are parallel (independent) and which ones are sequential (dependent). For parallel mechanisms, strain rates are additive, namely,

$$\dot{\varepsilon} = \sum \dot{\varepsilon}_i \tag{9.66}$$

thus the mechanism that yields the highest strain rate under a given condition will be dominant. Diffusional and dislocation creep are independent mechanisms because they involve different defects. For sequential mechanisms (or dependent mechanisms), the inverse strain rate is additive, namely,

$$\dot{\varepsilon} = \frac{1}{\sum(1/\dot{\varepsilon}_i)}. \tag{9.67}$$

Typical dependent mechanisms are dislocation glide and climb in high-temperature, power-law creep. There are some exceptions to this classification. For example, when creep is due to dislocation motion, at high-stress levels, dislocation creep is either controlled by recovery (or glide) or by glide without pinning due to dislocation interaction. In such a case, the two processes are independent, and the actual rate of deformation is the faster of the two, but not the sum of the two (in practice these two cases give similar results although the two cases are different theoretically).
3. Choose appropriate variables (grain size and temperature, temperature and stress etc.) to map the dominant mechanisms. Note that in most cases, a two-dimensional diagram is used for convenience in which case one needs to recall that some other parameters are fixed.
4. Estimate reasonable values of these variables relevant to a given geological process.
5. Draw conclusions as to the dominant deformation mechanism(s).

Examples of a deformation mechanism map are illustrated in Fig. 9.17.

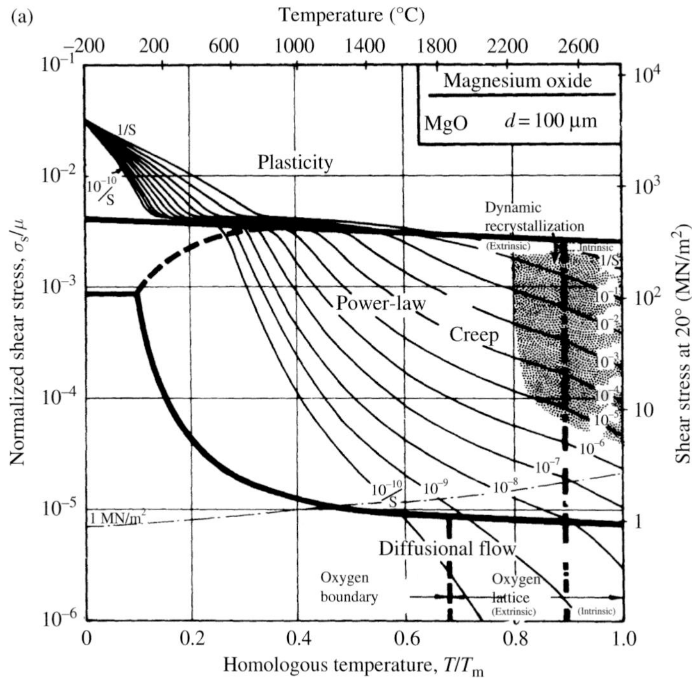

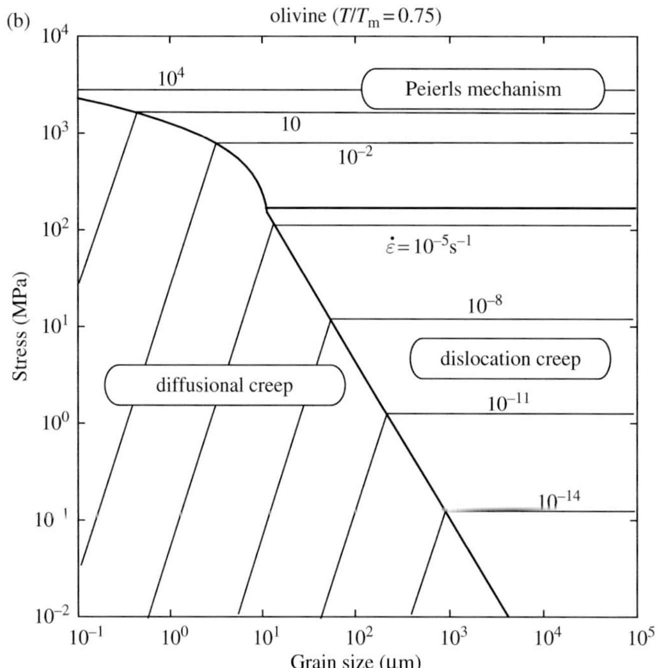

FIGURE 9.17 (a) A deformation mechanism map for MgO in the temperature–stress space for a grain size of 100 μm showing competition among power-law dislocation creep, exponential flow law ("plasticity") and diffusional creep (from FROST and ASHBY, 1982). (b) A deformation mechanism map for olivine in the grain size versus stress space at $T/T_m = 0.75$ under "dry" conditions (Coble creep, i.e., $m = 3$ is assumed for diffusional creep).

General conclusions from these maps are:

1. For a range of conditions that might be relevant to geological processes (we consider ductile deformation only), the following mechanisms appear to be potentially important: power-law creep, diffusional creep and low-temperature plasticity (the Peierls mechanism).
2. The Peierls mechanism is important at high stresses and/or low temperatures. Therefore it may be important in some portions of cold slabs.

3. Diffusion creep–dislocation creep boundaries are close to typical conditions in most of the Earth's interior. This implies that if there is any mechanism of grain-size reduction, then significant rheological weakening should occur (see also Chapter 16). This also implies that in boundary layers (lithosphere etc.) where stress is high, one might expect dislocation creep, whereas in the stagnant core of a convection cell, one might expect diffusional creep (see also KARATO, 1998d).

Applications of deformation maps will be discussed in Chapter 19 in relation to some specific geological questions. I should note that a deformation mechanism map is a good guide to infer possible mechanisms of deformation under a variety of conditions. However, in many geological or geophysical applications, it is the observations on deformation microstructures that provide a more robust conclusion on the dominant deformation mechanisms. A combination of microstructural observations (including observations on seismic anisotropy (see Chapter 21)) and a deformation mechanism map is a good way to understand the dominant deformation mechanisms. Finally I must quote a cautious note: *Both the equations used to describe creep behaviour, and the maps constructed from them, must be regarded as a first approximation only. The maps are no better (no worse) than the equations and data used to construct them* (FROST AND ASHBY, 1982).

10 Effects of pressure and water

Pressure has important effects on plastic flow. Pressure affects the rate of plastic flow through two distinct mechanisms. First, the height of the potential barrier for atomic motion changes with pressure, which usually makes plastic flow more difficult at higher pressures. This effect is mainly characterized by a parameter called activation volume. Experimental observations and theoretical models of activation volume are summarized including the pressure dependence of activation volume. For a constant stress, the rate of deformation changes with pressure by as much as ten orders of magnitude in Earth's interior. Second, pressure modifies the chemical environment particularly the fugacity of water that controls the concentration of point defects and changes the rate of plastic deformation. The fugacity of water becomes higher at higher pressures, which enhances the rate of plastic deformation by several orders of magnitude. Experimental observations and theoretical models of effects of pressure and water on plastic deformation are reviewed including empirical correlations such as the homologous temperature scaling and some atomistic models of defects.

Key words activation volume, homologous temperature, Keyes model, water fugacity, hydrogen-related defects, FT–IR spectroscopy.

10.1. Introduction

It is often argued that the strength of a rock in a brittle regime increases strongly with pressure, but the strength in the ductile (plastic flow) regime is relatively insensitive to pressure. This is misleading, however, and in fact, the effects of pressure on plastic deformation can be large and complicated under the conditions of Earth's deep interior. The large effect of pressure is obviously very important in the deep mantle where pressure becomes very large, but it is already important in the depth range exceeding ~30 km (mid-lithosphere and below). For instance, the effects of pressure can change the viscosity of the deep continental lithosphere (~200 km depth) by a factor of $\sim 10^3$–10^5.

There are three different ways by which pressure affects plastic deformation. First, pressure affects all the fundamental physical properties such as molar volume and elastic modulus. The *driving force for defect motion* is dependent on these properties and hence the rate of deformation becomes pressure-dependent through its effect on driving forces for defect motion. For example, the effects of stress to cause plastic deformation can be scaled as σ/μ (σ, stress; μ, shear modulus), thus the changes in the elastic modulus cause a change in plastic properties. This effect is, however, relatively small (elastic moduli change with pressure by a factor of ~3–5 in the Earth's mantle; see Chapter 4). A second, and more important effect, is the effect through the change in *defect mobility*. The defect mobility may change with pressure due to (1) the pressure dependence of the fugacity of some chemical species such as water and/or to (2) the pressure dependence of activation free energy (i.e., activation volume). These effects are large and in most cases act in opposite ways: the effects of water for instance are to enhance deformation whereas the effects of the activation volume are usually to suppress deformation. The effect of water is also pressure-dependent. The water effect is large only at high pressures (high water fugacity). Third, pressure causes phase transformations that modify plastic properties. The effects of phase transformations are discussed in Chapter 15, and in this chapter, I will focus on the first two effects.

Since the physical mechanisms are quite different between the pressure effects without the influence of water and the (pressure-dependent) water-weakening effects, I will treat them separately in this chapter. I will first (in section 10.2) focus on the effects of pressure on plastic properties without water-weakening effects (effects at the absence of water or at a constant water content) and then discuss the effects of water in section 10.3. However, as we will learn in this chapter, a proper quantitative analysis of the influence of water (hydrogen) can only be made when the influences of water and pressure are analyzed together.

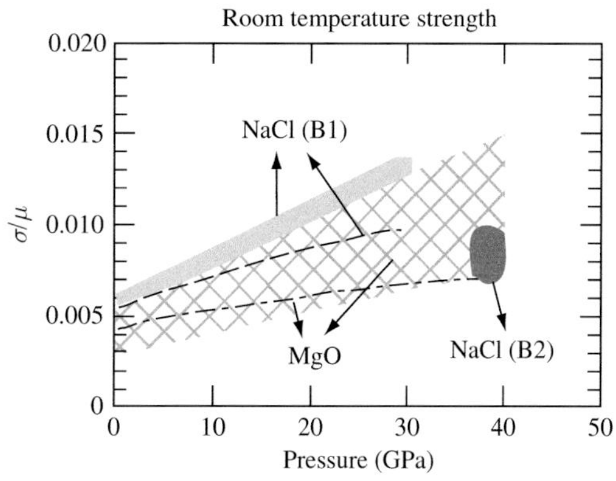

FIGURE 10.1 The pressure dependence of (normalized) strength, σ/μ, of various materials determined at room temperature (after KARATO, 1998b).

10.2. Intrinsic effects of pressure

In this section, I will summarize experimental observations and models on the effects of pressure on plastic deformation under the conditions where the effects of the chemical environment are negligible. In the geological or geophysical context, the most important pressure-sensitive chemical environment that affects plastic flow is the fugacity of water. Therefore the results summarized in this section would apply to Earth materials for constant water content (water fugacity) or for very small water content.

10.2.1. Experimental observations

Introduction

Two different causes for this type of pressure dependence can be distinguished. First, plastic deformation in most cases involves the thermally activated motion of defects, and the barrier for defect motion depends on pressure. This effect can be expressed by the pressure dependence of free energy for motion of defects, namely the *activation volume* (V^*) for defect motion (this may include the activation volume for formation of defects). For example when the power-law creep formula is assumed, the pressure dependence of strain rate due to this effect can be expressed as $\dot{\varepsilon} = A(\sigma^n/L^m)\cdot \exp(-(E^* + PV^*)/RT)$ ($\dot{\varepsilon}$ is strain rate, σ is stress, L is grain size, E^* is activation energy, V^* is activation volume, R is the gas constant, T is temperature, P is pressure A, n and m are constants), see Fig. 10.1. Second, pressure changes elastic constants and the dimension of unit cell and hence changes the stress field around a defect. This effect can be incorporated in the power-law creep equation as $\dot{\varepsilon} = A'(P, T)\cdot (b(P, T)^m/L^m)(\sigma^n/\mu(P, T)^n)\exp(-(E^* + PV^*)/RT)$ where b is the length of the Burgers vector and μ is the shear modulus. The length of the Burgers vector changes little with T and P (the change throughout the mantle is less than ~10%) but shear modulus increases significantly with pressure (the change throughout the mantle is a factor of ~2–3). Therefore this effect is mostly from the pressure dependence of shear modulus. This effect is through $b(P, T)^m/\mu(P, T)^n$ term and will result in the hardening effects. For example, when pressure is increased by ~15 (100) GPa, then $|\Delta b/b| \approx 0.03$ (~0.1) and $|\Delta\mu/\mu| \approx 0.3(\sim 3)$. Therefore the magnitude of this effect will be to reduce the strain rate by a factor of ~2 (~30). We conclude that the effects of pressure through this mechanism are modest. The most important effect is the effect through the activation volume. Because this effect is through an exponential term, i.e., $\exp(-PV^*/RT)$, this effect is small at low pressures and cannot be determined accurately at low pressures, but its effects become large at higher pressures. For example, let us consider a typical case of activation volume of $15\times 10^{-6}\,\mathrm{m^3/mol}$. At the temperature of 1600 K, this would result in a change of strain rate by a factor of ~2 when the pressure change is $\Delta P = 0.5$ GPa (this is the maximum pressure range that can be explored by a gas-medium deformation apparatus). If the pressure change is $\Delta P = 15$ GPa, the strain rate will be reduced by a factor of $\sim 10^8$. The effects of pressure on strain rate through the activation volume are shown in Fig. 5.2b. It is seen that the effect through $\exp(-PV^*/RT)$ term can be very large, but the magnitude of this effect depends strongly on the value of activation volume. It should also be noted that because the rate of deformation changes with pressure following the exponential function, $\exp(-PV^*/RT)$, the pressure effect is large only at high pressures.

Consequently, it is important to investigate the pressure effects in a wide range of pressures. Although the resolution of a low-pressure, gas-medium deformation apparatus is high, it is difficult to obtain a tight constraint on the pressure dependence of plastic deformation using a gas-medium apparatus (see Problem 10.1).

Problem 10.1*

Show that the error of the activation volume determined from the experiments at $P = P_1$ and $P = P_2$ corresponding to a generic power-law creep ($\dot{\varepsilon} = A(\sigma^n/L^m)\exp(-(E^* + PV^*)/RT)$) is

$$\left|\frac{\delta V^*}{V^*}\right| = \frac{RT}{|P_1 - P_2|V^*}\left(\left|\frac{\delta\dot{\varepsilon}}{\dot{\varepsilon}}\right| + n\left|\frac{\delta\sigma}{\sigma}\right| + m\left|\frac{\delta L}{L}\right| + \frac{PV^*}{RT}\left|\frac{\delta P}{P}\right| + \frac{H^*}{RT}\left|\frac{\delta T}{T}\right|\right)$$

where $\delta X/X$ is the error in determining the parameter X and $H^* = E^* + PV^*$ is the activation enthalpy ($n = 1$–5, $m = 0$–3; see Chapters 9 and 10). Typical values of activation volume and activation enthalpy in silicates or oxides are $(5\text{–}20) \times 10^{-6}\,\mathrm{m^3/mol}$ and (400–800) kJ/mol. Assuming the following values of errors in determining various parameters, discuss the relative merit of high-resolution, low-pressure experiments and low-resolution, high-pressure experiments in determining the activation volume.

Pressure range	0.1 MPa to 0.5 GPa	0.1 MPa to 15 GPa
Error in pressure	1%	5%
Error in stress	1%	10%
Error in strain rate	5%	10%
Error in temperature	0.1%	3%
Error in grain size	10%	10%

Solution

From the power-law constitutive equation, uncertainties in several measured parameters are related as

$$\log\dot{\varepsilon} + \frac{\delta\dot{\varepsilon}}{\dot{\varepsilon}} = \log A + n\left(\log\sigma + \frac{\delta\sigma}{\sigma}\right) - m\left(\log L + \frac{\delta L}{L}\right) - \frac{H^*}{RT}\left(1 - \frac{\delta T}{T}\right)\left(1 + \frac{PV^*}{H^*}\left(\frac{\delta P}{P} + \frac{\delta V^*}{V^*}\right)\right)$$
$$\approx \log A + n\left(\log\sigma + \frac{\delta\sigma}{\sigma}\right) - m\left(\log L + \frac{\delta L}{L}\right) - \frac{H^*}{RT}\left(1 - \frac{\delta T}{T} + \frac{PV^*}{H^*}\frac{\delta P}{P} + \frac{PV^*}{H^*}\frac{\delta V^*}{V^*}\right). \quad (1)$$

Writing this equation for $P = P_1$ and $P = P_2$, and subtracting each other, one obtains

$$\log\frac{\dot{\varepsilon}_1}{\dot{\varepsilon}_2} + \frac{\delta\dot{\varepsilon}}{\dot{\varepsilon}} = n\frac{\delta\sigma}{\sigma} - m\frac{\delta L}{L} - \frac{(P_1 - P_2)V^*}{RT} + \frac{H^*}{RT}\frac{\delta T}{T} - \frac{PV^*}{RT}\frac{\delta P}{P} - \frac{(P_1 - P_2)V^*}{RT}\frac{\delta V^*}{V^*} \quad (2)$$

where one notes that errors occur in a random fashion in each experiment so errors at $P = P_1$ and $P = P_2$ do not cancel. Now $\log(\dot{\varepsilon}_1/\dot{\varepsilon}_2) = -(P_1 - P_2)V^*/RT$, so

$$\frac{\delta\dot{\varepsilon}}{\dot{\varepsilon}} = n\frac{\delta\sigma}{\sigma} - m\frac{\delta L}{L} + \frac{H^*}{RT}\frac{\delta T}{T} - \frac{PV^*}{RT}\frac{\delta P}{P} - \frac{(P_1 - P_2)V^*}{RT}\frac{\delta V^*}{V^*} \quad (3)$$

one obtains

$$\left|\frac{\delta V^*}{V^*}\right| = \frac{RT}{|P_1 - P_2|V^*}\left(\left|\frac{\delta\dot{\varepsilon}}{\dot{\varepsilon}}\right| + n\left|\frac{\delta\sigma}{\sigma}\right| + m\left|\frac{\delta L}{L}\right| + \frac{PV^*}{RT}\left|\frac{\delta P}{P}\right| + \frac{H^*}{RT}\left|\frac{\delta T}{T}\right|\right). \quad (4)$$

With the assumed values of errors and parameters, the errors in activation volume determined from low-pressure experiments are dominated by the errors in grain size and/or pressure measurements and will be 20–60%. In contrast, the errors in activation volume determined from high-pressure experiments are dominated by those in pressure and/or temperature measurements and the error in activation volume will be 3–10%. This large difference comes mostly from the difference in the span of pressure, i.e., the difference in $RT/|P_1 - P_2|V^*$. Therefore although the errors in the measurement of various parameters (i.e., $|\delta\dot{\varepsilon}/\dot{\varepsilon}| + n|\delta\sigma/\sigma| + m|\delta L/L| + (PV^*/RT)|\delta P/P| + (H^*/RT)|\delta T/T|$) are large in high-pressure experiments, overall errors in the activation volume are significantly smaller than those in low-pressure experiments.

Some representative results

There are a wide variety of experimental observations on the pressure effects on plastic deformation. In general, experimental studies on plastic deformation at high pressures are difficult and many experimental studies have been performed under limited conditions. Consequently, physical significance of experimental results on pressure dependence of plastic properties can be very different among different sets of experiments. It is critical to identify the microscopic processes controlling the plastic flow in each experimental condition to obtain sensible information on the effects of pressure on plastic deformation from experimental studies. Also, the interpretation of results under high pressures is complicated due to the influence of water. Under high pressures, it is very difficult to avoid contamination of a sample with water that has a large effect on plastic deformation. Effects of water must be carefully corrected in order to determine the true effects of pressure.

(1) Low-temperature plasticity

Effects of pressure on low-temperature plasticity have been investigated for many materials including metals (e.g., Jesser and Kuhlmann-Wilsdorf, 1972), alkali halides (e.g., Davis and Gordon, 1968; Auten *et al.*, 1973; Meade and Jeanloz, 1988) and oxides and silicates (e.g., Auten *et al.*, 1976; Meade and Jeanloz, 1990; Mao *et al.*, 1998; Merkel *et al.*, 2003). Under these conditions, the plastic flow in materials is controlled by dislocation glide and the thermal activation plays relatively minor roles. Consequently, the "strength" determined in these experiments is insensitive to the strain rate and reflects largely the stress needed to move dislocations on their glide planes – such as the Peierls stress. *At high temperatures (most of Earth's interior), the rate-controlling processes for plastic flow are different, and consequently, results at low temperatures cannot be extrapolated to high temperatures and have little relevance to Earth science.* Fig. 10.1 shows some of these results (on ionic crystals). The pressure effects on strength are rather modest and the strength increases nearly linearly with pressure.

(2) High-temperature plasticity (Table 10.1)

Heard and Kirby (1981) determined the pressure dependence of creep strength of CsCl in high-temperature, power-law dislocation creep regime by conducting constant strain-rate deformation experiments at pressures up to 0.4 GPa and temperature of 423–673 K (0.45–0.73 T/T_m). This material has a large activation volume and its activation volume for creep was well determined by these low-pressure measurements ($V^* = 53 \times 10^{-6}\,\mathrm{m^3/mol}$). Similarly large values of activation volume were obtained for the formation processes of Schottky defects in alkali halides.

Pressure dependence of high-temperature dislocation creep in olivine has been investigated by many groups (Fig. 10.2). Ross *et al.* (1979) were the first to attempt to determine the activation volume for Earth material by conducting deformation experiments at high pressures. They used the solid-medium Griggs apparatus to $P = 1.5$ GPa at $T = 1373$–1623 K (0.64–0.74 T/T_m) and determined the activation volume for creep in olivine as $V^* = 13 \times 10^{-6}\,\mathrm{m^3/mol}$. Borch and Green (1987, 1989; Green and Borch, 1987) conducted deformation experiments on olivine to 2.5 GPa (at 1473 K) using an improved Griggs-type apparatus (see Chapter 6). They obtained $V^* = 27 \times 10^{-6}\,\mathrm{m^3/mol}$. Deformation experiments on olivine were conducted to much higher pressures using a multi-anvil apparatus (Bussod *et al.*, 1993; Karato and Rubie, 1997). Bussod *et al.* (1993) used grain size as a stress indicator to determine $V^* = 5$–$10 \times 10^{-6}\,\mathrm{m^3/mol}$, whereas Karato and Jung (2003) and Karato and Rubie (1997) obtained $V^* = 14 \times 10^{-6}\,\mathrm{m^3/mol}$ for "dry" conditions by estimating the strength from elastic constants using a theoretical model and from the dislocation densities respectively. This result agrees well with a result of theoretical calculation ($V^* = 13.9 \times 10^{-6}\,\mathrm{m^3/mol}$, Karato, 1977). Using a recently developed deformation apparatus, D-DIA, Li *et al.* (2003b) reported an even smaller activation volume, $V^* < 3 \times 10^{-6}\,\mathrm{m^3/mol}$. Fig. 10.2 compares these results. The discrepancy among several studies is large showing the difficulties in obtaining reliable results on pressure effects on plastic deformation. Important issues to be noted in evaluating these results include: (1) the resolution of (or the uncertainties in) stress measurements, (2) evaluation of the effects of water (hydrogen), (3) the flow-law formula used (power-law creep versus exponential flow law, i.e., deformation mechanisms) and (4) the influence of transient creep. In addition, in some cases defects introduced during the pressurization process are not well annealed, and that can potentially influence the mechanical behavior and introduce an extra degree of complication in interpreting the data. In order to resolve the cause for this discrepancy, a careful analysis of the issues listed above will be needed.

Although a large scatter in the reported values of activation volume for olivine is alarming, there are some robust observations for other materials. For instance, the activation volumes for diffusion (and creep) in alkali halides and MgO, which are well-constrained experimentally, can be explained by a theoretical model rather well (Karato, 1981a, 1981b) (see also the later part of this chapter). Also, some

TABLE 10.1 Experimental data on activation volumes (units: V^* ($\times 10^{-6}\ m^3/mol$), P (GPa), T (K)). Only results from high-temperature conditions ($T/T_m > 0.5$) are shown.

	V^*	P	T	property	remarks	ref.
CsCl	53 ± 6	10^{-4}–0.40	423–673	Dislocation creep		(1)
NaCl	50 ± 5	0.02–0.65	900–1000	Schottky formation	From conductivity	(2)
	7 ± 1	0.02–0.65	900–1000	Na migration	From conductivity	(2)
NaBr	44 ± 5	0.02–0.65	900–1000	Schottky formation	From conductivity	(2)
	9 ± 1	0.02–0.65	900–1000	Na migration	From conductivity	(2)
KCl	59 ± 5	0.02–0.65	900–1000	Schottky formation	From conductivity	(2)
	8 ± 1	0.02–0.65	900–1000	K migration	From conductivity	(2)
KBr	54 ± 5	0.02–0.65	900–1000	Schottky formation	From conductivity	(2)
	11 ± 1	0.02–0.65	900–1000	K migration	From conductivity	(2)
Garnet[1]	5.3 ± 3	1.3–4.3	1373–1753	Diffusion (Mg)	"Dry"[5]	(3)
	5.6 ± 3	1.3–4.3	1373–1753	Diffusion (Fe)	"Dry"[5]	(3)
	6.0 ± 3	1.3–4.3	1373–1753	Diffusion (Mn)	"Dry"[5]	(3)
Olivine	5.5	1.0–3.5	1173–1373	Diffusion (Mg–Fe)	"Dry"[5]	(4)
	5.4 ± 4	1.0–4.0	1473	Diffusion (Mg–Fe)	"Dry" (~200 ppm H/Si)	(5)
	1 ± 1	0.5–9	873–1173	Diffusion (Mg–Fe)	"Dry"[5]	(6)
	16 ± 6	0.3–6	1373–1450	Diffusion (Mg–Fe)	Water-saturated[6]	(7)
	0.7 ± 2.3	4.0–9.0	1773	Diffusion (Si)	"Dry" (< 1 ppm H/Si)	(8)
	10.6 ± 0.3	0.3–12	1373–1623	Hydrogen solubility	Water-saturated[6]	(9)
	13 ± 3	0.5–1.5	1373–1623	Dislocation creep	"Wet"[7]	(10)
	5–10	6–13.5	1723–1873	Dislocation creep	"Dry"[5,8]	(11)
	14 ± 1	0.3–15	1873	Dislocation creep	"Dry"[9] (<100 ppm H/Si)	(12)
	15 ± 5[3]	0.1–0.3	1523	Diffusional creep	"Dry"[5]	(13)
	24 ± 3	0.1–2.2	1473	Dislocation creep	Water-saturated[6]	(14)
	27 ± 5	1.2–2.6	1473	Dislocation creep	"Dry"[5]	(15)
	0–5	0.3–8	618–1780	Dislocation creep	"Dry"[5]	(16)
MgO	3.0 ± 0.4	15–25	2273	Diffusion of Mg	"Dry"[5]	(17)
	3.3 ± 2.4	15–25	2273	Diffusion of O	"Dry"[5]	(17)
	1.8 ± 1.2	7–35	1573–1973	Diffusion of Mg–Fe	"Dry"[5]	(18)
Anorthite[2]	24 ± 21	0.1–0.4	1273–1423	Diffusional creep	"Dry"[10]	(19)
	38[4]	0.1–0.4	1273–1423	Diffusional creep	"Wet" (water-saturated)	(19)

(1): Heard and Kirby (1981).
(2): Yoon and Lazarus (1972).
(3): Chakraborty and Ganguly (1992).
(4): Misener (1974).
(5): Farber *et al.* (2000).
(6): Jaoul *et al.* (1991).
(7): Hier-Majumder *et al.* (2005a).
(8): Béjina *et al.* (1999).
(9): Kohlstedt *et al.* (1996).
(10): Ross *et al.* (1979).
(11): Bussod *et al.* (1993).
(12): Karato and Rubie (1997).
(13): Mei and Kohlstedt (2000a).
(14): Karato and Jung (2003).
(15): Green and Borch (1987).
(16): Li *et al.* (2004b).

Notes to table 10.1 (cont.)

(17): VAN ORMAN *et al.* (2003).

(18): YAMAZAKI and IRIFUNE (2003).

(19): RYBACKI *et al.* (2006).

[1] Almandine–spessartine–pyrope

[2] Synthetic hot-pressed anorthite

[3] $\pm 5 \times 10^{-6}$ m^3/mol is the reported error bar. Because the pressure range explored in this study is very small, the actual error bar is likely much larger.

[4] No error bar is reported

[5] No water was added, but no report on water content after the experiment.

[6] The partial pressure of water was ~ total pressure. The activation volume was calculated after the correction for the fugacity effect.

[7] Water was supplied by the decomposition of talc. The stress measurements were made using the external load cell with a solid pressure medium. The results have very large uncertainties due to the influence of friction (see Chapter 6).

[8] Results from stress-relaxation tests. Recrystallized grain size was used to infer the stress.

[9] Results from stress-relaxation tests. The initial stress was estimated from the elastic constants of the components in the sample assembly.

[10] Water content ranges from 380 to 3100 ppm H/Si for "dry" samples.

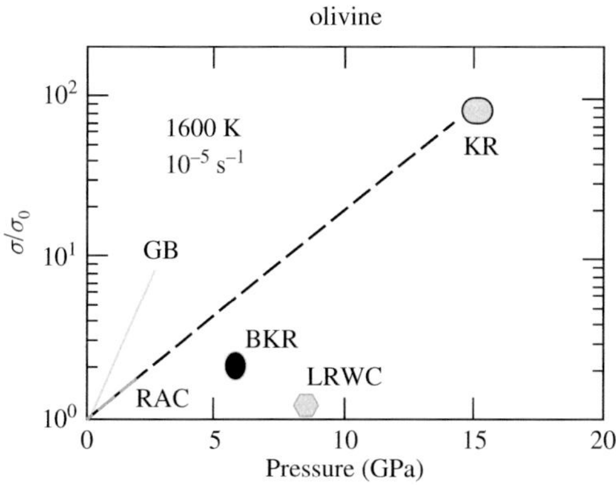

FIGURE 10.2 Experimental results for the pressure dependence of the creep strength of olivine at high temperatures ($T/T_m > 0.5$) (after KARATO, 1998b).

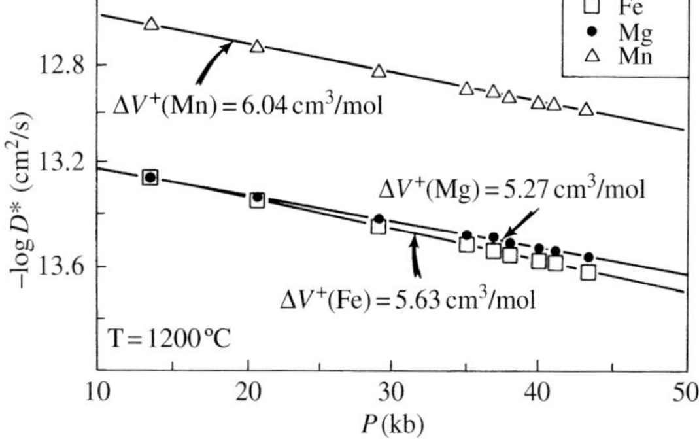

FIGURE 10.3 The pressure dependence of diffusion coefficients in garnet (after CHAKRABORTY and GANGULY, 1992).

well-defined experimental data on olivine such as the data reported by KOHLSTEDT *et al.* (1996), KARATO and JUNG (2003), and HIER-MAJUMDER *et al.* (2005a) under water-saturated conditions for a broad range of pressure can be explained by a combination of the influence of water and pressure (see section 10.3.6).

(3) *Diffusion (Table 10.1)*

Diffusion of atoms is one of the important elementary processes for high-temperature plastic deformation. Diffusion controls the rate of diffusional creep. Diffusion also plays an important role in high-temperature dislocation creep in many materials. An experimental set-up for diffusion experiments is relatively simple compared to that for plastic deformation. Consequently, a large number of experimental results are already available for diffusion under high pressures (for review see e.g., SAMMIS *et al.*, 1977; 1981). In many cases, the rate of diffusion decreases with pressure and the pressure dependence of diffusion can be characterized by the activation volume, namely, $D = D_0 \exp(-(E^* + PV^*)/RT)$ (Fig. 10.3).

Some caution must be exercised in interpreting high-pressure diffusion data, however. When diffusion coefficients are measured by high-pressure annealing using a high-pressure apparatus with solid media, effects of plastic deformation (i.e., effects of dislocations) could be important. This is particularly important because a very thin layer near the surface (interface) is used in diffusion measurements. Any "damage" of the surface (interface) during pressurization will seriously affect the results. Also, as noted before, a larger amount of water tends to be dissolved in many silicates and oxides at higher pressures that enhances diffusion. These two effects would reduce the observed pressure effects. Some of the observed unusually small effects of pressure on diffusion may be attributed to these effects.

10.2.2. Models for pressure dependence of plastic deformation

Some cautious notes must be made here because the concept of *activation volume* has not always been correctly understood. In many literatures, the activation volume of diffusion or deformation was identified directly with that of an atomic species that is involved in a relevant process. Also, a significant decrease of activation volume with pressure was frequently invoked to explain the modest depth variation of viscosity of Earth's mantle inferred from geodynamic considerations. Neither of these notions is supported by the physical models of pressure effects on deformation (or diffusion) as will be clear in this chapter.

Due to the difficulties in the experimental studies of plastic deformation at high pressures, experimental data on the pressure effects are sparse and many of them are conducted under limited conditions. As we will learn in this chapter, the physical significance of observed pressure dependence of plastic properties can be different among different processes of deformation. For instance, in the low-temperature regime where plastic deformation is controlled by dislocation glide (e.g., the Peierls mechanism), the effects of pressure on the strength are more or less linear (i.e., the strength is roughly proportional to the elastic modulus and hence increases approximately linearly with pressure) and only modest (up to a factor of 3–5 increase with pressure in the whole mantle). In contrast, in the high-temperature, power-law creep regime or in the diffusional creep regime, pressure can change the strength by ten orders of magnitude in the whole mantle. The results obtained for one particular mechanism of deformation cannot be extrapolated to the strength of materials under conditions where some other mechanisms control the strength. Therefore it is very important to identify the microscopic mechanisms of deformation before one can formulate the experimental results to apply them to Earth's interior.

General considerations

Pressure effects on plastic deformation can be analyzed either by measuring the influence of confining pressure on strain rate at constant stress, or by measuring the stress needed to deform a material at constant strain rate. Both techniques should give an identical result if deformation is steady state although there are some differences in the nature of uncertainties (see Chapter 6).

When the rate of deformation is controlled by thermally activated processes and when deformation occurs in the power-law regime, then, for "steady-state" deformation,[1] one can write

$$\dot{\varepsilon} = A \cdot \exp\left(-\frac{G^*}{RT}\right) \cdot \sigma^n = A \cdot \exp\left(-\frac{S^*}{R}\right) \exp\left(-\frac{E^* + PV^*}{RT}\right) \cdot \sigma^n. \tag{10.1}$$

Activation volume, V^*, is a partial derivative of the Gibbs free energy of activation with respect to pressure (see Chapter 2), i.e.,

$$V^* = \left(\frac{\partial G^*}{\partial P}\right)_T = \left(\frac{\partial H^*}{\partial P}\right)_T \tag{10.2}$$

and activation entropy S^* is given by $S^* = (-\partial G^*/\partial T)_P$. Therefore activation volume can be determined by the pressure dependence of activation enthalpy. Alternatively, activation volume V^* can be determined either from[2]

$$V^* = nRT\left(\frac{\partial \log \sigma}{\partial P}\right)_{T,\dot{\varepsilon}} \quad \text{for constant strain-rate tests} \tag{10.3a}$$

or

$$V^* = -RT\left(\frac{\partial \log \dot{\varepsilon}}{\partial P}\right)_{T,\sigma} \quad \text{for constant stress tests} \tag{10.3b}$$

if the pressure dependence of $A \cdot \exp(S^*/R)$ is negligible. The main contribution to the activation entropy

[1] In order to determine activation volume using this formula, deformation must be "steady state." Deviation from steady-state deformation can lead to a systematic error in the estimated activation volume.

[2] Green and Borch (1987) reported that V^* determined from $V^* = nRT \cdot (\partial \log \sigma/\partial P)_T$ does not agree with the value calculated from $V^* = (\partial H^*/\partial P)_T$ with $H^* = nR(\partial \log \sigma/\partial(1/T))_P$ (the former gives positive V^* whereas the latter gives a negative V^*). They argued the importance of the second derivatives of Gibbs free energy of activation such as $(\partial V^*/\partial P)_T = (\partial^2 G^*/\partial P^2)_T$ and $(\partial S^*/\partial T)_P = -(\partial^2 G^*/\partial T^2)_P$. However, the contributions from the second derivatives are usually small (see discussions in the later section) and the discrepancy between two results is likely due to the variation of n with pressure (n will increase with pressure if the increase in stress results in the change in the rheological regime from the power-law to the Peierls regime, which would lead to the decrease in apparent H^* with pressure).

is the change in vibrational frequencies (see Chapter 2), so the pressure dependence of $\exp(S^*/R)$ is small compared to $\exp(-PV^*/RT)$. The pressure dependence of A is usually small as I discussed above. Only exception is the case where A contains water fugacity. In this case, the pressure dependence of A is strong. Such a case will be discussed separately in section 10.3. It must also be noted that the determination of activation volume by equation (10.3a) with a constant n is valid only for the power-law regime. Because a material will become stronger at high pressures, it often happens that the stress level increases with pressure and the deformation mechanism changes to the Peierls mechanism. In such a case, the use of equation (10.3a) gives an apparent activation free energy that decreases with pressure (negative activation volume).

(1) Diffusional creep

For Nabarro–Herring diffusional creep, the constitutive law is (Chapter 8)

$$\begin{aligned}\dot{\varepsilon} &= A_1 \frac{\sigma b^3}{RT}\frac{D_0}{L^2}\exp\left(-\frac{E_1^* + PV_1^*}{RT}\right)\\ &= A_1 \nu_D \frac{\sigma b^3}{RT}\frac{b^2}{L^2}\exp\left(-\frac{E_1^* + PV_1^*}{RT}\right)\end{aligned} \tag{10.4}$$

where I used a relation $D_0 \approx b^2 \nu_D$ (b, the length of the Burgers vector; ν_D, the Debye frequency). Therefore the pressure dependence of creep rate comes from that of the Debye frequency, shear modulus and the activation volume term. When stress is kept constant, then,

$$\frac{\partial \log \dot{\varepsilon}}{\partial P} = \frac{\partial \log \nu_D}{\partial P} - \frac{1}{K} - \frac{V_1^*}{RT}. \tag{10.5}$$

Using the definition of the Grüneisen parameter (see Chapter 4), one obtains

$$\frac{\partial \log \dot{\varepsilon}}{\partial P} = \frac{\gamma - 1}{K} - \frac{V_1^*}{RT}. \tag{10.6}$$

For typical crustal or mantle minerals, the second term dominates (for $K = 100$ GPa, $\gamma = 1.5$, $V_1^* = 10 \times 10^{-6}$ m^3/mol, (second term)/(first term) ~ 50). Since the flow law is linear between stress and strain rate, a case of constant strain rate is the same as the case of constant stress.

(2) Power-law dislocation creep

A generic constitutive equation for power-law dislocation creep is (Chapter 9)

$$\dot{\varepsilon} = A_2 \nu_D \left(\frac{\sigma}{\mu}\right)^n \exp\left(-\frac{E_2^* + PV_2^*}{RT}\right). \tag{10.7}$$

Therefore for a constant stress,

$$\begin{aligned}\frac{\partial \log \dot{\varepsilon}}{\partial P} &= \frac{\partial \log \nu_D}{\partial P} - n\frac{\partial \log \mu}{\partial P} - \frac{V_2^*}{RT}\\ &= -\frac{(2n-1)\gamma + \frac{1}{3}n}{K} - \frac{V_2^*}{RT}.\end{aligned} \tag{10.8a}$$

Using the analysis similar to the diffusional creep, it can be shown that the second term dominates the pressure dependence.

For a constant strain rate,

$$\begin{aligned}\frac{\partial \log \sigma}{\partial P} &= \frac{1}{n}\left[n\frac{\partial \log \mu}{\partial P} - \frac{\partial \log \nu_D}{\partial P} + \frac{V_2^*}{RT}\right]\\ &= \frac{[(2n-1)/n]\gamma + \frac{1}{3}}{K} + \frac{V_2^*}{nRT}.\end{aligned} \tag{10.8b}$$

(3) The Peierls mechanism

The constitutive equation for the Peierls mechanism is given by

$$\dot{\varepsilon} = A_3 \nu_D \left(\frac{\sigma}{\mu}\right)^2 \exp\left(-\frac{H^*(P)(1-(\sigma/\sigma_P(P))^q)^s}{RT}\right). \tag{10.9}$$

Hence

$$\begin{aligned}\frac{\partial \log \dot{\varepsilon}}{\partial P} &= \frac{3\gamma + \frac{2}{3}}{K} - \frac{H^*[1-(\sigma/\sigma_P)^q]^s}{RT}\\ &\left\{\frac{\partial \log H^*}{\partial P} + \frac{sq(\sigma/\sigma_P)^q}{1-(\sigma/\sigma_P)^q}\frac{\partial \log \sigma_P}{\partial P}\right\}.\end{aligned} \tag{10.10}$$

The pressure dependence of creep comes from that of $\nu_D(P)$, H^* (P) and $\sigma_P(P)$ but similar to other mechanisms, the pressure dependence of $\nu_D(P)$ is small and the main effects are from $H^*(P)$ and $\sigma_P(P)$. The relative importance of these two terms depends on the stress level. At high temperatures and/or low strain rates, the stress is small and hence $\partial \log \dot{\varepsilon}/\partial P \approx -\bar{V}_2^*/RT$ where $\bar{V}_2^* \equiv \partial H^*/\partial P$, a result similar to power-law dislocation creep. At low temperatures and/or high strain rates, stress will be high, and $\partial \log \dot{\varepsilon}/\partial P \approx -\frac{H^*[1-(\sigma/\sigma_P)^q]^{s-1} sq(\sigma/\sigma_P)^q}{RT}\frac{\partial \log \sigma_P}{\partial P}$. The pressure dependence of strain rate in the Peierls mechanism is small (for a constant stress $\partial \log \dot{\varepsilon}/\partial P \propto -\partial \log \sigma_P/\partial P \approx -(2\gamma + \frac{1}{3})/K$. Note that the apparent activation volume (pressure dependence of creep) for this mechanism is significantly lower than that for the power-law creep. In many ultrahigh-pressure rheology experiments, a high stress is applied to a sample, and plastic deformation likely occurs by the Peierls mechanism or similar exponential flow-law mechanisms. A great care must be exercised when results from such studies are applied to Earth where power-law dislocation creep (or diffusional creep) is likely important.

Models for activation volume

As we learned above, activation volume is the most important parameter that determines the pressure dependence of plastic deformation. Models for activation volume can be classified into two categories, namely atomistic models and phenomenological models. Atomistic models are based on some microscopic models for thermally activated processes that control the rate of deformation. Some important concepts in this type of models will be reviewed. Phenomenological models refer to any models that are based on an empirical relationship without any atomistic details. The most important one in this category is the model of *homologous temperature scaling*. Physical basis for this model will be discussed in some detail.

(1) Microscopic models for activation volume

By definition, activation volume is the partial derivative of the Gibbs free energy associated with a process of thermal activation with respect to pressure. The pressure dependence of activation free energy is made of two distinct terms. First, when there is an explicit volume change associated with thermal activation, then there is a $P \cdot \Delta V$ term in the activation free energy where ΔV is the explicit volume change associated with thermal activation. Second, any work to create an activated state depends on pressure due to the dependence of physical constants such as elastic constants (in addition to the volume change effect). Both terms yield an activation volume, and hence in general,

$$V^* = \Delta V + \bar{V}^* \tag{10.11}$$

where ΔV is the explicit volume change of the system due to thermal activation without elastic distortion of a crystal and $\bar{V}^*$ is the activation volume due to the pressure dependence of physical properties that determines the activation enthalpy (e.g., elastic constants). It is important to distinguish these two effects because the origin of pressure dependence of activation volume is quite different between the two terms.

The presence of these two distinct terms can be understood easily by considering the following two hypothetical cases. First let us consider a case in which the inter-atomic potential is so weak that a change in atomic configuration between the ground state and the activated state does not change the free energy of the system (i.e., $\bar{V}^* = 0$). Even in such a case, when one moves an atom from the inside to the surface of a crystal (to form a vacancy), one does extra work corresponding to $P \cdot \Delta V$. So there is ΔV associated with this process. Second is a case where there is no explicit volume change associated with thermal activation (i.e., $\Delta V = 0$). Even in this case, changes in atomic configuration between the ground state and the activated state result in excess energy due to the change in inter-atomic potential energy. This energy depends on some physical properties that in general depend on pressure. Therefore even in this case, there must be some activation volume ($\bar{V}^*$). Note that this $\bar{V}^*$ may also include an explicit volume change. For example, formation of a vacancy may results in elastic singularity that results in an extra volume change, volume relaxation, in addition to the explicit change in volume due to the removal of an atom from the interior to the surface of a crystal (Eshelby, 1956).[3]

Let us consider the physical models for ΔV and $\bar{V}^*$. The origin of ΔV is straightforward in some cases. When formation of a vacancy is considered, this term is related to the volume change of a crystal associated with the formation of a vacancy. Note, however, that even in a case where the thermally activated process in consideration involves vacancy formation, ΔV is not necessarily the volume of the single vacancy. Vacancy formation in an ionic solid occurs as the formation of a group of vacancies to maintain the charge neutrality condition (a Schottky group), so that (Chapter 5),

$$\Delta V = \frac{\psi \Omega}{\alpha + \beta + \gamma + \cdots} \tag{10.12}$$

where ψ is a constant representing the atomic relaxation around a vacancy ($\psi \sim 1$), Ω is the volume of a Schottky group, and $\alpha, \beta, \gamma, \ldots$ are the stoichiometric coefficients, i.e., for olivine Mg_2SiO_4, $\alpha = 2$, $\beta = 1$ and $\gamma = 4$ (for olivine (or MgO) this gives $\Delta V \approx 6.5\ (5.5) \times 10^{-6}\ \mathrm{m^3/mol}$ for $\psi = 1$). Similarly, if a process involves an explicit volume change due to some chemical reaction (e.g., if deformation involves water-related species), then the rate of deformation is proportional to the concentration of relevant water-related species. The concentration of water-related species depends on pressure through the volume change. See the later part of this chapter for more detail.

[3] In terms of the notation in Chapter 5, this part of activation volume corresponds to $\delta V^* = (1 - \psi)\Omega$ where Ω is the atomic volume of species involved in the defect (see also equation (10.12)).

The origin of $\bar{V}^*$ may be investigated through atomistic calculations (e.g., GILMAN, 1981; LIDIARD, 1981). But here I will review two continuum models of activation volume. A continuum model of a thermally activated state is obviously a simplified model of actual processes, but a major advantage of such a simplified model is that some part of physics can be more easily understood than a more rigorous numerical approach. In a continuum model of defect, the difference in free energy between the activated and the ground state is modeled as a free energy change in a continuum. If an activated state has no excess electric charge, then the change in energy associated with the activated state may be modeled as elastic strain energy (e.g., ZENER, 1942; KEYES, 1963). If, on the other hand, an activated state is associated with an excess electric charge, then the electrostatic energy will also be changed by the formation of an activated state.

(1.i) Elastic strain energy model for $\bar{V}^*$

Activation energy for deformation or diffusion may be approximately treated as strain energy. Under this assumption, from the dimensional analysis, it is clear that one can write (KEYES, 1963),

$$G^*_{\text{ela}}(P) = \beta \cdot C(P) \cdot \Omega(P) \tag{10.13}$$

where β is a non-dimensional parameter, C is some combination of elastic constants and Ω is the volume of defects that are involved in the given process. By differentiating equation (10.13) with P, one obtains,

$$\bar{V}^*_{\text{ela}} = \left(\frac{\partial G^*_{\text{ela}}}{\partial P}\right)_0 = \frac{H^*_{\text{ela}}}{K}\left(K\frac{\partial \log C}{\partial P} - 1\right). \tag{10.14}$$

Now we recall $\partial \log C/\partial P = (2\gamma + \frac{1}{3})/K$ (Chapter 4), and hence,

$$\bar{V}^*_{\text{ela}} = \frac{H^*_{\text{ela}}}{K}\left(2\gamma - \frac{2}{3}\right). \tag{10.15}$$

For most materials, $\gamma = 1.5 \pm 0.5$ (Chapter 4), so $\bar{V}^*_{\text{ela}} \approx (2.3 \pm 1)H^*_{\text{ela}}/K$. For a typical case of $H^*_{\text{ela}} = 200-500\,\text{kJ/mol}$ and $K = 120\,\text{GPa}$, we get $\bar{V}^*_{\text{ela}} = (4\text{–}10) \times 10^{-6}\,\text{m}^3/\text{mol}$. This type of model was first proposed by KEYES (1963) and applied to Earth science by KUMAZAWA (1974), SAMMIS *et al.* (1977, 1981), POIRIER and LIEBERMANN (1984). However, the validity of this model is not obvious. For example, the process of formation of vacancies involves the explicit volume expansion simply because of the removal of atoms from the interior to the surface of a crystal. The volume expansion associated with this process is related to the atomic volume (Ω), $P \cdot \Omega$, which is not incorporated in the strain-energy model. Also, the formation of a charged defect includes a significant change in electrostatic energy that is not included in this model. The importance of explicit volume change and the electrostatic energy in formation of point defects in ionic solids was discussed by KARATO (1977, 1981a).

(1.ii) Effects of electrostatic energy on $\bar{V}^*$

In addition to the strain energy, a thermally activated state may also have excess energy associated with electrostatic charge. The enthalpy change associated with creating a defect with excess charge, Ze, is given by (e.g., FLYNN, 1972)

$$H^*_{\text{ele}} = \frac{Z^2 e^2}{2\varepsilon_0(P) \cdot r(P)} \tag{10.16}$$

where $\varepsilon_0(P)$ is the static dielectric constant and $r(P)$ is the "radius" of the defect at the activated state configuration. By differentiating (10.16) with pressure assuming that the radius of defect changes with pressure in the same way as the mean inter-atomic distance, one obtains

$$\bar{V}^*_{\text{ele}} = \frac{H^*_{\text{ele}}}{K}\left(\frac{\partial \log \varepsilon_0}{\partial \log V} + \frac{1}{3}\right). \tag{10.17}$$

According to KEYES (1963) and KARATO (1977, 1981a), $\partial \log \varepsilon_0/\partial \log V = 2.5 \pm 1.0$, and $\bar{V}^*_{\text{ele}} = (2.8 \pm 1.0) \cdot (H^*_{\text{ele}}/K)$. Therefore, numerically elastic and dielectric models give similar activation volumes. In fact, if one assumes that the pressure dependence of static dielectric constant is mainly through the vibration of optical mode frequency with pressure (the Szigetti relation, see KITTEL, 1986) then $\partial \log \varepsilon_0/\partial \log V = 2\gamma - 1$ and equation (10.17) becomes identical to equation (10.15) (only difference is the meaning of the Grüneisen parameter: it is the acoustic mode Grüneisen parameter for the elastic model and the optical mode Grüneisen parameter for the dielectric model).

(1.iii) Homologous temperature scaling

The temperature and pressure variation of rheological properties including diffusion coefficient of atoms is often scaled using the melting temperature, namely,

$$A(P, T) = A_0 \exp\left(-\beta' \frac{T_m(P)}{T}\right) \tag{10.18}$$

where $T_m(P)$ is the "melting temperature" of the material that changes with pressure, P. Such a relation has been observed for a large number of materials including metals and ionic solids (e.g., WEERTMAN, 1968, see Fig. 10.4).

SHERBY *et al.* (1970) and POIRIER and LIEBERMANN (1984) provided some theoretical explanation for the homologous temperature scaling. Here I will summarize POIRIER and LIEBERMANN'S (1984) discussion. They assumed, following KEYES (1963), that the activation enthalpy for deformation (or diffusion) can be expressed by elastic strain energy and used Lindemann's law for melting that assumes that melting occurs when the amplitude of lattice vibration reaches a critical value. LINDEMANN (1910) postulated that melting occurs when the amplitude of atomic vibration exceeds some critical value. Using the equipartition law and assuming that a solid behaves like a collection of harmonic oscillators, one obtains,

$$E = k\langle x\rangle^2 \tag{10.19}$$

where E is the energy of the crystal, k is a spring constant and $\langle x\rangle$ is the amplitude of atomic displacement due to lattice vibration. According to Lindemann, melting takes place when the amplitude of lattice vibration becomes a critical value, $\langle x\rangle_c = \delta \cdot a$. Then, at a melting temperature, we must have

$$E = k\delta^2 \cdot a^2 = RT_m = m\omega^2\delta^2 \cdot a^2 \tag{10.20}$$

where we used a relation $\omega = \sqrt{k/m}$ (ω, frequency of lattice vibration) and hence,

$$T_m = \frac{\delta^2 m\Omega^{2/3}\omega^2}{R} = \frac{\delta^2 C\Omega}{R} \tag{10.21}$$

where the relationship $\omega = V/a = \sqrt{C/\rho}(1/a) = \sqrt{C\Omega}m^{-1/2}\Omega^{-1/3} = C^{1/2}m^{-1/2}\Omega^{1/6}$ (Chapter 4) is used. Using the elastic strain-energy model for defect formation or migration, i.e., equation (10.13), one finds

$$G^* = \frac{\beta RT_m}{\delta^2} = \beta' RT_m \tag{10.22}$$

with

$$\beta' = \frac{\beta}{\delta^2}. \tag{10.23}$$

Given the relation (10.22), assuming β' does not change with pressure, one can calculate the activation volume,

$$V^* = \frac{\partial G^*}{\partial P} = \frac{H^*}{T_m}\frac{dT_m}{dP}. \tag{10.24}$$

This is a useful model because pressure dependence of melting temperature is better known than the pressure dependence of plastic deformation. However, theoretical basis of the assumptions is not very secure. In particular the validity of the Lindemann's law of melting for silicates can be questioned (e.g., WOLF and JEANLOZ, 1984). Also the appropriate choice of "melting temperature" is not obvious. A question is often raised as to what T_m we should use when evaluating the rheological properties of a multi-component system. In a multi-component system, melting occurs gradually from the solidus and is completed at the liquidus. The difference between the solidus and the liquidus can be large: for a peridotite the difference is $\sim$600 K at room pressure. It is often argued that the solidus should be used as T_m in rheological studies (e.g., GREEN and BORCH, 1987), but this notion is inconsistent with the observation that the rheological properties of pure olivine polycrystals (solidus is $\sim$2100 K) is virtually indistinguishable from those of peridotites (solidus is $\sim$1500 K) (see e.g., ZIMMERMAN and KOHLSTEDT, 2004). A large reduction of solidus temperature from the melting temperature of component solids is due to the influence of mixing entropy to reduce the free energy of liquid. Since the plastic properties of solid are our concern, solidus temperature that is affected by the properties of liquid should not be used. Rather than the solidus of the system, the melting temperatures of each phase must be used and some appropriate average must be taken (see also Chapter 12). One of such an analysis is given by YAMAZAKI and KARATO (2001b) for a two-phase mixture of (Mg, Fe)SiO_3–(Mg, Fe)O in the lower mantle. Also one should remember that the homologous temperature scaling is only approximately true.

Activation volume for the homologous temperature model can be compared with those of other models. If one uses the Lindemann's law, then from (10.24) one has $\partial \log T_m/\partial \log V = -2\gamma + \frac{2}{3} = -K(\partial \log T_m/\partial P)$, and one obtains,

$$V^* = \frac{H^*}{K}\left(2\gamma - \frac{2}{3}\right). \tag{10.25}$$

Therefore, all of these semi-empirical models ((10.14), (10.17) and (10.24)) essentially give an identical result, (10.25). For $H^* = 300-600\,\text{kJ/mol}$ and $K = 100$–$200\,\text{GPa}$, we obtain $V = (6–15) \times 10^{-6}\,\text{m}^3/\text{mol}$. Two points need to be emphasized. First, these relations are only semi-quantitative: they give a rough estimate of V^*, but not the precise value of V^*. Second, the relationships such as (10.25) are applicable only for $\bar{V}^*$. The actual activation volume should include ΔV as ($V^* = \Delta V + \bar{V}^*$).

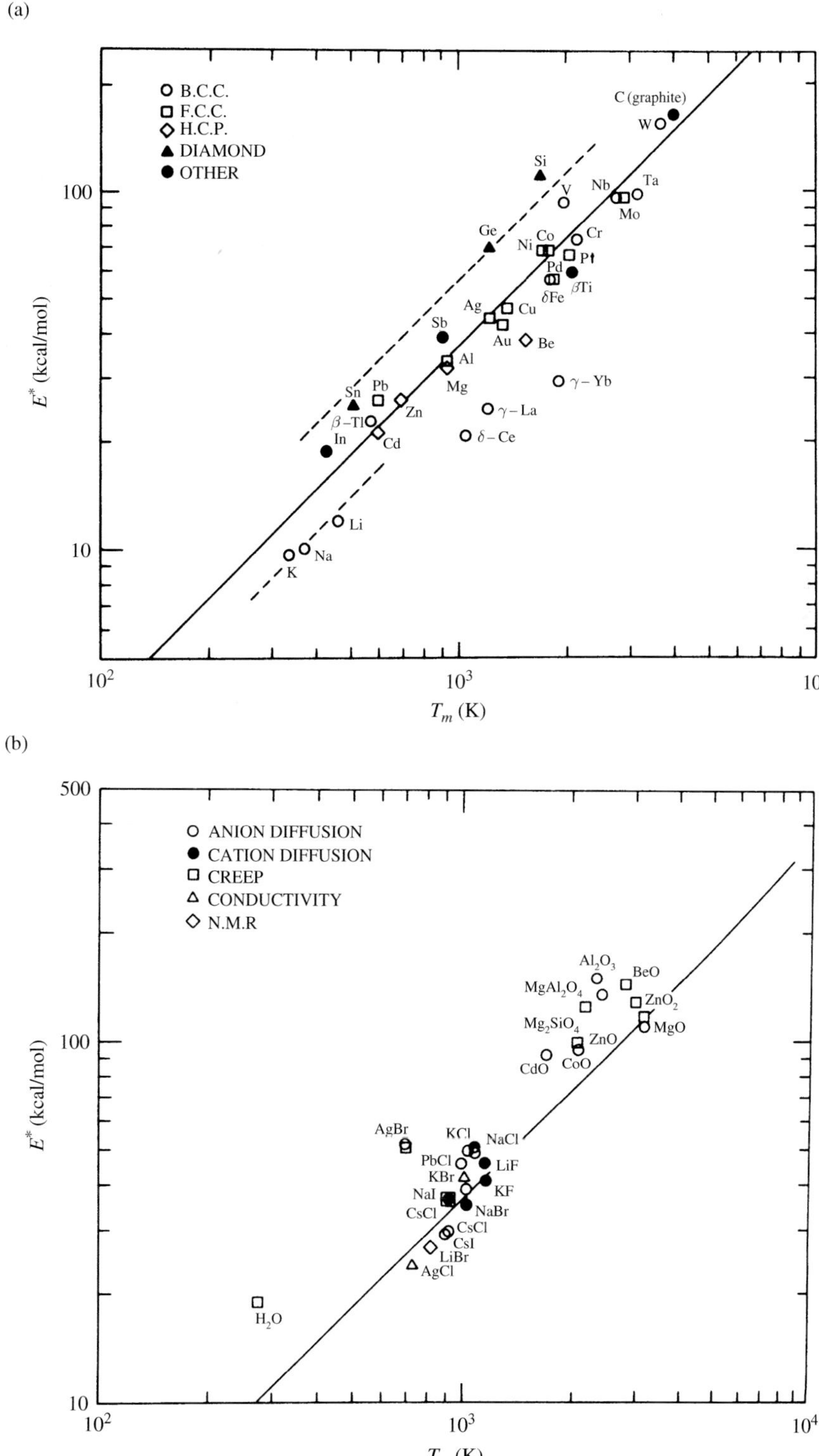

FIGURE 10.4 Correlation between melting temperatures, T_m, and activation energies, E^*, of creep and diffusion. (a) Data for metals, (b) data for non-metals (from SAMMIS *et al.*, 1981).

(2) Atomistic models

Several more atomistic models of activation volume have been developed. Each model has to include specific aspects of atomistic processes involved in a given process. For example, when diffusion occurs through vacancy mechanism, the activation free energy includes the energy change associated with cutting bonds as well as the work done against the volume change. KARATO (1977, 1981a) developed such a model that includes the effect of explicit volume change (ΔV), elastic strain energy and electrostatic energy associated with point-defect formation and migration ($\bar{V}^*$).

More detailed calculations using an empirical interatomic potential or first-principle approach have also been made (e.g., KARATO, 1978; WALL and PRICE, 1989; WRIGHT and PRICE, 1993; ITA and COHEN, 1997; BRODHOLT and REFSON, 2000; BRAITHWAITE *et al.*, 2003). In calculating the energy of formation of defects in ionic crystals, a neutral group of defects such as the Schottky defects must be considered as a unit process rather than the energy of an isolated charged defect. However, this means that such a calculation must incorporate a large number of atoms, exceeding a number of atoms for one defect ($\sim 10^5$–10^6). Alternatively, defect energy can be calculated for a smaller number of atoms if the long-range interaction is incorporated by a continuum approximation such as the method of MOTT and LITTLETON (1938). In this model, atomistic details are incorporated only in the vicinity of a defect and the long-range interaction of atoms is treated through the elastic and dielectric distortion of a crystal. The defect-energy calculations published so far involved only a small number of atoms (< 1000 atoms), and the validity of these results is questionable. However, due to the rapid progress in the computational power, it is likely that a realistic calculation of defect energy can be made for complicated materials such as silicates in the near future.

Pressure dependence of activation volume

In its simplest form, the equation $H^* = E^* + PV^*$ is a first-order approximation in which the activation enthalpy is considered to increase linearly with pressure. Under certain conditions this may not be a good approximation. In fact several authors suggested the importance of pressure dependence of activation volume (O'CONNELL, 1977; POIRIER and LIEBERMANN, 1984; BORCH and GREEN, 1987; HIRTH and KOHLSTEDT, 2003). Two types of argument have been presented. The first one is based on the elastic strain-energy model. Assuming this model, one obtains (Problem 10.2),

$$\frac{\partial \log \bar{V}^*}{\partial P} = -\frac{1}{K} + \frac{1}{\gamma - \frac{1}{3}} \frac{\partial \gamma}{\partial P}. \tag{10.26}$$

Therefore, the pressure dependence of activation volume critically depends on the pressure dependence of the Grüneisen parameter. If the Grüneisen parameter is independent of pressure, then $\partial \log V^*/\partial P = -1/K$, which is the same as the pressure dependence of a molar volume and the pressure dependence of activation volume is small. If, however, the Grüneisen parameter decreases significantly with pressure, then the activation volume decreases significantly with pressure. For example, it is often found that $\gamma\rho^q =$ constant (Chapter 4). In such a case, $\partial\gamma/\partial P = -q\gamma/K$ and $\partial \log V^*/\partial P = -(1/K)\frac{(1+q)\gamma - \frac{1}{3}}{\gamma - \frac{1}{3}} \approx -2.4/K$ for $\gamma = 1.5$ and $q = 1$. This means that V^* decreases with pressure more than the (molar) volume. O'CONNELL (1977) obtained a similar result assuming that the activation volume is a volume of a vacant sphere in a continuum but such a model is not supported by an atomistic theory of point defects (GILMAN, 1981).

Problem 10.2

Derive the relation (10.26).

Solution

From (10.25), one has

$$\frac{\partial \log \bar{V}^*}{\partial P} = \frac{\partial \log H^*}{\partial P} - \frac{\partial \log K}{\partial P} + \frac{\partial \log(2\gamma - \frac{2}{3})}{\partial P} = \frac{2\gamma - \frac{2}{3}}{K} - \frac{\partial \log K}{\partial P} + \frac{1}{\gamma - \frac{1}{3}} \frac{\partial \gamma}{\partial P}.$$

Now, $\partial \log K/\partial P = (2\gamma + \frac{1}{3})/K$ and hence $\partial \log \bar{V}^*/\partial P = -1/K + [1/(\gamma - \frac{1}{3})](\partial\gamma/\partial P)$.

Another argument is that activation volume depends on the pressure dependence of melting temperature and dT_m/dP decreases significantly with pressure if T_m is the solidus (BORCH and GREEN, 1987). However, T_m in relation (10.25) is not the solidus as discussed before. It should also be noted that a significant fraction of activation volume is the explicit volume change, ΔV, and in such a case there should be very small pressure dependence of ΔV. This can be seen in the results of KOHLSTEDT *et al.* (1996) where the activation volume associated with the dissolution of water in olivine is nearly constant for a pressure range of

0.3–13 GPa. In summary, there is no clear evidence for significant decrease of activation volume with pressure.

In estimating the depth variation of viscosity in Earth's mantle, the pressure dependence of activation volume is often invoked. O'Connell (1977) and Poirier and Liebermann (1984), for example, argued that the mantle viscosity is nearly depth-independent because of the large reduction of activation volume at high pressures. As we learned above, this notion is neither supported by theory nor by experimental observations, and if indeed mantle viscosity is nearly depth-independent, an alternative explanation must be found. Yamazaki and Karato (2001b) suggested that a weak depth dependence of viscosity through the lower mantle of Earth is due simply to the small activation energy and volume of diffusion in lower mantle minerals such as (Mg, Fe)O (Chapter 19).

Pressure-induced change in the mechanism of defect motion

Activation volume for defect motion depends on the atomistic mechanism. Defect motion involving vacancies involves a larger activation volume than defect motion involving interstitial atoms. Consequently, it is expected that the mechanism of defect motion changes from vacancy mechanism to interstitial atom mechanism as pressure increases. This issue was examined by Karato (1978) based on the theoretical calculation of formation free energy of vacancies and interstitial atoms in a model crystal following the method by Kanzaki (1957). Karato (1978) found that a change in the dominant type of defects occurs as pressure increases, and that the transition pressure depends strongly on the nature of inter-atomic potential: the "harder" the repulsive potential, the higher the transition pressure. Above this transition pressure, the activation volume is negative. Ito and Toriumi (2007) obtained similar results by the molecular dynamics simulation of diffusion in MgO.

10.3. Effects of water

10.3.1. General introduction

The effects of water on plastic deformation are large and deserve special attention. Griggs and Blacic (1965) and Griggs (1967) discovered that high-temperature plastic deformation of quartz is significantly enhanced by a small amount of water dissolved in quartz. Similar effects were also found for olivine (Blacic, 1972; Chopra and Paterson, 1981, 1984; Karato *et al.*, 1986) and for other minerals. In all cases, the effects of water are larger at higher pressures. It is also well documented that the solubility of water in minerals increases with pressure (at least at low pressures). Therefore, weakening effects of water are due to water-related species that are dissolved in minerals. Consequently, in order to understand the effects of water (hydrogen) on plastic deformation in Earth, we need to understand (1) the microscopic mechanisms of dissolution of water, (2) the microscopic mechanisms by which water-related species (defects) enhance deformation of minerals and (3) the mechanisms by which amounts of water in minerals are controlled in Earth. Important progress toward understanding of atomistic mechanism of water dissolution has been made through experimental observations and thermodynamic modeling of point defects. A combination of experimental observations and thermodynamic as well as atomistic modeling of defects (both point defects and dislocations) has provided some insights into the mechanisms of water weakening.

In the following, I will first summarize the mechanisms of water (hydrogen) dissolution in minerals, followed by the experimental observations on water weakening and its theoretical interpretations. Finally, a brief review will be given on the processes by which water contents in minerals may be controlled in Earth's interior.

10.3.2. Mechanisms of dissolution of water (hydrogen)

When the water (hydrogen) budget in solid Earth is discussed, often only hydrous minerals such as phlogopite, serpentine and some other hydrous minerals are considered. A major challenge to this notion was made in the early 1970s by Martin and Donnay (1972) who pointed out the importance of *nominally anhydrous minerals* (minerals that do not have H in their chemical formula, e.g., quartz (SiO_2) and olivine ($(Mg, Fe)_2SiO_4$) as a water (hydrogen) reservoir in Earth's interior. In fact, the pioneering studies by Griggs and his colleagues (Griggs and Blacic, 1965; Blacic, 1972) already showed that a certain amount of hydrogen can be dissolved in nominally anhydrous minerals such as quartz and olivine. Detailed experimental studies in the laboratory of Paterson at ANU (Australian National University) supported this notion for quartz and olivine (for review of earlier works see Karato, 1989a; Paterson, 1989), and the subsequent studies at other laboratories contributed to the understanding of the nature of hydrogen in olivine and other silicates. Reviews on this topic are also available (Bell and Rossman, 1992; Thompson, 1992; Kohlstedt and Mackwell, 1999; Ingrin and Skogby, 2000; Williams and Hemley, 2001; Bolfan-Casanova, 2005).

These studies have established that nominally anhydrous silicate minerals such as quartz (SiO_2) or olivine ($(Mg, Fe)_2SiO_4$) which do not have H (hydrogen) in their chemical formula can dissolve a significant amount of hydrogen. The amount of hydrogen that can be dissolved in these minerals depends on the thermodynamic conditions and increases with the fugacity of water. Consequently, the effects of water on kinetic processes (such as plastic deformation and diffusion) increase with the fugacity of water. Although the amount of water that can be dissolved in most nominally anhydrous minerals ($\sim 10^{-4}$–10^{-1} mole fraction, $\sim 10^{-3}$–1 wt%) is smaller than that in hydrous minerals (2.7 wt% in phlogopite, 4.3 wt% in serpentine), the concentration of hydrogen in these minerals is much larger than that of point defects under anhydrous conditions ($\sim 10^{-5}$–10^{-6} mole fraction). This is the fundamental reason why hydrogen enhances various kinetic processes in silicates.

It must also be noted that although the amount of hydrogen that can be dissolved into nominally anhydrous minerals is relatively small (compared to the concentration of water in hydrous mineral), the solubility is large enough to significantly affect water (hydrogen) budget in Earth's interior. For example, if the upper mantle is saturated with hydrogen, the total amount of water that goes to nominally anhydrous minerals will be $\sim 10^{21}$ kg, which is about the same as the amount of the sea water. The situation is somewhat different for minerals in the transition zone such as wadsleyite and ringwoodite. In these minerals (particularly in wadsleyite), the maximum solubility of H is very high and in wadsleyite, it reaches the point that 1/8 of Mg is replaced with H to change its chemical formula from Mg_2SiO_4 to $Mg_{1.75}H_{0.25}SiO_4$ (Inoue *et al.*, 1995). The amount of water that can be dissolved in wadsleyite (and ringwoodite) far exceeds the amount of sea water (~ 10 times of sea water, i.e., $\sim 10^{22}$ kg).

Let us consider atomistic mechanisms of dissolution of water in silicate minerals. Water, H_2O, can be dissolved in silicate minerals in a variety of ways. Water provides hydrogen and oxygen ions through the reaction,

$$H_2O = H_2 + \frac{1}{2}O_2 = 2H^+ + O^{2-}. \tag{10.27}$$

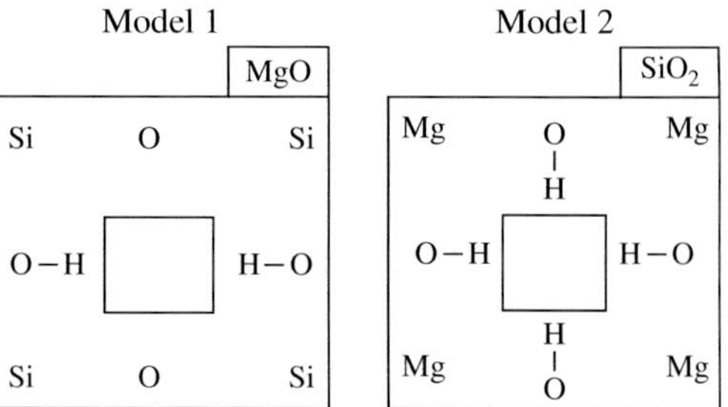

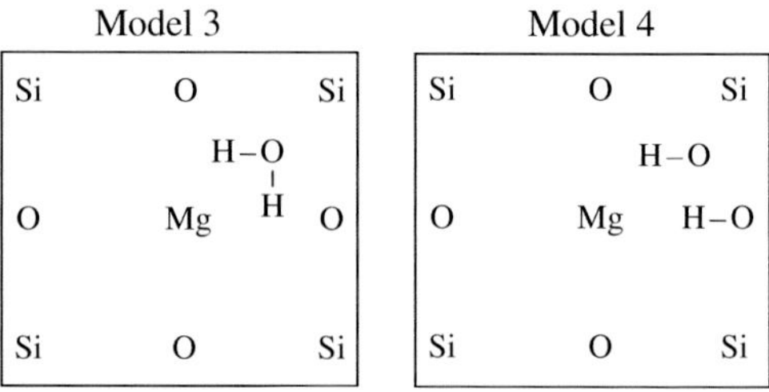

FIGURE 10.5 Mechanisms of dissolution of water in silicate minerals. Hydrogen (proton) and nearby oxygen form an electric dipole. The interaction between an electric dipole and the electromagnetic field (light) causes absorption of light. Note that each mechanism is associated with a specific volume change of a crystal.

Now both proton H^+ and oxygen ions O^{2-} go to some sites in a mineral. For an oxygen ion, there are two possibilities. An oxygen ion can either go into a lattice site that is usually occupied by oxygen, or it may go into an interstitial site. A hydrogen ion, namely proton may go into various sites that have negative charges. Therefore for minerals containing two types of cations such as olivine, wadsleyite or ringwoodite ($(Mg, Fe)_2SiO_4$), we will consider three possible sites that are negatively charged: (1) *M*-site (*M* represents either Mg or Fe) vacancy, (2) Si-site vacancy and (3) oxygen ion. Consequently, likely mechanisms of dissolution of water (hydrogen) in silicate minerals are (Fig. 10.5):

(i) two protons at *M*-site vacancies + oxygen at O-site
(ii) four protons at Si-site vacancies + oxygen at O-site
(iii) two protons at an interstitial oxygen
(iv) two proton at a regular O-site + at an interstitial oxygen

The mechanism (iii) is identical to the dissolution of "molecular" water. In minerals such as (Mg, Fe)O, only (i) or (iii) (or (iv)) may be considered. In minerals such as SiO_2, on the other hand, only mechanisms (ii), (iii) or (iv) can be considered. (One can consider a modified version of mechanism (iv) in which one proton is associated with an interstitial oxygen.)

Two points may be noted. First, in the first two cases shown in Fig. 10.5, hydrogen atoms completely neutralize the defect. However, partial neutralization is also possible. That is if two protons go to a M-site vacancy (V''_M) (for the notation of defects see Box 5.1), then the defect is completely neutralized ($(2H)^{\times}_M$, this defect has the same charge as in the perfect lattice site) but if only one proton goes to the M-site, it is only partially neutralized (H'_M, this defect has –1 effective charge compared to the perfect lattice). Similarly, if four protons go to the Si-site vacancy (V''''_{Si}), then one has a neutral defect ($(4H)^{\times}_{Si}$) whereas if only one proton goes to a Si-vacancy then one has a charged defect H'''_{Si}. The distinction between fully charge-compensated defects such as $(2H)^{\times}_M$ and partially charge-compensated defects such as H'_M is important because the latter affects the concentrations of other charged defects whereas the former does not. Consequently, the latter type of defects has a more important role in modifying physical properties. Note, however, even a neutral defect such as $(2H)^{\times}_M$ has an important effect on diffusion of atoms at the same site, i.e., the M-site in this case (e.g., Hier-Majumder *et al.*, 2005a).

Second, note that even in cases (i) and (ii) where protons are associated with cation vacancies, I draw Fig. 10.5 in such a way that protons are close to the nearby oxygen to form OH dipoles, $(OH)^{\bullet}_O$. It is an OH dipole that causes absorption of an infrared beam (see Box 10.1). To emphasize this fact, a neutral defect such as $(2H)^{\times}_M$ may be considered as a defect complex $\{(OH)^{\bullet}_O - V''_M - (OH)^{\bullet}_O\}^{\times}$ (e.g., Kohlstedt *et al.*, 1996). However, the dependence of concentrations of $(2H)^{\times}_M$ and $\{(OH)^{\bullet}_O - V''_M - (OH)^{\bullet}_O\}^{\times}$ on chemical environment is identical.

Equation of chemical reaction corresponding to model (i) is

$$H_2O(\text{fluid}) + M^{\times}_M \Leftrightarrow (2H)^{\times}_M + MO(\text{surface}). \quad (10.28)$$

This relation means that, by adding water, one M (Mg or Fe) is removed from an M-site and two protons occupy an M-site to form a neutral defect, $(2H)^{\times}_M$. The M atom removed from the M-site will be combined with an oxygen atom to create an excess MO (MgO) (at the surface). This implies that the solid will undergo a volume expansion of $\Delta v \approx V_{MO}$ where V_{MO} is the molar volume of MO. Applying the law of mass action one obtains

$$f_{H_2O} = K_{29}(T,P) \cdot [(2H)^{\times}_M] \cdot a_{MO}. \quad (10.29)$$

Hence

$$[(2H)^{\times}_M] = f_{H_2O}(T,P) \cdot a^{-1}_{MO} \cdot K^{-1}_{29}. \quad (10.30)$$

The cases (ii) and (iii) can be treated similarly, and one obtains,

$$[(4H)^{\times}_{Si}] = f^2_{H_2O}(T,P) \cdot a^2_{MO} \cdot K^{-1}_{31} \quad (10.31)$$

and

$$[(H_2O)^{\times}_I] = f_{H_2O}(T,P) \cdot K^{-1}_{32} \quad (10.32)$$

respectively where $K_{29,31,32}$ are the relevant equilibrium constants given by

$$K^{-1}_{29}(T,P) = \exp\left(-\frac{\mu^{\otimes}_{H_2O} + \Delta u_{29} + P\Delta v_{29} - T\Delta s_{29}}{RT}\right) \quad (10.33a)$$

$$K^{-1}_{31}(T,P) = \exp\left(-\frac{2\mu^{\otimes}_{H_2O} + \Delta u_{31} + P\Delta v_{31} - T\Delta s_{31}}{RT}\right) \quad (10.33b)$$

$$K^{-1}_{32}(T,P) = \exp\left(-\frac{\mu^{\otimes}_{H_2O} + \Delta u_{32} + P\Delta v_{32} - T\Delta s_{32}}{RT}\right) \quad (10.33c)$$

where $\mu^{\otimes}_{H_2O}$ is the chemical potential of water at the reference state, Δu is the energy needed to create the hydrogen-related defect ($(2H)^{\times}_M$, $(4H)^{\times}_{Si}$ and $(H_2O)^{\times}_I$), Δv and Δs are the volume and entropy change associated with the formation of hydrogen-related defect respectively. With the reaction shown by (10.28), the volume change of the system Δv_{29} is composed of the volume change due to the addition of MgO and the volume change of M-site by replacing Mg with two protons. The latter is only a fraction of the volume of Mg^{2+} ($\sim 1.4 \times 10^{-6}\,m^3/mol$) and is much smaller than the volume of MgO ($\sim 11 \times 10^{-6}\,m^3/mol$) so that Δv_{29} must be similar to the molar volume of MgO. For $(4H)^{\times}_{Si}$, Δv_{31} is approximately that of SiO_2 ($\sim 22 \times 10^{-6}\,m^3/mol$) and for the volume change of Si-site (the molar volume of SiO_2 is approximately twice that of MgO. This is due to the fact that the molar volumes of oxides are dominated by those of oxygen ions: SiO_2 has two and MgO has one oxygen).

It is likely that for a given mineral there are a few possible sites in which hydrogen can be dissolved. The relative fraction of hydrogen at various sites will change with physical and chemical conditions. For example, the relations (10.30) and (10.31) indicate

Box 10.1. Infrared spectroscopy and hydrogen in minerals

Infrared spectroscopy is a powerful tool with which the nature of hydrogen (water) in minerals can be investigated (Fig. 10.6). A light beam (electromagnetic wave) interacts with the optical modes of lattice vibration that involve the relative motion of cations and anions. Both stretching mode of OH bonds and bending mode of H–O–H interact with light with $\sim 3\,\mu m$ wavelength range and energy levels of electrons and phonons can change due to this interaction. Consequently, light is absorbed by a substance that contains OH-related species. By measuring the absorption of light by a substance, one can determine the concentration of OH, and the possible sites at which OH dipoles are located. Three types of information (data) are useful to investigate the nature of OH-related defects. First, from the intensity of absorption, one can determine the concentration of OH. Second, from the wavenumber of absorption peaks, one can place constraints as to the plausible sites where OH may be present. Third, from the anisotropy of absorption, one can infer the orientation of OH dipoles in a given crystal that provides a useful constraint on the crystallographic positions where hydrogen atoms are located. A useful calibration needed for the determination of OH content from infrared absorption was provided by Paterson (1982), but the validity of this equation is debated (e.g., Bell *et al.*, 2003). It should be noted that when infrared absorption is measured at ambient pressure and temperature, results may not reflect the nature of OH-related species at high pressure and temperature: it is difficult to quench the atomic sites because of the fast diffusion of hydrogen. *In-situ* measurements of infrared absorption at high pressure and temperature will be needed to obtain definitive data to specify the likely lattice sites where hydrogen may sit in minerals at high pressures and temperatures.

Other techniques of measuring hydrogen content in minerals include secondary-ion-mass-spectrometer (SIMS) and nuclear-reaction-analysis (NRA) technique. Both techniques are available only at ambient conditions. Since NRA technique relies on the reaction of hydrogen with some atom such as ^{15}N, this technique is considered to provide the most robust results of hydrogen content in a mineral. However, IR spectroscopy has important advantages in that it gives constraints as to the speciation of hydrogen from the wave-number dependence of absorption and anisotropy in absorption. Such information cannot be obtained from SIMS or NRA measurements.

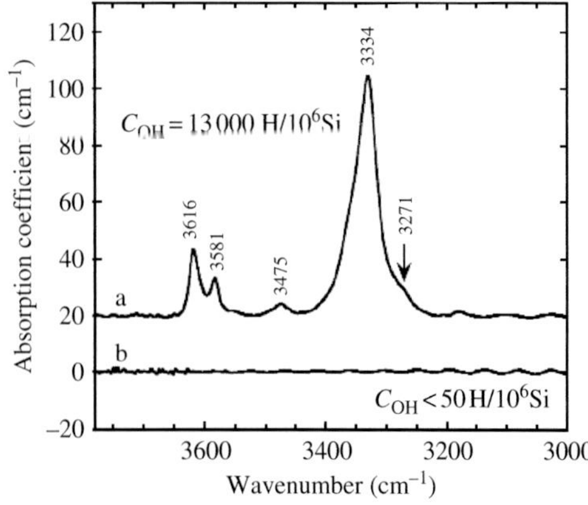

FIGURE 10.6 Infrared absorption spectra for wadsleyite samples cooked at $P = 15$ GPa, $T = 1670$ K under water-saturated conditions (Nishihara *et al.*, 2006).

that $(2H)_M^{\times}$ dominates at relatively low a_{MO}, but at high a_{MO}, $(4H)_{Si}^{\times}$ will dominate. This was confirmed by Lemaire *et al.* (2004). It should also be noted that the kinetics of dissolution of water (hydrogen) are different among different sites. The kinetics of dissolution of water (hydrogen) at M-site involves diffusion of V_M'' but the kinetics of dissolution of water (hydrogen) at Si-site involves diffusion of V_{Si}''''. Consequently, it is possible that for an experimental time-scale one only sees incorporation of hydrogen at M-site whereas in the natural environment (at longer time-scales), incorporation of hydrogen at Si-site might also occur.

Neutral defects do not have direct influence on the concentration of other defects. Therefore defects such as $(2H)_M^{\times}$ will not have direct influence on the defects at other sites such as oxygen or silicon-related defects. In contrast to neutral defects, charged defects have strong influence on the concentrations of other charged defects and consequently have important effects on a number of physical properties.

Several mechanisms can be considered for the formation of charged defects. For instance, a neutral defect $(2H)_M^{\times}$ may change to a charged defect through the ionization reaction

$$(2H)_M^{\times} = H_M' + H^{\bullet}. \tag{10.34}$$

A free proton $H^\bullet$ will react with a negatively charged defect such as V''_M to yield

$$H^\bullet + V''_M = H'_M \tag{10.35}$$

thus

$$2H'_M = (2H)^\times_M + V''_M. \tag{10.36}$$

Therefore, applying the law of mass action, one obtains

$$[(2H)^\times_M] = [H'_M][H^\bullet]K_{34} \tag{10.37a}$$

$$[V''_M][H^\bullet] = [H'_M]K_{35} \tag{10.37b}$$

$$[H'_M]^2 = [(2H)^\times_M][V''_M]K_{36}. \tag{10.37c}$$

This means that the concentration of a charged defect, H'_M, depends on the concentration of neutral defect, $(2H)^\times_M$, as well as that of M-site vacancy, V''_M. Obviously $K_{36}^{-1} = K_{34}K_{35}$. The importance of free proton, $H^\bullet$ (or $OH^\bullet_o$), has been shown by the study of electric conductivity in olivine, wadsleyite and ringwoodite (Huang *et al.*, 2005; Wang *et al.*, 2006).

Let us now consider how the dominant types of point defects change with water fugacity. At water-free ("dry") conditions, the dominant types of point defects in olivine and other ferro-magnesian silicates are M-site vacancy and ferric iron. The concentration of M-site vacancy depends on the oxygen fugacity as (see Chapter 5)

$$f_{O_2}^{1/2} = K_{38}[V''_M][Fe^\bullet_M]^2 a_{MO}. \tag{10.38}$$

At dry conditions, the dominant defects in olivine are M-site vacancy, V''_M, and ferric iron, $Fe^\bullet_M$, and the charge neutrality condition is $[Fe^\bullet_M] = 2[V''_M]$ and hence

$$[Fe^\bullet_M] = 2[V''_M] = 2^{1/3} f_{O_2}^{1/6} a_{MO}^{-1/3} K_{38}^{-1/3}. \tag{10.39}$$

When one inserts this relation and (10.30) into (10.37c) and (10.37b), one finds that $[H'_M] \propto f_{H_2O}^{1/2}$ and $[H^\bullet] \propto f_{H_2O}^{1/2}$. The dependence of concentrations of other defects on chemical environment can be determined using the law of mass action (Table 10.2).

Note that in this regime, the concentration of H'_M increases with water fugacity as $[H'_M] \propto f_{H_2O}^{1/2}$ but the concentration of V''_M is independent of f_{H_2O}. Therefore at certain f_{H_2O}, the dominant negatively charged defect will change from V''_M to H'_M (Fig. 10.7), and consequently the charge neutrality condition becomes

$$[Fe^\bullet_M] = [H'_M]. \tag{10.40}$$

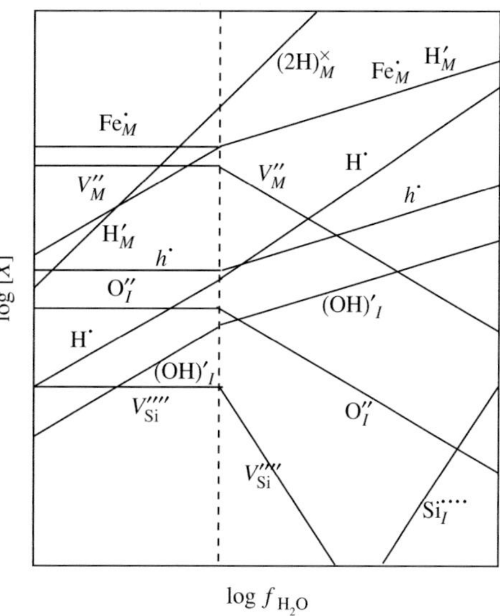

FIGURE 10.7 A schematic diagram showing the concentrations of various defects as a function of water fugacity in olivine. At low water fugacity conditions, the dominant charged defects are ferric iron $Fe^\bullet_M$ and M-site vacancy V''_M, whereas when water fugacity increases they change to $Fe^\bullet_M$ and H'_M.

Using this charge neutrality condition and the relations (10.29), (10.37c) and (10.38), one has

$$[H'_M] = f_{H_2O}^{1/4} f_{O_2}^{1/8} a_{MO}^{-1/2} K_{29}^{1/4} K_{36}^{1/4} K_{38}^{-1/4}. \tag{10.41}$$

Concentration of all other charged defects can be calculated from (10.41) using the law of mass action for relevant reactions. The results are summarized in Table 10.2 (for comparison, this table also includes defect concentrations corresponding to the charge neutrality condition of $[Fe^\bullet_M] = 2[V''_M]$).

Similar analyses can be made for other types of defects (charge neutrality conditions such as $[(OH)'_I] = [Fe^\bullet_M]$). Note that different defects have different dependence on thermodynamic parameters. Consequently, the data on the dependence of physical properties (such as the rate of deformation or diffusion) on thermodynamic parameters provide useful constraints on the dominant defect that controls the process.

Problem 10.3

Diffusion of Mg(Fe) in olivine occurs through the vacancy mechanism. Compare the dependence of Mg(Fe)-related defects on water fugacity and infer which defect controls diffusion of Mg under hydrous conditions using the experimental observation that $D_{Mg-Fe} \propto f_{H_2O}$. (Assume that the charge neutrality

TABLE 10.2 Dependence of concentration of point defects in $(Mg, Fe)_2SiO_4$ or $(Mg, Fe)SiO_3$ on chemical environment ($[X] \propto f^{p}_{H_2O} f^{q}_{O_2} a^{r}_{MO}$) for various charge neutrality conditions.

Defect	$[Fe^{\bullet}_M] = 2[V''_M]$			$[Fe^{\bullet}_M] = [H'_M]$		
	p	q	r	p	q	r
$[(2H)^{\times}_M]$	1	0	-1	1	0	-1
$[H'_M]$	$\frac{1}{2}$	$\frac{1}{12}$	$-\frac{2}{3}$	$\frac{1}{4}$	$\frac{1}{8}$	$-\frac{1}{2}$
$[Fe^{\bullet}_M]$	0	$\frac{1}{6}$	$-\frac{1}{3}$	$\frac{1}{4}$	$\frac{1}{8}$	$-\frac{1}{2}$
$[V''_M]$	0	$\frac{1}{6}$	$-\frac{1}{3}$	$-\frac{1}{2}$	$\frac{1}{4}$	0
$[M^{\bullet\bullet}_I]$	0	$-\frac{1}{6}$	$\frac{1}{3}$	$\frac{1}{2}$	$-\frac{1}{4}$	0
$[O''_I]$	0	$\frac{1}{6}$	$\frac{2}{3}$	$-\frac{1}{2}$	$\frac{1}{4}$	1
$[Si^{\bullet\bullet\bullet\bullet}_I]$	0	$-\frac{1}{3}$	$-\frac{10}{3}$	1	$-\frac{1}{2}$	-4
$[(OH)'_I]$	$\frac{1}{2}$	$\frac{1}{12}$	$\frac{1}{3}$	$\frac{1}{4}$	$\frac{1}{8}$	$\frac{1}{2}$
$[(H_2O)^{\times}_I]$	1	0	0	1	0	0
$[V^{\bullet\bullet}_O]$	0	$-\frac{1}{6}$	$-\frac{2}{3}$	$\frac{1}{2}$	$-\frac{1}{4}$	-1
$[(OH)^{\bullet}_O]$	$\frac{1}{2}$	$-\frac{1}{12}$	$-\frac{1}{3}$	$\frac{3}{4}$	$-\frac{1}{8}$	$-\frac{1}{2}$
$[V''''_{Si}]$	0	$\frac{1}{3}$	$\frac{10}{3}$	-1	$\frac{1}{2}$	4
$[H'''_{Si}]$	$\frac{1}{2}$	$\frac{1}{4}$	3	$-\frac{1}{4}$	$\frac{3}{8}$	$\frac{7}{2}$
$[(2H)''_{Si}]$	1	$\frac{1}{6}$	$\frac{8}{3}$	$\frac{1}{2}$	$\frac{1}{4}$	3
$[(3H)'_{Si}]$	$\frac{3}{2}$	$\frac{1}{12}$	$\frac{7}{3}$	$\frac{5}{4}$	$\frac{1}{8}$	$\frac{5}{2}$
$[(4H)^{\times}_{Si}]$	2	0	2	2	0	2
$[H^{\bullet}]$	$\frac{1}{2}$	$-\frac{1}{12}$	$-\frac{1}{3}$	$\frac{3}{4}$	$-\frac{1}{8}$	$-\frac{1}{2}$
$[h^{\bullet}]$	0	$\frac{1}{6}$	$-\frac{1}{3}$	$\frac{1}{4}$	$\frac{1}{8}$	$-\frac{1}{2}$
$[e']$	0	$-\frac{1}{6}$	$\frac{1}{3}$	$-\frac{1}{4}$	$-\frac{1}{8}$	$\frac{1}{2}$

conditions are either $[Fe^{\bullet}_M] = [H'_M]$ or $[Fe^{\bullet}_M] = 2[V''_M]$.)

Solution

Diffusion coefficient of Mg(Fe) is proportional to the defect concentration related to Mg(Fe), i.e., $(2H)^{\times}_M$, H'_M, V''_M or $M^{\bullet\bullet}_I$. From Table 10.2, $[V''_M] \propto f^{-1/2}_{H_2O}$, $[H'_M] \propto f^{1/4}_{H_2O}$, $[(2H)^{\times}_M] \propto f_{H_2O}$ and $[M^{\bullet\bullet}_I] \propto f^{1/2}_{H_2O}$ for the charge neutrality condition of $[Fe^{\bullet}_M] = [H'_M]$, or $[V''_M] \propto f^{0}_{H_2O}$, $[H'_M] \propto f^{1/2}_{H_2O}$, $[(2H)^{\times}_M] \propto f_{H_2O}$ and $[M^{\bullet\bullet}_I] \propto f^{0}_{H_2O}$ for the charge neutrality condition of $[Fe^{\bullet}_M] = 2[V''_M]$. The experimental observation by Hier-Majumder *et al.* (2005a) shows that $D_{Mg-Fe} \propto f_{H_2O}$. Therefore we conclude that diffusion of Mg(Fe) in olivine under hydrogen-rich conditions is due to $(2H)^{\times}_M$ (recall that diffusion of Mg(Fe) in olivine under hydrogen-poor conditions is due to V''_M).

Based on the comparison of model predictions (Table 10.2) with experimental observations, dominant mechanisms of incorporation of water in various minerals have been investigated. Existing experimental observations suggest that the dominant mechanism of water (hydrogen) dissolution in quartz and olivine (and wadsleyite) is through the neutral defects $(4H)^{\times}_{Si}$ and $(2H)^{\times}_M$ respectively (Doukhan and Paterson, 1986; Kohlstedt *et al.*, 1996). Also the experimental observations suggest that the charge neutrality condition is either $[Fe^{\bullet}_M] = [H'_M]$ or $[Fe^{\bullet}_M] = 2[V''_M]$. Less extensive studies have been made for other minerals, but for garnets both $(4H)^{\times}_{Si}$ and $(OH)'_I$ are proposed (Aines and Rossman, 1984; Lu and Keppler, 1997; Withers *et al.*, 1998; Katayama *et al.*, 2003), for orthopyroxene the results by Rauch and Keppler (2002) suggest $(2H)^{\times}_M$ as a dominant type of defect, for clinopyroxene defects associated with cation vacancies were suggested by Smyth *et al.* (1991), Skogby (1994), Katayama and Nakashima (2003), Bromiley *et al.* (2004), and for (Mg, Fe)O, a charged defect H'_M is proposed to be the dominant hydrogen-related defects (Bolfan-Casanova *et al.*, 2002). The results on (Mg, Fe)O are at odds in comparison to other minerals. The absorption spectra reported by Bolfan-Casanova *et al.* (2002) show evidence of a large amount of free water (and brucite) suggesting that a large fraction of hydrogen dissolved in their samples was not quenched as OH-related defects. Similarly, the incorporation of carbon in olivine is likely through a neutral defect $C_{Si}^{\times}$.

Special attention should be paid to a case in which water (hydrogen) is dissolved into a mineral in combination with another species. For example, the water solubility in orthopyroxene is known to correlate with the concentration of Al (RAUCH and KEPPLER, 2002). In this case, the solubility is controlled by a reaction,

$$\mathrm{Si}_{Si}^{\times} + \tfrac{1}{2}\mathrm{H_2O(fluid)} + \tfrac{1}{2}\mathrm{Al_2O_3} = (\mathrm{H}\cdot\mathrm{Al})_{Si}^{\times} + \mathrm{SiO_2}. \qquad (10.42)$$

Alternatively, Al and H may replace two M (Mg or Fe) as

$$2M_M^{\times} + \tfrac{1}{2}\mathrm{H_2O(fluid)} + \tfrac{1}{2}\mathrm{Al_2O_3} = \left(\mathrm{H}'_M\cdot\mathrm{Al}_M^{\bullet}\right) + 2M\mathrm{O}. \qquad (10.43)$$

In such cases, the concentration of water (hydrogen) in a mineral depends not only on water fugacity but also the activity of Al_2O_3,

$$\left[(\mathrm{H}\cdot\mathrm{Al})_{Si}^{\times}\right] \propto f_{\mathrm{H_2O}}^{1/2} a_{\mathrm{Al_2O_3}}^{1/2} a_{\mathrm{SiO_2}}^{-1} \qquad (10.44)$$

or

$$\left[\left(\mathrm{H}'_M\cdot\mathrm{Al}_M^{\bullet}\right)\right] \propto f_{\mathrm{H_2O}}^{1/2} a_{\mathrm{Al_2O_3}}^{1/2} a_{\mathrm{SiO_2}}. \qquad (10.45)$$

The kinetics of water (hydrogen) dissolution into (or escape from) a mineral involves diffusion of Al ion that is slow compared to the diffusion of hydrogen and cation vacancy (or electron holes). Therefore hydrogen contents in these minerals may be better preserved than hydrogen in olivine (or wadsleyite).

It is important to note that the solubility of these defects depends on both pressure and temperature. In particular, the pressure dependence of solubility is large because both water fugacity and the equilibrium constants strongly depend on pressure. At low pressures, water fugacity is low and the equilibrium concentration of water-related defects (hydrogen-related defects) is low. However, the solubility of these defects increases strongly with pressure and the effects of water on physical properties become more important at high-pressures. Temperature also has important influence on solubility of volatile species (hydrogen, carbon). For olivine, the existing study indicates strong positive dependence of hydrogen solubility on temperature (ZHAO *et al.*, 2004) whereas the data on ringwoodite indicate a negative dependence (OHTANI *et al.*, 2000). DEMOUCHY *et al.* (2005) discussed that the temperature dependence of chemical composition of coexisting fluid phase affects the temperature dependence of water (hydrogen) solubility at high temperatures.

It is likely that grain boundaries can dissolve some water. It is often reported that the water content in a polycrystalline sample exceeds significantly that of a single crystal (e.g., MEI and KOHLSTEDT, 2000a), which suggest an important role of grain boundaries. However, the nature of "water" on grain boundaries is not well understood. The relative contribution between grain boundary and the concentration of intragranular hydrogen (water) may change with pressure. Also it is possible that some of the polycrystalline samples have a small amount of melt (that is quenched to a glass upon cooling) that contains a large amount of water (even $\sim$0.1 % of melt with $\sim$10% water gives $\sim$1500 ppm H/Si). The role of grain boundaries in the dissolution of water remains unconstrained and deserves a detailed study.

10.3.3. Experimental observations on the role of water on plastic deformation

The effects of water to enhance high-temperature creep have been documented in a number of minerals (quartz, olivine, plagioclase, halite). Here I will focus on two minerals, quartz and olivine, and review some essential observations.

Quartz

Historically, quartz was the first material in which water-weakening effect was found (GRIGGS and BLACIC, 1965; GRIGGS, 1967; BLACIC, 1975). Consequently a large number of experimental studies have been conducted since the mid-60s but the essence of this remarkable effect is not yet fully understood (for a review, see PATERSON, 1989).

There are two reasons why the mechanisms of water weakening in quartz are difficult to study (compared to those in olivine). First, diffusion coefficient (chemical diffusion coefficient; see Chapter 8 about diffusion) of water in quartz is very low particularly when the water content is low (KEKULAWALA *et al.*, 1981; KRONENBERG *et al.*, 1986). Therefore the concentration of water (or hydrogen-related defects) in quartz is likely not in equilibrium with the given thermochemical conditions in many experiments. Consequently, small-scale features such as water-containing bubbles play an important role to modify the effective concentration of water-related defects near dislocation lines (e.g., MCLAREN *et al.*, 1983; DOUKHAN and TRÉPIED, 1985; PATERSON, 1989) causing complications in interpreting the experimental data. Second, as a consequence of the first point, deviation from thermochemical equilibrium is large at lower pressures and hence the majority of the mechanical data were obtained at higher pressures (>1 GPa) using a solid-medium deformation apparatus. Interpretation of mechanical data from a solid-medium apparatus is difficult due to the large error in stress measurements caused by friction (see Chapter 6). Some high-resolution mechanical tests have also been made at low pressures (typically <0.3 GPa) (Kekulawala *et al.*, 1978, 1981; Paterson and Kekulawala, 1979; Mainprice and Paterson, 1984; Mackwell and Paterson, 1985; Paterson, 1989) or room pressure (Linker and Kirby, 1981; Linker *et al.*, 1984) but the interpretation of these results are difficult because of a large deviation from thermochemical equilibrium.

Another complication for quartz is that since dislocation mobility is so low (due to strong covalent bonding of Si–O), deformation (in the dislocation-creep regime) is not always steady state: a whole regime of deformation ranging from transient to steady-state creep involving dislocation multiplication (work hardening or work softening) and annihilation (recovery) must be considered to fully understand the role of water in quartz deformation (e.g., Griggs, 1974; Paterson, 1989). The possible importance of dislocation multiplication in quartz deformation is emphasized by (McLaren *et al.*, 1989).

Although the exact details of the way in which water affects quartz deformation are not clearly understood, the main observations can be summarized as follows. The effect of water on plastic deformation of quartz is very large. Quartz crystals with little water (<80 ppm H/Si)[4] are very strong. At a confining pressure of 0.3 GPa , even at 1573 K and stress of ~1 GPa , deformation is largely due to brittle fracture and plastic deformation due to dislocations occurs only near crack tips (Doukhan and Trépied, 1985). However, with a larger amount of water (>500 ppm H/Si), plastic deformation becomes much easier: at ~850 K, strain rate of $10^{-5}\ s^{-1}$ can be achieved at ~100 MPa deviatoric stress (if extrapolated to 1573 K, the flow stress for strain rate of $10^{-5}\ s^{-1}$ will be ~4 MPa). The easiness of plastic deformation is correlated with the water content in quartz crystals (e.g., Hobbs *et al.*, 1972; Doukhan and Trépied, 1985). A similar observation was made by Kronenberg and Tullis (1984) who showed that the creep strength of quartz decreases with pressure when water is present. They interpreted their results as due to the increased solubility of water in quartz with pressure. Similarly, Mackwell *et al.* (1985) found that even with the presence of water, quartz is strong at low pressures (0.3 GPa), whereas once annealed at higher pressures (1.5 GPa) with the presence of water, quartz becomes weak even at a low pressure (0.3 GPa), a result presumably due to the pressure effect on water solubility (see also Cordier and Doukhan, 1989). Using a liquid-cell assembly with the Griggs apparatus, quantitative studies on the influence of water on deformation of quartzite have been conducted by Gleason and Tullis (1995) and Post *et al.* (1996). When the influence of water is parameterized as $\dot{\varepsilon} \propto f_{H_2O}^r$, $r = 1$–2. The value of r is not well constrained by the previous studies (Fig. 10.8). The reported value of r should be considered as a lower bound since the important effect of the influence of the $\exp(-PV^*/RT)$ term is not included in the currently available literature (for discussion of this point see section 10.3.6). Blacic (1975) noted anisotropic effects of water weakening.

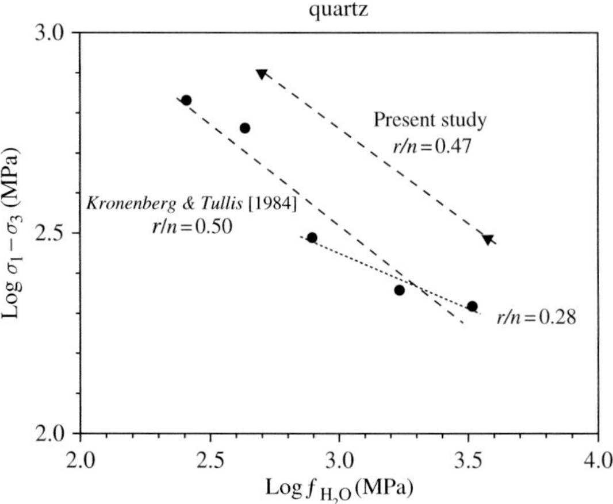

FIGURE 10.8 Dependence of the creep strength of quartz on water fugacity (Post *et al.*, 1996). A constitutive relation of $\dot{\varepsilon} \propto f_{H_2O}^r(P,T) \cdot \sigma^n \cdot \exp\left(-\frac{E^* + PV^*}{RT}\right)$ ($n = 4$) is assumed. The water fugacity is changed by changing the pressure, but the influence of the $\exp(-PV^*/RT)$ term is not included in this analysis (i.e., effectively it is assumed that $V^* = 0$).

Water enhances many kinetic processes of quartz including diffusion (Dennis, 1984; Farver and Yund, 1991), dynamic recrystallization (Hobbs, 1968), dislocation recovery (Tullis, 2002) and grain growth (Tullis and Yund, 1982).

[4] Two units are used to define the water (hydrogen) content. H/Si is the ratio of number of hydrogen to silicon atoms, whereas wt% is the weight percent (of water, H_2O). 1 wt% of water in Mg_2SiO_4 (SiO_2) corresponds to 156 000 ppm H/Si (67 000 ppm H/Si). Both are non-dimensional units, but unit must be specified to clearly indicate the water (hydrogen) content.

All the published results on the role of water on deformation of quartz are for dislocation creep. However, diffusion coefficients (of oxygen) in quartz increase with water content (DENNIS, 1984; FARVER and YUND, 1991), and hence diffusional creep is likely enhanced by water.

Olivine

The effects of water to enhance plastic deformation of olivine single crystals were first observed by BLACIC (1972) using the Griggs apparatus. CHOPRA and PATERSON (1981, 1984) performed the first high-resolution experimental studies on deformation of natural dunites using a gas-medium deformation apparatus at 0.3 GPa. When a natural dunite was deformed *as is*, a large amount of water supplied from the dehydration of accessory hydrous minerals, and a dunite was weaker compared to a sample that was dried at high temperatures (at room pressure). Studies using synthetic samples were initiated by KARATO *et al.* (1986) and follow-up studies were conducted in Kohlstedt's lab (e.g., HIRTH and KOHLSTEDT, 1995a, 1995b; MEI and KOHLSTEDT, 2000a, 2000b). These early studies were conducted at low pressures, < 0.45 GPa.

The kinetics of chemical diffusion of water (hydrogen) in olivine is fast (MACKWELL and KOHLSTEDT, 1990; KOHLSTEDT and MACKWELL, 1998) and most of the experiments can be conducted under equilibrium conditions even at low pressures (say ~ 0.3 GPa) where precise mechanical measurements are possible. Consequently the influence of water on plastic deformation in olivine is much better understood than that in quartz. However it should be noted that the knowledge obtained at low pressures (<0.5 GPa) is limited and cannot be applied to Earth's upper mantle below ~ 20 km (see KARATO and JUNG, 2003).

The effect of water on plastic deformation of olivine is modest compared to that on quartz at least at low pressures. At $P = 0.3$ GPa and at $T = 1573$ K, the difference in stress to achieve deformation of water-free and water-saturated olivine at $10^{-5}\ s^{-1}$ strain rate is a factor of 1.5 to 3 (e.g., CHOPRA and PATERSON, 1984; MACKWELL *et al.*, 1985; KARATO *et al.*, 1986). Also MACKWELL *et al.* (1985) found that the effects of water are anisotropic: deformation due to dislocations with $\boldsymbol{b} = [001]$ is enhanced more than deformation by $\boldsymbol{b} = [100]$ dislocations. Similarly, YAN (1992) found water enhances dislocation recovery in olivine and its effect is anisotropic: stronger effect for $\boldsymbol{b} = [001]$ dislocations than $\boldsymbol{b} = [100]$ dislocations. KARATO *et al.* (1986) investigated the influence of water on plastic deformation using synthetic olivine aggregates in both diffusional and dislocation creep regimes and found that in both diffusional and dislocation creep regimes, water enhances plastic deformation of olivine. They also found that the strength of olivine is negatively correlated with the amount of dissolved water. Similar studies were conducted by MEI and KOHLSTEDT (2000a, 2000b).

MEI and KOHLSTEDT (2000a, 2000b) and KARATO and JUNG (2003) investigated the influence of pressure on water-weakening effect in olivine in detail. At low pressures (< 0.45 GPa), deformation is progressively more enhanced with increasing pressure (MEI and KOHLSTEDT, 2000a, 2000b) which is interpreted as due to the increase in water solubility with pressure (Fig. 10.9). However, at higher pressures both the influence of increasing water solubility and decreasing dislocation mobility are important, which results in the non-monotonic dependence of strength on pressure (KARATO and JUNG, 2003). The maximum effect of water in the deep upper mantle (say $\sim$300–400 km depth) is quite large: for a given stress, strain rate under water-saturated conditions is a factor of $\sim 10^4$ larger than that under water-free conditions. When the

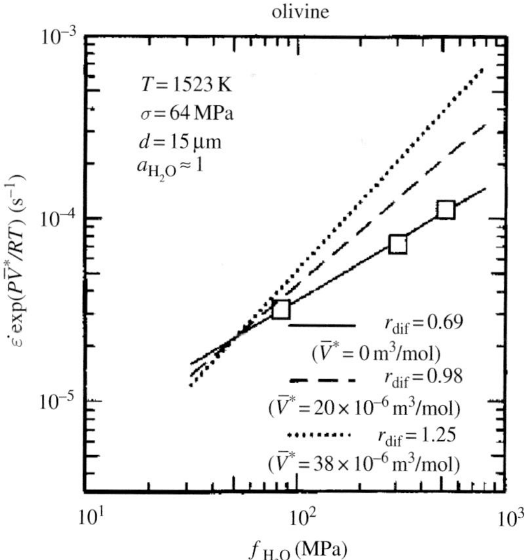

FIGURE 10.9 Influence of water fugacity on the rate of deformation in olivine in the dislocation-creep regime (MEI and KOHLSTEDT, 2000b). In the experiments by MEI and KOHLSTEDT (2000b), the water fugacity was changed by changing the total confining pressure. Consequently, the value of the water fugacity exponent, r, cannot be determined uniquely because it depends on the unknown activation volume $\bar{V}^*$ (for details see the later part of this chapter).

influence of water is parameterized as $\dot{\varepsilon} \propto f_{H_2O}^{r}$, one obtains $r \sim 1.0$–1.2.

Water also has important influence on other processes in olivine such as diffusion (of Fe–Mg) (HIER-MAJUMDER *et al.*, 2005a), grain-growth kinetics (KARATO, 1989b), dynamic recrystallization (AVÉ LALLEMANT and CARTER, 1970; POST, 1977; JUNG and KARATO, 2001a), fabric development (JUNG and KARATO, 2001b; KATAYAMA *et al.*, 2004; JUNG *et al.*, 2006), electrical conductivity (KARATO, 1990; WANG *et al.*, 2006) and KARATO (1995) suggested that seismic wave attenuation may also be enhanced by water.

Problem 10.4

Explain why chemical diffusion of water in quartz is more sluggish than that in olivine.

Solution

Note that by diffusion of water, we mean the rate of chemical reaction between water and mineral that is controlled by *chemical diffusion* (Chapter 8). When water reacts with a silicate mineral, hydrogen diffuses into a silicate, but another charged defect must also diffuse to maintain the charge neutrality. In olivine, M-site vacancies (or electron holes) can be this charge-compensating defect whose diffusion coefficient is large. In contrast, in quartz, the charge-compensating defect is either Si-site or O-site defect diffusion of both is more sluggish than that of M-site vacancies in olivine. This is a basic reason for the vast difference in chemical diffusion rate between two minerals. Note that this model suggests that the currently available experimental results on olivine may correspond to *partial equilibrium* (with respect to the M-site) and not to complete equilibrium. In the natural setting or in experimental studies with longer durations (or at higher temperatures), a larger fraction of hydrogen may go to the Si-site (that requires diffusion of Si) than observed in the currently available experimental data.

Other minerals

The role of water on plastic deformation in other minerals is less extensively studied. RUTTER (1972) and RUTTER *et al.* (1994) reported weakening effects of water on plastic deformation of calcite. TULLIS and YUND (1982) showed the enhanced grain growth in calcite by the presence of water. In both cases the effects are less pronounced than in quartz. Water-induced enhancement of plastic deformation and recovery in feldspar was investigated by TULLIS *et al.* (1979), TULLIS and YUND (1980) and STÜNITZ *et al.* (2003). AVÉ LALLEMANT (1978), BOLAND and TULLIS (1986) and HIER-MAJUMDER *et al.* (2005b) studied the role of water on deformation in clinopyroxene. The influence of water on creep in feldspar and clinopyroxene is stronger than that in olivine but weaker than that in quartz. SKOGBY and ROSSMAN (1989), SKOGBY *et al.* (1990), SKOGBY (1994), RAUCH and KEPPLER (2002) and KATAYAMA and NAKASHIMA (2003) investigated the water dissolution mechanisms in pyroxenes. Little is known about the effect of water on deformation of garnets (for preliminary results see JIN *et al.*, 2001). KATAYAMA and KARATO (2007) investigated the influence of water on plastic deformation of pyrope garnet and found that the rate of deformation is sensitive to water content. LAGER *et al.* (1989), ROSSMAN *et al.* (1989), WITHERS *et al.* (1998), ROSSMAN and AINES (1991) and KATAYAMA *et al.* (2003) investigated the water dissolution mechanisms in garnets.

Solubility of water (hydrogen) in wadsleyite has been studied by INOUE *et al.* (1995), KOHLSTEDT *et al.* (1996), and NISHIHARA *et al.* (2007a). These studies showed that the solubility could reach ~3 wt%, and at this limit, $\frac{1}{8}$ of M-site is completely replaced with two protons. NISHIHARA *et al.* (2007a) investigated the influence of oxygen fugacity on solubility of water in wadsleyite and showed very weak dependence on oxygen fugacity for the dominant absorption peaks. These observations suggest that the mechanism of water dissolution in wadsleyite is similar to that of olivine, two protons occupy the M-site vacancy, $(2H)_M^{\times}$, but at the same time, there are other hydrogen-related species whose concentration depends on chemical environment in a different manner. KUBO *et al.* (1998) showed that water enhances dislocation recovery in wadsleyite and enhances transformation kinetics. However, CHEN *et al.* (1998) reported small effects of water on the strength of wadsleyite at relatively low temperatures. NISHIHARA *et al.* (2006) found that water enhances grain-growth kinetics in wadsleyite. HUANG *et al.* (2005) showed water enhances electrical conductivity in wadsleyite and ringwoodite through diffusion of free protons, $H^{\bullet}$ (a similar result has been obtained for olivine (WANG *et al.*, 2006)). Nothing is known about the effects of water on plastic deformation of perovskite. The solubility of water (hydrogen) in $MgSiO_3$ was investigated by a few groups but the results show large discrepancy. MEADE *et al.* (1993) reported a solubility of ~700 ppm H/Si and MURAKAMI *et al.* (2002) reported a high water solubility, ~30 000 ppm H/Si, LITASOV *et al.* (2003) reported ~1000 ppm H/Si for Al-free perovskite and ~14 000–18 000 ppm H/Si for Al-bearing perovskite, whereas BOLFAN-CASANOVA *et al.* (2000, 2003) reported much smaller solubility (<10 ppm H/Si). The solubility of hydrogen in perovskite is not well constrained at this time. FREUND and WENGELER (1982) and BOLFAN-CASANOVA *et al.* (2002) showed a finite solubility of water in MgO. But no experimental data are available on the effects of water on the physical properties of MgO or magnesiowüstite.

10.3.4. Models for water (hydrogen) weakening

Diffusional creep

The model for water-weakening effects in diffusional creep is straightforward. Given the flow law for diffusional creep, $\dot{\varepsilon} = \alpha(D/L^2)(\sigma\Omega/RT)$ (equation (8.44)), the enhancement of diffusional creep for a fixed grain-size should be due to the enhancement of diffusion.[5] Since diffusion coefficient is determined by the concentration, Γ_f, and the mobility, Γ_m, of a relevant point defect (see Chapter 8), the influence of water on diffusion must be evaluated both for defect concentration and on mobility. As has been shown in section 10.3.3, the influence of water on defect concentration can be written as $\Gamma_f \propto f_{H_2O}^r$,[6] whereas the influence of water on defect mobility is through the change in the activation enthalpy for mobility, namely, $\Gamma_m \propto \exp(-H^*_{m,\text{wet}}/RT)$ where $H^*_{m,\text{wet}}$ is the activation enthalpy for defect migration under "wet" conditions. $H^*_{m,\text{wet}}$ is different from $H^*_{m,\text{dry}}$ (activation enthalpy for defect migration under "dry" conditions) because the relevant defect can be different between the two cases (e.g., $(2\text{H})^\times_\text{M}$ versus V''_M for Mg diffusion). Thus

$$\dot{\varepsilon}_\text{wet} = \alpha\frac{D_\text{eff,wet}}{L^2}\frac{\sigma\Omega}{RT} \propto f^r_{H_2O}\cdot\exp\left(-\frac{H^*_\text{diff,wet}}{RT}\right) \qquad (10.46)$$

where $H^*_\text{diff,wet}$ is the activation enthalpy for diffusional creep under "wet" conditions.

Dislocation creep

The situation is more complicated for dislocation creep. The rate of plastic deformation by dislocation motion is determined by the density of (mobile) dislocations and their velocity as $\dot{\varepsilon} = \rho b v$ (ρ is dislocation density, b is the length of the Burgers vector and v is the velocity of dislocation motion; see Chapter 9). The length of the Burgers vector does not change with water so much and therefore water effects should be either through its effect on dislocation density or dislocation velocity (or both). At steady state, dislocation density is controlled by force balance, $\rho \approx b^{-2}(\sigma/\mu)^2$ (see Chapter 5), and is nearly independent of water content and the main effect of water must be through its effects on dislocation velocity corresponding to a constant dislocation density. However, when a crystal has a low initial dislocation density and dislocation multiplication is difficult, then water effects could be through the enhancement of dislocation multiplication.

The latter situation is likely the case of experimental deformation of quartz (at least at modest temperatures) where intrinsic dislocation mobility is low due to the strong covalent Si–O bond (HOBBS *et al.*, 1972; GRIGGS, 1974) (see also PATERSON, 1989). Many of the experimental observations on quartz at modest temperatures are likely due to the effects of water on dislocation multiplication that determine the density of mobile dislocations. Dislocation multiplication is easy in olivine (or quartz at high temperatures), and in this case, most of the water-weakening effect is from its influence on dislocation mobility at a constant dislocation density.

Since both dislocation density and velocity are ultimately controlled by dislocation mobility, let us now consider how water might affect dislocation mobility. In crystals with relatively strong covalent bonding, important factors to control dislocation mobility are (1) kink nucleation and migration and (2) jog nucleation and migration. The jog migration rate is controlled by atomic diffusion whereas other processes (kink formation/migration and jog formation) are controlled mostly by dislocation core structures (Chapter 9).

A generic equation for dislocation velocity in a crystal with relatively high intrinsic resistance to dislocation motion (high Peierls stress) is (see Chapter 9)

$$v_{c,g} = c_{j,k} b v_{j,k} \qquad (10.47)$$

where $v_{c,g}$ is climb (or glide) velocity, $c_{j,k}$ is jog (or kink) concentration and $v_{j,k}$ is jog (or kink) velocity. Water effects may be through $c_{j,k}$ or $v_{j,k}$ (or both). HOBBS *et al.* (1972) and GRIGGS (1974) proposed that in most experimental studies in quartz, dislocation multiplication plays an important role in determining the resistance of quartz for plastic deformation. In this model, water

[5] In real Earth, grain size under "wet" conditions could be larger than that under "dry" conditions which could counteract the influence of enhancement of diffusion.

[6] The analysis in section 10.3.3 is for defects in the crystal. Defects at grain boundaries may show different dependence on water.

is considered to enhance glide mobility of dislocations by the increase in kink density, c_k, and/or kink mobility, v_k, by water (hydrogen). In order for this process of weakening to occur, hydrogen-related defects must diffuse to a dislocation line, and the formation and/or migration energy of kinks must be reduced by hydrogen. HEGGIE and JONES (1986) performed theoretical calculations of dislocation energies in quartz. They showed that hydrogen defects have large binding energies with a dislocation line and therefore tend to be bound along the dislocation line. They also showed that both nucleation and migration energies of double kinks are considerably reduced by the interaction with hydrogen defects. Consequently, their results support the GRIGGS (1974) and HOBBS *et al.* (1972) model for hydrolytic weakening. It is likely that similar effects are present in other silicates, although the extent to which nucleation and migration of kinks controls dislocation creep (i.e., the Peierls mechanism) depends on the bond strength. Such an effect is expected to be stronger in materials where a strong Si–O bond plays important roles. CORDIER and DOUKHAN (1995) discussed the importance of dislocation splitting on water weakening in quartz.

When dislocation recovery controlled by dislocation climb is the rate-controlling step, then $\dot{\varepsilon} \propto v_c$ where v_c is the velocity of dislocation climb. Now, the climb velocity is controlled by the jog velocity ($v_j \propto D, D$, diffusion coefficient) and jog concentration (c_j), $v_c = c_j b v_j$. Therefore the enhancement of diffusion or the increase in jog concentration by hydrogen will lead to enhanced rate of deformation. If one writes $v_j \propto f_{H_2O}^{r_D}$ and $c_j \propto f_{H_2O}^{r_j}$, then $\dot{\varepsilon} \propto v_c \propto f_{H_2O}^{r_c}$ with $r_c = r_D + r_j$. In olivine, a recovery-controlled model is proposed (e.g., MEI and KOHLSTEDT, 2000b; KARATO and JUNG, 2003). A jog may be hydrated, in which case the jog concentration is a function of hydrogen content (water fugacity). Consider the following reaction

$$\frac{n_j}{2} \cdot H_2O + (\text{jog}) \Longleftrightarrow (\text{hydrated jog}) \tag{10.48}$$

where n_j is the number of hydrogen atoms at a hydrated jog. Applying the law of mass action to reaction (10.48),

$$\tilde{c}_j = K_{48} \cdot f_{H_2O}^{n_j/2} \cdot c_j \tag{10.49}$$

where $\tilde{c}_j$ is the concentration of hydrated jogs and K_{48} is the equilibrium constant for the reaction (10.48) (which includes other factors such as oxide activity, see section 10.3.2). Consequently,

$$\dot{\varepsilon} \propto f_{H_2O}^{n_j/2} \cdot c_j \cdot v_j \propto f_{H_2O}^{n_j/2} \cdot D(C_H) \propto f_{H_2O}^{(n_j/2)+p_D} \tag{10.50}$$

where D is the diffusion coefficient and $D \propto f_{H_2O}^{p_D}$ (see Table 10.1). The observed anisotropy may be due to the anisotropy in the degree of hydration, K_{48}: for one type of dislocation, hydration affects the kink energy more than for other types of dislocation.

Alternatively hydrogen weakening in the dislocation-creep regime may be due to the enhancement of dislocation glide. In analogy with the theoretical study by HEGGIE and JONES (1986) on quartz, it is possible that the dislocation energy (such as the Peierls energy) in olivine is reduced by hydrogen thereby increasing the density and mobility of kinks. Therefore, within a glide-controlled model, one way to explain the enhancement of deformation by hydrogen is to assume (1) the dislocation velocity is controlled by $v_g = c_k b v_k$, and (2) the concentration (and mobility) of kinks increases by hydration. Similar to equation (10.48), the chemical reaction relevant to the hydration of a kink is

$$\frac{n_k}{2} \cdot H_2O + (\text{kink}) \Leftrightarrow (\text{hydrated kink}) \tag{10.51}$$

where n_k is the number of hydrogen atoms at a hydrated kink. Applying the law of mass action to equation (10.51), one has

$$\bar{c}_k = K_{51} \cdot f_{H_2O}^{n_k/2} \cdot c_k \tag{10.52}$$

where $\bar{c}_k$ is the concentration of hydrated kinks and K_{51} is the reaction constant for reaction (10.51). Therefore,

$$\dot{\varepsilon} = \bar{c}_k b \bar{v}_k = K_{51} \cdot f_{H_2O}^{n_k/2} \cdot c_k \cdot \bar{v}_k \tag{10.53}$$

where $\bar{v}_k$ is the velocity of hydrated kinks. Again, in this case, enhancement of creep by water can be anisotropic because both K_{51} and $\bar{v}_k$ will be anisotropic. If we assume that K_{51} and $\bar{v}_k$ are not dependent on the concentration of hydrogen (or water fugacity), the water fugacity dependence of creep rate comes mainly from the dependence of concentration of hydrated kinks on water fugacity thus

$$\dot{\varepsilon} \propto f_{H_2O}^{n_k/2}. \tag{10.54}$$

In a case where a hydrated kink contains a $(2H)_M^{\times}$, $n_k = 2$ ($\dot{\varepsilon} \propto f_{H_2O}$), whereas if a hydrated kink contains a $(4H)_{Si}^{\times}$, then $n_k = 4$ ($\dot{\varepsilon} \propto f_{H_2O}^2$).

In summary, the way in which the influence of water on the strain rate for dislocation creep can be expressed is by an equation similar to equation (10.46), namely,

$$\dot{\varepsilon}_{wet} \propto f_{H_2O}^r \cdot \exp\left(-\frac{H_{disl,wet}^*}{RT}\right) \qquad (10.55)$$

where $H_{disl,wet}^*$ is the activation enthalpy for dislocation creep that may include jog (kink) formation and migration enthalpy, and r is related to the influence of water on the concentration of hydrated jogs (or kinks).

One remarkable observation in olivine is the highly anisotropic effect of water weakening: deformation by $\boldsymbol{b}=[001]$ dislocations is enhanced by water more than $\boldsymbol{b}=[100]$ dislocations (MACKWELL *et al.*, 1985). Diffusion of Si and O in olivine is not strongly anisotropic (JAOUL *et al.*, 1981; JAOUL and HOULIER, 1983; HOULIER *et al.*, 1988, 1990), so a possible mechanism of anisotropy in the hydrogen-weakening effect is the anisotropic enhancement of jog formation. Alternatively, control by kink nucleation and migration may be a cause of anisotropy, because by its very nature the nucleation and migration of kinks is strongly controlled by the crystal structure.

It should be noted that the nature of transition between "dry" condition to "wet" condition is different from the nature of transition between diffusional and dislocation creep. In the latter case, the two deformation mechanisms are independent and both processes operate following the same constitutive relation under most conditions, and consequently the total strain rate is simply given by $\dot{\varepsilon} = \dot{\varepsilon}_{disl} + \dot{\varepsilon}_{diff}$ and each term ($\dot{\varepsilon}_{wet}$ or $\dot{\varepsilon}_{dry}$) has the same expression corresponding to a case where each of them operates in isolation. In contrast, when we consider a transition in deformation behavior between "wet" and "dry" conditions, a process that operates at one condition (e.g., "wet" condition) may not operate at another condition in the same way as it operates in isolation so that the constitutive relations may not be additive. For example, dominant point defect may change from vacancy to interstitial atom as water fugacity changes, then $\dot{\varepsilon}_{wet}$ and $\dot{\varepsilon}_{dry}$ are not necessarily additive (see Fig. 10.7). The analysis of experimental data spanning both "wet" and "dry" conditions needs to be done with a great care.

Other volatiles

Water (or hydrogen) is not the only volatile species in Earth, but there are other volatile species including carbon (e.g., WOOD *et al.*, 1996). Very little has been investigated on the influence of volatile elements on plastic deformation other than hydrogen. However, the solubility of carbon in olivine was investigated. KEPPLER *et al.* (2003) showed much smaller solubility of carbon in olivine than that of hydrogen (at $P=3.5$ GPa, $T=1400$ K, solubility of carbon is ~6 ppm C/Si (~0.5 wt ppm) as compared to that of hydrogen, ~2000 ppm H/Si). KEPPLER *et al.* (2003) also proposed that carbon is dissolved in olivine as $C_{Si}^{\times}$. From an atomistic point of view, the following points should be noted. The influence of volatile elements on plastic properties is through dissolved defect containing the volatile elements. The solubility of carbon in olivine is, for example, much smaller than that of hydrogen. However, carbon is dissolved in olivine at Si-site. Consequently, it is likely that all the dissolved carbon will directly influence plastic deformation that is related to the Si-site defects. Effects of both water and carbon dioxides (as well as other oxides such as P_2O_5) on melting of silicate has been studied (KUSHIRO, 1975; WYLLIE and HUANG, 1976). The results showed a clear trend between the melting temperature and melt composition and the concentration of these volatile species. Because there is a close link between melting temperature and plastic properties (i.e., homologous temperature scaling), one might expect some important effects of other volatiles. However, very little is known about the influence of carbon or other volatile elements on plastic properties.

10.3.5. Interplay between pressure and water (hydrogen) effects

Experimental studies on the influence of water on plastic deformation (and related properties) are often conducted by changing the water (hydrogen) content of a sample through the change in pressure. For example, one can change the water content of olivine by a factor of ~500 by changing pressure from 0.1 to 3 GPa. By plotting the strain rate as a function of water content, one could determine a parameter r ($\dot{\varepsilon} \propto f_{H_2O}^r$). However, one should recall that strain rate changes also with pressure due to the activation volume effect with the same water content, namely,

$$\dot{\varepsilon} \propto f_{H_2O}^r(P,T) \cdot \exp\left(-\frac{E^* + PV^*}{RT}\right). \qquad (10.56)$$

Therefore the influence of activation volume, $\exp(-(E^* + PV^*)/RT)$, needs to be corrected appropriately in order to determine r. Otherwise, the inference of a parameter r will be incorrect (see a discussion by KARATO, 2006a).

In order to evaluate the influence of the $\exp(-(E^* + PV^*)/RT)$ term, let us consider how water fugacity changes with pressure. As we learned

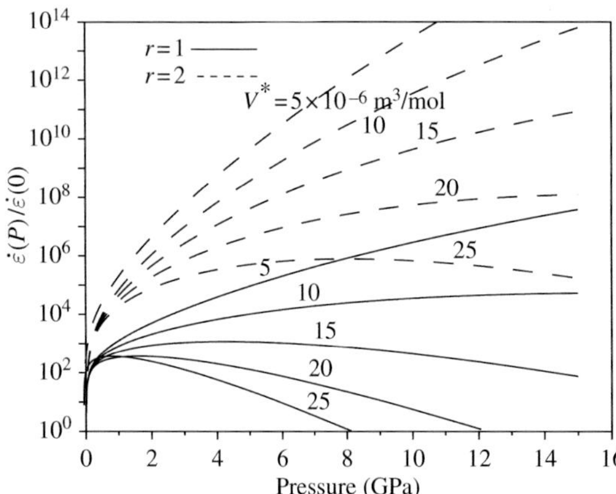

FIGURE 10.10 The variation of strain rate with pressure for constant stress and temperature using equation (10.56).

in Chapter 2, water is nearly an ideal gas at high temperatures (above ~1300 K) at low pressures (<0.5 GPa), but above a few GPa, fugacity of water changes with pressure approximately exponentially ($f_{H_2O}(P,T) \propto \exp(P\bar{V}^*/RT)$ with $\bar{V}^* \approx 11 \times 10^{-6}\,\mathrm{m}^3/\mathrm{mol}$, which is close to the molar volume of oxygen). Consequently, when a system is in equilibrium with a water reservoir, strain-rate changes with pressure as shown in Fig. 10.10. At low pressures (say <0.5 GPa), strain rate increases with pressure due to the increase in f_{H_2O}, whereas at higher pressures, the competition between the f_{H_2O} term and the $\exp(-PV^*/RT)$ term becomes important and often strain-rate decreases with pressure. These two effects are different and it is critical to characterize these two effects separately in order to determine the constitutive relationship (flow law) that can be extrapolated to Earth's interior. As seen from Fig. 10.10, it is important to realize that the full characterization of these two effects can only be made through the experimental studies under a broad pressure range spanning below and above ~0.5 GPa, at which the behavior of water changes from nearly ideal gas to highly non-ideal gas (see Chapter 2). An example of such an analysis that incorporates both the effects of $f_{H_2O}(P,T)$ and $\exp(-PV^*/RT)$ is shown in Fig. 10.11.

Problem 10.5*

Discuss how $\exp(-(E^* + PV^*)/RT)$ term affects the apparent water fugacity exponent determined from $r^* \equiv d\log\dot{\varepsilon}/d\log f_{H_2O}$ as compared to the true water fugacity exponent defined by $r \equiv (\partial \log\dot{\varepsilon}/\partial \log f_{H_2O})_{P,T}$.

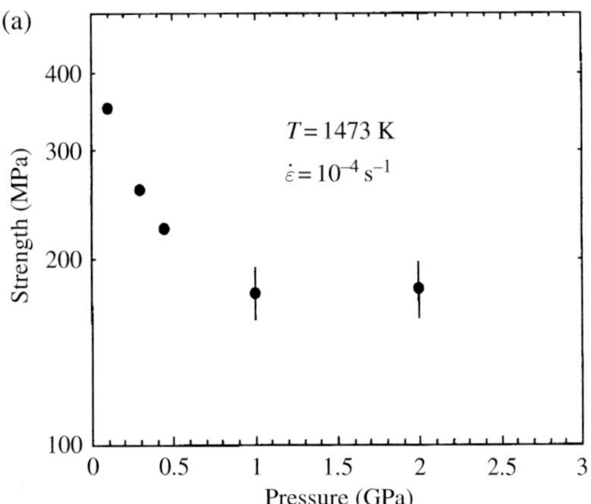

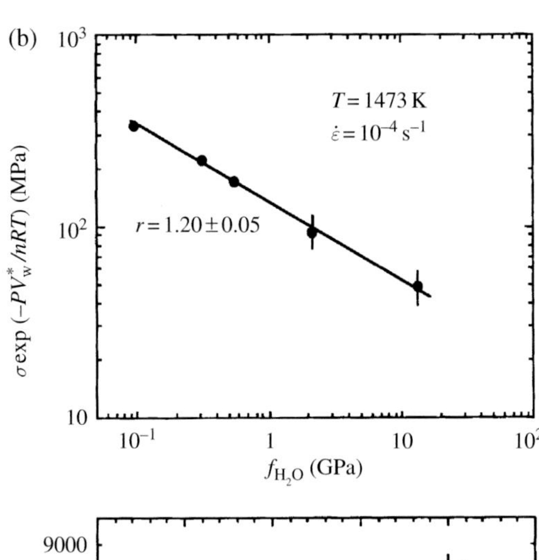

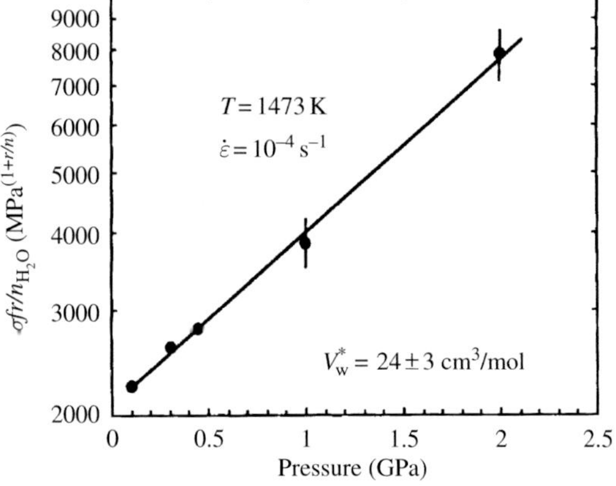

FIGURE 10.11 (a) The pressure dependence of the creep strength of olivine under water-saturated conditions and (b) the analysis of the data in terms of $\dot{\varepsilon} \propto f^r_{H_2O}(P,T) \cdot \sigma^n \cdot \exp\left(-\frac{E^*+PV^*}{RT}\right)$, where both r and V^* are determined together from the data (KARATO and JUNG, 2003).

Solution

From $\dot{\varepsilon} \propto f_{H_2O}^{r}(P,T)\cdot\exp(-(E^* + PV^*)/RT)$, one has $r^* \equiv d\log\dot{\varepsilon}/d\log f_{H_2O} = (\partial\log\dot{\varepsilon}/\partial\log f_{H_2O})_{P,T} - \frac{V^*}{RT}(\partial P/\partial\log f_{H_2O})_T = r - \frac{V^*}{RT}(\partial P/\partial\log f_{H_2O})_T$. Therefore $r = r^* + \frac{V^*}{RT}\left(\frac{\partial P}{\partial\log f_{H2O}}\right)_T$. Now, at low pressures and high temperatures where water is nearly an ideal gas, $(\partial P/\partial\log f_{H_2O})_T \approx P$ so that $r = r^* + PV^*/RT$. At higher pressures where $f_{H_2O}(P,T) \propto \exp(P\bar{V}^*/RT)$, then $(\partial P/\partial\log f_{H_2O})_T \approx RT/\bar{V}^*$ and hence $r \approx r^* + V^*/\bar{V}^*$. For a typical value of V^* ($\sim 10\times10^{-6}\,m^3/mol$), $PV^*/RT \approx 0.3$ (for $P = 0.5\,GPa$ and $T = 1600\,K$) and $V^*/\bar{V}^* \approx 1$. In either case, the difference between r and r^* is large.

An equation similar to (10.56) may apply to the rate of kinetic processes that are assisted by hydrogen, namely,

$$Y \propto \exp\left(-\frac{E_Y^* + PV_Y^*}{RT}\right) \propto C_{OH}^{\tilde{r}}(P,T)\cdot\exp\left(-\frac{\tilde{E}^* + P\tilde{V}^*}{RT}\right) \qquad (10.57)$$

where Y is the rate of some process such as electric conductivity or diffusion coefficient, $C_{OH}(P,T)$ is the concentration of hydrogen in the mineral, and $\tilde{r}$ is an appropriate constant. The concentration of hydrogen depends on pressure and temperature as (see section 10.3.2)

$$C_{OH}(P,T) \propto f_{H_2O}^{r_{OH}}(P,T)\exp\left(-\frac{E_{OH}^* + PV_{OH}^*}{RT}\right). \qquad (10.58)$$

For diffusion of Mg–Fe in olivine, $\tilde{r} = 1$ (Hier-Majumder *et al.*, 2005a), and for olivine $r_{OH} = 1$ (Kohlstedt *et al.*, 1996). Consequently, $V_Y^* = V_{OH}^* + \tilde{V}^*$. Hier-Majumder *et al.* (2005a) found $= 16\times10^{-6}\,m^3/mol$ for olivine under water-saturated condition, and $V_{OH}^* \sim 10\times10^{-6}\,m^3/mol$ for olivine (Kohlstedt *et al.*, 1996). This leads to $\tilde{V}^* \sim 6\times10^{-6}\,m^3/mol$, which is consistent with the observation of $V_{diff}^* \sim 5–6\times10^{-6}\,m^3/mol$ for Mg–Fe diffusion in olivine under "dry" conditions (e.g., Misener, 1974; Farber *et al.*, 2000) assuming $V_{diff}^*(wet) \approx V_{OH}^* + V_{diff}^*(dry)$.[7] Similarly, Karato and Jung (2003) obtained $V_{creep}^* = 24\times10^{-6}\,m^3/mol$ for olivine under water-saturated conditions, and $V_{creep}^* = 14\times10^{-6}\,m^3/mol$ under water-free conditions. These observations can be interpreted by noting $V_{creep}^*(wet) \approx V_{OH}^* + V_{creep}^*(dry)$.

[7] I used "≈" rather than "=" is to emphasize the fact that the process of thermal activation under water-saturated conditions may involve a different defect (hydrated defect) from that under dry (water-free) conditions and therefore the corresponding activation volume is likely somewhat different.

10.3.6. Some issues in investigating water (hydrogen) in minerals

Experimental studies on the nature of hydrogen are challenging for several reasons. First, hydrogen is so mobile that it could easily move into a sample from the surroundings or move out from a sample to the surroundings. Second, the hydrogen can occur as various forms in a given material depending on the pressure and temperature, and therefore the form of hydrogen that one observes at ambient conditions may not represent the form of hydrogen-related species in Earth's interior.

Let us first discuss the diffusional loss or gain of hydrogen. The characteristic time for a given reaction depends on the length-scale at which the reaction occurs and the chemical diffusion coefficient as,

$$\tau_{diff} \approx \frac{d^2}{\pi^2 D} \qquad (10.59)$$

where d is the length-scale at which diffusion takes place and D is the chemical diffusion coefficient of hydrogen in a given material. If the time-scale involved in a given process significantly exceeds τ_{diff}, then the process will occur to completion. On the other hand if the time-scale is much less than τ_{diff}, then the process does not alter the concentration of defects. Fig. 10.12 shows a diagram in which the characteristic times for diffusion-controlled reactions are compared with the length- and time-scales of various processes. For example, consider what could happen during a typical quenching process in the lab. The quenching time-scale is ~1–100 second (depending on the type of apparatus), and the sample size is ~1–5 mm and the mean distance among the various atomic sites is ~1 nm. From Fig. 10.12, it is seen $\tau \leq \tau_{diff}$ (τ is the time-scale of quenching) if d is the sample size. However, if d is the distance among different lattice sites, then $\tau \gg \tau_{diff}$. Therefore we conclude that the *total hydrogen content* of a sample is preserved in most cases during quenching but the *atomic sites* that hydrogen prefers at high pressure and temperature may not be preserved during quenching. Consequently, if the preferred sites of hydrogen-related defects change with pressure and temperature, there is no guarantee that the atomic sites

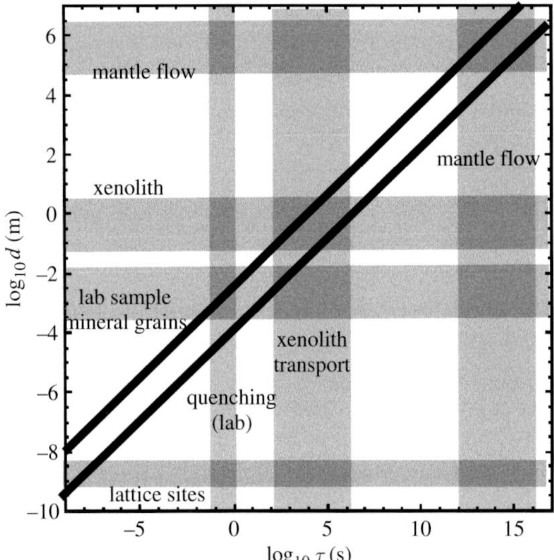

FIGURE 10.12 A characteristic distance (d) and time-scale (τ) diagram for the equilibration of hydrogen-related defects. The (chemical) diffusion coefficients are from KOHLSTEDT and MACKWELL (1998) for temperatures of ~1000–1500 K. The solid lines correspond to the relation (10-59). If the characteristic length and time for a given process (d, τ) fall above the solid lines, then the hydrogen-related species will be frozen during that process. If the characteristic length and time for a given process (d, τ) fall below the solid lines, then the nature of hydrogen-related species will be reset during the process (after KARATO, 2006a).

of hydrogen-related defects inferred from observations at ambient pressure and temperature are identical to those at high pressure and temperature.

A similar analysis applies to a natural sample. Recall that the hydrogen solubility in a given mineral is a strong function of pressure and temperature. At high pressures and temperatures in Earth's interior, the hydrogen solubility is high, whereas at low pressures and temperatures, the solubility is low. Imagine a mineral that occurs in a hydrous environment in Earth's deep interior. Then a large amount of hydrogen should be dissolved in the mineral. Suppose that this mineral is transported to Earth's surface. The pressure and temperature are now reduced, and the solubility of hydrogen is reduced. If we assume that the sample is embedded in a large rock so that diffusion is inefficient (i.e., $t \ll \tau_{\text{diff}}$ where t is the time-scale of depressurization) then *the total hydrogen content* in a mineral will be preserved, but the sample will be supersaturated with hydrogen. In such a case, the excess hydrogen will precipitate either as water-filled bubbles or hydrous minerals (depending on the pressure–temperature conditions). Consequently, a natural sample that one can investigate in the lab at ambient pressure and temperature will contain "frozen" hydrogen-related defects as well as precipitated water-filled bubbles and/or hydrous minerals. An FT–IR absorption spectrum from such a sample includes absorption peaks due to dissolved OH (due to frozen-in OH) in addition to those due to free water and hydrous minerals (BERAN and LIBOWITZKY, 2006). Therefore if the hydrogen content of a sample is estimated only from the absorption peaks corresponding to dissolved OH, then it will provide an underestimate of a true hydrogen content in a sample in Earth's interior. The amount of hydrogen in water-filled bubbles and hydrous inclusions must be added in order to infer the water (hydrogen) content of a sample in Earth's interior. Submicron-size hydrous minerals are frequently detected in natural samples by TEM studies (e.g., KITAMURA *et al.*, 1987; KHISINA *et al.*, 2001). They may represent the excess hydrogen that was dissolved in a mineral in Earth's deep interior. However, these minerals or fluid inclusions could have been formed as a result of later stage metasomatism. Origin of various hydrogen-related species must be examined carefully in order to obtain insight into the hydrogen (water) content in Earth from the analysis of samples from Earth.

10.3.7. Principles governing the distribution of water in minerals in Earth

The effects of water on plastic properties are large. By laboratory experiments, we can determine how water affects plastic properties, and when the results are properly parameterized based on a sound physical model, then the results can be extrapolated to Earth and planetary interiors. Given the strong effects of water, it is important to understand the principles by which the distribution of water is controlled.

Water content in any portion of Earth is controlled by a complex history of water budget including the water supply and loss. At a macroscopic scale (say more than ~1 km), diffusive transport of water is ineffective, and in most cases water is transported associated with macroscopic (advective) transport of material. This includes water transport by the migration of liquids (melts) containing water as well as water transport by convective motion of water-bearing minerals. In this regard, the stability of hydrous phases and their spatial distribution in subducting slabs play a critical role (e.g., RÜPKE *et al.*, 2004). Upwelling of relatively water-rich plumes will also modify the water distribution.

Box 10.2 Partial melting

Melting in a multi-component system occurs gradually when the temperature reaches a certain value called the *solidus*, and the degree of melting increases with further increase in temperature, and at a certain temperature (called the *liquidus*) the whole material will be molten. So in a broad range of temperature, a material is composed of a mixture of melt and solid. This is called *partial melting*.

Partial melting modifies the distribution of various elements such as hydrogen (water). The way in which an incompatible element such as hydrogen (water) is redistributed upon partial melting depends on the processes of melt transport. When melt moves easily, then upon melting, the melt will be removed from the system by gravity before the bulk of melt becomes in chemical equilibrium with the original solid. In this case, chemical equilibrium between melt and solid occurs only locally upon incremental melting. This process of melting is referred to as *fractional melting*. In contrast, if the melt transport is very slow, all the melt formed may stay in chemical equilibrium with the original solid. This mode of melting is referred to as *batch melting* (by batch melting alone, one can not separate the melt from the solid. In order to do so, one needs to invoke fractional melting at a later stage). The removal of an incompatible element is more efficient for the fractional melting. The experimental and observational studies suggest that the melt extraction in the shallow upper mantle is easy and partial melting in the upper mantle occurs in most cases as fractional melting.

In many cases, melting occurs as a progressive process. When melting occurs as fractional melting, then at any time, the total amount of melt that has been produced will be larger than the amount of melt at a given time. The *integrated (total) amount* of melt removed from the solid is referred to as the *degree of melting*. The degree of melting at mid-ocean ridges can be estimated from the volume ratio of the oceanic crust and the oceanic lithosphere and is ~10%. The *fraction of melt* in a partially molten material is the same as the degree of melting only when melting occurs as batch melting. Although the degree of melting is ~10% beneath the mid-ocean ridges, the melt fraction in a partially molten column beneath them can be much smaller than this value (conversely the melt fraction in a magma chamber will be, by definition!, always very high despite the fact that the melt fraction in the melting column is small). The degree of melting and the fraction of melt should not be confused.

Let us now consider more microscopic processes of water re-distribution. In a general multi-component system, water (hydrogen) at equilibrium is distributed among co-existing phases according to the partition coefficients (relative solubility limits) and the amount of each phase. If a system is open with respect to water, then all phases have water contents corresponding to the solubility for a given chemical environment (water fugacity etc.). Therefore whatever processes may operate, water contents in each phase will always be determined by the chemical equilibrium with the environment. However, when the system is closed with respect to water, then the formation of a new phase that has largely different water solubility will result in a marked re-distribution of water (hydrogen). An important example is partial melting (KARATO, 1986; HIRTH and KOHLSTEDT, 1996). Consider for simplicity, a case where melt formed stays with solid (this type of melting process is referred to as *batch melting*; see Box 10.2). In such a case, there must be chemical equilibrium between all the melt and solid in the system. Therefore if C_{s0} is the total water (hydrogen) content and $C_{m,s}$ are the concentration of water in melt and solid respectively then the conservation of water demands

$$\phi C_m + (1-\phi)C_s = C_{s0} \tag{10.60}$$

where ϕ is the fraction of melt.[8] Now the condition of chemical equilibrium (with respect to water) is

$$\frac{C_s}{C_m} = k \tag{10.61}$$

where k is a partition coefficient that is controlled by the relative solubility of water. From equations (10.60) and (10.61), one has

$$\frac{C_s}{C_{s0}} = \frac{k}{\phi + (1-\phi)k} \approx 1 - \frac{1-k}{k}\phi \qquad \text{for } \phi \ll 1 \tag{10.62}$$

where C_{s0} is the concentration of water in the solid before melting. Note that this relation also applies to

[8] If the fraction is volume (mass) fraction, then concentration must be measured per volume (mass).

any reaction involving two phases with different water solubility, such as the formation of a hydrous mineral. We conclude that the formation of a phase with a high water solubility such as melt or hydrous mineral in a *closed system* will decrease the concentration of water in an anhydrous mineral.

The above argument requires some modifications when one phase is easily removed from the system. Consider a case where melting occurs progressively (say by the decrease in pressure). Let us assume that melt has different density from surrounding solids and is easy to move. In such a case, as melt is formed it is removed from the system before the melt becomes in chemical equilibrium with the majority of the solids (this process of melting is referred to as *fractional melting*). In such a case, chemical equilibrium between melt and solid is achieved only at the time of melting between a small amount of melt and solid, say at grain boundaries. Let us imagine a small mass of solid, dm, is melted, this mass of material transforms to melt. Let dx be the amount of water in this newly formed melt. The concentration of water in the small amount of melt is dx/dm, and the concentration of water in the solid is x/m. By assumption, a small amount of melt and solid are in chemical equilibrium during this incremental process. Therefore

$$x/m = k(dx/dm). \tag{10.63}$$

Integrating this one obtains $(x/x_0)^k = (m/m_0)$ and hence,

$$\frac{x/m}{x_0/m_0} = \left(\frac{m}{m_0}\right)^{(1-k)/k}. \tag{10.64}$$

Inserting the relations $C_s = x/m$, $C_{s0} = x_0/m_0$, and $\phi = (m_0 - m)/m_0$ where ϕ is the total fraction of melt removed from the solid (i.e., the degree of melting) into (10.64), one finds,

$$\frac{C_s}{C_{s0}} = (1-\phi)^{(1-k)/k} \approx \exp\left(-\frac{1-k}{k}\phi\right) \approx 1 - \frac{1-k}{k}\phi \qquad \text{for } \phi \ll 1. \tag{10.65}$$

It can be seen that a significant depletion of water occurs when the degree of melting exceeds a few times

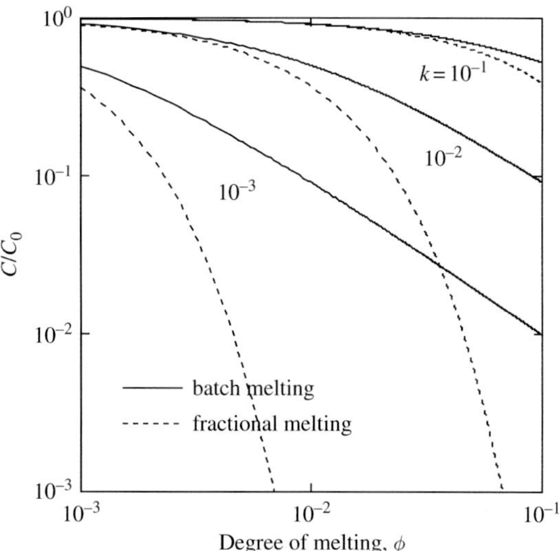

FIGURE 10.13 Depletion of water by partial melting as a function of the degree of melting, ϕ, and the partition coefficient, k.

the partition coefficient (95% of water is removed if $\phi/k = 3$). Fig. 10.13 shows plots of relations (10.62) and (10.65). Although the degree of removal of water in both processes is the same for a small degree of melting, the two models yield largely different results at a larger degree of melting: the degree of removal of water from solid by partial melting is larger for fractional melting than for batch melting. Note that the appearance of a hydrophilic (water-loving) phase (melt or hydrous minerals) will always reduce the water (hydrogen) content in co-existing nominally anhydrous minerals. Therefore the presence of a hydrous mineral does not mean that a rock is rheologically soft.

Plausible processes of water circulation in Earth have been investigated by McGovern and Schubert (1989), Richard *et al.* (2002), Bercovici and Karato (2003), Rüpke *et al.* (2004), and Karato *et al.* (2006). The large uncertainties in these models include the solubility of water (hydrogen) in deep mantle (particularly lower mantle) minerals, and the macroscopic and microscopic processes of water transport. "Inverse" approach to infer water content from geophysical observations is presented by Karato (2003b, 2006b).

11 Physical mechanisms of seismic wave attenuation

All materials show some deviation from perfect elasticity particularly at low frequencies and/or high temperatures. In geophysics, this effect can be seen as the attenuation of seismic waves, but it can also be seen as a reduction in seismic wave velocities. Many of the physical processes causing this effect involve the motion of crystalline defects (dislocations, grain boundaries) and are sensitive to temperature, grain size and the chemical environment such as the fugacity of water. This chapter provides a review of the solid-state mechanisms of seismic wave attenuation as well as those involving partial melt. Understanding the physical mechanisms of anelasticity (or visco-elasticity) is critical for the inference of distribution of temperature, grain size, water content and other variables describing physico-chemical environment such as the presence of partial melt from seismic wave attenuation (Chapters 18, 20).

Key words anelasticity, visco-elasticity, Q, velocity dispersion, Maxwell model, Voigt model, standard linear solid, Burgers model, absorption band model, melt squirt, Walsh model, Zener mechanism (thermoelasticity), Bordoni peak, Snoek peak, Kê peak, bulk attenuation.

11.1. Introduction

Understanding the microscopic mechanisms for anelasticity is critical for the interpretation of seismic wave attenuation in terms of the physical and chemical state of Earth's interior (e.g., JACKSON, 2000) and for the study of tidal dissipation that controls the thermal history and orbital evolution of planets (e.g., KAULA, 1964). The phenomenological theory of anelastic (or visco-elastic) behavior[1] is reviewed in Chapter 3. This chapter provides a materials science basis for anelasticity (or visco-elasticity) including a brief description of experimental techniques (section 11.2), experimental and theoretical results on solid-state mechanisms of anelasticity (section 11.3) and mechanisms unique to partially molten materials (section 11.4).

11.2. Experimental techniques of anelasticity measurements

11.2.1. General introduction

Experimental techniques of anelasticity (or visco-elasticity) measurements can be divided into two categories: (quasi-) static and dynamic measurements. In comparison to elasticity measurements, anelasticity measurements are challenging because small signals must be measured. Consider, for example, a static method of elasticity measurements (see Chapter 4). We can determine the elastic constants by measuring the strain caused by a known stress. In anelasticity measurements, we need to determine a small deviation from elastic deformation, so the

[1] *Anelasticity* is often distinguished from *visco-elasticity*, but when energy dissipation follows "high-temperature background" type behavior (see Chapter 3) where attenuation is related to frequency as $Q^{-1} \propto \omega^{-\alpha}$, this distinction becomes unclear. I therefore use a term "anelasticity" in a broad sense to describe deformation associated with energy loss in general.

resolution of strain measurements required for static measurements is high and such experiments are challenging. This is particularly the case because we need to confirm that anelastic response of a sample is linear with respect to stress. The linear behavior is usually observed below the strain of $\sim 10^{-6}$. If one wants to determine anelasticity (say a parameter $Q \sim 100$) with $\sim 10\%$ resolution, then one would need to determine strain with a resolution of $10^{-6}/100/10 = 10^{-9}$. For a 10 mm sample, this means that one needs to be able to measure the displacement of $\sim 10^{-2}$ nm! (using geometrical amplification, the displacement can be amplified to make the measurement more feasible: see Problem 11.1). In addition, anelastic behavior is sensitive to frequencies as well as temperature and pressure, chemical environment and microstructures such as grain size. Consequently, high-resolution measurements of displacement or amplitude of waves must be carried out under a well-controlled environment.

Problem 11.1

Estimate the resolution of strain measurements needed for static anelasticity measurements in which Q is measured by the deviation of elastic modulus from (high-frequency) unrelaxed modulus. Assume $Q \sim 100$, and calculate the resolution of strain measurements needed to determine Q within 10% error below the total strain of 10^{-6}.

Solution

Anelasticity affects the elastic modulus. Therefore, Q can be determined by the measurements of elastic modulus. Let us recall $Q^{-1} = \Delta M(\omega)/M(\infty)$ (see Chapter 3) where $\Delta M(\omega) = M(\infty) - M(\omega)$ ($M(\omega)$ is elastic modulus at frequency ω). Now elastic strain for a given stress is $\varepsilon = \sigma/M$. Therefore the change in strain due to anelasticity is $\Delta\varepsilon = (\sigma/M)\,(\Delta M/M) \approx (\sigma/M)Q^{-1} \approx 10^{-8}$. The corresponding displacement on the sample surface with a diameter r is $\Delta x = r\Delta\varepsilon \sim 0.1$ nm for $r = 1$ cm. Using a long arm (lever) R, one can amplify the displacement to $\Delta x = R\Delta\varepsilon \sim 10$ nm for $R = 1$ m. Therefore one needs to measure the displacement with a resolution of ~ 1 nm to determine $Q \sim 100$ with 10% accuracy.

A few points should be noted. First, almost all mechanisms of anelasticity are frequency-dependent. Consequently, the measurements are best made in the seismic frequency range (1–10^{-3} Hz). Second, most of the mechanisms for anelasticity are sensitive to temperature (and pressure). Therefore temperatures similar to those in Earth's interior must be used in the measurements or the temperature dependence of anelasticity must be determined with sufficient accuracy. The effects of pressure are two-fold. First, many anelasticity mechanisms are sensitive to the presence of grain boundaries. Therefore a polycrystalline sample is often used in these studies. Because of the anisotropy in thermal expansion (in most minerals), a polycrystalline sample tends to develop grain-boundary cracking when it is heated without confining pressure. Although this trend can be suppressed by using a fine-grained sample, the use of a high confining pressure, exceeding the magnitude of local stress at grain boundaries is recommended to minimize the unwanted effects of cracking (effects of cracking can be identified by investigating the pressure effects on attenuation). Finally, the stress magnitude must also be relatively small in order to assure that mechanisms that operate under laboratory conditions are linear anelasticity (or visco-elasticity).

11.2.2. Wave-propagation method

Attenuation of elastic waves can be determined by the change in amplitude of elastic waves. Similar to the ultrasonic wave velocity measurements, one generates ultrasonic waves using a piezoelectric transducer and measures the waveform of received elastic waves. The observed waveform includes waves with different frequencies. Using the Fourier analysis, one calculates the amplitude of waves with a range of frequencies at least at two positions. The amplitude of waves, $X(x, t)$, changes as the wave propagates,

$$X(x,t) = A_0 \exp(-\alpha x)\exp\left[i\omega\left(t - \frac{x}{V}\right)\right] \tag{11.1}$$

where α is the attenuation coefficient, ω is (angular) frequency, x is distance, t is time and V is the velocity of elastic wave. Using the definition of Q, one can show that

$$Q^{-1} = \frac{2aV}{\omega} \tag{11.2}$$

hence

$$X(x,t) = A_0 \exp\left(-\frac{\omega}{2VQ}x\right)\exp\left[i\omega\left(t - \frac{x}{V}\right)\right]. \tag{11.3}$$

Therefore by measuring the amplitude of elastic waves at a given position for a variety of frequencies or by measuring the amplitude of wave with a fixed frequency at different positions, one can determine Q. It is usually assumed that Q is independent of frequency, but one can also infer the frequency dependence of Q from the measured amplitude versus frequency results. Since the wavelength must be significantly shorter than the laboratory sample size, a high frequency is needed. For example, SATO *et al.* (1989) used 5 MHz in their study of attenuation using this technique.

Problem 11.2

Derive equation (11.2).

Solution

Recall that the energy of a wave is proportional to the square of its amplitude and the definition of Q is $Q^{-1} \equiv \Delta E/2\pi E$ where ΔE is energy loss per one cycle. For one cycle, a wave will travel a distance of one wavelength, so that

$$Q^{-1} = \frac{X^2(x)-X^2(x+\lambda)}{2\pi X^2(x)} = \frac{\exp(-2ax)-\exp(-2a(x+\lambda))}{2\pi\exp(-2\alpha x)} = \frac{1-\exp(-2a\lambda)}{2\pi} \approx \frac{a\lambda}{\pi}.$$

Now $V = \lambda\omega/2\pi$, therefore $Q^{-1} = 2aV/\omega$.

11.2.3. Quasi-static measurements

One of the major limitations of the wave-propagation method described above is that only high-frequency waves can be used. Also, the dependence of attenuation on frequency is not easy to determine because the frequency range that can be explored is limited. In fact, in many of the studies in which the spectra method is used, it is assumed that the attenuation (Q) is independent of frequency (e.g., SATO *et al.*, 1989). Seismic wave attenuation and its frequency dependence can be best determined by quasi-static methods. In the quasi-static method, one determines the change in amplitude and phase-lag of the displacement corresponding to a periodic force. If the specimen is elastic, then the amplitude of displacement is determined by the (unrelaxed) elastic constant that is determined by high-frequency measurements and there is no phase-lag: deformation is instantaneous. However, when non-elastic deformation is involved, then the amplitude will be smaller than expected from unrelaxed elastic modulus, and there must be a phase-lag in response. In short, one measures the frequency-dependent modulus, $M(\omega)$, and frequency-dependent phase-lag of displacement, $\delta(\omega)$. These two quantities are related to in-phase response, $M_1(\omega)$, and the out-of-phase response, $M_2(\omega)$, of a material as (see Chapter 3),

$$M(\omega) = \sqrt{M_1^2(\omega) + M_2^2(\omega)} = \frac{1}{\sqrt{J_1^2(\omega) + J_2^2(\omega)}} \tag{11.4}$$

and

$$\tan\delta(\omega) = \frac{M_2(\omega)}{M_1(\omega)} = \frac{J_2(\omega)}{J_1(\omega)} = Q^{-1}(\omega). \tag{11.5}$$

$M_1(\omega)$ and $M_2(\omega)$ are related by the Kramers–Kronig relation (Chapter 3), and therefore if either $M(\omega)$ or $\delta(\omega)$ is determined for an infinite frequency range, one set of results (either $M(\omega)$ or $\delta(\omega)$) is enough to completely characterize the non-elastic behavior of a material. Although measurements for an infinite frequency range are impossible in practice, the Kramers–Kronig relation is useful to assure the internal consistency of a measurement when results from a wide range of frequencies are available.

11.2.4. Low-frequency oscillation methods

KE (1947) developed an apparatus to measure low-frequency anelastic measurements. In this apparatus, a sample is attached to a "pendulum" and a force is initially applied to a specimen + pendulum to cause free-oscillation (at the frequency determined by the geometry and the elastic stiffness of the sample and supporting rod). The amplitude of this free-oscillation is measured by a sensor (usually by an optical encoder: see also WOIRGARD *et al.*, 1981). The decay of amplitude with time is measured from which attenuation (Q^{-1}) can be determined. GUEGUEN *et al.* (1989) applied this technique to determine Q in single crystals of forsterite.

BRENNAN and STACEY (1977) were among the first to use a forced-oscillation apparatus to determine low-frequency attenuation in geological materials. BERCKHEMER *et al.* (1982) and KAMPFMANN and BERCKHEMER (1985) applied this technique to high temperature (but at room pressure) to determine seismic wave attenuation in Earth materials. GRIBB and COOPER (1998) used a similar apparatus. In this type of apparatus, a sinusoidal stress is applied to one end of a sample and the displacement of another end of the sample is measured. The effects of attenuation can be determined by the delay in displacement and the decay in amplitude. In these studies anelasticity of polycrystalline samples was studied without confining pressure. Consequently, the possible role of grain-boundary cracking is not clearly ruled out. JACKSON and PATERSON (1987, 1993) further developed this type of apparatus to allow measurements of attenuation under a confining pressure and high temperature. With the current apparatus, measurements of seismic wave attenuation can be made to ~0.3 GPa and ~1600 K. The measurements of attenuation at high confining pressure are important for two reasons. First, cracking at grain boundaries probably occurs in most geological materials if a deviatoric stress is applied without confining pressure. A confining pressure of ~ 0.2 GPa is high enough to close most cracks if the magnitude of deviatoric stress is low (for attenuation measurements, a stress of 0.1–1 MPa is usually used). Second, some effects such as the effects of water can only be measured under high water fugacity conditions. Extension of such measurements to higher pressures is desirable but highly challenging.

11.3. Solid-state mechanisms of anelasticity

Mechanisms of seismic wave attenuation (anelasticity or visco-elasticity) can be conveniently divided into solid-state mechanisms and mechanisms involving partial melt. This section reviews the solid-state mechanisms of anelasticity (or visco-elasticity; see also KARATO and SPETZLER, 1990). When a small stress is applied then a perfect crystal will deform only elastically. Restoring force for atomic displacement is large and atoms will go back to their original positions as soon as the stress is removed. Non-recoverable or time-dependent strain will occur usually through the (thermally activated) motion of crystalline defects. Since all of these dissipative processes are time-dependent, their contribution to energy loss can be evaluated by the relative strain magnitude, $\varepsilon_{diss}/\varepsilon_{ela}$, where ε_{diss} is the strain associated with dissipative process(es) and ε_{ela} is the elastic strain. When dissipative deformation is characterized by viscosity, η, then $\varepsilon_{diss}/\varepsilon_{ela} = tM/\eta = t/\tau_M$ where $\tau_M\ (\equiv \eta/M)$ is the Maxwell time (Chapter 3). Typical values of viscosity of solid Earth materials is $\sim 10^{18}$–10^{22} Pa s, and $M \sim 10^{11}$ Pa so, $\varepsilon_{diss}/\varepsilon_{ela} \sim t(\mathrm{s})/(10^7$–$10^{11})$. Therefore we conclude that typical viscous behavior associated with a long-term viscosity does not contribute significantly to seismic wave attenuation except for very long-period phenomena such as the Chandler wobble ($\sim 10^7$ s). It is the viscous (dissipative) deformation at shorter time-scales that causes seismic wave attenuation. One should also note that the frequency dependence of seismic wave attenuation is in most cases $Q^{-1} \propto \omega^{-\alpha}$ with $\alpha = 0.2$–0.4 (e.g., ANDERSON and MINSTER, 1979; KARATO and SPETZLER, 1990; SHITO *et al.*, 2004) which is different from what is expected from simple viscous motion of defects (attenuation due to unlimited viscous motion of defects would yield $Q^{-1} \propto \omega^{-1}$, or constrained motion of a defect will lead to $Q^{-1} \propto \omega\tau/(1 + \omega^2\tau^2)$, see Chapter 3). Such weak frequency dependence requires a distribution of relaxation times or some transient creep behavior characterized by $\varepsilon \propto t^{\alpha}$ (Chapter 3).

11.3.1. Point defect mechanisms

Mechanisms of elastic wave attenuation due to motion of point defects may be classified into two categories. The first is anelastic relaxation due to local motion of point defects (motion with a scale of a few atoms), and the second is viscous (visco-elastic) relaxation due to grain-scale motion of point defects.

Anelastic relaxation due to local motion of point defects

Point defects are associated with strain (stress) field (see Chapter 5). Consequently when stress is applied, there will be interaction between a point defect and the applied stress. In considering the interaction between the applied stress and point defect, the symmetry of strain field plays a key role. Consider a point defect such as a vacancy in a cubic crystal located at regular atomic sites. Then the strain field associated with this defect has the same symmetry as an atom and all other possible positions for this type of defect are identical. In this case, applied (deviatoric) stress does not interact with the defect. In contrast, when a point defect is located at a site that has a lower symmetry than the host crystal, then the symmetry is broken by the defect and there are several non-equivalent positions in a crystal that have different energy. Such low-symmetry defects include an interstitial atom and a defect pair (see Fig. 11.1). In these cases, when deviatoric stress is

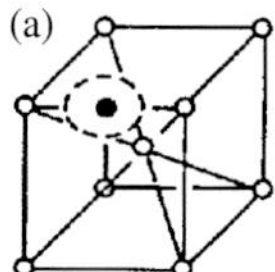

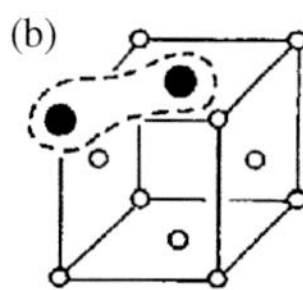

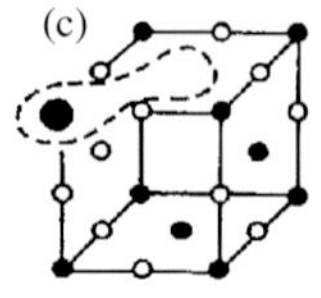

FIGURE 11.1 Point defects that can cause anelasticity (after KARATO and SPETZLER, 1990).

TABLE 11.1 Relaxation strength and relaxation times for point-defect mechanisms in silicate minerals such as Mg_2SiO_4 olivine.

	Relaxation strength	Relaxation time (s)
M-site defects		
Mg–Fe pair	$\sim 10^{-3}$	$\sim 10^{-5}$–10^{-4}
Hydrogen-related defects	$\sim 10^{-4}$–10^{-2}	$\sim 10^{-6}$–10^{-4}
Si-site defects		
Hydrogen-related defects	$\sim 10^{-4}$	$\sim$1–10

Modified from Karato and Spetzler (1990).

applied to a crystal, a defect (or a defect complex) will move from one configuration to another lower energy configuration through atomic diffusion causing *anelastic* relaxation. Anelastic behavior due to the motion of point defects is well documented in metals (Nowick and Berry, 1972). One is the *Snoek peak* due to the motion of interstitial impurity atoms in bcc (body-centered-cubic) metals, and another is the *Zener peak* caused by the motion of pairs of atoms in solid-solution alloys. At a larger scale, point defects at one side of a crystal have different energy than those of another orientation, which causes diffusive mass transport from one grain boundary to another. This mechanism will be discussed in Section 11.3.3 (grain-boundary mechanisms).

Anelastic relaxation by point defects can be characterized by relaxation time, τ, and relaxation strength, Δ (Chapter 3). The characteristic time for defect motion is the time for diffusion of an atom from one position to another. If the distance of two atomic positions is a then the characteristic time for this atomic motion is given by

$$\tau \approx \frac{a^2}{\pi^2 D} \tag{11.6}$$

where D is the diffusion coefficient and a is the distance of defect motion. The relaxation strength is proportional to the strain associated with a defect and the defect concentration as

$$\Delta = \frac{\varepsilon_{\text{diss}}}{\varepsilon_{\text{ela}}} = \frac{\varepsilon_{\text{an}}}{\varepsilon_{\text{ela}}} \approx \frac{\delta\varepsilon \cdot (C_0 - C)}{\sigma/M} = \frac{C_0(\delta\varepsilon)^2 \cdot M\Omega}{RT} \tag{11.7}$$

where C_0 is average concentration of the point defect, $\delta\varepsilon$ is strain due to the defect, Ω is the volume of the defect, M is the elastic modulus of the crystal, R is the gas constant and T is temperature, and we used the relation $C = C_0 \exp(-\sigma\Omega \cdot \delta\varepsilon/RT) \approx C_0(1 - \sigma\Omega \cdot \delta\varepsilon/RT)$. The relaxation time and the relaxation strength for various point defect mechanisms of anelastic relaxation are given in Table 11.1. In order for this mechanism to be important, the relaxation strength must be large ($>10^{-3}$) and relaxation time must be within the seismic period (inverse frequency, 1–10^{-3} Hz). Point defects that have large enough relaxation strength (i.e., defects with high concentrations) are hydrogen-related defects (concentration up to $\sim$0.1) and some solid-solution pairs such as Fe–Mg or Si–Al pairs. However, the diffusion coefficients for such defects are generally too high to be compatible with significant seismic wave attenuation, or the relaxation strength is too low.

Transient diffusional creep

Point defects such as vacancies have different energies at grain boundaries of different orientations with respect to the applied shear stress. This causes the flow of vacancies (and hence atoms) from grain boundary to grain boundary, causing a large-scale plastic strain. This type of plastic flow is called *diffusional creep* (Chapter 8). Diffusional creep is an obvious energy dissipation process, and therefore a possible mechanism for seismic wave attenuation. However, the time-scale of deformation associated with steady-state diffusional creep is too long to cause any appreciable seismic wave attenuation. Gribb and Cooper 1998; see also Cooper, 2002), however, proposed that transient diffusional creep may be responsible for some of the seismic wave attenuation. Upon the application of deviatoric stress, stress concentration occurs at grain boundaries which is to be relaxed by diffusional flow. Consequently, higher rate of deformation occurs at the initial stage of diffusional creep.

Lifshitz and Shikin (1965) showed that the rate of deformation in the transient diffusional creep is given by $\varepsilon \propto t^{\alpha}$ (with $\alpha \sim 0.5$). This leads to $Q^{-1} \propto \omega^{-\alpha}$ (see Problem 3.6, Chapter 3). The characteristic time of deformation for this mechanism is that of diffusional creep and hence strongly depends on grain size, namely,

$$\tau \propto L^{m} \tag{11.8}$$

with $m = 2$–3 (see Chapter 8). Consequently, for a power-law regime, attenuation would depend on grain size as

$$Q^{-1} \propto (\omega\tau)^{-\alpha} \propto L^{-\alpha m}. \tag{11.9}$$

A related mechanism of anelasticity caused by grain-boundary sliding is discussed in section 11.3.3.

11.3.2. Dislocation mechanisms

Dislocations are line defects (Chapter 5) whose motion results in non-elastic deformation. Dislocation motion is in most cases thermally activated and hence dissipates energy. There are numerous observations both in metals (e.g., Nowick and Berry, 1972; Fantozzi *et al.*, 1982) and minerals (e.g., Getting *et al.*, 1997; Gueguen *et al.*, 1989; Karato and Spetzler, 1990) that show the importance of dislocation mechanisms of anelasticity or visco-elasticity.

Dislocation motion results in long-term creep (Chapter 9). Therefore viscous motion of dislocations is an obvious mechanism for seismic wave attenuation. However, there are a number of issues that need to be examined in order to understand how dislocation motion leads to elastic wave (seismic wave) attenuation. First, the characteristic time corresponding to steady-state dislocation creep is too long compared to the periods of seismic waves as discussed before. Therefore if a dislocation motion can cause significant elastic wave attenuation at seismic time-scales (frequencies), there must be some mechanisms by which dislocations can move much faster at a short time-scale than their motion in steady-state creep, but such fast dislocation motion must terminate at longer time-scales (otherwise one would have unacceptably small long-term viscosity). Here we are looking for a possible *transient dislocation creep* wherein mobility of dislocations is much higher than that of steady-state creep. Second, the frequency-dependence of seismic wave attenuation ($Q^{-1} \propto \omega^{-\alpha}$ with $\alpha = 0.2$–0.4 (e.g., Karato and Spetzler, 1990; Shito *et al.*, 2004) is different from what is expected from simple viscous motion of dislocations (attenuation due to unlimited viscous motion of dislocations would yield $Q^{-1} \propto \omega^{-1}$, or constrained motion of a dislocation segment will lead to $Q^{-1} \propto \omega\tau/(1+\omega^2\tau^2)$). Therefore important points to be clarified are (1) the mechanisms for faster dislocation motion for short time-scale deformation and (2) the physical explanation for a limited dislocation motion (anelastic behavior as opposed to viscoelastic behavior). The understanding of these microscopic bases for attenuation is critical to connect seismic wave attenuation to long-term creep.

It is not difficult to find mechanisms of short-term fast dislocation motion because there are a range of mechanisms for dislocation motion and usually the most difficult process controls the rate of steady-state deformation (see Chapter 9). Among a variety of dislocation mechanisms for elastic wave attenuation, I will review two models that apply to different classes of materials.

In a material with a small Peierls stress (e.g., fcc metals, simple ionic solids such as alkali halides or MgO), resistance to dislocation motion by chemical bonding of a crystal is small (see Chapters 5 and 9). In these cases, the Peierls stress is small and dislocation glide is easy. Therefore a dislocation can be treated as a continuum string with a line tension, μb^2, where μ is the shear modulus and b is the length of the Burgers vector (Chapter 5). When a deviatoric stress is applied, a dislocation moves causing plastic strain and associated energy loss. However, in many cases, dislocation motion is not uniform due to the interaction with impurities or with other dislocations. Consequently, dislocations are often *pinned* at certain points. In such a case, applied stress will cause an expansion of dislocation line (Fig. 11.2a) until the applied stress balances with the stress caused by the line tension and dislocation motion stops. Therefore such a small-scale motion of dislocations causes *anelastic* relaxation. The relaxation strength and time for such a process are determined by dislocation density and mobility respectively as

$$\Delta = \frac{\varepsilon_{\mathrm{diss}}}{\varepsilon_{\mathrm{ela}}} = \frac{\varepsilon_{\mathrm{an}}}{\varepsilon_{\mathrm{ela}}} = \frac{1}{12}\left(\frac{l}{L}\right)^{3} = \frac{1}{12} l^2 \rho \tag{11.10}$$

and

$$\tau = \frac{l^2}{8\mu B} \tag{11.11}$$

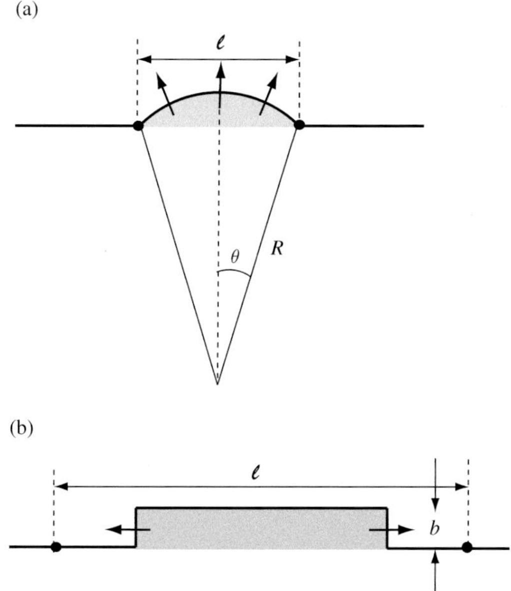

FIGURE 11.2 Geometry of pinned dislocations under stress. (a) The case where the dislocation geometry is not controlled by crystallographic orientation (low Peierls stress) and (b) the case of high Peierls stress.

where l is the distance between the pinning points, L is the mean distance between pinned dislocation segments, B is the mobility of a dislocation ($v = \sigma b B$; v, velocity of a dislocation) and we used a relation $\rho = l/L^3$ (ρ: dislocation density). Note that the strength of attenuation due to this mechanism depends on the geometry of dislocation pinning. The most common mechanism of pinning is due to jogs formed by the intersection of dislocations. In this case, $l \approx L$ (or $\rho \approx 1/l^2$), so one has large attenuation, $\Delta \approx 0.1$. Therefore if the relaxation time (equation (11.11)) is within or close to the periods of seismic waves (1–10^3 s), then dislocation motion can result in strong seismic wave attenuation. The mobility of a dislocation relevant to seismic wave attenuation is not known but it is likely higher than that relevant to long-term deformation. One possibility is that dislocation mobility relevant to seismic wave attenuation is that corresponding to dislocation glide, and dislocation mobility that controls long-term deformation corresponds to that for dislocation climb (e.g., MINSTER and ANDERSON, 1980, 1981). This would be a case for material in which the resistance for dislocation glide is small (i.e., pure materials with a low Peierls stress such as Cu or MgO), but most minerals do not belong to materials of this category.

Problem 11.3*

Derive relations (11.10) and (11.11).

Solution

The final equilibrium shape of a dislocation segment pinned at two points is determined by the balance of line tension and the force acting on a dislocation line. From the geometry of curved dislocation line, the force due to the line tension (per unit length of dislocation) is given by $\mu b^2/R$ where R is the radius of curvature. The force per unit length on a dislocation due to the applied stress is σb (Chapter 5). Therefore $R = \mu b/\sigma$. Now consider a block of crystal with linear dimension of L that contains a dislocation. When a complete slip occurs along the glide plane of a dislocation then the strain would be $\varepsilon = b/L$. When slip occurs only a fraction of glide plane, due to pinning, then the strain in a volume of L^3 will be $\delta\varepsilon = (b/L)\ (\delta s/s) = (b/L^3)\ \delta s \equiv \varepsilon_{an}$. When a dislocation segment moves from its initial position to the final equilibrium position, the area that a dislocation has swept is $\delta s = R^2\theta - \frac{1}{2}R^2 \sin 2\theta \approx \frac{2}{3}R^2\theta^3 = \frac{1}{12}(l^3/R) = \frac{1}{12}(l^3\sigma/\mu b)$ where we used the relations $2R\theta \approx l$ (see Fig. 11.2a) and $R = \mu b/\sigma$. Therefore $\Delta = \varepsilon_{\text{an}}/\varepsilon_{\text{ela}} = \frac{1}{12}(l/L)^3$. The characteristic time for dislocation motion is $\tau \simeq R(1-\cos\theta)/v \approx R\theta^2/2\sigma b B = l^2/8B\mu$ where v is the dislocation velocity and we used the relations $v = \sigma b B$, $2R\theta \approx l$ and $R = \mu b/\sigma$.

For materials with a high Peierls stress (i.e., most minerals), an alternative model for dislocation motion applies. In this case, dislocation motion (glide) occurs through the (nucleation and) migration of kinks (Chapters 5 and 9), and dislocation motion involves more discrete characteristics and is strongly controlled by the crystal structure. In such a case, a dislocation line does not behave like a string, but its geometry is controlled by the crystallographic orientation. When dislocations are pinned by some mechanism (such as jogs) then when a stress is applied, a dislocation will glide along a certain crystallographic direction until it reaches a pinned position (Fig. 11.2b). Therefore such dislocation motion will lead to anelastic behavior. The relaxation strength and time for this mechanism are given by SEEGER and SCHILLER (1966)

$$\Delta \approx \frac{b^4 \mu \rho l}{RT} \exp\left(-\frac{2H_k^*}{RT}\right) \tag{11.12}$$

and

$$\tau \approx \frac{1}{\nu} \exp\left(\frac{H_k^*}{RT}\right) \tag{11.13}$$

respectively, where l is the distance between two pinning points, ρ is dislocation density, H_k^* is the energy of a kink ($\approx \mu b^3$, see equation (5.71), Chapter 5; note that this energy may also depend on the Peierls stress, see Chapter 9), and ν is the characteristic frequency of a dislocation. The mobility of dislocations in this case is controlled by the nucleation and growth of a pair of kinks. An attenuation peak due to this mechanism is called the *Bordoni peak*. A somewhat different case can occur in materials that are deformed by long-term plastic deformation. In these materials, dislocations are already generated and they assume certain microstructures. Some dislocations already have kinks due to dislocation–dislocation interactions or the interaction with the applied (long-term) stress. A small stress associated with seismic wave propagation can cause small-scale migration of pre-existing kinks (*geometrical kinks*) that requires kink migration but not kink nucleation. This provides a possible explanation of a vast difference in dislocation mobility between steady-state creep and seismic wave attenuation (KARATO, 1998a).

In both cases, dislocation mechanisms likely have a range of relaxation times, because of the presence of different slip systems and a variety of crystal orientations as well as the distribution of pinning point spacings. This leads to a weak frequency dependence of attenuation (e.g., MINSTER and ANDERSON, 1981; KARATO, 1998a).

Consideration of microscopic physics of dislocation motion also provides a plausible mechanism for transition from anelastic to viscous (visco elastic) behavior (e.g., KARATO and SPETZLER, 1990; KARATO, 1998a). At high temperatures and/or low frequencies, pinning will no longer be effective allowing large-scale motion of dislocations. Similarly, when geometrical kinks (pre-existing kinks) are exhausted, then kink nucleation will become needed leading to large-scale motion of dislocations. Therefore there will be a gradual transition from anelastic behavior (in a strict sense) to visco-elastic behavior as temperature increases and/or as frequency decreases. LAKKI *et al.* (1998) report experimental observations on gradual transition from anelastic to viscous behavior.

11.3.3. Grain-boundary mechanisms

Grain-boundary sliding

Grain boundaries have structures different from the bulk of grains, and usually have weaker resistance to deformation. Consequently, upon the application of (shear) stress, grains across a boundary slide across each other. This sliding will create strain mismatch at grain corners that builds up back-stress that prevents further grain-boundary sliding. Sliding eventually stops when the back-stress balances applied stress. Therefore viscous grain-boundary sliding causes *anelastic* behavior. Assuming that the grain boundary can be treated as a viscous thin film (η, viscosity; δ, thickness), the force balance equation for viscous deformation associated with the displacement, x, is

$$\frac{1}{\delta} \frac{dx}{dt} = \frac{1}{\eta}\left(\sigma - M\frac{x}{L}\right) \tag{11.14}$$

where σ is the applied stress and $M(x/L)$ is the back-stress (M, elastic modulus). Solving this equation, and noting that anelastic strain due to grain-boundary sliding is given by $\varepsilon_{an} = x/L$, one gets,

$$\varepsilon_{an} = \frac{\sigma}{M}\left[1 - \exp\left(-\frac{t}{\tau_{gs}}\right)\right]. \tag{11.15}$$

Therefore grain-boundary sliding mechanism yields anelastic relaxation characterized by

$$\Delta \approx 1 \tag{11.16}$$

and

$$\tau_{gs} = \frac{\eta L}{M\delta}. \tag{11.17}$$

Note that the maximum magnitude of this effect is large (more precise calculation yields $\Delta = 0.2$–0.3 e.g., NOWICK and BERRY, 1972). The frequency dependence of attenuation is therefore given by $Q^{-1} = \Delta[\omega\tau/(1+\omega^2\tau^2)]$ with a well-defined relaxation peak at $\omega = 1/\tau$. However, when grain size, L, grain-boundary viscosity, η, or grain-boundary width, δ, have distribution, then attenuation will show more modest dependence on frequency. Experimental studies by TAN *et al.* (1997, 2001); GRIBB and COOPER (1998) and JACKSON *et al.* (2002), show evidence of grain-boundary mechanism(s) with modest frequency dependence ($Q^{-1} \propto \omega^{-\alpha}$ with $\alpha \sim 0.2$–0.3). In these cases, attenuation due to grain-boundary sliding depends on grain size as

$$Q^{-1} \propto (\omega\tau)^{-\alpha} \propto L^{-\alpha}. \tag{11.18}$$

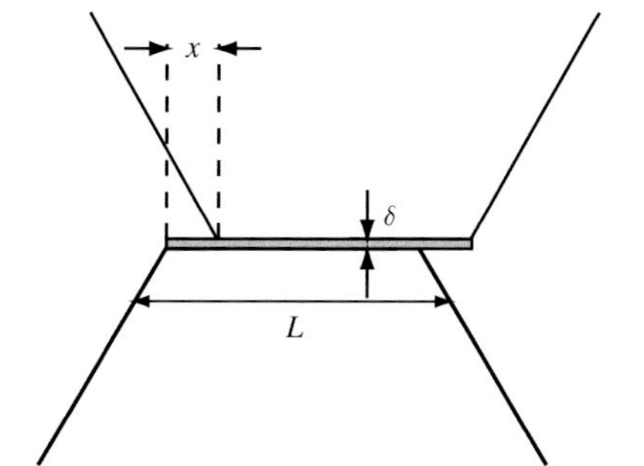

FIGURE 11.3 A schematic diagram showing grain-boundary sliding.

Note that the grain-size dependence of attenuation for this mechanism is different from that of transient diffusional creep (equation (11.9)). Therefore by comparing the frequency dependence and grain-size dependence of attenuation, one can determine which model (grain-boundary sliding or transient diffusional creep) is more important for seismic wave attenuation.

The experimental observation of weak frequency dependence for a wide range of frequency ($\sim$1–10^{-3} Hz) cited above would imply a wide distribution of relaxation time. Such a wide distribution is difficult to attribute to the distribution of grain size alone. The viscosity and/or the width of grain boundaries should have a wide range of values.

Anelastic behavior due to grain-boundary sliding (see Fig. 11.3) will gradually change to viscous (visco-elastic) behavior when the elastic back stress is reduced at higher temperatures and/or at lower frequencies by diffusion (Raj and Ashby, 1971).

Experimental observations on grain-boundary mechanisms of attenuation are reviewed by Nowick and Berry (1972). Gribb and Cooper (1998) and Jackson *et al.* (2002) report evidence for grain-boundary mechanisms of seismic wave attenuation in olivine aggregates.

Problem 11.4

Show how to distinguish the grain-boundary sliding model for attenuation from the transient diffusional creep model from the frequency and grain-size dependence of attenuation.

Solution

From equation (11.9) for the transient diffusional creep model and (11.18) for the grain-boundary sliding model, we have $(\partial \log Q^{-1}/\partial\omega)/(\partial \log Q^{-1}/\partial L) = 1$ for the grain-boundary sliding model, whereas $(\partial \log Q^{-1}/\partial\omega)/(\partial \log Q^{-1}/\partial L) = 1/m$ where m is the grain-size exponent for diffusional creep, i.e., $\dot{\varepsilon} \propto L^{-m}$ ($m = 2$–3).

Problem 11.5

Derive equation (11.15).

Solution

The strain rate of viscous materials on grain boundaries is given by $(1/\delta)\ (dx/dt)$, which is proportional to the applied shear-stress minus back-stress due to elastic deformation, and inversely proportional to viscosity. Therefore $(1/\delta)\ (dx/dt) = \ (1/\eta)\left(\sigma - M\frac{x}{L}\right)$. If we define $y \equiv x - \sigma L/M$, then this differential equation is reduced to $dy/dt = -(1/\tau_{gs})y$ with $\tau_{gs} = \eta L/M\delta$. Therefore we have $y = y_0 \exp(-t/\tau_{gs})$. From the initial condition that $x = 0$ for $t = 0$, we get $\varepsilon_{an} = x/L = (\sigma/M)\left[1 - \exp(-(t/\tau_{gs}))\right]$.

Anelasticity due to motion of twin boundaries, subboundaries

In materials that are subject to easy mechanical twinning, it is possible that twin boundaries move under external stress causing attenuation. Among various minerals, calcite and perovskite are unique in their tendency for mechanical twinning. However, there is no well-documented report on this effect (see, however, Harrison and Redfern, 2002).

Sliding along subgrain boundaries is a potential source of anelasticity (Gribb and Cooper, 1998). However, the magnitude of this effect is likely to be small because of the small misorientation between neighboring subgrains.

Grain-scale thermoelasticity

When a stress is applied to a polycrystalline aggregate, grains in an aggregate will deform but in general strain in each grain will be different. Let us consider an instantaneous response of a polycrystalline aggregate. The difference in volumetric strain in each grain will result in difference in the degree of adiabatic heating (or cooling). Consequently, each grain will develop a different temperature that will eventually be homogenized by thermal diffusion. Since a change in temperature causes a change in strain due to thermal expansion, strain distribution in a polycrystalline aggregate will be time-dependent, and is associated with energy dissipation. The mechanical response of a polycrystalline aggregate changes from adiabatic to isothermal response at sufficiently long time-scales. This mechanism therefore causes *anelastic* behavior and is referred to as *thermoelasticity*. A full account of thermoelasticity is given by ZENER (1948a). The characteristic time for this process is the time for thermal diffusion between grains and is given by

$$\tau \approx \frac{L^2}{\pi^2 \kappa} \tag{11.19}$$

where L is the grain size and κ is the thermal diffusivity. The strength of relaxation is determined by the difference between the isothermal and adiabatic elastic modulus, and hence on the Grüneisen parameter (Chapter 4), γ, and elastic anisotropy of crystals,

$$\Delta = \frac{\pi \rho C_P T \gamma^2 \tilde{R}}{M} \tag{11.20}$$

where ρ is density, C_P is specific heat, T is temperature, M is elastic modulus and $\tilde{R}$ is a factor that depends on the anisotropy or heterogeneity of elastic moduli,

$$\tilde{R} \approx \left(\frac{\Delta M}{M}\right)^2. \tag{11.21}$$

Problem 11.6

Show that when energy loss occurs only for shear deformation but not for bulk deformation, then $Q_P/Q_S \approx 3V_P^2/4V_S^2$ for small dissipation.

Solution

P- and S-wave velocities are related to shear and bulk modulus as $V_P = \sqrt{(K + \frac{4}{3}\mu)/\rho}$ and $V_S = \sqrt{\mu/\rho}$. Recall that the energy loss for deformation characterized by an elastic constant $M = M_1 + iM_2$ is given by $Q^{-1} = M_2/M_1$ (see Chapter 3). Therefore if one writes $K = K_1 + iK_2$ and $\mu = \mu_1 + i\mu_2$, then $Q_K^{-1} = K_2/K_1$ and $Q_\mu^{-1} = \mu_2/\mu_1 = Q_S^{-1}$. Now $Q_P^{-1} = (K_2 + \frac{4}{3}\mu_2)/(K_1 + \frac{4}{3}\mu_1) = [K_1/(K_1 + \frac{4}{3}\mu_1)]\,(K_2/K_1) + \frac{4}{3}\,[\mu_1/(K_1 + \frac{4}{3}\mu_1)](\mu_2/\mu_1) \approx \left(1 - \frac{4}{3}(V_S^2/V_P^2)\right)Q_K^{-1} + \frac{4}{3}(V_S^2/V_P^2)Q_S^{-1}$ where we used the assumption of small dissipation (i.e., $V_P \approx \sqrt{(K_1 + \frac{4}{3}\mu_1)/\rho}$ and $V_S \approx \sqrt{\mu_1/\rho}$). If $Q_K = \infty$, then $Q_P/Q_S \approx 3V_P^2/4V_S^2$.

For a reasonable range of grain size (1–10 mm) and thermal diffusivity ($\sim 10^{-6}\,\mathrm{m^2/s}$), we have $\tau \approx 10^{-1}$–10 s. This is exactly the time-scale of seismic (body) wave propagation therefore thermoelasticity is potentially an important mechanism of seismic wave attenuation. However, the importance of this effect strongly depends on the anisotropy or heterogeneity factor, $\tilde{R}$. $\tilde{R}$ for a single-phase material can be estimated from elastic anisotropy and in most cases $Q^{-1} \approx 10^{-3}$–10^{-4}. Similarly in most mineral aggregates, the contrast in elastic moduli is less than 10% so attenuation is $Q^{-1} \approx 10^{-3}$–10^{-4}. Therefore in most cases the magnitude of anelasticity due to this mechanism is not very large (e.g., KARATO, 1977; HEINZ *et al.*, 1982). However, the magnitude of this effect becomes large when heterogeneity in elastic constants becomes large. KARATO (1977) investigated the thermoelasticity in a partially molten material and showed that its magnitude become $\Delta \approx 10^{-2}$ and this could be an important mechanism of seismic wave attenuation (see section 11.3).

One of the important aspects of this mechanism is that since this mechanism involves differential heating due to *volumetric strain*, it results in significant bulk attenuation and hence a small value of Q_P/Q_S which is distinct from the Q_P/Q_S of most other mechanisms (for many other mechanisms there is no bulk attenuation ($Q_K = \infty$) and $Q_P/Q_S = 3V_P^2/4V_S^2 \sim 2.3$).

11.3.4. Experimental studies on solid-state mechanisms of anelasticity in Earth materials

Experimental studies on anelasticity in solids are reviewed by CHANG (1961), FANTOZZI *et al.* (1982) and KARATO and SPETZLER (1990). Generally, dominant peaks of attenuation are found at relatively low temperatures, but the attenuation behavior becomes more diffused and a power-law behavior $Q^{-1} \propto \omega^{-\alpha}\exp(-\alpha E^*/RT)$ is commonly observed at high temperatures.

Two sets of experimental observations on olivine (or olivine-rich rocks) deserve special attention. First, GUEGUEN *et al.* (1989) conducted laboratory experiments (at room pressure) to measure the attenuation of elastic waves in forsterite single crystals at seismic frequency at high temperatures. They measured Q values of undeformed and deformed forsterite and noted a distinct increase in attenuation by deformation. This is strong evidence for the importance of dislocations on anelasticity. Their results show a power-law behavior, $Q^{-1} \propto \omega^{-\alpha}\exp(-\alpha E^*/RT)$, with $\alpha \sim 0.2$–0.3 and $E^* = 440$ kJ/mol. Second, two groups reported

TABLE 11.2 Typical experimental results on anelasticity in minerals. The data follow the constitutive relation $Q^{-1} = A \cdot \omega^{-\alpha} \cdot L^{-m} \cdot \exp(-\alpha H^*/RT)$. For polycrystalline samples, only the results from high-pressure experiments are listed. Units: A ($s^{-\alpha}(\mu m)^m$), H^* (kJ/mol), P (GPa), T (K).

	A	α	m	H^*	P	T	Remarks[2]	ref.
Olivine	8.1×10^2	0.2	–	440	RP[1]	1273–1673	S, forsterite	(1)
Olivine	7.5×10^2	0.26	0.26	424	0.2	1273–1573	P, Fo_{90}	(2)
$CaTiO_3$	6.3×10^2	0.37	–	500	0.2	1173–1573	P	(3)
$SrTiO_3$	5.0×10^3	0.37	–	590	0.2	1173–1573	P	(3)
MgO	2.3×10^2	0.30	–	233	RP	1000–1500	S	(4)
MgO	1.5×10^2	0.30	–	250	0.2	973–1173	P	(5)
bcc Fe	8.1×10^1	0.23	–	270	0.2	873–1073	Mild steel	(6)

[1] RP indicates room pressure (10^{-4} GPa).
[2] S indicates a single crystal, P indicates a polycrystal.
(1) Gueguen *et al.* (1989).
(2) Jackson *et al.* (2002).
(3) Webb *et al.* (1999).
(4) Getting *et al.* (1997).
(5) Webb and Jackson (2003).
(6) Jackson *et al.* (2000a).

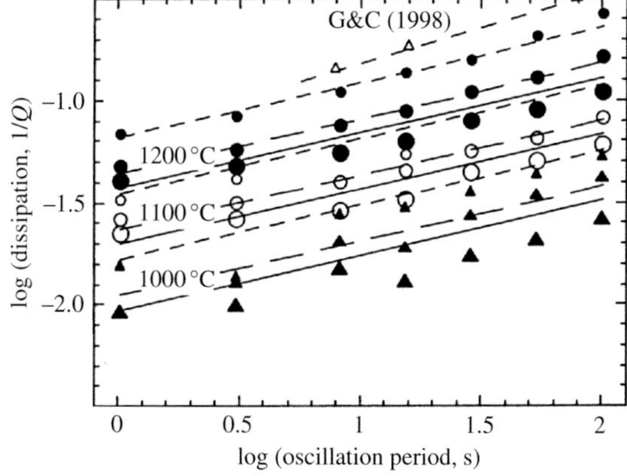

FIGURE 11.4 Attenuation of seismic waves in a polycrystalline olivine (after Jackson *et al.*, 2002).

grain-size sensitive attenuation in olivine (or olivine-rich) aggregates (Tan *et al.*, 1997, 2001; Gribb and Cooper, 1998; Jackson *et al.*, 2002), see Fig. 11.4. Samples with smaller grain size show systematically higher attenuation.

Evaluation of the relative importance of these two mechanisms is difficult because the relationship between dislocation density and attenuation has not been investigated quantitatively. Also, there are no experimental studies on the pressure effects on attenuation (except for the indirect effects due to cracking). An extrapolation of laboratory data to Earth suggests that grain-boundary mechanisms marginally explain some of the seismologically observed attenuation (Jackson *et al.*, 2002). Gribb and Cooper (1998) interpreted their results in terms of a model of transient diffusional creep that implies a stronger grain-size effect than Jackson *et al.* (2002) and concluded that grain-boundary effects do not contribute significantly to seismic wave attenuation unless subgrain boundaries play a similar role as grain boundaries.

Another important issue that has not been investigated in detail is the influence of water (hydrogen). Almost all solid-state kinetic processes in silicates are enhanced by water (hydrogen) and consequently, anelasticity is likely to be enhanced by water. An earlier study by Jackson *et al.* (1992) reported a difference in anelasticity between *as-is* dunite and dried dunite. *As-is* dunite contains more water than dried dunite (Chopra and Paterson, 1984) and consequently, such results are likely to be due to enhanced anelasticity by water. However, there have been no systematic studies on this effect at this time (2007).

Experimental studies on anelasticity were also performed on calcite (Jackson and Paterson, 1987), MgO (Getting *et al.*, 1997; Webb and Jackson, 2003), titanate perovskite (Webb *et al.*, 1999) and iron (Jackson *et al.*, 2000a). In most cases, the attenuation behavior follows the power-law, $Q^{-1} \propto \omega^{-\alpha} \exp(-\alpha E^*/RT)$, with $\alpha = 0.2$–0.4 (Table 11.2). It should be noted that, although the influence of pressure and water on any defect-related processes is expected to be very large (see Chapter 10), experimental studies on these effects are difficult and there have been no systematic experimental studies to investigate these effects.

11.4. Anelasticity in a partially molten material

11.4.1. General introduction

Partial melting affects elastic and anelastic properties of material. At infinite frequency (immediately after the application of stress), a partial melt will have unrelaxed elastic moduli that are smaller than those of melt-free materials. In addition to this reduction of elastic constant for unrelaxed state, anelastic relaxation will occur at finite frequency (or finite time) due to viscous motion of the melt. The transition from unrelaxed to relaxed state occurs with certain characteristic times. Consequently, this results in anelastic relaxation. Anelastic relaxation results in frequency-dependent elastic moduli and the attenuation of elastic waves (Chapter 3). Unrelaxed moduli are determined by the contrast in elastic constants between the melt and the solid, and the fraction and geometry of melt. In addition melt viscosity will also play an important role for anelastic properties. In this section I review theoretical models and experimental observations on the anelastic properties of a partially molten material. Mechanisms of anelasticity unique to a partially molten material include: (1) energy dissipation due to the viscous motion of melt, (2) thermoelasticity caused by inhomogeneous stress distribution (Karato, 1977). Both of them are sensitive to the geometry as well as the viscosity (and the bulk modulus) of melt. Briefly, when melt completely wets grain boundaries, then the effects of partial melting to modify anelastic behavior are large whereas if melt assumes isolated pockets, then the effects are small. Geometry of partial melt is controlled by interfacial energy and stress and is discussed in Chapter 12.

11.4.2. Attenuation due to viscous motion of melt

When viscous deformation occurs in the melt pocket, deformation becomes time-dependent and therefore there is energy loss. In most cases where melt is confined in a material, viscous motion of melt will terminate at a certain point, causing anelastic relaxation. The anelasticity is therefore characterized by the relaxation strength and the relaxation time. The relaxation strength is proportional to the melt fraction, ϕ, and melt geometry,

$$\Delta = A\frac{\phi}{\xi} \tag{11.22}$$

where A is a non-dimensional constant of order ~ 0.1, ϕ is the melt fraction, $\xi = \delta/L$ (δ, thickness of a film; L, the size of a film, i.e., grain size) for melt film and $\xi = r/2\pi L$ (L, grain size; r, the radius of a tube) for a melt tube. The relaxation time depends on the viscosity of melt, η, elastic constant of solid matrix, M, and the aspect ratio of the melt pocket as

$$\tau = \frac{B\eta}{M\xi^n} \tag{11.23}$$

where B (~ 1) is a constant that depends on melt geometry and n is a constant that depends on the mechanism of flow in the melt pocket. Note that the value of n has a strong influence on the relaxation time.

There are two distinct mechanisms of flow in the melt pocket (Fig. 11.5). First, when a melt is confined to an inclusion in a solid matrix (i.e., an isolated melt pocket), viscous motion within a melt inclusion (melt pocket) dissipates energy and causes attenuation. Walsh (1968, 1969) developed a theory to relate energy dissipation with viscosity of melt and geometry of melt pocket. Upon the application of a force, traction is transmitted to an isolated melt pocket that causes a viscous motion. Viscous motion gradually relaxes the stress eventually leading to another equilibrium (relaxed state). Therefore this mechanism leads to *anelastic* behavior. In both cases, energy loss occurs for both shear and bulk deformation. For this mechanism, $n = 1$ and $\tau \approx (10^{-10}\text{–}10^{-11})(1/\xi)$ where I used the viscosity of 1–10 Pa s (Kushiro, 1976) and the elastic modulus $M \sim 10^{11}$ Pa. Therefore in order to bring this time-scale towards the seismic time-scale, one would need an unacceptably small aspect ratio ($\xi = 10^{-10}\text{–}10^{-13}$). Consequently, attenuation due to this mechanism is unlikely to be important for the upper mantle, although it might be important in the crust where melt viscosity can be higher. It is noted, however, that since the peak frequency of this mechanism is likely to be higher than seismic frequencies, a partially molten material will show a relaxed behavior for seismic wave propagation. Consequently, the seismic wave velocity will be reduced by this mechanism although attenuation is unlikely to be significant

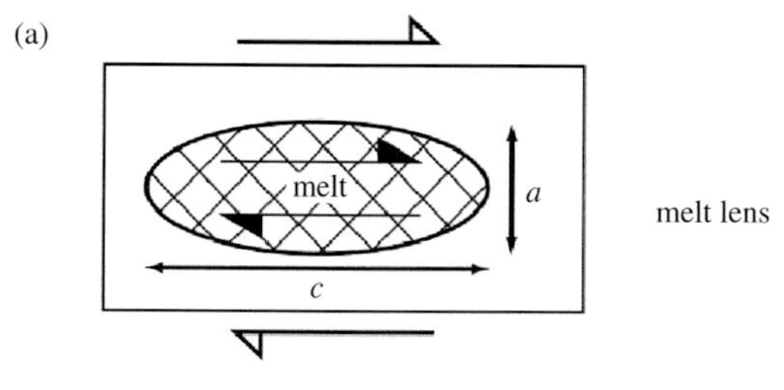

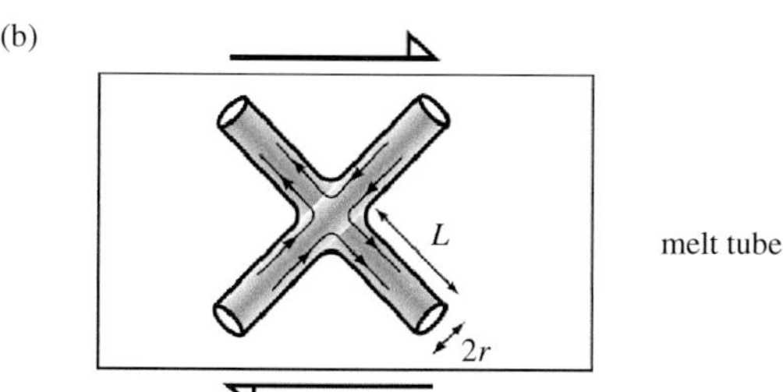

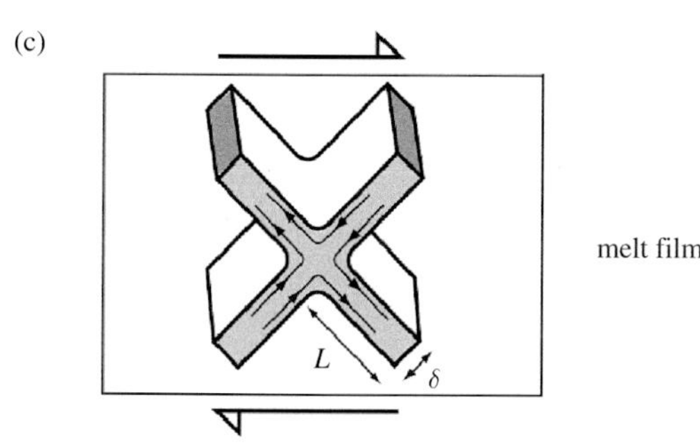

FIGURE 11.5 The geometry of a melt pocket in a partially molten material and the nature of energy dissipation in a partial melt. (a) An isolated melt pocket. Shear deformation of a fluid in the pocket can cause energy dissipation. (b) A melt tube. Differential pressure in different regions of connected tubes causes a flow of fluid, resulting in energy dissipation. (c) Connected melt films (melt squirt). Differential pressure causes fluid flow similar to a melt tube.

(Karato, 1977). Since viscous motion of fluid occurs due to shear traction, this mechanism causes attenuation mostly for shear deformation and not for bulk deformation ($Q_K \gg Q_\mu$, $Q_S < Q_P$).

Problem 11.7

Calculate the magnitude of velocity reduction by the Walsh mechanism.

Solution

The velocity reduction by an anelastic relaxation with the Zener model is given by $\Delta V/V \approx \frac{1}{2}\Delta = A\phi/2\xi$ (Table 3.3). For typical mantle materials $A \sim 0.2$, so $\Delta V/V \approx 0.1(\phi/\xi)$. For a tube model, the volume of melt per cube with dimension l is $3\pi r^2 l \cdot X(\theta)$ where $X(\theta)$ is a geometrical factor that depends on the dihedral angle θ ($X(\theta) \sim 0.2$–0.4 for $30^\circ < \theta < 60^\circ$) and the volume of a cube is l^3, so that $\xi \equiv r/2\pi l \approx \sqrt{\phi/12\pi^2 X(\theta)}$ and $\Delta V/V \approx 0.6\sqrt{\phi}$. Therefore for $\phi = 10^{-2}$–10^{-3}, $\Delta V/V \approx 2$–6%. The velocity reduction by this mechanism is large although this mechanism is unlikely to be important for attenuation. For a melt film, $3\xi = \phi$, and $\Delta V/V \approx 30\%$.

Another mechanism was proposed by O'Connell and Budianski (1974, 1977); Mavko and Nur (1975) and Mavko (1980). When a melt forms some connected network, then the applied stress can cause a flow of melt from one region to another. Consider a simple case where melt pockets have a geometry shown in Fig. 11.5c. Since the pressure in the melt pocket depends on the geometry (orientation of surfaces) of a melt pocket, different pressures will develop in melt pockets with different orientations. Consequently, flow of melt will occur to relax this pressure gradient. The flow in this case is a viscous flow in a tube or thin layer and hence the time-scale of flow is highly sensitive to the melt geometry. In a polycrystalline material, the space-scale at which pressure gradient is relaxed is determined by the space-scale of melt geometry that is roughly the size of grains (1–10 mm). Thus this mechanism results in anelastic behavior, but the characteristic frequency is different from that of the Walsh mechanism. Due to the difference in flow geometry, the value of n in these mechanisms is different from that of the Walsh mechanism ($n = 1$), $n = 2$ for melt tube and $n = 3$ for melt film.

Remembering that if melt has a single geometry, ϕ (melt fraction) and β (melt geometry) must have a simple relation and consequently (see Problem 11.7),

$$\Delta = 3A' \quad \text{for a film} \tag{11.24a}$$

and

$$\Delta = (2\pi)^{3/2} A' \phi^{1/2} \quad \text{for a tube.} \tag{11.24b}$$

The magnitude of relaxation, Δ, can be very high, but for these mechanisms to be important in attenuation, the characteristic frequencies must be in the seismic frequency band. The characteristic frequencies are highly sensitive to the geometry of melt, and in order for the characteristic frequencies to be in the seismic frequency band, the aspect ratio needs to be $\xi = 10^{-3}$–10^{-4} for a film geometry and $\xi = 10^{-5}$–10^{-6} for a melt tube.

SHANKLAND *et al.* (1981) analyzed various cases for the distribution of aspect ratios, and HAMMOND and HUMPHREYS (2000a, 2000b) studied the influence of geometry of melt in experimental samples. HAMMOND and HUMPHREYS (2000a, 2000b) concluded that this mechanism has characteristic frequencies higher than seismic frequencies. However, there are two issues that need attention. First, the experimental study by JACKSON *et al.* (2004a) shows that the peak frequencies of anelasticity of partially molten peridotite are in the seismic frequency range. If one interprets the experimentally observed peak frequencies by this mechanism, then one would need to assume either the effective viscosity of a thin layer of melt is much higher than that of a bulk fluid or the flow geometry might be more complicated than a simple model. NAKASHIMA *et al.* (2004) reviewed the literature and showed that the effective viscosity of a thin fluid layer can indeed be much higher that that of the bulk of a fluid. FAUL *et al.* (2004) proposed an alternative model that a change in grain-boundary geometry by partial melting may cause enhanced anelasticity due to grain-boundary sliding.

It should be noted that because the geometry of melt is likely to be distributed as emphasized by SHANKLAND *et al.* (1981), it is highly likely that some portion of melt pockets have characteristic frequencies higher than seismic frequency so that they do not contribute to attenuation but reduce seismic wave velocities. If the melt fraction is $\sim$1%, the velocity reduction will be a few percent (a tube model is used, see Problem 11.7).

One of the important characteristics of this mechanism is that it causes non-negligible bulk attenuation, $Q_K \neq \infty$. Consequently, the Q_P/Q_S ratio tends to be small for this mechanism. This is particularly true when a material contains melt-filled pores with a variety of shapes (SCHMELING, 1985); in these cases, melt pockets with different shapes will develop different pressures upon compression that causes flow of melt between different melt pockets. DUREK and EKSTRÖM (1995) reported evidence for bulk attenuation in the asthenosphere, which could be attributed to the presence of partial melt.

11.4.3. Thermoelasticity in a partially molten material

The magnitude of grain-scale thermoelasticity depends strongly on the contrast in elastic constants of each grain. Consequently, thermoelasticity can be significant in a partially molten material. KARATO (1977) analyzed the thermoelasticity in partially molten materials assuming some simplified melt geometries. The characteristics of this mechanism are common to that of thermoelasticity in a solid polycrystal. The characteristic frequency is similar to (11.19). In particular, this mechanism provides bulk attenuation and hence the Q_P/Q_S ratio tends to be small.

11.4.4. Experimental studies on anelasticity in partial melts

Partial melting is often assumed to be the cause of low velocity and high attenuation in Earth's interior (e.g., SHANKLAND *et al.*, 1981). MIZUTANI and KANAMORI (1964) and SPETZLER and ANDERSON (1968) conducted experimental studies on elastic wave attenuation in partially molten systems using analog materials. Both of these studies reported a large increase in attenuation upon partial melting. SATO *et al.* (1989) investigated elastic wave attenuation of partially molten peridotites at high frequencies (5 MHz) and noted that partial melting does not cause appreciable change in attenuation. However, they did observe significant reduction in elastic wave velocities, which implies that the characteristic frequencies for anelastic relaxation associated with partial melting are higher than the frequency range that they used. GRIBB and COOPER (2000) conducted room-pressure, high-temperature and seismic-frequency experiments on an olivine + melt mixture. JACKSON *et al.* (2004a) performed anelasticity measurements on an olivine + melt system at high pressure, high temperature and seismic frequencies (see Fig. 11.6). A detailed study by JACKSON *et al.* (2004a) for a broad range of frequencies showed that the

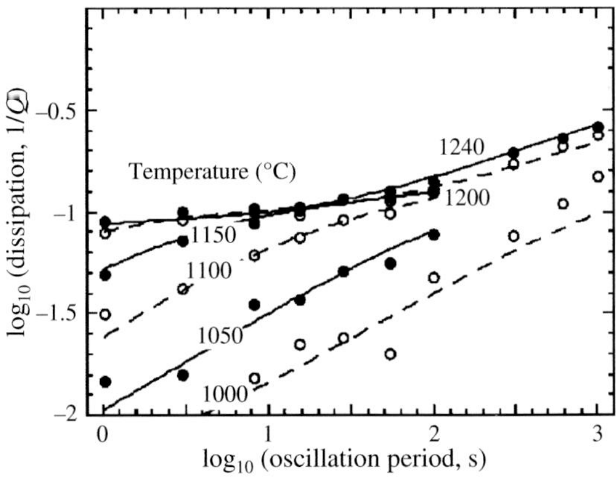

FIGURE 11.6 The results of an experimental study on attenuation of seismic waves in the olivine + basalt melt (JACKSON *et al.*, 2004a).

presence of melt causes a broad absorption peak in addition to the high-temperature background.

11.4.5. Seismological observations

Seismological studies on the presence of partial melting are controversial. Historically, many of the observations on low velocity (and/or high attenuation) were attributed to the presence of partial melting. However, the importance of solid-state relaxation mechanisms was recognized in the 1970s and early 1980s based on theoretical argument (e.g., Gueguen and Mercier, 1973; Karato, 1977; Minster and Anderson, 1980), and the experimental studies support this notion (e.g., Karato, 1993a; Jackson *et al.*, 2000a). Briefly, the observed velocity reduction in the low-velocity zone can be attributed to a combined effect of anharmonicity and solid-state anelasticity. However, current seismological observations provide an inconclusive answer as to the importance of partial melting to cause anomalies in seismic wave attenuation and velocities. First, a detailed search for evidence of partial melting by seismic anisotropy near an ocean ridge (East Pacific Rise) failed to detect evidence for partial melting in seismic anisotropy (Wolfe and Solomon, 1998). Second, the frequency dependence of attenuation in the wedge mantle in the Philippine Sea region is consistent with the solid-state mechanisms but inconsistent with partial melt mechanisms (Shito *et al.*, 2004). Third, however, (Durek and Ekström, 1995) reported significant bulk attenuation in the asthenosphere that may be due to the presence of partial melting.

In summary, currently there is no strong evidence for the influence of partial melting on either velocity or attenuation anomalies in Earth's upper mantle. A majority of the seismological observations can be attributed to solid-state processes. Experimental, theoretical and geochemical studies suggest that a region of a significant melt fraction in Earth is limited because of effective compaction and fast melt transport. However, petrological and geochemical studies also suggest that there is a broad region of small melt fraction, $\sim 0.1\%$, corresponding to the asthenosphere (e.g., Plank and Langmuir, 1992; see also Faul, 2001 who suggested a larger value). A small amount of melt could cause velocity reduction without causing large enhancement of attenuation. There are some reports showing a large, sharp decrease in seismic wave velocities at the lithosphere–asthenosphere boundary (e.g., Rychert *et al.*, 2005). Observed ultra-low velocity regions in the D'' layer are usually attributed to partial melting (e.g., Williams and Garnero, 1996; Lay *et al.*, 2004), but an alternative model invoking chemical heterogeneity is also possible (e.g., Garnero and Jeanloz, 2000).

12 Deformation of multi-phase materials

Most parts of the Earth are made of multi-phase materials. The physical mechanisms of deformation of a multi-phase material differ from those of a homogeneous material in several ways. A key issue here is how the rate of deformation (or the strength of a polycrystalline material) is related to those of individual materials (or crystals) and their volume fraction, orientation and geometry. Experimental observations and theoretical models of the deformation of a multi-phase material are reviewed. It is shown that the plastic deformation of a multi-phase material is controlled not only by the rheological contrast (the contrast in effective viscosity) and the volume fraction of each phase but also by the stress–strain distribution that depends on the geometry of each phase. The case of partial melt deserves special attention. Mechanical contrast is large in this case and as a consequence the properties of a partial melt depend strongly on the fraction and geometry of the melt. Principles that determine the geometry of melt in a partially molten material are discussed.

Key words Reuss average (model), Voigt average (model), Hill average (model), Hoff's analogy, Taylor average (model), Sachs average (model), variational principle, self-consistent approach, percolation, partial melt, dihedral angle.

12.1. Introduction

So far, we have considered the deformation of single-phase, homogeneous materials. This is a natural starting point, but there are several fundamental differences in deformation and microstructural development between a single- and a multi-phase material. Obviously, for the most part Earth is made of multi-phase materials and therefore an understanding of the deformation of a multi-phase material is essential in Earth science. A heterogeneous microstructure may occur at various scales, but we will consider, in most of this chapter, a multi-phase material where heterogeneity of mechanical properties occurs at the scale of grains or a group of grains. Deformation of a single-phase polycrystalline material can be treated in a similar way: each single crystal with a different orientation can be considered as a material with different mechanical properties.

The relation between the mechanical properties of multi-phase materials and the properties of individual materials is important for several reasons. The seismic wave velocities of a multi-phase aggregate depend on the stress–strain distribution as well as the volume fraction and geometry of each phase (e.g., WATT *et al.*, 1976). The effects of the presence of multi-phases are more important in plastic properties because the contrast in plastic properties of various phases (or crystals with different orientations) is generally much larger than that of elastic properties. Consequently, the focus in this chapter is the plastic properties of multi-phase materials, although I will also briefly discuss the average elastic properties of multi-phase aggregates. Many of the formal aspects of the theory are common between the elastic and plastic properties.

The plastic properties of multi-phase aggregates may be addressed from two different end-member viewpoints. If the volumetrically dominant phase is a weaker phase, then one can ask how the addition of a strong material increases the aggregate strength (e.g., YOON and CHEN, 1990; RYBACKI *et al.*, 2003). In contrast, if the volumetrically dominant phase is stronger than a minor component, then the question is how the addition of a weaker material reduces the overall strength (e.g., BLOOMFIELD and COVEY-CRUMP, 1993; KOHLSTEDT, 2002). As we will learn, in both cases, the key question is the distribution of stress and strain in co-existing phases that depends strongly on the *geometry* (morphology) of the two phases, which evolves with strain. It must also be emphasized that a multi-phase mixture tends to assume a small grain size due to the sluggish grain growth that *indirectly* influences the plastic properties. The direct effects due to mechanical interactions of two phases and the indirect effects due to grain size must be clearly distinguished in analyzing the experimental data.

Unique features of plastic deformation of multi-phase rocks have been noticed by a number of geological observations. Upper crust rocks are made of quartz, feldspar and micas whose rheological properties are different, resulting in markedly contrasting deformation microstructures in these rocks (BEHRMANN and MAINPRICE, 1987). Upper mantle rocks are made of olivine, orthopyroxene and garnet (in the deep portion) and again their rheological contrast can be large (e.g., KARATO *et al.*, 1995a). Similarly, in the lower mantle where the major constituent minerals are magnesiowüstite and perovskite, the rheological contrast in co-existing phases is likely to be large (YAMAZAKI and KARATO, 2001b). In these cases, the question of how to estimate the rheological properties of a multi-phase aggregate from the rheological properties of component minerals becomes important.

In this chapter, I will discuss some of the important characteristics of the deformation of a multi-phase material. To simplify the discussion, I will mostly treat deformation of a two-phase material but the generalization to deformation of a material with more than three-components is straightforward. A fundamental issue in treating the deformation of multi-phase composite materials is that the stress–strain distribution in these materials is heterogeneous and difficult to characterize precisely. Consequently, in most cases, properties of multi-phase materials can only be calculated approximately. It should also be noted that the stress–strain distribution is sensitive to the geometry of two phases that may evolve during deformation.

12.2. Some simple examples

Let us consider two simple cases shown in Fig. 12.1. In Fig. 12.1a there are two materials that are connected in parallel whereas in Fig. 12.1b, two materials are connected in series. When external (compressional) stress is applied, these materials will deform, but the stress–strain distribution between the two elements is different. In Fig. 12.1a the strain (rate) of the two elements is identical whereas the stress in each element is different, in contrast in Fig. 12.1b the stress in each element is identical whereas the strain (rate) is different.

For case (a) in which two elements are connected in parallel along the compression direction, the strain in each element must be the same. If a linear constitutive relation is assumed, then,

$$\sigma_1 = M_1\varepsilon, \quad \sigma_2 = M_2\varepsilon. \tag{12.1}$$

Note that the following analysis applies to elastic as well as viscous deformation. In the case of elastic deformation, ε is strain, and $M_{1,2}$ are the elastic constants, and in the case of viscous deformation, ε should be read as strain rate and $M_{1,2}$ will be the relevant viscosity. This correspondence between elastic and viscous deformation works also for non-linear constitutive relations as far as viscous behavior at steady state is considered and is referred to as *Hoff's analogy* (HOFF, 1954). Now the total stress is given by

$$\sigma = x_1\sigma_1 + x_2\sigma_2 = (x_1M_1 + x_2M_2)\varepsilon \equiv M_a\varepsilon \tag{12.2}$$

where $x_{1,2}$ is the volume fraction of 1- (2-) material ($x_1 + x_2 = 1$) with

$$M_a = x_1M_1 + x_2M_2 \tag{12.3}$$

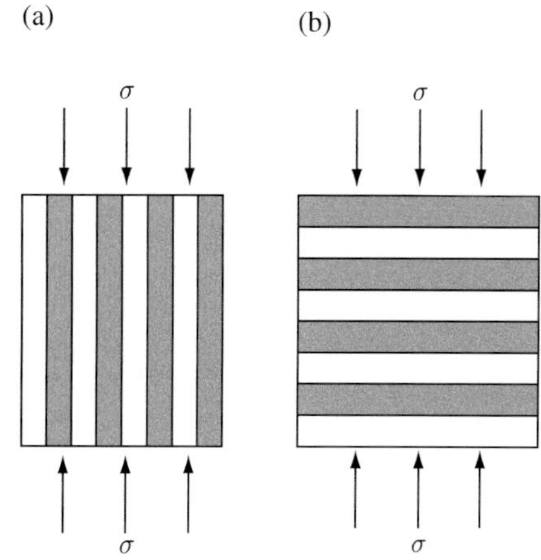

FIGURE 12.1 Two simple models of a combination of two materials: (a) homogeneous strain and (b) homogeneous stress.

where M_a is the effective modulus for case (a).

Similarly for case (b), the stress in each element must be same but the strain is the sum of each strain. Thus

$$\sigma = M_1\varepsilon_1 = M_2\varepsilon_2 \tag{12.4}$$

and

$$\varepsilon = x_1\varepsilon_1 + x_2\varepsilon_2 = \left(\frac{x_1}{M_1} + \frac{x_2}{M_2}\right)\sigma \equiv \frac{\sigma}{M_b} \tag{12.5}$$

with

$$M_b = \frac{1}{x_1/M_1 + x_2/M_2} \tag{12.6}$$

where M_b is the effective modulus for case (b). The actual modulus of a two-phase mixture must be in between these two end-member cases, so that

$$M_a \geq M \geq M_b. \tag{12.7}$$

The average modulus for a composite corresponding to case (a) (homogeneous strain) is referred to as the *Voigt average* for elastic deformation, and as the *Taylor average* for viscous (plastic) deformation, and case (b) (homogeneous stress) as the *Reuss average* for elastic deformation, and as *Sachs average* for viscous (plastic) deformation.

Problem 12.1

Show $M_a \geq M_b$.

Solution

From equations (12.3) and (12.6), one has $M_a - M_b = x_1x_2(M_1 - M_2)^2/(M_1M_2(x_1/M_1 + x_2/M_2)) \geq 0$.

Problem 12.2

Consider the case of a 1:1 mixture of two phases. Calculate the ratio of the upper and the lower bound as a function of ratio of moduli.

Solution

For $x_1 = x_2 = \frac{1}{2}$, equations (12.3) and (12.6) yield $\frac{M_a}{M_b} = (x_1M_1 + x_2M_2)(\frac{x_1}{M_1} + \frac{x_2}{M_2}) = \frac{1}{4}(2 + \xi + \frac{1}{\xi})$ where $\xi \equiv M_1/M_2$. As $\xi \equiv M_1/M_2$ changes from 1 to ∞, M_a/M_b changes from 1 to $\sim \frac{1}{4}\xi$. The modulus ratio of two cases is on the order of modulus ratio of the two components.

The elastic modulus or the viscosity for a composite material for case (a) (parallel combination) gives an upper bound and that for case (b) (serial combination) yields the lower bound (HILL, 1952). These two bounds yield very different properties if the contrast in properties (elastic constants of viscosities) is large (see equations (12.3) or (12.6)) (see Problem 12.2). HILL (1952) suggested that some mean values of the upper and the lower bounds ($M_{H1} = \frac{1}{2}(M_a + M_b)$ or $M_{H2} = \sqrt{M_aM_b}$) would give a reasonable estimate of real aggregate properties. The arithmetic mean ($M_{H1} = \frac{1}{2}(M_a + M_b)$) is often called the *Hill average*. There is no theoretical reason to justify this model but it has been frequently used for practical applications.

12.3. More general considerations

12.3.1. Variational principle

The above analysis showed that one can calculate the bounds for the average properties of a multi-phase aggregates by considering some specific distribution of stress or strain. Such a result is a special case of a more general result derived from the *variational principle* (see e.g., HASHIN and SHTRIKMAN, 1963; ASHBY *et al.*, 1978). Briefly, the variational principle states that among the various possible stress–strain distributions in a system that satisfy the boundary conditions, the actual state is the one in which the energy (energy dissipation if ε is strain rate) or complementary energy (dissipation) is the minimum. When applied to a material with linear constitutive relation, the energy and the complementary energy density can be defined as

$$U = \int \sigma \, d\varepsilon = \frac{1}{2}M^*\varepsilon^2 = \frac{1}{2}\sigma\varepsilon \tag{12.11}$$

and

$$\bar{U} = \int \varepsilon \, d\sigma = \frac{1}{2}\frac{1}{M^*}\sigma^2 = \frac{1}{2}\sigma\varepsilon \tag{12.12}$$

respectively, where we define the macroscopic (average) modulus, M^*, by $\sigma \equiv M^*\varepsilon$. If strain (rate) is homogeneous in a multi-phase material, then

$$U_0 = \frac{1}{2}\left(\sum_i x_iM_i\right)\varepsilon^2 \tag{12.13}$$

where x_i is the volume fraction of the ith phase and each phase has the following constitutive relation, $\sigma_i = M_i\varepsilon_i$. Similarly for homogeneous stress,

$$\bar{U}_0 = \frac{1}{2}\left(\sum_i \frac{x_i}{M_i}\right)\sigma^2. \tag{12.14}$$

The variational principle states that $U_0 \geq U$ and $\bar{U}_0 \geq \bar{U}$. Consequently, we have

$$\sum_i x_i M_i \geq M^* \geq \frac{1}{\sum_i (x_i/M_i)}. \tag{12.15}$$

Problem 12.3

Show that the relation (12.15) can be generalized for a non-linear constitutive relation, $\varepsilon \equiv (1/M^*)\sigma^n$, to $\left(\sum_i x_i M_i^{1/n}\right)^n \geq M^* \geq \frac{1}{\sum_i (x_i/M_i)}$.

Solution

Using a non-linear constitutive relation, $\varepsilon \equiv (1/M^*)\cdot\sigma^n$, one obtains

$$U = \int \sigma\, d\varepsilon = \frac{n}{n+1} M^{*\,1/n} \varepsilon^{(n+1)/n} = \frac{n}{n+1}\sigma\varepsilon$$

and

$$\bar{U} = \int \varepsilon\, d\sigma = \frac{1}{n+1}\frac{1}{M^*}\sigma^{n+1} = \frac{1}{n+1}\sigma\varepsilon.$$

Equation (12.13) becomes $U_0 = [n/(n+1)](\sum_i x_i \cdot M_i^{1/n})\varepsilon^{(n+1)/n}$ and (12.14) becomes $\bar{U}_0 = [1/(n+1)]\cdot(\sum_i (x_i/M_i))\sigma^{n+1}$. Consequently, using the relation $U_0 \geq U$ and $\bar{U}_0 \geq \bar{U}$, we get $(\sum_i x_i M_i^{1/n})^n \geq M^* \geq \frac{1}{\sum_i (x_i/M_i)}$.

12.3.2. A more general stress–strain distribution

General introduction

Let us consider a more general case where both stress and strain distribute inhomogeneously. In such a case, one may write

$$\sigma = x_1\sigma_1 + x_2\sigma_2 \tag{12.16}$$

and

$$\varepsilon = x_1\varepsilon_1 + x_2\varepsilon_2 \tag{12.17}$$

where $x_{1,2}$ is the volume fraction of the phase 1 and 2, and

$$x_1 + x_2 = 1. \tag{12.18}$$

Note that in using the relation such as equations (12.16) and (12.17), I made an approximation that the properties of a mixture are represented by volume fraction. This is often called *coarse microstructure approximation*. This approximation is valid when the characteristic length-scale of each component is larger than the length-scale of microstructure such as grain size (e.g., Hill, 1965). In such a case, interaction of two phases can be treated by a continuum mechanics. Now define the following parameters, A_1 and A_2, to treat the non-uniform distribution of strain as

$$\varepsilon_1 = A_1\varepsilon, \qquad \varepsilon_2 = A_2\varepsilon. \tag{12.19}$$

From equations (12.17) and (12.19), we find

$$x_1A_1 + x_2A_2 = 1. \tag{12.20}$$

If strain is uniform then

$$A_1 = A_2 = 1. \tag{12.21}$$

Now let us introduce linear constitutive relations for individual phases,

$$\sigma_i = M_i\varepsilon_i, \qquad \varepsilon_i = S_i\sigma_i \quad (\text{for } i = 1, 2). \tag{12.22}$$

Then

$$\varepsilon = x_1S_1\sigma_1 + x_2S_2\sigma_2 \tag{12.23a}$$

and

$$\sigma = x_1M_1\varepsilon_1 + x_2M_2\varepsilon_2. \tag{12.23b}$$

If one defines the aggregate modulus by

$$\sigma \equiv M^*\varepsilon \tag{12.24}$$

then

$$M^* = x_1M_1A_1 + x_2M_2A_2. \tag{12.25}$$

One can obtain similar relations for compliance, S, by defining

$$\sigma_1 = B_1\sigma, \qquad \sigma_2 = B_2\sigma \tag{12.26}$$

to get

$$x_1B_1 + x_2B_2 = 1 \tag{12.27}$$

and

$$S^* = x_1S_1B_1 + x_2S_2B_2. \tag{12.28}$$

From equations (12.25) and (12.28), one obtains

$$M^* = M_1 + x_2A_2(M_2 - M_1) \tag{12.29a}$$

and

$$S^* = S_1 + x_2B_2(S_2 - S_1). \tag{12.29b}$$

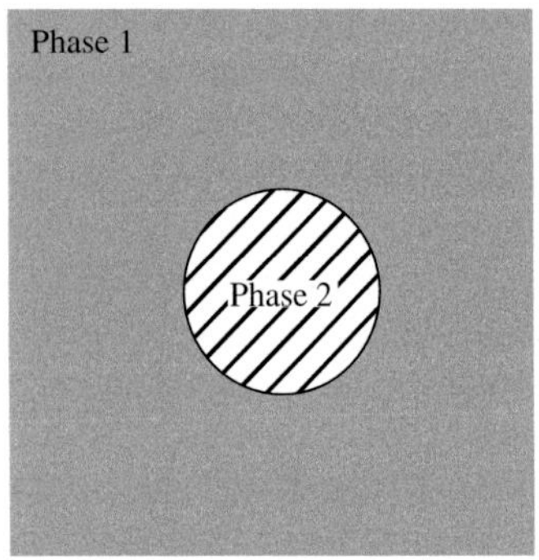

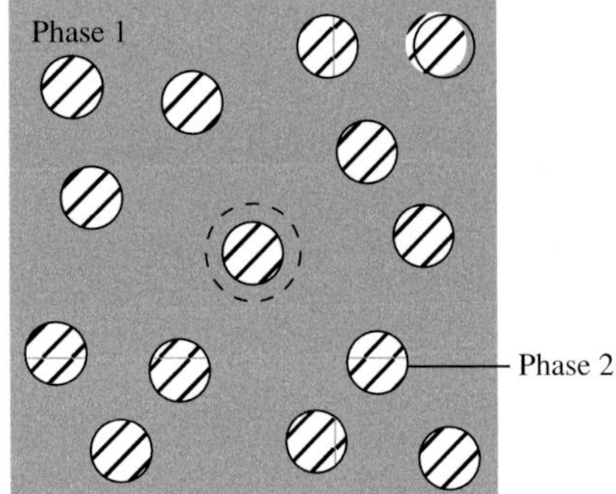

FIGURE 12.2 (a) An inclusion of phase 2 in phase 1 and (b) an inclusion of phase 2 in a mixture of phase 1 and 2. In the self-consistent approach, the matrix made of phases 1 and 2 is treated as a homogeneous medium with one (unknown) modulus.

Equations (12.29a) and (12.29b) mean that the mechanical properties of a two-phase aggregate can be calculated only when one knows A_2 or B_2 (i.e., strain or stress distribution). The Voigt average (or Taylor model), i.e., homogeneous strain corresponds to the assumption of $A_1 = A_2 = 1$, and the Reuss average (or Sachs model), i.e., homogeneous stress corresponds to the assumption of $B_1 = B_2 = 1$. When strain is homogeneous, stress is heterogeneous, $\sigma_1/\sigma_2 = C_1/C_2$, and when stress is homogeneous, then strain is heterogeneous, $\varepsilon_1/\varepsilon_2 = S_1/S_2 = C_2/C_1$.

Direct calculation

When the concentration of a secondary phase is small and has a simple shape such as a sphere, then the strain in the secondary phase, i.e., A_2, can be calculated directly (Fig. 12.2a). The solution for A_2 (sometimes called the concentration factor) for a spherical inclusion, $A_2(M_2, M_1)$, becomes a scalar and is given by

$$A_2(M_2, M_1) = \frac{M_1}{(1-\beta)M_1+\beta M_2} \tag{12.30}$$

with $\beta = \frac{2}{15}[(4-5\nu)/(1-\nu)]$ for shear and $\beta = \frac{1}{3}[(1+\nu)/(1-\nu)]$ $(0<\beta<1)$ for dilatation (ν, Poisson's ratio) (for a viscous fluid for shear $\beta = \frac{2}{5}$) (ESHELBY, 1957). Therefore

$$M^* = M_1\left[1 + \frac{M_2-M_1}{(1-\beta)M_1+\beta M_2}x_2\right]. \tag{12.31}$$

If the secondary phase has a smaller (larger) modulus, it will reduce (increase) the modulus of a composite.

Self-consistent approach

The approximation of dilute mixture does not work in many practical cases where the fraction of the second component is large. In these cases, one needs to include the effects of interaction of each component to evaluate M^*. One of these models is a so-called *self-consistent approach*, in which the interaction of two components is treated as follows. For simplicity, let us consider a two-phase mixture. Let us consider deformation of a piece of component 2 embedded in a matrix that is made of both components (1 and 2) (Fig. 12.2b). In contrast to the previous case where deformation of component 2 in the matrix made of component 1 is treated, we consider deformation of component 2 in a matrix that is made of a mixture of components 1 and 2 whose property M^* is unknown. We make an assumption (approximation) that the properties of the matrix are homogeneous and characterized by a single quantity, M^*. In this approach, one calculates A_2 as a function of M^* and inserts this A_2 into equation (12.29a), and demands that the M^* thus calculated must be the bulk property of an aggregate, M^*. Briefly, in this approach, one has $A_2 = A_2(M_2, M^*)$, and hence equation (12.29a) becomes,

$$M^* = M_1 + x_2A_2(M_2, M^*)\cdot(M_2 - M_1). \tag{12.32}$$

This equation is an implicit equation by which M^* can be determined from the known volume fraction and mechanical properties of each phase. Because the requirement (12.32) is the self-consistency of the procedure, this approach is called the self-consistent approach (e.g., HILL, 1965). Extension of such an approach to non-linear materials is discussed by CHEN and ARGON (1979) and PONTE CASTAÑEDA and WILLIS (1988).

Let us apply the self-consistent approach to a simple case where the second phase has the spherical shape. For a spherical inclusion, the concentration factor, $A_2(M_2, M^*)$, becomes a scalar and is given by (e.g., ESHELBY, 1957), $A_2(M_2, M^*) = \frac{M^*}{(1-\beta)M^*+\beta M_2}$, and hence

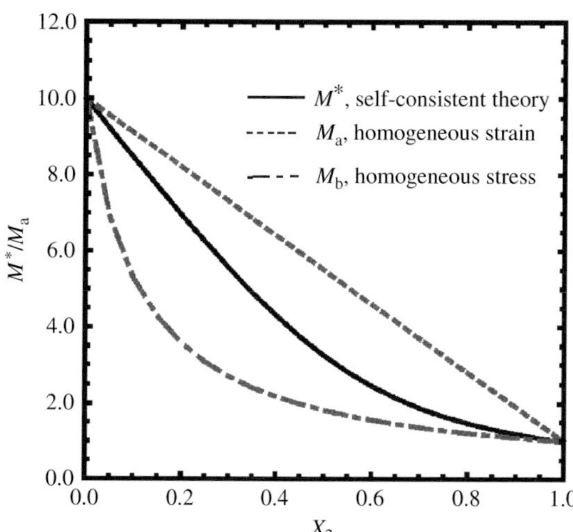

FIGURE 12.3 The dependence of the modulus (elastic modulus, viscosity) of a composite on the fraction of each phase for a case of $\xi \equiv M_1/M_2 = 10$. The homogeneous strain (rate) upper bound, M_a, homogeneous stress lower bound, M_b, and the results of a self-consistent approach, M^*, are shown.

$$(1-\beta)M^{*2} + [(M_1+M_2)\beta - (M_1x_1+M_2x_2)]M^* - \beta M_1 M_2 = 0 \quad (12.33a)$$

or

$$f(y) \equiv (1-\beta)y^2 + [(\beta + x_2 - 1)\xi + \beta - x_2]y - \beta\xi = 0 \quad (12.33b)$$

with $y \equiv M^*/M_2$ and $\xi \equiv M_1/M_2$. Because $1-\beta>0$ and $\beta\xi>0$, this equation has a positive and a negative root. Remembering that the moduli for homogeneous strain (rate) and stress are given by $M_a = x_1M_1 + x_2M_2$ and $M_b = ((x_1/M_1)+(x_2/M_2))^{-1}$ respectively, it is easy to show that $f(M_a/M_2)\cdot f(M_b/M_2)<0$ (Problem 12.4). Consequently, the solution of equation (12.33) must be between the moduli corresponding to homogeneous stress (M_b) and strain (rate) (M_a) (see Fig. 12.3), i.e.,

$$M_a \geq M^* \geq M_b. \quad (12.34)$$

Results of a self-consistent approach are compared with the upper and the lower bounds in Fig. 12.3. The results approach the lower bound at a large amount of a weaker phase and the upper bound at a small amount of a weaker phase. This does not agree with the experimental results such as Bloomfield and Covey-Crump (1993) showing that a weaker phase plays an important role even at a relatively small volume fraction.

The above approach can be extended to a case where the shape of the secondary phase is not spherical. Such a generalization can be made if the concentration factor is calculated as a function of the shape of an inclusion, S, $A_2(M_2, M^*, S)$.

Treagus (2002) extended the above analysis to include anisotropic inclusions. She used the Eshelby solution for elliptic inclusions and assumed an isotropic modulus to obtain an effective modulus of a two-phase mixture with anisotropic shape of secondary phase material. She further assumed that the results can be applied to a case of non-random distribution of oriented inclusions and discussed the influence of secondary phase shape on viscous anisotropy (Treagus, 2003). However, Treagus (2002, 2003) did not treat the matrix as an anisotropic medium, and therefore her results apply to the case for randomly oriented inclusions but not for a material with anisotropic plasticity.

As one can see from Fig. 12.1, the upper and the lower bounds of modulus can be interpreted as the anisotropy in modulus (viscosity or elastic constant).

Problem 12.4

Derive equation (12.33a,b) and show $f(M_a/M_2)\cdot f(M_b/M_2)<0$.

Solution

Inserting equation (12.32) into equation (12.26), one obtains

$$(1-\beta)M^{*2} + [(M_1+M_2)\beta - (M_1x_1+M_2x_2)]M^* - \beta M_1 M_2 = 0.$$

Dividing this equation by M_2^2 and using $x_1 + x_2 = 1$ and the definitions $y \equiv M^*/M_2$ and $\xi \equiv M_1/M_2$, one obtains the relation (12.33b). Inserting $y = M_a/M_2$ into (12.33b), $f(M_a/M_2) = -\beta(y-\xi)(y-1)>0$ $(0<\xi<1)$. Similarly, $f(M_b/M_2) = -\beta[\xi^2/(1+\xi)^2] - y_b(y_a - y_b)<0$ $(y_a>y_b>0)$.

Jordan–Handy model

Variation of stress–strain distribution in two-phase mixture with strength contrasts and phase geometries is discussed qualitatively by Jordan (1988) using a concept of gradual transition from nearly homogeneous strain to localized deformation as a volume fraction of a weak phase increases. He employed a model of deformation of porous materials to formulate a flow law for homogeneous strain following THARP (1983). HANDY (1994) incorporated this idea into a more elaborated mathematical formula. He starts from an energy dissipation relation,

$$\dot{E} = x_1\sigma_1\dot{\varepsilon}_1 + x_2\sigma_2\dot{\varepsilon}_2 \tag{12.35}$$

where $\dot{E}$ is the energy dissipation per unit volume. He distinguishes two cases. (1) A case where much of the load is supported by a stronger phase. This case is called the LBF (load-bearing frame). He postulates that strain is homogeneous in this case (Voigt or Taylor model), so equation (12.35) is modified to

$$\dot{E} = x_1\sigma_1\dot{\varepsilon} + x_2\sigma_2\dot{\varepsilon} \equiv \sigma_{LBF}\dot{\varepsilon}. \tag{12.36}$$

Therefore

$$\sigma_{LBF} = x_1\sigma_1 + x_2\sigma_2. \tag{12.37}$$

By defining the strength ratio by $\xi \equiv \sigma_1/\sigma_2$ (1, strong phase; 2, weak phase), this equation can be rewritten as

$$\frac{\sigma_{LBF}}{\sigma_2} = (1 - x_2)\xi + x_2. \tag{12.38}$$

(2) Another is a case where much of the strain is due to a weaker phase (phase 2). This is called the IWL (interconnected weak layers). HANDY (1994) postulates that the strain in this case is completely concentrated in a weaker phase for a limit of $\xi \to \infty$ (i.e., the phase 2 is extremely weak), namely, $\dot{\varepsilon} \approx x_2\dot{\varepsilon}_2$ as $\xi \to \infty$. Also strain must be homogeneous when the two phases have identical properties, namely, $\dot{\varepsilon} = \dot{\varepsilon}_1 = \dot{\varepsilon}_2$ for $\xi = 1$. HANDY (1994) chooses the following functional form that satisfies these requirements,

$$\dot{\varepsilon} = x_2^{\beta(\xi)}\dot{\varepsilon}_2 \quad \text{with } \beta(\xi) = 1 - \tfrac{1}{\xi} \tag{12.39}$$

With this assumption and from $\dot{\varepsilon} = x_1\dot{\varepsilon}_1 + x_2\dot{\varepsilon}_2$, one has

$$\dot{\varepsilon}_1 = \frac{1 - x_2^{1-\beta}}{1 - x_2}\dot{\varepsilon} \tag{12.40}$$

and from (12.35)

$$\frac{\sigma_{IWL}}{\sigma_2} = \left(1 - x_2^{1-\beta}\right)\xi + x_2^{1-\beta} = \left(1 - x_2^{1/\xi}\right)\xi + x_2^{1/\xi}. \tag{12.41}$$

$\sigma_{IWL} \to (1 - \log x_2)\sigma_2$ as $\xi \to \infty$. HANDY (1994) suggests that between these two cases, the one with lower energy dissipation will be a stable one. The IWL is similar to a case of Reuss or Sachs model, but stress is not homogeneous in this case (see Problem 12.5).

Handy's model provides a practical means to calculate the strength of a two-phase material as a function of strength ratio and volume fraction. However, the choice of the functional form, (12.39), is arbitrary and there is no strong theoretical basis behind this model. In fact the assumption that the stress–strain partitioning depends only on the strength ratio is not supported by the experimental study by BLOOMFIELD and COVEY-CRUMP (1993).

Problem 12.5

Calculate the stress distribution in two phases for the IWL model and show that the result approaches to the LBF if $\beta \to 0$ ($\xi \to 1$).

Solution

For the IWL model, one has $\dot{\varepsilon} = x_2^{\beta}\dot{\varepsilon}_2 = x_2^{\beta}S_2\sigma_2^n = x_2^{\beta}M_2^{-1}\sigma_2^n$ from equation (12.39) where we use the relation $\dot{\varepsilon}_i = S_i\sigma_i^n \equiv (1/M_i)\sigma_i^n$. From equation (12.40), $\dot{\varepsilon}_1 = [(1 - x_2^{1-\beta})/(1 - x_2)]\dot{\varepsilon} = S_1\sigma_1^n = M_1^{-1}\sigma_1^n$. Therefore $\sigma_2 = x_2^{-\beta/n}M_2^{1/n}\dot{\varepsilon}^{1/n}$ and $\sigma_1 = ((1 - x_2^{1-\beta})/(1 - x_2))^{1/n}M_1^{1/n}\dot{\varepsilon}^{1/n}$ and $\sigma_1/\sigma_2 = ((x_2^{\beta} - x_2)/(1 - x_2))^{1/n}(M_1/M_2)^{1/n}$. So the stress in the IWL model is not homogeneous (different from Reuss or Sachs model). For a limiting case of $\beta \to 1$ ($\xi \to \infty$), we have $\sigma_1/\sigma_2 = 0$, i.e., the load is supported entirely by a weaker phase, 2. If $\beta \to 0$ ($\xi \to 1$), $\sigma_1/\sigma_2 \to (M_1/M_2)^{1/n}$ and $\dot{\varepsilon}_1 = \dot{\varepsilon}_2 = \dot{\varepsilon}$, a result for a homogeneous strain rate (LBF model).

Numerical modeling

TULLIS *et al.* (1991) conducted numerical calculations of plastic deformation of two-phase materials using a finite element approach. They chose a plagioclase and clinopyroxene mixture and investigated the deformation behavior of an aggregate in the dislocation creep regime. They explored the conditions of a modest strength contrast, i.e., the relative strength of the two phases changes by a factor of ~1–10. They found that both stress and strain are heterogeneous and that the relation between the average strength of an aggregate and the strength of an individual phase depends on the relative strength and geometry of the two phases as follows:

(i) If a volumetrically small phase is relatively weak, the average strength of a mixture is approximately the same as that of a homogeneous strain rate.

(ii) If a volumetrically small phase is relatively strong, the average strength is close to that of homogeneous strain.

(iii) However, if the geometry of a weak phase is such that most of the strain is taken by the weak phase then the average strength is close to that of homogeneous stress.

The trends (i) and (ii) are similar to the results of a self-consistent model (see section (12.3.2)). As we will see later, experimental observations show that when the strength contrast is very large, a significant deviation from the above trend can be seen. An important case is a two-phase mixture with a very weak minor phase. In such a case, strain tends to be partitioned largely into a weaker phase and even for a small volume fraction, rheological behavior is similar to the case of homogeneous stress (see (12.7)). Note also that the above theoretical models do not consider the effects of evolving geometry of two-phases.

12.3.3. Effective stress exponent and activation enthalpy

Let us now consider a case of a mixture of two materials that deform plastically by the power-law creep with different stress exponents and activation enthalpies. What is the effective stress exponent and the activation enthalpy of a two-phase mixture?

Let us define the flow-law parameters of each phase as

$$\dot{\varepsilon}_i = A_i \exp\left(-\frac{H_i^*}{RT}\right)\sigma^{n_i} \quad i = 1, 2. \tag{12.42}$$

Now let us define the composite flow law as

$$\dot{\varepsilon}_c = A_c \exp\left(-\frac{H_c^*}{RT}\right)\sigma_c^{n_c}. \tag{12.43}$$

Consider two cases: homogeneous strain (rate) and homogeneous stress. For homogeneous strain rate, one has

$$\sigma_c = x_1\sigma_1 + x_2\sigma_2 = x_1\left(\frac{\dot{\varepsilon}_c}{A_1\exp(-H_1^*/RT)}\right)^{1/n_1} + x_2\left(\frac{\dot{\varepsilon}_c}{A_2\exp(H_2^*/RT)}\right)^{1/n_2} \tag{12.44}$$

Similarly for homogeneous stress,

$$\dot{\varepsilon}_c = x_1\dot{\varepsilon}_1 + x_2\dot{\varepsilon}_2 = x_1A_1\exp\left(-\frac{H_1^*}{RT}\right)\sigma_c^{n_1} + x_2A_2\exp\left(-\frac{H_2^*}{RT}\right)\sigma_c^{n_2}. \tag{12.45}$$

From equations (12.44) and (12.45), one obtains (Problem 12.6; French *et al.*, 1994)

$$n_c = x_1\frac{\dot{\varepsilon}_1}{\dot{\varepsilon}_c}n_1 + x_2\frac{\dot{\varepsilon}_2}{\dot{\varepsilon}_c}n_2 \quad \text{for homogeneous stress} \tag{12.46a}$$

$$\frac{1}{n_c} = x_1\frac{\sigma_1}{\sigma_c}\frac{1}{n_1} + x_2\frac{\sigma_2}{\sigma_c}\frac{1}{n_2} \quad \text{for homogeneous strain rate} \tag{12.46b}$$

and

$$H_c^* = x_1\frac{\dot{\varepsilon}_1}{\dot{\varepsilon}_c}H_1^* + x_2\frac{\dot{\varepsilon}_2}{\dot{\varepsilon}_c}H_2^* \quad \text{for homogeneous stress} \tag{12.47a}$$

$$H_c^* = x_1\frac{n_c\,\sigma_1}{n_1\,\sigma_c}H_1^* + x_2\frac{n_c\,\sigma_2}{n_2\,\sigma_c}H_2^* \quad \text{for homogeneous strain rate.} \tag{12.47b}$$

The corresponding pre-exponential factors of an aggregate can also be related to properties of individual phases namely,

$$A_c = \left[x_1A_1\exp\left(-\frac{H_1^*}{RT}\right)\sigma_c^{n_1} + x_2A_2\exp\left(-\frac{H_2^*}{RT}\right)\sigma_c^{n_2}\right]\cdot\exp\left(\frac{H_c^*}{RT}\right)\sigma_c^{-n_c} \quad \text{for homogeneous stress} \tag{12.48a}$$

$$A_c = \left[x_1A_1^{-1/n_1}\exp\left(\frac{H_1^*}{n_1RT}\right) + x_2A_2^{-1/n_2}\exp\left(\frac{H_2^*}{n_2RT}\right)\right]^{-n_c}\cdot\exp\left(\frac{H_c^*}{RT}\right) \quad \text{for homogeneous strain rate.} \tag{12.48b}$$

These results mean that the effective stress exponent or activation enthalpy of a two-phase mixture depends on the distribution of stress or strain rate as well as the stress exponent, the activation enthalpy and the pre-exponential factor in each phase.

Problem 12.6

Derive equations (12.46) and (12.47).

Solution

For homogeneous stress, taking the logarithmic derivative of equation (12.45) $\partial \log \dot{\varepsilon}_c / \partial \log \sigma_c = n_c = x_1(\dot{\varepsilon}_1/\dot{\varepsilon}_c)n_1 + x_2(\dot{\varepsilon}_2/\dot{\varepsilon}_c)n_2$ and $\partial \log \dot{\varepsilon}_c / \partial(-1/RT) = H_c^* = x_1(\dot{\varepsilon}_1/\dot{\varepsilon}_c)H_1^* + x_2(\dot{\varepsilon}_2/\dot{\varepsilon}_c)H_2^*$. Similarly, for homogeneous strain rate, from equation (12.44),

$$\frac{\partial \log \sigma_c}{\partial \log \dot{\varepsilon}_c} = \frac{1}{n_c} = x_1 \frac{\sigma_1}{\sigma_c}\frac{1}{n_1} + x_2 \frac{\sigma_2}{\sigma_c}\frac{1}{n_2}$$

and

$$\frac{\partial \log \sigma_c}{\partial(1/RT)} = \frac{H_c^*}{n_c} = x_1 \frac{\sigma_1}{\sigma_c}\frac{H_1^*}{n_1} + x_2 \frac{\sigma_2}{\sigma_c}\frac{H_2^*}{n_2}.$$

Ji and Zhao (1993) presented a similar model in which the *logarithmic* stress or strain rate is assumed to be additive, namely, $\log \sigma_c = x_1 \log \sigma_1 + x_2 \log \sigma_2$ and $\log \dot{\varepsilon}_c = x_1 \log \dot{\varepsilon}_1 + x_2 \log \dot{\varepsilon}_2$. However, there is no obvious reason to assume that the logarithmic stress or strain is additive. Chen and Argon (1979) studied the relationship between n_c and the stress exponents of individual phases using the self-consistent approach.

12.4. Percolation

12.4.1. General introduction

The above models are applicable to the case where the contrast in properties among co-existing phases is small. When the contrast in properties is large, some other approaches are needed, and it is in such a case where the issue of deformation of two-phase materials becomes particularly important. However, theoretical understanding of deformation of such a material (i.e., deformation of a two-phase mixture with highly contrasting rheological properties) is not well developed. *Percolation theory* is one of the models in which one can treat behavior of a two-phase mixture with a large contrast in property. In this theory, one investigates how the overall properties of a mixture depend on the mixing ratio (volume fraction of each phase). In cases where the properties are vastly different, the properties of a mixture depend strongly on the geometry of a mixture, particularly the *interconnectedness* of a weak phase. To see this point, consider a mixture of an insulator and a metal, or a liquid and a solid. In these cases, when the volume fraction of a "weaker" phase (a phase with low viscosity or high conductivity) reaches a critical value, then a drastic change in bulk property will occur. For example, in the case of a mixture of an insulator and a metal, when metallic particles are distributed in an isolated manner, then the overall electrical conductivity is dominated by the insulator, and hence is zero. However, at a certain critical volume fraction of a metallic phase, a metallic phase will be connected from one end of the sample to another, and then conductivity will increase dramatically. This sharp transition between isolated to connected geometry is called the *percolation* transition. The purpose of a model is to find this critical volume fraction, *percolation threshold*, and to understand how the connectivity of a weaker phase evolves with its volume fraction. For a review of percolation theory one can consult Stauffer and Aharony (1992). A succinct discussion of percolation theory and its applications to the properties of heterogeneous rocks are given in Gueguen and Palciauskas (1994).

In this approach, the connectivity of a weaker phase is a key. Consider a square lattice where each site can be occupied by a weaker phase or empty (i.e., occupied by a stronger phase). In Fig. 12.4, we define a group of weaker phase materials that are connected each other. Such a group is called a *cluster*. The central issue of a percolation theory is to calculate the size of cluster, S (i.e., the number of sites that are contained in a cluster), and define the condition at which S becomes infinite. As we will find, the cluster size approaches infinity at a certain volume fraction of a weaker phase as

$$S \propto \frac{1}{(p_c - p)^x} \qquad (p < p_c) \tag{12.49}$$

where x is a constant that depends on the mechanisms of percolation, p is the probability of finding a site occupied by a weaker phase (i.e., the volume fraction of a weaker phase), and p_c is the critical volume

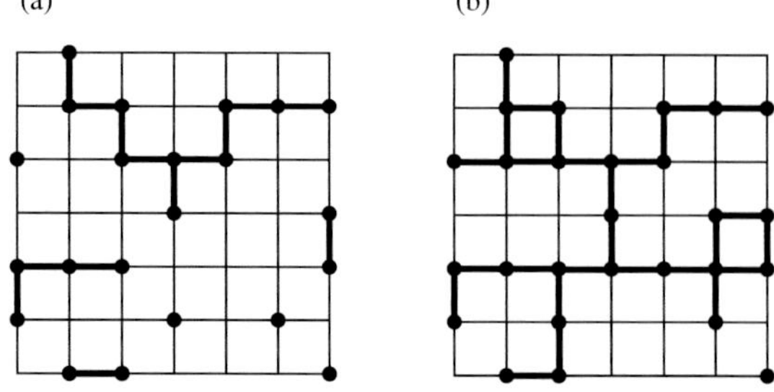

FIGURE 12.4 Cluster and percolation.

TABLE 12.1 The percolation threshold for various three-dimensional lattices.

Lattice	Bond percolation	Site percolation
Diamond	0.388	0.428
Simple cubic	0.249	0.312
bcc	0.179	0.245
fcc	0.119	0.198

fraction of a weaker phase (the percolation threshold). In addition to S (the size of cluster), the fraction of sites that belongs to infinite clusters, P, is also an important quantity. This quantity determines the properties of a two-phase mixture above the percolation threshold.

12.4.2. The Bethe lattice

Exact calculations of properties of a random mixture of two materials are difficult, and most results have been obtained by numerical calculations (see Stauffer and Aharony, 1992). However, an analytical result can be obtained in one particular model. In this model, we consider sites that are connected to z neighboring sites (z: coordination number). Consider a site from which bonds are formed. There are z bonds from each site. From each of these new sites (A_i) that are bonded to this original site, there will be $z-1$ bonds to the neighboring sites extending to "outside" (A_{i+1}), and so on. This network of objects involving *infinite* number of objects is called the *Bethe lattice* (Bethe, 1935) (Fig. 12.5). Now let us assume that each site is occupied by an object with a probability p (p is the fraction of a weak phase). Then for a given site (A_i), the probability that this site is bonded to the neighboring sites (A_{i+1}) is $(z-1)p$. Repeating this process to the next neighboring sites, one finds that the probability to find a connected bond at the n-th step is $[(z-1)p]^n$. If $(z-1)p<1$, then $[(z-1)p]^n \to 0$ as $n \to \infty$. Consequently, if one were to have an infinite cluster, $(z-1)p = 1$ and hence the condition for percolation is given by

$$p_c = \frac{1}{z-1}. \tag{12.50}$$

More sophisticated calculations have been made for the percolation threshold for various lattice. The results are summarized in Table 12.1. Note that there are two values of percolation threshold: one for *bond percolation* and another for *site percolation*. Bond

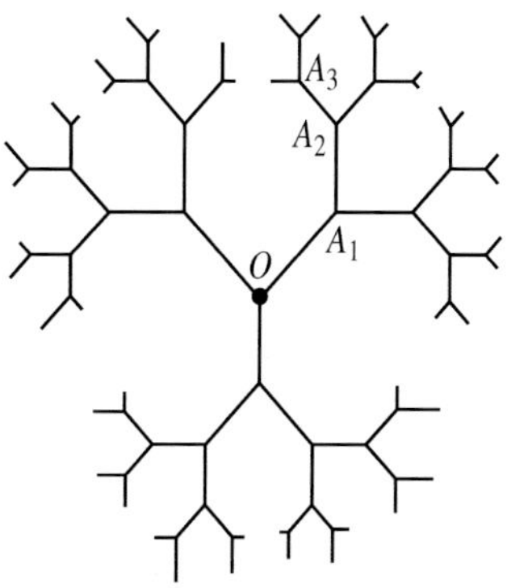

FIGURE 12.5 The Bethe lattice. O is origin, which is connected to $A_1, A_2, \cdots$. From each site there are z branches.

percolation refers to percolation that occurs in a system in which all sites are occupied with a weaker phase, but the connection (bond) between each site is made only with certain probability, p. Site percolation occurs in a system in which each site is occupied by a weaker phase with probability, p. It is interesting to compare this value with the value for the critical melt fraction at which a solid-like behavior changes to a liquid-like behavior. van der Molen and Paterson (1979) determined that this values is ~ 0.2–0.3 for plastic deformation of a partially molten granite (see also Scott and Kohlstedt, 2006). It is also noted that this critical volume fraction is close to the fraction of a weaker phase ((Mg, Fe) O) in Earth's lower mantle (e.g., Karato, 1981b; Yamazaki and Karato, 2001b).

Let us now calculate how the size of cluster, S, increases as p approaches a critical value, p_c, and how the fraction of sites that belong to an infinite cluster changes if p exceeds the critical value, p_c. Consider the Bethe lattice made of infinite sites (Fig. 12.5). z branches emanate from the origin. Let us define, T, as the average size of each branch, i.e., the average number of sites that belong to one branch. From one point, say A_i, there are $z-1$ neighboring sites, A_{i+1}, that will extend further away. They are either unoccupied sites or occupied sites. Since the network is infinite, one can calculate the average size of a cluster by counting the number of sites in the connected network starting from any site. The chosen site may either be occupied or unoccupied with a weak phase. The unoccupied sites contribute 0 to the size of cluster, whereas the occupied sites contribute 1 from the site itself and $(z-1)T'$ from the extending cluster where T' is the size of a cluster that starts from this particular site. However, since we consider an *infinite* network, the average size of subbranch of cluster is also T, i.e., $T = T'$. Therefore

$$T = (1-p)\cdot 0 + p(1+(z-1)T) \tag{12.51}$$

hence

$$T = \frac{p}{1-(z-1)p} = \frac{1}{z-1}\frac{p}{p_c-p}. \tag{12.52}$$

The size of a cluster emanating from the origin is 0 for an unoccupied case and $1+zT$ for an occupied case, hence

$$S = (1-p)\cdot 0 + p(1+zT) \tag{12.53}$$

and

$$S = \frac{p(1+p)}{1-(z-1)p} = \frac{1}{z-1}\frac{p(1+p)}{p_c-p}. \tag{12.54}$$

The size of a cluster goes to infinity as $p \to p_c$.

Now let us calculate the average fraction of material points, P, that belong to infinite clusters when p exceeds a critical value, p_c. As P increases, the strength of a mixture will decrease. To calculate the value of P, let us first define the probability, Q, that a path from the origin is interrupted somewhere. This can happen either by (i) having an occupied point but all the out-going branches are not connected to infinity or (ii) by having this point unoccupied $(1-p)$. Since there are $z-1$ branches, the first probability is pQ^{z-1}, and obviously the probability for (ii) is $1-p$. Therefore

$$Q = pQ^{z-1} + 1 - p. \tag{12.55}$$

Rearranging equation (12.55), one finds

$$(Q-1)\left[1-p(Q^{z-2}+Q^{z-3}+\cdots+Q+1)\right] = 0. \tag{12.56}$$

Now the probability that the origin is occupied but not connected to infinity is pQ^z. The probability that the origin is occupied and connected to infinity is P. Therefore

$$p = pQ^z + P. \tag{12.57}$$

Equations (12.55) and (12.57) can be solved to obtain,

$$\left(\frac{p-P}{p}\right)^{1/z} - p\left(\frac{p-P}{p}\right)^{(z-1)/z} + p - 1 = 0. \tag{12.58}$$

This equation gives a relation $P = P(p)$ (Fig. 12.6). The solution has a general behavior such that $P \to 1$ as $p \to 1$ and $P \to 0$ $(P \to (p-p_c)^\beta$ $(\beta = 1$ for $z = 3))$ as $p \to p_c$ $(p > p_c)$ (Problem 12.7).

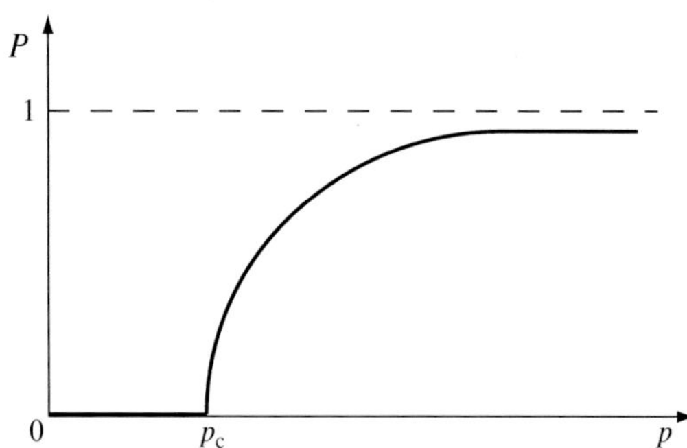

FIGURE 12.6 Plots of P (the fraction of sites occupied by a weak phase) versus p (the fraction of a weak phase). $P = 0$ for $p < p_c$ and $P \to 1$ as $p \to 1$ and $P \to (p - p_c)$ as $p \to p_c$ $(p > p_c)$.

Problem 12.7

Show that $P \to 1$ as $p \to 1$ and $P \to (p-p_c)^\beta$ $(\beta = 1$ for $z = 3)$ as $p \to p_c (p > p_c)$.

Solution

For any value of z, let $p \to 1$ in equation (12.58), then one gets $(1-P) = (1-P)^{z-1}$. Thus $(P-1)P\cdot\left\{1+(1-P)+\cdots+(1-P)^{z-3}\right\} = 0$ since $1+(1-P)+\cdots+(1-P)^{z-3} > 0$, $P = 1$. From equation (12.58), as $p \to p_c$, $(1-P/p_c)^{p_c/(p_c+1)}$ $-p_c(1-P/p_c)^{1/(p_c+1)} + p_c - 1 = 0$ and hence $P \to 0$. Therefore $P \propto (p-p_c)^\beta$. For $z = 3$, equation (12.58) can be written as $X - pX^2 + p - 1 = -p(X-1)\cdot(X-(1-p)/p) = 0$ with $X \equiv ((p-P)/P)^{1/3}$. Since $X = 1$ corresponds to $P = 0$, a physically meaningful solution is $X = (1-p)/p$, hence $P = p-(1-p)^3/p^2 = 2\left(p-\frac{1}{2}\right)\left(p^2-p+1\right)/p^2 \to 6\left(p-\frac{1}{2}\right) = 6(p-p_c)$ as $p \to p_c$ (Fig. 12.6).

The above analytical results are for a special case of the Bethe lattice. For a more general case, numerical modeling must be employed. However, a general relationship is that S diverges as $p \to p_c$, and $P \to 1$ as $p \to 1$ and $P \to (p-p_c)^\beta$ as $p \to p_c$ $(p > p_c)$. Rheological behavior above the percolation threshold is likely to be dominated by a weaker phase. This approach will be useful when the contrast in property between two phases is very large. However, this approach has not been applied to long-term creep.

12.5. Chemical effects

Presence of a secondary phase has another obvious effect. Consider an aggregate of olivine, $(Mg, Fe)_2SiO_4$. When a very small amount of (Mg, Fe)O (or $(Mg, Fe)SiO_3$) (say 1%) is added, it will not have a direct mechanical effect (other than controlling the grain size through Zener pinning; see Chapter 13). However, addition of (Mg, Fe)O (or $(Mg, Fe)SiO_3$) changes the oxide activity and hence indirectly changes the rheological properties of the aggregate (Bai *et al.*, 1991).

Also, the presence of a secondary phase may influence the rheological properties through its influence on the properties of interfacial boundaries. Consider a mixture of A and B. Unlike a pure A (or B) phase aggregates, a mixture of two phases has a large fraction of A–B boundary. Diffusional creep involving these interphase boundaries may be largely different from those of grain boundaries in a pure phase. Chen (1982) and Wheeler (1992) proposed a model of diffusional (or pressure-solution) creep in which the role of interphase boundary is emphasized. These authors analyzed the problem of diffusional creep involving interfacial boundaries and found that enhancement of creep could occur if deviation from stoichiometry is allowed in these phases. However the deviation from stoichiometry likely occurs only under very high stress (see Chapter 8). Under relatively small stress conditions that are normally the case for diffusional creep, the stoichiometry must be preserved and hence the enhanced deformation will not be realized. In fact the experimental observation by Hitchings *et al.* (1989) of higher deformation rate of a mixture of olivine with other phases quoted by Wheeler (1992) turned out to be an artifact caused by the systematically smaller grain size of these two-phase mixtures (the importance of systematic variation in grain size in two-phase mixture was noted by McDonnell *et al.* (2000) and Ji *et al.* (2001)). Therefore I conclude that there is no evidence of enhanced deformation due to the chemical effect as proposed by Wheeler (1992).

12.6. Deformation of a single-phase polycrystalline material

Deformation (elastic or plastic deformation) of a single-phase polycrystalline material can be treated in the same way as deformation of a multi-phase material. In this case, each grain has anisotropic mechanical properties, M_{ijkl}, and the question is how to take an average of anisotropic mechanical properties to obtain average properties of a polycrystalline aggregate. Such a problem has been solved for two end-member cases (homogeneous stress and homogeneous strain (rate)) for isotropic aggregates with linear mechanical properties. Here I simply summarize the results (for more detail see Hill, 1952; Watt *et al.*, 1976). For the upper bound (homogeneous strain (rate): Voigt, Taylor model),

$$\tilde{M}_a^D = \frac{A+2B}{3} \quad \text{for dilatational deformation} \tag{12.59a}$$

$$\tilde{M}_a^S = \frac{A-B+3C}{5} \quad \text{for shear deformation} \tag{12.59b}$$

with $A = \frac{1}{3}(M_{11} + M_{22} + M_{33})$, $B = \frac{1}{3}(M_{12} + M_{23} + M_{31})$, $C = \frac{1}{3}(M_{44} + M_{55} + M_{66})$ where Voigt notation of moduli is used (i.e., $M_{11} = M_{1111}$, $M_{12} = M_{1122}$, $M_{44} = M_{2323}$ etc.: see Chapter 4). And for the lower bound model (homogeneous stress: Reuss, Sachs model),

$$\tilde{M}_b^D = [3(X + 2Y)]^{-1} \quad \text{for dilatational deformation} \tag{12.60a}$$

and

$$\tilde{M}_b^S = \left[\frac{4X-4Y+3Z}{5}\right]^{-1} \quad \text{for shear deformation} \tag{12.60b}$$

with $X = \frac{1}{3}(S_{11} + S_{22} + S_{33})$, $Y = \frac{1}{3}(S_{12} + S_{23} + S_{31})$, $Z = \frac{1}{3}(S_{44} + S_{55} + S_{66})$ where S_{ij} is the abbreviated notation of compliance defined by $\varepsilon_{ij} = \sum_{kl} S_{ijkl}\sigma_{kl}$.

I should emphasize that the degree of anisotropy is usually much larger for plastic deformation than for elastic deformation. Consequently, the upper and the lower bounds give largely different effective moduli. The relation between single crystal deformation and the strength of a polycrystalline aggregate was investigated by Kocks (1970) and Hutchinson (1976, 1977). When homogeneous deformation is assumed, then the overall strength of a polycrystalline aggregate is dominated by that of a crystal with a hard orientation (see Chapter 9).

12.7. Experimental observations

Deformation of a multi-phase aggregate involves complex features of stress and strain distribution and therefore careful experimental studies are required to understand the nature of deformation of a multi-phase aggregate. In analyzing or conducting experimental studies, two issues must be noted. (i) Mechanical behavior of a two-phase material is very sensitive to microstructures including the geometry of each phase and the grain-size. Therefore the microstructures of deformed samples must be analyzed carefully in conjunction with the mechanical data. The morphology of each phase is important in estimating strain partitioning. Grain size is also critical, because grain size is likely to be affected systematically by the mixture of various phases (Chapter 13). Importance of grain-size sensitivity in plastic deformation two-phase aggregates has been noted by McDonnell *et al.* (2000) and Ji *et al.* (2001). (ii) Microstructures often evolve with strain, and therefore large-strain deformation experiments are needed to fully understand the mechanical behavior of a two-phase aggregate.

Experimental studies on plastic deformation of multi-phase aggregates involve the determination of mechanical properties of aggregates in relation to strain partitioning. Strain partitioning can be inferred from the microstructural analysis of a deformed sample. This provides $\varepsilon_1 = A_1\varepsilon$ and $\varepsilon_2 = A_2\varepsilon$. Stress partitioning can also be investigated using the equation (12.16) which can be rewritten as $\sigma(\varepsilon) = x_1\sigma_1(A_1\varepsilon) + x_2\sigma_2(A_2\varepsilon)$ (this relation is valid for non-linear constitutive relationships as well). Given strain partitioning (i.e., A_1 and A_2) and the volume fraction of each phase (i.e., x_1 and x_2), this relation gives the stress partitioning. Note that the stress and strain distribution in co-existing phases $((\sigma_1, \varepsilon_1), (\sigma_2, \varepsilon_2))$ and the average stress–strain (rate) (σ, ε) are related by a linear tie line defined by Bloomfield and Covey-Crump (1993) (see Problem 12.8)

$$\frac{\sigma - \sigma_1}{\varepsilon - \varepsilon_1} = \frac{\sigma - \sigma_2}{\varepsilon - \varepsilon_2}. \tag{12.61}$$

This relation allows a graphical analysis of the stress–strain distribution in a deformed two-phase mixture within the coarse-microstructure approximation (Fig. 12.7a). Bloomfield and Covey-Crump (1993) performed detailed analysis of plastic deformation of a mixture of calcite (harder phase) and halite (softer phase) and found that a large fraction of strain is partitioned in a softer phase (halite), and that the degree of strain partitioning to a weaker phase increases with strain (Fig. 12.7b). A similar observation was made by Ross *et al.* (1987). In particular, Ross *et al.* (1987) showed that the stress–strain curve can develop a negative slope when a large fraction of strain becomes accommodated by a weaker phase. Such observations suggest that the progressive partitioning in strain into a weaker phase may lead to shear localization under some conditions (see Chapter 16).

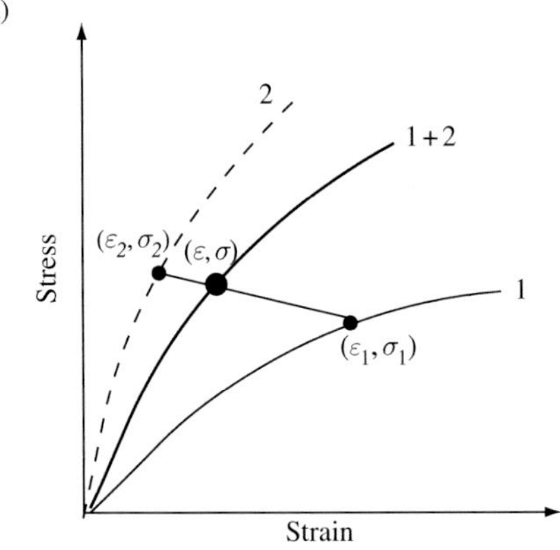

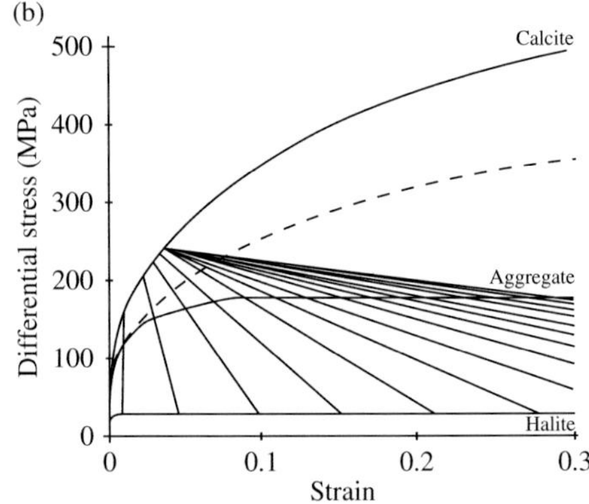

FIGURE 12.7 (a) A graphical representation of relation (12.61). (b) The stress–strain relationship in a deformed 30% halite + 70% calcite aggregate (after Bloomfield and Covey-Crump, 1993).

Rybacki *et al.* (2003) conducted a detailed experimental study on deformation of a mixture of calcite (soft phase) and quartz (hard phase) in a mixture where the volume fraction of a hard phase is small (1–30%). They found that as the fraction of a hard material increases both stress exponent and activation energy become closer to those of a hard material. This indicates that the mechanical behavior at larger fractions of a hard phase moves closer to homogeneous strain (see discussion in section 12.3.3) in which a large fraction of load is supported by a harder phase material. This concept is sometimes referred to as *load transfer* (e.g., Park and Mohamed, 1995; Li and Langdon, 1998). Other experimental works include Jordan (1987), Ross *et al.* (1987), Bons and Cox (1994), Bons and Urai (1994), Krajewski *et al.* (1995), Bruhn and Casey

(1997), BRUHN *et al.* (1999), MCDONNELL *et al.* (2000), JI *et al.* (2001) and XIAO *et al.* (2002).

Problem 12.8

Derive equation (12.61).

Solution

From equation (12.16), $\sigma = x_1\sigma_1 + x_2\sigma_2$, so using $x_1 + x_2 = 1$ one has $\sigma - \sigma_2 = x_1(\sigma_1 - \sigma_2)$ and $\sigma - \sigma_1 = -x_2(\sigma_1 - \sigma_2)$. Similarly, $\varepsilon - \varepsilon_2 = x_1(\varepsilon_1 - \varepsilon_2)$ and $\varepsilon - \varepsilon_1 = -x_2(\varepsilon_1 - \varepsilon_2)$. Eliminating x_1 and x_2, one finds $(\sigma - \sigma_1)/(\varepsilon - \varepsilon_1) = (\sigma - \sigma_2)/(\varepsilon - \varepsilon_2)$.

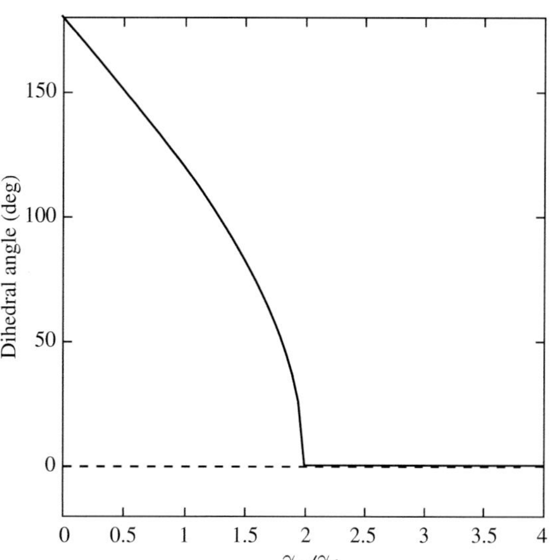

FIGURE 12.8 The relation between the dihedral angle and the ratio of interfacial energies.

12.8. Structure and plastic deformation of a partially molten material

The mechanical properties of a partially molten material have attracted much attention because partial melting is the most important process by which Earth (and other planets) evolves. Therefore the understanding of physical processes of evolution of Earth and terrestrial planets relies strongly on our understanding of physical properties of a partially molten material. In this book, we focus on the plastic deformation of a partially molten material. A partially molten material is one of the multi-phase materials, but the contrast in mechanical properties is large. Not only that the fluid phase has no rigidity, but also a fluid phase usually has a much lower bulk modulus. Consequently, the influence of partial melting on mechanical properties is large both for compressional properties (bulk modulus and compaction rate) and for shear properties (shear modulus and (effective) viscosity).

From a mechanical point of view, partial melt is a mixture of materials with largely contrasting properties. As we have learned above, the physical properties of a mixture (with large contrast in properties) are sensitive to the geometry of the phases involved. Therefore we will start with the geometry of partial melt, and then discuss the plastic flow properties of partial melt.

12.8.1. Geometry of partial melt

Consider a solid–liquid contact. There are two types of interface at this contact: a solid–solid interface and a solid–liquid interface. Each interface is associated with interfacial energy, γ_{ss} and γ_{sl} respectively. The force balance condition demands

$$\cos\frac{\theta}{2} = \frac{\gamma_{ss}}{2\gamma_{sl}} \tag{12.62}$$

where θ is the dihedral angle (Fig. 12.8). This equation means that the dihedral angle is controlled by the relative value of interfacial energies. The dihedral angle increases with the increase of γ_{sl}/γ_{ss} and $\theta = 0$ for $\gamma_{sl}/\gamma_{ss} \leq \frac{1}{2}$. Notice that $\theta = 0$ for a wide range of γ_{sl}/γ_{ss}, and as a consequence many materials show $\theta = 0$. It should also be noted that equation (12.62) is a simplified equation. In a more general case, interface energies are dependent on crystallographic orientation and hence the dihedral angle is in general orientation-dependent (e.g. FAUL *et al.*, 1994). Interfacial energies also change with the concentration and type of impurities (Chapter 5).

As shown in Fig. 12.9, the dihedral angle controls the morphology of melt. If $\theta = 0$, the melt completely wets grain boundaries. For $\frac{1}{3}\pi(60°) \geq \theta > 0$, the fluid pressure in the melt is smaller than outside and the tube of melt is a stable geometry. Whereas for $\theta > \frac{1}{3}\pi(60°)$, the fluid in a melt pocket will have a higher pressure than the surroundings, and the melt will assume isolated pockets (for a small melt fraction). For most silicate systems, dihedral angles of 30°–50° are often observed (e.g., WAFF and BLAU, 1979, 1982; COOPER and KOHLSTEDT, 1982; TORAMARU and FUJII, 1986),

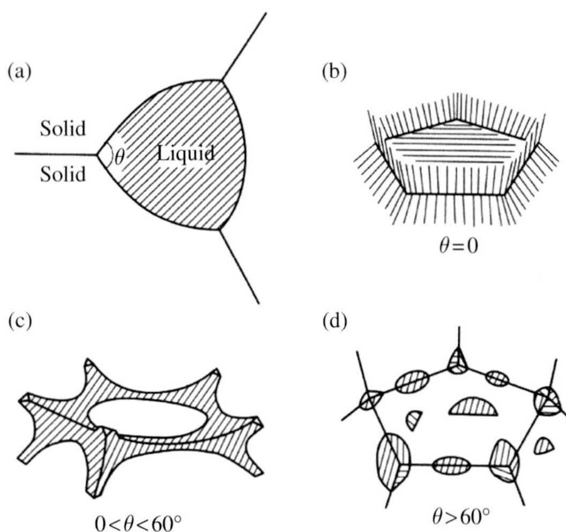

FIGURE 12.9 The geometry of partial melt as a function of the dihedral angle (after KARATO, 2003a).

whereas with silicate minerals and aqueous fluids $\theta > \frac{1}{3}\pi$ is observed at low pressures ($P < 2\,\text{GPa}$; WATSON and BRENAN, 1987). Note, however, that the dihedral angles depend strongly on the chemical composition of fluid that coexists with solid. When the compositions of fluid and solid are similar, dihedral angle is small. In fact, YOSHINO *et al.* (2007) showed that the dihedral angle in silicate melt and olivine changes systematically with pressure and reaches 0 degrees at $\sim$7 GPa. Fig. 12.8 shows that a small variation in γ_{sl}/γ_{ss} results in a large variation in dihedral angle. Therefore a wide range of dihedral angle is expected even in a simple one-component system. Wetting geometry is also affected by the presence of deviatoric stress. This can be understood by comparing the effective stress due to interfacial tension with the applied stress. For $\sim$1 μm radius of curvature, the interfacial tension will be $\sim$1 MPa (for interfacial energy of $\sim$1 J/m^2). So a stress larger than this will influence the melt geometry.

12.8.2. Deformation of a partially molten material

Deformation of a partially molten material may be classified into: (i) elastic deformation, (ii) anelastic deformation and (iii) plastic (viscous) deformation. Some aspects of anelastic deformation of a partially molten material have been discussed in Chapter 11. Here I will focus on (steady-state) plastic deformation of a partially molten material.

Deformation of a partially molten material occurs in two different modes. In one mode, when the relative motion of melt and solid is allowed, then volumetric deformation occurs leading to *compaction*. Physics of compaction was investigated by MCKENZIE (1984) and BERCOVICI *et al.* (2001a). Compaction is a process in which the fluid is squeezed out of a partially molten material due to the gravity force caused by density difference or to the external force. The compaction is characterized by a *compaction length* that is the characteristic length at which the porosity changes due to compaction. Since compaction involves viscous flow of fluid out of the solid skeleton that is accommodated by plastic deformation of solid, the compaction length depends on the permeability of fluid, k, as well as the viscosity of solid and melt, $\eta_{\text{solid,melt}}$, as,

$$\delta_c \equiv \sqrt{\frac{k\eta_{\text{solid}}}{\eta_{\text{melt}}}}. \tag{12.63}$$

Since deformation of the solid skeleton associated with compaction includes volumetric strain, the effective viscosity of solid, η_{solid}, includes not only shear viscosity but also *bulk viscosity*, $\eta_{\text{solid}} = \eta_{\text{solid}}^{\text{bulk}} + \frac{4}{3}\eta_{\text{solid}}^{\text{shear}}$ where $\eta_{\text{solid}}^{\text{bulk, shear}}$ is the bulk (or shear) viscosity of solid. The concept of a bulk viscosity has not always been made clear in the literature on fluid mechanics. A standard textbook such as LANDAU and LIFSHITZ (1987) does not give a physical explanation for the bulk viscosity. SCOTT and STEVENSON (1984) (see also RICARD *et al.*, 2001) provided a clear physical explanation for the bulk viscosity for deformation of a porous material: they showed that the bulk viscosity is in fact determined by the shear viscosity of solid and the porosity as $\eta_{\text{solid}}^{\text{bulk}} \sim \eta_{\text{solid}}^{\text{shear}}/\phi$ where ϕ is the porosity (note that $\eta_{\text{solid}}^{\text{bulk}} \to \infty$ as $\phi \to 0$ and $\eta_{\text{solid}}^{\text{bulk}} \to \eta_{\text{solid}}^{\text{shear}}$ as $\phi \to 1$). This relation implies that at a small porosity, $\eta_{\text{solid}}^{\text{bulk}} \gg \eta_{\text{solid}}^{\text{shear}}$ and squeezing melt will become progressively more difficult as the melt fraction (porosity) decreases.

Problem 12.9*

Derive the relation (12.63) for compaction length.

Solution

Recall that the compaction process involves fluid segregation due to the differential pressure that is controlled by permeability, k, namely, $J_{\text{melt}} = -(k/\eta_{\text{melt}})(dp/dx)$ (J_{melt}, melt flux; p, pressure). Now let the characteristic pressure difference be Δp and the characteristic length-scale at which pressure changes be δ_c. Then, $J_{\rm melt} = -(k/\eta_{\rm melt})(dp/dx) \approx -(k/\eta_{\rm melt})(\Delta p/\delta_c)$. Note that the pressure changes due to the change in porosity, and therefore the characteristic length at which the pressure changes is identical to the compaction length. Plastic deformation of solid must occur associated with the melt segregation. The mass flux of melt must be accommodated by the deformation of solid, so that $dV_{\rm solid}/dt \approx -dV_{\rm melt}/dt = -SJ_{\rm melt}$ (S, cross section area). Consequently, $dV_{\rm solid}/dt = V_{\rm solid}\dot{\varepsilon}_{\rm solid} = -SJ_{\rm melt} = S(k/\eta_{\rm melt})(\Delta p/\delta_c)$ where $\dot{\varepsilon}_{\rm solid} = (1/V_{\rm solid})\cdot(dV_{\rm solid}/dt)$. Now the volume change occurs in a column with a length of δ_c, so that $V_{\rm solid}/S \approx \delta_c$. Also the stress that causes deformation of solid is similar to the pressure difference, $\Delta p \approx \sigma$, and using the definition of solid viscosity, $\sigma = \eta_{\rm solid}\dot{\varepsilon}_{\rm solid}$, one obtains $\delta_c = \sqrt{k\eta_{\rm solid}/\eta_{\rm melt}}$.

Let us focus on shear (plastic) deformation of a partial melt in which a large-scale relative motion of melt and solid does not occur. In this case, we have a mixture of melt and solid. Two end-member cases can easily be distinguished. One is deformation of a mixture of a large amount of melt with a small amount of solid. In this case, the overall rheological property is controlled by that of a melt with a small influence of solid suspensions. Using the assumption that stress is supported by the melt but solid particles do not deform, $\dot{\varepsilon} \approx (1-\phi)\dot{\varepsilon}_{\rm melt}$, and one would get $\eta = \eta_0 \frac{1}{1-\phi} \approx \eta_0(1+\phi)$ where η_0 is the viscosity of pure liquid and ϕ ($\ll 1$) is the volume fraction of solid particles. However, some corrections must be made as to the distortion of flow lines by solid particles that yield the well-known Einstein formula (e.g., Einstein, 1906; Landau and Lifshitz, 1987),

$$\eta = \eta_0\left(1 + \frac{5}{2}\phi\right). \tag{12.64}$$

Experimental data and some models for viscosity of liquids with solid suspensions are summarized in McBirney and Murase (1984; see also Ryerson *et al.*, 1988). Interpreting (12.64) as an equation for a small change in viscosity for a small change in concentration of suspension, Roscoe (1952) obtained the following equation for the viscosity of a liquid with a relatively large amount of suspension,

$$\eta = \eta_0(1 - A\phi)^{-5/2} \tag{12.65}$$

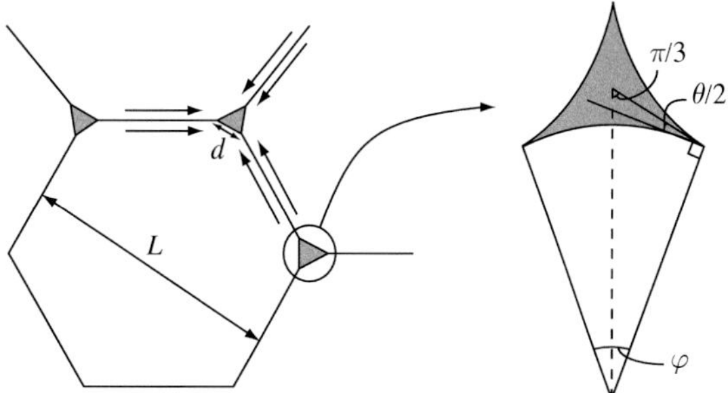

FIGURE 12.10 The geometry of diffusional mass transport along a grain boundary.

where A is a constant near unity (Roscoe, 1952). This formula gives an appropriate asymptotic limit of relation (12.64) for $\phi \to 0$, $\eta \approx \eta_0(1 + \frac{5}{2}A\phi)$, and also yields infinite viscosity as $\phi \to 1/A$.

Another end-member case is a case where a material is dominated by solid with a small amount of melt. In this case, deformation is similar to the LBF (load bearing framework) and the strength is given by $\sigma = (1-\phi)\sigma_s + \phi\sigma_m \approx (1-\phi)\sigma_s$ where ϕ is the melt fraction, and $\sigma_{s,m}$ is the strength of solid and melt respectively. Therefore the creep strength of a material will be modified through stress concentration as

$$\sigma_s \approx \frac{\sigma}{1-\phi}. \tag{12.66}$$

Hirth and Kohlstedt (1995b) showed that their data on deformation of olivine and basaltic melt in the dislocation creep regime roughly fit this relation.

For diffusional creep, additional effects will arise due to the reduction of diffusion path if diffusion occurs along grain boundaries, i.e., Coble creep (this effect will not be present in Nabarro–Herring creep). Diffusion distance will be shortened by the presence of melt that enhances the rate of deformation by $L_0/L = 1/(1 - B(\theta)\sqrt{\phi})$ where $B(\theta)$ is a geometrical factor that depends on the dihedral angle (Cooper and Kohlstedt, 1986; Cooper *et al.*, 1989) (see Fig. 12.10, Problem 12.10). The stress concentration effect given by equation (12.66) can also be cast in terms of the reduction of melt-free boundary, as

$$\sigma_s \approx \sigma\left(\frac{L_0}{L}\right)^2 = \frac{\sigma}{\left(1 - B(\theta)\sqrt{\phi}\right)^2}. \tag{12.67}$$

Therefore combining these two effects, one gets,[1]

[1] This formula is slightly different from the result proposed by Cooper *et al.* (1989), i.e., $\dot{\varepsilon} = \dot{\varepsilon}_0(L/L_0)^4 = \dot{\varepsilon}_0[1/(1 - B(\theta)\sqrt{\phi})^4]$. Different from the present discussion, Cooper *et al.* (1989) argued that the chemical potential gradient driving diffusional flow is enhanced by the presence of melt pocket.

$$\dot{\varepsilon} = \dot{\varepsilon}_0 \left(\frac{L}{L_0}\right)^3 = \dot{\varepsilon}_0 \frac{1}{\left(1 - B(\theta)\sqrt{\phi}\right)^3}. \tag{12.68}$$

Experimental results on grain-boundary diffusional creep (Coble creep) on olivine and basalt agree reasonably well with this model (see KOHLSTEDT, 2002).

KELEMEN *et al.* (1997) proposed the following relation that describes the rheological behavior of a partially molten rock for a broad range of melt content,

$$\dot{\varepsilon} = \dot{\varepsilon}_0 \exp(\alpha\phi) \tag{12.69}$$

where α is a non-dimensional constant ($\sim$20–40). However, there is no strong theoretical basis for this equation, and it should be considered as a practical means to fit the data. For example, the asymptotic behavior of (12.69) at $\phi \to 0$ does not agree with that of equation (12.68) and an extrapolation to $\phi \to 1$ would give $\dot{\varepsilon} = \dot{\varepsilon}_0 \exp(\alpha)$ which does not contain any physical constant for liquid phase which is physically unreasonable. A more appropriate formula that agrees with a relation (12.68) at $\phi \to 0$ is

$$\dot{\varepsilon} = \dot{\varepsilon}_0 \exp(\beta' \sqrt{\phi}) \tag{12.70}$$

with $\beta' = 3B(\theta)$ but this formula also yields an unreasonable limit for $\phi \to 1$. So both equations (12.69) and (12.70) must be considered as rough approximations for a small melt fraction.

The relation such as equations (12.67) and (12.68) indicates that the degree of enhancement of plastic deformation is highly sensitive to the wetting behavior through $B(\theta)$. For materials with smaller dihedral angles (larger $B(\theta)$), the degree of enhancement of deformation will be larger. The importance of wetting behavior in plastic deformation of partially molten materials was investigated by PHARR and ASHBY (1983).

JIN *et al.* (1994) conducted deformation experiments of partially molten peridotites and reported very large effects of partial melting in reducing the creep strength of peridotite at small melt fractions. These authors attributed this to near complete wetting of grain boundaries *during deformation*. The results by Green's group (JIN *et al.*, 1994) are in marked contrast to the results from Kohlstedt's group (see KOHLSTEDT, 2002) that show only limited wetting and hence modest weakening effects at small melt fractions (<5%). JIN *et al.* (1994) argued that complete wetting occurred during their experiments (at $P = 1.5$ GPa), but that after removing the stress melt geometry went back to equilibrium non-wetting geometry. If this is true then one needs to explain why this dynamic wetting did not occur in Kohlstedt group's experiments. The cause for this discrepancy is not resolved. Complete wetting is observed in static experiments in a peridotite system at high pressures (YOSHINO *et al.*, 2007), and in such a case, plastic deformation should be enhanced dramatically.

Rheological properties of a partially molten material evolve with strain. At high stresses and large strains, the morphology of melt changes from interface tension-controlled morphology to stress-controlled morphology leading to rheological weakening (e.g., ZIMMERMAN *et al.*, 1999). This phenomenon was first reported by BUSSOD and CHRISTIE (1991). The degree to which deformation affects the geometry of melt depends on the relative magnitude of applied deviatoric stress and the stress caused by interfacial tension. Briefly, the ratio of stress due to external force to the stress due to interface tension at melt–solid contact, $\sigma/(\gamma/r)$ (γ, interface energy; r, the radius of curvature of melt–solid interface), controls the influence of stress on melt geometry. I note that the stress at the melt–solid contact depends on the melt fraction. The radius of curvature is proportional to the melt fraction, ϕ, as $r \propto \sqrt{\phi}$, and consequently the critical stress changes with melt fraction as $\sigma_c \propto 1/\sqrt{\phi}$. This relation implies that the influence of applied stress on melt geometry is small for a small melt fraction, but the influence of stress becomes important at a large melt fraction. With increasing melt fraction, the radius of curvature will increase and the influence of applied stress will become more important. Some theoretical and experimental studies are available on this issue but much remains unexplored (NYE and MAE, 1972; RAJ, 1986; SROLOVITZ and DAVIS, 2001; MEI *et al.*, 2002; HOLTZMAN *et al.*, 2003a; HIER-MAJUMDER *et al.*, 2004).

Problem 12.10*

Consider a partially molten material with a melt pocket with a negative curvature (i.e., $0 < \theta \leq \frac{1}{3}\pi$). Show that the radius of curvature of a melt is proportional to $\sqrt{\phi}$ where ϕ is the melt fraction, and

$$B(\theta) = \left(\frac{8\sqrt{3}}{3}\right)^{1/2} \frac{\sin(\pi/6 - \theta/2)}{\sqrt{\frac{1}{3}\sqrt{3}(1+\cos\theta) + \sin\theta - (\pi/3 - \theta)}}.$$

Solution

Consider a two-dimensional section of a melt tube shown in Fig. 12.10. Let θ be the dihedral angle, and φ be the angle defined by the center of radius of curvature and the two corners of the melt pocket

(Fig. 12.10). From a geometrical consideration, one can show $\theta + \varphi = \pi/3$ and the cross sectional area of a melt tube is calculated to be $(3r^2/2)\cdot(\frac{1}{3}\sqrt{3}(1+\cos\theta)+\sin\theta-(\pi/3-\theta))$. A melt tube is contained in a solid with an area of $\frac{1}{4}\sqrt{3}L_0^2$ (L_0 the distance of two adjacent grains, Fig. 12.10). Therefore the porosity is $\phi = 2\sqrt{3}(r/L_0)^2(\frac{1}{3}\sqrt{3}(1+\cos\theta)+\sin\theta-(\pi/3-\theta))$, i.e., $r \propto \sqrt{\phi}$. The length of a grain boundary that is covered by a melt is $r(4\sqrt{3}/3)\sin(\pi/6-\theta/2)$. Using the definition of $B(\theta)$, $l/l_0 = 1 - B(\theta)\sqrt{\phi}$ where l_0 is the length of grain boundary and l is the length of grain boundary that is not wetted by melt, one obtains,

$$B(\theta) = \left(\frac{8\sqrt{3}}{3}\right)^{1/2} \frac{\sin(\pi/6-\theta/2)}{(\frac{1}{3}\sqrt{3}(1+\cos\theta)+\sin\theta-(\pi/3-\theta))}.$$

Rheological properties of partially molten materials at large melt fractions have been investigated by Arzi (1978), van der Molen and Paterson (1979), Mecklenburgh and Rutter (2003), and Scott and Kohlstedt (2006). In most of these studies, deformation involves grain–grain switching or grain crushing. Arzi (1978) defined a *rheologically critical melt fraction* at which the mechanical behavior of a partial melt changes from solid-like to liquid-like behavior. This critical melt fraction corresponds to the percolation threshold discussed in section 12.4. Rheological behavior near the critical melt fraction is important in the evolution of magma ocean (e.g., Abe, 1997).

In applying the results of experimental studies on the influence of partial melting on rheological properties, one needs to address how the melt fraction in a given rock is controlled. This problem not only involves the thermodynamics of melting but also the physics of compaction and melt transfer. Some aspects of these issues are discussed in Chapters 17 and 19.

13 Grain size

Grain size is one of the important microstructural parameters that control rheological properties. Conversely, grain size often reflects deformation conditions. This chapter provides a detailed account of the physical processes by which grain size is controlled. We start from the basic physics of grain-boundary migration, and discuss the kinetics of grain growth, mechanisms of dynamic recrystallization and finally the physics by which grain size is controlled by nucleation growth. Some examples of the application of basic physics to understand the grain-size distribution and rheology of Earth's interior are presented.

Key words grain boundary, grain-boundary migration, grain growth, Zener pinning, Ostwald ripening, dynamic recrystallization, nucleation growth in phase transformations, paleopiezometers.

13.1. Introduction

Grain size is an important parameter that controls the rheology of rocks (see Chapter 8). The grain size of rocks can vary from ~10 μm (ultra-mylonite, a fine-grained rock found in shear zones; e.g., BELL and ETHERIDGE, 1973; WHITE *et al.*, 1980; HANDY, 1989) to $\sim 10^2$–10^3 m (estimated grain size of the inner core; BERGMAN, 1998, see also Chapter 17 of this book) depending on its thermal and mechanical history. This large variation in grain size causes significant changes in rheological properties. For example, for a small grain size and low stress, a polycrystalline material tends to deform by grain-size sensitive creep mechanisms, such as diffusional creep (see Chapter 8). In these cases, the viscosity of a rock, η, changes with grain size, L, as $\eta \propto L^m$ where m is a constant between 2 and 3. Therefore a change in grain size by a factor of 100 (say 1 mm to 10 μm) in this regime would lead to a change in viscosity by a factor of 10^4 to 10^6. Thus the understanding the processes that may control the grain size of materials in Earth's interior is critical to the understanding of its rheological properties.

A close association of grain-size reduction and large (localized) strain is seen in ductile shear zones (*mylonite*) in Earth's crust (WHITE *et al.*, 1980; HANDY *et al.*, 1989). Grain-size reduction may lead to shear localization under some conditions (Chapter 16). The importance of grain size in controlling the style of mantle convection has also been studied by KARATO *et al.* (2001), SOLOMATOV (2001), HALL and PARMENTIER (2002) and KORENAGA (2005). Another example is the inner core. There is evidence for strong scattering of seismic waves in the inner core (CORMIER *et al.*, 1998; VIDALE and EARLE, 2000), suggesting that the grain size there is comparable to the wavelength of seismic waves (on the order of 10^3 m). This large grain size probably results in the relatively large viscosity that affects the dynamics of the inner core (Chapter 17). Therefore there is a strong indication of an important role for grain size on deformation, and it is crucial to understanding the physical processes that may control grain size in Earth's interior.

In this chapter, I will first summarize the fundamentals of grain-boundary migration and discuss three processes by which grain size may be controlled in

Earth: *grain growth*, *dynamic recrystallization* and *phase transformations* (*nucleation and growth*). Under static conditions, *grain growth* occurs to reduce the grain-boundary energy in a rock. The importance of grain growth has been inferred in several geological cases including the variation in grain size as a function of location in contact metamorphic rocks (e.g., Joesten, 1983; Covey-Crump and Rutter, 1989). The fundamental physics of grain growth has been well understood, although there are complications in the grain-growth processes in multi-phase materials. When deformation occurs by dislocation creep, heterogeneous distribution in dislocation densities often causes *dynamic recrystallization* by which grain size may be reduced. Evidence of this deformation-induced grain-size reduction is ubiquitous in many naturally deformed rocks. The physical processes of dynamic recrystallization are complicated and much less well understood than (static) grain growth. However, various models have been proposed, which will be summarized in this section. When a *phase transformation* (or chemical reaction) occurs, changes in grain size may also occur by *nucleation and growth* processes.

13.2. Grain-boundary migration

Grain-boundary migration is one of the key processes that control grain size. At a finite temperature, atoms are always moving around by thermal fluctuation and as a result an atom may change its position with time. Consequently, atoms near a boundary should be moving back and forth from one grain to another across the boundary. When the free energies are different between the two neighboring grains, then atoms tend to stay in a grain with a lower free energy. As a result, there is a migration of grain boundary.

13.2.1. Driving forces

The above picture of grain-boundary migration indicates that the driving force for grain-boundary migration is the difference in free energy across a grain boundary. There are several causes for the difference in free energies between neighboring grains. The first is the bulk free energy difference between adjacent grains. In this case, the driving force is simply,

$$F = \Delta g \tag{13.1}$$

where Δg is the difference in free energy per unit volume (the unit of the driving force is therefore the same as pressure). One of the most important types of energy difference is the difference in dislocation energy caused by the heterogeneity in dislocation density. In this case, equation (13.1) can be written as,

$$F = \mu b^2 \Delta\rho\,. \tag{13.2}$$

Heterogeneity of dislocation density could be caused by the orientation dependence of local stress, namely (see Chapter 5),

$$F_1 = \frac{\sigma^2}{\mu}\left|\Delta(S^2)\right| = 2\frac{\sigma^2 S^2}{\mu}\left|\frac{\Delta S}{S}\right| \tag{13.3}$$

where S is the Schmid factor (Chapter 5). Once a grain boundary starts to move due to heterogeneity in dislocation density, then a portion of a crystal behind the moving boundary will have zero dislocation density. Therefore the effective driving force will be the dislocation density of grains with a higher dislocation density.

Elastic strain energy difference is another driving force. In this case,

$$F_2 = |\Delta(\varepsilon\sigma)| = \frac{\sigma^2}{\mu}\left|\frac{\Delta\mu}{\mu}\right|. \tag{13.4}$$

Comparing F_1 and F_2, one has $F_1/F_2 = \left|\frac{2S^2(\Delta S)/S}{\Delta\mu/\mu}\right| \approx 2-10$ where we used the values of $S \sim 0.4$, $\Delta S/S = 0.5-1.0$ and $\Delta\mu/\mu = 0.1-0.2$. Therefore we conclude that the dislocation energy is in most cases more important than the elastic strain energy.

Surface (boundary) tension is another important driving force. A grain with a finite size has extra energy associated with its grain boundary. Consequently, grains with different grain size have different free energies. The energy per unit volume due to grain-boundary energy is $2(\gamma/L)$ (Problem 5.12). Therefore a difference in grain size causes a driving force for grain-boundary migration,

$$F_3 = 2\left|\Delta\frac{\gamma}{L}\right| = 2\frac{\gamma}{L}\left|\frac{\Delta L}{L}\right| \approx \frac{\gamma}{L} \tag{13.5}$$

where L is the grain size and we assumed, $\Delta L/L \sim \frac{1}{2}$. This driving force is large when grain size is small.

Problem 13.1

Assuming $S = 0.4$, $\mu = 10^{11}$ Pa, $\gamma = 1$ J/m^2, $\Delta S/S = 0.5$ and $\Delta L/L = 0.5$, construct a map on a stress and grain size space that shows the dominant driving force for grain-boundary migration.

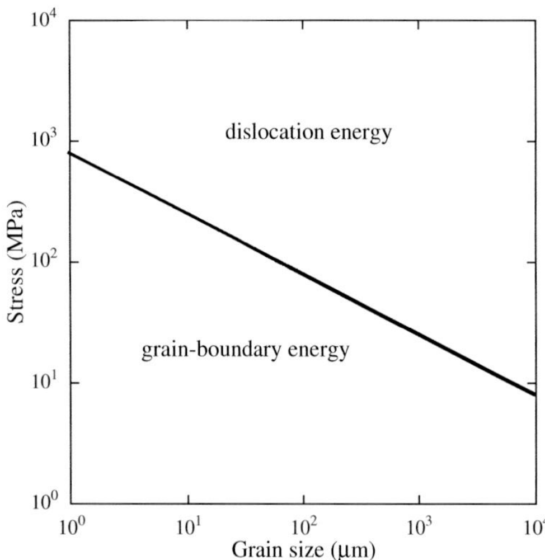

FIGURE 13.1 A diagram showing the stress–grain size region in which two different driving forces for grain-boundary migration dominate.

Solution

Using equations (13.3) and (13.5), the condition at which these two driving forces are equal is given by $|F_1/F_3| = \frac{(\sigma^2 S^2/\mu)|\Delta S/S|}{(\gamma/L)|\Delta L/L|} \approx \frac{\sigma^2 S^2/\mu}{\gamma/L} = S^2/\mu\gamma\sigma^2 L = 1$ (Fig. 13.1). Note that most of the conditions in laboratory experiments (σ =10–10^3 MPa, L= 10–10^2 μm), and in Earth (σ =1–10^2 MPa, L= 10^2–10^4 μm) are close to the boundary and therefore both driving forces play equally important roles.

13.2.2. Mobility of grain boundaries

Grain-boundary mobility is controlled by several factors. They can be classified into (1) intrinsic mobility, (2) impurity drag and (3) secondary phase drag (KINGERY *et al.*, 1976; YAN *et al.*, 1977).

Intrinsic mobility

This refers to the mobility of a clean, impurity-free boundary. It is controlled by the migration of grain-boundary ledges (steps). The mobility of grain boundary is proportional to the number of ledges and their mobility. The number of ledges is sensitive to the crystallographic orientation. Therefore, intrinsic mobility is often highly anisotropic leading to a faceted boundary. When a grain boundary is wetted by a liquid phase, other factors such as the effects of impurities tend to be less important because of their high mobility in the liquid phase, and the intrinsic mobility plays an important role (see DRURY and VAN ROERMUND, 1989).

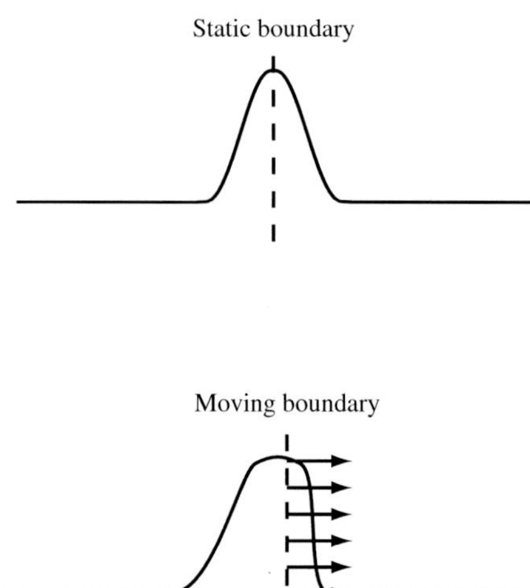

FIGURE 13.2 A schematic diagram showing the distribution of impurity atoms near a grain boundary.

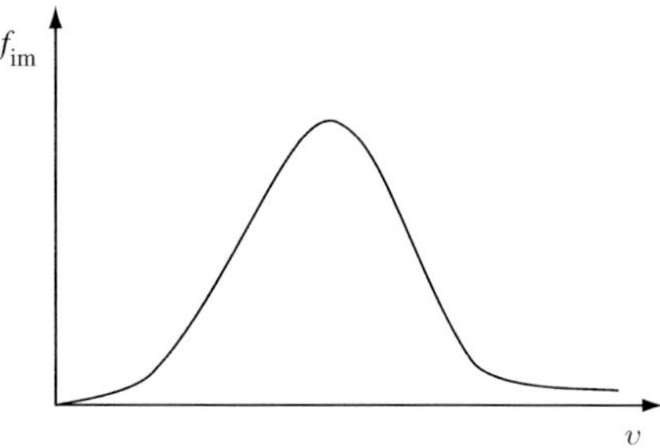

FIGURE 13.3 A relation between the drag force due to impurities on grain boundaries, $f_{im}(v)$ and grain-boundary velocity.

Impurity drag

Impurity atoms often occur on grain boundaries. Their distribution near boundaries is symmetric in the static situation, whereas it becomes asymmetric when boundaries move (Fig. 13.2). TORIUMI (1982) reported asymmetric distribution of impurities around grain boundaries in olivine.

This exerts a drag force that depends on the velocity, $v : f_{im}(v)$. This drag force must have the following properties: When velocity is small, the drag force increases with velocity, v. When velocity is very fast, most of the impurities will be left behind and no appreciable drag will occur. Thus, the force versus velocity curve should look like Fig. 13.4.

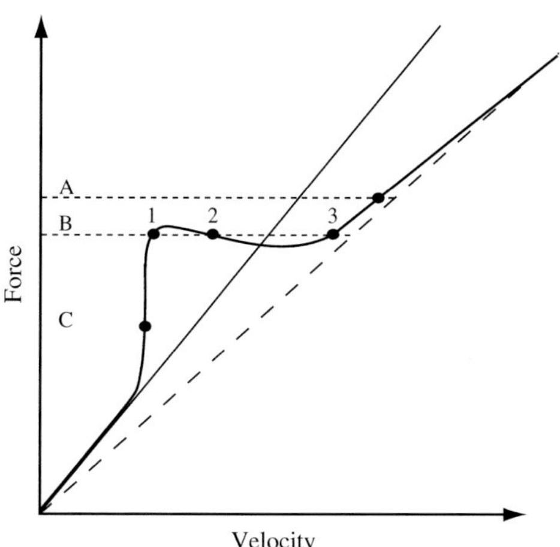

FIGURE 13.4 A force versus grain-boundary velocity curve.

At small velocities (or small driving forces), the grain-boundary mobility is controlled by the diffusion of impurity atoms and hence the drag force is inversely proportional to the diffusion coefficient and proportional to the concentration of impurity atoms. In contrast at high velocities (or large driving forces), impurity atoms tend to be broken away from the moving boundary and the degree to which impurities are broken away is controlled by the diffusion coefficient as well as the interaction potential. In this regime, for smaller diffusion coefficients, more break-away occurs and hence mobility is larger (force is smaller).

Therefore the total drag force is given by,

$$f = f_{im} + f_0 \tag{13.6}$$

where f_0 is the intrinsic drag (due to intrinsic mobility), i.e., $f_0 = v/M_0$. Thus, the force versus velocity diagram for a grain boundary looks like Fig. 13.4.

The behavior of grain-boundary migration can be understood from the graphical solution of equation (13.6) (see Fig. 13.4). Let us consider a case where we increase the force on a grain boundary from level C to A. At a small force, the level C or below, the diagram has only one solution, which is stable (see Problem 13.2). As one increases the force to the level of B, then one has three solutions (1, 2 and 3). Among these three, two are stable (1 and 3) and one (2) is unstable (see Problem 13.2). Again at a level A, there is only one stable solution. Physically, at a low level of force, the grain boundary moves with impurities, the velocity of boundary migration being determined by the mobility of impurities. In contrast, at a high level of force, the grain boundary moves without impurities (impurities are left behind the moving boundaries), the velocity of migration being controlled only by the intrinsic mobility. In between these two regimes, there is a transition from loaded boundary to free boundary (Fig. 13.4). When this occurs, there is a marked increase in grain-boundary mobility (Guillopé and Poirier, 1979).

Problem 13.2

Examine the stability of solutions of equation (13.6) using the graph (Figure 13.4). A solution is stable (unstable) if a small increase in velocity results in an increase (decrease) in force for a grain boundary.

Solution

The basic physics of stability of grain-boundary motion is the same as that of a dislocation that interacts with impurities (Problem 9.1). When $\partial F/\partial v > 0$ (F, force, v, velocity of a boundary), the motion is stable whereas when $\partial F/\partial v < 0$, it is unstable.

Effects of secondary phase particles

Secondary phase particles have a large influence on grain-boundary migration. Effects of secondary phases are similar to those of impurities. When secondary phase particles are present, then grain boundaries either move with them or leave them behind. When the mobility of secondary phases is high, then they move together with a boundary, but the velocity of the boundary is less than that without secondary phases. When the mobility of secondary phases is low and a large enough force acts on a boundary, then the boundary can move leaving secondary phase particles behind. Thus the interaction of secondary particles with grain boundaries is similar to that between the grain boundary and impurity atoms. A detailed account for the case of pore-boundary interaction is given by Brook (1969). An important difference between the role of secondary phase particles and impurities is that the secondary phase particles often change their distribution during annealing (Fan *et al.*, 1998). A detailed discussion of the role of secondary phase on grain growth will be made in section 13.3.2.

Effects of a fluid phase

Let us now consider the effects of a fluid phase. The role of a fluid phase depends on its geometry and chemical properties. Let us first consider a case where a fluid phase assumes a continuous film. Note that this

FIGURE 13.5 The geometry of mass transport through a fluid phase at a grain boundary.

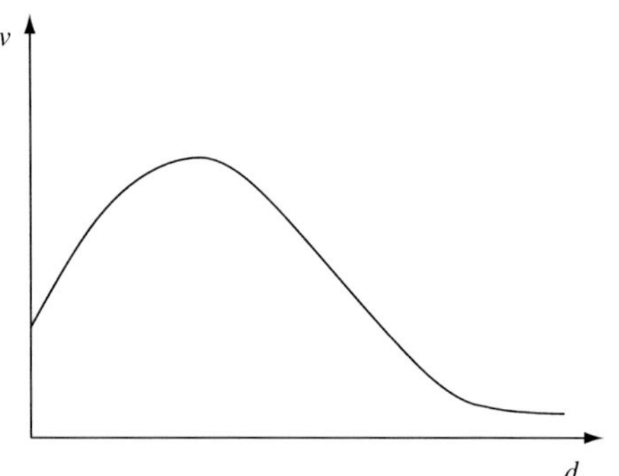

FIGURE 13.6 A plot of the relationship between the grain-boundary velocity and the thickness of a fluid phase, *d*.

geometry occurs only when the fluid phase completely wets the boundary.

In order for materials to be transferred across a grain boundary with a fluid phase, two successive processes are needed, namely the surface reaction to remove an atom from the solid to the fluid film (or vice versa) and diffusion through a fluid (Fig. 13.5). Thus,

$$v = \frac{d}{t_r + t_d} \tag{13.7}$$

where t_r is the reaction time for surface reaction to remove an atom from one boundary (or to attach an atom to another boundary) and t_d; time for diffusion. If $t_d \gg t_r$ then the boundary migration is controlled by diffusion (diffusion-controlled regime), thus $v \sim d/t_d$. Now $t_d \sim d^2/D_l c_l$ where D_l is the diffusion coefficient of material in the liquid phase and c_l is the solubility of solid material in the liquid hence,

$$v \sim \frac{D_l c_l}{d}. \tag{13.8}$$

If $t_r \gg t_d$ then the surface reaction controls the rate of mass transfer, and

$$v \sim \frac{d}{t_r}. \tag{13.9}$$

Thus the grain-boundary velocity versus fluid-layer thickness should look like Fig. 13.6.

A fluid phase may be present in a polycrystalline solid as an isolated pocket. In such a case, the role of a fluid phase in grain-boundary migration is similar to that of secondary phase particles. The difference is that a fluid phase is more mobile than a solid particle, and therefore its role to reduce the grain-boundary mobility is less pronounced.

In summary, a fluid phase enhances grain-boundary migration when it forms a thin continuous layer, whereas it will retard grain-boundary migration when it is thick or when it occurs as isolated pockets. A fluid phase may indirectly influence grain-boundary migration by helping the Ostwald ripening (see section 13.3) of the secondary phase or by dissolving the secondary phase.

13.3. Grain growth

Grain growth is a process whereby grain-boundary energy in a polycrystalline specimen is reduced by eliminating some grain boundaries. This occurs when grain-boundary energy is the main driving force for grain-boundary migration. Therefore grain growth is an important process at low dislocation densities (low stresses) and/or at small grain sizes (see Problem 13.1).

13.3.1. Normal grain growth

Basic theory

The actual process of grain growth is due to the imbalance of grain-boundary forces at grain-boundary junctions. Consider a case of isotropic grain-boundary energy. Let us also consider a two-dimensional model for the easiness of visualization (the essence is the same for three dimensions). Then, the interface tension will be balanced when three grain boundaries meet at 120°. However this is possible only for ideal grains that have a hexagonal shape (on two-dimensional projection; in three dimensions it is dodecahedron). If there are grains other than hexagons, then this force balance does not occur at triple junctions. This imbalance of interfacial forces causes large grains to grow and small grains to shrink, leading to grain growth. The kinetics of grain growth is therefore controlled by the rate of grain-boundary migration and the geometry of grains. Now the velocity of grain-boundary migration caused by grain-boundary energy imbalance is given by,

$$v = 3\gamma M\Delta\left(\frac{1}{L_1} + \frac{1}{L_2}\right) = 6M\gamma\Delta\frac{1}{L} \tag{13.10}$$

where $L_{1,2}$ is the curvature of grain boundary and L is the grain size. Now, let us assume that the rate at which grain size increases is proportional to the velocity of grain-boundary migration. More precisely, the velocity of grain-boundary migration in this case is proportional to the difference in the curvatures of neighboring grains, i.e.,

$$\frac{dL}{dt} = AM\gamma\left(\frac{1}{\bar{L}} - \frac{1}{L}\right) \tag{13.11}$$

where A is a constant of order unity and $\bar{L}$ is the average grain size. This equation means that grains whose sizes are larger than $\bar{L}$ grow at the expense of grains whose sizes are smaller than $\bar{L}$. Consequently, the average grain size $\bar{L}$ will increase with time. The evolution of average grain size can be determined from (13.11) together with a condition for the conservation of mass (e.g., HILLERT, 1965). Briefly, the rate at which average grain size increases is proportional to the rate at which smaller grains disappear. The rate at which small grains disappear is proportional to the velocity of grain-boundary migration that is proportional to $M\gamma(1/\bar{L} - 1/L)$. If the grain-size distribution remains the same, then $M\gamma(1/\bar{L} - 1/L) \propto M\gamma/\bar{L}$ and

$$\frac{d\bar{L}}{dt} = A'M\gamma\frac{1}{\bar{L}} \tag{13.12}$$

and hence

$$\bar{L}^2(t) - \bar{L}^2(0) = k_2 t \tag{13.13}$$

with

$$k_2 = \frac{1}{2}A'M\gamma. \tag{13.14}$$

Note that the material constant k_2 depends strongly on thermochemical conditions mainly through the mobility term, M.

The parabolic grain-growth law is a natural consequence of a process driven by grain-boundary curvature ($dL/dt \propto 1/L$) (e.g., HILLERT, 1965). However, the simple parabolic formula, (13.13), is not always observed experimentally (e.g., ATKINSON, 1988). More generally, the grain-growth kinetics can be written as,

$$L^n - L_0^n = k_n t \tag{13.15}$$

where $n \geq 2$ (hereafter I use L to represent an average grain size for simplicity). A review of experimental studies (ATKINSON, 1988) shows that even in very pure materials, grain-growth kinetics follows (13.13) with $n > 2$. The cause for deviation from a simple model ($n = 2$) is investigated by ATKINSON (1988) and others through numerical modeling (e.g. LOUAT and DUESBERY, 1994; COCKS and GILL, 1996; GILL and COCKS, 1996). The values of n large than 2 are attributed to the violation of the assumption (13.12). One obvious cause is the effect of secondary phase particles in which the grain-growth exponent is $n = 3$ or 4 (as will be discussed below).

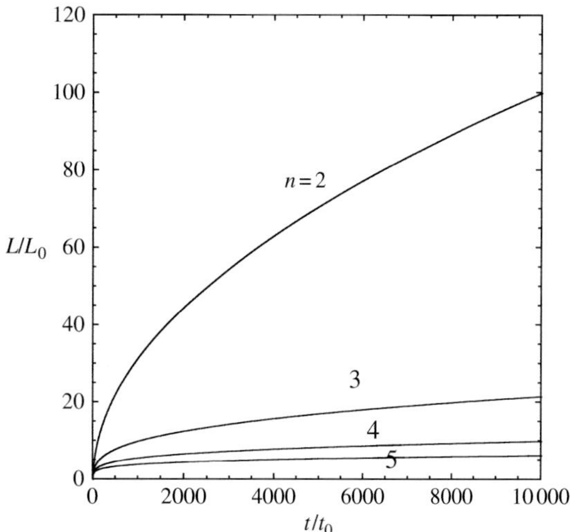

FIGURE 13.7 A plot of the grain size versus the time relation for various n ($t_0 \equiv L_0^n/k_n$).

Problem 13.3

Compare the time dependence of grain size expected from the grain-growth law with $n = 2$, 3, 4 and 5.

Solution

From equation (13.15), one obtains $(L/L_0)^n - 1 = k_n t/L_0^n \equiv t/t_0$, and hence $L/L_0 = (t/t_0 + 1)^{1/n}$. For $L^n \gg L_0^n$, the grain-growth kinetics is simplified to $L \propto t^{1/n}$. Grain growth is more sluggish with a higher value of n (Fig. 13.7).

13.3.2. Effects of secondary phase particles

Zener pinning

Grain growth is often inhibited by secondary phase particles. This effect is strong when the secondary phase particle is immobile. The essential reason for this is that when immobile particles are present at grain boundaries, moving boundaries have to increase their area in order to pass these particles (see Fig. 13.8). Therefore the excess energy is required to cause grain-boundary migration. When the driving force for grain-boundary migration is large (i.e., small grain size), moving boundaries "eat" particles to form mineral *inclusions*. However, when the driving force is small (coarse

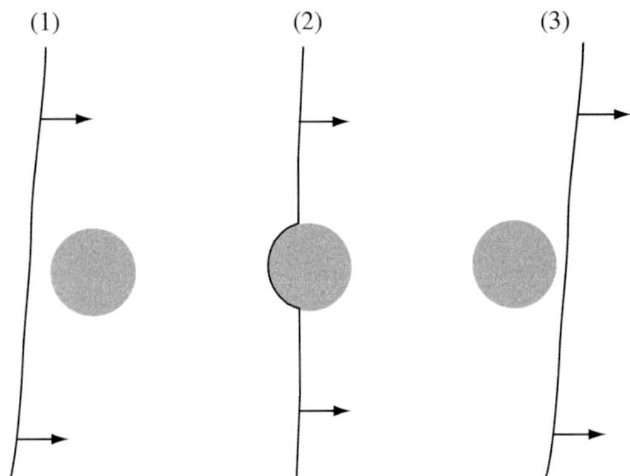

FIGURE 13.8 A grain boundary moving through a stationary particle.

grain size), then secondary phase particles are swept up by grain boundaries and move with them, if they have certain mobility. Finally, when secondary phase particles are totally immobile or the driving force is small, then grain-boundary migration (grain growth) will terminate when the driving force is balanced with the resistance force exerted by the presence of the secondary phase. The effect of the secondary phase particle to terminate the grain growth is called *Zener pinning*.

The terminal grain size controlled by the immobile secondary phase particles can be calculated as follows. The excess energy for a moving grain boundary to consume one particle is $2\pi a^2\gamma - \pi a^2\gamma = \pi a^2\gamma$ (a is the radius of the particle). If the mean distance of particles is H, then the volume fraction f is given by $f = \frac{4}{3}\pi a^3/H^3$. Therefore, the excess energy per unit volume due to secondary phase particles is $\pi a^2\gamma/H^3 = \frac{3}{4}f(\gamma/a)$. If this exceeds the driving force, γ/L, then grain growth will stop. Therefore the terminal grain size is given by,

$$L_Z = \frac{4a}{3f}. \tag{13.16}$$

In summary, the secondary phase particles (including a fluid phase) have important effects on grain-growth kinetics. When they are mobile, their effects are relatively minor. When they are immobile, their effects are large. The magnitude of the effects depends also on the volume fraction and the size of the secondary phase: the larger the volume fraction of secondary particles, the smaller the terminal grain size, and for a given volume fraction, the terminal grain size is smaller for the smaller particle size of the secondary phase.

Ostwald ripening

The average size of the secondary phase particles may, however, increase with time. Consider a case where the

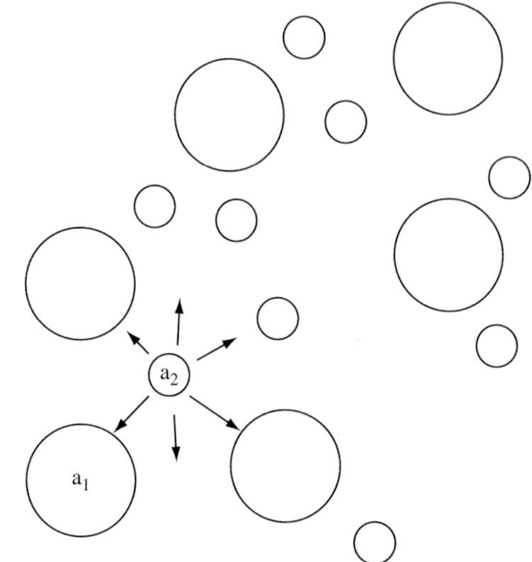

FIGURE 13.9 A schematic diagram showing the mechanism of Ostwald ripening.

secondary phase particles have a certain range of sizes and that the material of the particles has a finite solubility and diffusivity in the matrix. Recall that the solubility of materials in a particle into the matrix depends on the free energy difference of that material between the two phases. Furthermore note that the free energy of a particle depends on the size, because for a particle with a finite size there is a finite pressure caused by the surface tension, $F_3 = \Delta g/V = 4\pi\gamma a^2/\frac{4\pi}{3}a^3 = 3\gamma/a$ (a is the particle size). Therefore the solubility of the material of the particle in the matrix is dependent on the size of the particle as,

$$c(a) = c_0 \exp\left(\frac{3\gamma\Omega}{aRT}\right) \approx c_0\left(1 + \frac{3\gamma\Omega}{aRT}\right). \tag{13.17}$$

Consequently, there is a mass flux from smaller particles to larger ones leading to the growth of average size of secondary phase particles. This process is referred to as the *Ostwald ripening* (Fig. 13.9). The rate at which the size of the secondary particle grows can be calculated using a method similar to the one used in the theory of dislocation climb (see Chapter 10).

To simplify the analysis, let us make an approximation that the rate of grain coarsening is controlled by the rate at which the size of small particles shrinks (KINGERY *et al.*, 1976). This is a reasonable approximation because the change in particle size is more rapid for small particles. The steady-state total mass flux from small particles can be calculated from the steady-state concentration profile by $J^{total} = -4\pi r^2 D(dc/dr)$. The concentration profile must satisfy the Laplace equation in the spherical coordinates, namely,

$$\nabla^2 c = 0 = \frac{1}{r^2}\frac{d}{dr}r^2\frac{d}{dr}c \tag{13.18}$$

with the boundary condition (13.17) at $r=a$ and $c(\tilde{L}) = c_0$ where $\tilde{L}$ ($\gg a$) is the mean distance of the secondary phase particles. The solution of (13.18) with these boundary conditions is,

$$c(r) = c_0\left[1 + \frac{3\gamma\Omega}{RT}\left(\frac{1}{r} - \frac{1}{\tilde{L}}\right)\right]. \tag{13.19}$$

Noting that this flux must be equal to the flux of materials of growing large particles whose size is a, we obtain $J^{\mathrm{total}} = -4\pi r^2 D(dc/dr) = 4\pi a^2(da/dt)$. Therefore

$$a^2\frac{da}{dt} = \frac{3\gamma\Omega D c_0}{RT} \tag{13.20}$$

and hence

$$a^3(t) - a^3(0) = \frac{9\gamma\Omega D c_0}{RT}t. \tag{13.21}$$

Thus the size of secondary phase particles grows with time if there is finite solubility (c_0) and diffusivity (D) of materials of the secondary phase particles in the matrix. At the asymptotic limit ($a(t) \gg a(0)$), $a \propto t^{1/3}$, thus the size of grains controlled by (13.18) changes with time as $L \propto a \propto t^{1/3}$ (e.g., FAN *et al.*, 1998).

The kinetics of grain growth controlled by the Ostwald ripening is determined by the solubility and diffusivity of secondary phase material in the matrix. In contrast, the kinetics of grain growth in a single-phase material is controlled by diffusivity of material across grain boundaries. The latter is usually much faster than the former, and hence the grain-growth kinetics in a multi-phase material is significantly more sluggish than that in a single-phase material. Experimental studies on grain-growth kinetics in a two-phase mixture for Earth materials include (YAMAZAKI *et al.*, 1996; WANG *et al.*, 1999; ZIMMERMAN and KOHLSTEDT, 2004; OHUCHI and NAKAMURA, 2006).

Problem 13.4

Using equation (13.21) show that when the rate of growth of the secondary phase particles is controlled by grain boundary diffusion in the matrix, then $n=4$ in equation (13.15) whereas when it is controlled by volume diffusion $n=3$.

Solution

When grain-boundary diffusion dominates in the transport of secondary phase materials in the matrix, then $D = (\pi\delta/L)D^B$. If the size of grains is controlled by the Zener pinning, then $L = 4a/3f$. Therefore $L^3(dL/dt) = 128\pi\gamma\Omega\delta D^B c_0/27RTf^3$ and integrating one gets, $L^4(t) - L^4(0) = (32\pi\gamma\Omega\delta D^B c_0/27RTf^3)t$.

When volume diffusion dominates, one gets $L^3(t) - L^3(0) = (128\gamma\Omega D c_0/81RTf^3)t$.

The grain-size distribution in the steady-state, normal grain growth remains constant with time (by definition). A model by HILLERT (1965) leads to a steady-state grain-size distribution

$$P(x) = (2e)^3\frac{3x}{(2-x)^5}\exp\left(-\frac{6}{2-x}\right) \quad (0<x<2) \tag{13.22}$$

where $x \equiv L/L_c$, L is grain size, L_c is the critical grain size ($\sim$ average grain size) above which a grain will grow, $P(x)\,dx$ is the probability of finding a grain with size between x and $x+dx$. In this model, grains with a size larger than $2L_c$ do not exist during normal grain growth.

Problem 13.5

Grain-growth kinetics in wadsleyite (at $P = 15$–16 GPa) is described approximately by the following equation (NISHIHARA *et al.*, 2006)

$$L^2(t) - L^2(0) = A\cdot C_{\mathrm{OH}}^{1.7}\exp\left(-\frac{H^*}{RT}\right)t$$

where $A = 10^{-18}$ m^2/s, $H^* = 120$ kJ/mol and C_{OH} is the content of OH in ppm H/Si (at a fixed oxygen fugacity of 10^4 Pa). Calculate the critical temperature above which grain size grows to $\sim$1 mm in a typical slab in the transition zone. Consider a range of water content, $C_{\mathrm{OH}} = 100$–$100\,000$ ppm H/Si, and assume that the initial grain size is much smaller than 1 mm. Discuss what conditions are needed to maintain the grain size smaller than 1 mm for a significant portion (say larger than $\sim$100 km) of a slab subducting with a velocity of 10 cm/yr.

Solution

Because $L(t) \gg L(0)$, we can make an approximation $tC_{\mathrm{OH}}^{1.7}\exp(-H^*/RT) \approx L^2(t)/A$, thus $\log_{10} t = -1.7\cdot\log_{10} C_{\mathrm{OH}} + 0.434\frac{H^*}{RT} + 2\log_{10}L(t) - \log_{10}A$. For the subduction velocity of 10 cm/y, it takes 1 Myr for a slab to move 100 km. Therefore we choose $t = 1$ Myr, and calculate the temperature needed to increase the grain size to 1 mm in 1 Myr. From the above grain-growth law, the conditions under which grain size smaller than 1 mm is maintained are $-1.7 \cdot \log_{10} C_{OH}(\mathrm{ppm}\frac{H}{Si}) + \frac{6270}{T(K)} > 1.5$. For a relatively dry slab where $C_{OH} \sim 100$ ppm H/Si, the critical temperature is 1280 K, whereas for a wet slab where $C_{OH} \sim 100\,000$ ppm H/Si, the critical temperature is 630 K. The temperatures in subducting slabs depend on the initial temperature (before subduction) and the velocity of subduction. For cold, fast subducting slabs, the temperature at the coldest (central) region of the slab at ~500 km depth is ~900 K. Therefore if a slab is nearly saturated with water, grain growth is almost instantaneous and it is impossible to maintain small grain size whereas if a slab is dry (less than ~100 ppm H/Si), then a small grain size can survive for a considerable length in its cold regions ($T < 1200$ K).

13.3.3. Abnormal grain growth

In many cases, the statistical balance maintains the shape of grain-size distribution, and only the average grain size increases. This type of grain growth is referred to as *normal grain growth*. In contrast, under some conditions, a small number of grains grow at the expense of small ones and grain-size distribution changes with time. This is referred to as *abnormal grain growth*. In such a case, a material will be composed of bimodal grain-size distribution: a fine-grained matrix and coarse-grained crystals. HILLERT (1965) showed that abnormal grain growth occurs when the following conditions are met simultaneously: (1) normal grain growth cannot take place due to the presence of secondary phase particles, and (2) there is at least one grain much larger than the average grains.

13.3.4. Effect of stress: the limiting grain size

So far, we have neglected the effects of stress (deviatoric stress) on grain growth. If one considers the finite rate of Ostwald ripening, then there is no ultimate limit for grain size and consequently the grain size in a static environment will be controlled by the time and temperature (and other thermodynamic factors). However, when deviatoric stress is present, even when plastic flow and resultant dynamic recrystallization (see the next section) are ignored, there is a mechanism by which grain size is limited to a certain level. This is due to the effects of the stress field. The energy due to elastic deformation caused by strain mismatch is dependent on the size of secondary phase particles. Consequently stress can affect the grain-growth kinetics and could limit the grain size at a finite level. JOHNSON (1984) and JOHNSON and CAHN (1984) investigated a related subject, but the role of stress on grain-growth kinetics has not been investigated in any detail (see also VOORHEES, 1985, 1992).

13.3.5. Some examples

Some of the representative results are summarized in Table 13.1. An experimental study of grain growth is straightforward. One prepares a fine-grained starting material and anneals it at high temperature and hydrostatic conditions for a certain period. The change in grain size as a function of time, temperature and other physical and chemical conditions is determined that gives a functional relation of grain-growth kinetics. Note, however, that most of the results of grain-growth experiments at room pressure have a strong influence of cavitation (KARATO, 1989b) and are not applicable to geological problems. All the results shown in Table 13.1 are results of high-pressure experiments.

Fig. 13.10 shows a typical example of results of normal grain growth (for quartz; TULLIS and YUND, 1982). Grain size increases with annealing time and temperature. In quartz, the grain-growth kinetics increases with pressure. The authors interpreted this result to indicate that the grain-growth kinetics is controlled by the solubility of quartz in the interstitial fluid. Grain-growth kinetics is also sensitive to other parameters such as oxygen fugacity and water content. Fig. 13.11 shows results on wadsleyite in which the dependence of grain-growth kinetics on oxygen fugacity, temperature and water content was investigated in detail (NISHIHARA *et al.*, 2006b). Less detailed work was made for other minerals such as olivine (Fig. 13.12). Note that most of the studies are on single-phase materials and very little is known on the grain-growth kinetics in multi-phase systems. Also, although water is known to have a large effect on grain-growth kinetics, detailed studies on the influence of water are limited. Consequently, the applicability of the presently available experimental data to real geological problems is limited.

It is important to note that the results on grain-growth kinetics for single-phase materials probably

TABLE 13.1 Experimental results on grain-growth kinetics in some minerals. Parameters are for equation (13.15), $L^n - L_0^n = k_n t$, with $k_n = k_{n0} C_{OH}^r \exp(-H^*/RT)$ for "wet" case where the water content is known (C_{OH}, water content, in ppm H/Si), but $k_n = k_{n0} \exp(-H^*/RT)$ is used when water content is not known (the values of r was determined only in NISHIHARA *et al.* (2006b)). In all other studies under "wet" conditions, samples contained water close to the saturation limit at the experimental conditions, but the value of r was not determined. k_{n0} also depends on other chemical parameters such as oxygen fugacity. (The values of k_{n0} for wadsleyite are for the Ni-NiO buffer.) Units: T (K), P (GPa), k_{n0} ($(\mu m)^n s^{-1}$), H^* (kJ/mol).

Material	T	P	n	r	$\log_{10}k_{n0}$	H^*	Remarks	Ref.
Quartz	1073–1273	0.5–1.5	~2	–	2	80	"Wet"	(1)[3]
Anorthite	1373–1623	0.0001	2.6	–	11	365	"Dry"	(2)[4]
Olivine	1473–1573	0.3	~2	–	4.2	160	"Wet"	(3)
Olivine	1573	2	~2	–	4.6	200[1]	"Dry"	(3)
Wadsleyite	1450–1673	15	~2	1.7	–6	120	"Wet"	(4)
Wadsleyite	1773–2173	15	~2	–	7.5	410	"Dry"	(4)
Ringwoodite	1473–2023	21	4.5	–	–8	414	"Damp"[2]	(5)
Perovskite (+MgO)	1573–2173	25	10.6	–	–45	320	–	(6)[5]
Olivine + diopside	1473	1.2	4	–	–	–	"Dry"	(7)[6]
Lherzolite	1373–1523	0.3	3	–	5.4	700	"Dry"	(8)[7]

[1] Estimated from the activation energy for grain-boundary mobility.
[2] Samples contained variable amounts of water (~200–450 ppm H/Si).
[3] In reference (1), the authors analyzed the data using $L - L_0 = k_n t^n$ with $k_n - k_{n0} \exp(\ H^*/RT)$ rather than (13–15). The use of this form of grain-growth law is not appropriate, and I have recalculated the parameters from reference (1) using equation (13–15).
[4] Synthetic samples prepared from a glass.
[5] No report on water content.
[6] Synthetic samples. All experiments were conducted at $T = 1473$ K and therefore activation enthalpy was not determined. The kinetic parameters depend on the mixing ratio. The values shown here are for $Fo_{70}Di_{30}$. For $n = 4$, $\log_{10} k\left((\mu m)^4 s^{-1}\right) = 0.29$.
[7] 62% olivine, 26% orthopyroxene, 10% clinopyroxene, 2% spinel; less than 30 ppm H/Si.
(1): TULLIS and YUND (1982).
(2): DRESEN *et al.* (1996).
(3): KARATO (1989b).
(4): NISHIHARA *et al.* (2006b).
(5): YAMAZAKI *et al.* (2005).
(6): YAMAZAKI *et al.* (1996).
(7): OHUCHI and NAKAMURA (2006).
(8): ZIMMERMAN and KOHLSTEDT (2004).

FIGURE 13.10 The evolution of grain size of quartz with annealing time and temperature (TULLIS and YUND, 1982). (A), (B), (C): novaculite (A: initial sample, B: 172 hour, C: 796.5 hour annealing) (D), (E), (F): flint (D: initial sample, E: 172 hour, F: 796.5 hour annealing)

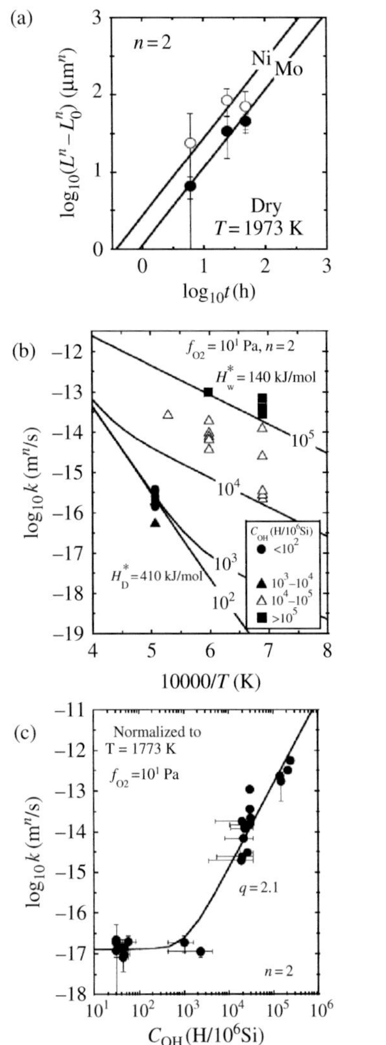

FIGURE 13.11 The grain-growth kinetics in wadsleyite. (a) The grain size versus time relation, (b) the temperature dependence of the grain-growth parameter, and (c) the water content dependence of the grain-growth parameter (NISHIHARA *et al.*, 2006).

provide a gross overestimate of actual grain growth in multi-phase materials. Also note that except for wadsleyite for which the influence of water is studied in detail, there is no experimental work in which the influence of water is quantified. Fig. 13.13 provides a rough estimate of the influence of secondary phase particles on grain-growth kinetics in peridotite. It is also noted that the influence of chemical environment such as water fugacity (water content) is not always well investigated, although effects of chemical environment, particularly that of water content, are likely to be very large (see NISHIHARA *et al.*, 2006).

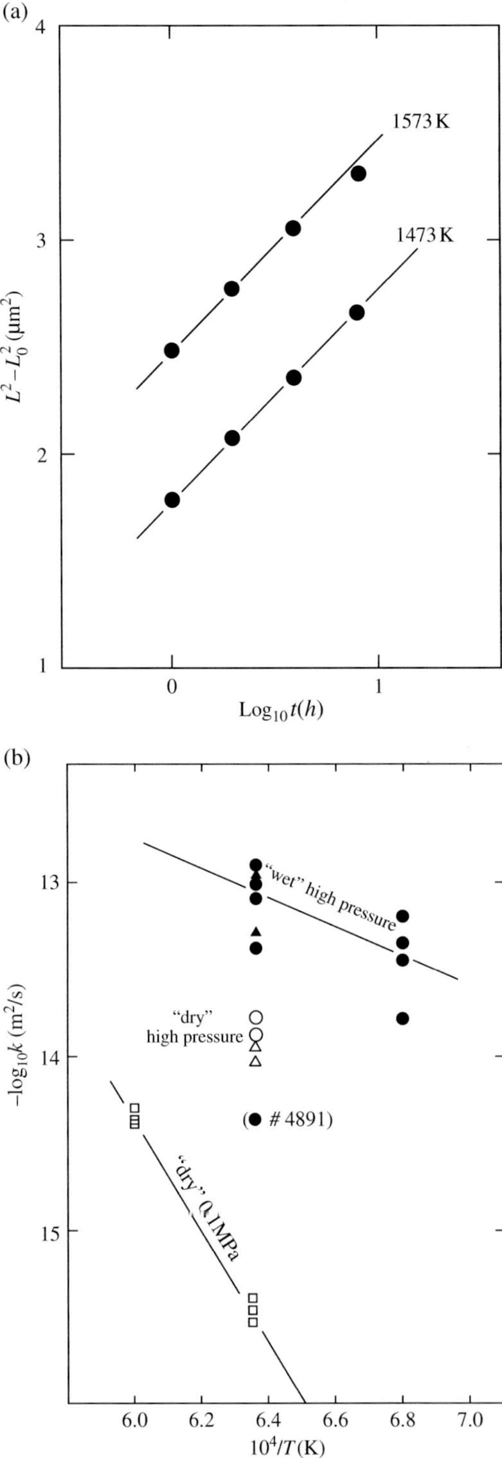

FIGURE 13.12 Grain-growth kinetics in olivine as a function of (a) time and (b) temperature (and water content) (KARATO, 1989b).

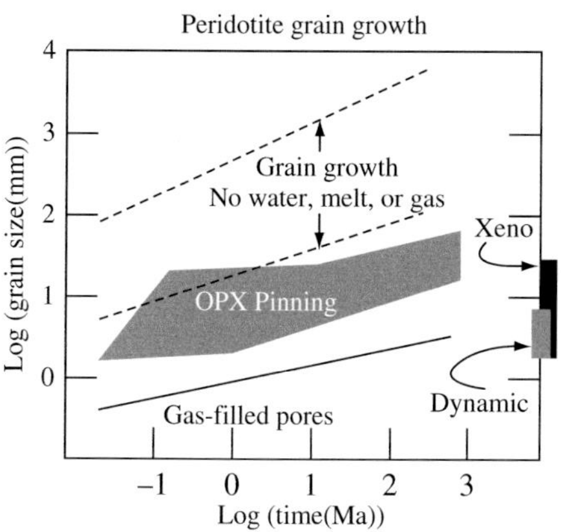

FIGURE 13.13 The grain size versus time relation for peridotite at 1523 K as inferred from laboratory data on grain-growth kinetics. The influence of pressure or water is not considered. The influence of Opx pinning is estimated through a model that assumes that the kinetics of grain growth is controlled by grain-boundary diffusion of Si on olivine boundaries (after EVANS *et al.*, 2001).

13.4. Dynamic recrystallization

Grain growth, as we have just discussed, occurs when the dominant driving force for grain-boundary migration is grain-boundary energy. This would be the case for low stress and/or small grain-size conditions (Problem 13.1). In contrast, under high stress and/or coarse grain-size conditions, dislocation energy dominates the driving force for grain-boundary migration. Grain-scale microstructural reorganization caused by dislocations is loosely called *dynamic recrystallization*. This includes formation of new (high-angle) grain boundaries and mobilization of pre-existing grain boundaries. As we will learn in this section, similar to grain growth where *heterogeneity* in grain size is the driving force, it is the *heterogeneity* in dislocation distribution that causes dynamic recrystallization: if dislocation distribution were perfectly homogeneous none of the processes referred to as dynamic recrystallization would occur. Consequently, dynamic recrystallization occurs easily in materials where plastic anisotropy is strong. In such a material, dislocation density tends to be highly heterogeneous because the dislocation density strongly depends on the orientation of individual crystals. Most minerals are highly anisotropic and therefore dynamic recrystallization is commonly observed in many minerals. For similar reasons, dynamic recrystallization occurs preferentially near grain boundaries where strain heterogeneity is large.

Dynamic recrystallization has important effects both in the strength of rocks and in the microstructure of rocks. This section provides a brief summary of experimental observations and theoretical models of dynamic recrystallization.

13.4.1. Experimental observations

Microstructural observations

(1) Subgrain rotation (Fig. 13.14)

In some materials, deformation by dislocation creep at high temperatures leads to the formation of subgrain boundaries (subboundaries) (Fig. 13.14a). The size of subgrains is uniquely determined by stress as,

$$\frac{L_S}{b} = K\frac{\mu}{\sigma} \tag{13.23}$$

where L_S is the subgrain size, K is a constant ($\sim$20). With progressive deformation, misfit angles of subboundaries may increase (Fig. 13.14b). When misfit angles reach a certain critical angle ($\sim$10–20 degrees), subboundaries evolve into normal grain boundaries because dislocations in subboundaries lose their identity when their distance becomes comparable to the length of the Burgers vector. In these cases, one can observe microstructural evolution from subboundaries (low-angle boundaries) to grain boundaries (high-angle boundaries) with progressive strain (Fig. 13.14a). Usually, this evolution can also be seen in a given specimen: evolution to recrystallized grains occurs more rapidly near existing grain boundaries (because of large strain gradient) than the central portion of pre-existing grains.

The size of subgrains is nearly the same as the size of recrystallized grains. But, more precisely, the size of recrystallized grains evolves with time: the size of the recrystallized grains usually increases with the progression of dynamic recrystallization and remains constant after a certain strain (or time).

(2) Grain-boundary migration

Polycrystalline materials deformed by dislocation creep often show grain-boundary migration. This occurs due to the heterogeneity of dislocation density. Boundaries migrate towards grains with higher dislocation densities and portions of crystals behind migrating boundaries have little dislocations. Dislocation distribution is highly heterogeneous in materials undergoing migration recrystallization. Grain boundaries are often serrated. Subgrain rotation and grain-boundary migration may

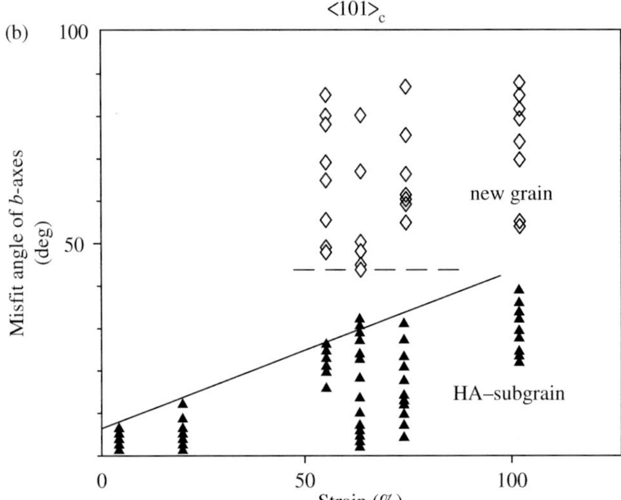

FIGURE 13.14 Recrystallization due to subgrain rotation. (a) Microstructure of a deformed peridotite (from Avachinsky Volcano, Kamchatka) showing the development of subgrains in olivine (seen as a slightly different color), (b) the relation between the misfit angles between subgrains and the strain (TORIUMI and KARATO, 1985) Plate 1.

not be totally independent processes. Subgrain rotation can increase the local misfit angles and hence increases the contrast in dislocation density in neighboring grains. At a certain point, the contrast in dislocation density will become large enough to initiate grain-boundary migration.

(3) *Recrystallized grain size versus stress relation*

After certain strain (usually ~10–100%), dynamically recrystallizing materials assume "steady-state" microstructure. The size of grains at this stage is determined primarily by applied stress and only weakly dependent on other parameters such as temperature, pressure or water fugacity (e.g., KARATO *et al.*, 1980; VAN DER WAL *et al.*, 1993). Therefore dynamically recrystallized grain size can be used to infer stress at which a given rock was deformed: such a relation is therefore called a *paleopiezometer*.

GUILLOPÉ and POIRIER (1979) showed, however, that the grain size versus stress relation depends on the mechanisms of recrystallization. Similarly, JUNG and KARATO (2001a) found that the size of dynamically recrystallized grains in olivine depends on water content.

Evolution of dynamically recrystallized grain size during deformation in dunite was studied by ROSS *et al.* (1980). According to ROSS *et al.* (1980), the size of recrystallized grains does not change significantly with progressive recrystallization. The size of grains at an early stage of recrystallization is similar to that at a later stage. However, SANDSTRÖM (1977) and KARATO *et al.* (1980) noticed some increase in grain size from the earlier stage to final steady state.

TABLE 13.2 The relation between the size of dynamically recrystallized grains and stress. A relation $L_g = A_g\sigma^{-r_g}$ is used. Units: T (K), P (GPa), A_g (μm · MPa^{r_g}).

	T	P	$\log_{10}A_g$	r_g	Remarks	Ref.
Ni	1035–1552	RP[1]	4.33	1.33	–	(1)
Mg	673–873	RP	2.74	1.28	SGR[2] + GBM[2]	(2)
NaCl	523–1063	RP	2.45	1.18	SGR	(3)
NaCl	523–1063	RP	3.95	1.28	GBM	(3)
$NaNO_3$	293–573	RP	2.82	1.54	SGR	(4)
$NaNO_3$	293–573	RP	3.40	1.54	GBM	(4)
Calcite	873–1273	0.3	3.22	1.11	SGR	(5)
Calcite	873–1273	0.3	3.75	1.11	GBM	(5)
Quartz	973–1173	1.5	1.89	0.61	Regime 1[3]	(6)
Quartz	1173–1373	1.5	3.56	1.26	Regime 2 & 3[3]	(6,7)[4]
Feldspar	1173	1.5	1.74	0.66	Regime 1[3]	(8)[5]
Olivine	1773–1923	RP	3.92	1.18	Dry, SGR	(9)[6]
Olivine	1473–1573	2.0	4.41	1.18	Wet, MR	(10)[6]
Olivine	1473–1573	0.3	4.14	1.33	Dry/wet	(11)[7]

(1): Luton and Sellars (1969).
(2): Drury *et al.* (1985).
(3): Guillopé and Poirier (1979).
(4): Tungatt and Humphreys (1984).
(5): Rutter (1995).
(6): Stipp and Tullis (2003).
(7): Stipp *et al.* (2006).
(5): Karato *et al.* (1980).
(6): van der Wal *et al.* (1993).
(7): Jung and Karato (2001a).
(8): Post and Tullis (1999).
(9): Karato *et al.* (1980).
(10): Jung and Karato (2001a).
(11): van der Wal *et al.* (1993).
[1] RP indicates room pressure.
[2] SGR refers to subgrain rotation recrystallization and MR migration recrystallization.
[3] "Regimes 1, 2 and 3" refer to three different deformation conditions defined by Hirth and Tullis (1992) based on microstructures.
[4] Stipp *et al.* (2006) investigated the influence of water for ~260 to ~640 ppm H/Si and found little water effect in this range.
[5] ~0.1 wt% water.
[6] "Dry" samples have less than ~10 ppm H/Si, whereas "wet" ones have ~800–1400 ppm H/Si.
[7] "Dry" samples have less than ~100 ppm H/Si, and "wet" ones ~300 ppm H/Si (in the olivine lattice), and no significant dependence on water content was observed.

At a steady state,

$$L_g = A_g\sigma^{-r_g} \tag{13.24}$$

where A and r are constants that are not very sensitive to temperature or pressure etc. It must be noted that the sensitivity of dynamically recrystallized grain size on parameters other than stress has not been investigated in any detail. Theoretical models (see later) suggest some dependence of recrystallized grain size on other thermodynamic parameters such as temperature and water fugacity. The representative results on minerals are summarized in Table 13.2.

(4) Other effects

Another obvious effect is resetting of grain shape. When dynamic recrystallization occurs, grain shape no longer follows macroscopic strain.

Dynamic recrystallization involving grain-boundary migration can modify lattice-preferred orientation (LPO) whereas dynamic recrystallization due to subgrain rotation does not have an important influence on LPO (KARATO, 1987b, 1988).

Mechanical aspects

(1) Weakening associated with grain-size reduction

When the degree of grain-size reduction is large enough, then a deformation mechanism could change to a grain-size sensitive one, causing weakening. The association of large shear with small grain sizes in many crustal shear zones (mylonites) suggests that this may be an important mechanism of shear localization.

There have been some reports showing weakening due to dynamic recrystallization in silicates presumably due to grain-size reduction (POST, 1977; ZEUCH, 1982, 1983; TULLIS and YUND, 1985).

(2) Weakening due to grain-boundary migration

There is clear evidence in metals (e.g., Ni) where the work hardening effect is eliminated by fast grain-boundary migration (LUTON and SELLARS, 1969). Single peak stress–strain curves or oscillation in stress strain curves are seen to be dependent upon deformation conditions (SAKAI and JONAS, 1984). This type of weakening occurs only when grain-boundary migration is significant and work hardening is important. The role of work hardening in minerals has not been well investigated, but is likely to be not as important as in metals because glide mobility of dislocations is much less in minerals than in metals as a result of high Peierls stresses.

13.4.2. Models for dynamic recrystallization

The experimental observations summarized above suggest the importance of two distinct processes: migration of grain boundaries and their mutual collisions, and the formation of subgrain boundaries and their evolution to high-angle grain boundaries.

Now having summarized some of the important experimental observations, let us discuss physical mechanisms by which this microstructural reorganization

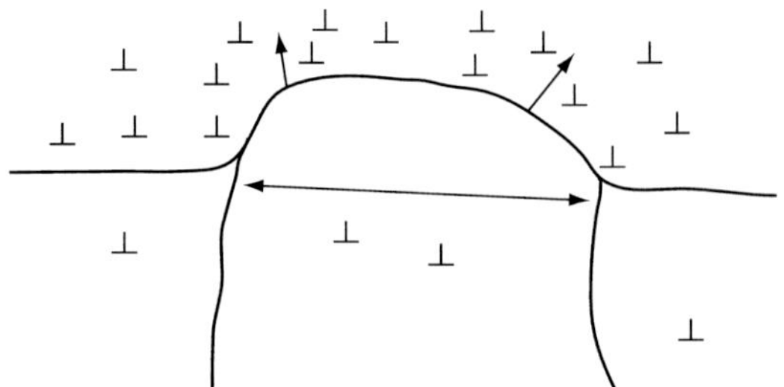

FIGURE 13.15 Bulging of a grain boundary.

occurs. Two processes are of particular importance: mobilization of a pre-existing grain boundary and the formation of subgrain boundaries.

Strain-induced grain-boundary migration and impingement

A contrast in dislocation density provides a driving force for grain-boundary migration. A grain boundary moves to a direction of higher dislocation density of grains. One point that needs to be noted is that grain-boundary migration in a real material occurs heterogeneously: migration occurs in a segment between pinning points. This means that a moving boundary must increase its boundary area which requires extra energy (see Fig. 13.15). Therefore one needs a finite amount of dislocation density contrast to overcome this energy barrier. Since the boundary energy is proportional to the square of the size of the bulge (Fig. 13.15), and the bulk dislocation energy is proportional to the volume of the bulge (l^3), the critical condition for the migration is

$$\Delta E = \mu b^2 \Delta\rho > \beta\frac{\gamma}{l} \quad (13.25)$$

where β is a constant of order unity. Consequently, deformation-induced grain-boundary migration (grain-boundary migration caused by dislocation density contrast) occurs only after a certain dislocation density contrast is established. Once this critical condition is met, a portion of grain boundary starts to move. After a grain boundary moves, then a region behind a moving boundary will be dislocation-free and therefore the driving force for boundary migration will increase. Also as the moving boundary bulges out further, its size gets larger hence the excess energy due to boundary energy will be reduced. Therefore the grain-boundary migration will be accelerated once it starts. Consequently moving boundaries keep moving until they collide with each other.

At a steady state, the dynamic balance between nucleation of new grains (i.e., the achievement of critical strain to establish the condition (13.25)) and grain growth determines the grain size. During steady state, the balance between nucleation and growth implies that in a time taken for a recrystallizing grain boundary to sweep a volume of a grain of diameter L, there will be on average one nucleation event per equivalent volume.

Let t_1 be the time to sweep out the volume of radius $L/2$. Then

$$t_1 = \frac{L}{2v} \tag{13.26}$$

where v is the velocity of grain-boundary migration. Now if $\dot{N}_{\bar{V}}$ is the nucleation rate (per unit volume), then the time interval of nucleation in a volume unit of radius $\bar{L}$, t_2, is given by,

$$t_2 = \frac{1}{\frac{4}{3}\pi(L/2)^3 \dot{N}_V} = \frac{6}{\pi L^3 \dot{N}_V} \quad \text{(for volume nucleation).} \tag{13.27a}$$

Similarly, for boundary nucleation ($\dot{N}_A$: the nucleation rate per unit area),

$$t_2 = \frac{1}{\pi(L/2)^2 \dot{N}_A} = \frac{4}{\pi L^2 \dot{N}_A} \quad \text{(for boundary nucleation).} \tag{13.27b}$$

At statistical balance, $t_1 = t_2$, and therefore,

$$L = \left(\frac{12v}{\pi \dot{N}_V}\right)^{1/4} \quad \text{(for volume nucleation)} \tag{13.28a}$$

and

$$L = \left(\frac{8v}{\pi \dot{N}_A}\right)^{1/3} \quad \text{(for boundary nucleation).} \tag{13.28b}$$

In this model, the size of dynamically recrystallized grains is related to two microscopic processes namely grain-boundary migration and nucleation of new grains. Up to this point, the model is quite general and as we will see in the next section, this type of model can also be applied to other processes involving nucleation and growth. However, in order to obtain a specific relationship between the size of grains and the conditions at which the nucleation and growth occur, we need a model specific to a particular physical process. Here let us consider the nucleation and growth associated with dynamic recrystallization involving grain-boundary migration. The velocity of grain-boundary migration is proportional to the mobility and driving force. Derby and Ashby (1987) considered that the major driving force is grain-boundary energy, namely,

$$v = MF \propto \frac{M}{L}. \tag{13.29}$$

The nucleation rate can be estimated by calculating the rate at which the condition (13.25) is met. The rate at which energy difference is created by deformation is obviously proportional to strain rate, but it may also depend on the grain size. Thus, the nucleation rate is given by,

$$\dot{N}_{A,V} \propto \frac{\dot{\varepsilon}}{L^q} \tag{13.30}$$

where a constant q represents the dependence of nucleation rate on grain size. Thus in general one has

$$L \propto \left(\frac{M}{\dot{\varepsilon}}\right)^{1/(m-q+1)} \tag{13.31}$$

where $m = 4$ for volume nucleation and $m = 3$ for boundary nucleation. Assuming the power-law creep, $\dot{\varepsilon} = \dot{\varepsilon}_0 \exp(-H^*_{\dot{\varepsilon}}/RT)\sigma^n$, and the temperature (and pressure) dependence of boundary mobility, $M = M_0 \exp(-H^*_M/RT)$, this equation can be translated to

$$L(T,P,C,\sigma) = A(T,P,C)\sigma^{-n/(m-q+1)} \tag{13.32}$$

where $A(T,P,C) \propto (M_0/\dot{\varepsilon}_0)^{1/(m-q+1)} \exp\left((H^*_{\dot{\varepsilon}} - H^*_M)/(m-q+1)RT\right)$ (T, temperature; P, pressure; C, parameters that characterize chemical environment, i.e., water fugacity). The stress exponent of grain size in this model is $-n/(m-q+1)$. In most cases, grain-boundary nucleation dominates, and therefore this translates to $L \propto \sigma^{-n/(4-q)}$. For olivine, the stress exponent is ~ -1.2 (Karato *et al.*, 1980; van der Wal *et al.*, 1993; Jung and Karato, 2001a) and $n = 3$–3.5 and therefore we get $q \sim 1.1$–1.5.

An important point to be noted is that according to this model, the size of dynamically recrystallized grains depends not only on stress but in general also on other thermodynamic variables that affect the kinetic processes (grain-boundary migration, plastic deformation) involved in dynamic recrystallization. In many cases, $H^*_{\dot{\varepsilon}} - H^*_M > 0$, and therefore grain size probably decreases with increasing temperature. This effect is observed in a magnesium alloy (De Bresser *et al.*, 1998). Similarly when water enhances grain-boundary migration more than deformation, then the size of a dynamically recrystallized grain increases with water content (Jung and Karato, 2001a). However, the details of the dependence of recrystallized grain size on temperature and other thermochemical parameters have not been well investigated.

A somewhat different model was proposed by De Bresser *et al.* (1998). Noting that grain growth occurs in the diffusion creep regime to bring grain size large enough for dislocation creep to occur, they postulated that the grain size in these cases is determined by the balance of strain rates in two deformation mechanisms, i.e., $\dot{\varepsilon}_{\text{diff}}(T, P, \sigma, L) = \dot{\varepsilon}_{\text{disl}}(T, P, \sigma)$. Although the interplay of diffusion and dislocation creep is important as first noted by Rutter and Brodie (1988), one can imagine a case in which dynamic recrystallization leads to a grain size and stress condition that is still definitely in the dislocation creep regime. In such a case, obviously this model does not apply. It is also important to recognize that the balance between diffusional and dislocation creep is different from the microscopic balance between nucleation and growth (impingement) as postulated by Derby and Ashby (1987). The latter occurs at the grain level that may determine the size of grains at the steady state of dynamic recrystallization, and has nothing to do with diffusional creep itself. The former type of balance concerns processes after the completion of dynamic recrystallization. It is important to distinguish these two levels of statistical balance. The difference in time-scale between these processes is critical when one evaluates the influence of these processes in strain-localization (see Chapter 16).

Subgrain rotation and subgrain growth

In an idealistic case, deformation by dislocation motion occurs through the motion of homogeneously distributed dislocations. However, because of strong long-range interaction between dislocations, homogeneous distribution is often unstable and dislocations tend to assume certain structures. The one that is often observed at high temperatures is subgrain-boundary (subboundary) formation in which dislocations with the same sign and type tend to arrange themselves into a planar well-organized array called subboundaries (Fig. 13.14a) and between these boundaries the density of dislocations is less. Dislocations in subboundaries are less mobile and hence they do not contribute to deformation. Subboundaries are characterized by the misfit angles that are determined by the density of dislocations in them. Under some conditions, the misfit angles of subboundaries increase with strain leading to the formation of high-angle grain boundaries. Subboundary formation is enhanced by heterogeneous deformation. In realistic heterogeneous deformation there will be regions where dislocations with one sign may occur more than others. These are called *geometrically necessary dislocations* (Ashby, 1970). Once one has an excess population of dislocations with one sign, then these dislocations interact each other to form subboundaries to reduce the free energy.

Holt (1970) considered the fluctuation in dislocation distribution and postulated that the size of fluctuation with the maximum growth rate corresponds to the size of subgrains. His model assumes heterogeneous distribution of (screw) dislocations without external force, and considers dislocation motion caused by the mutual interaction whose strength is inversely proportional to the distance of dislocations. It is also assumed that there is no dislocation multiplication or annihilation. When homogeneous distribution of dislocation is modified, then the energy caused by mutual interaction will also be modified which drives the dislocation flux. The growth of fluctuation of dislocation density is analyzed and shown to be dependent on the wavelength of fluctuation in dislocation density. He postulates that the wavelength at which the growth rate is the maximum corresponds to the size of subgrains. Kubin (1993) presented a similar but more sophisticated model in which the dislocation cell structure is treated as a result of self-organization in non-equilibrium processes. Shimizu (1998) considered a different situation in which the size of recrystallized grains is controlled by the growth of subgrains.

These models predict,

$$\frac{L}{b} = K' \left(\frac{\mu}{\sigma}\right)^{r_{sg}} \tag{13.33}$$

where K' and r_{sg} are constants depending on the details of the model. See also Sherby *et al.* (1977), Amadeo and Ghoniem (1988) and Selitser and Morris (1994). Since the physical basis for grain size is different between subgrain rotation and grain-boundary migration mechanisms of recrystallization, the stress versus grain size relationship is different between these cases (Guillopé and Poirier, 1979).

13.4.3. Application of paleopiezometers

Some microstructural parameters such as dynamically recrystallized grain size and dislocation density (see Chapter 5) are dependent mainly on (deviatoric) stress and can serve as stress indicators. Since these microstructures can be used to infer the magnitude of deviatoric stress in the geological past they are referred to as *paleopiezometers* (e.g., Nicolas, 1978). Three of them have been used in the geological community:

dislocation density, ρ, subgrain size, L_s, and dynamically recrystallized grain size, L_g, namely,

$$\rho = A_1 b^{-2}\left(\frac{\sigma}{\mu}\right)^{n_1} \tag{13.34a}$$

$$L_s = A_2 b\left(\frac{\sigma}{\mu}\right)^{-n_2} \tag{13.34b}$$

and

$$L_g = A_3 b\left(\frac{\sigma}{\mu}\right)^{-n_3} \tag{13.34c}$$

where $A_{1,2,3}$ and $n_{1,2,3}$ are constants. These microstructures can be measured on deformed rocks and using relations (13.34) one can easily determine the stress magnitude. However, caution must be exercised in using these paleopiezometers. Two points must be noted. First, strictly speaking these piezometers work only at steady-state deformation. No deformation in the Earth is steady state and therefore the effects of deformation history must be evaluated. One effective way is to compare several microstructural parameters. For example, if deformation that was responsible for microstructures was really steady state, then when one plots two different paleopiezometer values from a given rock (say dislocation density and recrystallized grain size), all of these microstructural parameters (dislocation density, recrystallized grain size etc.) must fall on a single line (Problem 13.6). The deviation of the data points from such a line suggests the effects of non-steady-state deformation. Second, the physical basis of paleopiezometers is not always clear and the extrapolation of laboratory data to the Earth may not always be justified. This is less of a problem for the dislocation density paleopiezometer than the recrystallized grain size paleopiezometer. Major problems in using the dynamically recrystallized grain size paleopiezometer include the effects of recrystallization mechanisms (one needs to identify the recrystallization mechanism in a given specimen and use the paleopiezometer corresponding to the same recrystallization mechanism) and the effects of thermochemical environment on recrystallized grain size.

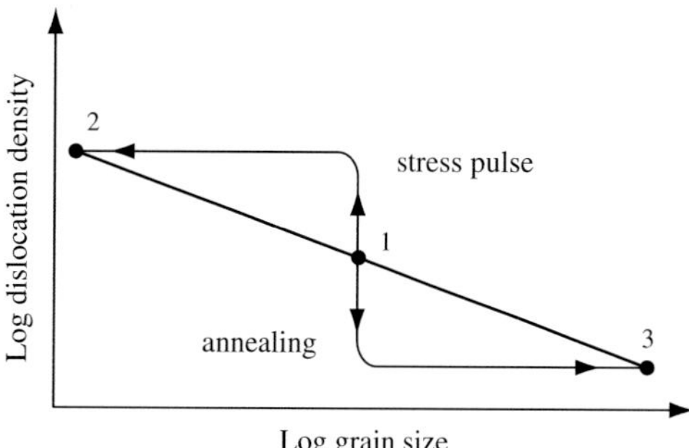

FIGURE 13.16 A dislocation density versus dynamically recrystallized grain size plot. If both dislocation density and recrystallized grain size correspond to a steady state, then the data should fall on a single line. Deviations from this line indicate non-steady-state deformation such as a stress pulse and/or static annealing.

Problem 13.6

Show that the log(dislocation density) versus log(recrystallized grain size) corresponding to steady-state deformation defines a line. Discuss where the dislocation density versus grain size data would fall if there is a stress pulse (or if there is post-deformation annealing). Note that a steady state can be achieved more easily for dislocation density than recrystallized grain size during a stress pulse and recovery is in most cases faster for dislocation density than for grain size (Fig. 13.16).

Solution

Eliminating the stress from equations (13.34a) and (13.34c), one gets $\rho^{n_3}\left(L_g\right)^{n_1} = A_3^{n_1} A_1^{n_3} b^{n_1-2n_3}$ and hence $n_3 \log \rho + n_1 \log L_g = n_1 \log A_3 + n_3 \log A_1 + (n_1 - 2n_3)\cdot \log b$. Consequently, if these microstructural parameters correspond to a steady state, then the data should fall on the line on a log(dislocation density) versus log(recrystallized grain size) plot. If a stress pulse and corresponding transient stage of deformation is responsible for these microstructures, then dislocation density will quickly increase whereas grain size will change more slowly. Consequently the data points would fall at the high dislocation density side of this line. In contrast, if post-deformation annealing is important, then the dislocation density will decrease quickly whereas grain size will adjust to a smaller stress more slowly and hence the data points would fall below that line.

13.5. Effects of phase transformations

In dynamic Earth, a piece of material may be brought into a new physical/chemical condition (a new temperature, pressure or the fugacity of relevant species) so that a previous crystal structure or mineral assembly becomes thermodynamically unstable. *Metamorphism* occurs in this case to make a new thermodynamically stable assembly.

Subduction of oceanic lithosphere is one of the most important cases in which such *metamorphism* occurs. In this case, the most important process is the change in crystal structure (*phase transformation*) of constituent minerals. These processes could affect rheology by various mechanisms including the effects of grain-size reduction (Chapter 15). In this chapter, we will focus on the effects of grain-size reduction.

13.5.1. Kinetics of a phase transformation

There are a variety of processes by which phase transformations or chemical reactions may change the grain size. For example, a phase transformation may occur through a collective displacement of atoms (e.g., *martensitic transformation*) or through more localized atomic motion involving *nucleation* and *growth* (AARONSON, 1990). Here we consider the grain-size change associated with *nucleation* and *growth*. Although *nucleation* and *growth* are not always the processes controlling the kinetics of a phase transformation, these processes occur in a wide range of geological conditions and are also the essence of chemical reactions.

The physical processes by which grain size is controlled during nucleation–growth associated with a phase transformation have some similarity with that of dynamic recrystallization. The concept of *nucleation* is, however, better defined for a phase transformation than dynamic recrystallization. Formation of a new phase occurs because the bulk free energy of one phase becomes smaller than another phase. A phase with a smaller free energy is more stable. However, the processes by which a material is transformed from one phase to another involve formation of new interfacial boundaries. Therefore there is a competition between bulk free energy and the boundary energy. Let us write a change in free energy associated with the formation of a new grain with grain size r. By forming a new grain the bulk free energy is reduced by $(4\pi/3)\, r^3(G_1 - G_2)$. However, by creating an interfacial boundary, we increase the free energy by $4\pi r^2\gamma$ where γ is the interfacial energy. Therefore the net change in the free energy is

$$\Delta G = 4\pi r^2\gamma - \frac{4\pi}{3} r^3(G_1 - G_2). \qquad (13.35)$$

This relation means that the change in free energy is the maximum at $r = 2\gamma/(G_1 - G_2)$ and the free energy difference for this r is $(16\pi/3)\,[\gamma^3/(G_1 - G_2)^2]$.

A phase transformation may occur through a variety of mechanisms. Two classes of mechanisms are often observed: nucleation–growth and martensitic mechanisms. In a nucleation–growth type mechanism, a new nucleus of new phase is formed in old phase (either inside a grain or at grain boundaries (or along a dislocation)) and they grow consuming old materials. In a martensitic mechanism, finite shear in an old crystal causes transformation to a new structure. The latter process is assisted by the external differential stress and therefore it tends to dominate at high stress conditions. In most conditions in the Earth, the nucleation–growth mechanism dominates.

Nucleation–growth

(1) Nucleation

When the thermodynamic condition change is large enough, then a material that was stable under the previous thermodynamic condition could become less thermodynamically stable than a new phase. This means that the free energy of a new phase will be smaller than the free energy of an old phase. Therefore the old phase tends to transform to a new phase. However, this does not happen all of a sudden. The transformation must start from making a "nucleus." However, when a new phase material is created not only a new phase (that has a lower free energy) but also a new interfacial boundary must be created that has excess energy. This excess free energy associated with the creation of a new nucleus depends on where the new nucleus is formed. The excess energy is, in general, smaller when a nucleus is formed on "defects" (such as grain boundaries or dislocations). However, the density of the possible nucleation sites will be smaller for these defects in comparison to the bulk. Therefore the nucleation at defects is favored when the P–T (pressure–temperature) conditions are close to equilibrium, whereas when P–T conditions are far from equilibrium then nucleation in the bulk will be favored. Nucleation in the bulk is often referred to as homogeneous nucleation and nucleation at grain boundaries is referred to as inhomogeneous nucleation or grain-boundary nucleation.

The net change in free energy associated with the formation of a new phase is given by,

$$\Delta G = \frac{4\pi}{3} r^3 \Delta G_V + 4\pi r^2\gamma \qquad (13.36)$$

where ΔG_V is the free energy difference between a new phase and an old phase (per volume) ($\Delta G_V < 0$ if a new phase is stable), r is the size of a new phase and γ is the interfacial energy of a boundary between a new phase and an old phase. Thus if a new phase is too small it will be unfavorable and will disappear. If a new phase happens to be large, then it will be stable and grow.

The critical size r_c can be given by $\partial\Delta G/\partial r = 0$, thus

$$r_c = \frac{2\gamma}{|\Delta G_V|} \tag{13.37}$$

and the corresponding energy is

$$\Delta G^* = \frac{16\pi\gamma^3}{3(\Delta G_V)^2}. \tag{13.38}$$

Free energy varies approximately linearly with T (temperature) and P (pressure), so that $\Delta G_V \propto (T_c - T), (P_c - P)$ where T_c or P_c is equilibrium T or P and therefore,

$$\Delta G^* \propto \frac{1}{(T_c - T)^2}, \qquad \frac{1}{(P_c - P)^2}. \tag{13.39}$$

Now the probability for nucleation is proportional to $\exp(-\Delta G^*/RT)$. When atomic diffusion is involved in nucleation (this will be the case in a phase transformation in a solid-solution), then the rate at which nucleation occurs is also proportional to the diffusion coefficient, $D = D_0 \exp(-H^*/RT)$, thus,

$$I = A \exp\left(-\frac{\Delta G^*}{RT}\right) \exp\left(-\frac{H^*}{RT}\right). \tag{13.40}$$

Note that because of the relation (13.43), the nucleation rate, I, depends on T or P very strongly, namely,

$$I = A \exp\left[-\frac{B}{(T - T_c)^2}\right] \exp\left(-\frac{H^*}{RT}\right) \tag{13.41}$$

or

$$I = A \exp\left[-\frac{C}{(P - P_c)^2}\right] \exp\left(-\frac{H^*}{RT}\right). \tag{13.42}$$

For example, if either T or P gets close to the equilibrium T or P, then the nucleation rate I becomes so small that nothing would happen. Measurable nucleation will occur only when a finite "overshoot" in T or P is made.

(2) Growth

After nucleation, each nucleus will grow. They impinge upon each other and the growth process stops. Further progress of transformation occurs by new grains that are formed in untransformed regions. Let us now calculate the rate of transformation as a function of nucleation rate I and growth rate G. For simplicity, we assume that both nucleation and growth rates are constant with time.

Now, consider one three-dimensional spherical new grain with a radius r. Its volume V_1 is given by,

$$V_1 = \frac{4\pi}{3} r^3 = \frac{4\pi}{3} G^3 (t - \tau)^3 \tag{13.43}$$

where t is time and τ is time when growth started. Thus,

$$dV_1 = 4\pi G^3 (t - \tau)^2 dt. \tag{13.44}$$

Now we consider a number of new grains that are growing in a transforming material. Recall that nucleation and growth occur in a volume fraction which has not been transformed, $u(t) = 1 - x(t)$ ($x(t)$ is the volume fraction of transformed material). Therefore the volume fraction of newly formed material during a time period $d\tau$ is given by,

$$dV_2 = dV_1 \cdot dn \cdot u(t) \tag{13.45}$$

where dn is the number of nuclei formed per time period $d\tau$. dn is related to the nucleation rate I,

$$I = \frac{dn}{d\tau}. \tag{13.46}$$

Thus,

$$\frac{dV_2}{dt} = 4\pi G^3 (t - \tau)^2 I \cdot u(t) \cdot d\tau \cdot \tag{13.47}$$

Now we must consider all the grains that have nucleated from $\tau = 0$ to $\tau = t$ to calculate the total rate of increase in the volume fraction of a new phase, dV_3/dt. Thus by integrating equation (13.47), one obtains,

$$\frac{dV_3}{dt} = 4\pi G^3 I \cdot u(t) \int_0^t (t - \tau)^2 \, d\tau = \frac{4\pi}{3} G^3 I \cdot u(t) t^3 \tag{13.48}$$

Remembering that $V_3(t) = x(t)$ so that $dV_3(t)/dt = dx(t)/dt = -du(t)/dt$, one finds,

$$-\frac{du(t)}{dt} = \frac{4\pi}{3} G^3 I \cdot u(t) t^3. \tag{13.49}$$

By integrating (13.49), one finally obtains,

$$u(t) = \exp\left(-\frac{\pi}{3} G^3 I \cdot t^4\right) \tag{13.50}$$

and

$$x(t) = 1 - \exp\left(-\frac{\pi}{3} G^3 I \cdot t^4\right). \tag{13.51a}$$

A similar analysis can be applied to a two-dimensional growth to get,

$$x(t) = 1 - \exp\left(-\frac{\pi}{3} G^2 I \cdot t^3\right). \tag{13.51b}$$

Problem 13.7

Derive equation (13.51b) for two-dimensional growth.

Solution

The only difference between two and three dimensions is the geometry of nuclei. Instead of a spherical nucleus, we have a circular nucleus. Thus $dV_1 = 2\pi G^2(t-\tau)\,dt$. And the rate at which new grains are formed is $dV_2/dt = (dV_1/dt)\, I \cdot u(t) = 2\pi G^2(t-\tau) I \cdot u(t)\, d\tau$. By integrating this formula from $\tau = 0$ to $\tau = t$, the total rate of formation of new grains is given by $dV_3/dt = dx(t)/dt = -du(t)/dt = 2\pi G^2 I \cdot u(t) \int_0^t (t-\tau)\, d\tau = \pi G^2 I \cdot u(t) \cdot t^2$. Therefore $-du(t)/dt = \pi G^2 I \cdot u(t) t^2$ and $u(t) = \exp\left(-\frac{1}{3}\pi G^2 I \cdot t^3\right)$ and hence $x(t) = 1 - \exp\left(-\frac{1}{3}\pi G^2 I \cdot t^3\right)$.

In general, the volume fraction of transformed grains is given by,

$$x(T,t) = 1 - \exp(-A(T) t^k). \tag{13.52}$$

This equation is referred to as the *Avrami equation*. From these relations, one can see that the volume fraction of a new phase $x(T, t)$ increases with time and when one can define a critical time above which a new phase will occupy a significant fraction of a given material. Thus,

$$A(T) t^k > \beta \tag{13.53}$$

(where β is a non-dimensional parameter ($\beta \sim 3$ for $\sim$95% transformation)) then the transformation is effectively complete. From this one can define the *Avrami time*

$$t_{Av} = \left(\frac{\beta}{A(T)}\right)^{1/k} \tag{13.54}$$

which is a measure of time-scale of a phase transformation.

For three-dimensional (homogeneous), nucleation

$$t_{Av} \propto \left(\frac{1}{G^{k-1} I}\right)^{1/k} \tag{13.55}$$

with $k = 3$ for two-dimensional nucleation (grain-boundary nucleation) and $k = 4$ for homogeneous nucleation.

Now consider the size of grains of a new phase. Fundamental physics is similar to the physics of dynamic recrystallization. In this case, grain size is controlled by the impingement. Therefore the size of grains after a phase transformation is determined by the equation similar to (13.29) or (13.30). Therefore, the size of grains is controlled by the competition between nucleation and growth. A useful measure of the size of the new grain is given by the *Avrami length* δ_{Av}

$$\delta_{Av} \equiv G \cdot t_{Av} \propto \left(\frac{G}{I}\right)^{1/k}. \tag{13.56}$$

If nucleation is very slow but growth rate is fast, grain size will be large. In contrast, if nucleation is easy and growth rate is slow, then grain size will be small. This simple physics has been well known and often applied in diamond manufacture.

Problem 13.8

We have defined two different types of grain size: one is the critical size for nucleation (equations (13.25) and (13.37)) and the other is the size determined by the balance of nucleation and growth (equations (13.28a,b) and (13.56)). Show that the former size is unstable whereas the latter size is stable.

Solution

In the former (nucleation), the competition is between bulk free energy and surface energy. Consider a small deviation in grain size from the critical size. If the grain size becomes larger (smaller) than the critical size, then the absolute value of the bulk energy (the value is negative) term increases (decreases) more than the surface energy term and then the system will obtain a lower (higher) energy state, so that the grain size will increase (decrease). Therefore the system is unstable for this perturbation. In contrast, if the grain size becomes larger (smaller) than the values determined by dynamic balance (equations (13.28) or (13.56)), then both growth and nucleation time-scales will increase (decrease) but the nucleation time-scale increases (decreases) more than the growth time. Consequently the grain size will decrease (increase). Therefore the grain size determined by this type of dynamic balance is stable.

In this sense, the model by Twiss (1977) gives an unstable grain size and physically unreasonable as correctly pointed out by Poirier (1985).

To illustrate how this physics works in real Earth, let us consider a case of a phase transformation in a downgoing slab. When temperature in the slab is warm (young slab and/or slow subduction rate), then nucleation is easy but growth rate will also be fast. In contrast, in a cold slab, both nucleation and growth are difficult. However, in this latter case, actual nucleation rate is not very small, because actual nucleation will occur after a very large overstep. Thus, the main difference between cold and warm slabs is the difference in growth rate. As a result, grain size after a phase transformation in a cold slab will be significantly smaller than that in a warm slab. Calculations by RIEDEL and KARATO (1997) based on kinetic parameters by RUBIE and ROSS (1994) showed that for a very cold slab, the grain size of spinel can be as small as ~1 μm. This will lead to a very significant reduction in strength. As a consequence, they predicted that cold slabs could be weaker than warm ones under some conditions. This unusual behavior provides a natural explanation for a seemingly strange result of seismic tomography that shows evidence of larger deformation for older slabs (in the western Pacific) than younger slabs (in the eastern Pacific).

Problem 13.9

When grain size is dependent on temperature, then the strain rate associated with grain-size sensitive creep will be temperature-dependent due to two different reasons: first due to the thermal activation for defect motion and second due to the temperature sensitivity of grain size. Show that under certain conditions, the rate of deformation of materials *decreases* with the increase of temperature. Discuss under which environment this unusual temperature sensitivity of creep might occur.

Solution

The flow law for grain-size sensitive creep can be written as $\dot{\varepsilon} = A\left(\sigma^n/L^m\right)\exp\left(-H_{\dot{\varepsilon}}^*/RT\right)$ where $H_{\dot{\varepsilon}}^*$ is the activation enthalpy for deformation. However, the grain size can also be temperature-dependent when it is controlled by dynamic balance of nucleation and growth. Under these conditions the grain size is dependent on temperature as $L = L_0 \exp(-H_L{}^*/RT)$. Therefore $\dot{\varepsilon} = A(\sigma^n/L_0^m)\exp\left(-(H_{\dot{\varepsilon}}^* - mH_L^*)/RT\right)$. When grain size is controlled by nucleation and growth of a new phase, then the activation energy for grain size is large (comparable to that of deformation) and $H_{\dot{\varepsilon}}^* - mH_L^* < 0$. In such a case, creep rate has a negative effective activation enthalpy. Such a situation probably occurs in subducting slabs, leading to a weaker slab in a colder region (KARATO *et al.*, 2001).

13.6. Grain size in Earth's interior

So far, we have reviewed elementary processes by which grain size may be controlled in Earth. We now consider what grain size would one expect in the dynamically evolving Earth. Let us consider, first, what would happen when a polycrystalline rock is deformed in Earth to large strain. As we learned before, most minerals are plastically anisotropic and prone to dynamic recrystallization. The grain size at the completion of dynamic recrystallization is determined by the local balance of nucleation and growth. This grain size is primarily controlled by stress, although it is also influenced by other parameters such as temperature and water content (water fugacity). Under certain conditions, grain size after the completion of dynamic recrystallization is large enough so that deformation still occurs in the dislocation creep regime. In this case, not much would happen.

An interesting case is the situation where the grain size after dynamic recrystallization is so small that deformation occurs in the diffusion creep (or other grain-size sensitive creep) regime. In this case, the new grain size is not stable for the long term. Grain growth should occur that brings grain size up to the size of the dislocation creep regime. In the dislocation creep regime, however, grain-size reduction will bring grain size back to a small size.

Such interplay between dislocation creep, dynamic recrystallization and grain growth was discussed by ZEUCH (1982, 1983). In particular, DE BRESSER *et al.* (1998) postulated that the grain size versus stress relationship corresponding to dynamic recrystallization may coincide with the boundary between dislocation and diffusion creep. Several points must be recognized. First, one can imagine a case where this hypothesis clearly does not work. This is a case where the size of grain size after dynamic recrystallization is larger than the critical size for diffusion creep. This can occur, because the microscopic processes by which the grain size is controlled by dynamic recrystallization is different from the competition between grain growth and grain-size refinement. In dynamic recrystallization, the governing microscopic process is a microscopic balance between nucleation and grain impingement.

Second, it is important to distinguish between two scales of microstructural balance (equilibrium). At a small scale, grain size can be controlled by the balance of nucleation and growth. Here the driving force for grain-boundary migration is dislocation density contrast or subboundary energy. (This is similar to the *primary recrystallization.*) After this stage, grain growth driven by grain-boundary energy will occur (similar to the *secondary recrystallization*). Once grain size becomes large enough for dislocation creep to dominate, then grain size will be reduced to a value that is controlled by the *microscopic* balance of nucleation and growth. Therefore grain size will oscillate and there will be a bimodal grain size in dynamically recrystallizing materials. The mechanical behavior of such materials will be similar to the one analyzed by SAKAI and JONAS (1984). Oscillatory behavior will appear under some conditions.

Now having reviewed some of the fundamental processes by which grain size may be controlled, what are the actual grain sizes of rocks in Earth? The grain size of rocks from the crust and upper mantle can be determined directly from the rock samples collected on Earth's surface. In the crust one can see a vast range of grain size reflecting their thermomechanical history. For example, basalts have very fine grain size almost glassy texture to ~10 μm grain size, which is due to rapid crystallization (fast nucleation at modest growth rate) caused by rapid cooling. As a basalt is annealed in the crust, its grain size increases to ~0.1 to 1 mm size (this *metamorphosed* basalt is called diabase and finally gabbro). For a similar reason, granites, which are usually formed in the deep crust and hence cooled slowly, show large grain size, 1–10 mm. Deformed crustal rocks show a range of grain size. The most intensive plastic deformation usually occurs in the lower crust (RUTTER and BRODIE, 1992). In the lower crust, the grain size has a wide variation. The smallest grain size occurs in highly sheared region and ultra-mylonites have grain size as small as ~10 μm.

A similar variation in grain size can be seen in the upper mantle rocks. Here grain size is mostly correlated with the magnitude of stress (AVÉ LALLEMANT *et al.*, 1980; MERCIER, 1980), but it can also vary from one mineral to another. Under most conditions, olivine, the most abundant mineral in the upper mantle, is the softest mineral and undergoes dynamic recrystallization readily. Orthopyroxene, the second common mineral in upper mantle rocks, is stronger than olivine in the dislocation creep regime (under water-poor conditions). Consequently, in many cases where deformation occurs initially by dislocation creep, orthopyroxene undergoes less deformation and recrystallizes later. However, after dynamic recrystallization, the grain size of orthopyroxene is significantly smaller than that of olivine (by a factor of ~5–10). Consequently, orthopyroxene after dynamic recrystallization often shows evidence of grain-size sensitive creep, and orthopyroxene under these cases can be significantly weaker than olivine. For example, in typical coarse-grained upper mantle xenoliths, the grain size of olivine, orthopyroxene and other minerals is a few mm (e.g., NICOLAS, 1978; MERCIER, 1980). However, in highly sheared upper mantle rocks, the grain size can be as small as ~10 μm.

The grain size in Earth deeper than the transition zone cannot be estimated from direct observations. There are some reports on mantle rocks or fragments of minerals deeper than ~300 km showing a few mm grain size (e.g., HAGGERTY and SAUTTER, 1990). A recent report on the presence of seismic anisotropy in the transition zone (TRAMPERT and VAN HEIJST, 2002) suggests deformation by dislocation creep implying a relatively large grain size, whereas the lack of seismic anisotropy in most of the lower mantle (MEADE *et al.*, 1995) suggests a relatively small grain size. By comparing the diffusion coefficients measured in the laboratory with the viscosity inferred from geophysical observations (see Chapter 19), YAMAZAKI and KARATO (2001b) estimated a grain size of ~2–3 mm at the top of the lower mantle. Based on a numerical simulation of grain-growth kinetics, SOLOMATOV *et al.* (2002) inferred a similar grain size for the lower mantle. The grain size in the inner core could be much larger. From the comparison of laboratory data on seismic wave attenuation with seismological observations, the grain size in the inner core is estimated to be ~10^2–10^3 m (Chapter 17). This inferred large grain size likely reflects a small deviatoric stress and a very high temperature (close to the melting temperature) in the inner core, which is likely to be made of nearly single phase.

14 Lattice-preferred orientation

Non-random distribution of crystallographic orientation, lattice-preferred orientation (LPO) (or crystallographic-preferred orientation: CPO), is one of the important microstructural characteristics of rocks. LPO in naturally deformed rocks can be measured in the laboratory or LPO in rocks deformed in Earth's interior can be inferred from seismic anisotropy. If the physical processes by which LPO is formed are understood, then one can use measured or seismologically inferred LPO to infer physical processes such as a flow pattern in Earth. This chapter provides a review of the various ways of representing LPO and the physical processes for the formation of LPO including deformation-induced LPO and crystallization- (or recrystallization-) induced LPO. LPO is controlled by both the microscopic physics of deformation or the nucleation–growth of crystals and the macroscopic geometry of field in which LPO develops (finite strain ellipsoid, stress, magnetic field, heat flow etc.). The relation between LPO and deformation geometry can change with physical (and chemical) conditions of deformation.

Key words Lattice-preferred orientation (LPO), Euler angles, orientation distribution function (ODF), pole figure, inverse pole figure, *J*-index, *M*-index, Schmid–Boas relation, Taylor–Bishop–Hill model, self-consistent model, dynamic recrystallization, fabric transition.

14.1. Introduction

Distribution of crystallographic orientation of grains in a polycrystalline material may be non-random. It can be measured on rock samples and it can also be inferred from geophysical observations of anisotropic properties such as seismic anisotropy (see Chapter 21). This non-random distribution of crystallographic axes is called *lattice-preferred orientation* (LPO) or *crystallographic-preferred orientation* (CPO) or simply *fabric*.[1] LPO reflects the processes by which a rock was formed and/or has been deformed in a given environment ("field"). Therefore by knowing the relationship between LPO and the anisotropic field in which a rock's LPO was established, one can infer the anisotropic field that would provide an important insight into the dynamic processes in Earth.

In this chapter we will first learn the experimental methods of LPO measurement and the representation of LPO (section 14.2). Then the physical mechanisms of formation of LPO will be discussed based on both theoretical models and experimental observations (section 14.3).

One of the most important concepts in studying LPO is that the nature of LPO is controlled by both *microscopic* processes and *macroscopic* anisotropic fields. When it is caused by deformation, LPO depends on deformation mechanisms as well as the geometry of deformation. Consequently, in order to obtain some

[1] Either the term *lattice-preferred orientation* (LPO) or *crystallographic preferred orientation* (CPO) is used in the literature. The term *fabric* is also used for LPO in geological literature, but the term *fabric* refers to grain shape, grain-size distribution in the materials science literature, which is called *texture* in geological literature.

information as to the macroscopic field from LPO, one needs to understand the microscopic processes of LPO development. A more extensive discussion of LPO can be found in books such as Kocks *et al.* (1998) and Wenk (1985) (for a new technique of LPO measurement, EBSD, see Randle, 2003).

14.2. Lattice-preferred orientation: definition, measurement and representation

14.2.1. Measurement of lattice-preferred orientation

Lattice-preferred orientation can be measured using a range of techniques. A classic technique is the optical microscope measurement using a *universal stage*. In this case, the anisotropic optical properties of crystals are used to determine the orientation of each crystal. Spatial resolution is determined by the sample thickness relative to grain size, and usually the orientations of grains smaller than ~20 μm are difficult to measure. X-ray measurements can be used to determine LPO. In this technique, one measures the intensity of various peaks of diffracted X-rays as a function of direction in a sample. From the orientation dependence of intensity of diffracted X-ray one can determine LPO. This technique has the advantage of being able to be applied to samples with fine grain size but the calculation of LPO from X-ray intensity is not straightforward because of the corrections for absorption and fluorescence. A recent and the most powerful technique is a scanning electron microscope (SEM)-based technique, called the electron backscatter diffraction technique (EBSD; Fig. 14.1; Dingley and Randle, 1992; Adams *et al.*, 1993; Prior and Wheeler, 1999; Randle, 2003). In this technique, a narrowly focused electron beam is scanned on a sample surface and one measures the distribution of diffracted electrons due to multiple scattering. Because of multiple scattering, the spatial pattern of diffracted electron beam reflects the orientation of crystal but not the direction of the incident electron beam. These diffracted electrons form a "Kikuchi pattern," which varies with crystal orientation (Fig. 14.1). The electron beam is usually obtained from a scanning electron microscope, and the diffraction pattern is captured by a high-sensitivity camera. In order to obtain a high quality Kikuchi pattern, the sample surface must be well polished. A common polishing procedure includes chemico-mechanical polishing using a colloidal silica. A high density of defects (dislocations) reduces the quality of patterns, but even in highly deformed mylonites (that have dislocation density of $\sim 10^{12}$–10^{13} m^{-2}), the quality of images is good enough to conduct high quality measurements with a recent version of the SEM-EBSD system.

The indexing of the patterns can be made by comparing the observed pattern with calculated ones for various orientations (commercially available software can be used for this purpose; Fig. 14.1b). This technique can be applied to almost all crystals and the determination of orientation is accurate to 1° or better. The spatial resolution depends on the nature of the electron beam. If a conventional tungsten filament is used, then the spatial resolution is ~2 μm. Using a high-density electron beam produced by a field-emission-gun, one can obtain a spatial resolution of ~0.2 μm. Another diffraction-based technique

(a)

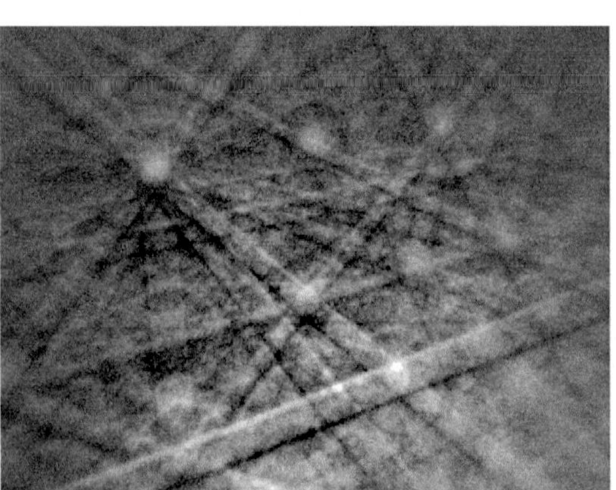

(b)

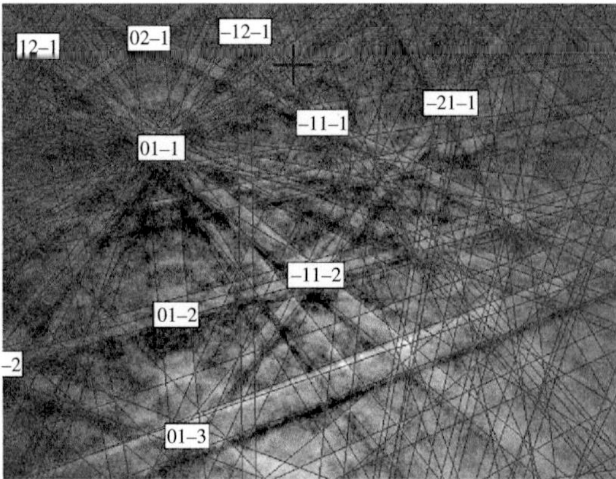

FIGURE 14.1 (a) A Kikuchi pattern of deformed olivine and (b) a comparison of a calculated Kikuchi pattern with the observed one. Plate 2.

for LPO measurement includes neutron diffraction measurements (e.g., Wenk, 2002). Neutrons interact with crystals including those containing hydrogen (X-rays do not interact with protons because a proton atom does not have an electron). One of the merits of this technique is that the absorption of neutrons by materials is much less than that of electrons and hence a large sample volume can be studied. Wenk (1985, 2002) contain reviews of preferred orientation measurements.

14.2.2. Representation of lattice-preferred orientation

Pole figure

Lattice-preferred orientation can be represented in several ways. In the Earth science community, pole figures are most commonly used. A pole figure is a representation of the distribution of crystallographic orientations in a sample coordinate. One chooses a particular coordinate fixed to a sample (shear plane–shear direction, foliation–lineation etc.), and plots the direction of a particular crystallographic orientation in this space. In order to show the orientation distribution in a two-dimensional diagram, a stereographic projection is used. Usually, the direction is projected to the lower hemisphere. Although this representation can cover all the orientations in the sample space, only a finite number of crystallographic orientations can be chosen. Also the relative orientations of the different crystallographic axes of each grain cannot be represented in a pole figure. Therefore the information in pole figures is incomplete. For example, when one chooses [100], [010], [001], then no information of orientation distribution of other orientations, say [110], can be represented.

An example of pole figures is shown in Fig. 14.2. In this representation, one can see how some of the different crystallographic orientations are distributed in the space relative to the deformation geometry (e.g., foliation, lineation).

Mathematically, a pole figure represents the angular distribution of a given crystallographic direction $c = [hkl]$ with respect to a sample coordinate system, and therefore it represents a volume fraction of grains having their crystallographic orientation parallel to the sample direction y and $y + dy$ ($y = \{\alpha, \beta\}; dy = \sin\alpha \cdot d\alpha \cdot d\beta, \alpha$, co-latitude; β, longitude),

$$\frac{dV}{V} = \frac{1}{4\pi} P_c(y)\, dy = \frac{1}{4\pi} P_c(\alpha, \beta) \sin\alpha \cdot d\alpha \cdot d\beta \quad (14.1)$$

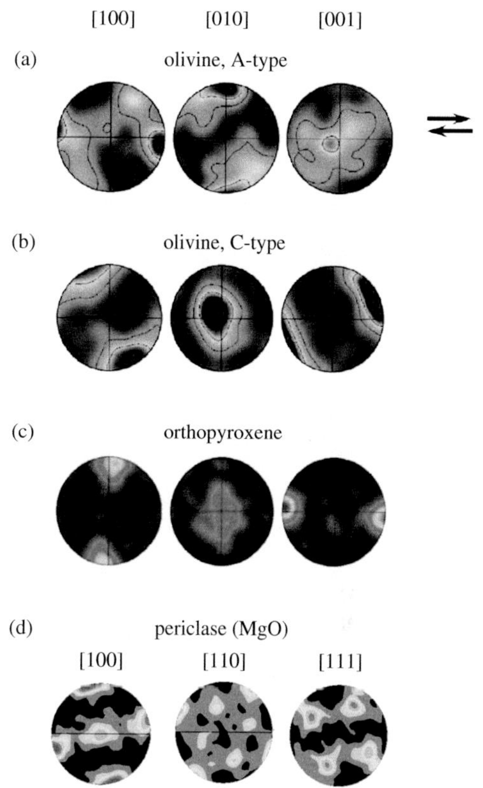

FIGURE 14.2 Some examples of pole figures showing the distribution of crystallographic orientations in the coordinate system defined with respect to the deformation geometry. The color code indicates the density of data (red, high; blue, low). (a) Olivine experimentally deformed in simple shear (water-poor, low-stress, high-temperature conditions), (b) olivine experimentally deformed in simple shear (water-rich, low-stress) (from Jung *et al.*, 2006), (c) naturally deformed orthopyroxene showing a girdle type LPO (from Skemer *et al.*, 2006), (d) MgO experimentally deformed in simple shear (from Long *et al.*, 2006). Plate 3.

where $P_c(y)\, dy$ is the fraction of grains whose crystallographic orientation is between y and $y + dy$. A pole figure is usually normalized as $P_c(y)_{\text{random}} = 1$.

Inverse pole figure

Another frequently used representation is the *inverse pole figure*. In an inverse pole figure, one plots the orientation of particular orientations in space (e.g., compression direction, shear direction, lineation) using a coordinate defined with respect to the crystallographic axes. This plot can cover all of the crystallographic orientations but only a small number of orientations in the sample space can be presented. An example of an inverse pole figure is shown in Fig. 14.3. Because all the crystallographic orientations are shown in this plot, one can easily find how crystallographic orientations are related to a given external reference frame such as

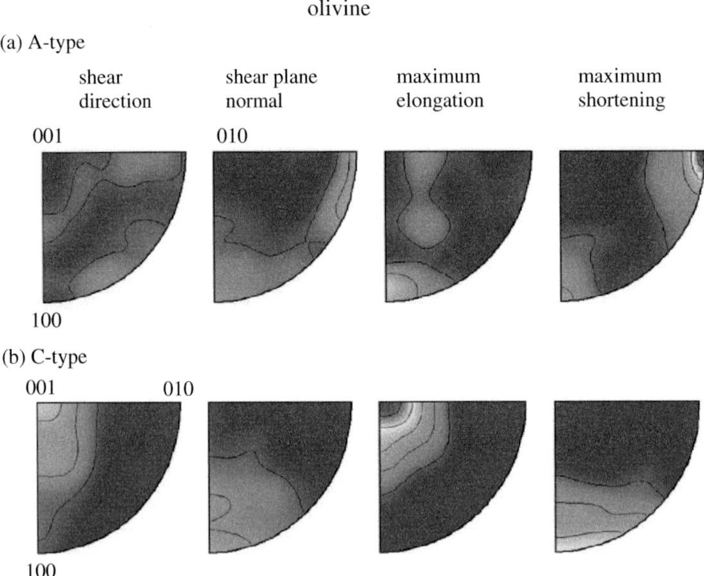

FIGURE 14.3 Examples of inverse pole figures showing the distribution of orientations in space relative to the crystallographic coordinate. (a) Olivine deformed in the A-type regime (the same sample as shown in Fig. 14.2a) and (b) olivine deformed in the C-type regime (the same sample as in Fig. 14.2b). The color code indicates the density of data (red, high; blue, low). Plate 4.

the shear direction (or shear plane). Therefore the inverse pole figure is very useful in inferring the dominant slip system(s).

Similar to the pole figure, the inverse pole figure is defined as,

$$\frac{dV}{V} = \frac{1}{4\pi} R_y(h)\, dh \tag{14.2}$$

where $R_y(h)dh$ is the fraction of grains for which a given orientation in space is between h and $h + dh$ of crystallographic orientation space. An inverse pole figure is usually normalized as $R_y(h)_{\text{random}} = 1$.

Because both pole figures and inverse pole figures are incomplete, information on LPO provided by these two diagrams are not redundant but complementary. Therefore use of both pole figures and inverse pole figures is recommended. For instance, it is easy to visualize the pattern of LPO in the sample space from pole figures, but it is often more instructive to use inverse pole figures to infer the dominant slip system(s).

Problem 14.1

Compare the pole figures shown in Fig. 14.2b and the corresponding inverse pole figures shown in Fig. 14.3b and discuss (i) the dominant slip system(s) and (ii) the controlling framework of LPO development.

Solution

In Fig. 14.2b, the olivine [001] axis has peaks nearly parallel to the shear direction and the [100] axis is nearly normal to the shear plane. However, there is a significant angle between the peak positions and these structural frameworks. Also the orientations of other crystallographic axes are not shown in these diagrams. In the inverse pole figure shown in Fig. 14.3b, the olivine [001] axis is clearly parallel to the maximum elongation direction and the olivine [100] axis is parallel to the maximum shortening direction, and no other crystallographic orientation has any correlation to the finite strain ellipsoid. Consequently, we conclude that [001](100) is the dominant slip system in this sample.

Orientation distribution function

A more complete representation of LPO is the *orientation distribution function* (ODF). The orientation of each crystal can be completely specified by three Euler angles (Fig. 14.4, Box 14.1). The Euler angles of one crystal are defined by three angles, so a given orientation of a three-dimensional object can be specified by a vector $\boldsymbol{g}$ in a three-dimensional space. The orientation distribution function, $f(\boldsymbol{g})$, is defined as $f(\boldsymbol{g})\, d\boldsymbol{g}$ being the volume fraction of materials whose Euler angle is between $\boldsymbol{g}$ and $\boldsymbol{g} + d\boldsymbol{g}$, namely,

$$\frac{dV}{V} = f(\boldsymbol{g})\, d\boldsymbol{g}. \tag{14.3}$$

Therefore the ODF contains the most complete information of LPO. The pole figures or the inverse pole figures are the subsets of ODF. For this reason,

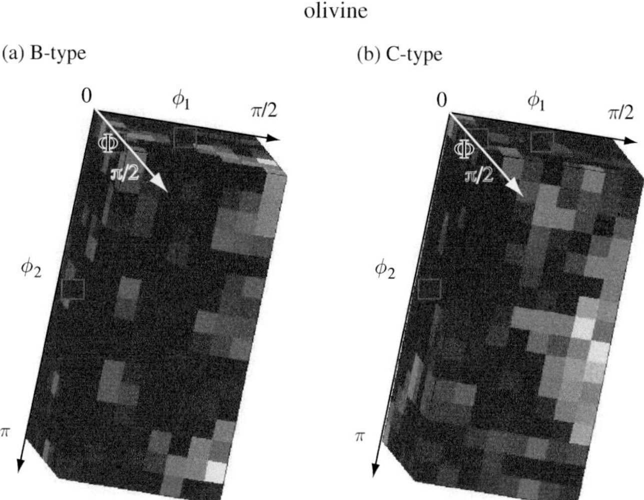

FIGURE 14.4 Some examples of the orientation distribution function (ODF). (a) Olivine deformed in the B-type regime and (b) olivine deformed in the C-type regime. The distribution of crystallographic orientation of individual grains is shown in the Euler angle space. The color code indicates the density of data (red, high; blue, low). Plate 5.

the ODF is commonly used in materials science. However, interpretation of ODF is not necessarily intuitive and it is still not very commonly used in the Earth science community. The orientation distribution function is normalized as

$$\int f(\mathbf{g})\, d\mathbf{g} = 1. \tag{14.4}$$

To understand how to interpret ODF, let us take a simple example. A case for single slip fabric is straightforward. Consider the A-type olivine fabric in which the olivine [100] axis is nearly parallel to the shear direction and the [010] axis is nearly normal to the shear plane. In this case, if the Euler angles are defined as shown in Box 14.1, then the data points corresponding to olivine A-type fabric will be plotted near $(\phi_1, \Phi, \phi_2) = \left(0, \frac{1}{2}\pi, 0\right)$. Schmid and Casey (1986) and Benn and Allard (1989) provide some examples of the application of ODF for geological materials (see also Mainprice and Nicolas, 1989).

None of these representations provide information about the spatial correlation of orientations of grains. Spatial distribution or correlation of orientations can be plotted or studied from digital EBSD measurements and can provide important information as to the processes of deformation (Fig. 14.5).

Another useful measure of the relative orientations of grains is *misorientation* of two grains. Given a pair of grains, one can define a common axis and the rotation of one crystal relative to another will make the orientations of two grains coincide. This angle is called misorientation angle. The distribution of misorientation angles reflects the nature of relative orientations in grains (e.g., Wheeler *et al.*, 2001). Skemer *et al.* (2005) used misorientation angles to quantify the strength of deformation fabrics (see the next section).

Problem 14.2

In addition to the A-type fabric discussed before, olivine assumes a range of LPO depending on the physical and chemical conditions. B-type fabric: the [001] axis is nearly parallel to the maximum elongation direction and the [010] axis is parallel to the maximum shortening direction, C-type fabric: the [001] axis is nearly parallel to the maximum elongation direction and the [100] axis is parallel to the maximum shortening direction, E-type fabric: the [100] axis is nearly parallel to the maximum elongation direction and the [001] axis is parallel to the maximum shortening direction, and D-type fabric: the [100] axis is nearly parallel to the maximum elongation direction and the [001] and [010] axes form a girdle. Show how olivine B-, C-, E- and D-type fabrics are plotted in the ODF (x' axis is [100] axis, y' axis is [010] axis and z' axis is [001] axis). Assume that the fabric symmetry follows the strain ellipsoid and the x axis is the maximum elongation direction, and the z axis is the maximum shortening direction.

Solution

B type: ODF is near $(\phi_1, \Phi, \phi_2) = \left(\frac{1}{2}\pi, \frac{1}{2}\pi, 0\right)$
C type: ODF is near $(\phi_1, \Phi, \phi_2) = \left(\frac{1}{2}\pi, \frac{1}{2}\pi, \frac{1}{2}\pi\right)$
E type: ODF is near $(\phi_1, \Phi, \phi_2) = (0, 0, 0)$
D type: ODF is near where $(\phi_1, \Phi, \phi_2) = (0, \Phi, 0)$ where Φ has any value $(0 \le \Phi \le \pi)$.

Box 14.1 Euler angles and the orientation distribution function (ODF)

The orientation of a three-dimensional object in a space can be defined by three angles. For example, one can define the orientation of a crystal by defining the orientation of a crystallographic plane (by two parameters) and by defining the orientation of one crystallographic direction on that plane (by one parameter). This can be done in the following way. Let us assume that a crystal and a sample coordinate (i.e., foliation–lineation) coincide. Starting from this, let us rotate the crystal coordinate (x', y', z') relative to a sample coordinate (x, y, z) using the following three steps. (i) Rotate the crystal coordinate across the sample axis z by ϕ_1 (0 to 2π). (ii) Then rotate the crystal coordinate around the x' axis by Φ (0 to π). (iii) Finally rotate the crystal axis around z' by ϕ_2 (0 to 2π). These three angles that define the crystal orientation in the sample coordinate are called the *Euler angles*. The Euler angles for a crystal completely define its orientation (relative to the sample space). The space made of three angles (ϕ_1, Φ, ϕ_2) is called the Euler space. Orientation of each crystal corresponds to a point in the Euler space. The distribution of these points in the Euler space can be described by a function that defines the density of these points. This function is called the *orientation distribution function, ODF*. In the Euler space, the direction along the Φ axis is the direction in which the crystal's x' axis is parallel to the x direction. Similarly the ϕ_1 direction is the direction in which the crystal's z' axis is parallel to the sample axis of z direction, and the ϕ_2 direction is the direction in which the crystal's y' axis is parallel to the sample axis of y direction.

FIGURE 14.5 Orientation mapping. Orientations of each grain can be color-coded and the spatial distribution of orientation is visualized (from JUNG *et al.*, 2006). Plate 6.

14.2.3. Statistics and fabric strength

The actual measurements of crystallographic orientations of grains by EBSD or similar techniques are made on individual points. From these measurements, we infer the geometry of flow and calculate seismic anisotropy and other anisotropic properties. One of the practical questions is "how many grains do we need to measure in order to obtain statistically meaningful results for fabrics?" WENK (2002) and SKEMER *et al.*, (2005b) investigated this issue. As we learned before the most complete information of LPO is contained in ODF, $f(\mathbf{g})$. This function is normalized as equation (14.4), so the simplest parameter to characterize the nature of LPO will be a *J-index* defined by (BUNGE, 1982)

$$J \equiv \int |f(\mathbf{g})|^2 \, d\mathbf{g}. \tag{14.5}$$

This parameter is a scalar, so it does not characterize the geometry of LPO, but it reflects the strength of LPO. The *J*-index increases with increasing the strength of LPO and for random distribution $J = 1$ and for a single crystal distribution it is infinite (Problem 14.3). The amplitude of seismic anisotropy has a positive correlation with the *J*-index (e.g., MAINPRICE and SILVER, 1993; SILVER *et al.*, 1999).

Problem 14.3

Show that $J=1$ for a random fabric and $J=\infty$ for a single crystal fabric.

Solution

For random distribution, $\int f(\mathbf{g})d\mathbf{g} = 1 = f_{\text{random}} \int d\mathbf{g} = f_{\text{random}}$ where we used a relation, $d\mathbf{g} = 1/8\pi^2 \sin\Phi \cdot d\varphi_1\, d\varphi_2\, d\Phi$ and hence $\int d\mathbf{g} = 1$. Consequently, $J_{\text{random}} = \int |f_{\text{random}}|^2 d\mathbf{g} = \int d\mathbf{g} = 1$. For a single crystal fabric, $f(\mathbf{g}) = \delta(\mathbf{g} - \mathbf{g}_0)$ where $\delta(x)$ is the Dirac delta function ($\delta(x) = 0$ for $x \neq 0$ and $\int_{-\infty}^{\infty} \delta(x)dx = 1$) and $J_{\text{single}} \equiv \int |\delta(\mathbf{g})|^2\, d\mathbf{g} = \infty$.

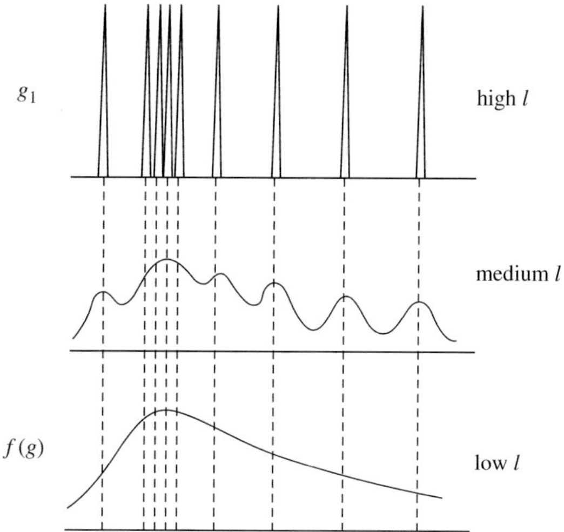

FIGURE 14.6 A schematic diagram showing the relation between discrete measurements of orientation and a continuous orientation distribution function (after BUNGE, 1982).

However, the J-index must be applied cautiously because actual measurements of fabric are often made on individual grains and some numerical artifacts can influence the calculation of the J-index. As shown before, the J-index for a continuous distribution of orientation is between one and infinity and should not depend on the number of data points (grains). However, as noted by MATTHIES and WAGNER (1996) and XIE *et al.* (2003), the J-index can be dependent on the number of data points when the smoothing factor is not chosen appropriately. In general, when we study the fabric, we seek to determine the anisotropic structure of the aggregate that is independent of the number of grains. However, any practical measurements are made on a finite number of grains. If an infinite number of spherical harmonics are used to represent the fabric, then the actual ODF would have discrete values (0 or 1) in the Euler angle space[2] (Fig. 14.6). In contrast, for an ideal case of an infinite number of grains, ODF should be a smooth function in the Euler angle space. Consequently, one must truncate the spherical harmonic order to capture the appropriate nature of ODF and obtain the J-index that is independent of the number of data points. SKEMER *et al.* (2005) investigated the influence of these numerical artifacts on the results of the J-index and found that in most cases the calculated J-index from a given sample depends strongly on these parameters (the maximum degree of spherical harmonics and the number of data points), and consequently, the J-index must be used with great caution.

SKEMER *et al.* (2005) introduced an alternative measure of fabric strength based on misorientation angles. For any pair of two grains, one can determine the *misorientation angle* (Chapter 5). A misorientation distribution function $R(\theta)$ is defined such that $R(\theta)\, d\theta$ is a fraction of pairs of grains for which the misorientation angle is between θ and $\theta + d\theta$. $R(\theta)$ for random fabric, $R_0(\theta)$, can be calculated from the symmetry of the crystal (e.g., GRIMMER, 1979), and one can use the difference between the actual $R(\theta)$ and $R_0(\theta)$ as a measure of fabric strength, namely,

$$M \equiv \frac{1}{2}\int_0^{\theta_{\max}} |R(\theta) - R_0(\theta)|\, d\theta. \tag{14.6}$$

This is called the *M-index*. The M-index is 0 for random fabric and is 1 for a single crystal. Misorientation angle can easily be calculated from the data of orientation distribution function using a commercial software package. One advantage of using misorientation angles is that there are a large number of misorientation angles for a given finite number of data points (${}_NC_2 = N(N-1)/2 \approx N^2/2$) so that there are fewer numerical artifacts than the J-index. In fact, SKEMER *et al.* (2005) showed that the values of the M-index are stable, for a typical fabric strength, for a number of grains exceeding ~100–200 (this number is higher for a crystal with high symmetry: for a cubic crystal it is ~500–1000 grains) (Fig. 14.7).

[2] The problem found by MATTHIES and WAGNER (1996) is due to this artifact. For a weak fabric, $\int f(g)\, dg = 1 \approx f'N$, where we use the approximation that ODF is nearly constant, and there are N data points in the Euler space. $J = \int |f|^2\, dg \approx f'^2 N \approx 1/N$ Therefore as found by MATTHIES and WAGNER (1996).

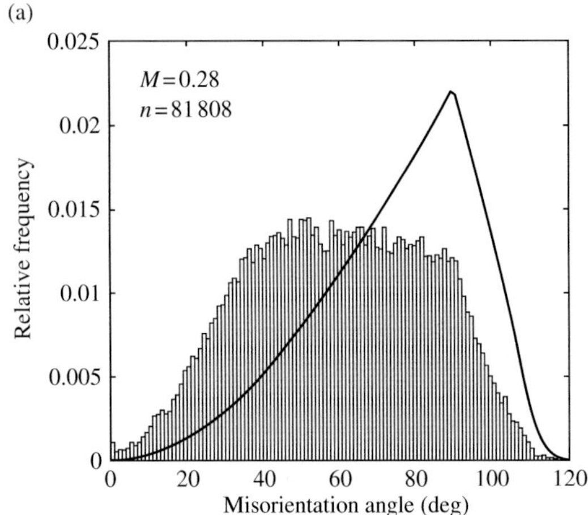

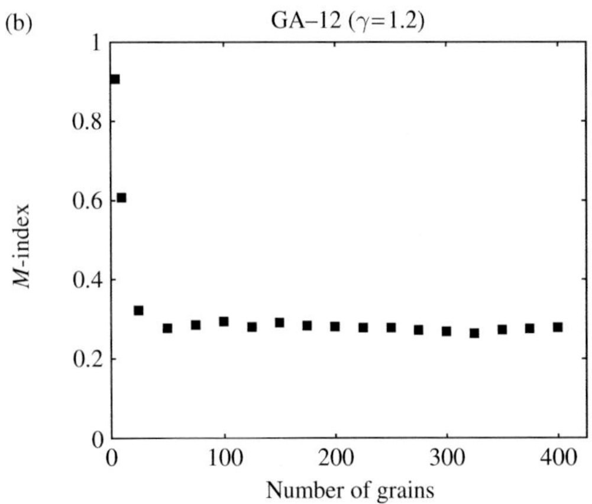

FIGURE 14.7 (a) The distribution of the misorientation angle in deformed olivine in the dislocation creep regime as compared with a theoretical distribution corresponding to a random distribution of orientations. (b) The M-index of deformed olivine as a function of the number of grains for which orientations were measured (after SKEMER *et al.*, 2005).

14.2.4. Symmetry of fabrics

The symmetry of LPO is controlled by the symmetry of the "field" (strain etc.) and the symmetry of crystals. Consequently, knowing the symmetry of crystals, one can deduce the symmetry of deformation from the LPO. Some detailed discussion of this issue is presented by PATERSON and WEISS (1961) and WEISS and WENK (1985). "Symmetry" is defined by an operation or a set of operations by which a given object (in this case pole figures) become identical. For instance [100] direction in a cubic crystal is equivalent to [010] and [001], and these orientations are denoted collectively as ⟨100⟩. Consequently, in a pole figure, there should be three equivalent points corresponding to the orientation of ⟨100⟩. The symmetry of deformation should also affect the symmetry of a pole figure. For instance, if deformation is uni-axial, then the pole figure should have uni-axial symmetry whereas simple shear deformation results in a pole figure that has the orthorhombic symmetry. Some of the examples are shown in Fig. 14.2.

14.3. Mechanisms of lattice-preferred orientation

Two microscopic processes can control the development of lattice-preferred orientation (fabric). In one mechanism, while individual atoms remain in each grain, the crystallographic orientation of individual crystal may rotate by deformation with respect to a certain external reference frame. Another mechanism is that grains with different orientations may grow at the expense of grains with other orientations. This latter mechanism involves grain-boundary migration. The former is called deformation-induced lattice rotation, and the latter (re)crystallization.

14.3.1. Fabric developments by deformation

Rotation of a crystal by constrained deformation

Consider a simple case where a single crystal with only one slip system is deformed experimentally by uni-axial compression. The end of the specimen is *constrained* by the surrounding materials. Let us assume that the deformation occurs by a slip using a single slip system. In this case deformation is simple shear, and if the crystal orientation is fixed with respect to the external reference frame, then there will be a gap between the surrounding materials and the specimen (Fig. 14.8). A crystal rotates to fill this gap. The amount of rotation can be calculated as follows. Note that the length of a material line along the slip direction is preserved. If we define the angle between the compression direction and the slip plane normal to be θ, then the length of a material line along the slip direction, x, is $x = l/\cos\theta$ where l is the thickness of the sample. Because this length does not change by deformation, $dx = dl/\cos\theta + l\sin\theta\, d\theta/\cos^2\theta = 0$ and one obtains,

$$\tan\theta\, d\theta = -dl/l. \tag{14.7}$$

Integrating this, one finds

$$\frac{\cos\theta}{\cos\theta_0} = \frac{l}{l_0} \tag{14.8}$$

where θ_0 is the initial angle and l_0 is the initial sample thickness. This equation means that upon compression, the slip plane rotates toward the normal to the

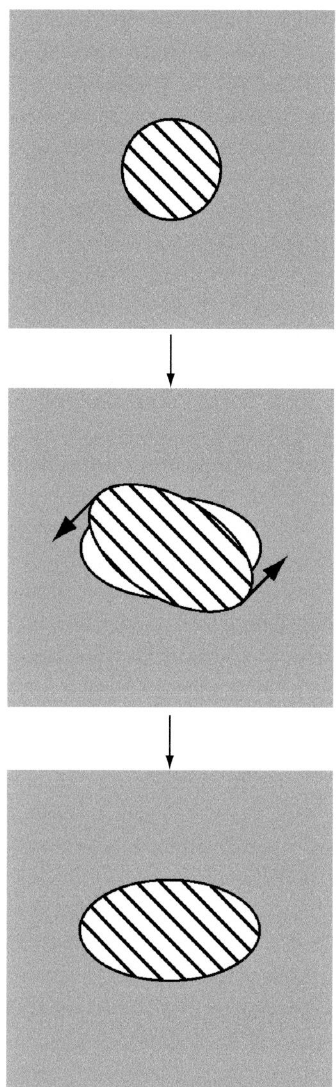

FIGURE 14.8 Constrained deformation of a single crystal with a single slip system (lines indicate the trace of slip planes).

compression direction ($\theta \to \frac{1}{2}\pi$ as $l \to 0$). This is the *Schmid–Boas relation* extended to a finite strain.

This simple case shows that the essence of lattice rotation is related to the rotational component of deformation. Deformation caused by slip (dislocation glide) is simple shear (Chapters 5 and 9), and therefore there is non-zero rigid body rotation, ω_{ij}. The essence of the Schmid–Boas relation is that a crystal rotates its orientation relative to the external frame in such a way that the rotational component of deformation due to slip (ω_{ij}^{ML}: rotation of material points relative to the lattice) and rigid-body rotation (ω_{ij}^{LX}: rotation of lattice relative to the external frame (microscopic rotation)) agrees with the imposed rotational component of deformation (ω_{ij}^{MX}: rotation of material points relative to the external frame (macroscopic rotation)).

Consequently, a general relation describing the rotation of a crystalline lattice relative to the external frame is given by

$$\omega_{ij}^{LX} = \omega_{ij}^{MX} - \omega_{ij}^{ML}. \tag{14.9}$$

Briefly, this equation means that the rotation of crystallographic axes relative to the external reference frame is controlled both by the geometry of *macroscopic* deformation, ω_{ij}^{MX}, and the geometry of *microscopic* deformation, ω_{ij}^{ML}. This equation contains the essence of the mechanism of LPO due to deformation.

During deformation, the orientations of individual grains continuously rotate relative to the external reference frame. When deformation involves rotational components (i.e., deformation due to dislocation glide or twinning), then all the crystals will rotate due to the mismatch of rotation due to microscopic deformation and macroscopic deformation. Among many crystals those in which $\omega_{ij}^{LX} \approx 0$ will rotate their orientation more slowly than others so that these crystals will dominate the LPO. It should be noted that there are no stable or equilibrium orientations toward which crystals rotate. Rather all crystals rotate their crystallographic orientations and the LPO is controlled by the relative rates of rotation as emphasized by Wenk and Christie (1991).

The rotation of material points relative to the crystalline lattice, ω_{ij}^{ML}, is the microscopic controlling factor for LPO. When deformation is due to crystalline slip (dislocation glide) or twinning, rotation of material points relative to the lattice, ω_{ij}^{ML}, is determined by the combination of rotations associated with individual slip systems (and/or twin systems) as

$$\delta\omega_{ij}^{ML} = \frac{1}{2}\sum_{S}\left(l_i^s n_j^s - l_j^s n_i^s\right)\delta\gamma^s \tag{14.10}$$

where $\delta\gamma^s$ is shear strain, $\boldsymbol{n}^s$ is the slip-plane normal and $\boldsymbol{l}^s$ is the slip direction for the *s*th slip system (Chapter 5). To determine the LPO from modeling, one needs to calculate $\delta\gamma^s$ for individual slip systems. This is a complicated issue and in general $\delta\gamma^s$ cannot be determined uniquely. However, some general conclusions can be drawn as follows.

A simple case is deformation due to diffusional mass transport, such as diffusional creep or dislocation creep due to dislocation climb (Nabarro, 1967a). In these cases, microscopic deformation does not have a rotational component, so $\omega_{ij}^{ML} = 0$ and hence $\omega_{ij}^{LX} = \omega_{ij}^{MX}$. This means that all the grains rotate their crystallographic axes in the same way as imposed by the macroscopic geometry of deformation. Consequently no LPO will develop due to deformation by diffusional creep (or dislocation climb) and the pre-existing LPO will rather be destroyed. The destruction of the pre-existing LPO by superplastic deformation (Chapter 8) was demonstrated in metals (e.g., Au–Cu eutectic alloy, Edington *et al.*, 1976) and in some minerals (e.g., orthopyroxene, Boullier and Gueguen, 1975).

When deformation is due to slip (dislocation glide) and/or twinning, then microscopic deformation includes rotational components. Consequently, the nature of LPO in these cases is controlled by the partitioning of deformation among various slip (twin) systems. Several theoretical models have been developed by which LPO by dislocation glide can be calculated, including the *Taylor–Bishop–Hill model* (e.g., Lister *et al.*, 1978; van Houtte and Wagner, 1985) and the *self-consistent approach* (e.g., Molinari *et al.*, 1987; Wenk *et al.*, 1991). In the Taylor–Bishop–Hill model, strain is assumed to be homogeneous and therefore more than five independent slip systems are needed. In general, the partitioning of strain among several slip systems cannot be uniquely calculated. In this model, strain in each slip system is determined using a criterion that the actual strain partitioning occurs in such a way that the energy dissipation is minimum (van Houtte and Wagner, 1985). However, in many of anisotropic minerals, the number of independent slip systems is less than five and the Taylor–Bishop–Hill model cannot be applied. The self-consistent approach is one of the models by which fabric development and strength of polycrystalline materials with less than five slip systems can be calculated (see Chapter 12). In this model, a crystal is considered to be an inclusion in a continuum matrix whose properties depend on the interaction between the inclusion and the matrix itself. The interaction equations between the matrix and the inclusion are solved such that the matrix properties are considered to be a function of the stress-strain field of an inclusion: hence the name "self-consistent approach" (Molinari *et al.*, 1987; Lebensohn and Tomé, 1993; also see Chapter 12). The issue of strain compatibility is circumvented by treating the matrix as a continuum. An alternative model was proposed by Etchecopar and Visseur (1987), Ribe (1989a), and Ribe and Yu (1991) in which the strain accommodation problem is solved approximately by minimizing the gaps and/or overlaps. Although the details are different among various models, the LPOs calculated for some minerals such as olivine are very similar. As a general rule, the largest strain is partitioned in the easiest (set of) slip systems, and has the largest influence on LPO.

The simplest case in this category is a case in which only one slip system dominates. This end-member case occurs when contributions from subordinate deformation mechanisms are significant (which provide strain components needed for strain accommodation) but does not directly alter the orientations of crystals. LPO developed by deformation under conditions close to the dislocation–diffusional creep boundary is such an example where both dislocation and diffusional creep contribute to total strain significantly. In this case, ω_{ij}^{ML} is for that particular slip system and the dominant orientations of crystals are such that deformation-induced microscopic rotation agrees with the rigid-body rotation imposed by the macroscopic deformation geometry. In simple shear, the orientation of the microscopic shear direction (plane) becomes sub-parallel to the macroscopic shear direction (plane). Since deformation conditions in most parts of Earth are close to the boundary between dislocation–diffusional creep (e.g., Karato *et al.*, 1986; De Bresser *et al.*, 1998), this end-member LPO is often observed in Earth materials.

A similar but somewhat different case is a case in which a single set of slip systems such as the $\langle 111 \rangle \{1\bar{1}0\}$ or $\langle 1\bar{1}0 \rangle \{111\}$ slip system for bcc or fcc crystals respectively provides enough strain components to accommodate arbitrary strain. In such a case, although a single set of slip systems controls the LPO, the single set of slip systems has a large number of equivalent slip systems and therefore ω_{ij}^{ML} must be calculated using various equivalent slip directions and slip planes. In a more general case, a combination of different slip systems must be considered to explain the LPO.

Inference of a dominant slip system from the inverse (or pole) figure is straightforward only in cases where a single slip dominates in low-symmetry materials. In more general cases, the crystallographic direction parallel to the shear direction in simple shear does not necessarily correspond to the crystallographic slip direction. For instance, in NaCl, the slip direction is $\langle 110 \rangle$, but in the inverse pole figure, the $\langle 111 \rangle$ and/or $\langle 100 \rangle$ directions are nearly parallel to the shear direction or extrusion direction (Fig. 14.3a).

This is due to the combination of several equivalent slip systems in this high-symmetry crystal (Wenk *et al.*, 1989).

Problem 14.4

Assuming only one slip system, [100](010), dominates in a crystal with orthorhombic symmetry (e.g., olivine), predict LPO for simple shear, uni-axial tension and compression (for infinite strain).

Solution

For simple shear, the [100] direction becomes parallel to the shear direction and the (010) plane becomes parallel to the shear plane. For uni-axial tension, the [100] direction becomes parallel to the tension direction and the other two axes are normal to the tension direction. For uni-axial compression, the [010] direction becomes parallel to the compression direction, and the other two axes are normal to it.

14.3.2. Fabric development due to oriented crystallization

LPO due to orientation-dependent crystallization is well known in metallurgy and in the study of ice sheets in oceans (e.g., Bennington, 1963; Nes *et al.*, 1984; Gandin *et al.*, 1995; Wettlaufer *et al.*, 1997; Kurz and Fischer, 1998). When a polycrystalline material is formed by crystallization from a liquid, grains with specific orientation may be selectively crystallized if the free energy of the grain depends on the orientation (*oriented crystallization*). Alternatively, the growth rate of grains with certain direction(s) could be faster than grains with other directions, causing LPO (*oriented growth*). The orientation dependence of free energy can occur when some anisotropic *field* is present. For example, Karato (1993b) proposed that when crystallization occurs in the presence of a magnetic field, then grains with low magnetization energy will grow faster than others. Similarly if the growth rate is controlled by heat flow then anisotropic growth could occur because thermal conductivity can be anisotropic (e.g., Bergman, 1997). For example, when the rate of solidification is controlled by the rate of removal of heat, then the crystal with the orientation that is most effective for the removal of latent heat tends to grow faster than the others. In this case, the LPO will be controlled by the anisotropy in thermal conductivity.

14.3.3. Effects of dynamic recrystallization on crystallographic fabric

In the previous discussion of deformation-induced LPO, we assumed that the crystallographic orientation of a group of atoms in a given grain remains the same during the LPO development, and LPO develops as a result of change in orientation of a group of atoms as a whole. This assumption can be invalid during high-temperature deformation. During high-temperature deformation, formation of new grains and/or growth and consumption of grains can occur as a result of heterogeneity of deformation. These processes are collectively referred to as *dynamic recrystallization* (Chapter 13). Dynamic recrystallization affects LPO in several ways (e.g., Gottstein and Mecking, 1985; Urai *et al.*, 1986a; Karato, 1987b; Jessel, 1988a, 1988b; Doherty *et al.*, 1997; Wenk *et al.*, 1997; Kaminski and Ribe, 2001; Lee *et al.*, 2002).

The essence of the processes by which LPO is affected by dynamic recrystallization can be summarized as follows. First, dynamic recrystallization often results in grain-size reduction. Grain-size reduction promotes grain-size sensitive deformation mechanisms, thereby reducing the number of necessary slip systems to achieve plastic deformation. This will promote the development of an end-member, single-crystal LPO. This process is important in both in Earth and in laboratory experiments, because deformation conditions are close to the boundary between dislocation and diffusional creep in both cases (see Chapters 9 and 19). Therefore materials in regions of a smaller grain size can often deform by grain-size sensitive creep such as diffusional creep that relaxes the constraints on deformation in coarse-grained regions that deform by dislocation creep (Lee *et al.*, 2002). Second, grain-boundary migration affects LPO when grains with certain orientations grow at the expense of grains with other orientations (*oriented growth*). Third, grains with particular orientations may be selectively nucleated (*oriented nucleation*) and if the subsequent growth process does not modify their orientations, then oriented nucleation would affect LPO. The subgrain rotation that occurs in most minerals during the early stage of dynamic recrystallization does not have a strong direct effect on crystal orientation. Its effect is mainly to make the orientation distribution more diffuse (Karato, 1988). It is the subsequent grain-boundary migration and/or resultant grain-size reduction that affects LPO.

There are two key issues that need to be addressed in examining the role of grain-boundary migration in LPO development. First, one needs to know how selective nucleation and/or growth result in a change in LPO. The grain-boundary migration associated with dynamic recrystallization occurs due to the contrast in dislocation density between neighboring grains. As deformation proceeds, the dislocation density becomes higher and in many cases, it becomes more heterogeneous due to the constraints imposed by neighboring grains (e.g., geometrically necessary dislocations, Ashby, 1970). Consequently, at some point, the difference in dislocation density becomes high enough to overcome the excess energy associated with creating new (portions of) grain boundaries. When this condition is met, then grains with lower dislocation densities will grow at the expense of grains with higher dislocation densities (Karato, 1987b). Therefore a key factor here is the orientation dependence of dislocation density. The orientation dependence of dislocation density depends critically on the stress–strain distribution in a deforming polycrystal. When stress is homogeneous, i.e., when individual grains deform like an isolated free grain, then crystals with soft orientations will have higher dislocation densities (because of a higher resolved shear stress on a dominant slip system). In contrast, when strain is homogeneous, i.e., when the constraints for deformation are tight, then crystals with hard-to-deform orientations must deform to the same strain as others, which implies that the local stress for crystals with hard orientations will be higher than crystals in other orientations. In the latter case, recrystallization would proceed first in grains with hard orientations, whereas in the former, recrystallization will proceed in grains with soft orientations. If these grains with high dislocation densities are selectively consumed, LPO would develop. Both cases have been observed in different materials and, obviously, the issue of the orientation dependence of dislocation density depends on the materials and the physical/chemical conditions of deformation (Urai *et al.*, 1986a; Karato, 1987b; Karato and Lee, 1999).

Second is the competition between recrystallization-induced LPO and deformation-induced LPO. Both processes occur concurrently and hence the relative importance of these two processes must be evaluated. Karato (1987b) introduced a non-dimensional parameter,

$$\xi \equiv \frac{L\dot{\varepsilon}}{v\varepsilon_1} \tag{14.11}$$

where L is grain size, $\dot{\varepsilon}$ is strain rate, v is the velocity of grain-boundary migration and ε_1 is the critical strain for recrystallization. For a smaller ξ, the effects of grain-boundary migration play a more important role than deformation-induced lattice rotation whereas for a large ξ effects of deformation dominates. A similar parameter was used by Kaminski and Ribe (2001).

In summary, the processes of LPO development during dynamic recrystallization are more complicated than those in deformation-induced LPO. The details of the formation of heterogeneity and the anisotropy of dislocation density and the competition of grain-boundary migration with concurrent deformation-induced lattice rotation need to be considered. Observations of deforming samples by "see-through" experiments on analog materials provide important insights into the complicated geometrical changes during dynamic recrystallization (e.g., Urai, 1983, 1987; Urai *et al.*, 1986a). However, the most critical information is the dislocation distribution in samples that have undergone dynamic recrystallization (see, e.g., Lee *et al.*, 2002). Obviously currently available models do not capture all of the essential physics of dynamic recrystallization and a lot more needs to be learned on the role of dynamic recrystallization in LPO development.

14.3.4. Fabric development in diffusional creep or superplasticity

Diffusional creep or superplasticity (Chapter 8) does not have an obvious process to rotate the crystal lattice. Deformation by diffusional mass transport does not have a rotational component. Consequently, diffusional creep does not result in LPO. In fact, Edington *et al.* (1976) reported that a pre-existing fabric was destroyed by superplastic deformation of an Au–Cu eutectic alloy. Similarly, large-strain experiments on $CaTiO_3$ perovskite in simple shear in a diffusional creep regime showed very weak LPO (Karato *et al.*, 1995b). However, some exceptional cases have also been reported. First is the experimental observation on calcite in the so-called superplastic regime. Rutter *et al.* (1994) conducted tri-axial compression and tension tests on calcite in the superplastic creep regime and found slow but significant LPO development (also Karato *et al.* (1995b) found significant LPO in a diffusion creep regime when the conditions are close to the boundary with dislocation creep in $CaTiO_3$ perovskite, $n \sim 1$). Similarly Pieri *et al.*

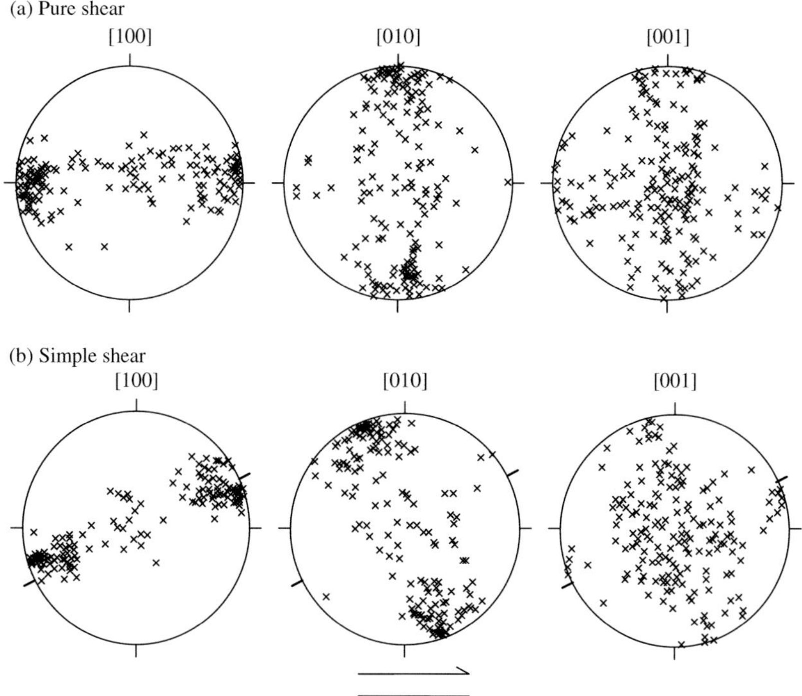

FIGURE 14.9 Pole figures of olivine (Ribe and Yu, 1991) based on numerical modeling (simple shear).

(2001) conducted torsion tests on calcite and found very strong fabric after large-strain deformation in a similar regime. Superplasticity is defined here as a deformation processes characterized by a small stress exponent ($n \sim 1.7$) and by the lack of significant grain elongation. These results are at odds in view of the common belief that superplastic deformation does not create significant fabrics (see 14.3.1). The current interpretation of these results is that a significant fraction of deformation in these cases is due to dislocation glide, and this is particularly the case of Pieri *et al.* (2001) where the grain-size reduction occurred as a result of dynamic recrystallization. In such a case, the grain size remains close to the critical grain size for transition to dislocation creep (see Chapter 13), and an end-member, single-slip fabric could develop. Second is the fabric development during pressure-solution creep (Chapter 8). Bons and Den Brok (2000) showed that the rate of dissolution and/or precipitation may depend on crystallographic orientation that could cause fabric development in the diffusion-controlled creep.

14.3.5. Controlling framework for LPO

The essence of LPO caused by slip is the geometrical mismatch of microscopic and macroscopic deformation (equation (14.8)). Therefore the LPO is controlled by the geometry of deformation and *not directly by the orientation of applied stress*. However, there is a controversy as to the exact controlling factor of LPO formed by crystalline slip. On the one hand, based on numerical modeling, the symmetry of LPO (pole figures) sometimes follows the shear direction and shear plane (quartz: Lister and Hobbs, 1980). In this case, the directions of the peaks in pole figures do not change with strain, when the shear plane–shear direction is used as a reference frame. On the other hand, other numerical models show that the symmetry of pole figures follows that of a finite-strain ellipsoid (olivine: Ribe and Yu, 1991; Wenk *et al.*, 1991; Tommasi *et al.*, 2000; Fig. 14.9). In this case, the directions of peaks in pole figures rotate when the shear plane–shear direction is used as a reference frame in pole figures. The distinction between these two cases is important because this issue is related to the inference of deformation geometry (such as sense of shear) from LPO (Bouchez *et al.*, 1983). If pole figures follow the shear plane and shear direction, then the pole figures plotted using the finite-strain ellipsoid as a reference frame will show a deviation of peak position with respect to the finite-strain ellipsoid. This deviation can be used as an indicator of the sense of shear (Bouchez *et al.*, 1983). On the other hand, if the pole figures follow the symmetry of the finite-strain ellipsoid, then the peak positions in a pole figure would coincide with the main axes of the pole figures (when pole figures are constructed using the finite-strain ellipsoid as a reference frame), and therefore there is no way of inferring the sense of shear from LPO (see also TAKESHITA *et al.*, 1990). The experimental study by ZHANG and KARATO (1995) on olivine under water-poor conditions showed that at relatively low strains where little dynamic recrystallization is observed, the symmetry of LPO follows closely that of a finite-strain ellipsoid, whereas after the onset of dynamic recrystallization, it follows close to the shear direction and shear plane.

In addition to these cases, JUNG *et al.* (2006) reported cases in which the pole figures' peak positions deviate significantly from both of these frameworks. The origin of such a deviation is not well understood, but one possibility is the significant influence of fast grain-boundary migration: when grain-boundary migration is fast, stress orientation could influence LPO (KARATO, 1987b). Also when two slip systems that have conjugate slip directions/slip planes (i.e., the [100](001) and [001](100)) play an important role, the position of peaks in pole figures depends on the partitioning of strain between the two slip systems and can deviate significantly from the strain ellipsoid or from shear plane–shear direction. Consequently, care must be exercised in inferring the sense of shear from the fabrics of this type (see KATAYAMA *et al.*, 2004).

14.4. A fabric diagram

Deformation fabrics in a given material depend on the deformation geometry and the physico-chemical conditions of deformation. A change in the fabric (LPO) due to the change in physico-chemical conditions of deformation is referred to as a *fabric transition*. If the conditions under which a particular type of fabric dominates are experimentally constrained then from the fabric one can infer the conditions under which a rock has attained the fabric (e.g., CARTER and AVÉ LALLEMANT, 1970; LISTER, 1979; LISTER and PATERSON, 1979; HOBBS, 1985; JUNG and KARATO, 2001b; KATAYAMA *et al.*, 2004; JUNG *et al.*, 2006). This is similar to the inference of physical conditions (such as temperature and pressure) from the chemical composition or existing stable phases in a rock. In the latter, the phase diagram (see Chapter 2) plays a key role. Similarly, when one wants to infer the physico-chemical conditions from deformation fabrics, one needs to know the relation between deformation fabrics and physico-chemical conditions of deformation. A diagram showing dominant deformation fabrics in a given parameter space is referred to as a *fabric diagram*. Understanding the nature of a fabric diagram is important in applying experimental results to Earth.

All of the laboratory studies of deformation experiments are conducted at much faster strain rates than those in Earth. Consequently, the LPO observed in laboratory experiments may not be an important one under geological conditions. In order to apply laboratory data on LPO to deformation in Earth, one needs to understand the *scaling laws* in fabric diagrams.

An obvious mechanism for a fabric transition is a change in the dominant (easiest) slip system(s). In this case, the fabric transition will occur when the strain rates of two competing slip systems become equal, namely,

$$\dot{\varepsilon}_1(T, P, \sigma, L; C) = \dot{\varepsilon}_2(T, P, \sigma, L; C) \tag{14.12}$$

where $\dot{\varepsilon}_{1,2}$ is the strain rate for slip systems 1 and 2, respectively, and T is temperature, P is pressure, σ is stress, L is grain size and C is a parameter that characterizes chemical environment such as water fugacity. Therefore the boundary between two fabrics is defined by a function

$$F(T, P, \sigma, L; C) = 0 \tag{14.13}$$

that is a solution of equation (14.12). Therefore the fabric boundaries are represented by a hyper-surface in the multi-dimensional space. Several properties of a fabric diagram can be immediately noted (KARATO, 2007).

(1) A fabric diagram does not explicitly depend on the strain rate. This is obvious from (14.13), which does not contain strain rate. This important property is due to the fact that the fabric boundaries are determined by the *relative* strain rates of two slip systems. Consequently, as far as strain rates are concerned, there is no need for extrapolation: experimental results obtained at much higher strain rates than in Earth can be directly applied to Earth where strain rates are much lower than those in laboratory experiments.
(2) Although the boundaries among various fabric types shown in a fabric diagram do not depend on the strain rates, experimental results obtained at high strain rates cover only some portion of the parameter space. For example, most experimental studies are conducted at much higher stresses than those in most of the Earth. Therefore proper

extrapolation is needed to construct a fabric diagram for a broad range of parameter space. This extrapolation is often made using equations such as (14.12) and requires the knowledge of flow-law equations.

(3) A fabric transition could occur not only by a change in the relative easiness of deformation by different slip systems, but also by the change in relative rates of lattice rotation by deformation and the rate of grain-boundary migration. In such a case, we need to compare rates of these processes to define the fabric boundary (KARATO, 1987b).

Fig. 14.10 shows two examples of such fabric diagrams (one for quartz and the other for olivine).

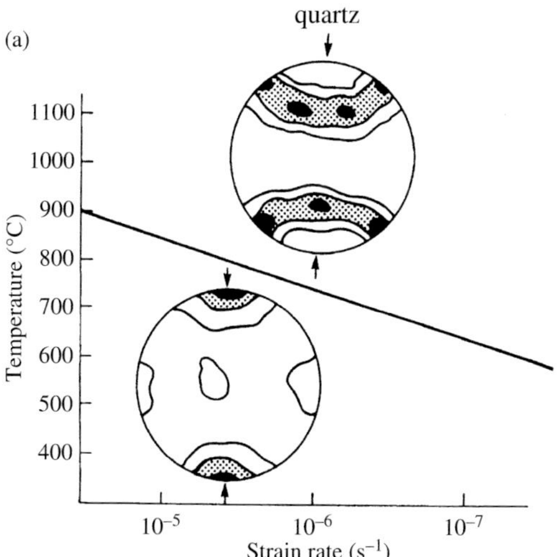

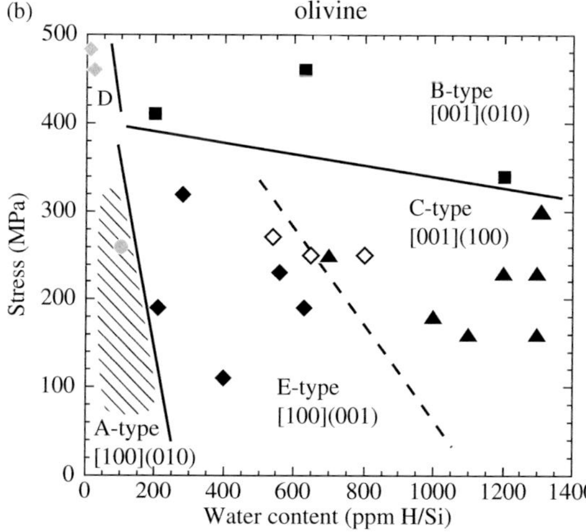

FIGURE 14.10 (a) A fabric diagram for quartz (TULLIS *et al.*, 1973) and (b) a fabric diagram for olivine (KATAYAMA *et al.*, 2004).

Let us consider some examples. Consider a case where a power-law equation is appropriate for all the slip systems involved,

$$\dot{\varepsilon}_{1,2} = A_{1,2}\sigma^{n_{1,2}} \exp\left(-\frac{H^*_{1,2}}{RT}\right). \quad (14.14)$$

The fabric boundary for this case is given by

$$\log\frac{A_1}{A_2} + (n_1 - n_2)\log\sigma - \frac{H^*_1 - H^*_2}{RT} = 0. \quad (14.15)$$

For an exponential flow law

$$\dot{\varepsilon}_{1,2} = A_{1,2}\sigma^2 \exp\left(-\frac{H^*_{1,2} - B_{1,2}\sigma}{RT}\right) \quad (14.16)$$

the fabric boundary is given by

$$RT\log\frac{A_1}{A_2} - (H^*_1 - H^*_2) + (B_1 - B_2)\sigma = 0. \quad (14.17)$$

If the dependence of the flow-law parameters on water content (water fugacity) is known, these relations can be written in terms of water content (or fugacity).

14.5. Summary

Significant progress has been made in the study of lattice-preferred orientation in rocks. Major driving forces for this progress are (1) the improved analytical techniques for measuring LPO (i.e., EBSD), (2) the improvement in experimental studies for large-strain, non-coaxial deformation experiments and (3) new developments in theoretical modeling.

It is now straightforward to measure the LPOs from naturally deformed rocks and calculate any structure-related properties such as ODF and seismic anisotropy. However, inferring geologically or geophysically significant conclusions from such measurements is not straightforward. This requires an understanding of the relationship between LPO and deformation conditions. In this chapter, I have reviewed some of the fundamentals of LPO development based on both experimental and theoretical studies.

The key factors in the materials science of LPO are (i) the dominant slip systems of dislocations and (ii) the deformation mechanism maps (boundary between dislocation and diffusional creep). Currently the only meaningful way to determine the dominant slip systems in a given crystal is to conduct an experimental study of plastic deformation under well-controlled physical and chemical conditions. Experimental studies on olivine by JUNG and KARATO (2001b), KATAYAMA *et al.* (2004), JUNG *et al.* (2006) and Katayama and Karato (2006) have revealed a rich variety in LPO in olivine based on large-strain deformation experiments in simple shear under a controlled chemical environment. However similar studies on deep Earth minerals are difficult and we currently have little knowledge on LPO of materials composing more than 90% of Earth. There have been some attempts to infer dominant slip systems in minerals and calculate LPO or the nature of seismic anisotropy from theoretical calculations (e.g., Thurel *et al.*, 2003b; Tommasi *et al.*, 2004; Durinck *et al.*, 2005a; Oganov *et al.*, 2005). This would be a useful approach because deformation experiments are difficult to conduct under deep Earth conditions (see Chapter 6). Unfortunately, the usefulness of this approach is highly limited at present since in all of the studies the strength was calculated for homogeneous shear and the concept of dislocation was not used in the calculations of strength.

I should note, however, that theoretical approaches have the advantage of being able to calculate LPO development for complicated or a wide range of geometries (e.g., Ribe, 1989b; Tommasi *et al.*, 2000; Kaminski *et al.*, 2004), but the range of predictability of current models is again limited. Beyond the qualitative prediction of LPO based on the (assumed) dominant slip systems, many details from such calculations are unconstrained from a materials science point of view. For instance, the role of dynamic recrystallization that controls the rate of LPO development and its geometry (e.g., Karato, 1987b; Zhang and Karato, 1995; Lee *et al.*, 2002) is not well incorporated in currently available models. Consequently, the evolution of LPO with strain predicted from these models has limited predictive power at this time.

In summary, the best approach for using LPO to infer dynamic processes in Earth is to combine materials science knowledge of LPO development (based on experimental results and theoretical analysis of scaling laws) with a range of plausible models of dynamic processes, and compare the results of various models with seismological observations. An example of such an approach is given by Kneller *et al.* (2005) on the subduction zone where the physical and chemical conditions are likely to change in a complicated fashion.

15 Effects of phase transformations

Phase transformations occur in many regions in Earth (particularly in the transition zone, 410–660 km depth in the mantle) and have several important effects on plastic properties. A phase transformation changes crystal structure and the nature of the chemical bonding that affects any physical properties including plastic properties. In addition, a first-order phase transformation results in a change in grain size and a redistribution of internal stress–strain that may influence the nature of plastic deformation. A second-order phase transformation will not result in these changes, but will result in the anomalous reduction of an elastic constant and/or the anomalous increase in the dielectric constant. All of these changes have certain effects on rheological properties. This chapter presents a summary of the experimental observations and theoretical models on which one can estimate the relative importance of various effects in geological processes.

Key words isomechanical group, transformation plasticity, Greenwood–Johnson model, Poirier model, grain-size reduction.

15.1. Introduction

Phase transformations of materials occur in Earth's interior. The most important ones are those that occur in the mantle transition zone (410–660 km; see Chapter 17), but phase transformations also occur in the crust and the upper mantle, and perhaps in the lower mantle. A phase transformation affects the rheological properties in a variety of ways. (1) A phase transformation results in a different crystal structure and chemical bonding, which will modify all the properties including the rheological properties (*crystal structure, chemical bonding effects*). (2) A first-order phase transformation (for a classification of phase transformations see Chapter 2) results in the volumetric strain that causes deviatoric stress and strain, which may affect the rheological behavior (*internal stress–strain effects*). (3) A (first-order) phase transformation results in the modifications of microstructures particularly grain size, which will cause a change in rheological properties (*grain-size effect*; also see Chapter 13). (4) A first-order transformation results in the release (or absorption) of heat, which results in a change in temperature and hence a change in rheological properties. (5) Second-order phase transformations can be associated with anomalies in elastic and dielectric properties (e.g., GHOSE, 1985). There are some reports indicating anomalous rheological behavior associated with second-order phase transformations.

15.2. Effects of crystal structure and chemical bonding: isomechanical groups

A phase transformation will change the geometry of the atomic arrangement (crystal structure) as well as the nature of chemical bonding. The bcc (body-centered-cubic) structure to hcp (hexagonal-close-pack) structure transformation in iron, for example, changes the crystal structure from cubic to hexagonal symmetry and hence causes a large change in anisotropy in many physical properties including plastic properties. The graphite to diamond transformation changes the crystal structure but in this case, the change in chemical bonding (from weak van der Waals bonding in graphite to strong covalent bonding in diamond) has a larger effect in changing the rheological properties. For some cases, the effects of phase transformations on plastic properties can be directly studied, whereas in many cases direct deformation experiments are difficult under the conditions where a phase transformation occurs. In the latter case, the use of analog materials that assume the same crystal structure and chemical bonding at accessible conditions helps to infer the plastic properties of a material for which direct mechanical tests have not yet been conducted. Frost and Ashby (1982) provide an extensive review of the effects of crystal structure and chemical bonding on rheological properties. Karato (1989c, 1997b) discusses some of the important effects in oxides and silicates. In particular, Karato (1989c) noted the important effect of crystal structure on the creep strength of oxides in the dislocation creep regime.

15.2.1. Direct mechanical tests on plastic properties of deep Earth materials

The effects of change in the crystal structure on the rheological properties can be investigated directly if deformation experiments are conducted under the conditions under which a phase transformation occurs. For example, for quartz, it is possible to conduct deformation experiments both for α- and β-quartz. In these cases, a direct comparison of the rheological properties is possible. However, for the majority of the phase transformations that occur in Earth, direct deformation experiments are difficult because many phase transformations occur at high pressures (e.g., the olivine to wadsleyite transformation occurs at $\sim$13–16 GPa depending on the temperature). Under these conditions, quantitative deformation experiments are challenging, although significant progress is being made using new types of deformation apparatus (Weidner *et al.*, 1998; Durham *et al.*, 2002; Xu *et al.*, 2005). I should emphasize that although there have been some reports on room- or low-temperature ($T/T_m < 0.3$) mechanical properties of deep Earth materials under pressures similar to deep Earth conditions (e.g., Meade and Jeanloz, 1990; Mao *et al.*, 1998, 2002; Merkel *et al.*, 2002, 2003, 2004; Wenk *et al.*, 2004), the relevance of these data to Earth is highly questionable because the deformation mechanisms at low temperatures are different from those under high temperatures.

Some experimental observations

- SiO_2 (α-quartz and β-quartz: $T_{tr} = 846$ K (at room P), quartz to coesite transformation: $P_{tr} = 3$–4 GPa at 700–1000 K, coesite to stishovite: $P_{tr} = 8$–9 GPa, at $T = 1000$–1500 K)

Quartz undergoes a series of phase transformations. The α- to β-quartz transformation that occurs at 846 K (at room P) is nearly a second-order transformation. Whereas most other transformations are first-order transformations. The high-temperature deformation of quartz in both the α- and β-quartz field was investigated by Kirby (1977) who conducted deformation experiments on synthetic quartz single crystals that contain $\sim$900 ppm H/Si at temperatures of 676–1047 K. The slip system $\{\bar{1}2\bar{1}0\}[0001]$ was activated in this study and no marked change in strain rate was observed at $T = T_{tr}$ (846 K), but a marked change in the activation energy ($E^* = 162$ kJ/mol in the α-quartz and $E^* = 60$ kJ/mol in the β-quartz field) was found (Fig. 15.1). The interpretation of this result is difficult because the sample was not in chemical equilibrium with the surrounding atmosphere. The solubility of water in quartz is pressure-dependent and therefore there must have been some degree of dehydration or precipitation of water during deformation experiments without confining pressure. In contrast to this work, Schmidt *et al.* (2003) concluded that there is a weakening near the $\alpha-\beta$ transformation. Their work is, however, indirect since they studied the deformation through the change in pressure in fluid inclusions.

The plastic deformation of coesite was studied by Renner *et al.* (2001). Their results show a higher strength of coesite than quartz at comparable conditions, but the role of water has not been well quantified. Shieh *et al.* (2002) investigated the strength of stishovite as well as its high-pressure polymorph ($CaCl_2$ structure) using a diamond anvil at room temperature.

- $(Mg,Fe)_2SiO_4$ (olivine, wadsleyite, ringwoodite, perovskite + magnesiowüstite)

The first experimental study on the effects of olivine to spinel transformation was made by Vaughan and Coe (1981) on Mg_2GeO_4 that undergoes the olivine to spinel transformation at low pressures (less than 1 GPa). This is an analog material of $(Mg,Fe)_2SiO_4$ that undergoes a similar transformation at much higher pressures (>17 GPa). Vaughan and Coe (1978) showed that the slip systems in Mg_2GeO_4 olivine are

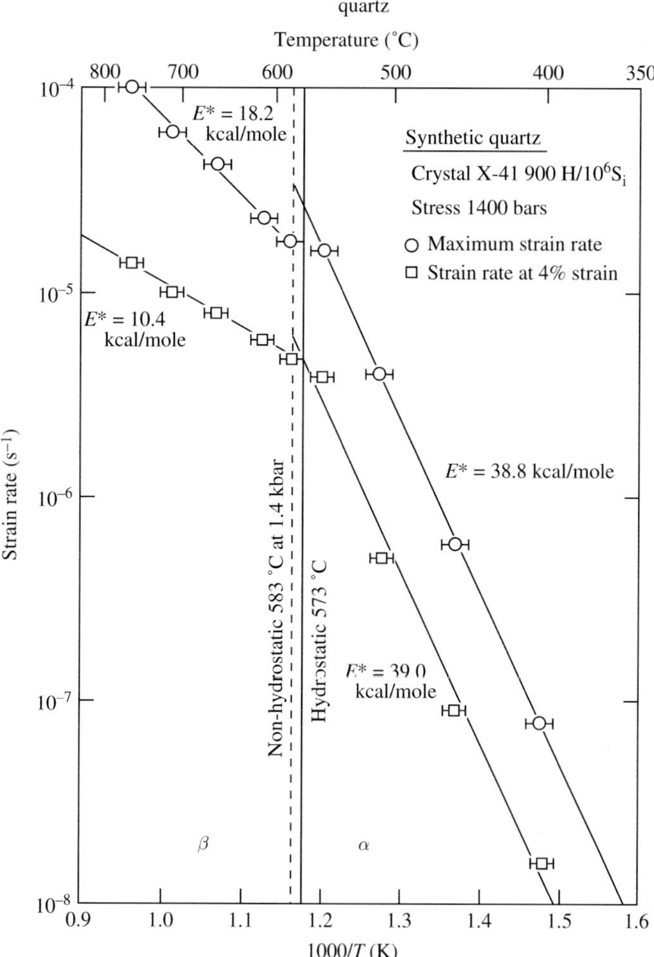

FIGURE 15.1 Effects of an α–β phase transformation on the deformation of quartz (Kirby, 1977).

similar to $(Mg,Fe)_2SiO_4$ olivine. Vaughan and Coe (1981) concluded that the main effect of this phase transformation is to change the strength through the change in grain size. However, the relative strength of olivine and spinel at the same normalized conditions was not determined. Karato *et al.* (1998) made deformation experiments on the spinel phase of $(Mg, Fe)_2SiO_4$ (i.e., ringwoodite) at transition-zone conditions. In both of these studies, the stress was not measured with enough accuracy, and therefore the results must be considered as semi-quantitative. Nevertheless, these two studies identified the importance of grain-size sensitive creep and Karato *et al.* (1998) determined the deformation mechanism map for ringwoodite under the limited conditions.

Deformation mechanisms in wadsleyite and ringwoodite have been studied using TEM analysis (Sharp *et al.*, 1994; Dupas-Bruzek *et al.*, 1998; Karato *et al.*, 1998; Cordier, 2002). At high stress and/or coarse grain size, dislocation creep dominates and the dominant slip systems have been identified. The slip systems in ringwoodite (silicate spinel) are similar to (nearly stoichiometric) oxide spinel. The $\langle 110\rangle\{1\bar{1}0\}$ and $\langle 110\rangle\{1\bar{1}1\}$ slip systems dominate with minor contribution from the $\langle 110\rangle\{001\}$ slip systems in both materials. The situation is more complicated in wadsleyite. Chen *et al.* (1998) investigated the creep properties of olivine, wadsleyite and ringwoodite at high pressures and modest temperatures and found that wadsleyite and ringwoodite are stronger than olivine, but the water-weakening effects are large only for olivine (see also Xu *et al.*, 2003). Kubo *et al.* (1998) investigated the kinetics of olivine to wadsleyite transformation as a function of water content and inferred the role of water on plastic deformation of wadsleyite. Under the assumption that the kinetics is controlled by the plastic deformation of wadsleyite, they concluded that water enhances plastic deformation of wadsleyite. They also show evidence of enhanced dislocation recovery in wadsleyite from TEM observations. The significance of these results to Earth's mantle is not clear, however, because both of them are indirect measurements: CHEN *et al.* (1998) investigated the deformation of powder samples at modest temperatures and therefore microscopic processes supporting the stress are unclear, and KUBO *et al.* (1998) did not directly investigate plastic deformation. NISHIHARA *et al.* (2007b) determined the creep strength of wadsleyite and olivine using a rotational Drickamer apparatus and concluded that at water-poor conditions, wadsleyite is significantly stronger than olivine.

POIRIER *et al.* (1986) investigated the microstructure of perovskite + magnesiowüstite aggregates after the transformation from ringwoodite and concluded that perovskite is stronger than magnesiowüstite. CHEN *et al.* (2002b) and KARATO *et al.* (1990) investigated the strength of perovskite at low to modest temperatures but at pressures outside its stability field. MEADE and JEANLOZ (1990), MEADE *et al.* (1995) and MERKEL *et al.* (2003) investigated the deformation of perovskite under high-pressure conditions at room temperature. Because the mechanisms of plastic deformation probably change with temperature, the relevance of low-temperature experiments to deformation in Earth's deep interior is questionable. CORDIER *et al.* (2004) investigated the plastic deformation of $MgSiO_3$ perovskite under upper lower mantle conditions using the stress-relaxation test. Slip systems are inferred from X-ray diffraction. The stress magnitude was not determined. KUBO *et al.* (2000) emphasized the importance of grain-size reduction associated with the formation of the perovskite and magnesiowüstite assembly that could result in rheological weakening. All of these studies are semi-quantitative and the plastic properties of wadsleyite, ringwoodite and silicate perovskite are largely unconstrained at this time (2007).

- NaCl, CsCl (B1 to B2)

MEADE and JEANLOZ (1988) investigated the plasticity of NaCl across the transformation from the B1 (NaCl) to B2 (CsCl) structure. The experimental study was made at room temperature using a diamond anvil cell. Samples were squeezed between two anvils and the differential stresses supported by the sample were determined by the lateral variation of pressure that was measured by the shift of fluorescence peak of ruby (for more details of the technique used see Chapter 6). These authors found that the room temperature strength of the B2 structure is significantly lower than that of the B1 structure. They proposed a model based on ionic radii consideration, but the results can

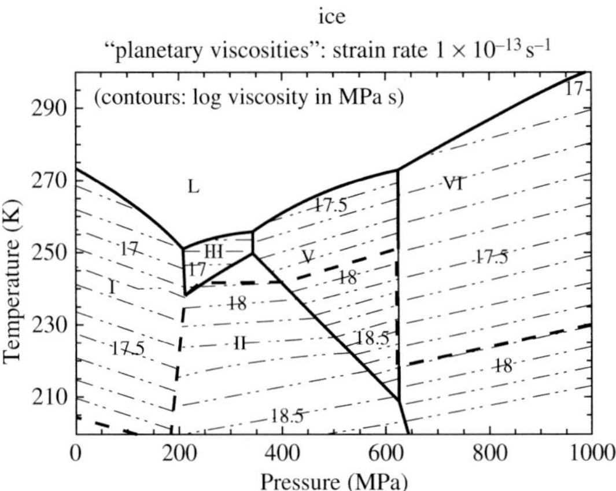

FIGURE 15.2 Effects of phase transformations on the deformation of ice (DURHAM and STERN, 2001).
I, II, III, V, VI indicate different crystal structures of ice (L is liquid water). A broken line shows the temperature–pressure conditions corresponding to a viscosity of $10^{17.9}$ MPa s (at the strain rate of $10^{-13}\,s^{-1}$).

be interpreted by the model of Peierls stress-controlled flow that implies that the strength is highly sensitive to the ratio of spacing of glide planes and the length of the Burgers vector (TAKEUCHI and SUZUKI, 1988). However, because the measurements were made at room temperature and no measurements were made for the dependence of strength on strain rate nor temperature, these results cannot be used to infer the high-temperature plasticity of this material across the B1 to B2 transformation.

SAMMIS and DEIN (1974) conducted deformation experiments of CsCl across the transformation temperature from CsCl (B2) structure to NaCl (B1) structure and inferred enhanced deformation.

- H_2O (ice)

H_2O undergoes a series of phase transformations under high pressures. POIRIER *et al.* (1981) and DURHAM and STERN (2001) conducted rheological measurements on H_2O in various phases. Rheology of high-pressure H_2O is a key to the understanding of dynamics and evolution of icy satellites. DURHAM and STERN (2001) provide an extensive review of this topic. They used a high-resolution creep apparatus to investigate creep properties of ice under the conditions of $T = 160$–270 K, $P = 0.1$–500 MPa. This range of temperature and pressure covers the stability fields of ice I, II, III, V and VI. Ice II and V have considerably higher creep strength than ice I and VI (Fig. 15.2). However, the possible importance of grain size documented for ice I (GOLDSBY and KOHLSTEDT, 2001) makes this comparison complicated. In cases where grain size plays an important role, changes in grain size across a phase boundary (see Chapter 13) must be evaluated in order to estimate the role of phase transformation.

- iron (bcc, *fcc*, hcp)

Iron undergoes a series of phase transformations. The influence of the phase transformation from low-temperature Fe (bcc, α-Fe) to high-temperature Fe (*fcc*, γ-Fe) ($T_{tr} = 1183$ K at room pressure) on plastic properties has been extensively studied (DE JONG and RATHENAU, 1959, 1961; CLINARD and SHERBY, 1964). The plastic properties of hcp Fe (ε-Fe) were not studied in any detail. MAO *et al.* (1998) and WENK *et al.* (2000) conducted room-temperature deformation experiments using a diamond anvil cell. They reported evidence for the importance of basal slip from the deformation fabrics. No experimental studies were conducted on ε-Fe at high temperatures.

- titanium (hcp to bcc: $T_{tr} = 1155$ K)

Titanium undergoes a phase transformation at room pressure from low-temperature hcp structure (α-Ti) to high-temperature bcc structure (β-Ti). There is a significant jump in self-diffusion coefficients as a phase changes from hcp (lower temperature) to bcc (higher temperature). CHAIX and LASALMONIE (1981) investigated deformation of titanium during the transformation between α-Ti and β-Ti and observed enhanced deformation during the transformation.

- cobalt (hcp to *fcc*: $T_{tr} = 722$ K)

Cobalt undergoes a phase transformation at room pressure from a low-temperature hcp structure (α-Co) to a high-temperature *fcc* structure (β-Co) at 722 K. ZAMORA and POIRIER (1983) observed enhanced plastic deformation when temperature is increased from below to above this transformation temperature (Fig. 15.3).

15.2.2. Isomechanical groups (classification of mechanical properties, scaling law, systematics)

For many geological applications, direct experimental tests on rheological properties or microstructural development are difficult because of the extreme conditions (such as high pressures) that are required for such experimental studies. For example, quantitative experimental studies of deformation of $(Mg,Fe)SiO_3$ perovskite under lower mantle conditions ($P > 24$ GPa, $T > 2000$ K) or hcp Fe or under inner core conditions ($P > 330$ GPa, $T > 5000$ K) respectively are extremely difficult and have not been performed to date. Although various attempts are being made to develop new technology to achieve such a goal (e.g., DURHAM *et al.*, 2002), quantitative studies on rheological properties under deep mantle conditions remain highly challenging and can be made only under a limited range of parameters (e.g., under relatively high stresses). Many of the earlier studies on the plastic properties of deep Earth materials were at low temperatures (KARATO *et al.*, 1990; MEADE and JEANLOZ, 1990; MAO *et al.*, 1998; MERKEL *et al.*, 2003). These studies provide some insights into the plastic properties at low temperatures, such as the magnitude of the Peierls stress (Chapter 9), but the relevance of such results to high-temperature rheology in most of Earth's interior is highly questionable.

A useful approach under these circumstances is to use analog materials and find some general trends. If one finds analog materials that are stable at low-pressure conditions, then one can conduct detailed quantitative studies on rheological properties to understand the basic physical mechanisms of deformation in that class of materials. If one finds that rheological properties of that class of materials are similar after a proper normalization, then one can use the results from such studies to infer rheological properties of deep Earth materials. Such an indirect approach was highly successful in elasticity (e.g., LIEBERMANN, 1982). The basic premise behind such a philosophy is the presence of *systematics* in a given property under consideration.

An *isomechanical group* refers to a group of materials that show similar mechanical behavior when properties are compared after appropriate *normalization*. Once an isomechanical group is established for materials that are geologically important, then a trend established for that group, based on studies on analog materials, can be applied to estimate properties of Earth materials using an appropriate scaling scheme.

Qualitative aspects (microstructures)

One use of the analog materials approach is to investigate microstructural development. Examples include the use of some organic transparent materials for a better understanding of deformation microstructures through *in-situ* observations of microstructural development during deformation ("see-through" experiments). In many cases, the main purpose of such an experiment is to get qualitative ideas about microstructural development and not much discussion has been made as to the validity of the use of analogs.

Examples include the see-through deformation experiments on sodium nitrate, octachloropropane

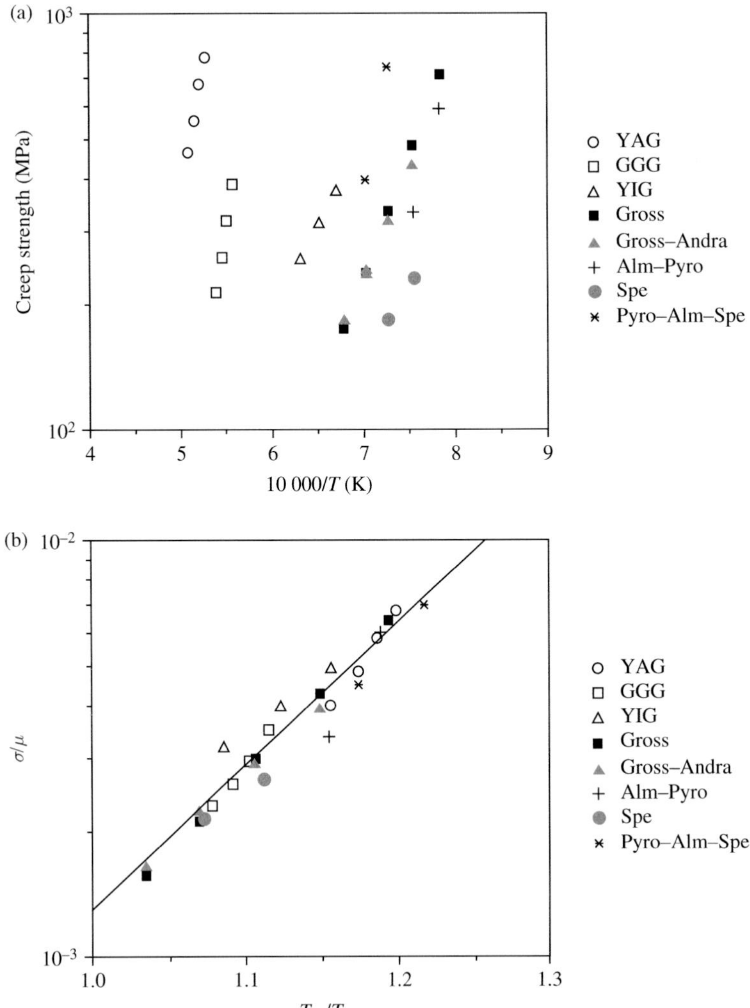

FIGURE 15.3 Systematics in high-temperature creep in garnets (Karato *et al.*, 1995a).
(a) The creep strength of various garnets at a strain rate of $10^{-5}\,s^{-1}$ as a function of temperature.
(b) A comparison of the creep strength versus temperature relationships after normalization.

and canalite (Urai, 1983). These *in-situ* observations on microstructural development have provided important new insights into the processes of *dynamic recrystallization* (e.g., Urai *et al.*, 1986a; Drury and Urai, 1990), such as the role of grain-boundary migration or of a fluid phase, which are difficult to investigate using conventional techniques where microstructures are observed after deformation. However, the relevance of these observations to real rocks must be evaluated with great caution. A physical understanding of the processes that will provide the scheme of scaling from analog materials to real rocks is necessary before any of these observations can be used to infer geological processes. Such studies (studies of a *scaling law*) have been performed for creep properties (Ashby and Brown, 1982; Frost and Ashby, 1982; Karato, 1989c), but have not yet been made for dynamic recrystallization-related microstructure developments.

Quantitative aspects (scaling laws for creep)

Similar to the case of elasticity, it would be useful if one could infer rheological properties of deep Earth materials from the studies on rheological properties of analog materials. For example, germanates are known to be a good analog for silicates in terms of phase relations and elastic properties. Since chemical properties of Ge^{4+} are similar to Si^{4+} (both have inert gas-type electron configuration, namely $2s^22p^6$ for Si^{4+} and $3s^23p^6$ for Ge^{4+}) but the ionic radius of Ge^{4+} (0.040 nm) is significantly larger than that of Si^{4+} (0.026 nm) (see Table 4.3). Crystal structures of ionic solids are primarily controlled by the ratio of ionic radii (e.g., Bloss, 1971). As pressure goes up, large oxygen ions are squeezed more than small cations. Therefore the ratio r_{cation}/r_{anion} will increase with pressure. Thus, germanates often show crystal structures that would be seen in silicates under higher pressures. For example, Mg_2GeO_4 is stable in spinel structure at room pressure (to ~1300 K), whereas the spinel structure is stable for Mg_2SiO_4 only above ~17 GPa. Similarly, $CaTiO_3$ crystallizes into perovskite structure at room pressure but $MgSiO_3$ crystallized into perovskite structure only above ~24 GPa. Thus it would be nice if we can learn something from the studies on these analogs. The most important issue here is to establish *scaling laws* for rheology.

Scaling laws and normalization of parameters (Ashby and Brown, 1982)

When one uses analog materials, results obtained at one condition for one specimen must be used to infer some properties of other materials at different conditions. For example, we might conduct deformation experiments for $CaTiO_3$ perovskite in a certain range of temperature and stress and obtain a flow law that describes a relation between T, stress and strain rate. How could one infer the rheological properties of $MgSiO_3$ perovskite from these data?

The first thing that one needs to establish is the presence (or not) of a "universal flow law" for a class of materials under consideration. If a universal flow law exists, then one can use it to predict the flow law of materials that belong to the same class (*an isomechanical group*) under other conditions. To see if a universal flow law exists or not, one has to compare the flow laws of various materials. To allow such a comparison, one needs to non-dimensionalize the flow law.

Creep constitutive law at steady state can in general be written as

$$\dot{\varepsilon} = f(\sigma, T, S; X) \tag{15.1}$$

where S is the structural parameters such as grain size and LPO (lattice-preferred orientation), X is a set of materials parameters such as T_m (melting temperature), n (stress exponent) and m (grain-size exponent). Now, let us choose a set of normalizing parameters σ^*, T^*, $\dot{\varepsilon}^*$ such that the parameters controlling deformation are normalized (non-dimensionalized) as, $\tilde{\sigma} = \sigma/\sigma^*$, $\tilde{T} = T/T^*$, $\tilde{\dot{\varepsilon}} = \dot{\varepsilon}/\dot{\varepsilon}^*$ and $\tilde{X} = X/X^*$ respectively. Thus, equation (15.1) becomes,

$$\tilde{\dot{\varepsilon}} = f(\tilde{\sigma}, \tilde{T}, \tilde{S}; \tilde{X}) \tag{15.2}$$

where $\tilde{S}$ and $\tilde{X}$ are non-dimensionalized structural and materials parameters. If the data on creep for various materials for equivalent $\tilde{S}$ converge to a single master curve $\tilde{\dot{\varepsilon}} = f(\tilde{\sigma}, \tilde{T}; \tilde{X})$, then non-dimensional materials parameters, $\tilde{X}$, must be constants for the members of that group.

The choice of a set of normalization parameters $(\sigma^*, T^*, \dot{\varepsilon}^*)$ is critical for this approach but is not unique. The following normalization scheme is often used (Frost and Ashby, 1982): $\sigma^* = \mu$ (μ, shear modulus), $T^* = T_m$ (T_m, melting temperature), $\dot{\varepsilon}^* = D_m \Omega^{-2/3}$ (D_m, diffusion coefficient at melting temperature). Such a normalization scheme has some theoretical basis. Creep laws suggest that stress affects creep through interaction with defects such as dislocations. The interaction of stress with dislocations is proportional to the elastic modulus such as the shear modulus (see Chapter 5). The temperature dependence of creep in many cases comes from that of diffusion for which homologous temperature (T/T_m) normalization works well (Chapter 10). Such an argument can be extended to include pressure and parameters characterizing chemical environment (such as oxygen fugacity and water fugacity). In most cases, the pressure effect can be expressed by the pressure effects of melting temperature. In such a case, pressure does not explicitly come into play but appears indirectly through $T_m(P)$ (see Chapter 10). The way to normalize the parameter characterizing the chemical environment is not well known. In the case of water fugacity that has a large effect on plasticity, one choice is again through its effect on melting temperature, i.e., $T_m(C)$. However, these choices are not based on any rigorous theory, and should be considered as semi-empirical. The normalization scheme for strain rate, for example, could be interpreted as a rate of a thermally activated process at melting temperature.

The choice of normalization discussed above is not always practical for applications to Earth sciences. The most problematic one is $\dot{\varepsilon}^* = D_m \Omega^{-2/3}$ because the diffusion coefficient at the melting temperature is not usually known for Earth materials or their analogs. Under these conditions, an alternative choice is $\dot{\varepsilon}^* = \nu_D$ where ν_D is the Debye frequency that can be calculated from the elastic moduli or seismic wave velocities (Chapter 4). Using an empirical relation $D \approx a^2 \nu_D \cdot \exp(-g(T_m/T))$ (see Chapter 10), it is seen that these two choices are equivalent (note that $D_m \approx a^2 \nu_D \exp(-g)$)

except for the factor of $\exp(-g)$. Because the variation in ν_D among various materials with the same crystal structure is small, normalization with respect to strain rate is not very important in most cases.

Let us take an example of diffusional creep. The flow law of diffusional creep (Nabarro–Herring creep) is (see (9.44)),

$$\dot{\varepsilon} = \alpha \frac{D_0 \sigma \Omega}{RTL^2} \exp\left(-\frac{H^*}{RT}\right). \tag{15.3}$$

This equation gives a relationship among macroscopic variables, $\dot{\varepsilon}, T, \sigma$, structural variable ($L$, grain size) and materials parameters, Ω, D_0, H^*. The equation (15.3) can be normalized to,

$$\tilde{\dot{\varepsilon}} = \alpha \frac{\tilde{X}_1}{\tilde{S}^2} \frac{\tilde{\sigma}}{\tilde{T}} \exp\left[-\tilde{X}_2\left(\frac{1}{\tilde{T}} - 1\right)\right] \tag{15.4a}$$

where $\tilde{\dot{\varepsilon}} = \dot{\varepsilon}/D_m\Omega^{-2/3}$, $\tilde{T} = T/T_m$, $\tilde{\sigma} = \sigma/\mu$, $\tilde{S} = L/\Omega^{1/3} = L/b$, $\tilde{X}_1 = \mu\Omega/RT_m$ and $\tilde{X}_2 = H^*/RT_m$.

Alternatively,

$$\tilde{\dot{\varepsilon}}' = \alpha \frac{\tilde{X}_1}{\tilde{S}^2} \frac{\tilde{\sigma}}{\tilde{T}} \exp\left(-\frac{\tilde{X}_2}{\tilde{T}}\right) \tag{15.4b}$$

with $\tilde{\dot{\varepsilon}}' = \dot{\varepsilon}/D_0\Omega^{-2/3} = \dot{\varepsilon}/\nu_D$. In either case, when a master curve can be defined for a group of materials, then $\tilde{X}_1 = \mu\Omega/RT_m$ and $\tilde{X}_2 = H^*/RT_m$ must be similar to all the materials in that group. The formula (15.4b) is preferred because all the normalization parameters are those that are easily determined. Similarly a non-dimensional creep constitutive equation can be obtained for power-law dislocation creep, $\dot{\varepsilon} = A(\sigma/\mu)^n \exp(-H^*/RT)$, namely,

$$\tilde{\dot{\varepsilon}}' = \tilde{X}_3 \tilde{\sigma}^n \exp\left(-\frac{\tilde{X}_2}{\tilde{T}}\right) \tag{15.5}$$

where $\tilde{\dot{\varepsilon}}' = \dot{\varepsilon}/\nu_D$ and $\tilde{X}_3 = A/\nu_D$. Since $\tilde{X}_3 = A/\nu_D$ (ν_D) is nearly constant for many materials, the relations (15.4b) or (15.5) indicate that the creep properties for the same class of materials are essentially controlled by the (normalized) grain size ($\tilde{S} = L/b$), normalized stress ($\tilde{\sigma} = \sigma/\mu$) and normalized temperature ($\tilde{T} = T/T_m$).

Problem 15.1

Derive equations (15.4a), (15.4b) and (15.5).

Solution

Using the definition $D_m = D_0 \exp(-H^*/RT_m)$, and using the relations $\tilde{T} = T/T_m$, $\tilde{\sigma} = \sigma/\mu$ and $\tilde{S} = L/\Omega^{1/3} = L/b$, the equation (15.3) becomes $\dot{\varepsilon} = \alpha(D_m\Omega^{-2/3}/\tilde{T}\tilde{S}^2)(\mu\Omega/RT_m)\tilde{\sigma}\exp((H^*/RT_{T_m})\cdot(1 - 1/\tilde{T}))$. Using the definitions, $\tilde{X}_1 = \mu\Omega/RT_m$, $\tilde{X}_2 = H^*/RT_m$, and $\tilde{\dot{\varepsilon}} = \dot{\varepsilon}/D_m\Omega^{-2/3}$, this is translated to (15.4a), $\tilde{\dot{\varepsilon}} = \alpha(\tilde{X}_1/\tilde{S}^2)(\tilde{\sigma}/\tilde{T})\cdot\exp[-\tilde{X}_2(1/\tilde{T} - 1)]$. If one uses $\tilde{\dot{\varepsilon}}' = \dot{\varepsilon}/D_0\Omega^{-2/3} = \dot{\varepsilon}/\nu_D$, then $\tilde{\dot{\varepsilon}} = \tilde{\dot{\varepsilon}}' \exp(H^*/RT_m)$ and hence $\tilde{\dot{\varepsilon}}' = \alpha(\tilde{X}_1/\tilde{S}^2)(\tilde{\sigma}/\tilde{T})\exp(-\tilde{X}_2/\tilde{T})$. Now the power-law creep equation, $\dot{\varepsilon} = A(\sigma/\mu)^n \exp(-H^*/RT)$, can be scaled by using $\tilde{X}_2 = H/RT_m$, $\tilde{X}_3 = A/\nu_D$, $\tilde{\sigma} = \sigma/\mu$ and $\tilde{T} = T/T_m$ as $\dot{\varepsilon} = \nu_D \tilde{X}_3 \tilde{\sigma}^n \exp(-\tilde{X}_2/\tilde{T})$, and using the definition $\tilde{\dot{\varepsilon}}' = \dot{\varepsilon}/D_0\Omega^{-2/3} = \dot{\varepsilon}/\nu_D$ one obtains (15.5).

Once a scaling law such as (15.4b) or (15.5) is established for a class of materials and we know non-dimensional material parameters such as $\tilde{X}_1, \tilde{X}_2, \tilde{X}_3$ for that class of materials, then the creep properties of materials that belong to this class can be predicted from the properties of those materials that are used for normalization. In the above scheme, these properties are ν_D (Debye frequency), μ (shear modulus), b (the length of the Burgers vector) and T_m (melting temperature). For most materials, these properties are well known from laboratory experiments including those materials that occur in Earth's deep interior. Therefore the creep properties in Earth's deep interior can be inferred from this approach. Karato *et al.* (1995a) studied the creep systematics in garnet (Fig. 15.3) and applied that result to infer the rheological stratification in the transition zone, and Karato (2000) inferred the viscosity of the inner core from the experimental data on hcp metals. However, the validity of such a normalization scheme is not always obvious. There are some complications in other materials such as perovskite (e.g., Drury and Fitz Gerald, 1998; Wang *et al.*, 1993).

An example of the search for systematics was made by Ashby and Brown (1982; see also Frost and Ashby, 1982). Frost and Ashby (1982) in particular showed that at a very general level, the nature of resistance of materials for plastic deformation is related to the nature of chemical bonding: the hierarchy of strength is covalent bonding > ionic bonding > metallic bonding ≫ van der Waals bonding.

In silicate minerals the Si–O bond is the strongest bond (Si–O bond is highly covalent when Si is surrounded by four oxygens). However, when Si is surrounded by six oxygens, then the bond length increases and the nature of bonding becomes more ionic. Consequently, some reduction of "strength" may occur when the coordination of Si changes from four to six. Shieh *et al.* (2002) determined room temperature strength of stishovite and found that the normalized stress of stishovite is significantly smaller than other silicates (with four coordination), suggesting an effect of weakened bond. Similar results were obtained for $MgSiO_3$ perovskite where Si is surrounded by six oxygens (Merkel *et al.*, 2003). However, the effects of coordination of Si on *high-temperature* plasticity of silicates have not been investigated.

Effects of crystal structure on plastic flow are also significant. For example, the bcc (body-centered-cubic) to hcp (hexagonal-closed-pack) structure transformation in Fe results in a large difference in plastic anisotropy. The transition from B1 structure (NaCl structure) to B2 (CsCl structure) results in a large difference in the spacing of glide planes that results in a significant weakening particularly at low temperatures. Also a transformation to the garnet structure results in significant strengthening, because of the unusually large unit cell structure of garnet that makes dislocation glide difficult.

Problem 15.2

The viscosity of the inner core plays a critical role in controlling the inner core rotation. But there is no direct measurement of high-temperature plasticity in iron in the inner core. Table 15.1 below shows the properties of some of the hcp metals (b, the length of the Burgers vector; c/a, the ratio of two lengths of unit cell; T_m, melting temperature; V_P, P-wave velocity; V_S, S-wave velocity; μ, shear modulus; D_0, pre-exponential term for diffusion coefficient; H^*, activation enthalpy of diffusion). The known properties of hcp Fe (ε–Fe) in the inner core are $b = 0.225$ nm, $T_m = 5800$ K, $V_P = 11.1$ km/s, $V_S = 3.6$ km/s and $\mu = 170$ GPa. Using the results shown in Table 15.1, predict the diffusion coefficient of Fe in the inner core and calculate the viscosity of the inner core appropriate for the Nabarro–Herring creep corresponding to the grain size of 1–1000 m (assume the temperature of 0.90 T_m) (grain size of ~1000 m is inferred: Chapter 17).

Solution

To infer yet to be determined diffusion coefficient, one needs to find the systematic trend in diffusion coefficient through normalization. Diffusion coefficients $D = D_0 \exp(-H^*/RT)$ contain two constants, D_0 and H^*. Normalize D_0 by $\tilde{D}_0 = D_0/b^2\nu_D$ and $\tilde{H} = H^*/RT_m$ to find the systematics and determine $\tilde{D}_0$ and $\tilde{H}$ for hcp metals, and use the relation (15.3) to calculate the viscosity, $\eta = RTL^2/\alpha D\Omega$ with $\alpha = 14$. The Debye frequency is related to V_P and V_S as $\nu_D = \left(9N_A/4\pi\Omega\left(1/V_P^3 + 2/V_S^3\right)\right)^{1/3}$ (see Chapter 4). Use $\Omega = (\sqrt{3}/2)b^3(c/a)N_A$ (N_A, the Avogadro number) for the molar volume.

TABLE 15.1 Properties of some hcp metals, b (the length of the Burgers vector, nm), T_m (melting temperature, K), V_P (velocity of P-wave, km/s), V_S (velocity of S-wave, km/s), μ (shear modulus, GPa), D_0 (pre-exponential term for diffusion coefficient, m^2/s), H^* (activation enthalpy for diffusion, kJ/mol).

	Zn	Cd	Mg
b	0.267	0.293	0.321
c/a	1.856	1.886	1.624
T_m	693	594	924
V_P	4.11	3.142	5.780
V_S	2.35	1.676	3.151
μ	49.3	27.8	16.4
D_0	1.3×10^{-5}	5.0×10^{-6}	1.0×10^{-4}
H^*	91.7	76.2	135

Systematics in high-temperature creep in oxides and silicates

The presence of systematics was first extensively studied by Ashby and his coworkers for a range of materials (Brown and Ashby, 1980; Ashby and Brown, 1982; Frost and Ashby, 1982). Systematics in high-temperature creep in Earth materials has been investigated by several groups (for a review see e.g., Karato, 1997b). In many cases, well-defined creep systematics have been demonstrated (e.g., garnet; Fig. 15.3), although some complications have been noted in other materials such as perovskite (e.g., Poirier *et al.*, 1983; Beauchesne and Poirier, 1989, 1990; Wang *et al.*, 1993).

Figure 15.4 summarizes the systematics in high-temperature creep in some of the oxides and silicates (Karato, 1989c). As shown, crystal structure has a large effect on high-temperature creep

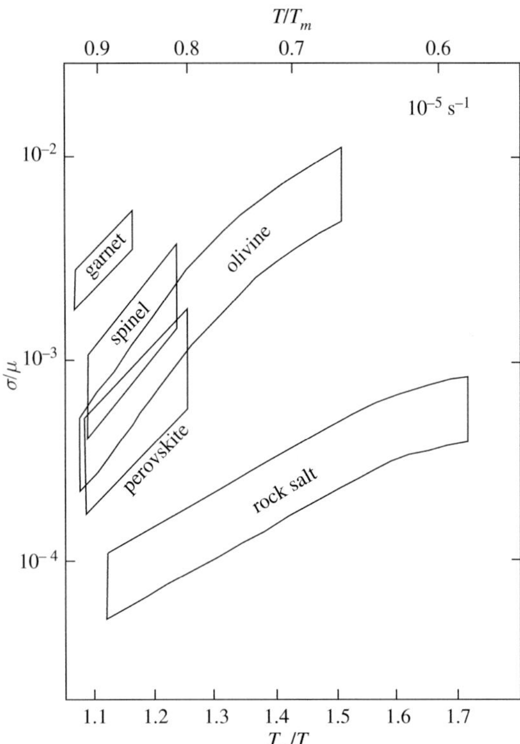

FIGURE 15.4 Systematics in high-temperature creep in dense oxides (KARATO, 1989c).

strength. Even at the same homologous temperature (T/T_m), the creep strength (effective viscosity) of materials with different crystal structures can differ as much as several orders of magnitude. Such a trend is markedly different from the trend in elasticity. In elasticity, density plays the major role: materials with higher densities tend to have higher elastic constants (higher seismic wave velocities). This is *Birch's law* (Chapter 4), which states that elastic properties are primarily controlled by density. Thus, *in plastic properties, Birch's law does not work*: density plays a minor role in plastic properties and denser materials can have smaller creep strengths than less dense materials. Crystal structure has a greater effect.

Some caution must be exercised in using the systematics in creep laws. The creep systematics provide a guideline for inferring still unknown creep properties of materials. However, such a systematic trend should not be considered as more than just a guideline. Important details such as the role of impurities (e.g., water) are not well incorporated in the normalization scheme.

15.3. Effects of transformation-induced stress–strain: transformation plasticity

A phase transformation causes a change in the shape and/or volume of a crystal. Consequently, a phase transformation may affect the plastic properties by two distinct mechanisms: (i) if the shape of a crystal changes in response to applied stress, then it results in plastic strain, or alternatively, (ii) if there is a finite volume change associated with a phase transformation, this volume change creates internal stress–strain that may affect the rheological behavior. The first case is similar to deformation by mechanical twinning that is discussed in Chapter 9 (section 9.6). I will focus on the second mechanism in this chapter. Unusual deformation behavior associated with transformation-induced internal stress–strain is often referred to as *transformation plasticity* and the first detailed study was made by GREENWOOD and JOHNSON (1965). GORDON (1971) pointed out the possible role of transformation plasticity in Earth's mantle (see also SAMMIS and DEIN, 1974 and PARMENTIER, 1981). POIRIER (1982) and PANASYUK and HAGER (1998) proposed theoretical models using specific assumptions (the details will be discussed in this section). This phenomenon occurs, in contrast to the effects of crystal structure, only during a phase transformation: internal stress or strain caused by a transformation will be relaxed by plastic flow, and after complete relaxation, enhanced deformation will no longer occur. However, if the degree of enhancement of deformation is large, it still has an important influence on geodynamics. The importance of this phenomenon depends on the magnitude of this effects and the duration of this process. As we will learn in this section, the current models provide a good estimate of total strain due to this effect. However the conditions under which transformation plasticity is important are likely to be limited to modest temperatures.

15.3.1. Experimental observations

Enhancement of plastic deformation during a first-order phase transformation can be observed in two different ways. First, when a small external stress is applied during a first-order phase transformation, permanent strain is created whose amount is proportional to the applied stress and increases linearly with the magnitude of volume change. When temperature cycling is made around a transformation temperature, this

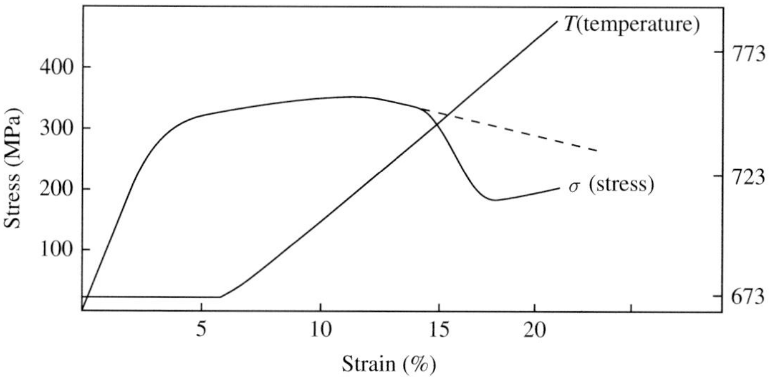

FIGURE 15.5 Experimental observations on transformation plasticity in Co (ZAMORA and POIRIER, 1983).

At a temperature of ~722 K, a phase transformation occurs from hcp to fcc structure that leads to weakening.

permanent strain is proportional to the number of cycles (COTTRELL, 1964). Similarly, when a constant strain rate is imposed during a temperature ramp that includes a transformation temperature, then an unusual weakening occurs after a transformation temperature is reached (CHAIX and LASALMONIE, 1981; ZAMORA and POIRIER, 1983; Fig. 15.5). Grain-size sensitivity in transformation plasticity was also observed in Bi_2WO_4 and Bi_2MoO_6 (WINGER *et al.*, 1980).

15.3.2. Models for transformation plasticity

Greenwood–Johnson model: an unrelaxed model

Models of transformation plasticity may be classified into two categories. The first is an *unrelaxed model* in which the internal *stress* caused by the volumetric strain is considered to modify the rheological behavior. This model was proposed by GREENWOOD and JOHNSON (1965) who assumed an isotropic flow law for non-linear rheology, namely (see Chapter 3),

$$\dot{\varepsilon}'_{ij} = B \cdot II_{\sigma'}^{(n-1)/2} \sigma'_{ij} \tag{15.6}$$

where $\dot{\varepsilon}'_{ij}$ is the deviatoric strain-rate, σ'_{ij} the deviatoric stress, n the stress exponent, B a constant and $II_{\sigma'}$ the second invariant of the deviatoric stress tensor (see Chapter 1). GREENWOOD and JOHNSON (1965) postulated that under the condition in which the magnitude of transformation-induced deviatoric stress, σ'_{tr}, is significantly larger than that of applied deviatoric stress, the second invariant of stress tensor is dominated by the internal stress and the rheology is linear with respect to the applied stress. However, the internal stress caused by transformation is macroscopically isotropic, so that it does not directly contribute to the deviatoric stress term, σ'_{ij}, in equation (15.6).

Integrating (15.6), one obtains

$$\varepsilon'_{ij} = B_1 \cdot II_{\sigma'}^{(n-1)/2} \sigma'_{ij} \tag{15.7}$$

where σ'_{ij} and $II_{\sigma'}$ are assumed to be constant with time and $B_1 = B \cdot \tau_{\mathrm{T}}$. From equation (15.7), one finds

$$II_{\varepsilon'} = B_1^2 \cdot II_{\sigma'}^n. \tag{15.8}$$

Thus

$$\varepsilon'_{ij} = B_1^{1/n} \cdot II_{\varepsilon'}^{(n-1)/2n} \sigma'_{ij}. \tag{15.9}$$

Now $II_{\varepsilon'}$ is related to the volume change associated with a phase transformation. The total (deviatoric) strain is the sum of imposed macroscopic strain (uniaxial deformation along the z axis (x_3 axis) is assumed here) and the (deviatoric) strain (strain from which the volumetric strain is subtracted) due to the volume change associated with a phase transformation, ξ_{ij}.

$$\varepsilon_{ij}^T = \begin{pmatrix} -\frac{1}{2}\varepsilon & 0 & 0 \\ 0 & -\frac{1}{2}\varepsilon & 0 \\ 0 & 0 & \varepsilon \end{pmatrix} + \xi_{ij}. \tag{15.10}$$

The magnitude of deviatoric strain caused by a phase transformation (ξ_{ij}) is proportional to the volume change associated with a phase transformation but its orientation is random if there is no preferred orientation of a new phase material. Consequently, we have

$$I_\xi = \xi_{11} + \xi_{22} + \xi_{33} = 0 \tag{15.11a}$$

and

$$II_\xi = -\xi_{11}\xi_{22} - \xi_{22}\xi_{33} - \xi_{33}\xi_{11} + \xi_{12}^2 + \xi_{23}^2 + \xi_{31}^2 = \beta\left(\frac{\Delta V}{V}\right)^2 \tag{15.11b}$$

where β is a constant of order unity. Therefore from the definition of $II_{\varepsilon'}$ one obtains,

$$II_{\varepsilon'} = \frac{3}{4}\varepsilon^2 + \frac{3}{2}\varepsilon\xi_{33} + \beta\left(\frac{\Delta V}{V}\right)^2. \tag{15.12}$$

Problem 15.3

Derive equation (15.12).

Solution

Using the definition of the second invariant of strain and equation (15.10), $II_{\varepsilon'} = -\left(-\frac{1}{2}\varepsilon + \xi_{11}\right)\left(-\frac{1}{2}\varepsilon + \xi_{22}\right) - \left(-\frac{1}{2}\varepsilon + \xi_{22}\right)(\varepsilon + \xi_{33}) - (\varepsilon + \xi_{33})\left(-\frac{1}{2}\varepsilon + \xi_{11}\right) + \xi_{12}^2 + \xi_{23}^2 + \xi_{31}^2 = \frac{3}{4}\varepsilon^2 - \frac{1}{2}\varepsilon(\xi_{11} + \xi_{22} - 2\xi_{33}) + \mathrm{II}_{\xi}$. Using (15.11), one gets $II_{\varepsilon'} = \frac{3}{4}\varepsilon^2 + \frac{3}{2}\varepsilon\xi_{33} + \beta(\Delta V/V)^2$.

From (15.9) and (15.12), one finds

$$\varepsilon + \xi_{33} = \sigma'_{33} B_1^{1/n} II_{\varepsilon'}^{(n-2)/2n}$$
$$= \sigma'_{33} B_1^{1/n} \left(\frac{3}{4}\varepsilon^2 + \frac{3}{2}\varepsilon\xi_{33} + \beta\left(\frac{\Delta V}{V}\right)^2\right)^{(n-1)/2n}. \tag{15.13}$$

Solving this equation for strain (assuming $\varepsilon \ll |\Delta V/V|$) and after taking the average over ξ_{33},

$$\varepsilon = \sigma'_{33} B_1^{1/n} \left(\frac{\Delta V}{V}\right)^{(n-1)/n} \beta^{(n-1)/2n} \left[1 - \frac{3(n-1)}{4n} \frac{\bar{\xi}_{33}^2}{\beta(\Delta V/V)^2}\right]$$
$$= \sigma'_{33} B_1^{1/n} (\Delta V/V)^{(n-1)/n} \beta^{(n-1)/2n} \frac{4n+1}{5n} \tag{15.14}$$

where we used the values of $\beta = \frac{1}{3}$ and $\bar{\xi}_{33}^2/(\Delta V/V)^2 = \frac{4}{45}$ (Greenwood and Johnson, 1965) and it was assumed that the orientation of strain due to the phase transformation is random, i.e., $\bar{\xi}_{33} = 0$ (where $\bar{\xi}_{33}$ is the average value of ξ_{33}). Thus this model predicts that the total strain during a phase transformation is linearly proportional to the applied stress. This prediction is borne out by the experimental observations.

Problem 15.4*

Derive equation (15.14).

Solution

Expanding the right-hand side of equation (15.13) assuming $\varepsilon \ll |\Delta V/V|$, one finds,

$$\varepsilon + \xi_{33} = C\left[1 + \frac{3(n-1)}{4n} \frac{\varepsilon\xi_{33}}{\beta(\Delta V/V)^2}\right]$$

with $C \equiv \sigma'_{33} B_1^{1/n} \beta^{(n-1)/2n} (\Delta V/V)^{(n-1)/n}$. Solving this for ε,

$$\varepsilon = \frac{C - \xi_{33}}{1 - C(3(n-1)/4n)\left[\xi_{33}/\beta(\Delta V/V)^2\right]}$$
$$\approx (C - \xi_{33})\left[1 + \frac{3(n-1)}{4n} \frac{C\xi_{33}}{\beta(\Delta V/V)^2}\right]$$
$$= C\left[1 - \frac{3(n-1)}{4n} \frac{\xi_{33}^2}{\beta(\Delta V/V)^2}\right]$$
$$+ \xi_{33}\left[\frac{3(n-1)}{4n} \frac{C^2}{\beta(\Delta V/V)^2} - 1\right].$$

Taking the average for all orientations, this leads to

$$\varepsilon = C\left[1 - \frac{3(n-1)}{4n} \frac{\bar{\xi}_{33}^2}{\beta(\Delta V/V)^2}\right] = \frac{C(4n+1)}{5n}$$

where we used the values of $\beta = \frac{1}{3}$ and $\frac{\bar{\xi}_{33}^2}{(\Delta V/V)^2} = \frac{4}{45}$ $(\bar{\xi}_{33} = 0)$.

Because of the assumption that $II_{\sigma'}$ is constant with time, this model does not give the precise value of *strain rate* during the transformation, but the average value of strain rate can be roughly calculated from (15.14) as

$$\dot{\varepsilon} \approx \frac{\varepsilon}{\tau_T} \tag{15.15}$$

where τ_T is the characteristic time for a phase transformation. The characteristic time for transformation depends both on nucleation and growth rate but in most cases in Earth, the rate at which a phase transformation occurs is largely controlled by the rate at which a material moves vertically. In such a case, $\tau_T \approx \Delta z/\nu$ where Δz is the depth interval in which a phase transformation is completed (~10–15 km for the olivine to wadsleyite transformation, ~3–5 km for the transformation from ringwoodite to perovskite and magnesiowüstite transformation) and ν is the vertical velocity of material motion (~1–10^2 mm/yr).

Note that for $n = 1$, (15.14) gives $\varepsilon = \sigma'_{33} B_1$ which is the strain without the effects of a phase transformation. Enhancement of deformation occurs only when $n > 1$ i.e., only when the deformation mechanism is non-linear dislocation creep (and not for diffusional creep).

A similar model was proposed by Cottrell (1964) to explain deformation in a polycrystalline material during thermal cycling. When a polycrystalline material is subjected to a temperature cycling under small applied stress, then plastic deformation is observed whose strain is proportional to $\Delta\alpha \cdot \Delta T$ where $\Delta\alpha$ is the anisotropy in thermal expansion and ΔT is the amplitude of temperature cycling. COTTRELL (1964) considered that this is due to large internal stress due to thermal expansion anisotropy that dominates the stress invariant in equation (15.6).

PANASYUK and HAGER (1998) analyzed a similar model but calculated the magnitude of internal stress directly from the estimated local strain rate due to the volume change associated with a phase transformation. Briefly, they assumed the non-linear rheology (equation (15.6)) and postulated that the second invariant of deviatoric stress tensor is dominated by the internal stress caused by the volumetric strain associated with a phase transformation. Under these assumptions $II_{\sigma'} \propto \sigma_{tr}'^2$ instead of $II_{\sigma'} \propto \sigma_0'^2$ (for a case without a transformation-induced stress; σ_0' is the background stress), and the magnitude of enhancement of deformation is roughly given by,

$$\frac{\dot{\varepsilon}}{\dot{\varepsilon}_0} \approx \left(\frac{\sigma_{tr}'}{\sigma_0'}\right)^{n-1}. \tag{15.16}$$

Note that similar to the Greenwood–Johnson model (these models are in fact equivalent; see Problem 15.5), this mechanism works only for non-linear rheology, $n > 1$ (i.e., dislocation creep), and that this mechanism is important only when the transformation-induced stress is significantly larger than the background stress.

The magnitude of internal stress can be estimated from the rate of transformation assuming that the materials surrounding the transformation materials must deform to accommodate the strain. Assuming this accommodation occurs via plastic flow, one can estimate the stress magnitude from the strain rates. For example, for a general upwelling flow, the rate of upwelling is estimated to be ~1 mm/yr for which the volumetric strain rate associated a phase transformation is on the order of $10^{-16}\,s^{-1}$, which would result in the stress on the order of $0.1-1$ MPa. This is the same stress level expected in the general area of Earth's mantle (see Chapter 9). Therefore in this case, the enhancement of deformation is negligible. However, in a subducting slab, the descending velocity is much faster, ~$1-10$ cm/yr (the asymmetry of vertical velocity is due to the symmetry of flow geometry), and hence the volumetric strain rate will be $\sim 10^{-14}-10^{-15}\,s^{-1}$. These strain rates are much faster than those expected at the low-temperature environment in slabs. Therefore the influence of transformation-induced internal stress can be significant in subducting slabs (although other effects such as the effects of grain-size reduction can also be large; see section 15.4).

Problem 15.5*

Show that the Greenwood–Johnson and the Panasyuk–Hager models are essentially identical.

Solution

From the relationships (15.14) and (15.15) for the Greenwood–Johnson model, the ratio of strain rate associated with a phase transformation to the background strain-rate is given by

$$\frac{\dot{\varepsilon}}{\dot{\varepsilon}_0} = \frac{(B\tau_T)^{1/n}(\Delta V/V)^{(n-1)/n}\beta^{(n-1)/2n}[(4n+1)/5n]\sigma_0'}{B\sigma_0'^n}$$
$$= \left\{\frac{(\Delta V/V)}{B\tau_T}\right\}^{(n-1)/n}\frac{\beta^{(n-1)/2n}[(4n+1)/5n]}{\sigma_0'^{n-1}}$$

where I used the relation $\dot{\varepsilon}_0 = B\sigma_0'^n$. Now the strain rate associated with a phase transformation is given by $\dot{\varepsilon}_{tr} = \frac{\xi(\Delta V/V)}{\tau_T}$ where ξ is a constant of order unity and this strain-rate is associated with the internal stress σ_{tr}' through the relation $\dot{\varepsilon}_{tr} = B\sigma_{tr}'^n$. Thus $\left\{\frac{(\Delta V/V)}{B\tau_T}\right\}^{(n-1)/n} = \left(\frac{\dot{\varepsilon}_{tr}}{B\xi}\right)^{(n-1)/n} = \sigma_{tr}'^{n-1}\xi^{-\frac{n-1}{n}}$. Therefore $\dot{\varepsilon}/\dot{\varepsilon}_0 = K\left(\sigma_{tr}'/\sigma_0'\right)^{n-1}$ with $K \equiv \frac{\left(\beta/\xi^2\right)^{(n-1)/2n}(4n+1)}{5n} = 0.95$ for $\beta = \frac{1}{3}$, $\xi = \frac{1}{2}$ and $n = 3$. This means that the Greenwood–Johnson and the Panasyuk–Hager models are identical except for a minor difference in the numerical factor.

Poirier model: a relaxed model

In contrast to these models in which the internal stress is assumed to be not relaxed, POIRIER (1982) considered a case where the internal stress caused by volumetric strain is relaxed by plastic flow. This would correspond to a case of phase transformation at high temperatures and/or a later stage of phase transformation. POIRIER (1982) writes

$$\dot{\varepsilon} = \dot{\varepsilon}_A X_A + \dot{\varepsilon}_B X_B + \alpha\dot{\varepsilon}_T + \dot{\varepsilon}_a(1 - X_B) \tag{15.17}$$

where in the first and the second term, $\dot{\varepsilon}_{A,B}$ and $X_{A,B}$ are the strain rate and volume fraction of A, B phase respectively ($X_A + X_B = 1$). The third and the fourth terms in equation (15.17) are unique to the phase transformation. The third term represents the direct contribution of transformation strain (associated with the volume change, $\Delta V/V$) to the total strain rate and is given by

$$\alpha\dot{\varepsilon}_T = \varsigma\left|\frac{\Delta V}{V}\right|\frac{dX_B}{dt} \quad (15.18)$$

where α and ς are constants of order unity. The fourth term is the contribution to the strain rate from the generation of dislocations caused by the volume change associated with the transformation. In writing an equation for this term, Poirier (1982) postulates that the deformation due to transformation-induced dislocations occurs only in the phase A (a weaker phase) and that volumetric strain associated with a phase transformation results in *geometrically necessary dislocations* (Ashby, 1970) and used the Orowan equation (see Chapter 6.3),

$$\dot{\varepsilon}_a = \rho_a b\nu \quad (15.19)$$

where ρ_a is the density of dislocations created due to a phase transformation, υ is dislocation velocity. It is further postulated that the velocity of dislocations is linearly proportional to the applied stress and that ρ_a is proportional to the absolute value of volume change $|\Delta V/V|$ and the volume fraction of a new phase X_B, i.e.,[1]

$$\rho_a(t) = \rho_T X_B(t), \quad (15.20)$$

with

$$\rho_T = C\left|\frac{\Delta V}{V}\right| \approx \frac{1}{bL}\left|\frac{\Delta V}{V}\right| \quad (15.21)$$

where L is the space-scale of heterogeneity of strain, i.e., ~grain size. Therefore

$$\dot{\varepsilon}_a(t) = bM(T)\rho_T\sigma X_B(T,t) \quad (15.22)$$

where $\upsilon = M\sigma$ (M is the mobility of dislocations) and C is a constant. When one considers a case where the first three terms are small, then equation (15.17) is reduced to

$$\dot{\varepsilon}(T,t) = bM(T)\rho_T\sigma X_B(T,t)(1 - X_B(T,t)). \quad (15.23)$$

This equation means that there is enhanced deformation during a phase transformation and the degree to which deformation is enhanced changes with time and is proportional to the volume change associated with the phase transformation. If we use the relation (13.52) ($X_B(T,t) = 1 - \exp(-A(T)t^k)$), then we can also calculate the amount of permanent strain due to a phase transformation, namely,

$$\varepsilon_\infty(T) = \frac{\rho_T M(T)b}{A(T)^{1/K}}\sigma F(K) \quad (15.24)$$

with

$$F(K) \equiv \frac{1 - 2^{-1/K}}{K}\Gamma\left(\tfrac{1}{K}\right) \quad K = 3, 4 \quad (15.25)$$

where $\Gamma(x) \equiv \int_0^\infty y^{x-1}\exp(-y)\,dy$ is the gamma function ($F(1) \equiv 0.5$, $F(2) \equiv 0.5192$, $F(3) \equiv 0.5527$, $F(4) \equiv 0.5768$ etc.: Problem 15.6).

Problem 15.6*

Derive equations (15.24) and (15.25).

Solution

Integrating equation (15.23), one has $\varepsilon_\infty(T) = \int_0^\infty \dot{\varepsilon}(T,t)\,dt = CbM(T)(\Delta V/V)\sigma \int_0^\infty X_B(T,t)(1 - X_B(T,t))\,dt$.

Now $X_B(T,t) = 1 - \exp(-A(T)t^k) \equiv 1 - \exp(-z^k)$ with $z \equiv A^{1/k}t$. Thus $dz = A^{1/k}dt$ and $\int_0^\infty X_B(t)(1 - X_B(t))\,dt = A^{-1/k}\int_0^\infty \exp(-z^k)(1 - \exp(-z^k))\,dz$.

Let us define $\int_0^\infty \exp(-z^k)(1 - \exp(-z^k))\,dz \equiv F(k)$. Then $F(k) = \int_0^\infty \exp(-z^k)\,dz - \int_0^\infty \exp(-2z^k)\,dz$. Putting $z^k \equiv y$, $dz = (1/k)y^{(1-k)/k}\,dy$ and one obtains $F(k) = [(1 - 2^{-1/k})/k]\int_0^\infty y^{1/k-1}\exp(-y)\,dy = [(1 - 2^{-1/k})/k]\Gamma(1/k)$.

Therefore $\varepsilon_\infty(T) = (CM(T)b/A(T)^{1/k})(\Delta V/V)\sigma F(k) = (CM(T)b/A(T)^{1/k})(\Delta V/V)\sigma\frac{1-2^{-1/k}}{k}\Gamma(\frac{1}{k})$.

When the third term in equation (15.17) is unimportant, the enhancement of strain rate by this mechanism is given by,

$$\frac{\dot{\varepsilon}_a}{\dot{\varepsilon}_0} \approx 1 + \frac{C(\Delta V/V)X_B(1 - X_B)}{\rho_0} \approx 1 + \frac{\rho_T}{\rho_0}. \quad (15.26)$$

In short, the degree of enhancement of deformation in this model is mainly determined by the ratio of transformation-induced dislocation density to background dislocation density (ρ_T/ρ_0). The dislocation density associated with accommodation strain due to a transformation is given roughly by $\rho_T \approx (1/b)(\partial\gamma/\partial x) \approx (1/bL)|\Delta V/V|$, and one would obtain $\rho_T \approx 10^{11}$–$10^{13}\ m^{-2}$ (where we assumed $L = 10^{-2}$–1 mm) for olivine to wadsleyite transformation, which is comparable to or higher than typical

[1] The density of dislocations generated by volume change associated with a phase transformation can be calculated from the theory of Ashby (1970) on *geometrically necessary dislocations*, which indicates that the dislocation density is related to the strain gradient ($\rho_T \approx 1/b\,|\,\partial\gamma/\partial x\,|$, γ: strain, $\gamma \approx |\,\Delta V/V\,|$) (see Chapter 5).

dislocation densities in Earth's mantle ($\sim 10^8 - 10^{12}\,\mathrm{m}^{-2}$: Chapter 5). However, as soon as these transformation-induced dislocations are exhausted (i.e., after newly formed grain boundaries have swept all the pre-existing grains), this mechanism will cease to operate. The essence of this model is the presumption that the transformation-induced dislocations enhance plastic deformation. In other words, in calculating the strain rate using the Orowan equation, it is assumed that the dislocation velocity is determined by the magnitude of applied stress and is not affected by the transformation-induced dislocations. This is not obvious, because one of the well-known effects of high-dislocation density on deformation is *work hardening* (see Chapter 9), which will *reduce* the strain rate. If work hardening plays an important role, the effects of transformation strain-induced dislocations on rheology will be small or even *reduce* as opposed to enhance the rate of deformation. In fact the concept of geometrically necessary dislocations was proposed to explain some aspects of work hardening (see e.g., Ashby, 1970). Therefore this effect is likely to be important only in a narrow range of temperature. Meike (1993) also discussed this issue and concluded that there is no convincing case for transformation-enhanced plasticity in minerals.

Also, the high-dislocation density that is considered to be a cause of enhanced plasticity can be maintained only for a certain period, if the role of dislocation recovery is considered (Paterson, 1983). Briefly, if the characteristic time for dislocation recovery is much shorter than that of a phase transformation, then the enhanced plastic deformation considered in the above model will work only for a limited period. Since the time-scale of a phase transformation is largely controlled by the vertical motion by convection, the relative time-scales of phase transformation and dislocation recovery are mainly determined by temperature. The enhancement of deformation would occur only at relatively low temperatures.

Problem 15.7*

Show that a high-dislocation density and resultant (possible) enhancement of deformation due to a phase transformation occurs only when $\tau_R \gg \tau_T$ where $\tau_{R,T}$ are the characteristic times of dislocation recovery and transformation respectively (Paterson, 1983).

Hint: assume the following relations for the kinetics of a phase transformation and of dislocation recovery:

$$\rho_a(t) = \rho_T\left[1 - \exp\left(-\frac{t}{\tau_T}\right)\right] \quad \text{dislocation generation}$$

and

$$\rho(t) = \rho_0 - (\rho_0 - \rho_s)\left[1 - \exp\left(-\frac{t}{\tau_R}\right)\right] \quad \text{dislocation recovery}$$

where $\rho_{0,s}$ are the initial and the steady-state dislocation density respectively.[2]

Solution

The time variation of dislocation when both a phase transformation and dislocation recovery occur can be calculated from the relation $d\rho/dt = (d\rho/dt)^+ + (d\rho/dt)^-$ where $(d\rho/dt)^+$ is the rate of generation of dislocations by a phase transformation and $(d\rho/dt)^-$ is the rate of dislocation recovery. From equation (15.20) we get $\rho_a(t) = \rho_T(1 - \exp(-t/\tau_T))$ and $(d\rho/dt)^+ = (\rho_T/\tau_T)\exp(-t/\tau_T)$. Also from $\rho(t) = \rho_0 - (\rho_0 - \rho_s)\cdot[1 - \exp(-t/\tau_R)]$, we get $(d\rho/dt)^- = -(\rho - \rho_s)/\tau_R$. Therefore $d\rho/dt + \rho/\tau_R = (\rho_T/\tau_T)\exp(-t/\tau_T) + \rho_s/\tau_R$. Integrating this equation, one gets $\rho(t) = \rho_s\{1 + (\rho_T/\rho_s)\ [\tau_R/(\tau_T - \tau_R)][\exp(-t/\tau_T) - \exp(-t/\tau_R)]\}$ and hence the strain rate due to the intrinsic deformation and deformation due to increased dislocation density is given by $\dot{\varepsilon}(t) = \dot{\varepsilon}_s\{1 + (\rho_T/\rho_s)\cdot[\tau_R/(\tau_T - \tau_R)][\exp(-t/\tau_T) - \exp(-t/\tau_R)]\}$. It can be shown that the strain rate has a maximum value of $\dot{\varepsilon}(t) = \dot{\varepsilon}_s(\rho_T/\rho_s)\exp(-[\tau_R/(\tau_T - \tau_R)]\log(\tau_T/\tau_R))$ at $t_m = [\tau_T\tau_R/(\tau_T - \tau_R)]\log(\tau_T/\tau_R)$. Therefore if $\tau_R \gg \tau_T$, $\dot{\varepsilon}(t) = \dot{\varepsilon}_s(1 + \rho_T/\rho_s)$, whereas if $\tau_R \ll \tau_T$, $\dot{\varepsilon}(t) = \dot{\varepsilon}_s$. The time-scale of transformation is mainly determined by the time-scale of convection, whereas the characteristic time for recovery is controlled by temperature, $\tau_R = \tau_R^0\exp(H_R^*/RT)$, where H_R^* is activation enthalpy for recovery. Therefore the condition $\tau_R \ll \tau_T$ means that the temperature should be lower than some critical value $(T < (H_R^*/R)\cdot\log(\tau_R^*/\tau_T))$.

Note that when we use a physical interpretation of the Levy–von Mises formulation of non-linear rheology, i.e., when the second invariant of stress ($\mathrm{II}_{\sigma'}$) in the constitutive relation for plastic flow is interpreted to be

[2] The use of first-order kinetics as opposed to second-order kinetics (see Chapter 10) is justified because we consider the annihilation of dislocations with equal signs.

dislocation density (Chapter 3), then the two classes of models (the Greenwood–Johnson and the Poirier model) just discussed appear to be equivalent. Therefore these models have common limitations as we have just discussed: enhanced deformation occurs only when a high-dislocation density is maintained for a sufficiently long time (i.e., low temperature) and when work hardening is not important (i.e., high temperature). These conditions are mutually exclusive and the conditions under which transformation plasticity is important are rather limited.

15.3.3. Experimental observations

Experimental observations of transformation-induced plasticity are reviewed by GREENWOOD and JOHNSON (1965), POIRIER (1985) and MEIKE (1993). The linear relationship between applied stress and strain, i.e., relation (15.24), is in most cases well documented. However, in most cases, the enhanced deformation associated with a phase transformation is reported only when a phase transformation occurs at relatively low homologous temperature. The interplay between work hardening and softening, as well as the influence of dislocation recovery need to be investigated in more detail.

15.4. Effects of grain-size reduction

A first-order phase transformation such as the olivine to wadsleyite transformation can lead to a significant change in grain size. VAUGHAN and COE (1981) were the first to suggest that the grain-size reduction associated with the olivine–spinel (ringwoodite) transformation might result in rheological weakening. RUBIE (1984) discussed the possible role of weakening for a range of slabs with different thermal structures. Grain-size reduction is often observed in laboratory experiments that have led to the idea of rheological weakening. However, the real question here is if the grain sizes observed in laboratory experiments are representative of the grain sizes in Earth. The time scale of a phase transformation in a laboratory is vastly different from that in Earth, and therefore an appropriate *scaling law* is needed to estimate the grain size associated with a phase transformation in Earth. In fact, grain size after a phase transformation can become smaller or larger dependent on the conditions of phase transformation (see Chapter 13).

The scaling law for the grain-size evolution associated with a phase transformation in Earth has been analyzed by RIEDEL and KARATO (1996, 1997) and KARATO *et al.* (2001), and they applied the results to calculate the grain-size reduction associated with phase transformations that occur in a subducting slab. Briefly, the degree to which grain-size reduction occurs is highly sensitive to the temperature at which a phase transformation occurs. The grain size after a transformation is roughly determined by (see Chapter 13),

$$L \approx G(T)\frac{\Delta Z}{\upsilon_{\mathrm{sub}}} \tag{15.27}$$

where $G(T) = G_0 \exp\left(-H_G^*/RT\right)$ is the growth rate that depends on temperature (H_G^* is the activation enthalpy for growth), υ_{sub} is the velocity of subduction and ΔZ is the overshoot depth for phase transformation. For cold slabs ($T \sim 800$–900 K), the grain size can be as small as $\sim$1 μm, leading to substantial weakening. In addition, at low temperatures, the grain-growth kinetics is sluggish. Consequently, a significant rheological weakening occurs that can last for a substantial amount of time in cold subducting slabs. In contrast, such effects would not be important in warm slabs where the grain size after a phase transformation is large and also the grain-growth kinetics is fast. KARATO *et al.* (2001) explained the large contrast in the nature of deformation of subducting slabs between the eastern and western Pacific by this mechanism.

RUBIE (1983) discussed the role of reaction-induced grain-size reduction on the rheology of Earth's crust. FURUSHO and KANAGAWA (1999) and NEWMAN *et al.* (1999) reported evidence for such an effect in Earth's upper mantle. STÜNITZ and TULLIS (2001) reported such an effect in experimentally deformed plagioclase aggregates.

Note that the effects of grain-size reduction are particularly important in cases where a phase transformation involves formation of two phases, e.g., ringwoodite to perovskite and magnesiowüstite transformation. In such a case, a small grain size will last longer and the weakening effects will be significant (e.g., KUBO *et al.*, 2000).

15.5. Anomalous rheology associated with a second-order phase transformation

A second-order phase transformation can be associated with anomalous elastic and dielectric behavior (e.g., GHOSE, 1985; CARPENTER, 2006). For example, one of the elastic constants vanishes and/or the dielectric constant becomes infinite at the thermodynamic condition at which a second-order phase transformation occurs (the former occurs when anomalous behavior is associated with the acoustic mode of lattice vibration and the latter behavior occurs when anomalous behavior occurs in the optical mode of lattice vibration; see Chapter 4, and GHOSE (1985) and CARPENTER (2006) for more detail).

Second-order phase transformations are rather uncommon in Earth's interior, but many of the structural phase transformations observed in materials with the perovskite structure are of second order. Some of the phase transformations in SiO_2 are nearly second order. Anomalies in these properties (elastic and dielectric properties) could cause drastic changes in rheological properties. CHAKLADER (1963), WHITE and KNIPE (1978) and SCHMIDT *et al.* (2003) reported on an anomalous mechanical behavior (enhanced deformation) of quartz associated with the α to β transformation (see however a conflicting result by KIRBY (1977)) which is nearly second-order transformation (almost no volume change). Room-temperature plasticity was investigated at around a second-order phase transformation from stishovite to $CaCl_2$ structure of SiO_2 (SHIEH *et al.*, 2002). Although these authors did not find any anomalous behavior in the measured strength–elastic modulus ratio, they concluded anomalous weakening near the (second-order) phase transformation by translating their results to strength using the Reuss-average elastic constant (their Figure 4). The validity of this conclusion is questionable since their results of calculation reflect mostly the anomaly in one elastic constant (in the Reuss-average scheme, the average elastic constant is dominated by the small elastic constants), and the question of how the anomaly in one elastic constant affects the plastic strength (which is the central issue here) was not investigated in their study. KARATO and LI (1992) and LI *et al.* (1996) investigated the high-temperature creep behavior of $CaTiO_3$ across a structural phase transformation (between orthorhombic and tetragonal structures) and found only a small change in diffusional creep behavior similar to those observed by KIRBY (1977). In summary, drastic effects of second-order phase transformations on (high-temperature) plasticity have not been well documented, although such an effect might result in anomalies in plastic properties. In particular, high-elastic anisotropy near the second-order phase transformation could cause large plastic anisotropy, but there have been no definitive studies on this subject.

15.6. Other effects

A first-order phase transformation also involves a change in the entropy. Consequently, there is a change in temperature associated with a first-order phase transformation. An exothermic (endothermic) phase transformation such as the olivine to spinel (or wadsleyite or ringwoodite) transformation will result in the increase (decrease) in temperature. The degree to which temperature is modified depends on the relative rate of a phase change and that of thermal diffusion as well as the magnitude of change in the entropy. The maximum value of temperature change for transformations in the transition zone is $\sim\pm 100$ K, which would result in a change in viscosity of about an order of magnitude.

Another effect includes a change in element partitioning associated with a phase transformation. For example, during a phase transformation from olivine to wadsleyite, Fe is partitioned more into wadsleyite leading to weakening of this phase. Similarly, the transformation from ringwoodite to perovskite and magnesiowüstite results in Fe-enrichment in magnesiowüstite in most cases that leads to weakening of magnesiowüstite relative to co-existing perovskite. A more drastic effect is the redistribution of water (hydrogen) associated with melting (KARATO, 1986; see also Chapter 19).

In some cases a phase transformation of a phase results in decomposition into two (or more) phases. In such cases, the plastic properties of the resultant aggregates are those of a two (or more)-phase mixture. One particularly interesting case is where one of the phases is a fluid. In such a case, dramatic effects are expected depending on the geometry of the fluid phase (see Chapter 12). When the fluid assumes a continuous film, then the effective pressure is significantly reduced and shear instability can occur.

16 Stability and localization of deformation

Deformation in Earth often occurs in a localized fashion particularly in the lithosphere. The most marked example of this is deformation associated with plate tectonics, in which deformation is localized at plate boundaries: plate tectonics would not operate without shear localization. This chapter reviews the physical mechanisms of the localization of deformation. The criteria for instability and localization are reviewed not only for infinitesimal deformation but also for finite amplitude deformation. Several mechanisms of the localization of plastic flow are discussed including thermal runaway instability, localization due to grain-size reduction, localization due to strain partitioning in a two-phase material and localization due to the intrinsic instability of dislocation motion. Many processes may play a role in shear localization in Earth, but grain-size reduction appears to play a key role in shear localization in the ductile regime.

Key words instability, localization, bifurcation, shear band, necking instability, adiabatic instability, mylonite, grain-size reduction.

16.1. Introduction

The plastic deformation of solids involves the motion of crystalline defects, so deformation is always heterogeneous (i.e., localized) and non-steady at the level of the individual defect (dislocations, grains etc.). However, when averaged over the space-scale of many defects and over the time-scale of the motion of a large number of defects, deformation in the ductile regime occurs in most cases homogeneously both in space and time. Deviation from such an idealized form of deformation occurs under some limited conditions, which leads to (spatially) localized, and (temporally) unstable deformation. Localized deformation is ubiquitous in the shallow portions of Earth (i.e., the upper crust) where brittle failure is the dominant mode of deformation. Localized deformation could extend to deeper portions, e.g., the entire lithosphere, and even in the deeper portions of Earth.

The fundamental causes for localization in the brittle regime are (1) the intrinsic instability of motion of a single crack, due to the interplay between stress concentration and crack motion, and (2) the strong interaction among cracks to cause positive feedback that enhances localized deformation (e.g., Paterson and Wong, 2005, see also Chapter 7). However, brittle deformation is suppressed by a large confining pressure, and one does not expect brittle failure at depths exceeding ~10–20 km. In this chapter, we will focus on the mechanisms of localization of deformation in the plastic regime. In this regime, deformation occurs through the motion of smaller scale defects such as dislocations and point defects (Chapters 8 and 9). Motion of these defects is usually viscous and intrinsically stable and the interaction among these defects is less strong than the interaction among cracks. Consequently, instability and localization are not often seen in the plastic regime. However, under some limited conditions, the interaction among defects or the interaction of defects with other "fields" (such as the temperature field) could lead to unstable localized deformation. Shear localization in the plastic regime is generally considered to be the cause of crustal shear zones (e.g., White *et al.*, 1980). Classical examples of instability (localization) in metals in the ductile regime include the propagation of the *Lüders band* (localized deformation observed in iron or Al–Mg alloy at modest temperatures through the propagation of shear zones) and the *Portevin–Le Chatelier effect* (unstable deformation of some alloys such as Al–Li and Cu–Sn that is characterized by the serrated stress–strain curves; see e.g., Hirth and Lothe, 1982). In geologic materials, localized deformation is often observed when a region of fine grain size develops (e.g., Post, 1977; White *et al.*, 1980; Furusho and Kanagawa, 1999). Also well known is the *adiabatic instability*, i.e., localized deformation due to localized shear heating (e.g., Rogers, 1979). These processes of localized deformation are found under limited conditions in experimental studies. When one wants to evaluate if some of these processes (or processes similar to them) might occur in Earth, one needs to understand the basic physics behind them because the applications of laboratory data to geological problems always involve a large degree of extrapolation.

Mechanisms of instability (and localization) of deformation in the plastic regime have been reviewed by Hill (1958), Hart (1967), Argon (1973), Rudnicki and Rice (1975), Rice (1976), Kocks *et al.* (1979), Poirier (1980), Bai (1982), Evans and Wong (1985), Fressengas and Molinari (1987), Estrin and Kubin (1988), Hobbs *et al.* (1990), Montési and Zuber (2002), and Regenauer-Lieb and Yuen (2003). Reviews with a geological context include Evans and Wong (1985), Handy (1989), Hobbs *et al.* (1990), Drury *et al.* (1991), Braun *et al.* (1999), Bercovici and Karato (2002), Montési and Hirth (2003), and Regenauer-Lieb and Yuen (2003). In this chapter, I will discuss some of the fundamental physics behind instability and localization during deformation in the plastic regime.

16.2. General principles of instability and localization

16.2.1. Criteria for instability and localization: infinitesimal amplitude analyses

Instability (and localization) of deformation occurs when the changes in properties of materials and/or local stress caused by a small excess deformation in a region cause further deviation from homogeneous deformation in that region. This occurs, for example, when a material has a property such that the resistance to deformation decreases with strain (or strain rate). In fact, in most laboratory experiments, the localization of deformation is associated with some sort of *weakening*. For example, localized deformation in the soft steel called the Lüders band propagation is associated with a sharp drop of stress in a constant strain-rate test (e.g., Nabarro, 1967b). As we have learned in previous chapters (Chapters 3, 8 and 9), in most cases, resistance to plastic deformation *increases* with strain (work hardening) and strain rate (positive strain-rate exponent). Therefore deformation in the plastic regime is usually stable and homogeneous and the instability and localization of deformation will occur only when a material has some atypical properties and/or when macroscopic deformation geometry satisfies some conditions. The main theme of this chapter is to provide a brief summary of the physical basis for unstable and localized deformation to identify the conditions under which unstable and localized deformation may occur.

The condition for instability mentioned above can be translated into an inequality

$$\frac{\delta \log \dot{\varepsilon}}{\delta \log \varepsilon} > 0. \tag{16.1}$$

When this condition is met, then in a region where more than average deformation occurred ($\delta \log \varepsilon > 0$), strain rate there will become larger ($\delta \log \dot{\varepsilon} > 0$). Therefore strain there will increase further and deformation will be localized in that region.

In the literature, the condition for localization (instability) was often defined in different ways. In a classical paper on the necking instability for tension tests, Hart (1967) defined the condition for unstable deformation in terms of cross sectional area A as

$$\frac{\delta \log \dot{A}}{\delta \log A} < 0 \text{ or } \frac{\delta \dot{A}}{\delta A} > 0 \tag{16.2}$$

where $\dot{A}$ is the time derivative of A. When this condition is met, an area where the cross section becomes smaller than average will keep shrinking that leads to the *necking instability*.

Poirier (1980) presumed that deformation is unstable when the load needed to deform a material is reduced by a small increase in strain, namely,

$$\frac{\delta \log F}{\delta \varepsilon} \leq 0 \tag{16.3}$$

where $F = \sigma A$ is the load (force). For simple shear where A is constant, this relation is equivalent to $\delta \log \sigma / \delta \varepsilon < 0$.

An alternative way to define the instability is to investigate the time dependence of strain. When strain is written as $\varepsilon = \varepsilon_0 \exp(\lambda t)$, then if

$$\lambda > 0, \tag{16.4}$$

strain increases exponentially with time and deformation is said to be unstable. This approach has been used in many linear stability analyses (e.g., ESTRIN and KUBIN, 1988).

MONTÉSI and ZUBER (2002) defined an effective stress exponent, $n_{\mathrm{eff}} \equiv \delta \log \dot{\varepsilon} / \delta \log \sigma$, and use the negative effective stress exponent,

$$n_{\mathrm{eff}} < 0, \tag{16.5}$$

as a condition for instability and localization.

Yet another way to define the condition for instability is to seek a condition under which a system can assume two different states corresponding to the same imposed boundary conditions (*bifurcation*; e.g., RUDNICKI and RICE, 1975; BERCOVICI and KARATO, 2002). For example, for a given stress, when a system can assume two different strain rates, then a system will evolve into *bifurcation* in which parts of the system will assume one strain (strain rate) and other parts will assume another strain (strain rate). This leads to localization if the strains (strain rates) in these regions are sufficiently different.

How are these different expressions related? The correspondence between (16.1) and (16.4) is straightforward. Note that if $\varepsilon = \varepsilon_0 \exp(\lambda t)$, then $\dot{\varepsilon} = \lambda \varepsilon$ and $\delta \dot{\varepsilon} = \lambda \delta \varepsilon$. Therefore $\delta \dot{\varepsilon} / \delta \varepsilon = \lambda$ and (16.1) and (16.4) are equivalent. To compare (16.1) with (16.2), let us note that the relations $\delta \varepsilon = -\delta \log A$ and $\dot{A}/A = -\dot{l}/l = -\dot{\varepsilon}$ hold for deformation of a cylindrical sample with no volume change. From $\dot{A}/A = -\dot{l}/l = -\dot{\varepsilon}$, one obtains $\delta \dot{A}/\delta A = \dot{A}/A = \delta \dot{\varepsilon}/\delta \varepsilon$. Note that (16.1) and (16.2) are not identical in general except for a case where $|\delta \dot{A}/\delta A| \gg |\dot{A}/A|$ (e.g., ESTRIN and KUBIN, 1986).

The correspondence of *bifurcation* to the condition of instability described above can be understood if one notes that for most cases $\delta \log \dot{\varepsilon} / \delta \log \varepsilon \leq 0$ at small strains (work hardening). In such a case, the condition $\delta \log \dot{\varepsilon} / \delta \log \varepsilon > 0$ means that $\delta \log \dot{\varepsilon} / \delta \log \varepsilon$ changes sign from negative to positive as strain increases. Consequently, a strain versus strain-rate curve, $\varepsilon(\dot{\varepsilon})$, for a constant stress, will not be a single-valued function, so there will be two strains corresponding to a given strain rate that leads to bifurcation. Bifurcation analysis can also be made for $\dot{\varepsilon}(\sigma)$. Recall $\delta \log \sigma / \delta \log \dot{\varepsilon} > 0$ in most materials for small strain rates (positive stress exponent), therefore if $\delta \log \sigma / \delta \log \dot{\varepsilon}$ becomes negative at a large strain rate, then $\dot{\varepsilon}(\sigma)$ will not be a single-valued function leading to bifurcation. Note that the conditions for localization correspond to $\delta \log \sigma / \delta \log \dot{\varepsilon} < 0$, i.e., negative effective stress exponent.

The instability condition by POIRIER (1980) (i.e., (16.3)) is not equivalent to $\delta \log \dot{\varepsilon} / \delta \log \varepsilon > 0$ (or equivalently to (16.2) and (16.4)). Equation (16.3) can be written as $\delta \log \sigma / \delta \varepsilon + \delta \log A / \delta \varepsilon < 0$. ($\delta \log A / \delta \varepsilon$) ($\equiv k$) is a constant that depends on the geometry of deformation (− 1 for uni-axial deformation, 0 for simple shear). Consider a case of simple shear ($k = 0$). Then (16.3) becomes $\delta \log \sigma / \delta \log \varepsilon = (\partial \log \sigma / \partial \log \varepsilon)_{\dot{\varepsilon}} + (\partial \log \sigma / \partial \log \dot{\varepsilon})(\partial \log \dot{\varepsilon} / \partial \log \varepsilon) < 0$. It is clear that the condition $\delta \log \sigma / \delta \log \dot{\varepsilon} < 0$ and $\delta \log \dot{\varepsilon} / \delta \log \varepsilon > 0$ are not connected directly. In fact, the condition of $\delta \log \sigma / \delta \log \varepsilon < 0$, i.e., the maximum stress in a stress–strain curve, is often observed well before shear instability occurs, indicating that it is not an appropriate criterion for shear instability (localization) (ARGON, 1973).

Similarly, the condition of negative effective stress exponent, (16.5), does not necessarily lead to shear instability (localization). This can be seen by writing $\delta \log \dot{\varepsilon} / \delta \log \varepsilon = (\partial \log \dot{\varepsilon} / \partial \log \varepsilon)_{\sigma} + (\partial \log \dot{\varepsilon} / \partial \log \sigma)(\partial \log \sigma / \partial \log \varepsilon)$. Even if $(\partial \log \sigma / \partial \log \dot{\varepsilon}) < 0$, if $(\partial \log \sigma / \partial \log \varepsilon) > 0$ (i.e., work hardening), then $(\partial \log \dot{\varepsilon} / \partial \log \sigma)(\partial \log \sigma / \partial \log \varepsilon) < 0$ and instability will not necessarily occur. Also the condition of a negative effective stress exponent does not apply to thermal runaway instability as will be shown later.

16.2.2. Development of strain localization: finite amplitude instability

Although most of the discussions in this chapter are on the conditions for instability at a small amplitude of perturbation in strain, it is important to discuss some issues of instability at finite amplitude. The condition $\delta \log \dot{\varepsilon} / \delta \log \varepsilon > 0$ means that deformation is unstable and that a material has the potential for strain localization, but this does not tell us how fast strain localization develops. Therefore this condition alone is not sufficient to evaluate if significant shear localization develops or not. Also important is the boundary condition or the property of the forcing "machine." As we will learn, certain conditions must be met about the properties of the "machine" (materials surrounding the region in which shear localization occurs), in order for significant localization to develop.

Influence of time dependence of materials parameters

The instability of deformation may not necessarily lead to significant shear localization unless the growth of instability is rapid. Figure 16.1 illustrates this point. Curve *A* shows the case where instability occurs and rapidly grows leading to significant shear localization. Curve *B* shows the case in which the instability occurs initially but the rate of growth of instability decreases rapidly, and significant localization will not occur. Case *C* is the case where instability occurs fast initially but the rate of instability growth decreases with time. However, the rate of decrease in the rate of instability is small compared to the initial rate of growth of instability and one obtains a sufficiently large degree of localization. Curve *D* shows yet another case, where the initial physical conditions do not cause instability, but eventually a material develops some conditions (e.g., microstructures) that promote instability.

In order to treat the issue of cases *A*, *B* and *C* shown in Fig. 16.1, Estrin and Kubin (1988) used the following criteria for the development of finite amplitude instability (localization),

$$\lambda_0 > 0 \text{ and } \left(\frac{d\lambda}{dt}\right)_0 > 0 \quad (\text{case } A) \tag{16.6}$$

or

$$\lambda_0 > 0, \left(\frac{d\lambda}{dt}\right)_0 < 0 \text{ and } \left|\left(\frac{d\lambda}{dt}\right)_0\right| \ll \lambda_0^2. \quad (\text{case } C) \tag{16.7}$$

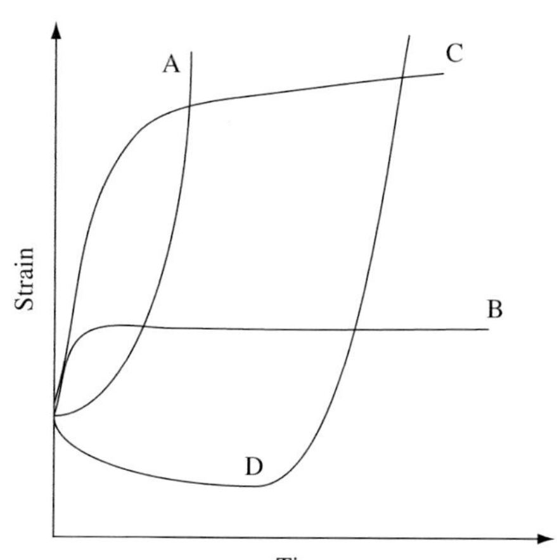

FIGURE 16.1 A schematic diagram showing the development of shear localization and instability (modified after Kocks *et al.*, 1979).

The physical meaning of (16.6) is obvious. The conditions (16.7) mean that the rate with which instability grows decreases with time, but the initial rate of growth is fast enough to establish significant degree of localization (note that the rate of initial growth is λ_0 and the rate of change in growth rate is $\left|(d\lambda/dt)_0\right|/\lambda_0$). The case *B* would correspond to

$$\lambda_0 > 0, \left(\frac{d\lambda}{dt}\right)_0 < 0 \text{ and } \left|\left(\frac{d\lambda}{dt}\right)_0\right| \geq \lambda_0^2. \quad (\text{case } B) \tag{16.8}$$

The evaluation of a situation *D*, i.e., $\lambda_0 < 0$ but $\lambda > 0$ at later stage, requires the analysis of instability for a large range of strain (time). Conditions for instability are often not met at small strains, but gradually evolve with strain. Analyses of $\delta \log \dot{\varepsilon}/\delta \log \varepsilon$ at various strains will be critical to address this point.

Although the following analyses will focus on linear stability analyses (analyses for infinitesimal strain), I will touch upon these issues when they play a critical role.

Problem 16.1*

Show that the conditions (16.7) lead to significant localization but (16.8) will not.

Solution

By integrating $\delta\varepsilon = \delta\varepsilon_0 \exp(\lambda t)$ with $\lambda = \lambda_0 + (d\lambda/dt)_0\, t$, one obtains

$$\begin{aligned}\frac{\varepsilon}{\delta\varepsilon_0} &= \int_0^\infty \exp\left(\lambda_0 t + \left(\frac{d\lambda}{dt}\right)_0 t^2\right) dt \\ &= \int_0^\infty \exp\left(\lambda_0 t - \left|\left(\frac{d\lambda}{dt}\right)_0\right| t^2\right) dt \\ &= \frac{\sqrt{\pi}}{2}\exp\left(\frac{\xi}{4}\right)\left[1 + \operatorname{erf}\left(\sqrt{\xi}\right)\right]\left(\xi \equiv \frac{\lambda_0^2}{\left|\left(\frac{d\lambda}{dt}\right)_0\right|}\right)\end{aligned}$$

where I used the definition $\operatorname{erf}(x) \equiv \frac{2}{\sqrt{\pi}}\int_0^x \exp(-y^2)\, dy$ ($\operatorname{erf}(\infty) = 1, \operatorname{erf}(0) = 0$). If the condition (16.7) is met (i.e., $\xi \gg 1$), then $\varepsilon/\delta\varepsilon_0 \approx \sqrt{\pi}\exp(\xi/4) \to \infty$, so significant localization will occur. If $\xi < 1$ (i.e., (16.8)), then $\varepsilon/\delta\varepsilon_0 \approx (\sqrt{\pi}/2)\exp(\xi/4) \approx \sqrt{\pi}/2$ (for $\xi \to 0$) and localization will be limited.

Influence of elasticity (machine stiffness)

Boundary conditions can also play an important role in the *growth* of instability. Consider a sample connected to a "machine" which is described by an elastic spring

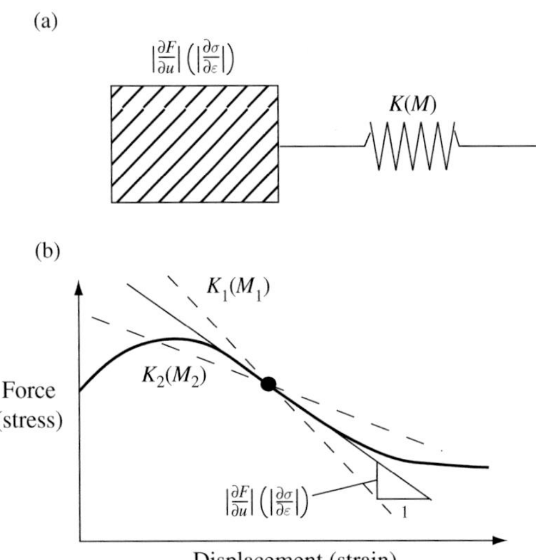

FIGURE 16.2 (a) A spring connected to a material that shows time-dependent deformation and (b) a strain (displacement) versus stress diagram showing the condition for development of a finite-amplitude instability.

When $|\frac{\partial\sigma}{\partial\varepsilon}| < M$ (or $|\frac{\partial F}{\partial u}| < K$), then instability does not occur even if there is strain softening, whereas when $|\frac{\partial\sigma}{\partial\varepsilon}| > M$ (or $|\frac{\partial F}{\partial u}| > K$), then instability does develop.

(Fig. 16.2a). Force is supplied from this machine to a sample so that when instability starts in the sample, the machine must provide energy in order to keep up with continuing *growth* of instability. If the machine is very stiff, then even though the properties of a sample become weak and satisfy the conditions for instability, no continuing development of localization is possible. The machine must be sufficiently soft to realize the localization. In other words, the degree of weakening must be sufficiently large in order to achieve a significant amount of localization (e.g., SCHOLZ, 2002).

To see this point, let us consider a stress (force) versus strain (displacement) diagram (Fig. 16.2b). When $\partial F/\partial u < 0$, a displacement will cause unloading of the machine. The unloading will follow the slope, $-K$, defined by the machine stiffness, K, where $F = Ku$ (F, force; u, displacement). If this unloading of the machine is not as large as the unloading from the sample, then instability does not develop to result in significant localization. So the condition $|\partial F/\partial u| > K$ must be satisfied in order for the machine to keep up with the progressive deformation. For shear localization in Earth, a similar formula can be obtained if regions outside a shear zone behave like an elastic body. This will be a good approximation when shear localization occurs in a short time-scale in a relatively cold region. In such a case equation $|\partial F/\partial u| > K$ may be cast into the equivalent form

$$\left|\frac{\partial\sigma}{\partial\varepsilon}\right| > M$$

where M is the elastic modulus of the material. In other words, when the finite amplitude instability is considered, the condition for instability is stronger than those discussed in 16.2.1. For example, instead of $\delta\sigma/\delta\varepsilon < 0$, one must have $\delta\sigma/\delta\varepsilon < -M$.

Problem 16.2

Show that $|\partial F/\partial u| > K$ and $|\partial\sigma/\partial\varepsilon| > M$ are equivalent.

Solution

From the definition of the stiffness $F = Ku$ and the definition of an elastic constant M ($\equiv \Delta\sigma/\Delta\varepsilon = \Delta F \cdot u/\Delta u \cdot A$), one gets $K = MA/u$. So $|\partial F/\partial u| > K = MA/u$ and hence $|\partial\sigma/\partial\varepsilon| > M$.

16.2.3. Orientation of shear zones

In most of the analyses of shear localization, deformation is assumed to be plane deformation and the plane of shear localization is assumed to be the same as the shear plane. This assumption is valid if only one component of stress (strain) plays an important role. However, when deformation involves volume change (volume expansion, in most cases), then such an assumption will no longer be valid. ANAND *et al.* (1987) extended shear localization analysis in the ductile regime to two dimensions and determined the orientation of a shear zone. They made a linear stability analysis involving the following properties of materials: the strain-rate hardening, strain hardening, thermal softening and pressure hardening coefficients. Conditions under which a small perturbation in shear strain grows are determined as well as the orientation along which the perturbation grows most rapidly. They found that shear bands are formed, in simple shear geometry, in two directions. Strain in such a case must be accommodated by the combination of deformation in two shear bands. The fastest growing shear band is formed in the direction (measured from the direction of the instantaneous maximum stretching direction) (Fig. 16.3)

$$\chi = \pm\left[\frac{\pi}{4} + \frac{\beta}{2}\right] \quad (16.9)$$

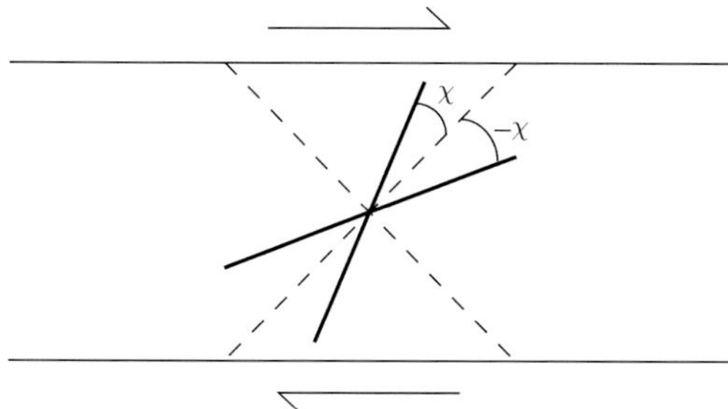

FIGURE 16.3 Orientations of shear zones (shown by thick lines). Broken lines indicate the direction of principal stress. The orientation of shear zones depends critically on the pressure sensitivity of the strength (i.e., volumetric strain). When there is no pressure sensitivity (i.e., no dilatation), shear bands will be formed on a plane parallel to the shear plane. When the strength has a significant pressure dependence (i.e., a large volumetric strain), then the shear bands will have a certain angle with respect to the shear plane (after ANAND *et al.*, 1987).

with

$$\beta = \sin^{-1}\left[\frac{1}{P_0}\left(1 - \sqrt{1 - P_0^2}\right)\right] \tag{16.10}$$

where P_0 is a non-dimensional parameter characterizing the pressure sensitivity of strength (i.e., $\tau = P_0 \cdot P$ where τ is the shear stress). If rheological properties are pressure-insensitive, then $P_0 = 0$, and $\beta = 0$. In such a case, a shear band will be formed parallel to the shear plane (another one would be normal to shear plane, but strain by this is small). According to this theory, there will be an inclination between the shear bands and the shear plane when the rheological properties are pressure-sensitive.[1] Note that the results of this theory are analogous to the theory of brittle fault (Chapter 7). In the brittle faulting, using Coulomb–Navier's law, the angle of a fault is controlled by the friction coefficient that is analogous to the parameter P_0. The observed tilted shear zones in the shear deformation of a partially molten material (e.g., ZIMMERMAN *et al.*, 1999; HOLTZMAN *et al.*, 2003b) suggest that the rheological behavior of material in their experiments is pressure-sensitive. SPIEGELMAN (2003) and KATZ *et al.* (2006) proposed a model to explain shear localization in partially molten materials that involves volume change during deformation.

[1] The pressure-insensitive rheology in this context means ductile rheology in which pressure-sensitivity is small. In terms of a non-dimensional parameter, P_0, brittle rheology would correspond to a large value of P_0 ($P_0 \approx 0.5$, P_0: friction coefficient). The ductile rheology is also pressure-sensitive, but $P_0 = \sigma V^*/RT \approx 10^{-4} - 10^{-2}$ (V^*: activation volume). In the context of shear band formation, ductile rheology is considered to be pressure-insensitive rheology.

16.3. Mechanisms of shear instability and localization

16.3.1. General considerations

In order to understand the physical meaning of conditions of instability (localization), let us consider a simple case where only one component of stress (or strain) plays an important role. In such a case, the relation between stress and various variables may be written as[2]

$$\dot{\varepsilon} = \dot{\varepsilon}(T, \varepsilon, \sigma;\ S;\ B) \tag{16.11}$$

where T is temperature, ε is strain, $\dot{\varepsilon}$ is strain rate, σ is stress, S is some microstructural parameter such as grain size, and a parameter B denotes a geometrical factor such as the cross sectional area which depends on sample and deformation geometry.

Now we consider how stress needed to deform a sample changes with a small variation of parameters, namely,

$$\begin{aligned}\delta \log \dot{\varepsilon} = &X_T \delta \log T + X_\varepsilon \delta \log \varepsilon + X_\sigma \delta \log \sigma \\ &+ X_S \delta \log S + X_B \delta \log B\end{aligned} \tag{16.12}$$

where $X_T = (\partial \log \dot{\varepsilon}/\partial \log T)_{\sigma, \varepsilon, S, B}$ etc. Each term represents the sensitivity of strain rate on various parameters that depends on the specific mechanisms (and processes) of deformation as well as deformation geometry.

Three of them correspond to intrinsic material properties. They are $X_T = (\partial \log \dot{\varepsilon}/\partial \log T)_{\sigma, \varepsilon, S, B}$, $X_\varepsilon = (\partial \log \dot{\varepsilon}/\partial \log \varepsilon)_{T, \sigma, S, B}$ and $X_\sigma = (\partial \log \dot{\varepsilon}/\partial \log \sigma)_{T, \varepsilon, S, B}$. $X_T = (\partial \log \dot{\varepsilon}/\partial \log T)_{\sigma, \varepsilon, S, B}$ represents the temperature sensitivity of strength, which is $X_T = H^*/RT$ for power-law rheology involving typical thermal activation processes. This term is positive, so a local increase in temperature by deformation tends to enhance instability and resultant localization. $X_\varepsilon = (\partial \log \dot{\varepsilon}/\partial \log \varepsilon)_{T, \sigma, S, B}$ represents the sensitivity to strain. This term is zero for steady-state rheology and has a finite value only for *transient* rheological behavior. In many cases, transient behavior is characterized by *work hardening* in such a case, this term may be represented by $\dot{\varepsilon} \propto \varepsilon^{-h}$ and hence $X_\varepsilon = -h$ where h is called the work hardening coefficient (*work softening* is also observed in some cases, in which

[2] Note that I have assumed that the plastic flow strength is pressure-insensitive. This is justifiable in most ductile deformation in which a change in pressure caused by applied deviatoric stress is small. When the strength is pressure-sensitive, i.e., deformation in the brittle regime, then at least two components of stress (strain) need to be considered.

case one can use a negative value of h). Work hardening will reduce the strain rate, so it has a stabilizing effect. the $X_\sigma = (\partial \log \dot{\varepsilon} / \partial \log \sigma)_{T,\varepsilon,S,B}$ term represents the stress sensitivity of strain rate, and is $X_\sigma = n$ for power-law rheology. The term $X_S = (\partial \log \dot{\varepsilon} / \partial \log S)_{T,\sigma,\varepsilon,B}$ represents the influence of some structural parameters such as dislocation density and grain size on the strength. The local variation in grain size is one of the important mechanisms of shear localization.

Finally, the term $X_B = (\partial \log \dot{\varepsilon} / \partial \log B)_{T,\sigma,\varepsilon,S}$ represents the influence of deformation geometry. For example, cross sectional area may change during uni-axial deformation that could lead to necking instability (e.g., Hart, 1967).

16.3.2 Specific mechanisms of instability and localization

Geometrical (necking) instability

Plastic deformation often causes shape change that may cause local variation in effective stress. Let us focus our attention to necking instability. The essence of necking instability is that once an area smaller than other area is formed during (uni-axial) tension, then stress is concentrated in that area causing further shrinkage of that area. Since the essence of this instability is the geometrical one, it is appropriate to use the instability criterion by Hart (1967), i.e., $\delta \log \dot{A} / \delta \log A < 0$. To write this inequality in terms of properties of materials, let us start from the relation $\delta \log \dot{\varepsilon} = n \cdot \delta \log \sigma - h \cdot \delta \log \varepsilon$ (we assumed a relation $\dot{\varepsilon} \propto \varepsilon^{-h} \sigma^n$). Now from $F = \sigma A$ where F is the applied force and A is the area through which force acts, the change in cross sectional area is related to the change in stress as $\delta \log \sigma + \delta \log A = 0$ and $\dot{\sigma}/\sigma + \dot{A}/A = 0$ (force is assumed to be constant). Also from $V = lA$ where V is volume and l is the length, we get $\delta \log l + \delta \log A = \delta\varepsilon + \delta \log A = 0$ ($\delta\varepsilon = \delta \log l$, elongation strain is taken positive here) and hence $\dot{\varepsilon} + \dot{A}/A = 0$ and $\delta\dot{\varepsilon} = -\delta\dot{A}/A + \dot{A}\delta A/A^2$ (volume is assumed to be constant). Inserting these relations into $\delta \log \dot{\varepsilon} = n\delta \log \sigma - h\delta \log \varepsilon$, one obtains $\delta \log \dot{A} / \delta \log A = n(h/\varepsilon n + 1/n - 1)$. Therefore the condition for instability, $\delta \log \dot{A} / \delta \log A < 0$, is

$$n > \frac{h}{\varepsilon} + 1. \qquad (16.13)$$

For $h = 0$, by integrating $\delta \log \dot{A} / \delta \log A = 1 - n$, one can show that the instability grows as (Hart, 1967)

$$\frac{\delta A}{\delta A_0} = \left(\frac{A}{A_0}\right)^{1-n} \quad (\text{for } h = 0). \qquad (16.14)$$

Note that the rate of growth of inhomogeneity is a strong function of stress exponent, n. If $n = 1$, there is no growth, but if $n \gg 1$, then the growth is rapid.

Problem 16.3

Derive equation (16.14).

Solution

From $d\log \dot{A} / d\log A = 1 - n$, one obtains $\dot{A}/\dot{A}_0 = (A/A_0)^{1-n}$. Now if the initial heterogeneity in cross section develops during a time δt, $\delta A_0 = \dot{A}_0\,\delta t$, then the corresponding heterogeneity in the final stage will be $\delta A = \dot{A}\,\delta t$ so that $\delta A/\delta A_0 = \dot{A}/\dot{A}_0 = (A/A_0)^{1-n}$.

A large stress exponent and/or small work hardening coefficient promote necking instability. For a small n, instability grows very slowly (see equation (16.14)) so that a material can deform uniformly. This is an explanation for a stable large extensional deformation of a material with a small stress exponent known as *superplasticity* (see Chapter 8). It should also be noted that this type of geometrical instability does not occur in simple shear in which the area to which stress acts is kept constant. Since most deformation geometry in Earth involves a large component of simple shear, necking instability does not usually play an important role (see, however, Tapponnier and Francheteau, 1978).

Adiabatic instability (thermal runaway instability)

Plastic deformation dissipates energy. Consequently, if there is a region in which a higher strain (strain rate) is created then more heat is generated in that region and hence that region is heated up and there will be a local increase in temperature (this often leads to melting). Increased temperature will enhance plastic deformation and therefore there is a positive feedback and potential for unstable, localized deformation due to deformation-induced heating. Processes that may counteract this include: (1) thermal diffusion and (2) work hardening. Therefore the essence of models of thermal (adiabatic) instability is to examine how these processes compete. The interplay between deformation-induced heating, work hardening and heat flow can be analyzed by solving the momentum balance equation and energy transport equation together. However, a simple analysis can be made if one focuses on infinitesimal amplitude instability. To do this let us assume a simple power-law constitutive relation with a work-hardening term, namely, $\dot{\varepsilon} \propto \exp(-\frac{H^*}{RT})\varepsilon^{-h}\sigma^n$.

We consider how small changes in controlling variables affect the strain rate. To evaluate the influence of temperature, we start from the energy conservation relation,

$$\frac{\partial T}{\partial t} = \kappa \frac{\partial^2 T}{\partial x^2} + \frac{\dot{\varepsilon}\sigma}{\rho C_P} \tag{16.15}$$

where κ is thermal diffusivity, $\dot{\varepsilon}\sigma$ is the rate of energy dissipation per unit volume, ρ is the density, C_P is the specific heat. When there is no heat diffusion, temperature will rise according to $\partial T/\partial t = \dot{\varepsilon}\sigma/\rho C_P$ whereas if temperature diffusion is fast, then energy balance is dominated by $\partial T/\partial t = \kappa(\partial^2 T/\partial x^2)$ and therefore the characteristic time of heat diffusion is $\tau_{\text{th}} \approx L^2/\kappa$. Consequently, the temperature increase due to mechanical work is roughly given by (e.g., Argon, 1973)

$$\delta T = \beta \frac{\dot{\varepsilon}\sigma}{\rho C_p} \delta t \tag{16.16}$$

with

$$\beta \equiv \frac{\tau_{\text{th}}}{\tau_{\text{def}}} = \frac{L^2 \dot{\varepsilon}}{\kappa \varepsilon} \tag{16.17}$$

which depends on the dimension of the region of interest, L, and thermal diffusivity, κ and hence

$$\delta T = \frac{\dot{\varepsilon}}{\varepsilon} \frac{L^2}{\kappa} \frac{\dot{\varepsilon}\sigma}{\rho C_p} \delta t = \frac{L^2}{\kappa} \frac{\dot{\varepsilon}\sigma}{\rho C_p} \delta \log \varepsilon. \tag{16.18}$$

Inserting these relations into the instability condition, $\delta \log \dot{\varepsilon}/\delta \log \varepsilon > 0$, we obtain

$$\frac{H^*}{RT^2} \frac{L^2}{\kappa} \frac{\dot{\varepsilon}\sigma}{h\rho C_P} > 1. \tag{16.19}$$

We find that the adiabatic instability occurs when (1) a material is deformed at a high energy dissipation rate, $\dot{\varepsilon}\sigma$ (fast strain rate and/or high stress), (2) a large region is subject to temperature heterogeneity (large L), and (3) when deformation occurs at relatively low temperatures. A large sample size and high strain rate are needed in order to develop localized high temperature regions without significant heat diffusion. Consequently, for small samples in a typical laboratory experiment, one needs a very high rate of deformation (e.g., Spray, 1987; Tsutsumi and Shimamoto, 1997) and the adiabatic instability is not usually observed in typical slow laboratory deformation experiments. However, for a large body, conditions for adiabatic instability are more easily met and evidence for shear heating is found in the rock called *pseudotachylyte*, a very fine-grained (sometimes glassy) rock observed on fault planes that has been quenched from the melt (Sibson, 1975, 1977).

Griggs and Baker (1969), Hobbs *et al.* (1986), Ogawa (1987), Hobbs and Ord (1988) and Karato *et al.* (2001), discussed the possible implications of adiabatic instability for the origin of deep (and intermediate depth) earthquakes.

Problem 16.4

Calculate the critical energy dissipation rate for adiabatic instability.

Solution

Rewriting (16.19), one gets $\dot{\varepsilon}\sigma > (\dot{\varepsilon}\sigma)_c \equiv h\rho C_P \kappa RT^2/H^* L^2$. If we plug in the following typical values ($h=1$, $\rho = 3000\,\text{kg/m}^3$, $C_P = 1200\,\text{J/kg K}$, $\kappa = 10^{-6}\,\text{m}^2/\text{s}$ and $H^* = 500\,\text{kJ/mol}$), then we get $(\dot{\varepsilon}\sigma)_c = 10^{-8}-10^{-10})\,\text{W/kg}$ for $T \sim 2000\,\text{K}$ and $L = 1$–$10\,\text{km}$. This value can be compared with the heat production rate from typical Earth material of 10^{-10}–$10^{-12}\,\text{W/kg}$ (e.g., Turcotte and Schubert, 1982). We conclude that the adiabatic instability likely occurs in Earth, but to obtain a high degree of localization (small L), atypically high energy dissipation is required.

Localization due to grain-size reduction

The association of a smaller grain size and shear localization is well known in mylonites (White, 1979; White *et al.*, 1980). Therefore the role of grain-size reduction in shear localization deserves special attention. When deformation occurs by the motion of crystalline defects, the reduction in grain size results in some degree of weakening if grain-size reduction is large. However, some processes could counteract this, including grain growth. Therefore the competition of these processes needs to be evaluated in order to understand the role of grain-size reduction in shear localization.

Details of the processes that control grain size are reviewed in Chapter 13. Grain-size reduction can occur through two different processes: nucleation–growth of new grains by phase transformation (or by chemical reaction) or by dynamic recrystallization. These processes may reduce the grain size. In contrast, grain growth driven by grain-boundary energy will increase the grain size.

Let us use a simple constitutive relation that includes the grain size sensitivity, $\dot{\varepsilon} \propto L^{-m}\varepsilon^{-h}\sigma^{n}$. Then noting that grain size changes with strain as well as time, we obtain (for a constant stress),

$$\frac{\delta \log \dot{\varepsilon}}{\delta \log \varepsilon} = -m\left[\left(\frac{\partial \log L}{\partial \log \varepsilon}\right)_t + \left(\frac{\partial \log L}{\partial t}\right)_{\varepsilon}\frac{\varepsilon}{\dot{\varepsilon}}\right] - h \qquad (16.20)$$

where $(\partial \log L/\partial \log \varepsilon)_t$ represents the rate at which grain size is reduced with strain (e.g., by dynamic recrystallization), and $(\partial \log L/\partial t)_{\varepsilon}$ represents the rate of change in grain size at a fixed strain (e.g., by grain growth). The condition for (infinitesimal strain) instability is therefore

$$\left(\frac{\partial \log L}{\partial \log \varepsilon}\right)_t + \left(\frac{\partial \log L}{\partial t}\right)_{\varepsilon}\frac{\varepsilon}{\dot{\varepsilon}} + \frac{h}{m} < 0. \qquad (16.21)$$

Essentially, this equation means that grain size must decreases with strain faster than the rate of grain growth and work hardening rate. This condition can be met for a certain range of physical parameters. One obvious case is a case where there is a large degree of grain-size reduction and grain growth is inhibited. Chemical reactions (i.e., metamorphism) under relatively low temperature conditions provide such an environment. Chemical reactions at relatively low temperatures occur only when a large departure from chemical equilibrium is achieved. In such a case, nucleation rate of new phases is relatively high because of a large driving force for the reaction. Yet the growth rate is low so that one tends to get a small grain size for the reaction product (see Chapter 13). Subsequent growth of grains is also slow particularly when the reaction products contain several different phases due to Zener pinning (see Chapter 13). Consequently, metamorphism provides a good opportunity to lead localized (enhanced) deformation (e.g., Rubie, 1983). Natural evidence for shear localization caused by grain-size reduction due to chemical reaction is reported by Drury *et al.* (1991) and Furusho and Kanagawa (1999).

The situation is more complicated when grain-size reduction occurs due to dynamic recrystallization. Dynamic recrystallization occurs only in the dislocation creep regime (Chapter 13), in which deformation is insensitive to grain size (Chapter 9). Consequently, $m=0$ so that there is no effect of grain-size reduction in this regime. Furthermore, when deformation occurs in grain-size sensitive mechanism such as diffusional creep, there is no mechanism to reduce the grain size.

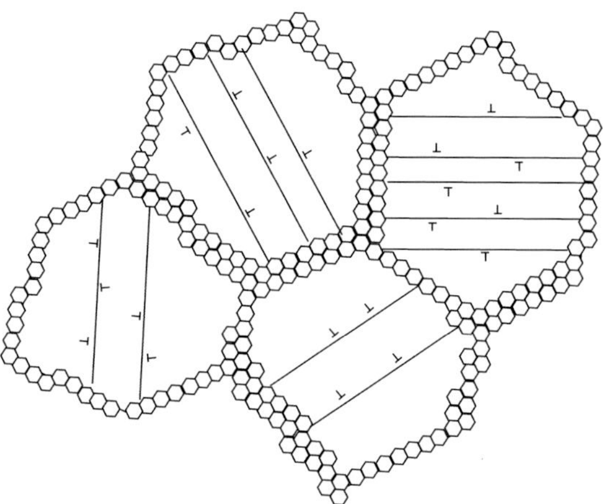

FIGURE 16.4 A microstructure of a material undergoing dynamic recrystallization. Large grains undergo deformation by dislocation creep which creates small recrystallized grains near grain boundaries.

Therefore it might appear that the concept of weakening and shear localization by dynamic recrystallization is fundamentally flawed. It is easy to see that such a simple view is not valid when one considers the *microstructural heterogeneity* in a dynamically recrystallizing material (Fig. 16.4). In most cases, dynamic recrystallization is initiated at or near existing grain boundaries (Chapter 13). Fine-grained regions are formed there whose thickness increases with strain. If the grain size in this region is below a threshold value, then the deformation mechanism there will be grain-size sensitive creep such as diffusional creep. During this stage, the "cores" of large grains are deformed by dislocation creep and hence keep producing heterogeneity in dislocation distribution that leads to the formation of fine-grained regions along the grain boundaries. Consequently, the thickness of fine-grained regions increases with time (strain) until the whole sample is completely recrystallized. In the intermediate stage of dynamic recrystallization therefore a material is composed of two regions: large grains deformed by dislocation creep, and fine-grained regions near the grain boundaries deformed by diffusional creep (or some sort of grain-size sensitive creep). It is in this *transient stage* where shear localization could occur.

Let us consider a model shown in Fig. 16.4. We assume that the grain size changes from L to L_0 ($L > L_0$) instantaneously and that the rate at which the volume fraction, ϕ, of a fine-grained region changes with time is proportional to the volume fraction of the coarse-grained region and to the rate at which that region deforms, namely,

$$\dot{\phi} = \alpha(1-\phi)\dot{\varepsilon}_1 \quad (16.22)$$

where α is a non-dimensional constant of order unity and $\dot{\varepsilon}_1$ is the strain rate of the coarse-grained region (Chapter 13). The solution of (16.22) with the initial condition of $\phi(0)=0$ is

$$\phi = 1 - \exp(-\alpha\dot{\varepsilon}_1 t) \quad (16.23)$$

Therefore the time-scale for completion of dynamic recrystallization is $\tau_{\rm recrys} \approx 3/\alpha\dot{\varepsilon}_1$ during which coarse-grained regions will deform to $\varepsilon_1 \approx 3/\alpha$.

In order for shear localization to occur by grain-size reduction, the fine-grained regions must be softer than the original material, i.e., $\dot{\varepsilon}_2(L_0) > \dot{\varepsilon}_1$ where $\dot{\varepsilon}_2(L_0) = A_0(\sigma/L_0^m)\exp(-H^*_{\rm diff}/RT)$ is the strain rate corresponding to grain-size sensitive creep (diffusional creep) for a grain size of L_0, and $\dot{\varepsilon}_1 = B_0\sigma^n\exp(-H^*_{\rm disl}/RT)$ is the strain rate corresponding to dislocation creep. By defining $\tilde{L}$ as a grain size at which the strain rate for diffusion creep is the same as that for dislocation creep, this condition can be written as

$$\zeta \equiv \left(\frac{\tilde{L}}{L_0}\right)^m > 1. \quad (16.24)$$

However, condition (16.24) itself does not guarantee if this leads to significant shear localization. Hardening will occur by grain growth and eventually shut off the localization. Strain by grain-size sensitive creep during grain growth needs to exceed strain by dislocation creep. In order to calculate the relative magnitude of strain by these two processes, let us assume, for simplicity, that the regions of small grain size are connected (above the percolation threshold, see Chapter 12). Then, one can assume that stress in these two regions is same and the strain in the fine-grained region during this period will be

$$\varepsilon_2 = \int_0^{\tau_{\rm recrys}} \dot{\varepsilon}_2\,dt = \frac{A}{k(q-m)/q} L_0^{q-m}\left[\left(1+\frac{3k}{\alpha\dot{\varepsilon}_1 L_0^q}\right)^{\frac{q-m}{q}} - 1\right] \quad (\text{for } q \neq m) \quad (16.25a)$$

$$= \frac{A}{k}\log\left(1+\frac{3k}{\alpha\dot{\varepsilon}_1 L_0^q}\right) \quad (\text{for } q = m) \quad (16.25b)$$

where the flow law of the fine-grained region is $\dot{\varepsilon}_2 = A/L^m$ and the grain-growth kinetics is $L^q - L_0^q = kt$. Therefore the necessary condition for significant localization, $\varepsilon_2/\varepsilon_1 \gg 1$, is approximately given by

$$\frac{\zeta}{\xi}\frac{(1+\xi)^\chi - 1}{\chi} \gg 1 \quad (\text{for } q \neq m) \quad (16.26a)$$

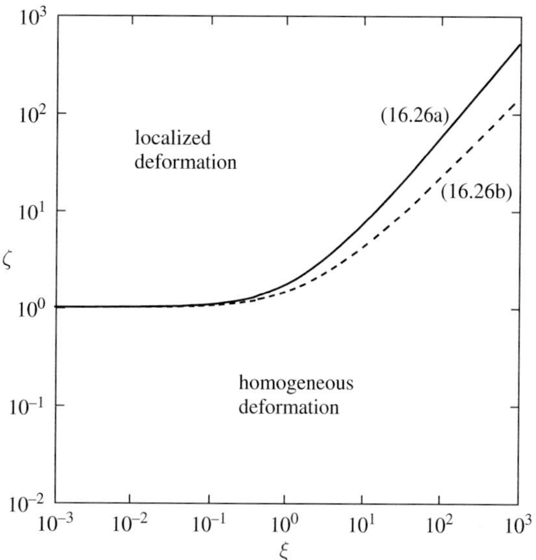

FIGURE 16.5 A diagram showing the conditions for shear localization described by relations (16.26) in the (ξ, ς) space.

$$\frac{\zeta}{\xi}\log(1+\xi) \gg 1 \quad (\text{for } q = m) \quad (16.26b)$$

with $\xi \equiv 3k\tilde{L}^m/\alpha A L_0^q$, $\chi \equiv 1 - m/q$ and $\tilde{L}$ is the grain size at which the deformation mechanism changes from dislocation creep to diffusional creep ($\dot{\varepsilon}_1 \equiv \dot{\varepsilon}_2 = A/\tilde{L}^m$). The parameter $\xi \equiv 3k\tilde{L}^m/\alpha A L_0^q$ represents the ratio of time-scale of grain growth and the time-scale of deformation by dislocation creep. The (ξ, ζ) space corresponding to the inequalities (16.26) is shown in Fig. 16.5. The conditions (16.26a,b) have two asymptotic branches. For $\xi \ll 1$, the relation (16.26) leads to $\zeta \gg 1$, and for $\xi \gg 1$, it leads to $\zeta \gg |\chi|\xi$ (for (16.26a) with $\chi < 0$; for (16.26b) this equation will be $\zeta \gg \xi/\log(1+\xi)$ but these relations are similar as can be seen in Fig. 16.5). The inequality $\zeta > 1$ or $\zeta < 1$ represents the competition between dislocation and diffusional creep and is identical to the relation (16.24) and the inequality $\zeta > |\chi|\xi$ or $\zeta < |\chi|\xi$ represents the competition between deformation by diffusional creep and grain growth (Problem 16.5). Therefore the conditions for localization given by inequality (16.26) correspond to a large degree of initial grain-size reduction (a large ζ, (16.24)), and a slow grain-growth rate relative to the rate of deformation by diffusional creep (a small ξ relative to ζ, (16.26)). These conclusions are similar to the one obtained in section 16.2.2 (relation (16.7)), i.e., the fast initial growth of instability and the slow rate of decrease in the rate of growth of instability.

The physical conditions (such as stress and temperature) corresponding to these localization conditions may be inferred if the relevant physical parameters are known. Those parameters are well constrained for diffusional and dislocation creep. Both activation enthalpy and the stress exponent for dislocation creep are higher than those for diffusional creep. Consequently, the condition (16.24) ($\zeta > 1$) corresponds to high stress and low temperature (see JIN *et al.*, 1998). The physical conditions corresponding to the inequality $\zeta > |\chi|\xi$ are less obvious because of the complications in the kinetics of grain growth. If grain-growth kinetics in pure olivine is used, then instability conditions are hardly satisfied because the kinetics of grain growth is much faster than that of deformation (i.e., $\xi \gg 1$). Influence of secondary phases on grain-growth kinetics needs to be included if grain-size reduction is responsible for shear localization in the upper mantle. However, not much is known about the effects of secondary phases on grain-growth kinetics (see EVANS *et al.*, 2001).

Problem 16.5

Show that the inequality $\zeta > |\chi|\xi$ or $\zeta < |\chi|\xi$ represents the competition between deformation by diffusional creep and grain growth.

Solution

From the definition, $\zeta/\xi = \left(\tilde{L}/L_0\right)^m / (3k\tilde{L}^m/\alpha A L_0^q) = \alpha A L_0^{q-m}/3k$. Recall the grain-growth law, $L^q - L_0^q = kt$, from which one can define the characteristic time for grain growth, τ_{gg} as a time needed to increase the grain size to twice the initial size. Then $\tau_{gg} \equiv (2^q - 1)L_0^q/k = 3L_0^q/k$ for $q = 2$. Now the characteristic time for deformation by diffusion creep is $\tau_{diff} \equiv 1/\dot{\varepsilon} = L_0^m/A$. Therefore $\zeta/\xi = \tau_{gg}/\tau_{diff}$ which means that the inequality $\zeta > |\chi|\xi$ or $\zeta < |\chi|\xi$ represents the competition between deformation by diffusional creep and grain growth ($|\chi|$ is a non-dimensional constant of order unity).

In their analysis of conditions for shear localization, JIN *et al.* (1998) considered only the competition between dislocation and diffusional creep (i.e., the parameter ζ). In other words, their treatment was limited to the conditions defined by equation (16.24). However, as can be seen in Fig. 16.5, this condition is valid only when $\xi \equiv 3k\tilde{L}^m/\alpha A L_0^q < 1$. When grain-growth rate is fast, this treatment is not adequate: the influence of a finite value of $\xi \equiv 3k\tilde{L}^m/\alpha A L_0^q$ needs to be addressed in addition to the effects of $\zeta \equiv \left(\tilde{L}/L_0\right)^m$. DE BRESSER *et al.* (1998) proposed that the grain size of a material undergoing dynamic recrystallization is controlled by the balance of grain-size refinement in the dislocation creep regime and grain growth in the diffusional creep regime (see Chapter 13). Using this model, DE BRESSER *et al.* (2001) argued that the grain size of a material cannot be far from the critical grain size between dislocation and diffusional creep, and therefore the degree of weakening by this mechanism is small. In essence they looked at the strain-rate enhancement at long-term equilibrium, but not the integrated strain. Consequently, the contribution from the initial stage deformation at a small grain-size is ignored and the conclusion is not necessarily valid (see 16.2.2): even though the growth rate of instability may become low in a later stage of growth of instability, if the rate of growth of instability is fast at the beginning, then significant localization can result.

The above analysis does not explicitly include the influence of evolution of volume fraction of recrystallized region, ϕ, on strain partitioning. A more complete analysis will have to include this effect, as well as the evolution of stress–strain partitioning.

Localization due to intrinsic instability of dislocation motion

Motion of defects at a microscopic scale can be unstable in some cases. This can happen either at a level of individual defect or at the level of a group of defects. For example, the Portevin–Le Chatelier effect seen in some alloys is interpreted to be due to unstable motion of individual dislocations due to the interaction with impurities (solute atoms): when break-away of solute atmosphere occurs, then dislocation velocity can be accelerated in an unstable fashion (e.g., NABARRO, 1967b; ESTRIN and KUBIN, 1991). Similarly, the propagation of the Lüders band is considered to be due to the weakening caused by the creation of a large number of mobile dislocations that is seen as a second yield point in a stress–strain curve (e.g., COTTRELL, 1953; NABARRO, 1967b). These cases are the examples in which either the work-hardening coefficient or stress exponent is negative ($h < 0$ or $n_{eff} < 0$). The relevance of these processes to deformation in Earth is not clear. Although some evidence of dislocation-point defects interaction is reported in minerals (e.g., KITAMURA *et al.*, 1986; ANDO *et al.*, 2001), its role in shear localization in Earth's interior has not been investigated in any detail.

Localization due to grain-boundary migration associated with dynamic recrystallization

A somewhat analogous mechanism is unstable (periodic) deformation associated with dynamic recrystallization involving grain-boundary migration (LUTON and SELLARS, 1969; SAKAI and JONAS, 1984). When dynamic recrystallization occurs in some metals such as Ni, the stress–strain curve often shows oscillation. This phenomenon is interpreted to be due to the temporal weakening caused by grain-boundary migration that has eliminated dislocations (SAKAI and JONAS, 1984). In these materials increasing dislocation density by deformation causes work hardening. Grain-boundary migration eliminates some of the dislocations causing weakening. Consequently, if weakening occurs coherently in a deforming polycrystal, significant weakening occurs. WHITE *et al.* (1980) considered this mechanism is important for mylonite formation and ZEUCH (1983) discussed that this mechanism may operate in olivine, but the role of grain-boundary migration on mechanical properties of silicate is less distinct than those of metals because work hardening is less important in silicates than in metals (Chapter 9).

Localization due to anisotropic microstructure (LPO) development

Development of crystallographic texture (LPO, Chapter 14) causes plastic anisotropy and hence changes the creep strength. If LPO develops in a certain region of a material, then this region will have different rheological properties. This rheological change due to LPO development is often characterized by a *Taylor factor* (see Chapter 14). This effect can be incorporated through the Taylor factor, M, defined by,

$$\dot{\varepsilon} \propto \varepsilon^{-h}\left(\frac{\sigma}{M(\varepsilon)}\right)^{n}. \tag{16.27}$$

For constant stress deformation,

$$\frac{\delta \log \dot{\varepsilon}}{\delta \log \varepsilon} = -h - n\frac{\delta \log M}{\delta \log \varepsilon} > 0. \tag{16.28}$$

Therefore if the rate of geometrical weakening is fast compared to the work-hardening rate, then localization due to LPO development will occur. Such a phenomenon is well-documented in metallurgy. WHITE *et al.* (1980) and HARREN *et al.* (1988) listed this mechanism as one of the most important mechanisms of shear localization in mylonites. However, not much is known about geometrical hardening (or softening) in geological materials, and currently available knowledge is mostly based on modeling that involves gross simplifications (e.g., TAKESHITA and WENK, 1988; TAKESHITA, 1989). Furthermore, LPO development in some minerals such as olivine causes hardening (e.g., WENK *et al.*, 1991) rather than softening. In such a case LPO development cannot cause shear localization.

Instability and localization in deformation of a two-phase material

Due to the inherent mechanical heterogeneity, deformation of a two-phase mixture has a high potential for shear localization. Deformation in a two-phase material depends critically on how strain (or strain rate) and stress are partitioned between two phases (Chapter 12). Deformation tends to be partitioned more in the weaker phase, and if a small increase in strain (strain rate) partitioning in the weaker phase promotes further increase in strain (strain rate) partitioning, then this will lead to instability (localization). A classic example is deformation in the brittle regime in which *stress concentration* at the crack tip causes this positive feedback leading to fracture along a fault plane (see Chapter 7 and PATERSON and WONG, 2005). When a fluid phase is formed due to a dehydration reaction or partial melting, then localized deformation occurs along a pore-rich (or melt-rich) region due to the increased pore pressure (e.g., RALEIGH and PATERSON, 1965; HOLTZMAN *et al.*, 2003b). A similar case was found in the experimental study of BLOOMFIELD and COVEY-CRUMP (1993) in which they found that strain is progressively partitioned more into a weaker phase. Shear localization associated with the formation of fine-grained regions by dynamic recrystallization can be viewed as shear localization in a two-phase material if fine-grained regions and coarse-grained regions are considered as two different phases with different rheological properties. An *S–P mylonite* is a natural example of this process in which strain is partitioned largely in the fine-grained recrystallized mineral (orthopyroxene) (e.g., BOULLIER and GUEGUEN, 1975).

Instability (localization) of deformation is enhanced in a two-phase material if a weaker phase coagulates. Local increase in the volume fraction of a weaker phase will lead to instability and localization. STEVENSON (1989) discussed the instability of deformation involving coagulation of a weaker phase (melt) in a partially molten material. BERCOVICI *et al.* (2001b) and BERCOVICI and RICARD (2003, 2005) developed a sophisticated theory of deformation of a two-phase mixture with the emphasis on the continuum mechanics aspects and applied it to shear localization and generation of plate boundaries. Deformation in the brittle field involves cracks and (elastic) solid and hence as far as cracks and solids are considered as two phases, their theory has some resemblance to a standard theory for brittle fracture. However, their theory differs significantly from a standard theory of fracture involving cracks and their interaction. Although the role of surface energy is included in both theories in a similar way, the role of *stress concentration* that plays a key role in the classical model of brittle fracture (Chapter 7) is not included in Bercovici and Ricard (2003) and Bercovici *et al.* (2001b). Instead the interaction of weak phases is incorporated through the interface energy and through an assumed rule of the effect of volume fraction of weak phase on strength. Bercovici and Ricard (2005) extended their analysis to include the influence of "fineness" (small grain size). Their model provides new insights into the processes of grain-size reduction and distribution of grain size but some of the important aspects of micro-physics such as the role of deformation mechanisms (see Fig. 16.4) and the processes of dynamic recrystallization (see Chapter 13) are not included in detail.

16.4. Long-term behavior of a shear zone

Localization of deformation is by definition a transient phenomenon. Once unstable deformation occurs, then a certain amount of strain is concentrated in a shear zone but the enhanced deformation will eventually stop. After being formed, how will a shear zone be deformed in the subsequent period? In order for a shear zone to continuously work as a weak zone for a geologically long period (e.g., a plate boundary), there must be some mechanisms to keep that zone weak.

The formation of a shear zone is associated with the formation of some types of defects (e.g., gouge in the brittle regime, fine-grained materials in the ductile regime). The long-term behavior of a shear zone depends critically on how these defects are *annealed* (*healed*).

Consider a shear zone formed by the adiabatic instability. In this case, after the formation of a shear zone, it will be made of quenched glass or very fine-grained materials (pseudotachylyte). Consequently, if a shear zone is made at modest depth (modest temperatures), then this zone will be a weak zone (Chapter 8). The same is true for a shear zone that was formed by grain-size reduction. In these cases, the long-term behavior of a shear zone is largely controlled by grain-growth kinetics. Therefore, the strength of a shear zone will increase with time. If fine-grained regions are made of a single phase, then the kinetics of grain growth is fast (growth to $\sim$100 μm will take $\sim 10^5$ yr for pure olivine at $T = 800$ K ($\sim$1–3 GPa) at water-poor conditions, see Chapter 13), but if several minerals co-exist, then growth kinetics will be much more sluggish (grain growth in a multi-phase material occurs by a slow process called Ostwald ripening, see Chapter 13). In the latter case, a shear zone could maintain a weak state for a geological time-scale ($\sim 10^8$ yr).

Another way to maintain a shear zone is to have repeated instabilities. A well-known example in the brittle field is earthquakes along plate boundaries. There are some possible processes by which repeated instabilities occur on a shear zone in the ductile regime including the adiabatic instability in a fine-grained soft shear zone. However, mechanisms of repeated instabilities (such as the "period" of instability) are not well understood.

16.5. Localization of deformation in Earth

There are a number of observations on localized deformation in the ductile regime in Earth. The most important observation is a *mylonite* which is commonly observed in the continental lower crust (White *et al.*, 1980) (mylonite is also observed in the oceanic upper mantle along a transform fault (Jaroslow *et al.*, 1996)). Mylonite is a fine-grained rock observed normally on or near a fault. Microstructural studies indicated that grain-size reduction is syntectonic (i.e., it occurred at the same time as tectonic deformation) and is due to ductile processes as opposed to fracture. Therefore mylonite presents a clear case for shear localization due to grain-size reduction in the plastic flow regime (see Chapter 7).

Despite the clear evidence for the formation of mylonite by grain-size reduction there has been some confusion as to the origin of mylonite. One of the puzzling observations is the fact that lattice-preferred orientation (LPO; for details of lattice-preferred orientation see Chapter 14) of mylonites is not always weak. If mylonite is formed as a result of grain-size sensitive creep such as diffusional creep, then one expects a weak LPO for mylonites (Chapter 14). In contrast to this expectation, the LPO in the mylonites is not always weak although in ultra-mylonites (rocks with very small grain size, $\sim$10 μm or less) it tends to be nearly random (Berthe *et al.*, 1979). Based on the absence of clear indication of random fabric, White *et al.* (1980) argued that grain-size reduction is not the major cause for shear localization, and that the main causes for shear localization are anisotropic microstructure (LPO) development and/or softening due to grain-boundary migration associated with dynamic recrystallization. In addition to the conceptual problem that a clear correlation between small grain size and shear zone is not readily explained by this model, there is a problem with this model from the view point of properties of materials. As I have reviewed in this chapter, these two processes, i.e., LPO-induced weakening and grain-boundary migration-induced weakening are not well documented in silicate minerals and unlikely to be important in Earth's interior. Furthermore, the recent experimental studies of deformation fabrics show that significant LPO develops even though when grain size is reduced and a sample is deformed by grain-size sensitive creep, as far as grain size is concerned reduction is not very large (Karato *et al.*, 1995b; Pieri *et al.*, 2001). Also Jin *et al.* (1998) demonstrated a clear correlation between grain-size reduction and shear localization (see also White, 1979; Handy, 1989; Drury *et al.*, 1991). Jin *et al.* (1998) showed that the volume fraction of fine-grained regions increases systematically toward the shear zone and that LPO in the fine-grained regions is significantly weaker than that of coarse-grained regions (ultra-mylonites from the studied region by Jin *et al.* (1998) have almost completely random LPO). Based on these studies as well as studies on other processes reviewed in this chapter, I consider that grain-size reduction due to dynamic recrystallization (and/or chemical reactions) is indeed a likely cause for shear localization. When the degree of grain-size reduction is very large it leads to the regime in which deformation is completely by diffusional creep (hence no LPO), whereas if the degree of grain-size reduction is modest, there will be some LPO although there is significant weakening.

Obviously a combination with other processes is always a possibility. Enhanced deformation in fine-grained regions sometimes leads to significant shear heating that will further enhance localization (Sibson, 1975; Obata and Karato, 1995; Jin *et al.*, 1998). Evidence for localized fluid penetration is often observed in the semi-brittle regime (Etheridge *et al.*, 1983) as well as in the ductile regime (along grain boundaries; Cahn and Balluffi, 1979). Fluid flow helps shear localization (see e.g., Hippertt and Hongn, 1998). Reduced grain size will also enhance the reaction kinetics that will in turn enhance deformation. In addition, the "yielding" of one mineral such as orthopyroxene due to twinning above a critical stress leads to a large rheological contrast that promotes shear localization.

A large number of processes probably occur in the shear localization in the actual Earth that make the inference of operating mechanisms difficult. However, the existing observations on naturally deformed rocks from shear zones and the experimental data strongly suggest that grain-size reduction accompanied with some other processes is the key to the formation of shear zones.

Where could shear localization occur in Earth? The materials science based analysis presented in this chapter provides some insights. The important mechanisms for shear localization in Earth are (1) grain-size reduction, (2) shear heating (adiabatic instability) and (3) the formation of a weaker secondary phase. Both grain-size reduction and shear heating work efficiently under relatively low temperatures, namely in the relatively shallow regions (or in the cold subducting slabs). Together with the well-known mechanisms of shear localization due to brittle fracture, shear localization during plastic flow at relatively low temperatures is the ultimate cause of plate tectonics. Localized deformation is also likely to occur when a two-phase mixture with a large rheological contrast is deformed. A well-documented example is deformation of a partially molten material (Holtzman *et al.*, 2003b, 2005). Interplay of melt segregation and deformation is a likely cause of narrowly focused melt transport beneath mid-ocean ridges. A similar process of shear localization may occur in the lithosphere when one of the minerals "yields" at the condition in which other minerals have relatively large strengths.

When shear localization plays an important role, then the pattern of material circulation associated with mantle convection is different from the case of homogeneous deformation. The flow law should include strain-weakening (or strain-rate weakening) rheology to incorporate localization (e.g., Tackley, 2000a, 2000b; Bercovici, 2003) but such a flow law consistent with the materials science of deformation has not been formulated in any detail.

Part III
Geological and geophysical applications

17 Composition and structure of Earth's interior

The current status of our understanding of the structure and composition of Earth and other terrestrial planets is summarized with the emphasis on those issues that are critical to the rheological properties and the dynamics/evolution of terrestrial planets. A brief summary of the methods of inference of pressure, temperature and major element compositions is presented together with some major results. Current knowledge of the mineralogical composition of the crust, mantle and core is briefly reviewed together with a review of our knowledge on the water content and grain size.

Key words Adams–Williamson equation, adiabatic temperature gradient, Bullen parameter, crust, mantle, core, transition zone, lower mantle, D″ layer, inner core, pyrolite model, chondrite model.

17.1. Gross structure of Earth and other terrestrial planets

This chapter provides a brief review of the structure and composition of Earth and other planetary interiors. The emphasis is on the issues that are critical to the understanding of rheological properties, and hence the dynamics and evolution of terrestrial planets. Several methods are currently used to infer the interior of Earth and other planets. (1) The gravity field outside a planet and related geodetic measurements provide constraints on the average density and the mass distribution in that planet. (2) The magnetic field outside a planet provides some clue as to the dynamics of planetary interiors. The magnetic field caused by remnant magnetization of crusts provides hints as to the dynamics of a planet in the past. The magnetic field that is currently generated in a planetary interior provides an important clue as to the dynamics and structure of that planet particularly the core. (3) Where available, the propagation of elastic waves (seismic waves) in the planet provides information as to the elastic properties of a planet. When interpreted on the basis of knowledge of material properties under high pressure and temperatures, the nature of elastic wave propagation gives strong constraints on the composition of Earth and planetary interiors. (4) Samples (either rocks or atmospheric gases) collected from (the surfaces of) a planet provide direct information as to the composition of the planet with the help of the theory of chemical differentiation. (5) The knowledge of the chemical composition of the solar system (obtained from the chemical composition of meteorites) combined with the theory of planetary formation and the associated chemical evolution gives an important constraint on the composition of a planet.

For obvious reasons, far more details are known about Earth's interior than the interiors of other terrestrial planets, and therefore the main focus of this chapter is the structure and composition of Earth's interior. However a review of the studies on the interiors of other terrestrial planets is also included to place the studies on Earth in a broader planetary perspective.

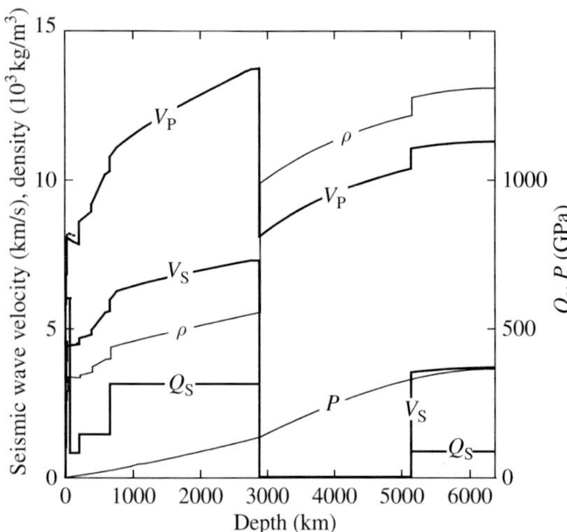

FIGURE 17.1 The PREM model for gross Earth structure (after Dziewonski and Anderson, 1981). V_P, P-wave velocity; V_S, S-wave velocity; ρ, density; P, pressure; Q_S, Q for the S-wave. Plate 7.

The distribution of elastic constants and density in Earth's interior can be determined from the combination of seismological observations with geodetic observations (observations of the topography and the gravity field). The seismological observations that are usually used are the velocities of P- and S-waves, but in some cases data on seismic wave attenuation and the anisotropy of seismic wave velocities are also included. One of such models is PREM (preliminary reference Earth model) developed by Dziewonski and Anderson (1981). This is a one-dimensional model in which all the physical parameters are assumed to change only with depth. Fig. 17.1 shows some of the basic features of PREM (see also Table 17.1). In this model, Earth is divided into several layers, crust, upper mantle (to ~400 km), transition zone (~400–670 km), lower mantle (~670–2890 km) and the core (~2890–6370 km). There are significant sharp jumps in seismic wave velocities and densities at the boundaries between these layers indicating that there must be a change in crystal structure of major constituent minerals and/or chemical composition.

Although much more limited, some information concerning the internal structure of other terrestrial planets (or satellites) is also obtained mostly from gravity and magnetic field observations. These data provide important constraints on the composition and internal structure of these planetary bodies particularly the gross chemical composition and the presence or absence of metallic cores.

17.2. Physical conditions of Earth's interior

17.2.1. Pressure distribution

When hydrostatic equilibrium is attained, the depth variation of pressure is determined by the hydrostatic equilibrium, namely,

$$dP = \rho g\, dz = -\rho g\, dr \tag{17.1}$$

where P is pressure, g is the acceleration due to gravity, ρ is density, z is depth and r is the distance from the center of Earth. The condition of hydrostatic equilibrium is very well satisfied in most portions of Earth because Earth materials are soft enough and cannot support large non-hydrostatic stress. The maximum non-hydrostatic stress that Earth materials can support for a geological time-scale is on the order of 1–100 MPa in most regions, which is far less than the magnitude of hydrostatic pressure (to ~135 GPa in the mantle).

Note that the acceleration due to gravity in turn depends on density,

$$g(r) = \frac{G \cdot M(r)}{r^2} = 4\pi G \frac{\int_0^r \rho(\xi)\xi^2\, d\xi}{r^2} \tag{17.2}$$

where G is the gravity constant and $M(r)$ is the mass of planetary materials within the radius r. Using equations (17.1) and (17.2), the density and hence pressure distribution in a planet can be calculated from the total mass $M(R)$ and the equation of state (i.e., the relationship between density and pressure). The equation of state can be inferred from seismological observations, because $V_\phi^2 \equiv V_P^2 - \frac{4}{3}V_S^2 = K_S/\rho = (\partial P/\partial\rho)_S$. Therefore the density distribution in Earth can be calculated from seismological observations. To see how this can be made, let us combine equations (17.1) and (17.2) to obtain,

$$\frac{d\rho}{dr} = -\frac{g(r)\rho^2(r)}{K_S(r)} = -\frac{4\pi G\rho(r)}{V_\phi^2(r)} \frac{\int_0^r \rho(\xi)\xi^2\, d\xi}{r^2} \tag{17.3}$$

where $K_S \equiv \rho\frac{dP}{d\rho}$. This is the *Adams–Williamson equation*. Given $V_\phi(r)$ from seismology, the density can be obtained by solving equation (17.3) numerically. This approach was used in the PREM model in estimating the density distribution in the lower mantle and the core. It is important to remember that the assumption behind this equation is that the depth variation of density is determined solely by adiabatic compression (see section 17.2.2) and that density changes only with depth (laterally homogeneous). Due to these

TABLE 17.1 Structure of Earth (PREM). Velocities and bulk modulus are those at the period of 1 s (after DZIEWONSKI and ANDERSON, 1981).

z	P	ρ	V_P*	V_S*	K**	μ**	η_B	Q_S***
0	0	1020	1.450, 1.450	0.000, 0.000	2.14	0.000	0.00	0
3	0.03	1020	1.450, 1.450	0.000, 0.000	2.14	0.000	0.00	0
3	0.03	2600	5.800, 5.800	3.200, 3.200	52.0	26.6	0.00	600
15	0.34	2600	5.800, 5.800	3.200, 3.200	52.0	26.6	0.00	600
15	0.34	2900	6.800, 6.800	3.900, 3.900	75.3	43.9	0.00	600
24.4	0.60	2900	6.800, 6.800	3.900, 3.900	75.3	43.9	0.00	600
24.4	0.60	3381	8.022, 8.019	4.396, 4.612	131.4	68.2	−0.13	600
40	1.12	3379	8.004, 8.179	4.400, 4.598	131.1	68.0	−0.13	600
60	1.79	3377	7.982, 8.165	4.404, 4.580	130.7	67.7	−0.13	600
80	2.45	3375	7.959, 8.150	4.409, 4.562	130.2	67.4	−0.13	600
80	2.45	3375	7.959, 8.150	4.409, 4.562	130.0	67.4	−0.13	80
115	3.62	3371	7.920, 8.124	4.417, 4.529	129.3	66.9	−0.13	80
150	4.78	3367	7.876, 8.102	4.425, 4.502	128.5	66.5	−0.12	80
185	5.94	3363	7.840, 8.066	4.433, 4.469	127.7	66.0	−0.12	80
220	7.11	3360	7.801, 8.049	4.441, 4.436	126.9	65.6	−0.12	80
220	7.11	3446	8.559	4.644	152.9	74.1	0.78	143
265	8.65	3463	8.656	4.675	157.9	75.7	0.79	143
310	10.2	3490	8.732	4.707	163.0	77.3	0.80	143
355	11.8	3516	8.819	4.738	168.2	79.0	0.82	143
400	13.4	3543	8.905	4.770	173.5	80.6	0.83	143
400	13.4	3724	9.134	4.933	189.9	90.6	1.73	143
450	15.2	3787	9.390	5.078	203.7	97.7	1.79	143
500	17.1	3850	9.646	5.224	218.1	105.1	1.86	143
550	19.1	3913	9.902	5.370	233.2	112.8	1.92	143
600	21.0	3976	10.158	5.516	248.9	121.0	1.98	143
635	22.4	3984	10.212	5.543	252.3	122.4	0.37	143
670	23.8	3992	10.266	5.570	255.6	123.9	0.37	143
670	23.8	4381	10.751	5.945	299.9	154.8	0.98	312
721	26.1	4412	10.910	6.094	306.7	163.9	0.97	312
771	28.3	4443	11.066	6.240	313.3	173.0	0.97	312
871	32.8	4504	11.245	6.311	330.3	179.4	0.97	312
971	37.3	4563	11.416	6.378	347.1	185.6	0.98	312
1071	44.9	4621	11.578	6.442	363.8	191.8	0.98	312
1171	46.5	4678	11.734	6.504	380.3	197.8	0.99	312
1271	51.2	4735	11.882	6.563	396.6	203.9	0.99	312
1371	55.9	4790	12.024	6.619	412.8	209.8	0.99	312
1471	60.7	4844	12.161	6.673	428.8	215.7	0.99	312
1571	65.5	4898	12.293	6.725	444.8	221.5	0.99	312
1671	70.4	4951	12.421	6.776	460.7	227.3	0.99	312
1771	75.4	5002	12.545	6.825	476.6	233.1	0.99	312
1871	80.4	5047	12.666	6.873	492.5	238.8	1.00	312
1971	85.4	5106	12.784	6.919	508.5	244.5	1.00	312
2071	90.6	5157	12.900	6.965	524.6	250.2	1.00	312
2171	95.8	5207	13.016	7.011	540.9	255.9	1.00	312
2271	101.0	5257	13.131	7.055	557.5	261.7	1.00	312
2371	106.4	5307	13.245	7.099	574.4	267.5	1.00	312
2471	111.8	5357	13.361	7.144	591.7	273.5	1.00	312
2571	117.3	5407	13.477	7.189	609.5	279.4	1.01	312
2671	123.0	5457	13.596	7.234	627.9	285.5	1.01	312

TABLE 17.1 (cont.)

z	P	ρ	V_P*	V_S*	K**	μ**	η_B	Q_S***
2771	128.7	5506	13.688	7.266	644.0	290.7	1.00	312
2871	134.6	5556	13.711	7.265	653.7	293.3	1.00	312
2891	135.8	5566	13.717	7.265	655.6	293.8	0.99	312
2891	135.8	9903	8.065	0	644.1	0	0.98	0
2971	144.2	10029	8.199	0	674.3	0	0.99	0
3071	154.7	10181	8.360	0	711.6	0	0.99	0
3171	165.1	10327	8.513	0	748.4	0	1.00	0
3271	175.4	10467	8.658	0	784.6	0	1.00	0
3371	185.6	10602	8.796	0	820.2	0	1.00	0
3471	195.7	10730	8.926	0	855.0	0	1.00	0
3571	205.6	10853	9.050	0	888.9	0	1.00	0
3671	215.3	10971	9.168	0	922.0	0	1.00	0
3771	224.8	11083	9.279	0	954.2	0	1.00	0
3871	234.2	11191	9.384	0	985.5	0	1.00	0
3971	243.2	11293	9.484	0	1015.8	0	1.00	0
4071	252.1	11390	9.579	0	1045.1	0	1.00	0
4171	260.7	11483	9.669	0	1073.5	0	1.00	0
4271	269.0	11571	9.754	0	1100.9	0	1.00	0
4371	277.0	11655	9.835	0	1127.3	0	1.00	0
4471	284.8	11734	9.912	0	1152.9	0	1.00	0
4571	292.2	11809	9.986	0	1177.5	0	1.00	0
4671	299.3	11880	10.056	0	1201.3	0	1.00	0
4771	306.1	11947	10.123	0	1224.2	0	1.00	0
4871	312.6	12010	10.187	0	1246.4	0	1.01	0
4971	318.7	12069	10.249	0	1267.9	0	1.01	0
5071	324.5	12125	10.310	0	1288.8	0	1.02	0
5150	328.9	12166	10.356	0	1304.7	0	1.03	0
5150	328.5	12764	11.028	3.504	1343.3	156.7	1.00	85
5171	330.0	12774	11.036	3.510	1346.2	157.4	1.00	85
5271	335.4	12825	11.072	3.535	1358.6	160.3	1.00	85
5371	340.3	12870	11.105	3.558	1370.1	163.0	0.99	85
5471	344.7	12912	11.135	3.579	1380.5	165.4	0.99	85
5571	348.7	12949	11.162	3.598	1389.8	167.6	0.99	85
5671	352.2	12982	11.185	3.614	1398.1	169.6	0.99	85
5771	355.3	13010	11.206	3.628	1405.3	171.3	0.99	85
5871	357.9	13034	11.223	3.640	1411.4	172.7	0.99	85
5971	360.0	13054	11.237	3.650	1416.4	173.9	0.99	85
6071	361.7	13069	11.248	3.658	1420.3	174.9	0.99	85
6171	362.9	13080	11.256	3.663	1423.1	175.5	0.99	85
6271	363.6	13086	11.261	3.667	1424.8	175.9	0.99	85
6371	363.8	13088	11.262	3.668	1425.3	176.1	0.99	85

z (km): depth, P (GPa): pressure, ρ (kg/m^3): density, V_P (km/s): compressional wave velocity, V_S (km/s): shear wave velocity, K (GPa): bulk modulus, μ (GPa): shear modulus, η_B: the Bullen parameter, Q_S: Q-factor for shear waves.
*In the top 220 km, the elastic properties are assumed to have transverse isotropy in PREM. The numbers in the first column of velocity correspond to those for vertical deformation (e.g., $V_{PV,SV}$) and the numbers in the second column to those for horizontal deformation (e.g., $V_{PH,SH}$).
**Bulk modulus and shear modulus in the top 220 km are for the average values.
***Different Q-models have also been published. For discussions see text of this chapter as well as Chapter 18.

assumptions, which are needed to infer density distribution, density distribution is, in general, less well constrained than seismic wave velocities. However, the integrated density is well constrained, and the pressure distribution is known with a high degree of accuracy (better than ~1%).

Problem 17.1

Derive the Adams–Williamson equation, (17.3).

Solution

Combing equations (17.1) and (17.2), one obtains $dP/dr = -4\pi G\rho(r)\left[\int_0^r \rho(\xi)\xi^2\, d\xi/r^2\right]$. Now $dP/dr = (\partial P/\partial\rho)_S\, d\rho/dr$, and $\rho(\partial P/\partial\rho)_S = K_S$ where we assumed the adiabatic compression. Inserting these relations into $dP/dr = -4\pi G\rho(r)\left[\int_0^r \rho(\xi)\xi^2\, d\xi/r^2\right]$, one finds equation (17.3).

17.2.2. Temperature distribution

In contrast to pressure, the distribution of temperature is less well constrained. This is because neither the distribution of heat sources nor the material properties that control heat transfer (i.e., viscosity, thermal conductivity) are well known. However some basic features of temperature distribution are well established. In this section, an overview of temperature distribution is given. The governing equations are *Fourier's law* of conductive heat transfer

$$J_i = -k\frac{\partial T}{\partial x_i} \tag{17.4}$$

where k is thermal conductivity (for simplicity we assume that thermal conductivity is isotropic) and the energy conservation equation for a moving material, i.e.,

$$\rho C_P\left(\frac{\partial T}{\partial t} + \sum_{i=1}^{3} u_i\frac{\partial T}{\partial x_i}\right) = -\mathrm{div}\boldsymbol{J} + A \tag{17.5}$$

where $\boldsymbol{u}$ is the velocity of motion of materials, C_P is the specific heat and A is the heat generation per unit volume. The left-hand side of equation (17.5) contains two terms, representing the energy gain in a parcel of material: the first term represents the increase in heat content and the second term corresponds to the advective transfer of heat. The right-hand side represents the energy gain due to the heat flux by conduction ($-\mathrm{div}\boldsymbol{J}$) and the energy gain due to heat generation (A).

Temperature distribution in a given region of Earth is controlled by the relative importance of various terms in equation (17.5) and the boundary and the initial conditions. Let us first compare the two terms, $\rho C_P \sum_{i=1}^{3} u_i(\partial T/\partial x_i)$ and $\mathrm{div}\boldsymbol{J}$, i.e., the advective heat transport and the conductive heat transport. The ratio of these two terms, a *Peclet number*, is given by[1]

$$\mathrm{Pe} \equiv \frac{\rho C_P \sum_{i=1}^{3} u_i(\partial T/\partial x_i)}{k\sum_{i=1}^{3}(\partial^2 T/\partial x_i^2)} \approx \frac{\rho C_P u T/L}{k(T/L^2)} = \frac{\rho C_P u L}{k} = \frac{uL}{\kappa}$$

where κ ($\equiv k/\rho C_P$) is thermal diffusivity (~10^{-6} m^2/s for most silicates) and L is the length-scale at which temperature changes. When considering the vertical temperature profile of a near-surface lithosphere, the vertical velocity of motion is <10 m/Myr and $uL/\kappa < 0.3$, and therefore the advective heat transfer is small (except in cases where hydrothermal fluid plays an important role). In contrast, for hot general regions of the mantle, $u = 10^{-3}$–10^{-1} m/yr and $L \sim 1000$ km, so $uL/\kappa \sim 3 \times (10-10^3)$ and the advection effect dominates.

The temperature distribution in a subducting slab is governed by a similar principle. Assuming a certain initial temperature profile before subduction, the temperature profile of a slab in the mantle can be calculated by solving the equation of thermal conduction.

The term A includes the heat generation by radioactive elements, the heating or cooling by the latent heat associated with a phase transformation and the frictional heating. The importance of this term relative to the actual heat flux in Earth can be estimated as follows. If a steady state is assumed, integrating the right-hand side of equation (17.5) for a one-dimensional problem, one obtains

$$k\left(\frac{\partial T}{\partial z}\right)_h - k\left(\frac{\partial T}{\partial z}\right)_0 + \int_0^h A\, dz = J_h - J_0 + \int_0^h A\, dz \tag{17.6}$$

where J_h and J_0 are the heat flux (positive upward) at depth h and at the surface respectively. The surface heat flux on Earth is on the order of 30–100 mW/m^2 (e.g., SCLATER *et al.*, 1980). The amounts of heat generation due to radioactive elements in various rocks are

[1] The velocity of materials in a convecting fluid is given by $u \approx (\kappa/L)Ra^{\beta}$ with $\beta \sim \frac{2}{3}$ and Ra is the Rayleigh number defined as $Ra \equiv \rho g \alpha_{th} L^3/\eta\kappa$. Therefore $uL/\kappa \approx Ra^{\beta}$.

TABLE 17.2 Concentration of radioactive elements in various portions of silicate Earth (the values of heat generation are those at present). Heat production rate in the core is not well known but is much less than that in silicate Earth.

	Continental crust	Oceanic crust	Depleted mantle	Undepleted mantle
Concentration				
U (wt ppm)	1.1	0.9	0.006	0.02
Th (wt ppm)	4.2	2.7	0.04	0.1
K (wt%)	1.3	0.4	0.01	0.04
Heat generation				
(10^{-10} W/kg)	2.7	1.7	0.02	0.057
(10^{-6} W/m^3)	0.7	0.5	0.006	0.02

summarized in Table 17.2. Radioactive heat generation in the mantle is ~10^{-12} W/kg, whereas radioactive heat generation in the crust is ~10^{-10} W/kg. From this table, one can find that the radioactive heat generation from the oceanic crust and the upper mantle makes a negligible contribution to the total heat flux. However, radioactive heat production is an important part of heat generation in the continental regions. For the whole Earth, the total amount of radioactive heat generation is only ~30–50% of the total energy loss (~44 TW) (the ratio of heat production by radioactive elements to the total heat loss from the surface is called the *Urey ratio* and the current estimate of the Urey ratio is ~0.3–0.5 (Schubert *et al.*, 2001)). This indicates that Earth is being cooled at the present time.

The contribution from latent heat may be estimated by estimating the temperature change associated with a phase transformation, $\Delta T = L/\rho C_P \sim \pm 50-100$ K (L, latent heat). The contribution from frictional heating can be calculated as

$$A_{\text{fric}} = \dot{\varepsilon}\sigma = \eta\dot{\varepsilon}^2 = \frac{\sigma^2}{\eta}. \tag{17.7}$$

The strain rate in a general region of convecting mantle is estimated to be 10^{-15} s^{-1} ($\dot{\varepsilon} \approx u/H$, u, velocity of plate motion; H, the depth of mantle). From (17.7), it is immediately obvious that in order for frictional heating to be important (in comparison to radioactive heating which yields $A \approx 10^{-8}$ W/m^3 for the mantle) for a strain rate of 10^{-15} s^{-1}, the stress magnitude must exceed ~10 MPa or viscosity must be larger than ~10^{22} Pa s. Therefore frictional heating plays relatively small roles in general portions of convecting mantle where the viscosity is $<10^{22}$ Pa s (see Chapters 18 and 19). However, in regions of localized deformation, this effect can be large. For instance, in shear zones, strains of order ~1 can occur in ~1 Myr with stress levels of ~100 MPa. In such a case, strain rate is ~10^{-13} s^{-1} and $A_{\text{fric}} \sim 10^{-5}$ W/m^3 that significantly exceeds radioactive heating and therefore frictional heating plays an important role in these regions. Note that both latent heat and frictional heating do not affect the net energy budget of a convecting system because these heat sources are ultimately from convecting mass transport itself. The main role of these processes is to redistribute the energy.

When the effect of convection (i.e., $u_i(\partial T/\partial x_i)$ term) and heat generation is negligible and when thermal conductivity is constant, then (17.4) is reduced to,

$$\frac{\partial T}{\partial t} = \kappa \sum_{i=1}^{3} \frac{\partial^2 T}{\partial x_i^2}. \tag{17.8a}$$

The assumption of ignoring heat generation is appropriate for the oceanic upper mantle. When we consider the depth variation of temperature in the oceanic upper mantle, lateral variation can be neglected ($\partial/\partial z \gg \partial/\partial x$, $\partial/\partial y$, z: vertical axis). If one makes another assumption that thermal conductivity is constant, then temperature distribution can be determined by solving equation (17.8a) with appropriate initial and boundary conditions assuming that temperature changes only with one direction (z, i.e., depth, in this case), namely,

$$\frac{\partial T}{\partial t} = \kappa \frac{\partial^2 T}{\partial z^2}. \tag{17.8b}$$

Let us consider the temperature profile of a column of materials that were injected from the deeper portion at a mid-oceanic ridge. These hot materials will be cooled from above as they spread horizontally. Therefore the initial condition for this problem is $T(z, 0) = T_\infty$ where T_∞ is the temperature of hot, deep materials (where we assumed that the temperature is independent of depth). The obvious boundary condition at the surface is $T(0, t) = T_0$ (surface temperature). When thermal diffusivity is constant, the solution of equation (17.8b) for these boundary and initial conditions is given by

$$T(z,t) - T_0 = (T_\infty - T_0)\,\mathrm{erf}\left(\frac{z}{2\sqrt{\kappa t}}\right) \qquad (17.9)$$

where t is the time measured from the time of injection of hot materials to the surface at a mid-ocean ridge (this is the "age" of the lithosphere) and erf(y) is the error function defined by $\mathrm{erf}(y) = (2/\sqrt{\pi})\int_0^y \exp(-z^2)\,dz$ (see Chapter 8, (8.11)).

Problem 17.2

Show that equation (17.9) is the solution of equation (17.8b) with appropriate boundary and initial conditions.

Solution

Using the definition of error function erf(y), one finds $d\,\mathrm{erf}(y)/dy = (2/\sqrt{\pi})\exp(-y^2)$ and $d^2\mathrm{erf}(y)/dy^2 = -(4y/\sqrt{\pi})\exp(-y^2)$. Therefore $\frac{\partial T}{\partial t} = (T_\infty - T_0)\cdot(d\mathrm{erf}\xi/d\xi)(\partial\xi/\partial t) = -[(T_\infty - T_0)z/2\sqrt{\pi\kappa}]\exp(-\xi^2)t^{-3/2}$ where $\xi \equiv z/2\sqrt{\kappa t}$. Similarly $\partial^2 T/\partial z^2 = -[(T_\infty - T_0)\cdot z/2\kappa\sqrt{\pi\kappa}]t^{-3/2}\exp(-\xi^2)$. Therefore $\partial T/\partial t = \kappa(\partial^2 T/\partial z^2)$. Noting that $\mathrm{erf}(0) = 0$ and $\mathrm{erf}(\infty) = 1$ it is easy to show that both initial and boundary conditions are met.

Equation (17.9) provides a good approximation for the temperature–depth profile for the oceanic upper mantle. Other boundary conditions involving latent heat release associated with partial melting at the base of the lithosphere were proposed (e.g., McKenzie, 1969; Yoshii *et al.*, 1976), but the presence of a large degree of partial melting at the base of the oceanic lithosphere is not supported by the later petrological studies (e.g., Plank and Langmuir, 1992). The basic feature of this model is that for the shallow region, $z \ll 2\sqrt{\kappa t}$, the temperature profile is linear with respect to depth, $T - T_0 = [(T_\infty - T_0)/\pi\sqrt{\kappa t}]z$ $\left((dT/dz)_0 \sim 8\,\mathrm{K/km}\right.$ for $t = 100$ Myr $\left.(T_\infty = 1600\,\mathrm{K})\right)$ whereas the temperature approaches an asymptotic value, T_∞, for $z/2\sqrt{\kappa t} \gg 1$. Therefore $z = 2\sqrt{\kappa t}$ gives a rough measure of the thickness of the lithosphere (for $\kappa = 10^{-6}$ m^2/s and $t = 100$ (30) Myr, $z \sim 110$ (60) km). Fig. 17.2 shows the temperature–depth relationship in a typical Earth's mantle where the near-surface temperature profile corresponds to the error function solution. Some modifications are needed when a more realistic behavior of thermal diffusivity is included. At near-surface conditions (i.e., at relatively low temperatures), thermal diffusivity is controlled mostly by phonon conduction. Thermal conductivity (diffusivity) due to phonons decreases with temperature (e.g., Kittel, 1986). Consequently, the temperature gradient increases with depth leading to somewhat higher temperatures at the deep portion of lithosphere than in a simple model where temperature dependence of thermal diffusivity is not included.

(a)
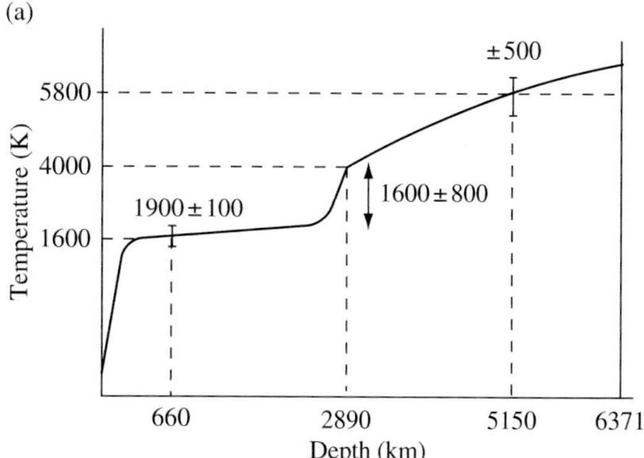

(b)
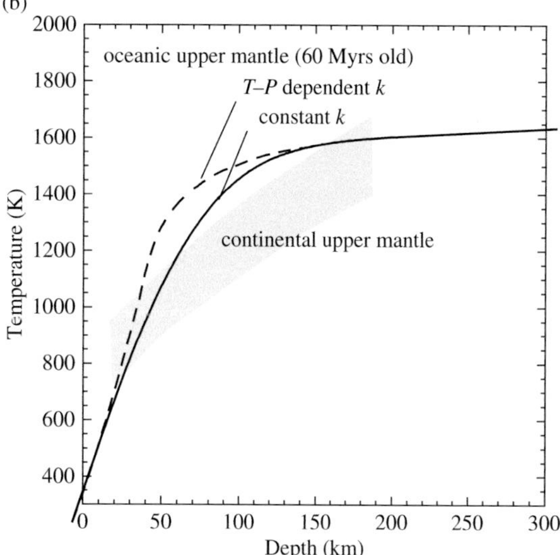

FIGURE 17.2 (a) A typical temperature–depth profile of Earth's mantle. (b) A close-up of temperature–depth profiles in the near-surface region.

Being thicker than the oceanic lithosphere, the continental lithosphere is colder, but the difference in temperatures between the oceanic and continental upper mantle becomes small at deeper portions ($>$200 km). In the continental lithosphere, the concentration of radioactive elements is larger (particularly in the upper crust) and the crust is thicker than in the oceanic regions. Therefore the effects of radioactive heating cannot be ignored. SCLATER *et al.* (1980, 1981) made a detailed comparison of temperatures between oceanic and continental lithosphere (see also ARTEMIEVA, 2006). The difference in temperatures is one important factor that determines the difference in creep strength between oceanic and continental upper mantle. The difference in water content also contributes to the ocean versus continent difference in strength (see Chapter 19).

A well-defined temperature gradient in Earth is *the adiabatic temperature gradient*. This is a temperature gradient in a material that moves vertically in the gravity field without the exchange of heat with surroundings. If the vertical motion is fast enough compared to the time-scale of thermal diffusion, then adiabatic gradient would be achieved. In reality, the temperature in the deep mantle is more or less adiabatic rather than a constant temperature assumed above. When material motion occurs adiabatically, then there is no change in entropy, i.e.,

$$dS = \left(\frac{\partial S}{\partial P}\right)_T dP + \left(\frac{\partial S}{\partial T}\right)_P dT = 0. \tag{17.10}$$

Thus, using the equation of hydrostatic equilibrium ($dP = \rho g\, dz$, equation (17.1))

$$\frac{dT}{dz} = -\frac{(\partial S/\partial P)_T}{(\partial S/\partial T)_P}\rho g = \frac{T g \alpha_{\text{th}}}{C_P} \tag{17.11}$$

where we used the thermodynamic relationships, $(\partial S/\partial T)_P = C_P/T$ and $(\partial S/\partial P)_T = -\alpha_{\text{th}} V$ (see Chapter 2), and the definition of thermal expansion, $\alpha_{\text{th}} = (1/V)(\partial V/\partial T)_P$. For typical values of relevant parameters ($T = 2000$ K, $g = 10$ m/s^2, $\alpha_{\text{th}} = 2 \times 10^{-5}$ K^{-1}, $C_P = 1200$ J/K $\cdot$ kg), the adiabatic temperature gradient is $\sim$0.4 K/km.

Temperature in the deeper regions can be inferred from seismological observations combined with mineral physics constraints. The seismological observations show that Earth's mantle is solid (except for a small amount of melt locally). Therefore the temperature in the mantle must be below the melting temperature (the liquidus) of mantle materials. Simply from this constraint, we can see that the temperature gradient, dT/dz, must decrease significantly with depth, otherwise temperature in the deep portions (say $z > 200$ km) of the mantle would significantly exceed the melting temperature. Such a notion is consistent with the inference of mantle temperatures from the depth of seismic discontinuity. If these discontinuities (at 410 and 660 km, for example) are caused by phase transformations, then their depths (i.e., pressure) must depend on the temperature. By comparing the actual depth of these discontinuities with the phase diagrams of corresponding materials, temperatures at these depths can be inferred (AKIMOTO *et al.*, 1976; AKAOGI *et al.*, 1989; ITO and KATSURA, 1989). In this way the temperature gradient in the transition zone has been estimated to be $dT/dz = 0.5 \pm 0.2$ K/km. More direct estimate of temperature and other thermal properties (such as thermal expansion) is possible from seismological observations when the effects of thermal properties on seismic wave velocities are taken into account. BROWN and SHANKLAND (1981) conducted such a study and concluded that in much of the Earth's mantle the temperature gradient is nearly adiabatic (a similar, extensive study was made by SHIMAZU (1954)).

Similarly, the fact that the outer core is molten indicates that the temperature there exceeds the melting temperature of the core materials (Fe-rich alloy), but the fact that the inner core is solid demands that the temperature at the inner core boundary (5150 km), ICB, must be the melting temperature of the core materials. In fact this latter constraint, in principle, gives a tight estimate of temperature, but due to the large uncertainties in experimental determination of melting temperature, the temperature estimate at the ICB is poorly constrained. SHEN *et al.* (1998) determined the melting temperature of pure Fe at the ICB pressure to be 5800 ± 1000 K. This should be considered to be the upper limit of the temperature at ICB, because alloying materials likely reduce the melting temperature significantly. ALFÉ *et al.* (2002a) calculated the melting temperature of core materials including the effects of secondary components. They estimated a similar melting temperature at the ICB ($\sim$5800 K).

Analysis of depth variation of density and elastic properties also provides some constraints on temperature distribution. To understand how it works, let us consider a simple case of compression of a homogeneous material. When density changes with depth through the change in temperature and pressure, we have,

$$\frac{d\rho}{dz} = \left(\frac{\partial \rho}{\partial P}\right)_T \frac{dP}{dz} + \left(\frac{\partial \rho}{\partial T}\right)_P \frac{dT}{dz}. \tag{17.12}$$

Inserting equation (17.1) and using the definitions of bulk modulus and thermal expansion, one has,

$$\frac{d\rho}{dz} = \frac{\rho^2 g}{K_T} - \rho\alpha_{th}\frac{dT}{dz}. \tag{17.13}$$

This relation means that if the density changes through compression and thermal expansion, then the actual depth variation in density depends on the depth variation in temperature.

Let us consider a case where the temperature–depth relation follows the adiabatic gradient, namely, $(dT/dz)_{ad} = Tg\alpha_{th}/C_P$. Inserting this relation into (17.13), one obtains,

$$\begin{aligned}\left(\frac{d\rho}{dz}\right)_{ad} &= \frac{\rho^2 g}{K_T} - \rho\alpha_{th}\frac{Tg\alpha_{th}}{C_P} \\ &= \frac{\rho^2 g}{K_T}\left(1 - \frac{T\alpha_{th}^2 K_T}{\rho C_P}\right) \approx \frac{\rho^2 g}{K_S} = \frac{\rho g}{V_\phi^2}\end{aligned} \tag{17.14}$$

where $V_\phi^2 \equiv V_P^2 - \frac{4}{3}V_S^2 = \frac{K_S}{\rho}$ and use has been made of the relation $K_T = K_S(1 - (\alpha_{th}^2 K_T T/\rho C_V)) \approx K_S \cdot (1 - (\alpha_{th}^2 K_T T/\rho C_P))$ (equation (4.21)).

Although the concept of adiabatic temperature gradient is well defined and nearly adiabatic temperature gradient is often assumed for much of the deep Earth, deviation from adiabatic temperature gradient in convecting fluids can occur. One of the causes for this is the effect of heating by radioactive elements. When a significant amount of radioactive elements are present, then heat is generated in a moving material and the assumption of adiabaticity will not be satisfied (e.g., BUNGE *et al.*, 2001).

The depth variation of density can also be inferred from seismological (plus gravity) observations. A ratio of actually inferred depth gradient of density to theoretical values corresponding to adiabatic compression (i.e., equation (17.11)) is known as the *Bullen parameter*,

$$\eta_B \equiv V_\phi^2(d\rho/dz)/\rho g = \frac{d\rho/dz}{(d\rho/dz)_{ad}}. \tag{17.15}$$

By definition, the Bullen parameter should be one for regions where density is determined by compression in the adiabatic temperature gradient. In contrast, in regions of super-adiabatic gradient, the density increases with depth less than in the case of adiabatic gradient. Therefore the Bullen parameter will be less than one in such a case. When the density increases with depth more than adiabatic compression due to phase transformations or the change in chemical composition (i.e., increase in Fe content with depth), then the Bullen parameter will be larger than one. Fig. 17.3 shows Bullen parameter as a function of depth. Note that the Bullen parameter is nearly one in most of the core and the lower mantle, whereas it is less than one in the shallow upper mantle and higher than one in the transition zone. It should be noted, however, the Bullen parameter is determined with some uncertainties and therefore some deviations from adiabaticity are allowed by the data.

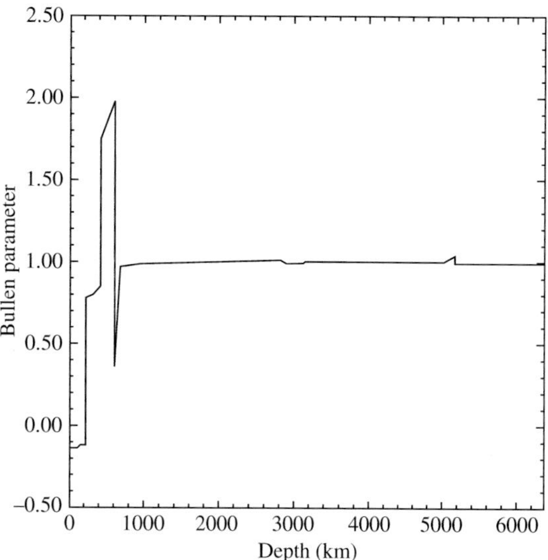

FIGURE 17.3 Distribution of the Bullen parameter in the Earth.

Problem 17.3

Assuming that the error bar for the Bullen parameter is ±10%, estimate the range of temperature gradients allowed by the data for regions where adiabatic temperature gradient is inferred ($\eta_B = 1$).

Solution

From (17.15),

$$\begin{aligned}\eta_B &\equiv V_\phi^2(d\rho/dz)/\rho g = \frac{d\rho/dz}{(d\rho/dz)_{ad}} \\ &= \frac{(d\rho/dz)_{ad} + \Delta(d\rho/dz)}{(d\rho/dz)_{ad}} = 1 + \frac{\Delta(d\rho/dz)}{(d\rho/dz)_{ad}}\end{aligned}$$

where $(d\rho/dz)$ is the variation in depth dependence of density due to the deviation in temperature gradient from adiabatic gradient. Thus $\Delta(d\rho/dz) = (\partial\rho/\partial T)\cdot\Delta(dT/dz) = -\rho\alpha_{th}\cdot\Delta(dT/dz)$ and using equation (17.14), one gets $\eta_B = 1 - (\alpha_{th}K_S/\rho g)\Delta(dT/dz) \equiv 1 + \eta_B$. Therefore, the errors of the Bullen parameter translate into the errors of temperature gradient as

$\Delta(dT/dz) = -(\rho g/\alpha_{th} K_S)\Delta\eta_B \sim \pm 0.5\,\text{K/km}$ for the errors of 10% of the Bullen parameter respectively. Note that quite a large range of non-adiabatic temperature gradient is allowed even for the lower mantle and the core by the data.

In summary, the gross structure of temperature–depth profile of Earth's interior can be understood based on the adiabatic temperature gradient plus conductive profiles in the horizontal boundary layers. Among the horizontal boundary layers, the thermal structure of the lithosphere at the surface is well known. In contrast, thermal structure of other possible boundary layers is poorly known. One of them is the D″ layer at the bottom of the mantle. Two lines of argument can be used to constrain the thermal structure of the D″ layer. One can extrapolate temperature from a shallow region assuming the adiabatic gradient. If one assumes $T \sim 1900$ K at 660 km and uses the adiabatic gradient of 0.4 K/km, then the temperature at the bottom of the mantle would be ~2800 K. Similarly, one can also extrapolate temperature from the inner core boundary (at 5150 km). Assuming the temperature there to be ~5800 K, and using the adiabatic temperature gradient of ~0.7 K/km in the core, then the temperature at the core–mantle boundary would be ~4600 K. Therefore there would be ~2000 K temperature change near the core–mantle boundary implying a well-defined thermal boundary layer. However, this estimate of a temperature increase has large uncertainties. The largest uncertainty is due to the fact that the estimated temperature at the inner core boundary is the upper bound because of the effects of light elements to reduce the melting temperature. The temperature change across the D″ layer is likely to be substantially smaller than 2000 K.

The temperature profile in the D″ layer can also be estimated from the heat flux from the core, the thermal conductivity of D″ layer materials and the thickness of the layer. Ignoring the contribution from radioactive elements and from convective heat transport, the temperature change across the D″ layer is given by $\Delta T = h(J/k)$ where J is the heat flux from the core (per unit area), k is the thermal conductivity (~10 W/mK) and h is the thickness (~100 km) of the D″ layer. Assuming the total heat flux from the core is 2×10^{12} W (i.e., $J = 1.3 \times 10^{-2}\,\text{W/m}^2$) which is ~5% of the total heat flux from Earth, one gets $\Delta T = 1300$ K. One notes that all the assumed parameters are rather poorly constrained (uncertainty of each parameter is likely to be as much as a factor of ~2) and we must conclude that the temperature profile in the D″ layer is not well known at this time.

17.3. Composition of Earth and other terrestrial planets

The gross chemical composition of a planet can be inferred from the mean density. However, more detailed knowledge of composition is often needed to investigate the dynamics and evolution of a planet. This section provides an overview of the current status of our knowledge of the chemical composition of Earth and other terrestrial planets. Emphasis is on the composition of Earth for which we have far more detailed constraints than the compositions of other planets.

17.3.1. Planetary atmospheres and the volatiles in Earth and terrestrial planets

In a typical terrestrial planet like Earth, the total mass is dominated by non-volatile materials such as silicates and metallic iron. However, at the surface there are layers composed primarily of fluids (atmosphere and hydrosphere (oceans)). There is strong evidence that a significant amount of aqueous fluid is present in Earth's crust, and water (and other volatile materials) is also present in various forms in the deep Earth (and other planets). These volatile materials have weak chemical bonding and they have important influence on mechanical properties. Table 17.3 summarizes the composition of the atmosphere (hydrosphere) of various terrestrial planets. The atmosphere is made primarily of N_2 and O_2 with a small amount of CO_2 and H_2O and other molecules. The hydrosphere (oceans) is mostly made of H_2O. Our main concern in this book is the mechanical properties of solid Earth, but as will be seen later, interaction of solid Earth with surface fluid layers (particularly with hydrosphere (H_2O)) has important influence on the properties of solid Earth. Note that the concentration of water in the atmosphere of Venus is much lower than that of Earth.

TABLE 17.3 Composition of planetary atmosphere/ocean (unit ppb kg/kg of total mass) (data from Hartmann, 1999).

	Venus	Earth	Mars
H_2O	6×10	2.5×10^5	$\sim 5 \times 10^3$
CO_2	10^5	0.4	$\sim 10^3$
N_2	2×10^3	2×10^3	3×10^2

17.3.2. The composition and structure of Earth

Models of Earth's composition

Among the rocks in the crust, sedimentary rocks are formed by transportation and compaction of pre-existing minerals and therefore they are of secondary origin: their formation involves small chemical processes. In contrast, volcanic rocks such as basalts are formed by partial melting of some parent rocks and they represent the fundamental chemical differentiation processes in Earth. Among the various volcanic rocks, basalt is the most abundant on Earth and on many of the other terrestrial planets and the formation of basalt, particularly the basalt at mid-ocean ridges (mid-ocean ridge basalt, MORB in short) is considered to be the most important volcanic activity on Earth. The mid-ocean ridges have a total length of $\sim 8 \times 10^4$ km, and $\sim 27\,km^3$ (or $\sim 8 \times 10^{13}$ kg) of basalt is created in a year at mid-ocean ridges. This volcanism is due to the partial melting caused by the upwelling of hot materials by convection. Consequently, the upper mantle materials should have chemical composition that can produce mid-ocean ridge basalt upon partial melting. Based on the chemical compositions of various rock types on Earth, Ringwood proposed a model of composition of Earth (e.g., Ringwood, 1975; Green and Falloon, 1998). Basic assumptions in this model are (1) basaltic volcanism at mid-ocean ridges is the most fundamental volcanism that is due to the partial melting of rocks in the deep upper mantle, (2) Mg- and Fe-rich rocks (peridotite) occasionally found as xenoliths are the residue of partial melting. Based on these assumptions, Ringwood proposed that mid-ocean ridge basalt is formed by partial melting of materials in the deep upper mantle (deeper than ~100 km) whose composition is similar to most peridotites but contains a larger amount of Ca-, Al-rich minerals than most peridotites (these minor elements are needed to form basalt). He called this hypothetical rock in the source region *pyrolite* which is a mixture of mid-ocean ridge basalt and (depleted, i.e., differentiated) peridotite such as harzburgite (Table 17.4). Chemical composition of some peridotite xenoliths is close to the pyrolite (e.g., xenolith from Kilbourne Hole (KLB1)

TABLE 17.4 Composition of the silicate portion (crust and mantle) of Earth (wt% of oxides).

For Py-1, Py-2 and CI model, data from McDonough and Sun (1995).
For MORB, granite and harzburgite, data from Carmichael *et al.* (1974).

	Py-1[1]	Py-2[2]	CI[3]	MORB[4]	Harzburgite[5]	Granite
SiO_2	45.0	45.0	49.9	49.27	43.37	71.32
TiO_2	0.201	0.17	0.16	1.26	0.15	0.27
Al_2O_3	4.45	4.4	3.65	15.91	1.19	14.31
Cr_2O_3	0.384	0.45	0.44	—	0.56	—
MnO	0.135	0.11	0.13	0.13	0.14	0.07
FeO	8.05	7.6	8.0	10.36	7.59	2.06
NiO	0.25	0.26	0.25	—	0.14	—
MgO	37.8	38.8	35.15	8.49	45.35	0.52
CaO	3.55	3.4	2.90	11.26	0.79	1.05
Na_2O	0.36	0.4	0.34	2.58	0.15	2.62
K_2O	0.029	0.003	0.022	0.19	—	6.87
P_2O_3	0.021	—	—	0.13	—	0.03
H_2O	—	—	—	0.35	—	0.65

[1] Primitive mantle based on the pyrolite model based on peridotites, komatiites and basalts.
[2] Primitive mantle based on the pyrolite model based on MORB and harzburgite. (These two models differ in the assumed process of chemical differentiation. Py-1 assumes differentiation not only by production of MORB but also by production of komatiite that occurred mostly in the Archean (older than ~2.7 Gyr). Py-2 assumes differentiation caused by the formation of MORB, a process that occurs at present.)
[3] The model composition based on the CI chondrite.
[4] Mid-ocean ridge basalt.
[5] Depleted peridotite made mostly of olivine and orthopyroxene.

in New Mexico in USA: (Jagoutz *et al.*, 1979)). The pyrolite model for mantle composition assumes that the mantle has a nearly homogeneous (with depth) composition.

There is another type of model of the chemical composition of Earth. This model is based on the composition of the solar system inferred from cosmochemistry. Two observations provide us with some ideas about the overall chemical composition of the solar system. One is the chemical composition of primitive meteorites, called *carbonaceous chondrite*. This is a type of chondrite that is a mixture of various minerals including iron, and a variety of silicate minerals. These materials are not in chemical equilibrium and the age of this type of chondrite is among the oldest in various materials in the solar system (4.57 Gyr). Consequently, carbonaceous chondrite is considered to be made of a mixture of various materials in the earliest stage of the history of the solar system. Another source of information as to chemical composition of the solar system comes from the analysis of chemical composition of the sun's outer convective layer from optical spectroscopy. The composition of the refractory elements of carbonaceous chondrite agrees well with that of the sun's convective layer, and therefore it is generally believed that the composition of carbonaceous chondrite is representative of that of the primitive solar system. One model for the chemical composition of Earth assumes that Earth's composition is the same as the composition of refractory elements of (type I) carbonaceous chondrite. This is called the *chondrite model* of Earth (Table 17.4). Chondrite composition has higher SiO_2 content and lower MgO content than the pyrolite model. As a consequence, the chemical composition of the upper mantle peridotites and that of pyrolite has significantly lower SiO_2 content than CI chondrite. This means that if Earth has a chondritic bulk composition, then there must be some regions deep in Earth where a large amount of Si occurs. If this region is the lower mantle (or some part of the deep mantle), it would imply that the upper mantle and the deep mantle are not mixed well throughout the history of Earth. Therefore mantle convection must be somewhat layered. In contrast, if the mantle is indeed more or less chemically homogeneous, it would be consistent with a whole mantle convection model of material circulation.

These two models (pyrolite and chondrite models) are the most important and well-defined models for the bulk composition of Earth. As you can understand from the above description, the pyrolite model does not have direct constraints on the composition of the deep mantle (below the region from which basalts are produced). The chondrite model provides an insight into the bulk chemical composition of Earth, but it provides only a loose constraint on the composition of the mantle (the precise composition of the mantle depends on the composition of the core and the details of chondrite composition).

Note that random sampling of surface rocks or meteorites does not give a good estimate of the bulk composition of Earth's interior. Rocks on Earth's surface or in space are there for some geological reasons, and therefore they represent a highly biased sampling of Earth's interior. In inferring the composition of Earth's interior, it is important to understand the chemical (and physical) processes by which rocks are distributed in Earth and in space.

The direct sampling of Earth materials is limited to the depth from which rocks are carried to the surface. The maximum depth from which most xenoliths (rock fragments) are transported is ~200 km. Although there are some unusual cases where rocks are considered to have been carried from more than 300 km deep in the mantle (e.g., Haggerty and Sautter, 1990), their volume fraction is extremely small and it is not obvious if these rocks are representative of these deep portions of Earth's mantle.

There are some geochemical constraints on the composition of the core. First, from the cosmochemical abundance, the only plausible material that is consistent with the density of the core is iron (plus some minor elements). The abundance of other plausible components of the cores of planets (such as FeS) can also be inferred from the theory of condensation assuming that materials that have been condensed at certain regions of solar nebulae are the source materials of a planet formed at that position of the solar system (Lewis, 1974) (see also Hartmann, 1999). The composition of the core, however, may also have been modified during Fe–silicate separation.

These petrological or geochemical models of Earth's composition must be consistent with geophysical observations. A detailed comparison of models with seismological observations will be made in the following sections. Before going into detail, I should note that the pyrolite model (homogeneous composition) is consistent with a simple whole mantle convection model whereas the chondrite model implies layering in major element composition (the deep mantle needs to be enriched with Si relative to the shallow mantle), and hence some dynamic processes need to be invoked to explain the layering in the chemical composition.

Crust

One of the most important observations about the crust is its heterogeneity. The crust is heterogeneous both laterally (variation with geological setting) and radially (variation with depth). The composition of the oceanic crust is different from that of the continental crust. Oceanic crust is made of a thin layer of sediments and a thicker basaltic layer. The thickness of the basaltic layer on the ocean floor is nearly homogeneous throughout the world and is ~7 km. In contrast, the continental crust varies its thickness from ~20 km to ~70 km and its composition also varies from granitic in the shallow portions to basaltic in the deeper portions. It is generally considered that crust materials are formed by a partial melting of upper mantle materials and that the difference in the composition and structure of crusts between the oceanic and continental regions is due to the difference in the nature of partial melting. There is evidence that the melting responsible for the formation of the continental crust involves a larger amount of water (e.g., Campbell and Taylor, 1983). Although a typical continental crust has two distinct layers (the upper and the lower continental crust), there is evidence that the lower continental crust has been removed (delaminated) in some regions (e.g., Meissner and Mooney, 1998).

The crust is made of silica-rich rocks such as basalt or granite. Important minerals in these rocks include quartz (SiO_2), plagioclase ($(Na, K)AlSi_3O_8$-$CaAl_2Si_2O_8$) and pyroxenes ($(Mg, Fe, Ca)SiO_3$)). Hydrous minerals such as muscovite ($(K, Na)Al_2AlSi_3(OH)_2O_{10}$), phlogopite ($KMg_3AlSi_3(OH)_2O_{10}$), amphibole ($(Mg, Fe, Ca)_2Mg_5Si_8(OH)_2O_{22}$) are also common in the crust. More silica (SiO_2)- and/or calcite ($CaCO_3$)-rich rocks are found as sedimentary rocks (e.g., chart or limestone (or marble)) in some regions of the crust. The densities of these rocks are ~2500–2900 kg/m^3.

Crustal rocks have a range of grain size and water content. Basalts have fine grain size ranging from sub-micron size in glassy, "fresh" basalts to a few millimeters for well-annealed basalts (gabbro). Granitic rocks usually have a large grain size of the order of mm to 10 mm. As a general rule, a newly formed volcanic rock that has been cooled quickly has small grain size and grain size increases by annealing. However, when these rocks are deformed at high stresses, then grain size becomes small (see Chapter 13). Highly deformed rocks in crustal shear zones (mylonites) have a grain size on the order of a few to several tens μm. The water content in crustal rocks varies from one to another. Many crustal rocks contain a large fraction of hydrous minerals, but some of the lower crustal rocks contain very few hydrous minerals (Hacker *et al.*, 2000).

Mantle

A layer below the Moho is called the mantle. Three distinct layers are identified in the mantle based mainly on seismological observations: (1) the upper mantle, (2) the transition zone and (3) the lower mantle.

(i) Upper mantle

The upper mantle is the layer below the Moho to ~410 km depth. At the Moho, seismic wave velocities (and densities) significantly increase. The seismological observations are consistent with the model of olivine-rich composition such as peridotite (Table 17.4) but some mixture with more pyroxene and/or garnet-rich rocks such as eclogite is also consistent with many geophysical and geochemical observations (e.g., Allègre and Turcotte, 1986). There is evidence for heterogeneous upper mantle from petrological and geochemical studies of basalts (e.g., Kogiso *et al.*, 1998).

One of the remarkable features of the upper mantle is its mechanical layering: the lithosphere–asthenosphere structure. There is a near-surface strong layer (~50–200 km thick) followed by a weak layer below. The strong layer is referred to as the *lithosphere*, and a weak layer the *asthenosphere*. Such a mechanical layering is inferred from the presence of a low velocity and high attenuation zone as well as the mechanical response of near-surface layers for some time-dependent loading (see Chapter 18). The asthenosphere in the shallow upper mantle (its exact depth depends on tectonic setting, but is generally from ~50 to ~200 km) was first recognized by Gutenberg (1926, 1948, 1954). Subsequent studies using mostly surface waves support this notion and further indicated that such structure changes from one region to another. As we learn in Chapters 11 and 19, the low seismic wave velocity and high attenuation imply that materials are mechanically weak. But the relationship between seismic properties and long-term rheological properties is not simple (see Chapters 11 and 19 for more detail).

Partial melting in the upper mantle produces basaltic magmas. There is no doubt that partial melting occurs in the upper mantle. Indeed estimated temperatures in the shallow upper mantle (~100 km) are close to or exceed the solidus of upper mantle rocks (see Fig. 17.2). Consequently, it was widely believed that significant partial melting occurs in the shallow upper mantle at the global scale causing low seismic waves and high attenuation (e.g., Gutenberg, 1926; Shimozuru, 1963; Solomon, 1972; Shankland *et al.*, 1981; Forsyth, 1992). However, recent experimental and theoretical studies of partial melting suggest that such a notion is not supported, and the significant partial melting is likely to be limited to the very vicinity of volcanoes (see Chapters 18–20).

Another important feature of the upper mantle is anisotropy. Seismic wave propagation in this layer is in most cases highly anisotropic (see Chapter 21). Strong anisotropy is consistent with a model containing a significant amount of olivine. A popular global Earth model such as PREM (Dziewonski and Anderson, 1981) includes anisotropy in the top 220 km. PREM is constructed based mainly on the data set of body wave travel times, free oscillation and surface waves that reflect the global average structure including mantle and crust below both oceanic and continental regions. However, the anisotropic structure is rather complicated and the inclusion of a discontinuity at 220 km as a global feature may not be appropriate. Another global model such as ak135 (Kennett *et al.*, 1995) does not have such a discontinuity at 220 km depth. However, the ak135 model is based mainly on travel time data of body waves and hence is biased toward the continental regions where most of the data are collected. The origin of discontinuities in the deep upper mantle (~200–300 km) is not well understood. High-resolution seismic studies using converted waves often detect discontinuities in this depth region (e.g., Gaherty and Jordan, 1995; Revenaugh and Jordan, 1991). There is growing appreciation that the structure of deep upper mantle is highly heterogeneous (e.g., Deuss and Woodhouse, 2002; Gu *et al.*, 2001). A phase transformation in orthopyroxene occurs at around this depth range (Shinmei *et al.*, 1999). The observed regional variation of these discontinuities may reflect a regional variation in composition such as the concentration of orthopyroxene (Matsukage *et al.*, 2005). An alternative model is that at least one of these discontinuities, i.e., the *Lehmann discontinuity*, is due to the change in anisotropy: from anisotropic structure in the shallow upper mantle to isotropic structure in the deep upper mantle (Karato, 1992). Another possibility includes the change in the nature of anisotropy due to the change in water content. Evidence for local/regional variation of water content in the deep upper mantle is recently reported (Shito *et al.*, 2006). The laboratory studies by Jung and Karato (2001b) suggest that a change in water content will result in a change in lattice-preferred orientation hence seismic anisotropy. Consequently, if such a fabric transition occurs in a narrow depth interval it will cause a seismic discontinuity.

Another important feature of the upper mantle is the possible presence of partial melting at the bottom of the upper mantle. Revenaugh and Sipkin (1994) suggested the presence of such a layer, and similar studies have been reported (e.g., Vinnik and Montagner, 1996; Song *et al.*, 2004). Partial melting at the bottom of the upper mantle is a distinct possibility because upwelling materials from the transition zone likely contain a large amount of water exceeding the water content in the upper mantle minerals at the solidus. In such a case, dehydration melting will occur as soon as materials come into the upper mantle just above the 410-km discontinuity (Bercovici and Karato, 2003; Karato *et al.*, 2006).

The composition of the upper mantle can be inferred from rocks at the surface. In addition to silica-rich rocks that compose the crust, we also find much denser rocks such as peridotites or eclogites. The densities of these rocks are ~3200–3400 kg/m^3 at ambient conditions. These rocks are made of minerals that contain a larger amount of (Mg, Fe)O and a smaller amount of SiO_2 than most crustal minerals and are often called *ultramafic* (i.e., very Mg, Fe-rich) rocks. They have more compact crystal structures than most crustal minerals. The important minerals in these rocks include olivine ($(Mg, Fe)_2SiO_4$), pyroxenes ($(Mg, Fe, Ca)SiO_3$) and garnets ($(Mg, Fe, Ca)_3(Al, Fe)_2Si_3O_{12}$). The ultramafic rocks occur as fragments in volcanic rocks (xenoliths) or in the deeply eroded fault zones in the collision zones in the continents or in the transform faults in the oceans. Hydrous minerals such as phlogopite and amphibole are less common in these rocks (except for serpentine ($Mg_3Si_2(OH)_4O_5$) as an alteration product). The densities and seismic wave velocities of these rocks are largely consistent with the seismological observations, and these rocks are considered to be the main constituent of the upper mantle.

The microstructures of upper mantle rocks can be investigated from xenoliths or from rocks carried to the surface by erosion or tectonic uplifting (exhumation). Upper mantle rocks have a range of grain size. Fine-grained rocks (~50 μm) can be found in shear zones or some deep xenoliths ("sheared lherzolites"; Nixon and Boyd, 1973) whereas the most typical grain size for upper mantle rocks is a few mm (e.g., Mercier, 1980; Karato, 1984).

Water contents of upper mantle rocks have been extensively studied (e.g., BELL and ROSSMAN, 1992; INGRIN and SKOGBY, 2000; BOLFAN-CASANOVA, 2005). The water content varies markedly from one rock to another. Spinel lherzolites (peridotite in the shallow upper mantle) contain up to ~30 ppm wt and garnet lherzolites (peridotites from deep upper mantle) up to ~180 ppm wt. There is some suggestion that water content in olivine increases more rapidly with depth than that in clinopyroxene and garnet. These water contents are well below the solubility limits of water in these minerals. Note, however, that these observations may not represent the water content of rocks in Earth's upper mantle. It is likely that some of the original water has been lost during the transfer of these rocks to the surface due to the fast diffusion of hydrogen or some water could have been added during the transport.

(ii) Transition zone

The layer between ~410 to ~660 km is separated from the layer above and below by the marked discontinuities in density and seismic wave velocities at these depths and both density and seismic wave velocities increase very rapidly with depth in this layer (in fact a steep gradient extends to ~900 km, but we use a convention that a layer below 660 km is defined as the lower mantle). Based on a number of studies pioneered by Birch and Ringwood (e.g., BIRCH, 1952; RINGWOOD, 1975), it is now well established that most of these features can be attributed to phase transformations that occur in major mantle minerals such as $(Mg, Fe)_2SiO_4$ and $(Mg, Fe)SiO_3$ (Fig. 17.4). However, some contribution from chemical layering cannot be ruled out (e.g., ANDERSON and BASS, 1986; OHTANI, 1988; AGEE, 1993).

The major constituents in this layer are wadsleyite (in the shallow transition zone; ~410 to ~520 km), ringwoodite (in the deep transition zone; ~520 to ~660 km) and majorite garnet. An important nature of these minerals, particularly wadsleyite and ringwoodite is that they can dissolve a significantly higher amount of impurities including "water (hydrogen)" (KAWAMOTO *et al.*, 1996; KOHLSTEDT *et al.*, 1996) and ferric iron (O'NEILL *et al.*, 1993). The maximum amount of water that can be stored in this layer is ~3 wt%, but the actual water content is not well known. HUANG *et al.* (2005) inferred water content in the transition zone in the Pacific to be ~0.1–0.2 wt% from observed electrical conductivity. However, regional variation in water content is highly likely (see BLUM and SHEN, 2004; TARITS *et al.*, 2004). Garnet is a unique mineral that has unusually high creep strength under water-poor conditions (KARATO *et al.*, 1995a). Consequently regions in which garnet controls the strength (such as the subducted oceanic crust in the transition zone; IRIFUNE and RINGWOOD, 1987) may have a relatively high viscosity compared to the surrounding regions if the water content is low. However, KATAYAMA and KARATO (2007) showed that the relative strength between garnet and olivine changes with water content: at high water contents, garnet will be weaker than olivine.

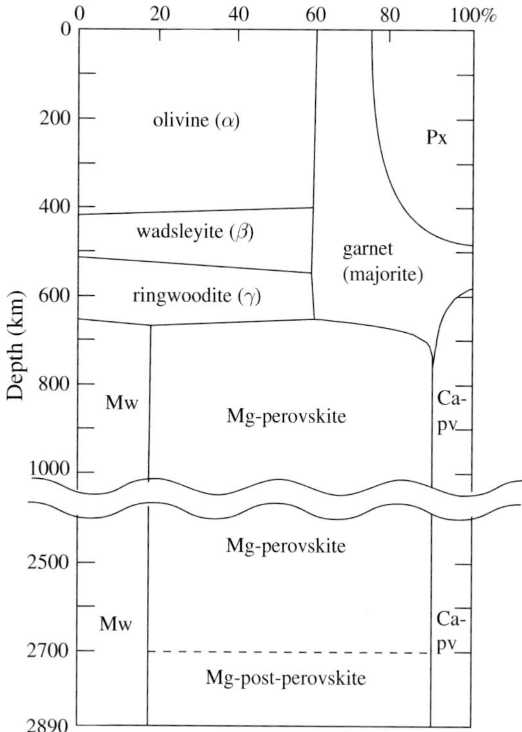

FIGURE 17.4 A major mineralogical constitution of Earth's mantle. Px, pyroxenes; Mw, magnesiowüstite.

A subtle, but important feature in geodynamics, is the topography on "410-" and "660-km" discontinuities. Although on average the depths of these discontinuities are 410 and 660 km respectively, the exact depths of the boundaries change regionally (e.g., SHEARER and MASTERS, 1992). This is considered to be due partly to the lateral temperature variation through the effects of temperature on the pressure at which the relevant phase transformation occurs (Chapter 2). Variation in chemical composition such as water content will also result in topography of these boundaries.

Due to the large density contrast across the transition zone, delamination of oceanic crust is possible in the transition zone if the rheological properties satisfy some conditions (Karato, 1997a). Shen and Blum (2003) reported evidence of subducted oceanic crust on top of the 660 km discontinuity.

There are no data to constrain the grain size of transition zone rocks. However, laboratory studies and mineral physics-based modeling suggest that the grain size in the transition zone is largely controlled by the kinetics of phase transformation and is likely to depend on the temperature at which a phase transformation occurs (Riedel and Karato, 1997; Karato *et al.*, 2001).

(iii) Lower mantle

The lower mantle is, in most part, remarkably homogeneous. This can be demonstrated by the Bullen parameter that is close to 1 (Fig. 17.3), indicating homogeneous chemical composition and nearly adiabatic temperature distribution. Also most of the lower mantle is nearly isotropic: the maximum amount of anisotropy is less than 1/100 of that of the upper mantle (Meade *et al.*, 1995).

The density and seismic wave velocities of the lower mantle are consistent with a model composition of a mixture of $(Mg, Fe)SiO_3$, $CaSiO_3$ perovskites and (Mg, Fe)O.

However, recent high-resolution seismological studies challenge this traditional view of featureless lower mantle. It is well established that the bottom ~200 km thick layer (D″ layer) is highly heterogeneous and anisotropic (e.g., Lay *et al.*, 1998; Garnero, 2004). Analysis of seismological observations strongly suggests that the deep lower mantle is chemically heterogeneous (see Chapter 20; Karato and Karki, 2001). There are some hints as to a more extensive chemical heterogeneity in the deep layer possibly starting from ~1600 km depth (Kellogg *et al.*, 1999; van der Hilst and Kárason, 1999). Evidence of (perhaps local or regional) seismic discontinuities was reported at around 900–1200 km depth (e.g., Kawakatsu and Niu, 1995, Kaneshima and Helffrich, 1999). Karato (1998d) proposed that the regional variation in anisotropy in the lower mantle is caused by the regional variation in the stress level (see also McNamara *et al.*, 2001, 2002, 2003). An alternative model is that the anisotropy is caused by layering (Kendall and Silver, 1996). The relative merits of these models are discussed in Chapter 21. Another notable structure of the D″ layer is the presence of an *ultra-low velocity layer* where shear wave velocity is lower than the average value by as much as 20–30% (Garnero *et al.*, 1998). Partial melting is a possible cause for the ultra-low velocity region (Williams and Garnero, 1996) but the enrichment in iron is an alternative explanation (Garnero and Jeanloz, 2000).

Phase transformations in major minerals such as $(Mg, Fe)SiO_3$ perovskite in the deep lower mantle have been reported (e.g., Wang *et al.*, 1992; Saxena *et al.*, 1996; Shim *et al.*, 2001; Murakami *et al.*, 2004; Oganov and Ono, 2004; Tsuchiya *et al.*, 2004b). If this is confirmed, they may explain some of the anomalies in the deep lower mantle.

The D″ layer is considered to be a thermal and chemical boundary layer and its structure reflects the dynamic processes of this planet including the formation of some plumes. There is a clear correlation between the low velocity anomalies in the lower mantle and the distribution of hotspots (e.g., Garnero, 2000). Lay *et al.* (1998), Garnero (2004) and Lay *et al.* (2005) provide excellent reviews of study of the D″ layer.

There are no direct observations on the grain size or water contents in lower mantle rocks. However, observation of no appreciable seismic anisotropy in the majority of the lower mantle implies that the grain size must be rather small (less than ~1 cm) in most of the lower mantle. Yamazaki and Karato (2001b) inferred that grain size in the shallow lower mantle is a few mm based on the comparison of viscosity inferred from post-glacial rebound with the laboratory data on diffusion coefficients. Kubo *et al.* (2000) investigated the grain size after the transformation to lower mantle mineralogy and suggested that the grain size of materials penetrating into the lower mantle may be rather small ($<10\ \mu m$).

Experimental studies show that the lower mantle minerals have much lower solubility of water than the transition zone minerals, although the actual values of solubility differ significantly among different studies (Bolfan-Casanova *et al.*, 2000, 2002 2003; Murakami *et al.*, 2002; Litasov *et al.*, 2003).

Core

Earth's core has much higher density than the mantle and is composed of two layers: the inner and the outer core. The outer core has no resistance to shear deformation and is considered to be made of molten iron. Based on Hf isotope analysis Lee and Halliday (1995) inferred that the core was formed some ~60 Myr after the formation of the solar system (for more recent results see Wood and Haliday, 2005). Therefore the formation of the core was almost contemporaneous with the formation of Earth. However, the age of the inner core is unknown. Based on a conjecture that strong dynamo action is only possible with the energy release due to the solidification of the inner core, Breuer and Spohn (1995) suggested that the inner core was formed initially at the Archean–Proterozoic boundary (~2.7 Gyr ago). However, the need for the presence of the inner core for geodynamo was recently questioned (e.g., Gubbins *et al.*, 2003). Some recent calculations suggest a rather young age of the inner core (e.g., Labrosse *et al.*, 2001), but such results are highly sensitive to the assumed amount of heat sources (e.g., potassium content) and the heat flux at the core–mantle boundary (e.g., Nimmo *et al.*, 2004).

The inner core has non-zero shear wave velocity and hence is largely solid. However, the density of the outer core is ~6–9% lower than that of molten iron as inferred from high-pressure studies (Birch, 1964; Brown and McQueen, 1980, 1982; Anderson, 2002). Consequently the presence of some light elements is inferred. However, the amount and speciation of light elements are not well constrained. Most of the candidate elements (H, O, C, K, Si, S) can be dissolved in the molten iron at high pressures and temperatures although the solubility of these elements depends on oxygen fugacity and hence oxygen and silicon, for instance, cannot be dissolved in a large amount simultaneously. Also based on the first-principles calculation Alfé *et al.* (2002b) showed that although the dissolution of oxygen results in a large density jump at the inner-core boundary (5150 km), silicon or sulfur do not. Consequently, the precise determination of the jump at ICB will provide some constraints as to the light elements in the outer core. The possible presence of potassium is important as an energy source for the core. The composition of the core is often inferred based on the consideration of chemical equilibrium between silicates and iron during core formation (e.g., Wood *et al.*, 2006). However, Karato and Murthy (1997) argued that a large fraction of core inherits the composition of core of accreted small planetesimals and the core composition is likely to be out of equilibrium with silicates, and the relevance of inference of core composition from the chemical equilibrium during core formation is questionable.

Buffett *et al.* (2000) pointed out that the amount of light elements in the outer core should have decreased with time as a result of secular growth of the inner core and these light elements are likely to have been deposited on top of the outer core. If the outer core is saturated with some of these light elements, then the growth of the inner core results in the supersaturation leading to the precipitation of light elements or compounds containing light elements. A part of the D″ layer including the ultra-low velocity layer might correspond to this "sediment" layer.

The inner core has a finite rigidity. Therefore the inner core–outer core boundary (sometimes called simply the inner core boundary, ICB) represents the solid–liquid transition. The density of the inner core is consistent with pure iron (and some Ni). Therefore the inner core is relatively pure compared to the outer core. The crystal structure of the inner core iron is not well understood but the *hcp* (hexagonal-close-pack) structure is most likely (Anderson, 2002).

Several points are noteworthy about the inner core:

(1) The inner core has most likely been formed by the solidification of the liquid outer core due to secular cooling. However, the time of inner core formation is not known.

(2) If the ICB represents the solid–liquid transition, and if the outer core contains some light elements, then the solid–liquid transition (melting) must occur with some temperature (pressure) interval. Consequently, a finite width is expected for the ICB. The width of the ICB depends on the melting phase relation as well as the dynamics of compaction (Sumita *et al.*, 1996). A detailed study of the structure of the ICB will provide useful constraints on the nature of light elements.

(3) The shear wave velocity in the inner core is rather slow and Poisson's ratio is consequently large (~0.44 as compared to ~0.28–0.30 for the mantle). It was sometimes considered that this is due to partial melting (e.g., Fearn *et al.*, 1981; Singh *et al.*, 2000). However, recent first-principles calculations of elastic properties of iron suggest that the large Poisson's ratio of the inner core can be explained without invoking partial melting (Alfé *et al.*, 2000; Steinle-Neumann *et al.*, 2001). In fact, the low viscosity of iron in the inner core conditions makes it difficult to retain a significant amount of melt (Sumita *et al.*, 1996). The observed seismic wave attenuation of the inner core is rather small (high Q) for a metallic material close to the melting point, and suggests that partial melting is not needed to explain seismic wave attenuation in the inner core. Instead, a large grain size is implied by the small attenuation.

(4) The inner core is anisotropic (Song, 1997; Tromp, 2001). Anisotropy is strong (~3% in P-wave velocity, faster along the rotation axis) and has roughly the axial symmetry. However, there appear to be some details about the anisotropic structure. First, there is evidence that the surface of the inner core is nearly isotropic and the thickness of the isotropic layer is different between two hemispheres (Song and Helmberger, 1998). Second, some authors suggest that in the very center of the inner core there is a region where the anisotropy is different from the shallower regions (Ishii and Dziewonski, 2002; Beghein and Trampert, 2003b).

(5) The Q values of $Q_P = 200–600$ (Bhattacharya *et al.*, 1993; Souriau and Roudil, 1995; Cormier and Li, 2002; Li and Cormier, 2002; Wen and Niu, 2002) which is surprisingly *large* (i.e., *low* attenuation) for a metal close to the melting temperature ($Q_S = 85$ for PREM). A plausible explanation is that the inner core has a large grain size (~100–1000 m: see Bergman, 1998 and also discussions in Chapter 19).

(6) The inner core is floating in the liquid outer core and therefore if there is any torque, it can rotate relative to the mantle. Some dynamo calculations predicted that the inner core might rotate relative to the mantle, and two papers published in 1996 presented evidence of inner core rotation from seismological observations (Song and Richards, 1996; Su *et al.*, 1996). However, later studies questioned the reliability of these works (e.g., Souriau, 1998; Souriau and Poupinet, 2002) and the issue of inner core rotation remains controversial (see also Creager, 2000; Tromp, 2001). Buffett (1997) showed that the inner core rotation is possible when the viscosity of the inner core is either very low ($<10^{16}$ Pa s) or very high ($>10^{20}$ Pa s). Rheology of the inner core plays an important role in the inner core dynamics.

The core is made of Fe–Ni alloy with some light elements (Birch, 1964). The exact nature of light elements is still controversial (Poirier, 1994). The comparison of the density estimated from seismological observations with high-pressure experiments or theoretical calculations shows that the outer core contains a significant amount of light elements, but the inner core density is consistent with nearly pure Fe (+Ni).

17.4. Summary: Earth structure related to rheological properties

The important factors that control the mechanical properties of Earth and other planets include (1) major element chemistry, (2) temperature (and pressure), (3) water content and (4) grain size. Among them the major element chemistry of Earth's interior is fairly well constrained. Temperature is not well known particularly in certain critical regions such as the D″ layer. However, the overall temperature distribution in Earth's mantle and core is now well constrained including the temperatures in the upper mantle. The largest uncertainties in terms of structure and composition that affect the mechanical properties (particularly rheological properties) are water content and grain size. Both of these factors could change the viscosity of Earth materials by more than a few orders of magnitude but neither of them is well known.

Variations of water content from one region to another might influence the rheological contrast between these regions. Similarly, water content in Earth is likely to be different from other terrestrial planets such as Venus. Venus is considered to have less water than Earth although Venus is otherwise very similar to Earth. The unique tectonic style on Earth (plate tectonics) and the strong magnetic field produced by vigorous convection in the metallic core may both be due to the presence of water in Earth. The grain size in Earth may differ from one region to another by a few orders of magnitude leading to large variation in strength in different regions. Although these factors (water content and grain size) are poorly constrained, much progress is being made through the interdisciplinary studies among mineral physics and geophysical, geochemical observations to constrain these parameters and thereby improving our understanding of dynamics of Earth and planetary interiors (e.g., Karato *et al.*, 2001; Solomatov, 2001; Hall and Parmentier, 2002; Karato, 2003b, 2006b; Blum and Shen, 2004; Shito *et al.*, 2006).

18 Inference of rheological structure of Earth from time-dependent deformation

Deformation of Earth can occur with a range of time-scales including attenuation of seismic waves (at a time-scale of ~ 1–10^3 s), post-glacial crustal rebound ($\sim 10^{10}$–10^{11} s), and deformation associated with mantle convection ($\sim 10^{14}$–10^{15} s). Theoretical analyses of observations on these processes of time-dependent deformation provide us with clues as to the non-elastic properties of Earth's interior. However, the time-scales, and hence the strain (and stress) magnitude involved in these processes are different, and therefore microscopic mechanisms can be different among the deformation involved in these processes. A brief review is presented on the essence of these processes including the nature of deformation and the methods of studying them, in addition to some of the representative results.

Key words seismic wave attenuation, Q, post-glacial rebound, relative sea level, geoid, dynamic topography, isostasy.

18.1. Time-dependent deformation and rheology of Earth's interior

Observations on time-dependent deformation of Earth provide important constraints on the rheological properties of Earth's interior. Time-dependent deformation of Earth can be directly observed from seismic or geodetic measurements or indirectly from gravity measurements. These phenomena have different time-scales and different strain (stress) amplitude. Consequently, different physical processes may operate for different time-dependent deformation processes.

For seismic wave attenuation (or Chandler wobble),[1] deformation occurs at short time-scales (1–10^3 s for seismic waves, $\sim 10^7$ s for Chandler wobble), low stress level ($< 10^3$ Pa), so deformation is linear but strain involved is small ($\sim 10^{-6}$–10^{-8}).[2] The distribution of non-elastic properties for this time-scale is well determined mostly from seismological observations. The best-studied time-dependent deformation of Earth is the one due to the surface loading after the melting of ice sheets. The main data related to post-glacial rebound include the sea-level change recorded on the coastlines as well as the change in Earth rotation due to the change in the moment of inertia caused by the change in mass distribution. The time-scale of this phenomenon is 10^{10}–10^{11} s. Deformation of Earth also occurs due to the internal loading associated with mantle convection. At these time-scales ($\sim 10^{14}$–10^{15} s), deformation is nearly steady state. Although deformation is not time-dependent, constraints on rheological structures can be

[1] A small variation in Earth's rotation axis that occurs at the time-scale of ~400 days. The period of this wobble can be calculated from the elasticity of Earth that involves an anelastic component.

[2] Tidal deformation of planets is another important type of deformation that is associated with energy loss. The strain magnitude associated with tidal deformation strongly depends on the distance of two bodies and can be high, and deformation could be non-linear.

obtained because steady-state deformation associated with the flow of materials is sensitive to the rheological contrasts in certain regions of Earth. The nature of deformation at this time-scale is inferred mainly from the anomalies of gravity field.

The main purpose of this chapter is to provide a theoretical framework of these methods, assumptions involved and the nature of deformation associated with these phenomena and some representative results on rheological structures. A more detailed description of theory and observations can be found in KARATO and SPETZLER (1990) and ROMANOWICZ and DUREK (2000) for seismic wave attenuation, NAKADA and LAMBECK (1987), PELTIER (1989) and LAMBECK and JOHNSTON (1998) for post-glacial rebound and HAGER (1984), RICHARDS and HAGER (1984) and HAGER and CLAYTON (1989) for the inference of rheological structure from the gravity field. As we will learn, these methods are complementary: each has its own advantages and limitations. Understanding the basic framework of the models used to infer rheological structures and the nature of deformation (stress level and the strain magnitude) is critical in assessing the results from these studies.

18.2. Seismic wave attenuation

Basics

Deformation associated with the propagation of seismic waves and other similar processes (free oscillation, Chandler wobble, tidal deformation) occurs with rather slow time-scales compared to some of the typical laboratory measurements of elasticity. Most laboratory measurements of elasticity use either ultrasonic waves or the interaction of light (laser beam) with acoustic phonons that use a frequency range of $\sim 10^7$–10^{10} Hz, whereas the propagation of seismic waves occurs at a time-scale of ~ 1–10^{-3} Hz (for body wave, surface waves and free oscillation) and $\sim 10^{-7}$ Hz for Chandler wobble. In addition, most of Earth's interior is at a high temperature (i.e., more than $\sim 50\%$ of melting temperature). Consequently, even in these relatively short-time-scale deformation processes (as compared to deformation with geological time-scale), deformation involves a significant non-elastic component.

Seismic wave attenuation can be either measured by the amplitude decay of propagating waves or through the broadening of peaks of standing waves. Seismic wave attenuation is usually characterized by a Q factor defined as (see Chapters 3 and 11)

$$Q^{-1} \equiv \frac{\Delta E}{2\pi E} \tag{18.1}$$

where ΔE is the energy loss per unit cycle, and E is the energy stored in a material. Details of seismological methods of determining Q can be found in some standard textbooks (e.g., LAY and WALLACE, 1995; AKI and RICHARDS, 2002). In measuring seismic wave attenuation from propagating waves, one uses a general expression for the amplitude of attenuating waves,

$$X(x,t) = A_0 \exp(-ax)\exp\left[i\omega\left(t - \frac{x}{V}\right)\right] \tag{18.2}$$

where a is the attenuation coefficient, ω is the (angular) frequency, x is distance, t is time and V is the velocity of the elastic wave. Using the definition of Q, one can show that

$$Q^{-1} = \frac{2aV}{\omega} \tag{18.3}$$

hence,

$$X(x,t) = A_0 \exp\left(-\frac{\omega}{2VQ}x\right)\exp\left[i\omega\left(t - \frac{x}{V}\right)\right]. \tag{18.4}$$

It is seen that attenuation, Q^{-1}, can be determined either from the frequency dependence of amplitude at a given propagation distance, x, or by the variation of amplitude with distance for a given frequency. Note that, if Q (and V) changes with distance, this equation needs some modification (Problem 18.1). Attenuation can also be measured from the broadening of peaks of free oscillation.

The spatial resolution of measurements of seismic wave attenuation (both depth variation and lateral variation of seismic wave attenuation) is better than other methods of inferring rheological properties such as the inference from post-glacial rebound or from geoid anomalies (see the later part of this chapter). However, the rheological properties inferred from seismic wave attenuation are only indirectly related to long-term rheological properties.

Problem 18.1

Show that when attenuation and velocity of seismic waves change (weakly) on the position, x, then equation (18.4) can be generalized to $X(x,t) = A_0 \cdot \exp(-\omega \cdot t^*/2)\exp[i\omega(t - x/V)]$ with $t^* = \int(dx/V \cdot Q)$.

Solution

In the case where the attenuation coefficient, α, changes as a function of position, we should interpret equation (18.2) as a local equation where $a(x)$ is a weak function of position. Then the amplitude of seismic wave is given by $A = A_0 \exp[-a(x)\cdot x]$. Therefore $dA/A = -x\cdot da - a\cdot dx \approx -a\cdot dx$ where we used the approximation, $|da/dx|x \ll a$. Integrating this equation and using $Q^{-1} = 2aV/\omega$, one obtains $A = A_0\exp(-\int a(x)\,dx) = A_0\exp(-\frac{\omega}{2}\int(dx/V(x)\cdot Q(x))) \equiv A_0\exp(-\omega\cdot t^*/2)$ with $t^* \equiv \int(dx/V(x)\cdot Q(x))$.

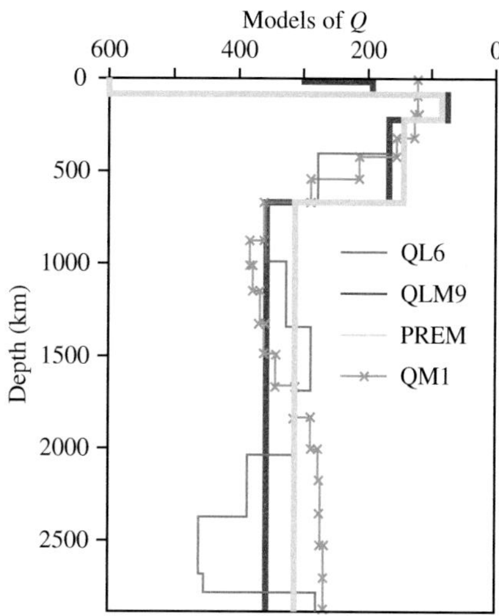

FIGURE 18.1 One-dimensional models of seismic wave attenuation in Earth's mantle (after LAWRENCE and WYSESSION, 2005).

Seismic wave attenuation and long-term rheology

Attenuation of seismic waves is sensitive to temperature and pressure (as well as other variables such as water content and grain size). Generally, the temperature and pressure dependence of seismic wave attenuation is complicated (see Chapter 11) but in most cases (high temperatures) its dependence on these variables can be expressed as,

$$Q^{-1}(T,P,C,L) \propto [\omega \cdot \tau(T,P,C,L)]^{-\alpha} \tag{18.5}$$

where T is temperature, P is pressure, C is some parameter characterizing chemical environment (e.g., water content), L is grain size, τ is the characteristic time of relaxation and α is a non-dimensional parameter with $\alpha = 0.3 \pm 0.1$ (see Chapter 11). In many cases, attenuation involves thermally activated processes, then $\tau(T,P,C,L) \propto \exp\left(H_Q^*(P)/RT\right)$, and

$$Q^{-1}(T,P,C,L) \propto \exp\left(-\frac{\alpha H_Q^*(P)}{RT}\right). \tag{18.6}$$

Therefore Q is a strong function of depth as well as geodynamic environment. Figure 18.1 shows a range of $Q(z)$ models where Q is assumed to change only with depth. Several points must be noted.

(i) Attenuation is small in the very shallow part (i.e., the lithosphere), whereas in a layer below the lithosphere, there is a layer in which attenuation is high, i.e., the asthenosphere.
(ii) Attenuation becomes small at deeper portions, but the increase in Q with depth is modest (Q in most of the lower mantle is ~300–400).
(iii) There is some suggestion that attenuation becomes somewhat higher (i.e., lower Q) toward the bottom of the lower mantle.

In almost all of the previous models of attenuation in Earth's mantle, Q is assumed to be independent of frequency. This is a common limitation of these models that is inconsistent with one of the most robust features of laboratory observations (see Chapter 11). This means that there is a systematic bias in the $Q(z)$ models such that the attenuation in the deeper portions tends to be over-estimated.

I should also mention that a one-dimensional model, $Q(z)$, is a gross simplification. In the real Earth, the lateral variation in Q can be as large as the depth variation (this is similar to the case of viscosity). There have been several attempts to determine the lateral variation in Q (e.g., FLANAGAN and WIENS, 1994; ROMANOWICZ, 1995; TSUMURA *et al.*, 2000; SHITO and SHIBUTANI, 2003a; GUNG and ROMANOWICZ, 2004; LAWRENCE and WYSESSION, 2006).

Note that to the extent that attenuation follows equation (18.6), attenuation and long-term creep are closely related. Long-term creep resistance (measured by effective viscosity that is defined as $\eta_{\mathrm{eff}} \equiv \sigma/2\dot{\varepsilon}$, see Chapters 3 and 19) depends on temperature and pressure as

$$\eta_{\mathrm{eff}}(T,P) \propto \exp\left(\frac{H^*_{\mathrm{creep}}}{RT}\right) \quad \text{(for constant stress)} \tag{18.7a}$$

and

$$\eta_{\mathrm{eff}}(T,P) \propto \exp\left(\frac{H^*_{\mathrm{creep}}}{nRT}\right) \quad \text{(for constant strain rate)} \tag{18.7b}$$

where H^*_{creep} is the activation enthalpy for creep and n is the stress exponent ($n = 1$–5). In many cases $H^*_Q = \xi H^*_{\text{creep}}$ (with $\xi = 0.8$–1.0). Consequently, there is a close connection between the spatial variation in attenuation and viscosity in these cases, namely,

$$\frac{\log[Q^{-1}(T,P)/Q_0^{-1}]}{\log[\eta(T,P)/\eta_0]} = \alpha n \xi \quad \text{(constant stress)} \tag{18.8a}$$

or

$$\frac{\log[Q^{-1}(T,P)/Q_0^{-1}]}{\log[\eta(T,P)/\eta_0]} = \alpha \xi \quad \text{(constant strain rate)} \tag{18.8b}$$

where Q_0^{-1} and η_0 are the seismic wave attenuation and effective viscosity at a reference state (infinite temperature).

As we will see later, the depth variation of Q summarized above (Fig. 18.1) correlates well with that of long-term viscosity (see Fig 18.9). Obviously, some differences are present in the microscopic processes of attenuation and long-term creep (see Chapters 8, 9 and 11) and consequently, relation (18.8) should be used with great caution.

Anderson and Given (1982) proposed that the frequency dependence of attenuation follows the power-law relation (18.5) in a certain frequency range, $\omega_m < \omega < \omega_M$, but $Q \propto \omega$ for $\omega > \omega_M$ and $Q \propto \omega^{-1}$ for $\omega < \omega_m$. They argued that seismic frequencies are inside the $\omega_m < \omega < \omega_M$. This is referred to as an *absorption band model* for seismic wave attenuation. Although the power-law relationship is very well established by the laboratory studies (see Chapter 11), and to a lesser extent by seismological observations (e.g., Anderson and Minster, 1979; Smith and Dahlen, 1981; Shito *et al.*, 2004), the frequency dependence of Q outside the absorption band is not well constrained either by experiments or seismological observations (there is a suggestion for $Q \propto \omega$ for $\omega > \omega_M$ by Sipkin and Jordan (1979), but later studies do not confirm this suggestion). I should also point out that in the absorption band model the energy dissipation would be negligible at the low-frequency limit, which is physically unreasonable. At the low-frequency limit ($\omega \to 0$), any solid material at high temperatures will behave like a viscous liquid and $Q \propto \omega$ as opposed to $Q \propto \omega^{-1}$.

18.3. Time-dependent deformation caused by a surface load: post-glacial isostatic crustal rebound

Slow crustal movement after the last glacial period has been one of the most important data sets to infer Earth's rheological structure. After the end of the last ice age, huge ice sheets that covered the polar regions melted rather quickly from $\sim 2 \times 10^4$ to $\sim 6 \times 10^3$ yr ago. The areas that were covered by the ice sheets tend to "rebound," but this rebound was not instantaneous as would have been the case if Earth behaved like an elastic body. In contrast, the crustal uplift (recorded by the change in the shore lines, i.e., the relative sea levels, RSL) has been occurring gradually with a characteristic time-scale of several 10^3 yr. This indicates that Earth behaves like a viscous body in this process. By analyzing this process through theoretical modeling, one can infer the viscosity of the Earth's mantle.

The physical conditions of viscous deformation associated with post-glacial rebound can be estimated as follows. The stress associated with this phenomenon is on the order of removed load, so that $\sigma \approx \Delta\rho g h \sim 60$ MPa ($\Delta\rho$, density difference between ice and the rock; h, the thickness of the ice sheet, $h \sim 3 \times 10^3$ m) just beneath the previous ice sheets, and $\sigma \approx \Delta\rho g h \sim 0.03$–$3$ MPa far from the previous ice sheet (in these areas the load is due to the change in sea level that is $h \sim 1$–10^2 m). The stress level for this phenomenon is similar to those for most tectonic processes (~ 0.1–100 MPa). However, strain associated with this phenomenon is on the order of vertical movement/horizontal scale $\sim 10^{-4}$–10^{-6}, which is much smaller than those for long-term tectonic processes and is similar to the elastic strain. Because of this small strain, issues of transient creep cannot be ignored in analyzing the post-glacial rebound (Karato, 1998c).

18.3.1. Basic theory

Let us consider a case where there is a thin layer (the asthenosphere) that has a significantly lower viscosity than other portions. In this case, viscous flow to relax stress occurs mostly in this thin channel.

Consider a two-dimensional model of relaxation corresponding to an initial topography of the surface (Fig. 18.2). Let us assume that the lithosphere is thin enough compared to the wavelength of the load so that

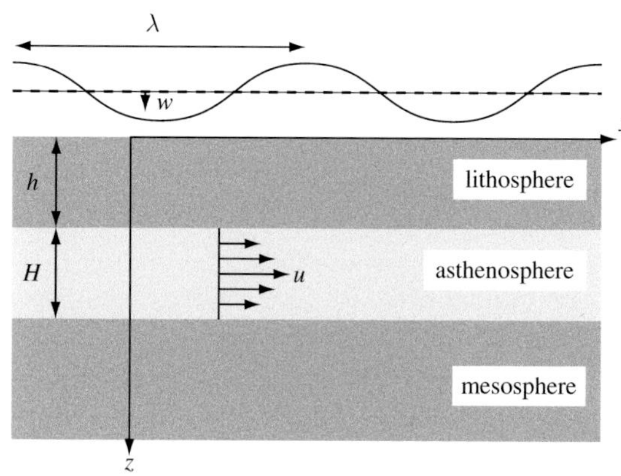

FIGURE 18.2 A schematic model for flow of materials associated with post-glacial rebound. Viscous flow occurs mostly in the low-viscosity layer (asthenosphere).

the stress caused by the lithosphere deformation is negligible compared to the pressure caused by the topography (a case where the elastic stress in the lithosphere is not negligible will be considered later). Under this assumption, the surface topography, $w(x, t)$, is directly related to the lateral variation in pressure $p(x, t)$,

$$p(x, t) = \rho g w(x, t). \tag{18.9}$$

This pressure gradient causes a viscous flow, $u(x, t)$, which is described by the Navier–Stokes equation,

$$0 = -\frac{\partial p}{\partial x} + \eta\left(\frac{\partial^2 u}{\partial x^2} + \frac{\partial^2 u}{\partial z^2}\right) \approx -\frac{\partial p}{\partial x} + \eta\frac{\partial^2 u}{\partial z^2} \tag{18.10}$$

where we assumed that $\partial^2 u/\partial x^2 \ll \partial^2 u/\partial z^2$ because $H \ll \lambda$ (H: thickness of a low viscosity layer ~100–200 km, λ: the size of ice sheet ~10^3–10^4 km). Solving, $-\partial p/\partial x + \eta(\partial^2 u/\partial z^2) = 0$, one obtains,

$$u = \frac{1}{2\eta}\frac{\partial p}{\partial x}\left[(z-h)^2 - H(z-h)\right] \tag{18.11}$$

where h is the thickness of the lithosphere.

Thus the mass flux q_x in the asthenosphere is given by,

$$q_x = \int_0^H u\, dz = -\frac{1}{12}\frac{\partial p}{\partial x}\frac{H^3}{\eta}. \tag{18.12}$$

Inserting equations (18.9) and (18.12) into the equation of continuity, $-\partial w/\partial t = \partial q_x/\partial x$, one obtains,

$$-\frac{\partial w}{\partial t} = -\frac{H^3}{12\eta}\frac{\partial^2 p}{\partial x^2} = -\frac{H^3 \rho g}{12\eta}\frac{\partial^2 w}{\partial x^2} \tag{18.13}$$

or

$$\frac{\partial w}{\partial t} = k\frac{\partial^2 w}{\partial x^2} \tag{18.14}$$

with

$$k = \frac{H^3 \rho g}{12\eta}. \tag{18.15}$$

Assuming $w \propto \exp(-t/\tau)\cos(2\pi x/\lambda)$, one finds,

$$\tau = \frac{3\lambda^2\eta}{\pi^2 \rho g H^3}. \tag{18.16}$$

From the observed relaxation time, τ, and the wavelength of the load, λ, one can estimate η/H^3. Note that $\tau \propto \lambda^2$ (a larger scale deflection relaxes more slowly than a small scale deflection) and that the viscosity can be estimated only when one knows the thickness of a low viscosity layer.

If viscosity does not change with depth, then $\lambda \ll H$ and the above analysis will not work. In such a case, one considers deformation of a half-infinite space loaded at the surface. There is no space-scale in this problem other than the length-scale of the load, λ. From the dimensional analysis, it can be shown that the characteristic time (relaxation time) in this case must be given by $\tau \propto \eta/\rho g \lambda$ (see Problem 18.2) and a more complete analysis shows (e.g., Turcotte and Schubert, 1982)

$$\tau = \frac{4\pi\eta}{\rho g \lambda}. \tag{18.17}$$

Note that the relaxation time is inversely proportional to the wavelength (compare this with equation (18.16)).

Problem 18.2

Using the dimensional analysis, show $\tau \propto \eta/\rho g \lambda$.

Solution

The characteristic time of isostatic adjustment must be related to the gravity force that is proportional to ρg, viscosity, η, and the wavelength of topography, λ. Therefore we assume that the relaxation time is given by $\tau \propto (\rho g)^\alpha \eta^\beta \lambda^\gamma$. The dimension of the left- and the right-hand sides of this relation must be the same, i.e., $[\tau] = [\rho g]^\alpha [\eta]^\beta [\lambda]^\gamma$, where $[X]$ is the dimension of the quantity X. Consequently, $1 = -2\alpha - \beta, 0 = \alpha + \beta$ and $0 = -2\alpha - \beta + \lambda$. Therefore $\alpha = -1$, $\beta = 1$ and $\gamma = -1$ and hence $\tau \propto \eta/\rho g \lambda$.

18.3.2. Effects of elastic lithosphere

In the previous analysis, we ignored the stress in the lithosphere. If the lithosphere is thick, then the deformation of the lithosphere due to surface load causes significant pressure variation. Consider the effects of a thin elastic layer on post-glacial rebound. We examine its effects using the theory of bending of a thin layer. The moment balance of an elastic thin layer reads (see e.g., Turcotte and Schubert, 1982),

$$D\frac{\partial^4 w}{\partial x^4} = f \tag{18.18}$$

where $D = Eh^3/12(1-\nu^2)$ is the flexural rigidity (E, Young's modulus; h, thickness of the lithosphere; ν, Poisson's ratio) and f is the vertical force per unit area acting on the plate. Thus

$$p = \rho g w + f = \rho g w + D\frac{\partial^4 w}{\partial x^4}. \tag{18.19}$$

Consequently, if we consider a thin channel model of flow of matter associated with a surface load, upon substituting (18.19) into (18.13), we get

$$\frac{\partial w}{\partial t} = k\frac{\partial^2 w}{\partial x^2} + \frac{k}{\rho g}\frac{\partial^2 f}{\partial x^2} = k\frac{\partial^2 w}{\partial x^2} + \frac{kD}{\rho g}\frac{\partial^6 w}{\partial x^6}. \tag{18.20}$$

Inserting $w \propto \exp(-t/\tau)\cos(2\pi x/\lambda)$,

$$\tau = \frac{(\lambda/2\pi)^2(1/k)}{1 + (D/\rho g)(2\pi/\lambda)^4}. \tag{18.21}$$

For $\lambda \gg \lambda_c \equiv 2\pi(D/\rho g)^{1/4}$, the effect of lithosphere is negligible and the result agrees with (18.21), but for $\lambda < \lambda_c$, much of the load is supported by elastic deformation of lithosphere and for $\lambda \ll \lambda_c$, $\tau \to \frac{9(1-\nu^2)}{4\pi^6}(\eta/E)(\lambda^6/H^3h^3)$.

Problem 18.3

Consider two types of loading; (1) loading by 100 km wide mountain range and (2) loading by 3000 km wide ice sheet in a region of lithosphere with thickness of $h = 50$ km, Young's modulus $E = 80$ GPa (the Poisson ratio is $\nu = 0.3$). Discuss how the load is supported in these two cases.

Solution

From $\lambda_c \equiv 2\pi(D/\rho g)^{1/4}$ the critical length for the elastic support in this case is $\lambda_c \sim 160$ km. Therefore a mountain range is largely supported by the elastic deformation of the lithosphere whereas the ice sheet will be supported by the fluid pressure or viscous stress.

Problem 18.4

Assume that there was a surface depression due to the sudden melting of an ice sheet (with a disk shape). Ignoring the effect of lithosphere, sketch how the surface topography will change with time for a thin channel model (see equation (18.16)) and for a depth-independent viscosity model (equation (18.17)).

Solution

For a thin channel model (equation (18.16)), the relaxation time is larger for longer wavelength topography. Therefore with time, topography will become smoother (Fig. 18.3a). In contrast, for a homogeneous viscosity model, the relaxation time is longer for short wavelength topography. Consequently, topography will become more rugged with time (Fig. 18.3b) which is not consistent with the observations. The wavelength dependence of relaxation times can provide a constraint on the depth variation on rheology including the depth variation of viscosity and the thickness of the lithosphere.

18.3.3. Importance of non-linear rheology

Most of the theoretical models of deformation associated with post-glacial rebound assume linear (Newtonian) rheology. This choice is mostly for mathematical simplicity, but in most of the laboratory deformation experiments, stress–strain-rate relations are non-linear. The non-linear rheology is a result of deformation by the motion of crystal dislocations

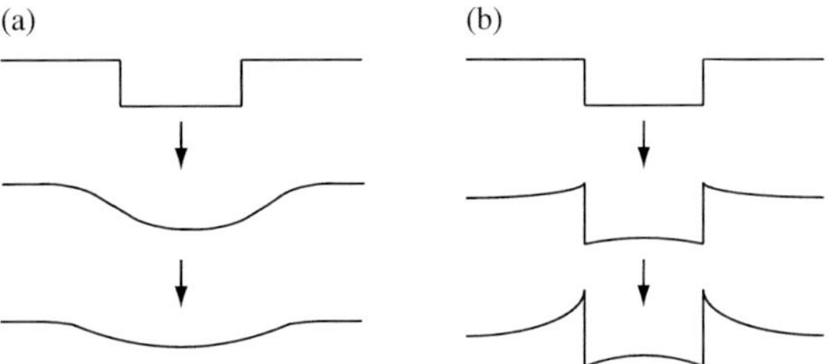

FIGURE 18.3 A schematic diagram showing the evolution of surface topography after the melting of an ice sheet corresponding to (a) a thin channel model and (b) a depth-independent viscosity model (the effect of elastic lithosphere is ignored).

(dislocation creep, see Chapter 9). Evidence for non-linear dislocation creep is strong for shallow upper mantle that includes the microstructures of deformed rocks from the upper mantle (Chapter 19) and the presence of strong seismic anisotropy (Chapter 21). However, based on laboratory studies of the deformation of rocks with a wide range of grain size, distribution of seismic anisotropy and also the analysis of post-glacial crustal uplift data, KARATO and WU (1993) concluded that linear rheology (diffusion creep) may dominate in the deep upper mantle. The observations on seismic anisotropy and fabric development in the lower mantle minerals also suggest that materials in the lower mantle may deform by diffusional creep (KARATO *et al.*, 1995b) except for the near D″regions (KARATO, 1998d).

It has often been argued (see e.g., TURCOTTE and SCHUBERT, 1982, p. 326; SCHMELING, 1987) that Earth materials may show apparent linear rheology for post-glacial rebound although long-term rheology for convection is non-linear. Discussions are as follows (TURCOTTE and SCHUBERT, 1982). Assume a non-linear rheology,

$$\dot{\varepsilon} = A\sigma^n. \tag{18.22}$$

Consider that stress has two components,

$$\sigma = \sigma_c + \sigma_g \tag{18.23}$$

where σ_c is convection-related stress and σ_g is glacial rebound-related stress. Thus,

$$\dot{\varepsilon} = A(\sigma_c + \sigma_g)^n \approx A\sigma_c^n\left(1 + n\frac{\sigma_g}{\sigma_c}\right) \approx \dot{\varepsilon}_c + \dot{\varepsilon}_g \tag{18.24}$$

where $\dot{\varepsilon}_c = A\sigma_c^n$ and $\dot{\varepsilon}_g = nA\sigma_c^{n-1}\sigma_g$. Therefore by defining $\eta_c \equiv \sigma_c/\dot{\varepsilon}_c$ and $\eta_g \equiv \sigma_g/\dot{\varepsilon}_g$, one gets

$$\frac{\eta_g}{\eta_c} = \frac{1}{n}. \tag{18.25}$$

(SCHMELING (1987) made a more sophisticated analysis using the Levy–von Mises formulation of non-linear rheology (see Chapter 3)). In this analysis, rheology for glacial rebound is linear because of the assumption, $\sigma_g \ll \sigma_c$. This is not true in most cases. The stress magnitude for convection is roughly the same as those for glacial rebound at least in regions near the paleo-ice sheets ($\sigma_g \approx \rho g w \sim 1$ MPa and $\sigma_c \approx \eta\dot{\varepsilon} \sim 1$ MPa). Furthermore, mathematical treatment such as the Taylor series expansion of non-linear stress terms may not be justified on the microscopic physical basis. The non-linearity comes mainly from the dependence of dislocation density on stress ($\dot{\varepsilon} = \rho b v$, with $\rho \approx b^{-2}(\sigma/\mu)^2$, v is the dislocation velocity, see Chapter 9). The Taylor expansion of this term therefore corresponds (physically) to a change in dislocation density. However, at the strain levels such as those associated with the post-glacial rebound, no appreciable dislocation multiplication occurs and hence the validity of such a mathematical treatment is dubious. In an end-member case where dislocation density is unchanged, then if dislocation velocity is a linear function of stress (see Chapter 9), then $\eta_g/\eta_c \approx 1$.

Problem 18.5*

A well-documented mechanism by which dislocation density changes is the multiplication through the Frank–Read source (see Chapter 9). In order for this mechanism of dislocation multiplication to work, a dislocation segment must move a distance that is significantly larger than the average distance of dislocations. Using the dislocation density versus stress relation, $\rho \approx b^{-2}(\sigma/\mu)^2$ (equation (5.74), ρ, dislocation density; b, the length of Burgers vector; μ, shear modulus; Chapter 5) and the Orowan relationship ($\dot{\varepsilon} = \rho b v$, v: dislocation velocity, Chapter 9), show that in order to change dislocation density strain must significantly exceed $\varepsilon_c \approx \sigma/\mu$.

Solution

Integrating the Orowan equation (9.3) with time, one gets $\varepsilon = \rho b l$ where l is the distance of dislocation motion. Now, the dislocation density is related to the mean dislocation distance, l', as $\rho \approx 1/l'^2$. Therefore, $l/l' \approx \sigma e/b\mu$. Inserting this into $\varepsilon = \rho b l$ with $l = l'$, one gets $\varepsilon_c \approx \sigma/\mu = \varepsilon_e$ where ε_e is the elastic strain. This equation means that in order for appreciable dislocation multiplication to occur, strain magnitude must significantly exceed the elastic strain.

18.3.4. Importance of transient creep

In post-glacial rebound, the initial elastic strain caused by the melting of ice sheets is slowly relaxed by plastic flow. Consequently, the strain magnitude is initially that of elastic strain corresponding to the initial load and the magnitude of plastic strain involved in this phenomenon is only a fraction of this elastic strain and is small ($\sim 10^{-4}$

in near the former ice sheets and $\sim 10^{-6}$ far from the melted ice sheets). Since all solids show some transient creep behavior at small strain level, transient creep may not be ignored. WEERTMAN (1978) PELTIER (1985a), SABADINI *et al.* (1987), and SMITH and CARPENTER (1987) discussed the role of transient creep in post-glacial rebound. However, the formulation of transient creep in these studies is not adequate, and furthermore the quality of the data they used was not high enough to reach clear conclusions as to the contribution of transient creep.

The analyses of transient creep presented in Chapters 8 and 9 indicate that in both diffusional and dislocation creep, significant transient creep occurs when applied stress is modified until steady-state internal stress distribution and/or microstructure is formed. The nature of transient creep is somewhat different between diffusional creep and dislocation creep, but the constitutive relation of transient creep behavior in both cases can generally be written as (KARATO, 1998c),

$$\dot{\varepsilon} = (\dot{\varepsilon}_0 - \dot{\varepsilon}_s)\exp\left(-\frac{t}{\tau}\right) + \dot{\varepsilon}_s \tag{18.26}$$

where $\dot{\varepsilon}_s$ is steady-state strain rate, $\dot{\varepsilon}_0$ is the initial strain rate and $\tau \approx \varepsilon_e/\dot{\varepsilon}_s$ is the characteristic time. The detailed studies on transient creep are missing but recent studies on micro-creep (e.g., JACKSON, 2000) provide some insight into micro-strain transient creep. These results suggest that the viscosity inferred from post-glacial rebound provides a lower bound for steady-state viscosity and the difference between the apparent viscosity inferred from post-glacial rebound and long-term ("steady-state") viscosity can be as much as a factor of ~ 10 (e.g., KARATO, 1998c).

18.3.5. Observations and some results

In order to determine the rheological properties from the time-dependent crustal motion after the melting of ice sheets, one needs to know (1) the volume of molten ice sheets (history of the ice sheet melting) and (2) the vertical crustal movement after the ice sheet melting. Actual observations on post-glacial rebound are the relative sea-level (RSL) changes that can be observed along the coastlines. The contribution to RSL can be classified into three sources,

$$\mathrm{RSL}(\theta, \lambda, t) = \varsigma_R(\theta, \lambda, t) + Z_1(\theta, \lambda, t) + Z_2(\theta, \lambda, t) \tag{18.27}$$

where θ and λ are latitude and longitude respectively, ς_R is RSL corresponding to a rigid Earth, Z_1 is the contribution from deformation of crust caused by the change in load in land due to melting of ice sheets and Z_2 is from deformation of crust caused by the change in load in the sea due to the melting of ice sheets (hydro-isostasy). The first effect reflects the history of ice sheet melting which is only imperfectly known. The third effect is sensitive to the geometry of coastline as well as the distance of the observational point from the ice sheet. Therefore it is important to separate (i) the uncertainties in the history of ice sheet melting and (ii) the effects of coastline to infer the rheology of the mantle from RSL data.

Due to the combination of these three effects, the RSL versus time curve depends on the location (Fig. 18.4). In the near-field, the first and the second effects dominate and hence the RSL falls with time due to crustal uplift. Note that the rate of crustal uplift in

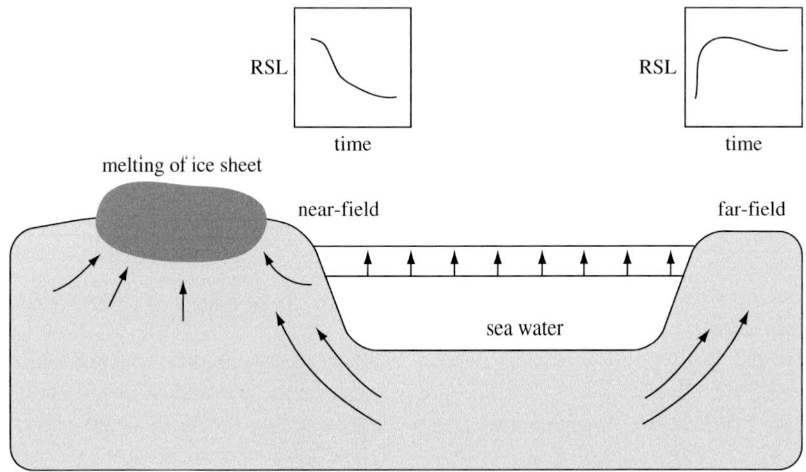

FIGURE 18.4 A cartoon showing the nature of the relative sea level (RSL) after the melting of an ice sheet. Note the difference between the near-field and far-field.

this area is highly sensitive to the history of melting of the ice sheet. In particular, the RSL curves in the periphery of previous ice sheets (e.g. Boston bay) are extremely sensitive to the exact geometry of ice sheets that is poorly constrained. Therefore, strong constraints on mantle rheology cannot be obtained from such data. Instead, these data can be used to place constraints on the history of ice sheet melting. The RSL curves from the central regions of previous ice sheets (e.g., Hudson bay) are less sensitive to the exact geometry of ice sheets, although the data are sensitive to the volume of ice sheets. These data can be used to place constraints on mantle rheology. The third effects dominate the RSL versus time curves in the far-field. In this case, up to $\sim 6 \times 10^3$ yr ago when the melting of the ice sheets was completed, the RSL curve is dominated by the effects of increase in seawater (*eustatic* sea level change), and after $\sim 6 \times 10^3$ yr ago, it is dominated by the delayed response of the solid Earth to seawater loading (hydro-isostasy).

The RSL curves in these regions (far from the ice sheet) are insensitive to the details of change in the geometry of ice sheets and therefore they provide useful constraints on the rheology of the mantle. The effects of uncertainties in melting history can further be minimized by considering the difference in RSL in two sites in the same region (NAKADA and LAMBECK, 1989). To understand this, we must understand the effects of the coastline geometry on RSL. Since the RSL in the far-field is dominated by the hydro-isostasy, the geometry of the coastline can play a significant role (NAKADA, 1986; NAKADA and LAMBECK, 1987, 1989). Consider two islands of different sizes. For a small island (small in comparison to the thickness of the lithosphere, asthenosphere), the change in load due to the change in the volume of seawater is nearly homogeneous and no mass flow will occur between the ocean and the island. Such an island acts as a passive marker of sea-level change and the RSL will be sensitive to the integrated viscosity of the mantle. In contrast, for a large island, increase in seawater volume causes mass flow from the ocean side to the island, causing crustal uplift and the RSL from these regions is sensitive mainly to the upper mantle viscosity. A similar argument applies to the RSL at a straight coastline and the RSL at a deep bay area. If one compares RSL records of these points in a similar area, then one can differentiate between the viscosity of deep mantle and that of the shallow (upper) mantle.

Time-dependent mass distribution associated with post-glacial rebound results in a time-dependence of the moment of inertia of Earth that can be detected by the time dependence of the gravity field and the resultant Earth rotation (PELTIER, 1985b; YUEN *et al.*, 1982). These data are particularly sensitive to the viscosity of the deep mantle that occupies a large fraction of Earth's mass.

Analysis of *RSL* provides very tight constraints as to the absolute value of "average" viscosity, $\langle\eta\rangle = 3 \times 10^{21}$ Pa s (the Haskell value, (HASKELL, 1937; MITROVICA, 1996)). However, the depth variation of viscosity is not well constrained. For the upper mantle, most studies show $\eta \sim (1\text{–}10) \times 10^{20}$ Pa s, with large regional variations. For the lower mantle, $\eta \sim 10^{22}$ Pa s (NAKADA and LAMBECK, 1989), but no constraints are obtained below ~1200 km from post-glacial rebound data (MITROVICA and PELTIER, 1991a, 1991b). Rotation data provide some constraints on the lower mantle viscosity, $\eta \sim 10^{22}$ Pa s (e.g., YUEN *et al.*, 1982; PELTIER, 1985b).

The remaining problems are given below.

1. Lateral variation in viscosity has not been well resolved although it is expected to be large (at least one order of magnitude) (e.g., NAKADA and LAMBECK, 1991).
2. Any fine viscosity–depth structures are hard to resolve from these studies, although complicated rheological stratification can be predicted from mineral physics studies (Chapter 19).
3. All of these (ice sheet melting-related) time-dependent deformation processes involve very small strains. Plastic properties of polycrystalline solids are usually dependent on strain magnitude. At small strains, transient creep as opposed to steady-state creep usually occurs. Therefore the viscosity that is inferred from these analyses may reflect the viscosity for transient creep, whereas the viscosity relevant to mantle convection is that for steady-state creep. The difference between them can be significant for some mechanisms of plastic flow.

18.4. Time-dependent deformation caused by an internal load and its gravitational signature

18.4.1. Basic theory

Excess mass *inside* Earth causes anomalies in the gravity field and the nature of the gravity field caused by the excess mass is sensitive to the mechanism by which excess mass is supported. Consequently measurements of the gravity field combined with the inference of excess mass distribution provide constraints on the mechanisms of support of excess mass and hence the rheological structure of Earth's interior.

The gravity between a point mass M and a unit mass is given by Newton's law,

$$g = -\frac{GM}{r^2} e_r \tag{18.28}$$

where G is the gravity constant ($G = 6.67 \times 10^{-11}\,\mathrm{m}^3\,\mathrm{kg}^{-1}\mathrm{s}^{-2}$), r is the distance between the two masses and e_r is the unit vector along the direction of r. Most of the important relationships on gravity can be derived from this law. The energy is conserved during the motion of matter by gravity. Consequently, one can define the *gravity potential*, U, from which the gravity force is derived[3]

$$g = \nabla U. \tag{18.29}$$

From (18.28) and (18.29), the potential for a point mass is given by

$$U = \frac{GM}{r}. \tag{18.30}$$

With this definition, the gravity potential is high toward the mass. When the mass distribution is modified, then the gravity potential will also be modified. The change in the equi-potential surface (*geoid*), δN, associated with mass anomaly, δM, can be calculated from (18.30) as $\Delta U = 0 = G\delta M/r - GM_0\delta N/r^2 = \delta U + g_r(r)\,\delta N$ ($g_r(r) = -GM_0/r^2$, $\delta U = G\delta M/r$: change in gravity potential due to the mass anomaly) and hence

$$\delta N = -\frac{\delta U}{g_r(r)} = \frac{r\delta M}{M_0}. \tag{18.31}$$

This means that the excess mass *increases* the height of the *geoid* (Fig. 18.5). However, it is important to note that actual anomalies in the gravity field associated with excess mass are caused not only by the presence of anomalous mass itself but also by the distortion of density boundaries due to the presence of mass anomalies (Hager, 1984; Richards and Hager, 1984). The nature of distortion of density boundaries depends on the way in which the mass anomaly is supported in Earth and hence is sensitive to the rheological structure. Therefore the gravity observations provide some constraints on the rheological structure of Earth's interior. Let us consider three cases.

[3] This sign convention is common in geophysical literature, but is opposite to a more common convention where potential energy, Φ, associated with a force f is $f = -\nabla\Phi$.

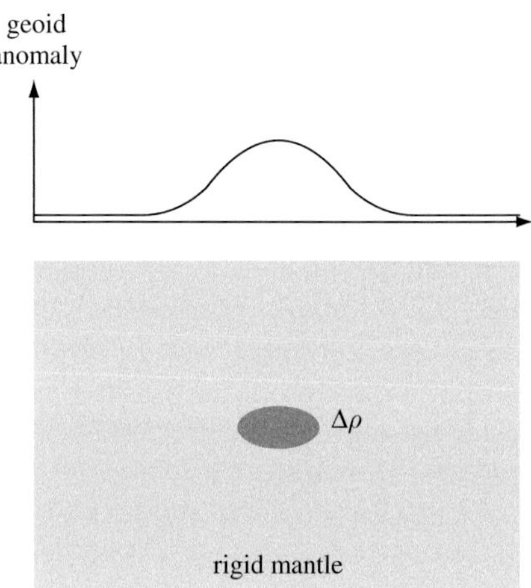

FIGURE 18.5 The excess mass and the geoid.

Elastic (or rigid body) support

When the size of the mass anomaly is small compared to the thickness of elastic lithosphere, then it will be supported by elastic distortion (see equation (18.21)). In this case, distortion is small and the geoid anomaly is caused mainly by the presence of a mass anomaly and the effect of deformation of the boundary is negligible. Therefore, a positive geoid anomaly is associated with the presence of an excess mass.

Static equilibrium (isostasy)

If the anomalous mass (density) is in a stratified fluid layer with different densities, then a static equilibrium is possible in which an anomalous mass is located at the boundary between two fluids, if the vertical position of the anomalous mass is such that the pressure at the bottom of the anomalous mass is the same as the pressure of the ambient fluid (Fig. 18.6). Such an equilibrium state is referred to as *isostatic equilibrium*. If we use a one-dimensional model (sheet mass anomaly), the condition for isostasy reads,

$$\int_{r_1}^{r_0} \delta\rho(y)\,dy = 0 \tag{18.32}$$

where $\delta\rho(y)$ is the density anomaly, r_1 is the distance of the bottom of the density anomaly from Earth's center and r_0 is the distance to the top of the density anomaly. Although the integrated mass anomaly for a column of materials is zero for this structure, there is a finite geoid anomaly. This is due to the fact that the effect of the mass anomaly on the gravity potential depends on the density anomaly as well as the distance between

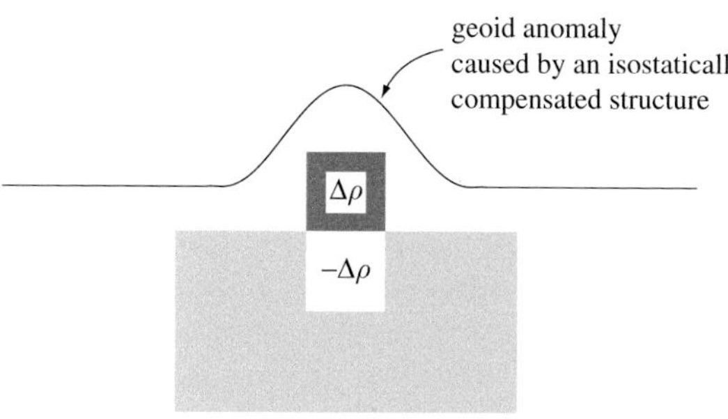

FIGURE 18.6 The mass distribution at a density boundary with isostatic equilibrium.

the mass anomaly and the observational point. A mass anomaly closer to the observation point (surface) makes a more important contribution to the geoid.

For more details, let us consider the gravity field caused by a sheet of mass. From the relation (18.28), it can be shown that the anomaly in the gravity field, $\delta g_r(r)$ (negative toward the center of Earth), due to the two-dimensional anomaly of density is given by (Problem 18.6)

$$\delta g_r(r) = -2\pi G \int_{r_0}^{r} \delta\rho(x)\, dx. \tag{18.33}$$

Problem 18.6

Derive equation (18.33) from (18.28).

Solution

Consider the gravity at a distance r from a thin layer of thickness dy with density anomaly $\delta\rho$. Using (18.28), the gravity due to the ring with a radius x to $x+dx$ is $-(2\pi Gx\, dx\, dy\, \delta\rho/(x^2+r^2)) \cdot (r/\sqrt{x^2+r^2})$. Integrating this with respect to x, the gravity anomaly due to this mass anomaly is $dg_r = -2\pi G\delta\rho(y)\, dy$ $\int_0^\infty (xr/(x^2+r^2)^{3/2})\, dx = -2\pi G\delta\rho(y)\, dy$. Integrating this with respect to y, one obtains $\delta g_r(r) = -2\pi G \int_{r_0}^{r} \delta\rho(x) dx$.

Using (18.29), the anomaly in gravity potential at the surface ($r = r_a$) is

$$\delta U(r_a) = -2\pi G \int_{r_0}^{r_a} \left(\int_{r_0}^{r} \delta\rho(x)\, dx \right) dr \tag{18.34}$$

where r_0 is the bottom of the mass anomaly. From equations (18.32) and (18.33), it is obvious that for a density distribution satisfying the isostasy (i.e., equation (18.32)), gravity anomaly is zero. However, there is a finite anomaly of the gravity potential and hence of geoid for an isostatically compensated structure. To see this point, we integrate equation (18.34) to get (Problem 18.7),

$$\delta U(r_a) = 2\pi G \int_{r_0}^{r_a} y \cdot \delta\rho(y)\, dy. \tag{18.35}$$

Using equation (18.31), the geoid anomaly δN is given by,

$$\delta N = -\frac{2\pi G}{g_r} \int_{r_0}^{r_a} y \cdot \delta\rho(y)\, dy. \tag{18.36}$$

Problem 18.7*

Derive equations (18.35) and (18.36) and show that when a positive mass anomaly is supported at a density boundary isostatically, then one would expect a positive geoid anomaly associated with a positive mass anomaly.

Solution

Applying (18.29) for a one-dimensional case, one gets $\delta U(r_a) = \int_{r_0}^{r_a} \delta g_r(y)\, dy$. Integrating this equation by parts, $\delta U(r_a) = (r_a\, \delta g_r(r_a) - r_0\, \delta g_r(r_0)) - \int_{r_0}^{r_a} (d\delta g_r(r_a)/dy) y \cdot dy$. Using the condition for isostatic equilibrium (e.g., (18.32)) and the relation $d\delta g_r/dy = -2\pi G\, \delta\rho(y)$ (from (18.33)), one gets $\delta U(r_a) = 2\pi G \int_{r_0}^{r_a} y \cdot \delta\rho(y)\, dy$. Using (18.32), $\delta N = -(2\pi G/g_r) \int_{r_0}^{r_a} y \cdot \delta\rho(y)\, dy$.

When a positive mass anomaly is supported at a density boundary (see Fig. 18.6), the higher density portion must be located in a less dense, upper layer. Let us consider a simple case where a density anomaly $\delta\rho$ occurs at the depth range r_b to $r_b + \xi$ and $-\delta\rho$ occurs at r_b to $r_b - \xi$. Then $\delta N = -(2\pi G/g_r) \int_{r_0}^{r_a} y \cdot \delta\rho \cdot dy$ $= -4\pi G\xi\delta\rho/g_r = 2\xi \cdot \delta\rho/\sigma > 0$ where we used a relation $g_r = -2\pi G \int d\rho = -2\pi G\sigma$ with σ, mass per unit area.

Dynamic topography

In a dynamic Earth, excess mass or mass deficit may be supported by the stress associated with fluid motion. The flow of matter caused by density anomalies will exert stress and hence deflect density discontinuities and results in additional geoid anomalies. The deflection of density discontinuities due to dynamic flow is referred to as *dynamic topography*. The effect of dynamic topography on gravity is large. For example, the surface will be depressed in the region of a down-welling current associated with excess mass (subduction zones) as much as an order of 1–10 km. If the mantle has constant viscosity, the effect of a surface depression will be larger than the effect of excess dense mass, so that one would see a *negative* gravity anomaly (*negative* geoid anomalies) over subduction zones. In reality, positive geoid anomalies are observed over the subduction zone. This means that the dynamic topography is less than expected for the simplest case of homogeneous viscosity. An increase in viscosity with depth is implied by this observation as we will learn in this section.

The dynamic topography is maintained by convective flow. Consequently, the rheological properties inferred from this observation must be identical to those for long-term deformation. The rheological properties inferred from this type of observation therefore do not have problems due to the contribution from transient creep that is an important issue in the application of results from post-glacial rebound.

In principle, the dynamic topography and related rheological stratification could be investigated through a study of the distortion of density boundaries. However, the resolution of such measurements is limited. Consequently, in most cases, geoid anomalies and density anomalies are used as data and we infer rheological stratification through the search for the rheological structure that best reproduces the observed geoid (or other gravity-related observations).

Because all the density anomalies, including those due to the distortion of density boundaries, are ultimately caused by the density anomalies associated with convecting materials, we assume that the density anomalies caused by the distortion of density boundaries (dynamic topography) are also proportional to the density anomalies of convecting materials. Consequently we extend the relations (18.36) (or (18.35)) to,

$$\delta U(r_a) = 2\pi G r_a \int_{r_0}^{r_a} K(y;\eta(y)) \cdot \delta\rho(y) \cdot dy \tag{18.37}$$

and

$$\delta N(r_a) = -\frac{2\pi G r_a}{g_r} \int_{r_0}^{r_a} K(y;\eta(y)) \cdot \delta\rho(y) \cdot dy \tag{18.38}$$

where $\delta\rho(r)$ is density anomaly, $K(r;\, \eta(r))$ is a non-dimensional function called the *geoid kernel* that depends on the rheological structure, $\eta(r)$, flow geometry and the density contrasts at the boundaries.

To see how $K(r;\, \eta(r))$ is related to the rheological structure, let us consider a simple model. The cause of dynamic topography is the distortion of a density boundary due to viscous stress. The stress associated with the deflection of a density boundary is $\approx (\delta\rho)_b g w$ ($(\delta\rho)_b$: density jump at the boundary, w: the topography on the density boundary). The stress associated with viscous flow is $\approx \eta_b(2\pi u/\lambda)$ where η_b is the viscosity of materials near the boundary, u is the flow velocity and λ is the space-scale at which flow pattern changes. Therefore at dynamic equilibrium, $w \approx (2\pi\eta_b u/(\delta\rho)_b g\lambda)$. Now if one uses the Stokes formula for homogeneous viscosity (i.e., $\eta_0 \Delta u - \rho_0 g = 0$), then $u \approx (\lambda^2(\delta\rho)_0 g/4\pi^2\eta_0)$ ($(\delta\rho)_0$: density contrast driving convection, η_0: average viscosity) and

$$(\delta\rho)_b w \approx (\delta\rho)_0 \frac{\eta_b}{\eta_0}\frac{\lambda}{2\pi}. \tag{18.39}$$

Note that the dynamic topography can be large. For $(\delta\rho)_0/(\delta\rho)_b \sim 10^{-3}$–$10^{-2}$, $\eta_b = \eta_0$ and $\lambda \sim 10^3$ km, the amplitude of topography is on the order of $w \sim 1$–10 km. It must also be noted that *the dynamic topography and hence the geoid kernel does not depend on the absolute value of viscosity but depends on the viscosity contrast (radial variation in viscosity).* Therefore knowing $(\delta\rho)_0$ and λ, one can calculate the geoid kernel for various η_b/η_0 to find a best-fit model. Since the most important contribution to the geoid is from the surface topography, η_b effectively corresponds to the viscosity of upper mantle and η_0 to that of average mantle that is dominated by deep mantle viscosity. For a more complete description, see HAGER (1984), RICHARDS and HAGER (1984) and HAGER and CLAYTON (1989).

Fig. 18.7 illustrates how different rheological structures cause different dynamic topography and hence different geoid anomalies at subduction zones. (a) If the viscosity is constant, then at both surface and internal density boundaries, the deflection of the boundary is given by $w \approx [(\delta\rho)_0/(\delta\rho)_b]\,(\lambda/2\pi)$. Although the topography of each boundary depends on the density contrasts at boundaries, $(\delta\rho)_b$, the integrated density anomaly from each boundary is the same and given by $(\delta\rho)_b w \approx (\delta\rho)_0(\lambda/2\pi)$. Consequently, the geoid anomaly caused by topography is dominated by that of the surface and is negative. The total geoid anomaly is the sum of this negative anomaly and the positive anomaly due directly to the presence of a positive mass anomaly of subducting slabs. Since surface topography has a strong effect, the net anomaly will be negative. (b) If the deep mantle has a higher viscosity, then the flow velocity is much smaller. Consequently the viscous drag by flow will now be smaller and the resultant

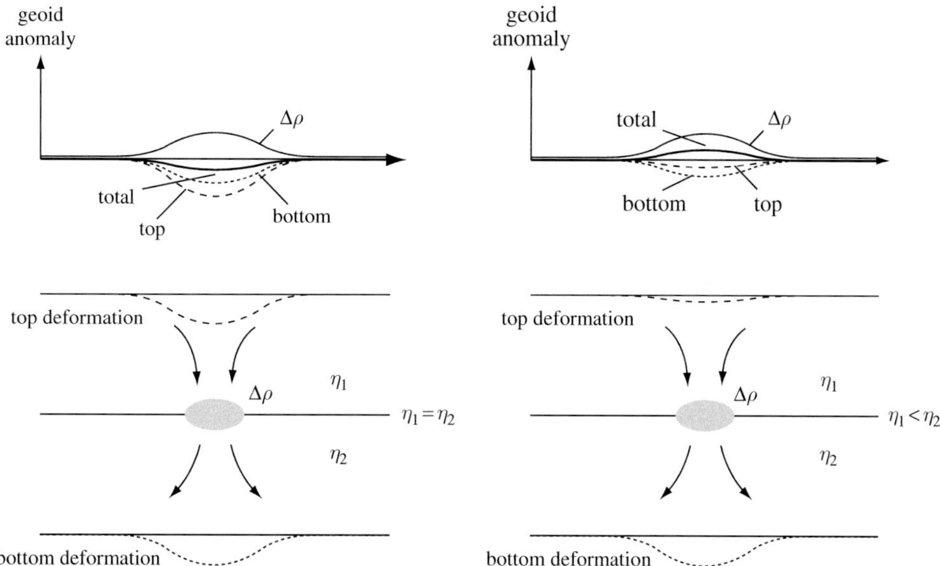

FIGURE 18.7 Schematic diagrams showing the deformation of density boundaries caused by steady-state flow corresponding to: (a) a case of constant viscosity and (b) a case where viscosity of the deep layer is much higher than that of a shallow layer (after HAGER, 1984).

dynamic topography will be less and the magnitude of the negative geoid anomaly due to topography will be small. As a result, the total geoid anomaly is dominated by the positive anomaly caused directly by the positive mass anomaly associated with subducting slabs.

In summary, geoid anomaly does not depend on the absolute values of viscosity, but it depends on the viscosity contrast. This is in marked contrast to the viscosity inferred from post-glacial rebound in which the data depend on the magnitude of viscosity but only weakly on the viscosity contrast.

One of the important input parameters for this exercise is the density anomaly, $\delta\rho$. $\delta\rho$ can be estimated from tomographic images of velocity anomalies using a conversion factor as

$$\frac{\delta\rho}{\rho} = \frac{\partial \log \rho}{\partial \log V_{P,S}} \frac{\delta V_{P,S}}{V_{P,S}} \tag{18.40}$$

where, $\delta V_{P,S}/V_{P,S}$ is velocity anomaly. Values of $\partial \log \rho / \partial \log V_{P,S}$ are critical in this exercise and the values of $\frac{\partial \log \rho}{\partial \log V_{P,S}}$ corresponding to temperature anomalies can be calculated from mineral physics (e.g., KARATO, 1993a; KARATO and KARKI, 2001, see Chapter 20). However, chemical heterogeneity and/or phase transformations can significantly modify these values.

Key issues in the use of geoids to infer viscosity include (i) the resolution of the seismic tomography model to infer $\delta\rho$ and (ii) the velocity to density conversion factor. The results also depend on the assumed convection pattern (whole mantle or layered). Recognizing the first point, KIDO and CADEK (1997) focused on oceanic regions where the effects of chemical heterogeneity are likely to be small. They found that a large class of models fit the data equally well including models with both a low- and a high-viscosity transition zone. Uncertainties are particularly large in the upper mantle where the largest velocity heterogeneity occurs. This is due to the effects of chemical heterogeneity. Uncertainties in density estimates associated with subducting slabs and those in the transition zone and the lower mantle are also large.

18.4.2. Observations and some results

As discussed in section 18.2, density anomalies with a short wavelength (<200 km or so) are supported elastically. It is the long wavelength geoid anomaly that can be used to constrain rheological structures. The key data in this exercise are geoid anomalies and density anomalies. The long wavelength geoid can be determined by satellite geodesy with great accuracy. The estimate of density anomalies in Earth is not straightforward. In most cases, density anomalies are inferred from observed anomalies of seismic wave velocities, but converting velocity anomalies to density anomalies is not trivial as discussed in Chapter 20. Furthermore, the geometry of mantle convection is also important. The observed plate motion on Earth's surface can be used to place some constraints on flow geometry. Large uncertainties still exist as to the nature of flow in the mantle, particularly around the transition zone (e.g. WEN and ANDERSON, 1997; FORTE, 2000).

The most remarkable long wavelength geoid anomaly is the circum Pacific high geoid highs. Starting from HAGER (1984) and RICHARDS and HAGER (1984) dynamic interpretation of this long wavelength geoid anomaly invariably showed evidence of a large increase in viscosity in the deep mantle. The subsequent studies, however, also showed that there is significant non-uniqueness of this type of modeling: the same data are equally well fit by different models (e.g., KING, 1995a, 1995b; KIDO and CADEK, 1997; CADEK and VAN DEN BERG, 1998).

These studies have shown that there is a large increase in viscosity with depth (average viscosity of the lower mantle is a factor of ~ 30–100 higher than that of the upper mantle). Some fine structures of rheological profiles have been proposed in these studies including high viscosity layers in the deep lower mantle (FORTE and MITROVICA, 2001) and a low viscosity channel below the 660-km discontinuity (KIDO and CADEK, 1997; FORTE, 2000). But due to the complications associated with the inference of density anomalies in these regions (the transition zone and the lower mantle), these features are not well resolved. One of the main issues in this type of study is the estimate of density anomalies. The effects of phase transformations (in the transition zone) and the possible effects of chemical heterogeneity on density (Chapter 20) make a robust estimate of density anomalies in these regions and hence rheological properties difficult. In addition, THORAVAL and RICHARDS (1997) showed the importance of boundary conditions on the inference of rheological structure from gravity signals. The dynamic topography, and hence inferred viscosity profiles, depends also on the geometry of convection that affects the stress magnitude on the density boundaries (e.g., WEN and ANDERSON, 1997; FORTE, 2000).

The inference of mantle viscosity has a long and twisted history. After the classic studies by Haskell (1935a, 1935b, 1937), extensive studies were made in the mid-1970s to mid-1980s, and the majority of these studies emphasized the constancy of viscosity with depth: the viscosity of the lower mantle was considered to be not higher than that of the upper mantle by a factor of ~2 (e.g., CATHLES, 1975; PELTIER, 1989). However, the later study by NAKADA and LAMBECK (1989) who paid attention to the details of hydro-isostatic response along the coastline and the separation of melting history from mantle response, indicated a significant increase in viscosity with depth in the deep mantle. Also the development of a new method of inferring rheological profiles from dynamic interpretation of geoid anomalies by HAGER (1984) and RICHARDS and HAGER (1984) improved our ability to infer deep mantle rheology. These studies showed a large increase in viscosity in the deep mantle (e.g., HAGER, 1984). The advantages and limitations of these methods are as summarized in Fig. 18.8.

Through the combination of these studies, we now have a good model for a gross one-dimensional viscosity–depth profile: the average viscosity of the mantle is $\sim 10^{21}$ Pa s, but the viscosity of the upper mantle is on average significantly lower than that of deep mantle (Fig. 18.9; HAGER, 1984; NAKADA and LAMBECK, 1989; PELTIER, 1998; FORTE and MITROVICA, 1996).

One should note that, beyond the general conclusion summarized above, most of the features of the

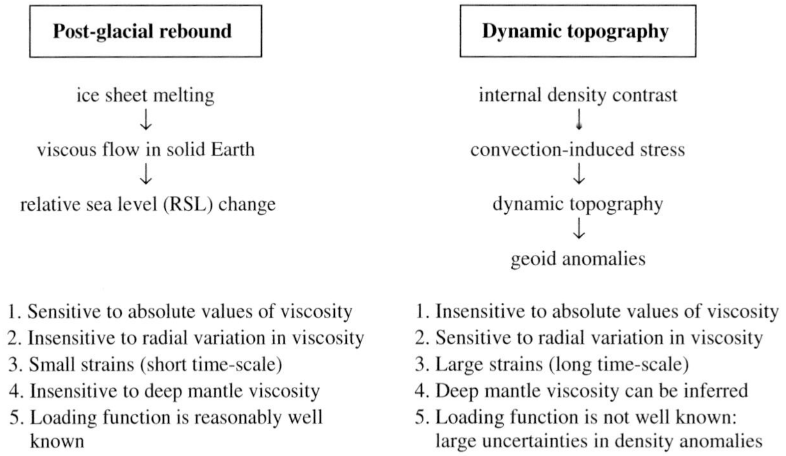

Post-glacial rebound	**Dynamic topography**
ice sheet melting ↓ viscous flow in solid Earth ↓ relative sea level (RSL) change	internal density contrast ↓ convection-induced stress ↓ dynamic topography ↓ geoid anomalies
1. Sensitive to absolute values of viscosity 2. Insensitive to radial variation in viscosity 3. Small strains (short time-scale) 4. Insensitive to deep mantle viscosity 5. Loading function is reasonably well known	1. Insensitive to absolute values of viscosity 2. Sensitive to radial variation in viscosity 3. Large strains (long time-scale) 4. Deep mantle viscosity can be inferred 5. Loading function is not well known: large uncertainties in density anomalies

FIGURE 18.8 A comparison of two methods of inferring mantle viscosity.

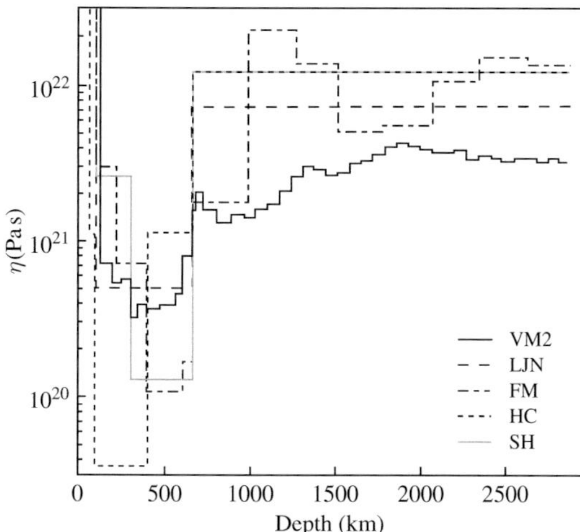

FIGURE 18.9 Viscosity–depth profiles inferred from geodynamic modeling (PELTIER, 1998).

rheological structure of Earth's interior remain poorly constrained. They include:

(1) Lateral variation in viscosity is expected to be large, but it is not well resolved from the post-glacial rebound or geoid anomalies (e.g., NAKADA and LAMBECK, 1991; ZHANG and CHRISTENSEN, 1993; CADEK and FLEITOUT, 2003). As noted above, the lateral variation in viscosity inferred from post-glacial rebound can be smaller than that for long-term deformation.
(2) There is some hint as to the presence of a low viscosity region in the subduction zone upper mantle due to the release of water (e.g., KARATO, 2003b). BILLEN and GURNIS (2001) and BILLEN *et al.* (2003) suggested such regions in the subduction zone from the analysis of gravity and topography.
(3) The distribution of seismic anisotropy provides strong evidence for non-linear rheology at least in the boundary layers (Chapter 21), but evidence for non-linear rheology is not well documented from the study of post-glacial rebound nor geoid anomalies (e.g., KARATO and WU, 1993).
(4) Some studies based on geoid anomalies suggest the presence of a low viscosity layer just below the 660-km discontinuity (KIDO and CADEK, 1997; FORTE, 2000) and very high viscosity layers in the deep lower mantle (FORTE and MITROVICA, 2001). Such conclusions are subject to large uncertainties due mainly to the uncertainties in density anomalies and the flow pattern in these regions.
(5) Mineral physics-based modeling suggests a highly complicated rheological structure (KARATO, 1997b; KARATO *et al.*, 2001) but such a small-scale rheological structure is beyond the resolution of current geodynamic study of mantle rheology.

18.5. Summary

Non-elastic properties of Earth can be inferred from a range of observations on time-dependent deformation. They include seismic wave attenuation, post-glacial crustal uplift and geoid anomalies associated with mass distribution in Earth. To a rough approximation, the results of the depth distribution of inferred non-elastic properties are similar (compare Fig. 18.1 and Fig. 18.9). The fundamental reason for this similarity is the fact that attenuation in Earth is described, in most cases, by the relation (18.5), $Q \propto \tau_Q^\alpha$, and viscosity is also related to the characteristic time, $\eta \propto \tau_\eta$. Both processes are thermally activated and therefore the functional relationships are similar. In the case of olivine and MgO for which some detailed experimental studies have been made on both long-term creep and seismic wave attenuation, even the activation enthalpies are similar between these two processes. Consequently, there is a close connection between long-term creep and seismic wave attenuation, and in fact, the depth profile of $Q^{1/\alpha}(z)$ ($\propto \tau_Q$) is similar to that of viscosity, $\eta(z)$.

However, some details can be different among the non-elastic properties inferred from different types of observations. It was discussed in Chapter 9, for example, that transient creep is likely to play an important role in the post-glacial rebound and in that case the influence of water on viscosity of the upper mantle will be weaker than that for a long-term (quasi-) steady-state creep. Similarly, the degree to which water affects attenuation can be different from that of long-term creep.

I should emphasize that the non-elastic properties inferred from seismic wave attenuation and from the post-glacial rebound probably involve small strain, and consequently, the mechanical response of materials to these short-term phenomena can be different from the rheological properties relevant to long-term, quasi steady-state deformation. Also it must be noted that the inference of the viscosity profile from the geoid (or any gravity signals) is subject to the uncertainties in the inferred density anomalies that cause geoid anomalies. It is important to appreciate the limitations and advantages of each method when comparing results from different approaches.

19 Inference of rheological structure of Earth from mineral physics

The current status of our understanding of the rheological structure of Earth's interior from a mineral physics point of view is reviewed. The rheological structure of the lithosphere–asthenosphere system is discussed in detail as well as the rheological structure of the deep mantle and the inner core. The inferred rheological structures depend critically on the temperature, water content, grain size and stress as well as the constitutive relations that control the dependence of rheological properties on these variables. These variables can be inferred for the shallow portions of Earth and some detailed analyses of the rheological structures there are presented including the regional variation in the rheological structure of the lithosphere. In contrast, temperature, water content and grain size are poorly known for Earth's deep interior. In addition, some of the important material parameters for deep Earth materials that determine the sensitivity of the rheological properties on these variables are not well constrained. Consequently, the inference of the rheological structure of Earth's deep interior can only be made with large uncertainties. In these cases, emphasis is placed on the basic framework by which rheological structures of Earth may be inferred from mineral physics rather than on specific models.

Keywords lithosphere, asthenosphere, partial melting, phase transformations, water, grain size.

19.1. Introduction

Similar to any other physical properties, the rheological properties in Earth and other terrestrial planets change both radially and laterally, and the manner in which the rheological properties change in space (and possibly in time) is highly dependent on the physical mechanisms of deformation. In this chapter, we will integrate the results summarized in previous chapters to present various models of rheological properties of Earth. As we have learned in the previous chapters, unlike elastic properties, rheological properties are highly sensitive to the variation in temperature as well as other factors such as water content and grain size. These key parameters can be inferred for the shallow portions of Earth from geophysical, geochemical or geological observations (see Chapter 17) and some detailed experimental studies on the dependence of rheological properties on these variables are available for upper mantle minerals such as olivine. Consequently, inference of rheological structures of the shallow portions of Earth's interior can be made with some confidence (e.g., Karato and Wu, 1993; Kohlstedt *et al.*, 1995; Hirth, 2002; Karato and Jung, 2003).

However, currently very little is known about the actual temperature, grain size and water content in the deep Earth (and in other terrestrial planets), and furthermore the dependence of rheological properties on these variables is known only for very few minerals such as olivine. Consequently, it is difficult to present any definitive models of rheological properties of deep

Earth (and other planets) at this time.[1] Even for the shallow regions where key parameters can be inferred with some confidence, possible variations in these parameters (such as temperature, grain size and water content) are so large that I believe that it is important to present a theoretical framework (or methodology) to infer rheological properties in addition to presenting some particular models. Therefore in many cases where I discuss the models of rheological structures, I have chosen to show a range of plausible structures for a range of parameters that are geologically or geophysically acceptable. In this way, the reader will obtain a solid physical basis behind any models of rheological structures to make her/his own judgment on the uncertainties in published models, or even preferably to develop her/his own new models.

In this chapter, I will first discuss some general principles for inferring rheological properties based on mineral physics (section 19.2). The rheological properties of shallow regions are summarized in section 19.3 in some detail with the emphasis on the lithosphere–asthenosphere structure. The rheological properties of the deep mantle (the transition zone and the lower mantle) are reviewed in section 19.4 and those in the core are reviewed in section 19.5.

19.2. General notes on inferring the rheological properties in Earth's interior from mineral physics

19.2.1. Constitutive relations

The most important concepts in developing models of rheological structures are:

(1) Strength of materials may be controlled either by brittle failure or plastic flow (Chapter 7). When strength is controlled by *brittle failure*, the strength is a function of various material parameters and variables that describe thermodynamic and mechanical conditions as (see Chapter 7),

$$\sigma = \tau_0 + \mu(\sigma_n - \alpha P_{\text{pore}}) \tag{19.1}$$

where σ is the strength, τ_0 is the "cohesion strength," μ is the friction coefficient, P_{pore} is the pore fluid pressure and α is a non-dimensional constant. The parameters τ_0, μ and α ($\tau_0 \sim 10$ MPa, $\mu \sim 0.7$, $\alpha \sim 1$) are nearly independent of materials and temperature and strain rate (see Chapter 7). However, the strength in this regime is sensitive to the pore fluid pressure, P_{pore}, as well as the stress state (tension, compression or shear) that changes the normal stress σ_n. The lithosphere in this depth range is weaker for tension than compression. A large amount of water (fluids) reduces the strength of lithosphere in this regime through the increase in pore pressure.

(2) Strength in the *plastic flow regime* is in most cases controlled by thermally activated processes. At low stresses ($\sigma/\mu < 10^{-3}$; σ: stress, μ: shear modulus), deformation by plastic flow is described by the following power-law formula (Chapters 9 and 10)

$$\dot{\varepsilon} = A\frac{\sigma^n}{L^m}C_{\text{OH}}^r \exp\left(-\frac{H^*(C_{\text{OH}}, P)}{RT}\right) \tag{19.2}$$

where $\dot{\varepsilon}$ is strain rate, A is a constant, L is grain size, C_{OH} is hydrogen (water) content, $H^* = E^* + PV^*$ is activation enthalpy for creep (E^*, activation energy; V^*, activation volume), R is the gas constant, T is temperature and P is pressure. In this regime, the strength may be defined through viscosity, $\sigma = 2\eta\dot{\varepsilon}$ (η, viscosity), namely,

$$\eta_\sigma = \frac{1}{2AC_{\text{OH}}^r}\frac{L^m}{\sigma^{n-1}}\exp\left(\frac{H^*(C_{\text{OH}}, P)}{RT}\right) \quad \text{for a given stress} \tag{19.3a}$$

or

$$\eta_{\dot{\varepsilon}} = \frac{L^{m/n}}{2A^{1/n}\cdot C_{\text{OH}}^{r/n}\cdot\dot{\varepsilon}^{(n-1)/n}}\exp\left(\frac{H^*(C_{\text{OH}}, P)}{nRT}\right) \quad \text{for a given strain rate} \tag{19.3b}$$

or

$$\eta_{\sigma\dot{\varepsilon}} = \frac{L^{2m/(n+1)}}{2A^{2/(n+1)}\cdot C_{\text{OH}}^{2r/(n+1)}}(\sigma\dot{\varepsilon})^{(1-n)/(1+n)} \exp\left(\frac{2H^*(C_{\text{OH}}, P)}{(n+1)RT}\right). \tag{19.3c}$$

for constant energy dissipation rate.

[1] The current status of research in deep Earth rheology is in much the same state of study as the elasticity of deep mantle minerals in the 1950s. One should recall Birch's "dictionary" for high-pressure research (Birch F., 1952).

High-pressure (rheological) form	Ordinary meaning
Undoubtedly	Perhaps
Certain	Dubious
Positive proof	Vague suggestion

I hope that some readers will make this "dictionary" outdated in the near future by making further advancements of this important, but currently undeveloped, area of Earth science.

Problem 19.1

Derive equation (19.3c).

Solution

From equation (19.2), one gets $\sigma = (1/2\eta)^{1/(n-1)} (L^m/AC_{OH}^r)^{1/(n-1)}$ and $\dot{\varepsilon} = (1/2\eta)^{n/(n-1)} (L^m/AC_{OH}^r)^{1/(n-1)}$. Therefore $\eta = \frac{1}{2} \frac{L^{2m/(n+1)}}{A^{2/(n+1)} \cdot C_{OH}^{2r/(n+1)}} (\sigma\dot{\varepsilon})^{(1-n)/(1+n)} \cdot \exp(2H^*(C_{OH}, P)/(n+1)RT)$.

Therefore the spatial variation of viscosity depends on how stress or strain-rate varies with space as well as the spatial variation of T, C_{OH} and L. Two points may be noted about water effects. (1) The water content C_{OH} is a function of temperature and pressure if a material is in chemical equilibrium with a water "reservoir" (open system). If a system is closed with respect to water (hydrogen), then C_{OH} is a constant. (2) The activation enthalpy, $H^*(C_{OH}, P)$, is a function of water content, but in many cases it is a discrete function of water content: activation enthalpy changes abruptly with water content and in a certain range of water content it remains a constant (Chapter 10). In these cases, the activation enthalpy can be assumed to be a constant within a certain range of water content.

In many cases, the grain size, L, is assumed to be a constant. However, grain size may vary from one place to another through various processes (see Chapter 13). For instance, when grain size is controlled by phase transformations (or chemical reactions), it will change with temperature–pressure history. When grain size is controlled by grain-growth kinetics, grain size will be temperature-dependent.

At higher stresses $\sigma/\mu > 10^{-3}$, strength determined by plastic flow is still thermally activated, but the influence of stress on thermal activation becomes significant, and the functional relationship between stress, strain rate and temperature is markedly different from that for a power-law creep. A generic flow law appropriate for this regime is (Chapter 9)

$$\dot{\varepsilon} = B(C_{OH})\,\sigma^{n'} \exp\left\{-\frac{H_0^*(C_{OH}, P)}{RT}\left[1 - \left(\frac{\sigma}{\hat{\sigma}(C_{OH}, P)}\right)^q\right]^s\right\} \tag{19.4}$$

where $B(C_{OH})$ is a constant, $n' \approx 2$, $H_0^* = E_0^* + PV_0^*$ is the activation enthalpy for creep (E_0^*, activation energy; V_0^*, activation volume), q and s are constants and other symbols are the same as before. This flow regime is called *exponential flow law* regime. Deformation by dislocation glide over the Peierls potential (Chapter 9) is an example. Deformation by twinning is another important case that occurs for some minerals including calcite. Similar to the power-law creep, H_0^*, $\hat{\sigma}$ can be a function of water content, but in many cases they are constant for a given mechanism (these parameters are likely to be different between water-poor and water-rich environments, but they remain constant in one regime). So these quantities can be left as constants. However, $B(C_{OH})$ is likely to be dependent on water content (or water fugacity) (see Chapter 10). Creep rate (or strength) in this regime is insensitive to grain size, but sensitive to chemical environment such as water content. The temperature and strain rate dependence of strength of materials for this regime can be written as

$$\frac{\sigma}{\hat{\sigma}(C_{OH}, P)} = \left[1 - \left(\frac{RT}{H_0^*(C_{OH}, P)}\right)^{1/s} \log\frac{\dot{\varepsilon}_0(C_{OH})}{\dot{\varepsilon}}\right]^{1/q} \tag{19.5}$$

where $\dot{\varepsilon}_0 \equiv B(C_{OH})\sigma^{n'}$. For a large range of stress and temperatures, $\log\dot{\varepsilon}_0$ does not change much. Therefore the variation of strength with temperature (or strain rate) in this regime is determined mostly by the $(RT/H_0^*(C_{OH}, P))^{1/s}$ term (or $\log\dot{\varepsilon}$ term) and hence the strength is much less sensitive to temperature and strain rate than in the power-law creep regime. Similarly pressure effect is through $(RT/H_0^*(C_{OH}, P))^{1/s}$ and $\hat{\sigma}(C_{OH}, P)$, and therefore it is much weaker than in the power-law creep regime. Importance of the exponential flow law (the Peierls mechanism of flow, for example) in geological processes was first emphasized by GOETZE and EVANS (1979).

Strength of Earth materials can be characterized by either equations (19.1), (19.3) or (19.5). Note that the constitutive relation (19.5) represents a mechanical behavior between brittle failure (19.1) and power-law creep (19.3). Equation (19.4) means that as temperature goes to $T \to 0$, deformation is possible only when $\sigma \approx \hat{\sigma}$ and $\hat{\sigma}$ is (nearly) independent of temperature. Figure 19.1 shows the variation of strength with temperature and strain rate represented by these three equations.

From equations (19.3a–c), the following points can be noted.

(i): The values of n and m depend on the dominant mechanisms of plastic flow, that can be inferred from laboratory experiments in combination with inferred physical and chemical conditions in Earth. For diffusional creep, $n = 1$ and $m = 2$–3

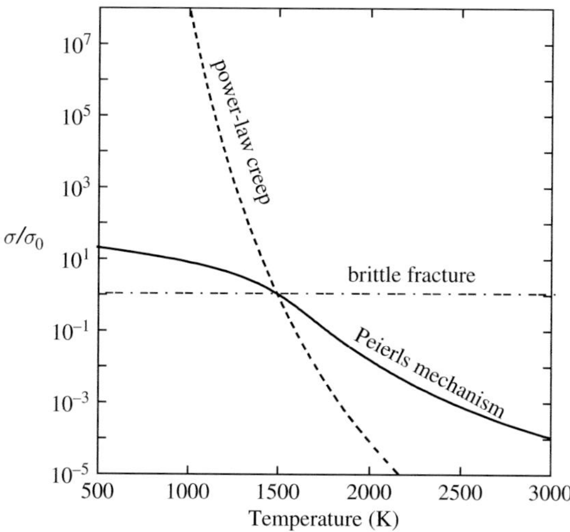

FIGURE 19.1 A schematic diagram showing the dependence of the (normalized) strength, σ/σ_0, on the temperature corresponding to three constitutive equations: brittle fracture (friction) (equation (19.1)), power-law creep (equation (19.3a)) and the exponential flow law (the Peierls mechanism, equation (19.5)). T_0 and σ_0 are the reference temperature (1500 K) and the reference strength (10 MPa), respectively. A strain rate of $10^{-15}\,s^{-1}$ is assumed.

(Chapter 8). For dislocation creep, $n = 3$–5 and $m = 0$ (Chapter 9). Presence and absence of seismic anisotropy can also be used to infer the dominant mechanisms of deformation (Chapters 14 and 21).

(ii): Temperature usually reduces the viscosity in the exponential fashion, $\eta \propto \exp(H^*/RT)$ for constant stress ($\eta \propto \exp(H^*/nRT)$ for constant strain rate, $\eta \propto \exp(2H^*/(n+1)RT)$ for constant energy dissipation rate).

(iii): Pressure has a large effect on rheology, and usually increases the viscosity as $\eta \propto \exp((E^* + PV^*)/RT)$ for constant stress ($\eta \propto \exp((E^* + PV^*)/nRT)$ for constant strain rate, $\eta \propto \exp(2(E^* + PV^*)/(n+1)RT)$ for constant energy dissipation rate) (see Chapter 10).

(iv): Viscosity is independent of stress for linear rheology ($n = 1$), but viscosity decreases with stress for non-linear rheology ($\eta \propto \sigma^{1-n}$ for constant stress in the power-law creep regime). This effect becomes very strong at high stress levels, and viscosity decreases exponentially with stress at very high stress levels (see Chapter 9). In the exponential regime, sensitivity of strength to temperature, pressure and water content is weaker than in the power-law creep regime (for details see below).

(v): Grain size has a large effect on viscosity at high temperature and low stress. Under these conditions grain size affects the viscosity as $\eta \propto L^m \exp(H^*/RT)$ for constant stress ($\eta \propto L^{m/n} \exp(H^*/nRT)$ for constant strain rate, $\eta \propto L^{2m/(n+1)} \exp(2H^*/(n+1)RT)$ for constant energy dissipation) (see Chapter 8; L, grain size; m, a constant ($m = 2$–3)).

(vi): Water content has a large effect on the viscosity. In many cases, water content affects the viscosity as $\eta \propto C_{OH}^{-r} \exp(H^*(C_{OH})/RT)$ for constant stress ($\eta \propto C_{OH}^{-r/n} \exp(H^*(C_{OH})/nRT)$ for constant strain rate, $\eta \propto C_{OH}^{-2r/(n+1)} \exp(2H^*(C_{OH})/(n+1)RT)$ for constant energy dissipation) (see Chapter 10; C_{OH}: water content, r: a constant ($r = 1$–2)). Note that the dependence of activation enthalpy on water content is usually discrete: $H^* = H_1^*$ for water-poor conditions, $H^* = H_2^*$ for water-rich conditions.

(vii): A phase transformation affects rheological properties in the plastic flow regime through its effects on chemical bonding and crystal structure as well as through the changes in grain size and water distribution and also through the redistribution of internal stress–strain (see Chapter 15).

(3) The overall creep strength of a multi-component aggregate is some average of strength of each component. The overall strength depends on the geometry of phases as well as the volume fraction of phases (see Chapter 12). The presence of a minor phase could have large effects if the minor phase has a much different strength than others and if the minor phase is connected each other. An important example is a partial melt (e.g., Kohlstedt, 2002), but this effect is also important in a solid–solid mixture with a large strength contrast (a likely example is the lower mantle material that is composed of a mixture of (Mg, Fe)SiO_3 perovskite + (Mg, Fe)O, e.g., Yamazaki and Karato, 2001b).

In addition to these effects, effects of varying the major element chemical composition and mineralogy will also control the rheological structure. Note that some of these effects are mutually related. For instance, the content of water (hydrogen) dissolved in minerals may increase with pressure leading to *pressure weakening* (as opposed to normally observed pressure hardening) in certain cases. Similarly, grain size of rocks increases with temperature when grain growth occurs. In such a case, viscosity of a rock may *increase* with temperature (as opposed to normally observed temperature weakening; SOLOMATOV, 1996).

In the following, I will illustrate how these factors may be integrated in developing models of rheological structures of Earth and planetary interiors. Although the emphasis in this book is ductile rheology, I will briefly touch upon the strength in the brittle regime for completeness (for brittle fracture, see also Chapter 7).

19.2.2. Spatial variation of stress and strain rate

As seen in section 19.2.1, the spatial variation in effective viscosity for non-linear rheology depends on the spatial variation in stress (strain rate). Consider two cases: a case of constant stress with depth and a case of constant strain rate with depth. For a power-law rheology, the effective activation enthalpy for the former case is H^*, while for the latter it is H^*/n. This causes a large difference in the depth variation of effective viscosity (e.g., KARATO, 1981b).

The spatial distribution of stress and strain rate is controlled by the nature of deformation. In a few cases, simple patterns of distribution of stress and strain (rate) can be inferred including the case of deformation of materials caused by the motion of surface plate and bending (folding) of a plate (Fig. 19.2). The two-dimensional flow pattern in a corner can also be calculated analytically if the stress and strain-rate relation is linear and the effective viscosity is homogeneous (see e.g., TURCOTTE and SCHUBERT, 1982). Both stress and strain (rate) are high near the corner because of the rapid change in flow geometry there (Fig. 19.3).

Problem 19.2

What is the depth variation of stress or strain rate in a two-dimensional flow beneath a moving plate driven by the motion of plate?

What is the variation of viscous strain rate in a bending plate across the plate (assume incompressible fluid behavior)?

Solution

For a two-dimensional flow, because of the condition for continuity of shear stress, the (shear) stress is constant with depth. strain rate at each depth depends on the local viscosity and changes with depth (Fig. 19.2a).

Bending (folding) causes tensional strain in the outer side and compressional strain in the inner side, the central plane being a neutral strain with no strain. If the volumetric strain is zero, then the corresponding strain distribution changes within a plate as shown by Fig. 19.2b. Both stress and strain is zero on the central neutral plane, and the strain (rate) increases as a distance from the central neutral plane. The distribution of stress depends on the local rheological properties.

19.2.3. Spatial variation in thermochemical variables and grain size

The principles governing the spatial variation of temperature and pressure are discussed in Chapter 17. Thermal conduction controls the temperature variation when the vertical motion is small. When the vertical motion is fast (compared to the conduction time-scale), then the *adiabatic change* in temperature becomes important. Pressure is in most cases controlled by the hydrostatic equilibrium.

Spatial variation in a composition such as water (hydrogen) content is controlled by the advective transport (transport by fluid flow) and diffusion. Diffusion is, however, effective only to a short distance. Consequently, chemical composition can be largely heterogeneous in many geological systems. An important example is the re-distribution of water (hydrogen) by partial melting (e.g., KARATO, 1986; HIRTH and KOHLSTEDT, 1996; see Chapter 10). This process will be discussed in section 19.3.3 in some detail.

The principles controlling the grain size are discussed in Chapter 13. The most important mechanisms are phase transformations, dynamic recrystallization and grain growth. Changes in grain size associated with phase transformations can be large. A specific case for subducting slabs will be discussed in section 19.4. Grain size can be quite small where high-stress deformation occurs. In typical hot regions of the upper mantle for which grain size is inferred from mantle xenoliths, grain size is on the order of a few mm.

19.3. Strength profile of the crust and the upper mantle

19.3.1. The brittle–plastic transition (see also Chapter 7)

In the shallow portions of Earth, the strength of materials is controlled by the brittle failure. The basic physics of the brittle failure is reviewed in Chapter 7. Assuming that the near-surface rocks contain a large number of

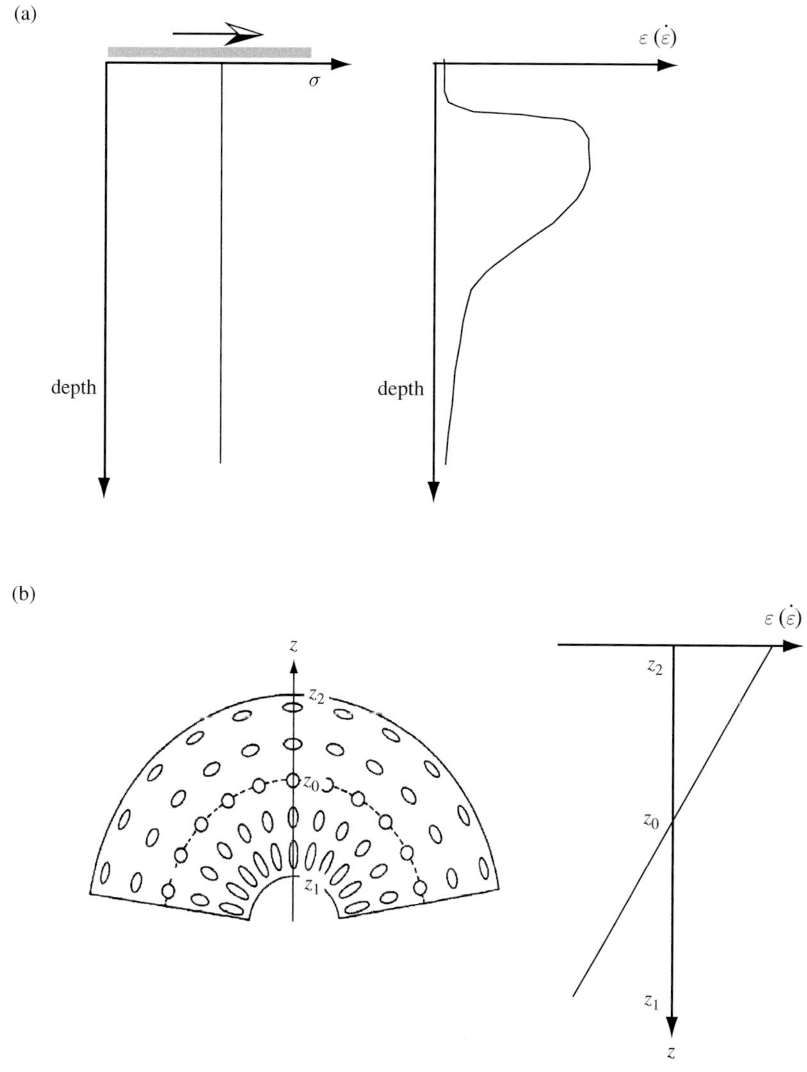

FIGURE 19.2 (a) The stress and strain (rate) distribution in deformation of materials beneath a moving plate (the strain distribution is controlled by the viscosity distribution). (b) The strain (rate) distribution in a bending (folding) plate.

pre-existing faults, the strength in this regime is determined by the strength against the propagation of faults. Under these conditions, the strength controlled by brittle failure is insensitive to rock type, strain rate, but sensitive to pore-fluid pressure and stress state (compression versus tension), and increases nearly linearly with effective confining pressure (see equation (19.1)). Therefore, the strength of rock at near-surface is very low and increases nearly linearly with depth in the brittle regime. In contrast, the strength controlled by plastic flow is sensitive to temperature and, in regions where temperature increases significantly with depth, the strength decreases with depth. In this regime (the plastic-flow regime), the strength of rocks is also sensitive to the strain rate and rock types (see equation (19.2)). Consequently, the strength versus depth profile in the shallow Earth has a maximum (or maxima) at a certain depth and hence much of the load in the lithosphere is supported in the region around the strength maximum (Fig. 19.4).

In constructing such a diagram, it is assumed that the two processes of deformation (i.e., brittle failure and plastic flow) are independent. Under this assumption, the smaller of the strengths calculated for each mechanism is chosen to be the strength of the material. An extensive discussion on the validity of this assumption (independence of brittle and plastic deformation) is given by Kohlstedt *et al.* (1995). This is a simplification of more complicated behavior of materials, and

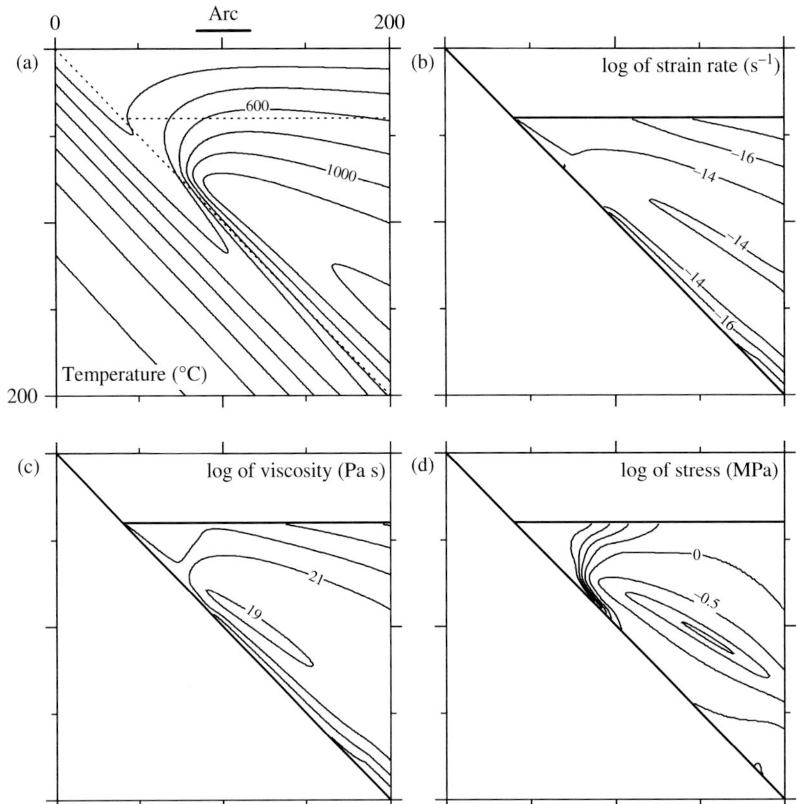

FIGURE 19.3 An example of stress and strain rate distributions in a convecting mantle (in a subduction zone) based on numerical modeling (after KNELLER *et al.*, 2005).

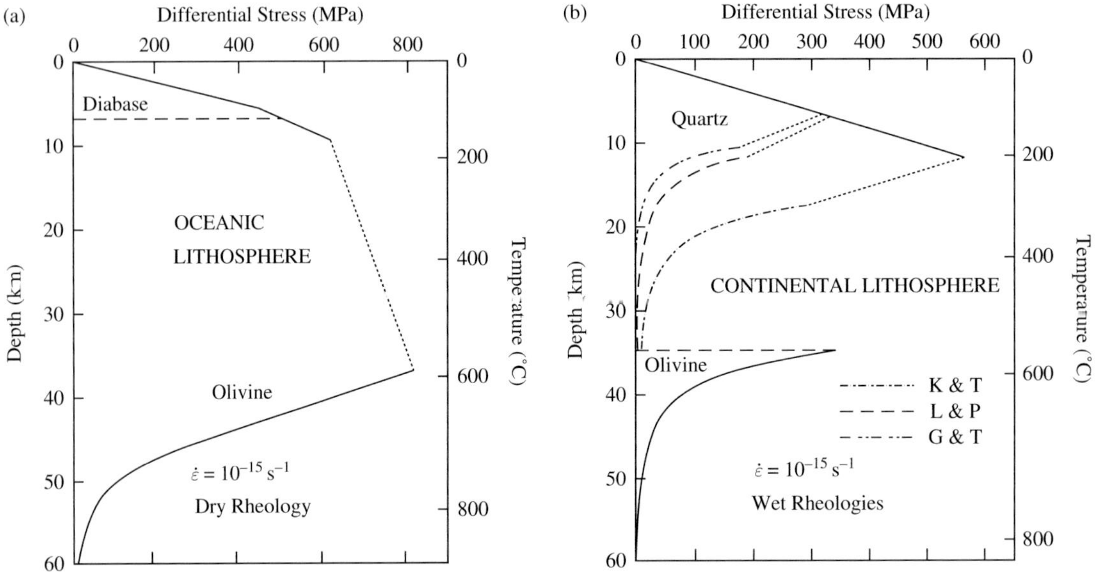

FIGURE 19.4 Strength–depth diagrams for (a) the oceanic upper mantle and (b) the continental upper mantle (after KOHLSTEDT *et al.*, 1995).

in reality a broad regime of *semi-brittle* behavior is observed. Microscopically, this semi-brittle behavior occurs when the interaction of cracks with dislocations becomes significant. Such an interaction reduces the stress concentration at the crack tip and thereby reduces the tendency for unstable localized deformation. However, a major limitation is that there is no well-accepted constitutive relation for the semi-brittle regime (Chapter 7). In addition, rheological properties of minerals at high-stress conditions are not appropriately treated in this work. For example, the role of exponential flow such as the Peierls mechanism or deformation by twinning (in orthopyroxene) is not included in this model. Consequently, the strength profile near the peak strength shown in Fig. 19.4 (after KOHLSTEDT *et al.*, 1995) has a large uncertainty and is likely to overestimate the actual strength substantially.

19.3.2. Crust

The crust of the oceanic region is made of basalt (or, in the deeper portions, diabase or gabbro that are rocks with the same chemical composition as basalt but with coarser grain size). Deformation there is in most cases by brittle fracture and plastic deformation occurs only in the close vicinity to the mid-ocean ridge where temperature is high (e.g., YOSHINOBU and HIRTH, 2002). Plastic deformation of the crust is important in the continental environment. The continental crust is made of two different layers: the upper crust is made of granitic rocks and the lower crust is made of basaltic rocks or some metamorphic rocks derived from basalt (see RUDNICK and FOUNTAIN, 1995). In most continents (and in island arcs), seismicity is confined to the upper crust. Seismicity in the lower crust is very rare (JACKSON, 2002a). The continental lower crust deforms by plastic flow and it plays an important role in continental tectonics (e.g., ROYDEN *et al.*, 1997; MEISSNER and MOONEY, 1998).

Rheological properties of crustal materials are more accessible than those of mantle materials for two obvious reasons. First, deformation of crustal materials can be studied directly through the studies of exposed crustal rocks (e.g., SIBSON, 1977; RUTTER and BRODIE, 1992; HIRTH *et al.*, 2001). Second, the experimental studies of deformation of crustal materials require less extreme conditions than those of mantle materials. However, rather surprisingly, the current status of experimental studies on deformation of crustal rocks is less complete than that of upper mantle materials (e.g., olivine). The main reasons for this include (i) the slow kinetics of chemical equilibrium with respect to water, and the resultant large ambiguity in interpreting the experimental observations (see PATERSON (1989) for a review on quartz), and (ii) the fact that at low temperatures relevant for the crust, stress levels are high at experimental conditions, and consequently a high-resolution, low-pressure apparatus such as the gas-medium Paterson apparatus (see Chapter 6) has limited applications (in order to investigate the flow laws in the ductile regime, the confining pressure must exceed the deviatoric stress; Chapter 6). Consequently, in many of the studies on deformation of crustal materials, a low-resolution, high-pressure solid-medium apparatus such as the Griggs apparatus was used, which results in large errors in mechanical data. To illustrate this point, I should point out that the functional relationship between water fugacity (water content) and rheological properties of crustal rocks or minerals has not been well constrained by the laboratory studies at this time. For quartz, for which a large number of experimental studies have been performed, some attempts have been made to quantify the influence of water fugacity on strain rate (e.g., HIRTH *et al.*, 2001; KOHLSTEDT *et al.*, 1995). These studies suggest $\dot{\varepsilon} \propto f_{H_2O}^r$ with $r \sim 1$. In contrast, POST *et al.* (1996) inferred $r \sim 2$. These estimates contain large uncertainties. These authors estimated the value of r by comparing the strength at different pressures without making corrections for the activation volume term. Consequently, the estimated value of r should be considered to be the lower bound. Second, the data used by these authors include low-precision tests using a solid-medium deformation apparatus such as the results by KRONENBERG and TULLIS (1984). For clinopyroxene HIER-MAJUMDER *et al.* (2005b) determined the water fugacity dependence and pressure dependence of creep (in the diffusional creep regime), but since the pressure range is limited (100–300 MPa), neither of these parameters characterizing water fugacity dependence (r) nor pressure dependence (V^*) was constrained well from that study. As KARATO (2006a) emphasized, there is a very strong trade-off between the values of r and V^* at low pressures (< 0.5 GPa), and these parameters cannot be determined with small enough errors from the experiments below 0.5 GPa (see Chapter 10).

Geological and geophysical observations on deformation in the crust are summarized by SIBSON (1977) (see also SCHOLZ, 2002). Deformation in shallow (upper) crust is mainly by brittle (highly localized) deformation. The mode of deformation gradually changes to ductile deformation at deeper portions. Localized deformation is common in the transitional regime, leading to the formation of mylonites where deformation occurs mainly in the ductile manner but in a localized fashion. In the even deeper portions (the lower crust), deformation is more distributed. Such a trend is consistent with the depth variation of seismicity (e.g., JACKSON, 2002a; SCHOLZ, 2002).

A simple model of rheological structure of the continental crust (and shallow upper mantle) is shown in Fig. 19.4. In such a model, the continental crust is assumed to be composed of two distinctive petrological units, granitic upper crust and more mafic lower crust (gabbro or diabase or metamorphic rocks with mafic composition). Since the strength of granitic rocks is significantly smaller than that of mafic rocks (under water-rich conditions), the strength versus depth profile across the continental crust has two distinct minima. The large uncertainties in this estimate include the water content and temperature in the actual crust (e.g., HACKER *et al.*, 2000), and the uncertainties in the experimental data on the influence of water content (water fugacity). For the review of this subject, see RUTTER and BRODIE (1992) and TULLIS (2002). Reviews from geophysical or geological points of view include ROYDEN *et al.* (1997), HACKER *et al.* (2000) and JACKSON (2002b).

In order to highlight the nature of the trade-off between the influence of temperature and water on the strength–depth profile, the strength versus depth profiles for a near-surface layer composed of a crustal rock were calculated for a few different temperature–depth profiles (corresponding to Venus and typical continental crust on Earth) and for either "dry" (water-free) or "wet" (water-saturated) conditions (Fig. 19.5). We consider a crust made of a coarse-grained basalt (diabase). The experimental results on diabase by CARISTAN (1982) and MACKWELL *et al.* (1998) are used. Note that both temperature and water content have large effects on the strength. The influence of water is very strong compared to that for upper mantle rocks (see Fig. 19.6). This is a result of the stronger influence of water (hydrogen) on the plastic deformation in silica-rich (crustal) rocks than in silica-poor (mantle) rocks (see Chapter 10). However, the influence of surface temperature is even more important. Because the surface temperature on Venus is much higher than that on Earth (by ~450 K), the strength of crust of Venus in the ductile power-law creep regime is likely to be lower than that of Earth's crust (for the same major element composition) even though Earth's crust is likely to have a larger water content. Indeed, the Magellan images show evidence of pervasive crustal deformation on Venus (SOLOMON *et al.*, 1991) suggesting a weak ductile crust on that planet. The influence of grain size on the strength of crust in the ductile regime is discussed by RYBACKI and DRESEN (2004).

(a)

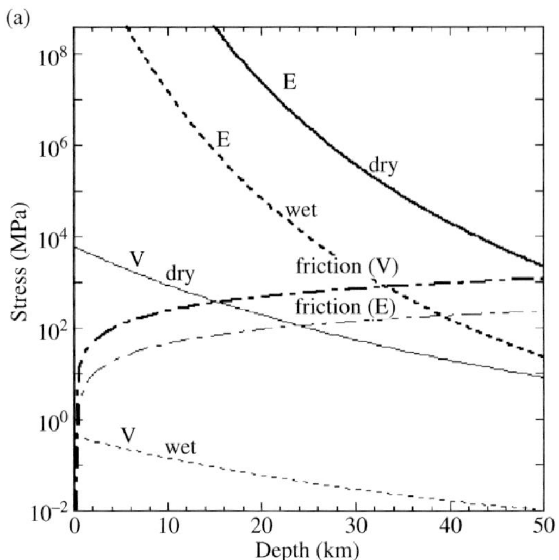

(b)

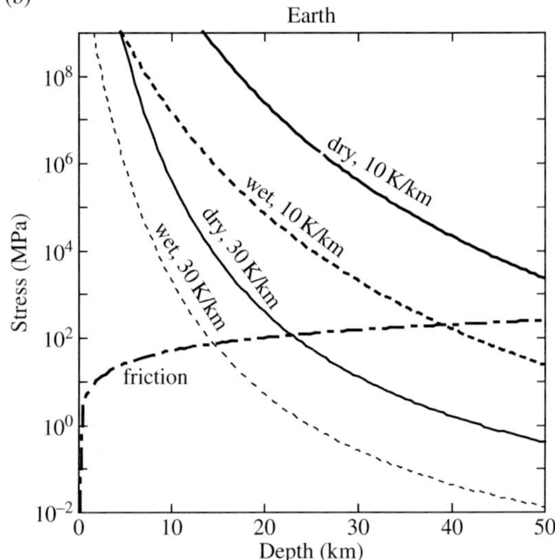

FIGURE 19.5 The strength–depth profiles for the crust. The crust is assumed to be composed of coarse-grained basalt (diabase) (similar results can be obtained for other silica-rich rocks). (a) Two temperature–depth profiles corresponding to Earth and Venus are considered. (b) Two temperature–depth profiles on Earth (corresponding to $dT/dz = 10$ and 30 K/km) are used. For each, the ductile creep strength for a strain rate of $10^{-15}\,s^{-1}$ is plotted for "dry" (water-free) and "wet" (water-saturated) conditions. The flow laws of diabase shown in Table 19.1 are used. Also shown are the strength profiles corresponding to frictional strength (zero pore-pressure (V), pore pressure $= 0.8 \times$pressure (E)).

TABLE 19.1 Flow law parameters of some crustal and upper mantle minerals and rocks A power-law formula, $\dot{\varepsilon} = A \cdot f_{H_2O}^r \cdot \sigma^n \cdot L^{-m} \exp(-(E^* + PV^*)/RT)$, is used. Units are: A ($s^{-1}(MPa)^{-r-n}(\mu m)^m$), E^* (kJ/mol), V^* (10^{-6} m^3/mol) (μm). Results from polycrystalline samples are selected.

	$\log_{10}A$	r	n	m	E^*	V^*	P	T	L	Apparatus	Water, melt etc.	Ref.
Quartz[1]	–	–	4	–	300	–	0.95–1.5	1223–1392	1.2–211	Solid-Griggs[26]	"Dried"	(1)[35]
Quartz[1]	–	–	2.6	–	130	–	0.35–1.6	973–1123	3.6–211	Solid-Griggs	Water added	(1)[35]
Quartz[2]	−4.0	1	4	–	223	–	1.5	1173–1373	100	Liquid-Griggs[27]	Without melt	(2)
Quartz[2]	−7.7	–	4	–	137	–	1.5	1173–1373	100	Liquid-Griggs	With 1–2% melt	(2)
Quartz[2]	–	2[22]	–	–	–	–	0.71–1.72	1173	100	Liquid-Griggs	No effect of f_{O_2}	(3)
Quartz[3]	–	–	4[23]	–	152	–	0.3	1100–1300	10–100	Paterson[28]	5000–10000 ppm H/Si	(4)
Quartz[4]	−0.4	–	1	2	220	–	0.3	1273–1473	0.4–4.5	Paterson	15 ppm H/Si (initial)	(5)
Quartz	−13	1	4	–	135	–	–	–	–	–	–	(6)[36]
Feldspar[5]	2.6	–	3	–	356	–	0.3	1270–1480	2.7–3.4	Paterson	"Wet" (~11500 ppm H/Si)	(7)
Feldspar[5]	13	–	3	–	648	–	0.3	1370–1480	2.7–3.4	Paterson	"Dry" (~640 ppm H/Si)	(7)
Feldspar[5]	1.7	–	1	3	170	–	0.3	1180–1480	2.7–3.4	Paterson	"Wet" (~11500 ppm H/Si)	(7)
Feldspar[5]	2.6	–	1	3	467	–	0.3	1370–1480	2.7–3.4	Paterson	"Dry" (~640 ppm H/Si)	(7)
Diopside[6]	9.8	–	4.7	–	760	–	0.3–0.43	1373–1523	5.2–330	Paterson	"Dry" (<10 ppm H/Si)	(8)
Diopside[6]	15	–	1	3	560	–	0.3–0.43	1373–1523	5.2–330	Paterson	"Dry" (<10 ppm H/Si)	(8)
Diopside[7]	0.09	1.4	1	3	340	14	0.1–0.3	1321–1421	6.6–10.5	Paterson	"Wet" (98–216 ppm H/Si)	(9)
Diopside[7]	17.5	–	1	3	760	–	0.1–0.3	1321–1421	6.6–10.5	Paterson	"Dry" (<27 ppm H/Si)	(9)
Enstatite[8]	−2.2	–	2.8	–	270	–	1	1273–1673	~1000	Solid-Griggs	"Wet" (water from talc)	(10)
Enstatite[9]	12	–	3.8	–	750	–	0.0001	1673–1723	–	Dead weight[29]	Dry	(11)
Enstatite[10]	8.8	–	2.9	–	600	–	0.45	1473–1523	~10	Paterson	"Dry" (see note [10])	(12)
Garnet[11]	13	–	3.2	–	270	–	4.3–6.8	1113–1573	2–10	D-DIA[30]	"Dry" (<5 ppm H/Si)	(13)
Garnet[12]	7.1	–	2.7	–	530	–	0.0001	1370–1430	–	MTS[31]	Dry	(14)
Garnet[13]	5.1	–	1.1	2.5	347	–	0.0001	1373–1543	2–6	Dead weight	"Dry" (~100 ppm H/Si)	(15)
Olivine[14]	5.4	–	3.5	–	540[24]	–	0.3	1573	25–60	Paterson	"Dry" (<50 ppm H/Si)	(16)
Olivine[14]	3.3	–	3	–	420[24]	–	0.3	1573	38–67	Paterson	"Wet" (~1000–1500 ppm H/Si)	(16)
Olivine[14]	4.9	–	1	2	290[24]	–	0.3	1573	7–18	Paterson	"Dry" (<50 ppm H/Si)	(16)
Olivine[14]	6.2	–	1	3	250[24]	–	0.3	1573	14–45	Paterson	"Wet" (water saturated)	(16)
Olivine[14]	3.2[21]	1[21]	3	–	470	20[21]	0.1–0.45	1393–1573	12–17	Paterson	"Wet" (water saturated)	(17)
Olivine[14]	4.7[21]	1[21]	1.1	3	295	20[21]	0.1–0.45	1473–1573	12–17	Paterson	"Wet" (water saturated)	(18)
Olivine[14]	6.8	–	1	3	315	–	0.3	1473–1523	10–14	Paterson	"Dry" (<50 ppm H/Si)	(19)
Olivine[14]	5.8	–	3	–	510	–	0.3	1473–1573	14–18	Paterson	"Dry" (<50 ppm H/Si)	(17)
Olivine[14]	6.1	–	3	–	510[24]	14	0.1–2.0[25]	1473	12–40	Solid-Griggs[32]	"Dry" (<100 ppm H/Si)	(20)

TABLE 19.1 (cont.)

	$\log_{10}A$	r	n	m	E^*	V^*	P	T	L	Apparatus	Water, melt etc.	Ref.
Olivine[14]	2.9	1.2	3	–	470[24]	24	0.1–2.0[25]	1473	12–40	Solid-Griggs[32]	"Wet" (water saturated)	(20)
Diabase[15]	−1.2	–	3.1	–	276	–	0.35–0.45	1073–1273	~50	Gas-apparatus[33]	"Wet" (saturated)	(21)
Diabase[15]	0.92	–	4.7	–	485	–	0.4–0.5	1213–1345	~50	Paterson	"Dry" (< 10 ppm H/Si)	(22)
Eclogite[16]	3.3	–	3.4	–	480	–	3	1450–1600	30–100	Liquid-Griggs	~1000 ppm H/Si in omphasite	(23)
Peridotite[17]	7.6	–	3.5	–	600	–	0.45	1473–1523	~10	Paterson	"Dry" (but see note [17])	(12)
Peridotite[18]	6.1	–	2.2	3	338	–	0.6	1173–1275	1–2	Gas-apparatus[34]	"Wet" (~0.5 wt%)	(24)
Peridotite[19]	8.8	–	1.7	–	538	–	0.0001	1473–1558	8–25	Dead weight	Dry	(25)
Peridotite[20]	9.1	–	1	3	370	–	0.3	1373–1573	8–34	Paterson	"Dry" (< 30 ppm H/Si)	(26)
Peridotite[20]	4.8	–	4.3	–	550	–	0.3	1373–1573	8–34	Paterson	"Dry" (< 30 ppm H/Si)	(26)

(1): Kronenberg and Tullis (1984), (2): Gleason and Tullis (1995), (3): Post *et al.* (1996), (4): Luan and Paterson (1992), (5): Rutter and Brodie (2004), (6): Hirth *et al.* (2001), (7): Rybacki and Dresen (2000), (8): Bystricky and Mackwell (2001), (9): Hier-Majumder *et al.* (2005b), (10): Ross and Nielsen (1978), (11): Mackwell (1991), (12): Lawlis (1998), (13): Li *et al.* (2006), (14): Karato *et al.* (1995a), (15): Wang and Ji (2000), (16): Karato *et al.* (1986), (17): Mei and Kohlstedt (2000b) , (18): Mei and Kohlstedt (2000a), (19): Hirth and Kohlstedt (1995a), (20): Karato and Jung (2003), (21): Caristan (1982), (22): Mackwell *et al.* (1998), (23): Jin *et al.* (2001), (24): McDonnell *et al.* (2000), (25): Ji *et al.* (2001), (26): Zimmerman and Kohlstedt (2004).

1 Heavitree quartitz, Arkansas novaculite, β-quartz field.
2 Black Hills quartzite, β-quartz field.
3 Synthetic quartz aggregate from sol-gel method, β-quartz field.
4 Hot-pressed samples prepared from Brazilian quartz, β-quartz field.
5 Synthetic hot-pressed anorthite ($CaAl_2Si_2O_8$).
6 Synthetic hot-pressed clinopyroxene ($Ca(Mg_{0.8}Fe_{0.2})\ Si_2O_6$).
7 Synthetic hot-pressed clinopyroxene ($Ca_{0.97}(Mg_{0.8}Fe_{0.2})\ Si_{1.99}O_6$).
8 ($Mg_{0.89}Fe_{0.8}Ca_{0.3}$)$SiO_3$.
9 Natural enstatite single crystals oriented for the [001](100) slip. Sample is in the proto-enstatite stability field in this study.
10 Synthetic hot-pressed sample (($Mg_{0.94}Fe_{0.04}Ca_{0.02})_2Si_2O_6$, ($Mg_{0.906}Fe_{0.091}Ni_{0.003})_2Si_2O_6$). No report on water content. P–T conditions are close to the boundary between ortho-enstatite and proto-enstatite fields.
11 Synthetic polycrystalline garnet (Py, $Py_{70}Alm_{16}Gr_{14}$).
12 High-temperature creep law was determined for eight different garnet single crystals (compression along the [100] orientation), and a universal creep law was derived. The parameters here are the estimated values corresponding to pyrope garnet.
13 Synthetic hot-pressed garnet ($Py_{88}Alm_{10}Gr_2$).

[14] Synthetic hot-pressed aggregate from San Carlos olivine.
[15] Maryland diabase.
[16] Synthetic aggregates from natural eclogite (~50% garnet, ~40% omphasite, ~10% quartz).
[17] Synthetic hot-pressed San Carlos olivine + orthopyroxene mixture (60:40), no report on water content. Enstatite is stronger than olivine.
[18] Synthetic hot-pressed aggregates of forsterite and Mg-enstatite (97:03 to 80:20). Samples are synthesized from sol-gel. The authors noted only weak dependence of creep strength on enstatite content above ~5 vol % of enstatite. The authors interpret the results in terms of grain-size sensitive creep.
[19] Synthetic hot-pressed polycrystalline aggregate of forsterite + Mg-enstatite (60:40). Enstatite is in the proto-enstatite stability field during deformation experiments. Cavitation occurred during creep whose effect is corrected assuming that the cavitation modifies the creep behavior through apparent elongation, $\Delta\varepsilon \approx \frac{1}{3}(\Delta\rho/\rho_0)$ (Raj, 1982). Strong dependence of creep strength on enstatite volume fraction was reported (enstatite is weaker than forsterite). Grain-size sensitivity was not investigated, but the authors interpreted the results in terms of power-law dislocation creep based on microstructural observations.
[20] Lherzolite (62% olivine, 26% orthopyroxene, 10% clinopyroxene, 2% spinel).
[21] The authors determined $r/n = 0.5$. If $n = 4$, then $r = 2$.
[22] The values of A, r and V^* are not independently constrained by this study. The values listed here are for the assumed value of $V^* = 20 \times 10^{-6}\,\mathrm{m^3/mol}$.
[23] This value is for "pure" quartz. With impurities more than ~0.1 wt%, n~2.3.
[24] The activation energy was not determined by this study. These are the values assumed based on other studies.
[25] The pressure range of 0.1–0.45 GPa corresponds to the study by Mei and Kohlstedt (2000b). The creep-law parameters are evaluated by the combination of low-pressure data by Mei and Kohlstedt (2000b) with the data at $P = 0.5$–2.0 GPa.
[26] The solid-medium deformation apparatus designed by Griggs. The stress reading has a large error due to friction.
[27] The solid-medium deformation apparatus with a liquid surrounding a sample. The stress reading is more reliable than the solid-medium Griggs apparatus (see Chapter 6).
[28] A gas-medium deformation apparatus designed by Paterson. The stress is measured by an internal load cell and therefore the stress measurements are with small errors (see Chapter 6).
[29] A room-pressure deformation apparatus with a dead weight as a stress generator.
[30] A deformation apparatus operated at high pressure (to ~10 GPa). The stress and strain are measured by X-ray (see Chapter 6).
[31] A room-pressure deformation apparatus driven by a servo-controlled hydraulic actuator.
[32] The deviatoric stress was estimated from the dislocation density measurements (see Chapter 6).
[33] A gas-medium deformation apparatus with an external load cell. The flow-law parameters are those cited by Caristan (1982).
[34] A gas-medium deformation apparatus with a semi-internal load cell.
[35] A complete flow law for all samples was not provided. The activation energy and stress exponent were determined for selected samples. The authors found remarkable pressure-weakening effects that are interpreted to be due to increase in water fugacity. The influence of water fugacity from this work was quantified later by Post *et al.* (1996). Weak grain-size sensitivity was observed ($m/n \sim 0.2$).
[36] The flow law of quartzite was estimated from a comparison of the microstructures of naturally deformed quartzites with laboratory observations.

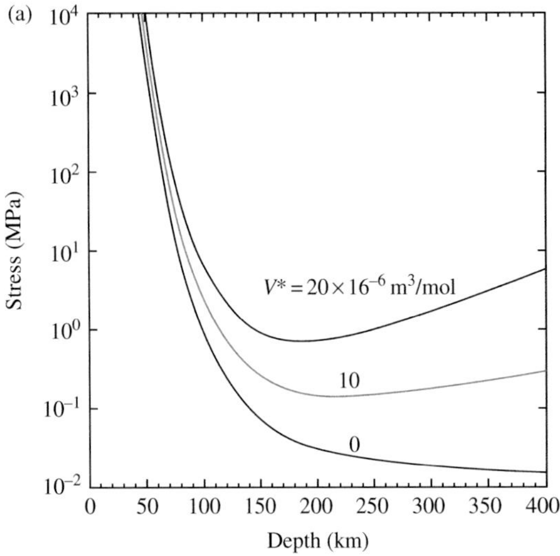

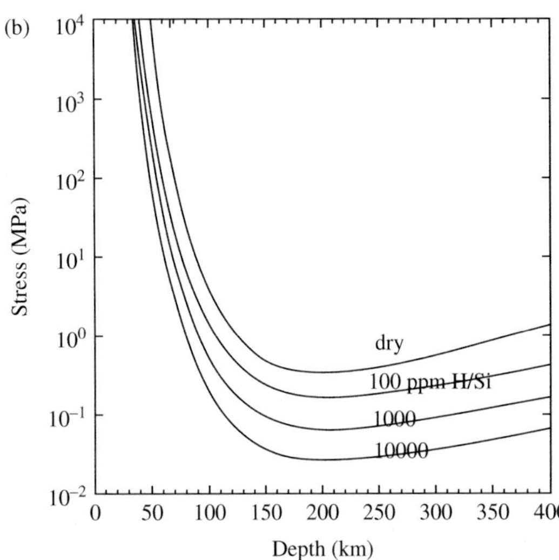

FIGURE 19.6 The strength profile of the upper mantle for a typical geotherm of the oceanic upper mantle based on the flow laws of olivine in the power-law dislocation creep regime. The strain rate is assumed to be constant with depth and the strength corresponding to a strain rate of 10^{-15} s^{-1} is calculated. Part (a) shows the strength–depth profiles for the water-free case for a range of activation volume. Part (b) shows the strength–depth profiles for a range of water content. The water content is assumed to be constant with depth (closed-system behavior). An activation volume of $V^* = 14 \times 10^{-6} m^3/mol$ is assumed. A jump in water content at some depth will lead to a jump in strength (e.g., HIRTH and KOHLSTEDT, 1996).

19.3.3. Upper mantle

The upper mantle is made of olivine, orthopyroxene and other minor phases. In most cases, olivine is weaker than other phases and volumetrically dominant, and therefore olivine is usually assumed to determine the rheological properties of the upper mantle. This assumption can be challenged in some cases where the grain size of orthopyroxene is much smaller than that of olivine (e.g., BOULLIER and GUEGUEN, 1975). It is also possible that the rheological contrast among different minerals (such as olivine and orthopyroxene) changes with water content. However, a large contrast in grain size is not often observed in naturally deformed rocks and the water effect on rheological properties of orthopyroxene has not been quantified. Consequently, I will make a conservative assumption that olivine controls the rheological properties of the upper mantle.[2] Based on this assumption, the effective viscosity of the upper mantle can be predicted in some detail, once temperature, pressure, stress (or strain rate), water content and grain size are given (Fig. 19.6). In this chapter I will discuss the following topics related to the rheological properties of the upper mantle: (1) The strength of the lithosphere, (2) the cause of the lithosphere–asthenosphere boundary, (3) the rheological contrast between the oceanic and the continental upper mantle, and (4) the rheological significance of the Lehmann discontinuity.

19.3.3.1. *The strength of the lithosphere*

The lithosphere is a strong near-surface portion of Earth. The strength of the lithosphere has an important influence on a number of geodynamic problems. A particularly important issue is the strength of the oceanic lithosphere that determines whether plate tectonics occurs or not. In order for subduction to occur at ocean trenches, which is a key element in plate tectonics, a lithosphere must be able to bend at a trench. If the lithosphere is too strong, then subduction is not possible, and hence the convection occurs only in the deeper portion of a planet and the surface layer does not participate in convection. This style of convection is called "stagnant-lid convection" (SOLOMATOV and MORESI, 1996). The transition between plate-tectonic type convection to stagnant-lid type convection occurs when the strength of the lithosphere reaches a certain value.

[2] Transient weakening effects due to small grain size of orthopyroxene may play an important role in short-term deformation events such as deformation associated with continental collision. Orthopyroxene could become weaker than olivine at high hydrogen contents. Also at stresses exceeding ~150 MPa, orthopyroxene deforms by transformation to clinopyroxene. Above this stress, orthopyroxene is much weaker than olivine. In these cases, the rheological properties estimated from coarse-grained olivine will be the upper limit for the actual creep strength.

In a simple model where the strength of the lithosphere is expressed by a "yield stress," the critical yield stress to allow a plate tectonic style of convection is ~200 MPa (e.g., Richards *et al.*, 2001). If one uses a model assuming truly plastic and truly brittle behavior, then the peak strength for an oceanic lithosphere will be ~800 MPa (see Fig. 19.4), that would exceed the critical strength for plate tectonics. Consequently, the cause of weakening that allows plate tectonics on Earth needs to be sought. There have been several models to explain the apparent weakness of the oceanic lithosphere. The first is to invoke the reduction of strength due to the interaction of cracks and dislocations (semi-brittle behavior). Deformation involving both dislocations and cracks has been identified both in laboratory samples and in naturally deformed rocks, but the limitation of this model is that there is no well-established microscopic model for semi-brittle behavior and a quantitative prediction of strength is not possible at this time (see Chapter 7). The second model is to attribute the strength reduction to localized deformation (e.g., Bercovici and Ricard, 2005). Localized deformation is ubiquitous at modest to low temperatures, which reduces the strength of materials (Chapter 16). Indeed, localized deformation such as brittle fracture is known to occur at a stress lower than the strength in the ductile regime (see Chapter 7). Again, the limitation is that there is no well-accepted model for localized deformation under the deep lithosphere conditions that allows us to predict the strength profiles (Chapters 7 and 16). Third, similar to the second model, Korenaga (2007) suggests that the macroscopic strength of the lithosphere of Earth is reduced by the strength of thin cracks containing serpentine formed by deep thermal cracking. Finally, stress-induced transformation of orthopyroxene to clinopyroxene could significantly reduce the strength of the lithosphere. Orthopyroxene cannot support deviatoric stress above the threshold stress for this transformation (~200 MPa) (Raleigh *et al.*, 1971). Above this stress, orthopyroxene behaves like a weak component, and hence deformation will be concentrated in orthopyroxene. However, the details of deformation caused by this phase transformation are not well constrained.

19.3.3.2. *The lithosphere–asthenosphere boundary*

Let us now consider the nature of the lithosphere–asthenosphere transition. Our focus is on the factors that control the strength in the ductile regime. Two issues are critical here: (1) the physical cause of the lithosphere–asthenosphere transition, and (2) the dependence of thickness of lithosphere on definitions. Obviously these two questions are related because the cause of the lithosphere–asthenosphere transition depends on the definitions.

I will first discuss several different definitions of lithosphere. The lithosphere is defined as a strong (near-surface) layer above a weak asthenosphere. However, as we have learned before the meaning of "strong" or "weak" depends on the physical conditions at which deformation occurs particularly the stress level and the time-scale. Consequently, the thickness of the lithosphere depends on the definition. In fact, the thickness of the lithosphere can vary as much as a factor of ~5 or more as we will see below.

Lithosphere thickness by various definitions

(i) Mechanical lithosphere: 1. From seismological observations

The seismic lithosphere may be defined as a layer of high seismic wave velocities on top of a seismic *low-velocity zone*. Because the structure of a low-velocity layer cannot be determined in any detail using the traditional body-wave travel-time analysis, the lithosphere–asthenosphere structure is usually studied using surface waves. These studies show that the thickness of the oceanic lithosphere is ~50–100 km (Kanamori and Press (1970) estimated it to be ~70 km) but varies with its age: the thickness changes from ~0 km at the mid-ocean ridges to ~50–100 km at ages older than ~80 Myr (Yoshii, 1973; Forsyth, 1975). Surface waves are sensitive to the depth variation of structures, but their resolution is poor because the wavelength is large (to investigate lithosphere–asthenosphere structure, one needs to use surface waves with ~500–1000 km wavelength). The recent studies using reflected body-waves show, however, that the thickness of a high-velocity lithosphere does not change much with the age and is ~65 km (Gaherty *et al.*, 1999). In this study, the thickness of the oceanic lithosphere is defined by the depth at which a reflection of body-waves comes from. Therefore that depth corresponds to a depth at which a *sharp* velocity jump occurs. Fig. 19.7 summarizes the thickness of seismic lithosphere determined by seismic wave velocities.

One important complication in estimating the thickness of the seismic lithosphere is the influence of *anisotropy* (Chapter 21). Regan and Anderson (1984) showed that the estimated thickness of the oceanic lithosphere depends strongly on anisotropy and the lithosphere thickness does not appear to increase with age so much when anisotropy is included in the analysis.

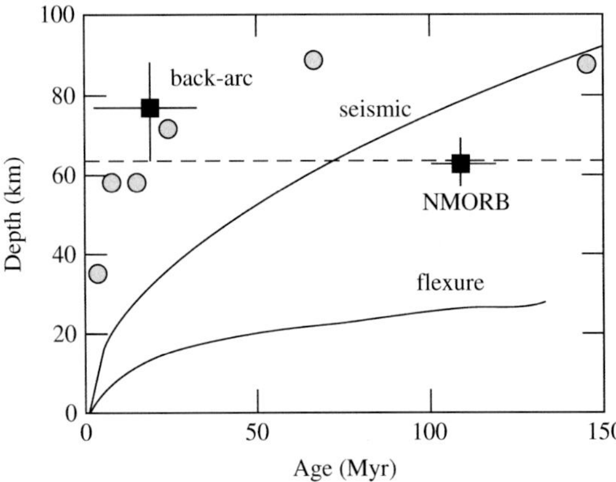

FIGURE 19.7 The relation between the thickness of the lithosphere versus the age of the lithosphere corresponding to various definitions of a lithosphere (modified from KARATO and JUNG, 1998). Circles are from surface-wave studies by FORSYTH (1975), solid squares are from body-wave studies by REVENAUGH and JORDAN (1991) and GAHERTY *et al.* (1996). The solid curve marked "seismic" denotes the results from YOSHII (1973) and the curve with "flexure" denotes elastic bending of the lithosphere. The seismic lithosphere is generally thicker than the lithosphere defined by bending (flexure).

ANDERSON (1979) suggested that the thickness of the continental lithosphere determined by JORDAN (1975) (~400 km) might be overestimated due to the influence of anisotropy. The recent studies including the influence of anisotropy show ~200–300 km thick seismic lithosphere for old cratons, and the deep continental roots are highly anisotropic (e.g., GUNG *et al.*, 2003). These studies clearly show that truly thick lithosphere is limited to the Archean cratons and the boundary between thick and thin lithosphere is sharp suggesting that chemical composition is different between Proterozoic (younger than ~2.7 Gyr) and Archean (older than ~2.7 Gyr) mantle.

(ii) Mechanical lithosphere: 2. From seismicity

Earthquakes in the oceanic lithosphere occur only in the relatively shallow, cold portions (CHEN and MOLNAR, 1983). The depth of this layer corresponds roughly to an isotherm of ~870 K. JACKSON (2002a) used the distribution of seismicity to infer the thickness (strength distribution) of the continental lithosphere but HANDY and BRUN (2004) pointed out that the relation between seismicity and strength is indirect.

(iii) Mechanical lithosphere: 3. From observations on bending (flexure)

The thickness of the lithosphere can be determined from the geometry of bending. In this approach, we assume that Earth is rheologically stratified as an elastic layer (the lithosphere) at the surface and a weak (low-viscosity) layer below (e.g., WALCOTT, 1970; PELTIER, 1984; FORSYTH, 1985; NAKADA and LAMBECK, 1989; MCKENZIE, 2003). When a load is applied at the surface (e.g., ice sheets, volcanoes), then there will be elastic deformation of the lithosphere. The bending of the lithosphere is characterized by a flexural rigidity, $D = Eh^3/12(1-\nu^2)$ (h, thickness of lithosphere; E, Young's modulus; ν, Poisson's ratio; Chapter 18), and by analyzing the topography (and gravity), one can determine D. Knowing the elastic properties (Young's modulus, E, and Poisson's ratio, ν) one can determine the effective thickness of the lithosphere, h. The thickness of the lithosphere determined by this method is also plotted in Fig. 19.7. In general, elastic thickness inferred from bending observation is much thinner than the thickness of seismic lithosphere (the results by PELTIER (1984) are an exception, but a later work by NAKADA and LAMBECK (1987) showed that the inference of lithosphere thickness from post-glacial rebound observation is highly non-unique due to strong trade-off between mantle viscosity and lithosphere thickness, and the influence of melting history of ice sheets, see Chapter 18). WALCOTT (1970) noted that the thickness of elastic lithosphere depends on the age of loading: longer is the age, thinner is the lithosphere.

(iv) Thermal lithosphere

The thickness of the lithosphere may be defined by the temperature profile. When temperature reaches a certain value, materials will become soft. Therefore the lithosphere–asthenosphere boundary may correspond to a certain temperature. Let us consider the oceanic upper mantle. When hot materials are carried to the surface at mid-ocean ridges and cooled from above while they spread (with velocity v), then the temperature profile across the near-surface layer is given by the error function, $T(z,t) - T_0 = (T_\infty - T_0)\,\mathrm{erf}\left(z/2\sqrt{\kappa t}\right)$, where T_∞ is the temperature of hot, deep materials, T_0 is the surface temperature, κ is thermal diffusivity (Chapter 17). One can define the bottom of the lithosphere as the depth at which temperature reaches a certain fraction of this asymptotic temperature. From this equation, it follows immediately that the thickness of the lithosphere with this definition increases with $\sqrt{t}$ where t is the age of the lithosphere,

$$d \propto \sqrt{\kappa t}. \tag{19.6}$$

The relation (19.6) is also shown in Fig. 19.7. This model works for the oceanic lithosphere (see Chapter 17). For the continental lithosphere, the definition of "age" becomes ambiguous, but if the "age" is interpreted to be the time from the last tectonic (volcanic) event, then the relation similar to (19.6) roughly holds for continents as well. The lithosphere thickness defined in this way could be inferred from the surface heat flow measurements (e.g., POLLACK *et al.*, 1993; ARTEMIEVA, 2006).

(v) Lithosphere defined by the flow pattern

The thickness of a lithosphere may be defined based on the flow pattern. When the surface plate moves horizontally, materials near the surface move with almost the same velocity as the surface plate, but materials deep in the mantle will have quite different velocities. One may define the lithosphere as a layer where materials move with a velocity higher than a certain fraction of the surface velocity (say 90%). For a simple two-dimensional flow driven by the surface motion of a rigid plate, one can get the depth variation of flow velocity of the mantle as (Problem 19.3),

$$u(z) = u(0)\left[1 - \frac{\int_0^z (dx/\eta(x))}{\int_0^L (dx/\eta(x))}\right] \tag{19.7}$$

where L is the depth at which horizontal flow is zero. The horizontal velocity of mantle materials is nearly the same as the surface where viscosity is large, but will deviate from the surface velocity when viscosity becomes low. Fig. 19.8 illustrates the depth variation of horizontal flow velocity. Note that the velocity pattern shows a sharp kink at a depth at which viscosity becomes low. A low viscosity layer *decouples* motion between a layer above and below.

If the depth variation of viscosity is written as $\eta(z) = \eta_0 \exp(-z/\tilde{L})$, then the lithosphere thickness defined this way is given by

$$z_L = \tilde{L} \log\left[(1-\xi)\exp\frac{L}{\tilde{L}} + \xi\right]. \tag{19.8}$$

where $\xi \equiv u(z_L)/u(0)$. For $\xi = 0$, $z_L = L$ and $\xi = 1$, $z_L = 0$ as it should be. The flow pattern depends on the boundary condition, i.e., L, and the depth variation of viscosity, $\tilde{L}$. Lithosphere defined by the flow pattern is not directly observable. However, distribution of seismic anisotropy, that is caused by finite-strain deformation, provides some clues as to the flow pattern in the mantle (see Chapter 21).

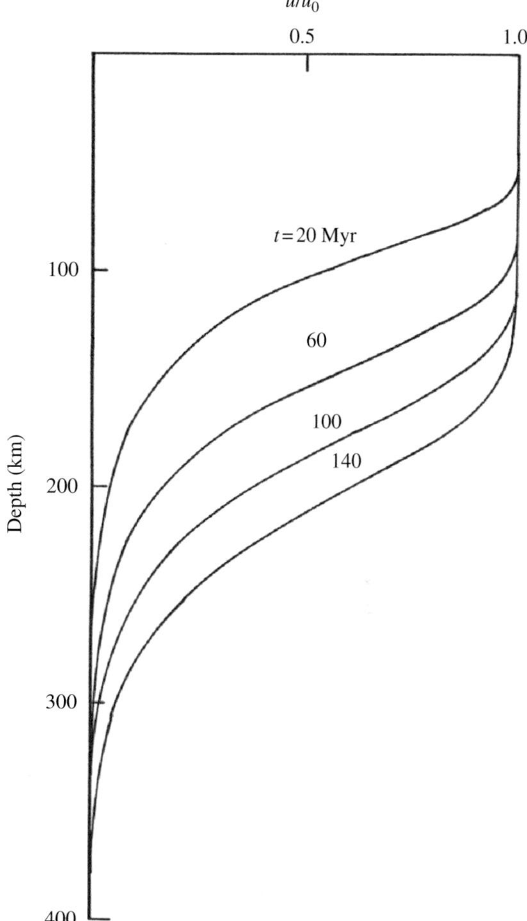

FIGURE 19.8 The flow pattern of material caused by the horizontal motion of a surface plate, where u is the horizontal velocity of the material, u_0 is the horizontal velocity of the surface plate and t is the age of the lithospheric plate.

Problem 19.3

Derive the relations (19.7) and (19.8).

Solution

In the two-dimensional model, the shear stress is independent of depth and therefore $\sigma = \eta(z)(\partial u/\partial z)$ where u is the horizontal velocity, and $\eta(z)$ is the depth-dependent viscosity. Therefore by integrating this (stress σ is constant with depth) with the boundary condition that the horizontal velocity is zero at $z = L$, one obtains $u(z) = u(0)\left[1 - \frac{\int_0^z (dx/\eta(x))}{\int_0^L (dx/\eta(x))}\right]$.

If $\eta(z) = \eta_0 \exp(-z/\tilde{L})$ is inserted into this, then after integration, $1 - u(z_L)/u(0) = \frac{\exp(z_L/\tilde{L})-1}{\exp(L/\tilde{L})-1}$ and hence (19.8).

Physical mechanisms for the lithosphere–asthenosphere boundary: generalities

The primary cause for the lithosphere–asthenosphere transition is temperature: asthenosphere is made of softer materials than lithosphere because of higher temperatures. This is obvious for the thermal definition of lithosphere, equation (19.6). Other definitions are on the basis of mechanical properties. To a first order, any mechanical properties, rheological and elastic properties, become "softer" at higher temperatures. Let us consider rheological properties. The lithosphere–asthenosphere transition occurs at a depth at which viscosity of materials becomes small enough. For example, the lithosphere defined by bending (flexure) is controlled by the elastic–plastic transition that is controlled by the Maxwell time (Chapter 3). If the Maxwell time is significantly larger than the time-scale of processes, then that region behaves like an elastic material and is regarded as a lithosphere, i.e.,

$$\tau_M(T, P, \sigma, C_{OH}, L) \equiv \frac{\eta(T, P, \sigma, C_{OH}, L)}{M(P, T, C_{OH})} > \tau_{geo} \qquad (19.9)$$

where τ_{geo} is the time-scale of the relevant geological phenomenon. Effective viscosity depends on a number of parameters including temperature, pressure, stress, water content and grain size (see section 19.2). Therefore the Maxwell time (viscosity) is a function of temperature, stress and water content, $\tau_M = \tau_M(T, P, \sigma, C_{OH}, L)$, hence the lithosphere–asthenosphere boundary is a function of these variables as well as the time-scale of the process, τ_{geo} (elastic constant, M, depends only weakly on pressure, temperature and water content; Chapters 4, 11).

Among these variables, the influence of grain size will be ignored because there is no reason to suppose any large variation in grain size across the lithosphere–asthenosphere boundary, and the dominant deformation mechanism in this region is likely to be dislocation creep in which the effective viscosity is insensitive to grain size. The temperature and pressure effects can be combined if one uses the homologous temperature scaling i.e., $\eta(T, P, C_{OH}) \approx \eta(T'(T, P, C_{OH}))$ with $T' = T/T_m(P, C_{OH})$ (see Chapter 10). So to a good approximation, equation (19.9) can be reduced to $\tau_M(T', \sigma) > \tau_{geo}$.

TABLE 19.2 Time-scales and stress magnitudes for various geophysical processes.

	Time-scale (s)	Stress (MPa)
Seismic waves	$1–10^3$	$10^{-4}–10^{-2}$
Bending	$10^{11}–10^{14}$	$10^2–10^3$
Mantle convection	$10^{14}–10^{15}$	1–10

Let us consider how the difference in the stress level and time-scale as summarized in Table 19.2 may affect the inferred lithosphere thickness. Other factors being the same, the increase in stress reduces viscosity and hence reduces the lithosphere thickness. Similarly, if the time-scale of deformation becomes larger, then critical viscosity to satisfy inequality (19.9) becomes larger hence lithosphere thickness will become smaller. Fig. 19.9 shows a relation between stress and the temperature at the base of the lithosphere for a range of time-scales of deformation, water content. In this calculation, I consider only dislocation creep, so the influence of grain size is ignored (this can be justified because there is strong seismic anisotropy in this region (see Chapter 21)). Note that the temperature (and hence the thickness) at the bottom of the lithosphere depends strongly on the stress magnitude. The lithosphere defined in terms of bending (flexure) corresponds to a stress of ~0.1–1 GPa, and corresponding temperature (thickness) is much lower (smaller) than that for lower stresses. The application of these results to a seismologically defined lithosphere–asthenosphere boundary is not straightforward because the physical mechanisms that control the temperature dependence of "softness" for seismic wave propagation are different from long-term plastic deformation (Chapters 11 and 20, see also Fig. 19.10 and related discussions).

The direct influence of partial melting is not considered because its effect is likely to be small (the fraction of melt in Earth's mantle is in most cases less than ~1%. The presence of ~1% melt will reduce the viscosity only by ~20%, see Chapter 12). An important effect of partial melting is its influence on re-distribution of water (hydrogen), so the influence of partial melting can be understood from Fig. 19.9 if one knows how water is distributed as a result of partial melting.

Influence of partial melting and water (hydrogen)

The depth at which the lithosphere–asthenosphere transition occurs roughly coincides with the depth at which geotherm is close to or above the solidus

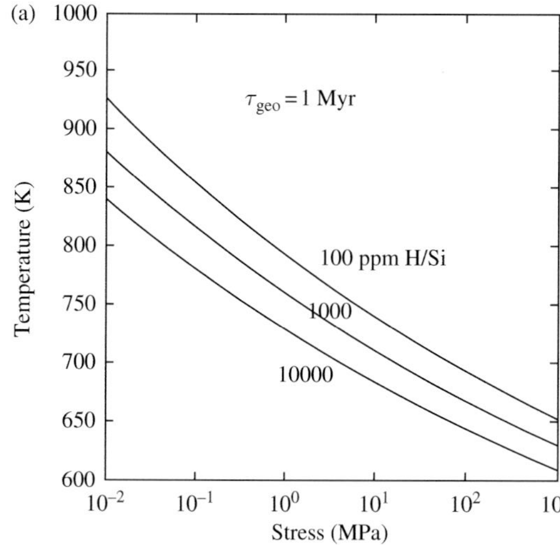

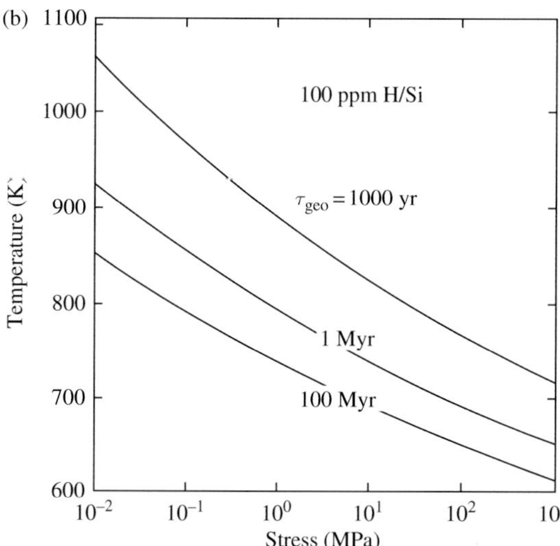

FIGURE 19.9 The relation between the temperature at the bottom of the lithosphere and the stress for a range of time-scales of deformation, τ_{geo}, water content. The experimental results from KARATO and JUNG (2003) are used. (a) The temperature at the bottom of the lithosphere as a function of stress and water content (for a time-scale of 1 Myr), (b) the temperature of the bottom of the lithosphere as a function of stress and time-scale (for a water content of 100 ppm H/Si).

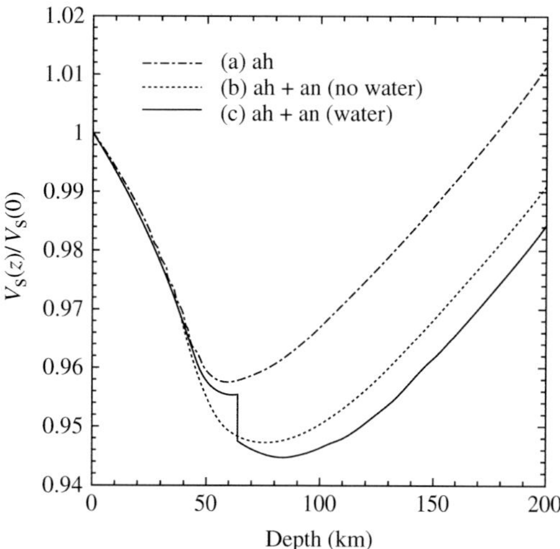

FIGURE 19.10 The seismic wave (shear wave) velocity–depth profile for the oceanic upper mantle for three models. (a) Anharmonicity (ah), (b) anharmonicity + anelasticity (temperature) (ah + an (no water)), (c) anharmonicity + anelasticity (temperature + water) (ah + an (water)).

(Chapter 17). Consequently, partial melting was often considered as a cause for the lithosphere–asthenosphere transition (e.g., SKINNER *et al.*, 2004; RINGWOOD, 1975). However, as discussed in Chapter 12, the direct effect of partial melting, i.e., the influence of the presence of melt phase on rheological properties is small for a typical upper mantle system (at low pressures) if the melt fraction is less than ~1%. The melt fraction in the asthenosphere (~50–150 km) is estimated to be less than 0.2% (e.g., LANGMUIR *et al.*, 1992). Therefore a classic idea that the presence of a small amount of melt is the cause of the asthenosphere is not supported by the mineral physics observations (e.g., KARATO, 1986; KARATO and JUNG, 1998).

A more important role of partial melting on rheological properties is likely to be through its effect on re-distributing water (KARATO, 1986). This idea is based on the experimental observations of a large difference in water solubility between melts and solid minerals, and of the strong effects of the presence of water on rheological properties. When partial melting occurs in a system that is closed with respect to water, then water is removed from the solid minerals that increases the creep strength of the solid part. At the same time, the presence of melt will reduce the creep strength of a rock (Chapter 12). The net effect will then be

$$\dot{\varepsilon} = \dot{\varepsilon}_0 \cdot F_1(\phi_1) \cdot F_2(\phi_2) \tag{19.10}$$

where $\dot{\varepsilon}$ is the strain rate of a partially molten material, $\dot{\varepsilon}_0$ is the strain rate before partial melting, $F_1(\phi_1)$ is a function representing the influence of removal of water by partial melting that depends on the *degree of melting*,[3] ϕ_1 , and $F_2(\phi_2)$ is a function representing the influence of the presence of melt that depends on the *fraction of melt*, ϕ_2, and the geometry of the melt pocket such as the dihedral angle (Chapter 12). In general, $F_1(\phi_1) < 1$ and $F_2(\phi_2) > 1$ and the net effect of partial melting depends

[3] The degree of melting should not be confused with the fraction of melt (Box 10.2).

on the competition of these two effects, i.e., either $F_1 \cdot F_2 > 1$ or $F_1 \cdot F_2 < 1$. When we consider rheological properties of a material after the removal of melt, then $F_2(\phi_2) = 1$, so we only consider the influence of water removal and the net effect is always hardening.

Let us consider some more details of $F_1(\phi_1)$ and $F_2(\phi_2)$. If the strain rate for a totally water-free sample is much less than the actual strain rate of a sample with some water,[4] and if the process of partial melting is fractional melting then (Chapter 10)

$$F_1(\phi_1) = (C_{OH}(\text{final})/C_{OH}(\text{initial}))^r \approx \exp\left(-\frac{r\phi_1}{k}\right) \tag{19.11}$$

where r is the water content exponent ($\sim 1.0 \pm 0.3$) and C_{OH}(initial, final) is the water content in solid minerals before or after partial melting respectively. The presence of melt enhances deformation following

$$F_2(\phi_2) \approx \exp\left(3B(\theta)\sqrt{\phi_2}\right) \tag{19.12}$$

where $B(\theta)$ is a function of the dihedral angle, θ (Chapter 12). So the net effect is

$$\dot{\varepsilon} = \dot{\varepsilon}_0 \cdot \exp(\Gamma) \tag{19.13}$$

with

$$\Gamma \equiv 3B(\theta)\sqrt{\phi_2} - \frac{r\phi_1}{k}. \tag{19.14}$$

If $\Gamma > 0$, there will be softening, whereas if $\Gamma < 0$, there will be hardening due to partial melting. Assuming fractional melting with a constant melt fraction, one can show that a partially molten material is weaker than the initial material at small melt fraction but becomes stronger than the unmelted material if the degree of melting exceeds a critical value, $\phi_1 > \phi_c$. From the relation (19.14), this critical value is given by,

$$\phi_c = 3B(\theta)\sqrt{\phi_2}\frac{k}{r}. \tag{19.15}$$

With reasonable values of relevant parameters ($3B(\theta) \sim 5$–10, $\phi_2 \sim 1\%$, $k \sim 10^{-2}$, $r \sim 1$), one has $\phi_c \sim 0.5$–1%.

Problem 19.4

Derive equation (19.15).

Solution

Let us assume fractional melting and assume that the melt fraction at any time is fixed to be ϕ_2. Let us also assume that the production rate of melt is constant so that $\phi_1 = Ax$, where x is a parameter to describe the progress of melting. Then from equation (19.14), $\Gamma = 0$ at $x = x_c \equiv (3B(\theta)\sqrt{\phi_2}/A)\ (k/r)$. Inserting this into $\phi_1 = Ax$, we get the critical degree of melting as $\phi_c \equiv 3B(\theta)\sqrt{\phi_2}\ (k/r)$. If the degree of melting is below this value, $\Gamma > 0$, so partial melting softens the material. Above this value of ϕ_1, $\Gamma < 0$ and there will be hardening.

In contrast, when one considers the rheological properties of materials after the completion of chemical differentiation due to partial melting or after cooling below the solidus, then the melt fraction is zero, and the only effect is through $F_1(\phi_1) \approx \exp(-r\phi_1/k)$, and the influence of partial melting is always hardening. In this case, significant hardening occurs when $r\phi_1/k > 3-4$ (for $r = 1$, $k = 10^{-2}$, $\phi_1 > 3-4\%$).

Let us apply this concept to the processes beneath a mid-ocean ridge. Beneath a mid-ocean ridge, hot materials ascend nearly adiabatically. Because the adiabatic temperature gradient is less than the pressure dependence of solidus (i.e., $(dT/dz)_{ad} < dT_s/dz$ where $(dT/dz)_{ad}$ is the adiabatic temperature gradient and T_s is the solidus), the temperature of ascending materials will reach the solidus at a certain depth. Then partial melting begins and water content of a rock will start to be reduced. Consequently, the solid part of the rock will be strengthened. So in a column of material beneath a ridge in which partial melting proceeds, the above analysis involving both $F_1(\phi_1)$ and $F_2(\phi_2)$ applies and we conclude that partial melting initially softens the material but when the degree of melting exceeds $\phi_c = 3B(\theta)\sqrt{\phi_2}(k/r)(\sim 0.5-1\%)$ the material starts to become harder than before. As the materials beneath a ridge ascend they eventually turn to move to the sub-horizontal direction and temperature will drop due to cooling from above. Consequently, at a distance far away from the ridge, the majority of materials there are free from melt. The rheological profile in this region is therefore determined mostly by the degree of melting that has removed water beneath the ridge (i.e., $F_1(\phi_1)$). The degree of melting beneath the ridge is controlled by the interplay between geotherm and solidus and increases as material ascends. When the geotherm becomes higher than the solidus for a given water content, partial melting starts. However, the degree of melting at this stage is controlled by the water content and is small (usually less than ~0.2 %). At a point where the geotherm becomes higher than the solidus for a water-free

[4] If this condition is not met, then the contribution from the strain rate of totally water-free material must be added to the net strain rate.

system, the degree of melting increases abruptly and exceeds a few %. For a typical oceanic upper mantle this occurs at ~60–70 km (e.g., LANGMUIR *et al.*, 1992). Consequently, a sharp decrease in $F_1(\phi_1)$ occurs at this depth and this depth makes a sharp change in viscosity (HIRTH and KOHLSTEDT, 1996).

Note that the above discussion holds for a typical upper mantle (and lower crust) where dihedral angles are non-zero. There are some experimental observations suggesting that the dihedral angle decreases with pressure (MIBE *et al.*, 1998). The small influence of partial melting cited above is based on low-pressure experiments on upper mantle systems. Evidence of complete wetting (i.e., zero dihedral angle) at high pressures is reported by Yoshino *et al.* (2007).

The extension of the above discussion to a seismologically defined lithosphere–asthenosphere boundary is not straightforward. The lithosphere–asthenosphere transition is characterized by a relatively sharp *decrease* in seismic wave velocity and an increase in attenuation with depth. Both of these features indicate that the asthenosphere is a layer where materials are *soft*. In understanding what soft means in seismological observations, it is important to recall that there are two distinct mechanisms by which seismic wave propagation is affected by the softness. These are *anharmonicity* (Chapter 4) and *anelasticity* (Chapter 11). Anharmonicity is essentially the effects of thermal expansion and hence reflects the temperature effect. Anharmonicity reduces the elastic constant almost linearly with temperature (Chapter 4). Since the temperature in the oceanic upper mantle changes with depth and age following roughly the error function formula (Chapter 17), the combination of temperature and pressure effects provides a *broad* minimum in velocity at the depth where the temperature–depth profile has a large change in slope as temperature approaches the adiabat (Fig. 19.10). Consequently, the depth at which this minimum occurs changes with age, t, following $z_s \propto \sqrt{\kappa t}$.

The influence of anelasticity was invoked in the late 1970s to early 1980s to explain the relatively sharp lithosphere–asthenosphere transition (e.g., GUEGUEN and MERCIER (1973) and MINSTER and ANDERSON (1980); for recent similar works see also FAUL and JACKSON (2005) and STIXRUDE and LITHGOW-BERTELLONI (2005a)). Anelasticity depends exponentially on temperature and hence a somewhat sharper transition is expected. However, the transition expected from thermal models is still gradual (Fig. 19.10). Therefore these models cannot explain some observations suggesting a sharp, although small, velocity reduction at the lithosphere–asthenosphere boundary (e.g., REVENAUGH and JORDAN, 1991; GAHERTY *et al.*, 1999). FAUL and JACKSON (2005) also included the influence of grain size, but its effect is small (Chapter 11). Incorporation of a sharp change in water content could make a relatively sharp change in the velocity–depth relation (KARATO and JUNG, 1998). However, the ability of this model to explain a sharp velocity reduction is limited to ~1% (corresponding to a Q_S of ~100). If there is more than a several percent sharp velocity drop at the lithosphere–asthenosphere boundary as suggested by RYCHERT *et al.* (2005), then some other mechanisms such as the direct effects of partial melting or a change in lithology will need to be invoked.

19.3.3.3. *Rheological contrast between the continental and oceanic upper mantle*

Seismological as well as geochemical studies showed that the contrast between the continental and the oceanic upper mantle to ~200–300 km is a long-lived structure, at least more than ~2 Gyr (CARLSON *et al.*, 1994; GUNG *et al.*, 2003). The rheological contrast between the oceanic and the continental upper mantle is a critical factor in understanding the long-term survival of the deep continental roots. DOIN *et al.* (1997) and LENARDIC and MORESI (1999) for example showed that in order to preserve thick (>200 km) continental roots more than ~1 Gyr, the viscosity of the continental roots must be significantly (by a factor of $\sim 10^3$ or more) larger than that of the oceanic upper mantle at a similar depth. Since the mineralogy is likely to be similar between the continental and the oceanic upper mantle (except for minor components such as clinopyroxene and garnet (e.g., JORDAN, 1981)), the main cause for the rheological contrast between the oceanic and the continental upper mantle must be either the difference in temperature or water content (or both). According to SCLATER *et al.* (1981), RUDNICK *et al.* (1998) and ARTEMIEVA (2006) however, the temperatures in the Archean craton (old continents) at 200 km depth are close to the mantle adiabat and hence similar to that of the oceanic mantle.

Therefore a likely cause for the rheological contrast is the contrast in water content. However, there are two conflicting observations about the water content in the continental upper mantle. On the one hand, most of the magmas found in the continents (e.g., kimberlites) contain a larger amount of volatiles including water (PASTERIS, 1984) and some authors assumed that the continental upper mantle is water-rich (e.g., KOHLSTEDT *et al.*, 1995). On the other hand, if the

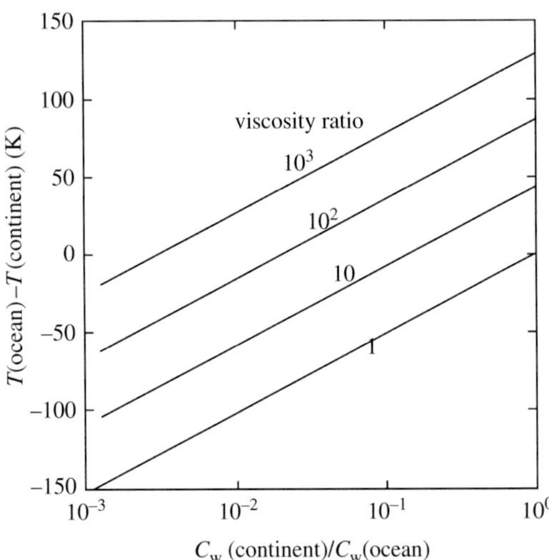

FIGURE 19.11 Viscosity contrasts (for a given stress) between continental and oceanic upper mantle as a function of contrasts in temperature and water content at ~200 km depth (C_w is the water content).

continental upper mantle has undergone a larger degree of depletion due to partial melting (e.g., JORDAN, 1981), then the continental upper mantle would have much less water than the oceanic upper mantle (e.g., POLLACK, 1986). If a large portion of the continental lithosphere is water-rich, then its viscosity will be *lower* than that of the oceanic upper mantle at the same depth (below ~200 km), and it will be difficult to understand the longevity of a thick continental lithosphere as demonstrated by geochemical studies (e.g., PEARSON, 1999). This apparent paradox can be solved if one accepts that the continental upper mantle is heterogeneous and the majority of the continental upper mantle is depleted with water but occasionally it is hydrated by local volcanic activities. However, it is not easy to obtain a factor of ~10^3 difference in viscosity by the contrast in water content alone: if the water content in the oceanic asthenosphere is ~1000 ppm H/Si, then the water content of the continental lithosphere will have to be ~1 ppm H/Si. A combination of a lower temperature and a lower water content of the continental lithosphere than those of the deep oceanic upper mantle could yield a large viscosity contrast needed to stabilize the continental deep roots (Fig. 19.11).

19.3.3.4. *The Lehmann discontinuity*

A seismic discontinuity is observed at around ~200–250 km depth mostly in the continental upper mantle. At this depth the seismic wave velocity increases slightly with depth. This is referred to as the *Lehmann discontinuity* (e.g., REVENAUGH and JORDAN, 1991; GU *et al.*, 2001; DEUSS and WOODHOUSE, 2004). The Lehmann discontinuity is enigmatic because there is no known phase transformation in major constituent minerals at this depth (Chapter 17). There have been some suggestions that the Lehmann discontinuity might be related to rheological behavior of upper mantle materials. REVENAUGH and JORDAN (1991) (see also GAHERTY and JORDAN, 1995) proposed that the layer above the Lehmann discontinuity corresponds to the deformed state and the layer below corresponds to the annealed state, implying no active deformation in the deep continents. However, it is difficult to understand the physical reason for the transition to annealed state. KARATO (1992) proposed that the Lehmann discontinuity might correspond to the change in deformation mechanisms from dislocation creep in the shallower mantle to diffusional creep in the deeper upper mantle. This hypothesis is based on the known large difference in activation volume between dislocation creep and diffusional creep in olivine that makes dislocation creep a favorable mechanism of deformation in the shallow portion. An implicit assumption here is that the grain size in the deep upper mantle is fixed by some mechanism such as the Zener pinning (see Chapter 13) so that grain size does not grow over the critical size for dislocation creep. DEUSS and WOODHOUSE (2004) discussed that the regional variation in the depth of the Lehmann discontinuity is consistent with this model.

19.4. Rheological properties of the deep mantle

The mantle below the 410-km discontinuity occupies ~80% of the mantle and the rheological properties of materials there have an important influence on the dynamics and evolution of this planet. However, the rheological properties of Earth materials below the transition zone are poorly constrained at this time because of the lack of reliable direct measurements of rheological properties under these conditions. The obvious reason is the difficulties in quantitative deformation experiments under deep mantle conditions (for details see Chapter 6). Consequently, the discussions below are necessarily more sketchy than the discussion on the rheological properties of the shallower portions of Earth. However, some estimates of rheological properties are possible by the combination of (1) systematics between rheological properties and crystal structure (see Chapter 15), (2) distribution of homologous temperature (T/T_m, T_m, melting temperature), (3) (limited) results of direct mechanical tests, (4) diffusion coefficients measured at deep mantle conditions and (5) models of grain-size distribution.

19.4.1. Transition zone

The transition zone of Earth's mantle is made mainly of high-pressure polymorphs of olivine and pyroxenes, i.e., wadsleyite, ringwoodite and majorite (see Chapter 17). The presence of seismic anisotropy in the transition zone (e.g., Trampert and van Heijst, 2002) suggests that the dominant mechanism of deformation is dislocation creep, although in some regions of small grain size (e.g., in a cold subducting slabs after phase transformation), diffusional creep will be important. Direct deformation experiments on majorite garnet have not been performed but some preliminary results are available for deformation of wadsleyite and ringwoodite (Sharp *et al.*, 1994; Chen *et al.*, 1998; Karato *et al.*, 1998; Kubo *et al.*, 1998; Mosenfelder *et al.*, 2000; Thurel and Cordier, 2003; Thurel *et al.*, 2003a, 2003b; Xu *et al.*, 2003; Nishihara *et al.*, 2007b). Most of these studies are for dislocation creep and they suggest that wadsleyite and ringwoodite are stronger than olivine compared at similar conditions. Karato *et al.* (1998) also reported evidence for the transition in deformation mechanisms between dislocation and diffusional creep (or superplasticity) in ringwoodite (Fig. 19.12). This transition was inferred by the microstructural observations including dislocation density and the morphology of grain boundaries. However, there is no consensus as to the role of water in deformation of these minerals. Based on the observed enhancement of the kinetics of olivine to wadsleyite phase transformation, Kubo *et al.* (1998) inferred that water significantly enhances plastic deformation of wadsleyite. In contrast, based on deformation experiments of powder samples, Chen *et al.* (1998) reported that water weakening effects are very small for wadsleyite and ringwoodite. Based on the systematics in creep and crystal structure (Karato, 1989c), Karato *et al.* (1995a) suggested that ringwoodite is somewhat stronger than olivine and garnet is significantly stronger than other co-existing minerals, and therefore a subducting slab in the transition zone is rheologically stratified having a relatively strong garnet-rich paleo-oceanic crust. Jin *et al.* (2001) reached the same conclusion. However, Katayama and Karato (2007) noted that the relative strength of garnet and olivine changes with water content.

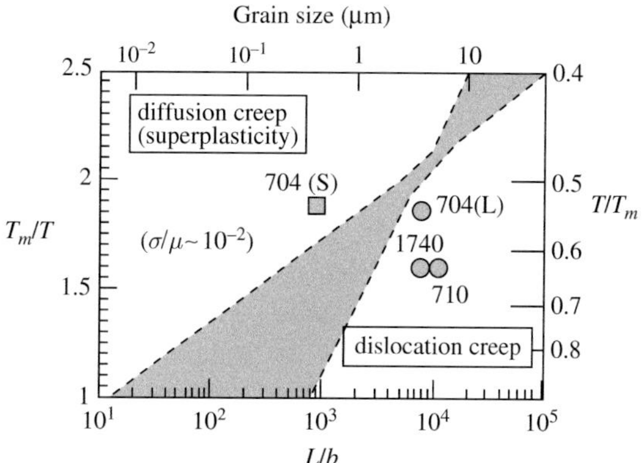

FIGURE 19.12 A deformation mechanism map for ringwoodite inferred from microstructural observations (after Karato *et al.*, 1998), where T is the temperature, T_m is melting temperature, L is grain size, and b is the length of the Burgers vector. Diffusional creep in fine-grained samples is inferred from the low dislocation density and the presence of many four-grain junctions that suggest significant grain-boundary sliding (see Chapter 8). Dislocation creep in coarse-grained samples is inferred from the high dislocation density.

The kinetics of phase transformation may affect the rheological properties (see Chapter 15). It was sometimes speculated that the internal stress (strain) associated with a phase transformation might enhance deformation (*transformational plasticity*) (Sammis and Dein, 1974; Parmentier, 1981; Panasyuk and Hager, 1998). These authors suggested that there may be a weak layer near the 410- or 660-km discontinuity due to transformation plasticity. However, the analysis of microscopic mechanisms of this process suggests that this effect is not important in most hot regions (Chapter 15).

Another better documented process is the weakening due to grain-size reduction. When a phase transformation occurs, the grain size of new phase material is sometimes reduced. Vaughan and Coe (1981) presented the first experimental data showing weakening due to grain-size reduction associated with the olivine–spinel phase transformation using a germanate analog (Mg_2GeO_4). Rubie (1984) discussed the possible role of grain-size reduction due to the olivine–spinel phase transformation in controlling the rheological properties of slabs. The physics of this process is discussed in Chapter 13. Briefly the grain size after a phase transformation is sensitive to the temperature at which the phase transformation occurs: when a phase transformation occurs at low temperatures, resultant grain size of a new phase tends to be small. Riedel and Karato (1997) and Karato *et al.* (2001) made a detailed analysis of this process for subducting slabs and suggested that significant weakening occurs in a cold slab but not in a warm one.

In summary, the transition zone minerals are somewhat stronger than the upper mantle minerals compared at similar conditions. The strength difference probably increases with the amount of garnet in the transition zone. Grain-size reduction in cold regions of subducting slabs due to phase transformations will reduce the strength locally. Water is likely to reduce the creep strength of transition zone minerals, but the quantitative experimental data are missing at this time.

19.4.2. Lower mantle

The lower mantle of Earth is composed mainly of (Mg, Fe)SiO_3 perovskite and (Mg, Fe)O. Based on the absence of seismic anisotropy, KARATO *et al.* (1995b) suggested that most of the lower mantle is deformed by diffusional creep (or superplasticity), although some regions of the bottom boundary layers may be deformed by dislocation creep. Flow laws of various perovskites were determined and a general trend in dislocation creep of crystals with perovskite structure was investigated (e.g., BEAUCHESNE and POIRIER, 1990; WANG *et al.*, 1993). These results can be compared with the extensive data set on deformation of (Mg, Fe)O ((Mg, Fe)O is stable at ambient conditions and many data are available for the rheological properties of this mineral at room pressure (for a review see FROST and ASHBY (1982)). The result suggests that (Mg, Fe)O is significantly weaker than perovskite in dislocation creep regime (see also Chapter 15).

Direct deformation experiments on (Mg, Fe)SiO_3 perovskite are very limited. KARATO *et al.* (1990) performed room-pressure indentation tests on single crystal $MgSiO_3$ perovskite and inferred the slip systems. MEADE and JEANLOZ (1990) and MERKEL *et al.* (2003) performed deformation experiments on silicate perovskite at lower mantle pressures at low temperatures. However, since rheological properties are sensitive to temperature, the relevance of these low-temperature experiments is questionable. CHEN *et al.* (2002b) performed deformation experiments of silicate perovskite under modest temperatures and high-pressure conditions and CORDIER *et al.* (2004) conducted similar experiments at higher pressures. CHEN *et al.* (2002b) obtained semi-quantitative results on the creep strength and suggested that silicate perovskite is stronger than other minerals in the mantle, but the flow law parameters are not determined and the reported very weak temperature dependence of strength is highly unusual. CORDIER *et al.* (2004) did not determine the strength but inferred the slip systems from X-ray studies. Both of these studies are for dislocation creep. However, almost nothing is known, at this time, about the quantitative rheological properties and dominant slip systems of silicate perovskite under the conditions relevant to Earth's lower mantle from direct deformation experiments.

The rheological properties of the lower mantle may be estimated from diffusion coefficients. This approach has merits for two reasons. First, diffusional creep is likely to be a dominant deformation mechanism in most portions of the lower mantle as KARATO *et al.* (1995b) suggested. In such a case, the viscosity of the lower mantle can be inferred from the diffusion coefficients. Second, the measurements of diffusion coefficients are easier than the measurements of creep strength under high-pressure conditions. In fact, there have been several experimental studies to determine the diffusion coefficients of oxygen or silicon under lower mantle conditions (e.g., YAMAZAKI *et al.*, 2001; YAMAZAKI and IRIFUNE, 2003; VAN ORMAN *et al.*, 2003). Given plausible estimates of temperatures and grain size, one can infer the viscosity profile from these studies (e.g., YAMAZAKI and KARATO, 2001b, Fig. 19.13). They found that a grain size of $\sim$2–3 mm gives a viscosity of $\sim 10^{22}$ Pa s as inferred from geophysical methods (Chapter 19). Note that the available experimental data on diffusion in the lower mantle minerals show relatively small activation energies and volumes leading to an only modest increase in viscosity with depth in the lower mantle.[5] YAMAZAKI and KARATO (2001b) noted that (Mg, Fe)O is likely to be much weaker than co-existing perovskite in the diffusion creep regime as well as in the dislocation creep regime. Consequently, the geometry of a weaker phase, (Mg, Fe)O, is likely to play an important role in controlling the creep strength of a two-phase aggregate (see Chapter 12). Another important point is a possible non-monotonic pressure dependence of diffusion coefficients. ITO and TORIUMI (2007) showed, based on molecular dynamics calculations, that the diffusion coefficient of oxygen in MgO has its minimum at a mid-lower mantle depth. If this result applies to Earth's

[5] There was a speculation that the activation volume decreases with pressure to explain the geophysically inferred modest depth variation in viscosity (see Chapter 10). Such a model is physically unreasonable and is not required to explain the observation.

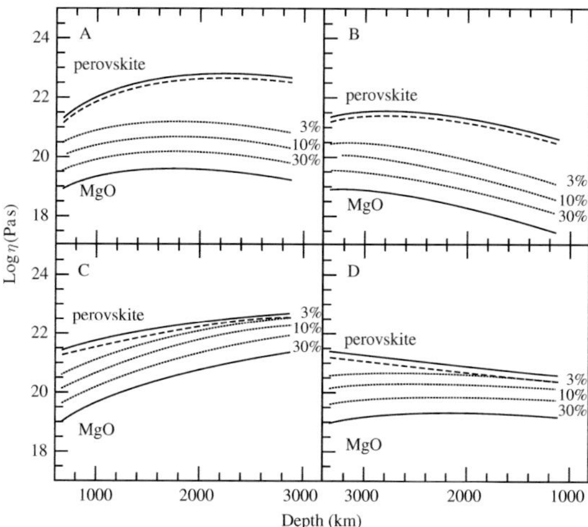

FIGURE 19.13 Viscosity–depth profiles for the lower mantle inferred from the diffusion coefficients in perovskite and (Mg, Fe)O (after YAMAZAKI and KARATO, 2001b). The viscosity was calculated from diffusion coefficients assuming a constant grain size for a range of geotherms. A modest depth variation in viscosity is predicted due to the small activation energy and volume.

lower mantle, then the viscosity of the lower mantle will have a maximum at a mid-lower mantle depth.

The role of water on plastic deformation was not investigated in any of these previous studies on the lower mantle phases although the solubility of water (hydrogen) was studied (e.g., BOLFAN-CASANOVA *et al.*, 2002, 2003; MURAKAMI *et al.*, 2002). The role of water on rheological properties is known to be large in all previous studies for low-pressure minerals (Chapter 10) and there is an urgent need to determine the influence of water on rheological properties of the lower mantle phases.

Recently a new phase (the post-perovskite phase) in silicate (Mg, Fe)SiO_3 was found and it is proposed that a significant fraction of the bottom part of the lower mantle might be made of this new phase (MURAKAMI *et al.*, 2004; OGANOV and ONO, 2004). Nothing is known about the rheological properties of this phase, but based on theoretical calculations, this mineral is likely to have large elastic anisotropy (TSUCHIYA *et al.*, 2004a). Therefore an understanding of the rheological properties, particularly the slip systems of this mineral, is important in interpreting seismic anisotropy in terms of flow pattern (see Chapter 21).

19.4.3. Some topics on deep mantle, related to rheological properties

A large number of important geodynamic issues are closely related to the rheological properties in the deep mantle minerals. They include (1) the origin of deep and intermediate earthquakes (e.g., GREEN and HOUSTON, 1995; KARATO *et al.*, 2001), (2) the rheological contrast between different geochemical reservoirs (e.g., MANGA, 1996; BECKER *et al.*, 1999) and (3) the structure and dynamics of plumes (e.g., KORENAGA, 2005). In all of them, an understanding of the rheological properties of Earth materials under deep Earth conditions is critical. For example, in investigating the origin of deep (and intermediate depth) earthquakes, we need to understand the rheological properties of subducting slabs, not only their "steady-state" rheological properties but also the instabilities of deformation (Chapter 16). The nature of mixing of different geochemical reservoirs depends on the rheological contrasts as well as on the density contrasts among them. Perhaps the most important cause for rheological contrast is the difference in water content. However currently almost nothing is known about the effect of water on the rheological properties of deep mantle minerals. Finally, the recent observations of structures of plumes by high-resolution tomography (e.g., MONTELLI *et al.*, 2004) show not-so-hot and thick plumes which are a marked contrast to the classic model of a plume with a big head and a hot and thin tail (e.g., WHITEHEAD and LUTHER, 1975; HILL *et al.*, 1992). The geodynamic significance of such observations has been investigated based on the basic physics of rheological properties (e.g., KORENAGA, 2005; KARATO, 2007).

19.5. Rheological properties of the core

19.5.1. Outer core

Seismic shear waves do not propagate through the outer core and therefore the outer core must be made mainly of liquid. The fundamental microscopic physics for the viscosity of a liquid is different from that of a solid and the viscosity of the outer core does not play a very important role in the core dynamics in comparison to the role of rheological properties in mantle dynamics. Consequently, I will only discuss this issue briefly for completeness. For a more detailed discussion on the physical properties of liquids, see POIRIER (2000), MARCH and TOSI (2002), and BARRAT and HANSEN (2003). The viscosity of the outer core may be inferred based on low-pressure measurements of viscosity of molten iron (alloy) (SECCO, 1995) with an extrapolation based on some theoretical models. POIRIER (1988) showed that the viscosity of a molten liquid follows a homologous temperature scaling law, i.e., $\eta \propto \exp(\beta T_m(P)/T)$, similar to solids. Viscosity on the order of 10^{-3}–10^{-2} Pa s is inferred (POIRIER, 1988). Since the viscosity of the outer core is so low, the outer core can be treated as an inviscid liquid except for a thin boundary layer (the Ekman layer) in most geodynamo modeling (e.g., MERRILL *et al.*, 1998).

19.5.2. Inner core

The inner core is solid, and is likely to be made of an iron-rich alloy with a hexagonal close pack (hcp) structure (Chapter 17). Recent seismological observations show surprisingly rich details of the central part of this planet. They include the discovery of strong anisotropy and possible relative rotation of the inner core with respect to the mantle (e.g., SONG, 1997; TROMP, 2001). An understanding of rheological properties of the inner core materials helps the geodynamic interpretation of these observations. First, the development of anisotropic structure depends on the rheological properties such as the deformation mechanism (dislocation versus diffusional creep) and the slip systems of dislocations (Chapter 14). Second, the nature of inner core rotation depends on its rheological properties. BUFFETT (1997) showed that the gravitational interaction between the inner core and the mantle can have an important control on the inner core rotation, and that the inner core rotation cannot occur when its viscosity is in the range of 10^{17}–10^{20} Pa s. Therefore two issues, namely the magnitude of viscosity and the mechanisms of deformation, need to be investigated to understand the geodynamic significance of these observations.

The temperature in the inner core is close to its melting temperature (0.96–0.99 T_m), and since thermal conductivity of metallic iron is very high, the magnitude of deviatoric stress is expected to be low. KARATO (1999, 2000) estimated that the deviatoric stress in the inner core is $\sim$10–10^4 Pa. Therefore the deformation mechanism in the inner core would be diffusional creep unless grain size is large. BERGMAN (1998) estimated the grain size of the inner core from the kinetics of crystallization and obtained $\sim 10^2$ m. The grain size of the inner core can also be inferred from the observed seismic wave attenuation. Attenuation in the inner core is characterized by the Q values of $Q_P = 200$–600 (BHATTACHARYA *et al.*, 1993; SOURIAU and ROUDIL, 1995; CORMIER and LI, 2002; LI and CORMIER, 2002; WEN and NIU, 2002) which is surprisingly *large* (i.e., *low* attenuation) for a metal close to the melting temperature. A plausible explanation for this low attenuation is to invoke a large grain size. Using the experimental data by JACKSON *et al.* (2000a), one can estimate the grain size of $\sim 10^2$–10^3 m. Such a large grain size is in harmony with the high-temperature, low-stress conditions in the inner core and is consistent with the observations of seismic wave scattering (e.g., CORMIER *et al.*, 1998; VIDALE *et al.*, 2000; VIDALE and EARLE, 2000). With this range of large grain size, deformation in the inner core may occur by dislocation creep that could result in lattice-preferred orientation. In fact, deformation at the stress estimated above would lead to a grain size of $\sim$1–10^4 m if grain size is controlled by dynamic recrystallization (Chapter 13). The order of magnitude of the effective viscosity of the inner core can be inferred from the crystal structure–plasticity systematics reviewed in Chapter 15. Such an analysis yields viscosities of $\sim 10^{15}$–10^{19} Pa s (KARATO, 2000). Interestingly, the viscosity values based on this rough estimate are close to the boundary between "locked" and "free" regimes for the mantle–inner core gravitational coupling calculated by BUFFETT (1997).

Some direct deformation experiments have been performed on hcp iron at high-pressure conditions (WENK *et al.*, 2000; MERKEL *et al.*, 2004), but the significance of these results to the inner core is questionable, because the stress and temperatures (in most cases room temperature) in these experiments are vastly different from those expected in the inner core.

It was sometimes considered that the inner core of Earth may contain a large fraction of melt (e.g., FEARN *et al.*, 1981; SINGH *et al.*, 2000). However, the presence of a large fraction of melt is unlikely because of efficient compaction (SUMITA *et al.*, 1996), and partial melting is not required to explain seismic attenuation. Most of the rheological properties of the inner core can be attributed to solid-state processes.

20 Heterogeneity of Earth structure and its geodynamic implications

Lateral heterogeneity in Earth exists at various scales. Such heterogeneity includes the lateral variation in seismic wave velocities and attenuation as well as the topography on seismic discontinuities. This chapter focuses on the seismological observations on large-scale (~ 100 km or larger) heterogeneity and provides a summary of (i) basic seismological observations on lateral heterogeneity of velocity and attenuation and the depth of seismic discontinuities, and (ii) their geodynamical interpretation. Interpretation of heterogeneity in seismic observations is not straightforward because a variety of factors could cause the lateral variation of seismological observables. However, some important conclusions have been obtained including (i) the large depth variation in the amplitude of velocity anomalies, and (ii) the fact that the ratio of anomalies in S- and P-wave velocities can be naturally interpreted by thermal anomalies, if the nature of anharmonicity and anelasticity is appropriately taken into account. In contrast, inferring chemical heterogeneity from seismological observations is challenging because of either low or poorly known sensitivity of observable parameters to chemical composition. Issues of identifying chemical heterogeneity are discussed including the heterogeneity in major element chemistry as well as heterogeneity in hydrogen (water) content.

Key words seismic tomography, topography on discontinuities, anharmonicity, Grüneisen parameter, anelasticity.

20.1. Introduction

Earth models discussed in Chapter 17 are characterized by laterally homogeneous and almost isotropic structure. One of the standard models, PREM (Dziewonski and Anderson, 1981), introduced an anisotropic layer in the topmost region (top 220 km), but the model is still highly idealized in that anisotropy is considered only in the top 220 km layer and more importantly, properties are assumed to be independent of the horizontal coordinate (i.e., the longitude or latitude). Such an ideal structure (i.e., a laterally homogeneous model) would, however, correspond to a static (dead) Earth, because there would be no driving forces for convection in such an idealized Earth.

In a real dynamic Earth, there must be deviations from such a structure corresponding to the lateral variation in density (temperature and/or chemical composition) that provides a driving force for convection. Also the structure in a dynamic Earth often becomes anisotropic due to flow-induced anisotropic structures. Therefore the deviations from an ideal structure, namely the lateral variation in seismic wave velocities and density, and the anisotropic structures, are closely related to the dynamics of the Earth's interior. High-resolution probing of the Earth's interior through seismology can provide crucial data to map out these deviations from an ideal structure.

In this chapter we focus on lateral heterogeneity (seismic anisotropy is discussed in Chapter 21). We will first learn the basic concepts behind *seismic tomography* and other high-resolution seismology from which the fine structures of the Earth can be determined. Then a brief review is provided on the basic results for lateral heterogeneity. The focus of this chapter is the interpretation of these observations based on mineral physics. Results from previous chapters (Chapters 4 and 11) are synthesized to provide a basis for interpretation of lateral heterogeneity in terms of physical and chemical anomalies in the Earth.

20.2. High-resolution seismology

20.2.1. Fundamentals of high-resolution seismology

Seismic tomography (velocity tomography)

The most commonly used technique to infer velocity heterogeneity is seismic tomography. *Tomography* is a powerful method to determine the three-dimensional structure of an object from the (two-dimensional) data outside of it. In seismology, a large number of data collected on Earth's surface are inverted for three-dimensional structures. The basic concept of seismic tomography is reviewed in LAY and WALLACE (1995) and some technical details can be found in NOLET (1987a) and IYER and HIRAHARA (1993).

The anomalies in seismic wave velocities can be inferred from (i) travel-time anomalies of body waves, (ii) anomalies in the dispersion of surface waves[1] and (iii) anomalies in the frequencies of free oscillations. The use of travel-time anomalies is the most straightforward technique, first developed by AKI *et al.* (1977) and DZIEWONSKI *et al.* (1977). In fact, in classic papers such as FUKAO *et al.* (1992) and VAN DER HILST *et al.* (1997), travel-time anomalies of body waves were used to obtain high-resolution images of fast velocity anomalies corresponding to subducted materials. Although simple, a major disadvantage of this technique is that it is sensitive to uncertainties in the location of earthquakes. To appreciate this, let us consider the influence of uncertainties in earthquake location of ~ 10 km (this is a typical value of uncertainty). With a distance of ~ 1000 km, this will yield the uncertainty in velocity of $\sim 1\%$. Improvement in earthquake locations played an important role in enhancing the resolution of body-wave tomography[2] (e.g., ENGDAHL *et al.*, 1998). An important improvement in the theoretical approach is the development of *waveform inversion* (e.g., NOLET, 1987b). In this technique, not only the travel-times of certain waves but also the whole data set of seismic records are used to constrain the structure of Earth. In addition, there is a complication due to the fact that there is a significant deviation from ray theory (i.e., Snell's law) in the propagation of actual seismic waves. The wavelength of seismic waves is often comparable to the scale of the lateral variation of the structure (for instance, an S-wave of frequency of 100 s has a wavelength of 400–600 km). Consequently the effects of a finite wavelength must be evaluated when structures whose characteristic length is comparable to the wavelength of seismic waves (NOLET, 2000; NOLET and DAHLEN, 2000) are to be determined. This is quite different from medical tomography, in which the wavelength of X-rays (~ 0.1 nm) is much shorter than the scale of heterogeneity (~ 1 cm).

A somewhat different approach is used when one uses surface waves or free oscillation data. The amplitude of surface waves decreases with distance less than body waves and consequently surface waves provide rich data on the structure of the upper mantle. The use of surface waves is critical for the study of upper mantle structure (e.g., ANDERSON *et al.*, 1992; ZHANG and TANIMOTO, 1993). The velocity of surface waves depends on the frequency as well as the depth variation of elastic properties. From the frequency dependence of the phase velocity of surface waves one can infer the structure of Earth's near-surface layers. The amplitude of surface waves decreases with depth and the information on the Earth's structure is limited to relatively shallow portions of the Earth (in most cases the upper mantle). In the analysis of free oscillation, one uses the slight deviation of free oscillation peak frequencies from those corresponding to an Earth with an ideal structure. Because the deviation of the real structure from the ideal structure often results in splitting of a peak, this distortion is often referred to as a "splitting function" (e.g., DAHLEN and TROMP, 1998). Results from the analysis of free oscillation are free from the uncertainties in the location of sources

[1] Seismic waves are classified into body waves and surface waves. Body waves are waves that propagate through a three-dimensional body without the influence of a surface boundary. When a free surface is present in otherwise infinite material, the displacement of the wave must satisfy the surface boundary conditions, and consequently, there will be special types of waves in which material displacement is confined to near-surface regions. These are called surface waves. The surface boundary conditions are satisfied by the combination of solutions with various "modes." Free oscillation is a special type of surface wave for which the boundary conditions at the surface are satisfied for waves propagating in a finite body.

[2] Note that to determine earthquake locations, we need to know the structure of Earth. Therefore the precise location of earthquakes is made through an iterative process.

but these results usually have low spatial resolution. Also these results are sensitive to the focal mechanisms (source processes such as the direction of propagation of rupture).

When one uses the data on the spheroidal modes of free oscillation, then one could infer the density anomalies as well as the velocity anomalies. This is because the spheroidal modes of free oscillation involve the radial motion of materials against gravity and hence they are sensitive to the density distribution. Anomalies in the gravitational field can also be used to obtain some insights into the density variation. Although the resolution of such an approach is limited, some intriguing results have been obtained from such an approach (Ishii and Tromp, 1999; Trampert *et al.*, 2004; see also Masters and Gubbins, 2003).

Thus, in seismic tomography, we are dealing with a less than ideal data distribution and because the scale of lateral heterogeneity is comparable to the wavelength of seismic waves, the seismological records are influenced by finite-frequency effects. Consequently, key to the success of seismic tomography are: (1) the development of better networks of seismic stations, (2) the improvement in computer technology to handle a large volume of data and (3) the development of better theoretical techniques to determine three-dimensional structure from noisy two-dimensional data, including the effects of finite wavelength.

Note that in most velocity tomography, elastic isotropy is assumed. However, as we will learn in Chapter 21, the amplitude of anisotropy is comparable to the amplitude of velocity heterogeneity in some cases. Some bias can be introduced by this assumption (e.g., Gung *et al.*, 2003).

Topography and the nature of discontinuities

Tomography is not the only technique to infer heterogeneous structures. Another equally useful technique is the mapping of topography on discontinuities. A simple Earth model such as PREM has several discontinuities such as the "410-km" or the "660-km" discontinuity. However, the actual depths of these discontinuities can deviate from these average values. Mapping the topography on these discontinuities can be made using signals caused by the conversion of seismic waves (from S to P or vice versa) or using the reflected waves. The topography on these boundaries has significant geodynamic implications. First, the density contrasts across these discontinuities are generally large (a few percent). Therefore, the topography on these discontinuities results in a large lateral variation in density which has an important influence on the convection pattern (e.g., Phipps Morgan and Shearer, 1993). Second, the topography on these boundaries may be caused by various factors including lateral variation in temperature and/or chemical composition (e.g., Bina and Helffrich, 1994; Helffrich, 2000; Blum and Shen, 2004). Consequently understanding the causes of the topography on these boundaries will improve our understanding of distribution of these parameters. Caution must be made in interpreting these results. The most important is that the topography obtained from these analyses depends on the wavelength of seismic waves. This is due to the fact that in a real Earth with a multi-component system, the structure of a discontinuity can be complicated. For example, the "660-km" discontinuity may be composed of various small-scale discontinuities at different depths each of which has a different sharpness and different magnitude of jumps in density/seismic wave velocity (e.g., Weidner and Wang, 2000). For a review of studies on the topography of discontinuities see Shearer and Masters (1992), Flanagan and Shearer (1998), Shearer (2000) and Gu *et al.* (2003).

The impedance contrast can also be determined when the amplitude of waves passing through a discontinuity is analyzed. When a seismic wave goes through a velocity (or density) discontinuity, reflection and refraction will occur and therefore a P-wave, for example, will be converted to both P- and S-waves. The amplitude of converted waves is sensitive to the impedance contrast at the discontinuity, and hence the velocity (or density) jump at a discontinuity can be determined from the observations of the amplitude of converted waves. Also from a detailed analysis of the waveforms, the sharpness of the boundary can be inferred (e.g., Benz and Vidale, 1993; van der Meijde *et al.*, 2003). The sharpness of a discontinuity contains some information as to the physical and chemical state of the Earth's interior (e.g., Wood, 1995; Karato, 2007).

Attenuation tomography

The sensitivity of anelastic properties to physical and chemical conditions in the Earth is quite different from that of seismic wave velocities (Chapter 11). Consequently, seismic wave attenuation provides important additional data to infer physical and chemical states in Earth's interior. Attenuation can be determined by (i) the measurements of amplitude decay with distance or (ii) from the broadening of peaks of free oscillation (or from the modifications to the waveform). Both amplitude and broadening measurements are subject to a number of complications. Among them, the most important source of uncertainty is the influence of geometrical spreading or focusing effects. One way to minimize the uncertainties in the body-wave attenuation measurements is to use the relative amplitude of P- and S-waves recorded for the same source–station pair (e.g., SHITO and SHIBUTANI, 2003a). GUNG and ROMANOWICZ (2004) discussed some of the technical issues in attenuation tomography with special reference to long-wavelength records such as surface wave data. In general, spatial resolution and errors in attenuation tomography are worse than those in velocity tomography (e.g., ROTH *et al.*, 1999). However, in regions with dense source–station coverage such as the Philippine Sea upper mantle (e.g., SHITO and SHIBUTANI, 2003a) and Tohoku, Japan (e.g., TAKANAMI *et al.*, 2000; TSUMURA *et al.*, 2000), relatively high-resolution images of attenuation can be obtained. SHITO and SHIBUTANI (2003b) and SHITO *et al.* (2006) used both velocity and attenuation tomography to infer the distribution of water in the upper mantle. LAWRENCE and WYSESSION (2006) inverted attenuation simultaneously with the velocity throughout the whole mantle using the body-wave data. The attenuation tomography is reviewed by ROMANOWICZ (1994) and ROMANOWICZ and DUREK (2000).

20.2.2. Some basic results

The first tomographic images of Earth's interior at the global scale were published in 1984 (DZIEWONSKI, 1984; WOODHOUSE and DZIEWONSKI, 1984). These pioneering studies revealed that there are large anomalies in seismic wave velocities at both the top and the bottom of the mantle. Low velocities along mid-ocean ridges were documented, but large-scale anomalies in velocities were also found near the bottom of the mantle. The pattern of anomalies in the D″ layer has a long-wavelength feature and is characterized by a circum-Pacific high-velocity anomaly and

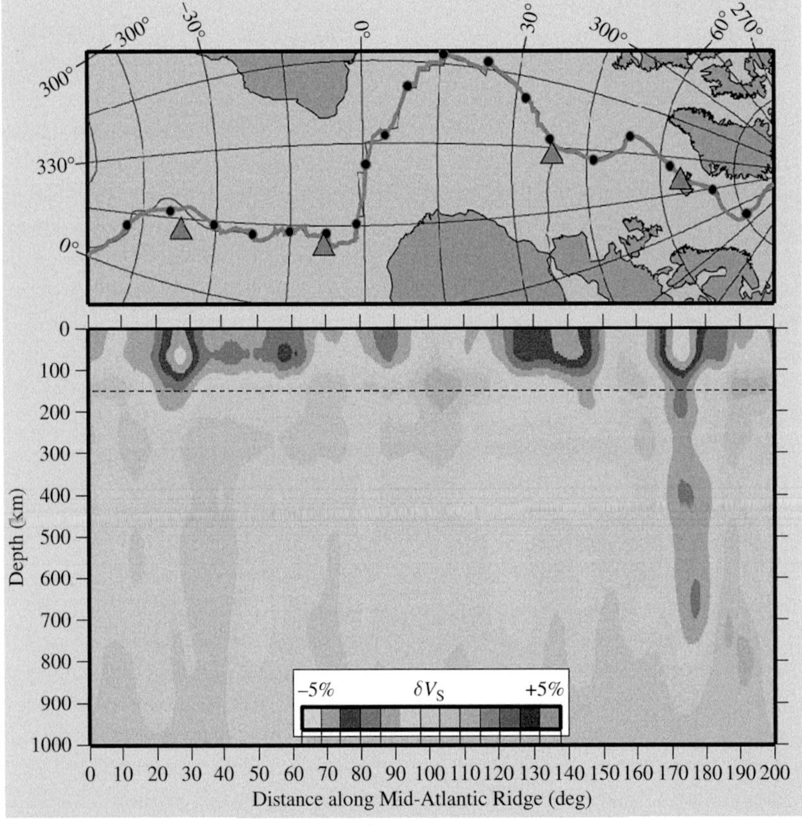

FIGURE 20.1 A result of high-resolution velocity tomography (S-wave velocity anomalies) in the upper mantle along the mid-Atlantic ridge (MONTAGNER and RITSEMA, 2001). Low-velocity anomalies beneath ridges are in most cases shallow (< 150 km), whereas in regions near hotspots, low-velocity anomalies occur in the deeper portions. Plate 8.

two broad low-velocity anomalies, one beneath the south Pacific and another beneath Africa. These long-wavelength features turned out to be robust features even after improvements in the resolution (e.g., RITSEMA *et al.*, 1999; MASTERS *et al.*, 2000). In addition, some detailed structures related to hot upwellings as well as cold down-going materials have been determined (e.g., FUKAO *et al.*, 2001; MONTELLI *et al.*, 2004). The main observations can be summarized as follows.

(1) There are well-developed low-velocity anomalies beneath mid-ocean ridges but their depth extent is limited to ~150 km (ZHANG and TANIMOTO, 1992, 1993; MONTAGNER and RITSEMA, 2001) (Fig. 20.1).
(2) There are high-velocity regions at the bottom of the mantle around the Pacific corresponding to the location of present and recent-past subduction zones (DZIEWONSKI, 1984; TANIMOTO and ANDERSON, 1990) (Fig. 20.2).
(3) There are two broad regions of low velocity at the bottom of the mantle, one beneath the south Pacific and another beneath Africa (DZIEWONSKI, 1984; TANIMOTO and ANDERSON, 1990) (Fig. 20.2). Some of the boundaries between these low-velocity regions and surrounding regions are sharp (e.g., WEN *et al.*, 2001).
(4) Low-velocity anomalies beneath "hot spots" are deep (> 200 km), and in some cases extend down to the core–mantle boundary but in other cases they extend only to ~660 km (ZHANG and TANIMOTO, 1992, 1993; MONTAGNER and RITSEMA, 2001; MONTELLI *et al.*, 2004; ZHAO, 2004) (Fig. 20.1).
(5) High-velocity anomalies beneath the continents are deep (~200–300 km) but are dependent on the age of continents. The very old Archean (older than ~2.7 Gyr) continents generally have a thicker root than young Proterozoic continents (e.g., JORDAN, 1975; ZHANG and TANIMOTO, 1993; GUNG *et al.*, 2003).

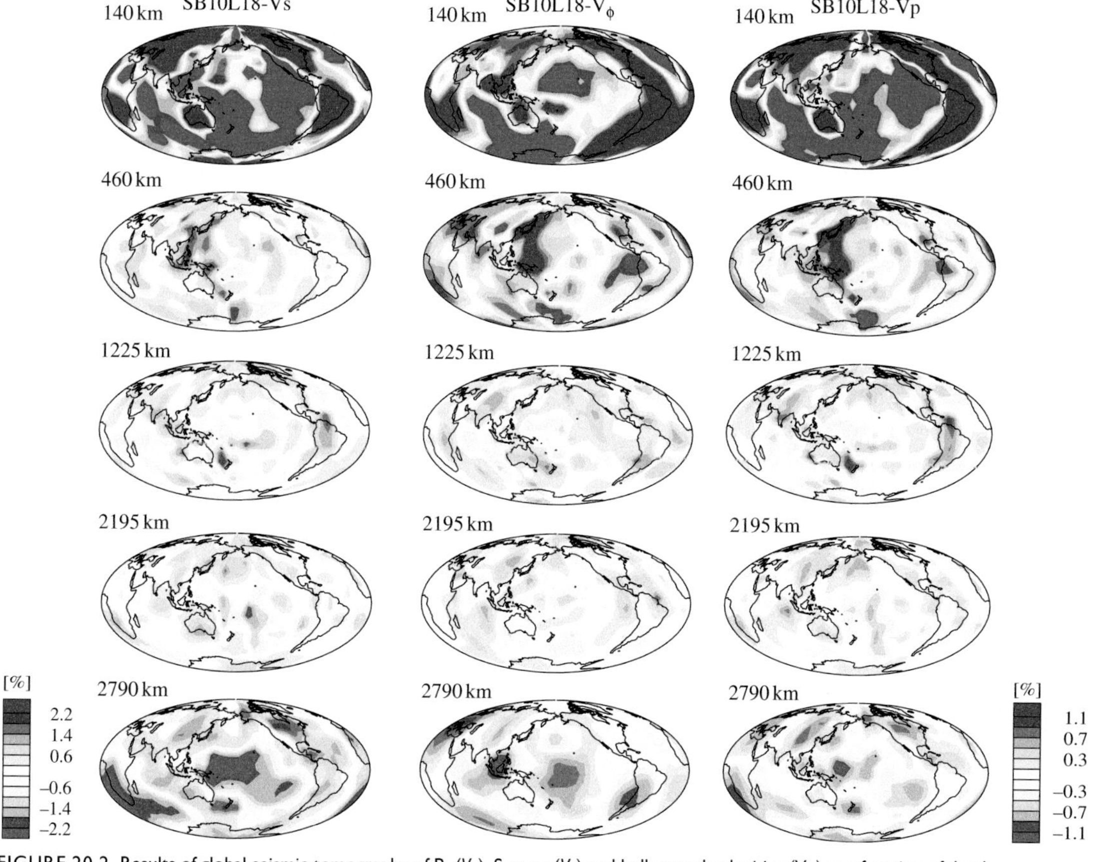

FIGURE 20.2 Results of global seismic tomography of P- (V_P), S-wave (V_S) and bulk sound velocities (V_ϕ) as a function of depth (after MASTERS *et al.*, 2000). Plate 9.

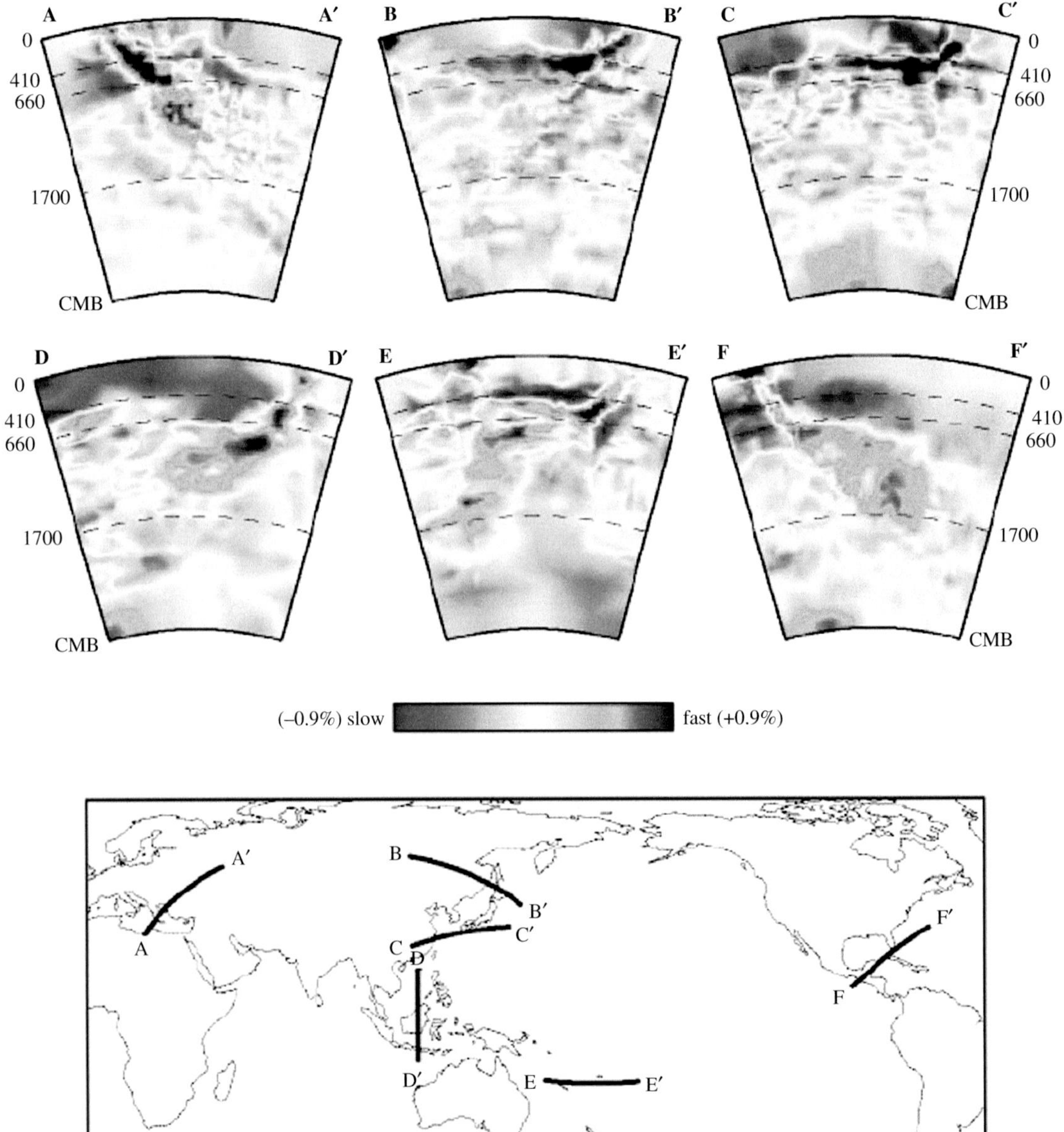

FIGURE 20.3 Examples of high-resolution velocity tomography for subduction zones. The pattern of fast velocity anomalies is modified in the transition zone in a variety of ways depending on regions (ROMANOWICZ, 2003). Plate 10.

(6) High-velocity anomalies associated with slabs continue to a greater depth. In some cases, one can trace the high-velocity anomalies to near the core–mantle boundary, but in many regions, there is significant distortion of the morphology of high-velocity anomalies around 500–800 km and at depths close to the core–mantle boundary, anomalies become fuzzy (e.g., FUKAO *et al.*, 1992, 2001, 2003; GRAND, 1994; VAN DER HILST *et al.*, 1997; ROMANOWICZ, 2003; ZHAO, 2004) (Fig. 20.3).

(7) The amplitude of velocity anomalies decreases significantly with depth until it becomes high again toward the bottom of the mantle (e.g., MASTERS *et al.*, 2000; ROMANOWICZ, 2003) (Fig. 20.4).

(8) The amplitude ratio of anomalies of S- to P-wave velocities, $R_{S/P}$, has a distinct depth variation. In the shallow upper mantle, this ratio is ~1–1.5, whereas it increases to ~3 in the deeper mantle (e.g., MASTERS *et al.*, 2000; ROMANOWICZ, 2003) (Fig. 20.5).

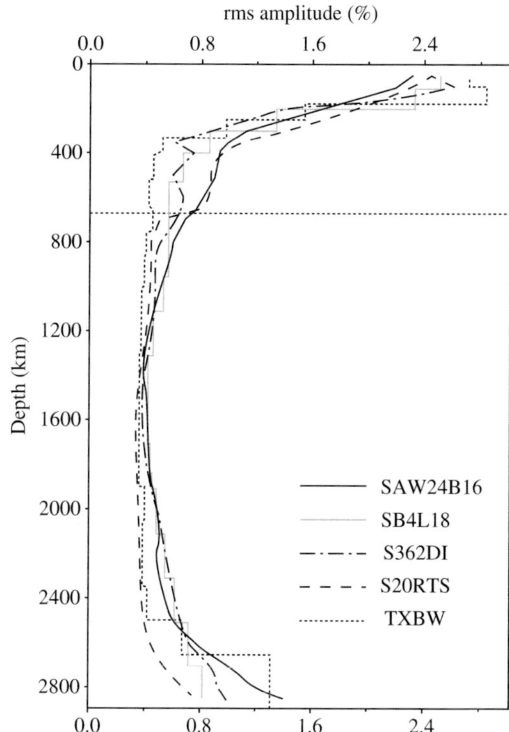

FIGURE 20.4 The rms (root-mean-square) values of S-wave velocity anomalies as a function of depth (ROMANOWICZ, 2003).

(9) The anomalies in S-wave velocities correlate with those of P-wave velocities, but the correlation between S-wave velocities and bulk sound wave velocities is weak and sometimes negative, particularly in the very deep lower mantle. Similarly, the correlation between S-wave velocity anomalies and density anomalies becomes negative in some regions, including the very deep lower mantle (ISHII and TROMP, 1999; MASTERS *et al.*, 2000) (Fig. 20.2).

(10) There is a distinct topography for some of the discontinuities (e.g., SHEARER and MASTERS, 1992; FLANAGAN and SHEARER, 1998; GU *et al.*, 2003). The "660-km" discontinuity tends to be deeper in regions near subduction zones. In contrast, the topography of the "410-km" discontinuity is sometimes anti-correlated with that of the 660-km discontinuity and sometimes not (Fig. 20.6).

(11) Three-dimensional mapping of attenuation structure has been performed (e.g., HASHIDA, 1989; ROMANOWICZ, 1995, 2003; ROTH *et al.*, 1999; BILLIEN *et al.*, 2000; TAKANAMI *et al.*, 2000; TSUMURA *et al.*, 2000; SHITO and SHIBUTANI, 2003a; GUNG and ROMANOWICZ, 2004). These studies show enhanced attenuation in the shallow upper mantle beneath mid-ocean ridges, deep into the mantle near hot spots and some parts of the subduction zone upper mantle (wedge mantle)

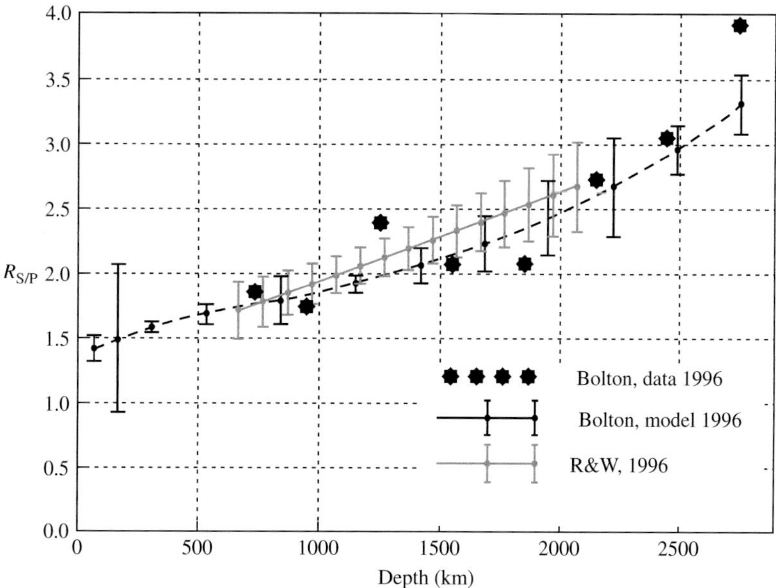

FIGURE 20.5 A plot of the ratio of the S-wave velocity anomaly versus the P-wave velocity anomaly ratio, $R_{S/P}$, as a function of depth (after MASTERS *et al.*, 2000).

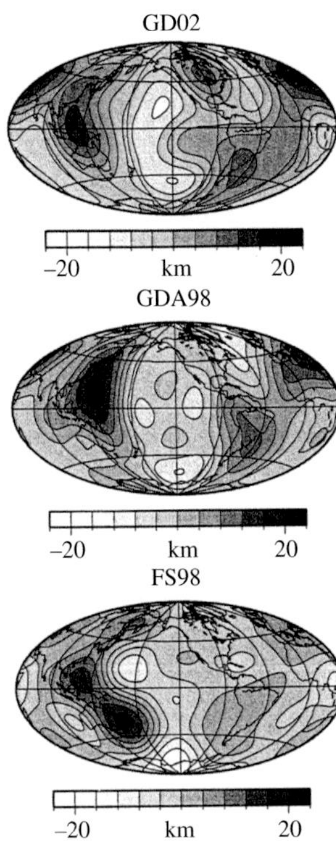

FIGURE 20.6 Topography of transition zone discontinuities (ROMANOWICZ, 2003).

(Fig. 20.7). Regional variation in attenuation in the lower mantle has also been documented by LAWRENCE and WYSESSION (2006).

20.3. Geodynamical interpretation of velocity (and attenuation) tomography

It is customary to interpret low- (high-) velocity regions as hot (cold) regions and materials there are subject to (negatively) buoyant forces. Obviously mid-ocean ridges are hot and subducting slabs are cold and these regions are seen as low- and high-velocity anomalies respectively. However, such an interpretation is not always secure because in many cases the influence of other factors such as the variation in chemical composition causes anomalies of magnitude comparable to the anomalies due to temperature variation. This is particularly true in the deep Earth where the amplitude of the temperature effect will decrease (due to high pressure), whereas the amplitude of the chemical effect may not decrease as we will learn below. If the variation in iron content is the major factor, for example, then low-velocity regions would have a higher density (higher Fe content) and would be sinking as opposed to rising. In fact, a non-thermal origin for velocity anomalies has been proposed in several cases. Based on a comparison of the amplitude of velocity anomalies due to plausible temperature variations and to the variation in chemical composition, STACEY (1992) argued that the majority of velocity anomalies in the lower mantle are due to chemical rather than thermal effects. Using the observations of both P- and S-wave velocity tomography, KARATO and KARKI (2001) and SALTZER *et al.* (2001) discussed evidence for chemical heterogeneity based on the comparison of P-and S-wave tomography in certain regions of the deep mantle. In most cases, chemical heterogeneity in major element composition such as Mg–Fe and Mg–Si is considered (e.g., FORTE and MITROVICA, 2001; TRAMPERT *et al.*, 2001; DESCHAMPS *et al.*, 2002), but heterogeneity in minor components such as calcium (KARATO and KARKI, 2001), aluminum (JACKSON *et al.*, 2004b) and hydrogen (water) (e.g., NOLET and ZIELHUIS, 1994; KARATO, 2003b; SHITO *et al.*, 2006; VAN DER MEIJDE *et al.*, 2003; BLUM and SHEN, 2004) was also proposed. Furthermore, on the basis of the expected small velocity variation due to the thermal effect, DUFFY and AHRENS (1992) proposed that the velocity anomalies in the majority of the lower mantle are due to the variation in the fraction of melt. In fact, interpretation of low-velocity anomalies in terms of partial melting is rather popular both for the shallow and the deep portions of the mantle (e.g., WILLIAMS and GARNERO, 1996; TOOMEY *et al.*, 1998), although such an interpretation is not often justifiable as we will learn in this chapter. Even in a case where a thermal interpretation of velocity anomalies is valid, it is not straightforward to obtain the correct conversion factor to infer temperature anomalies corresponding to velocity anomalies. If one uses the commonly reported temperature variation of velocity at high frequencies, one would obtain unreasonably high temperature and density anomalies. Subtle physics such as the influence of anelasticity needs to be incorporated in interpreting seismic tomography (KARATO, 1993a).

The results of seismic tomography have another direct geodynamic application. Velocity anomalies may be translated into density anomalies if the conversion factor is known. The density anomalies can then be used to calculate the flow pattern for a given viscosity profile. Such an approach has been used to infer the flow pattern and the rheological structure of Earth's interior (e.g., PHIPPS MORGAN and SHEARER, 1993; KING, 1995b; FORTE, 2000). These two problems, i.e., the cause of anomalies and the density to

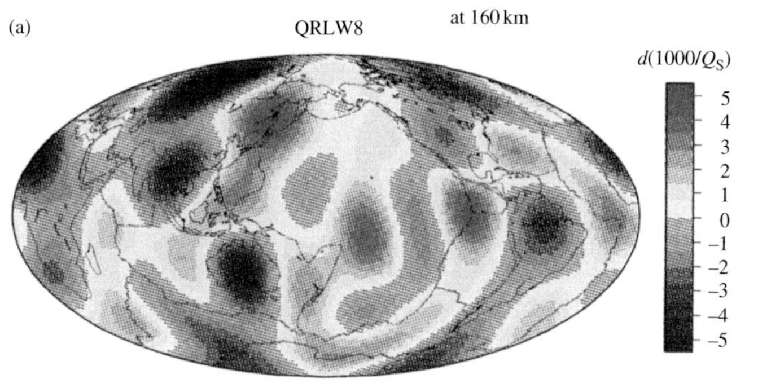

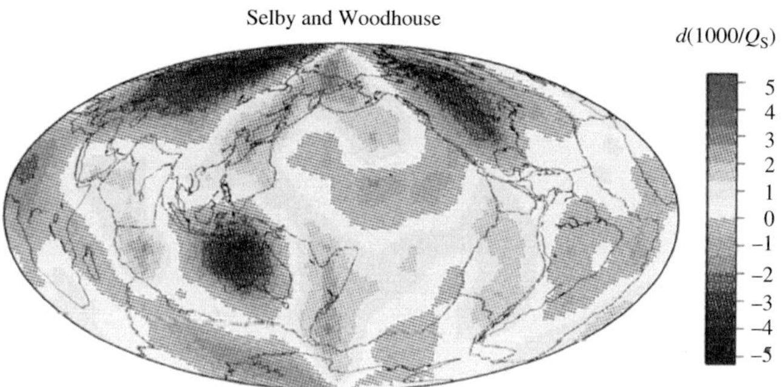

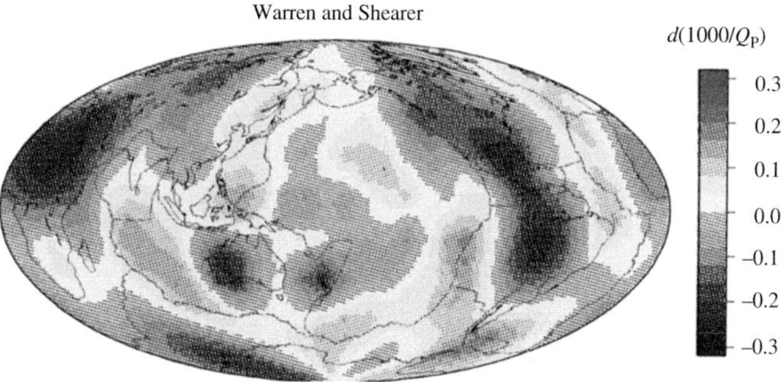

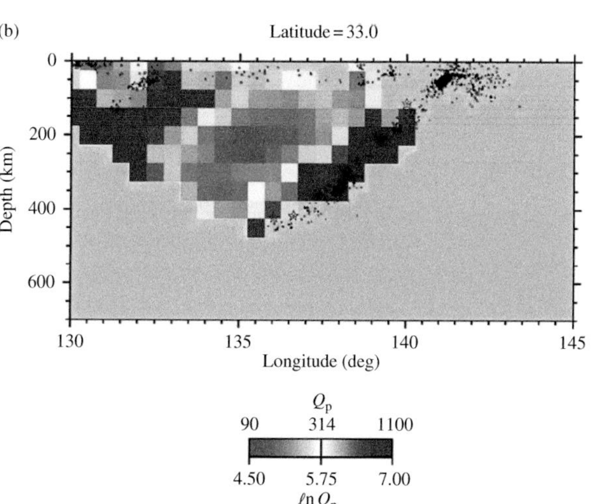

FIGURE 20.7 Some examples of attenuation tomography. (a) Several results of global attenuation tomography (ROMANOWICZ, 2003). (b) Results of a regional attenuation tomography based on the body-wave data (after SHITO and SHIBUTANI, 2003a).

velocity conversion factor are related. As I discussed above, if the cause of the velocity anomaly is the anomalies in iron content, then the density to velocity conversion factor will be *negative*, although the conversion factor is positive if there is a thermal origin for the velocity (and density) anomaly.

All of these issues are of great geodynamic significance, but the physical basis for these discussions involves subtle physics involving the mechanical properties of materials that need to be analyzed with great care. This chapter provides a summary of the mineral physics basis for interpretation of heterogeneity of seismic wave velocities and attenuation.

20.3.1. Origin of lateral heterogeneity: some fundamentals

Generalities: variation of seismic wave velocities with physical/chemical factors

In the most general case, seismological observations (assuming elastic isotropy) would provide us with the data on the lateral variation of P- and S-wave velocities (from which the lateral variation in bulk sound wave velocity can be calculated), the lateral variation of seismic wave attenuation and the lateral variation of density. As I discussed in Chapter 17, the lateral variation in pressure in Earth's mantle is so low that the lateral variation of these quantities should be due to the anomaly in temperature, T, and/or chemical composition, X_i, and/or the fraction of melt (or some other phases), ξ, and hence[3]

$$\delta \log V_{P(S,\phi)} = A_{V_{P(S,\phi)}T}\, \delta T + \sum_i A_{V_{P(S,\phi)}X_i}\, \delta \log X_i + A_{V_{P(S,\phi)}\xi}\, \delta\xi \tag{20.1a}$$

$$\delta \log Q^{-1}_{P(S)} = A_{Q^{-1}_{P(S)}T}\, \delta T + \sum_i A_{Q^{-1}_{P(S)}X_i}\, \delta \log X_i + A_{Q^{-1}_{P(S)}\xi}\, \delta\xi \tag{20.1b}$$

$$\delta \log \rho = A_{\rho T}\, \delta T + \sum_i A_{\rho X_i}\, \delta \log X_i + A_{\rho\xi}\, \delta\xi, \tag{20.1c}$$

where $\delta \log V_{P(S,\phi)}$ is the anomaly in P- (S- or bulk sound) velocity, $\delta \log Q^{-1}_{P(S)}$ is the anomaly in P- (S-) wave attenuation and $\delta \log \rho$ is the anomaly in density.

[3] Because $\delta \log W = (\partial \log W / \partial \log Z)\, \delta \log Z = (\partial \log W/\partial Z)\, \delta Z$ ($W = V_{P(S,\phi)}, Q^{-1}_{P(S,\phi)}, Z = T, X_i, \xi$), one can use either $A_{WZ} \equiv \partial \log W/\partial \log Z$ or $\tilde{A}_{WZ} \equiv \partial \log W/\partial Z$. These are equivalent.

$A_{V_{P(S,\phi)}T} \equiv \left(\partial \log V_{P(S,\phi)}/\partial T\right)_{X_i \xi}$ etc. are the parameters representing the sensitivity of the seismic wave velocity to a certain parameter such as temperature.

In writing these equations, I treat the fraction of melt as an independent variable, which is not rigorously true. The fraction of melt is controlled by the thermodynamic conditions (temperature, pressure and chemical composition) but also by the process of melt transport (note that the melt fraction is different from the degree of melting that is uniquely defined by the composition, temperature and pressure, see Chapter 10). Therefore ξ can be modified independently by P, T and X_i. Therefore it is convenient to treat the influence of melt separately from that of temperature, and chemical composition. In other words, the coefficients $A_{V_{P(S,\phi)}T}$, $A_{V_{P(S,\phi)}X_i}$ are the partial derivatives corresponding to a (hypothetical) fixed melt fraction.

When all of $(\delta \log V_P\ \delta \log V_S\ \delta \log Q_P^{-1}\ \delta \log Q_S^{-1}\ \delta \log \rho)$ are available as observed data, then equations (20.1a), (20.1b) and (20.1c) can be combined to yield,

$$\begin{pmatrix} \delta \log V_P \\ \delta \log V_S \\ \delta \log Q_P^{-1} \\ \delta \log Q_S^{-1} \\ \delta \log \rho \end{pmatrix} = \begin{pmatrix} \frac{\partial \log V_P}{\partial T} & \frac{\partial \log V_P}{\partial \log X_1} & \cdots & \frac{\partial \log V_P}{\partial \log X_n} & \frac{\partial \log V_P}{\partial \xi} \\ \frac{\partial \log V_S}{\partial T} & \frac{\partial \log V_S}{\partial \log X_1} & \cdots & \frac{\partial \log V_S}{\partial \log X_n} & \frac{\partial \log V_S}{\partial \xi} \\ \frac{\partial \log Q_P^{-1}}{\partial T} & \frac{\partial \log Q_P^{-1}}{\partial \log X_1} & \cdots & \frac{\partial \log Q_P^{-1}}{\partial \log X_n} & \frac{\partial \log Q_P^{-1}}{\partial \xi} \\ \frac{\partial \log Q_S^{-1}}{\partial T} & \frac{\partial \log Q_S^{-1}}{\partial \log X_1} & \cdots & \frac{\partial \log Q_S^{-1}}{\partial \log X_n} & \frac{\partial \log Q_S^{-1}}{\partial \xi} \\ \frac{\partial \log \rho}{\partial T} & \frac{\partial \log \rho}{\partial \log X_1} & \cdots & \frac{\partial \log \rho}{\partial X_n} & \frac{\partial \log \rho}{\partial \xi} \end{pmatrix} \times \begin{pmatrix} \delta T \\ \delta \log X_1 \\ \cdot \\ \cdot \\ \cdot \\ \delta \log X_n \\ \delta\xi \end{pmatrix}. \tag{20.2}$$

Note that this equation has a maximum of five data values at each point and has $n+2$ unknowns, where n is the number of chemical components. In many cases, not all of these data are available. Density anomalies are difficult to constrain, although some pioneering works for mapping density anomalies have been made (e.g., Ishii and Tromp, 1999; Deschamps *et al.*, 2001; Masters and Gubbins, 2003; Trampert *et al.*, 2004). In many cases, it is assumed that there is a unique relation between velocity and density anomalies, and the coefficient in such a relation is determined from gravity and seismic observations (e.g., Forte *et al.*, 1994). In these cases, density anomalies are not independent observations. Anomalies in seismic wave attenuation are also more difficult to determine than anomalies in seismic wave velocities, and in many cases, attenuation tomography is not available or is available only with much poorer resolution. So in general, we only have a limited number of observations to constrain anomalies in physical/chemical properties. Therefore the first challenge in using tomographic images to infer physical/chemical heterogeneity is to reduce the possible number of independent unknowns through the use of some models based on a theoretical understanding of material properties. The second challenge is to determine the values of matrix elements, A_{ij}, i.e., the partial derivatives of various physical properties with respect to temperature, composition and the degree of partial melting. In many cases, experimental data on the matrix elements are incomplete and one needs some theoretical models to extrapolate the experimental data to perform the inversion of equation (20.2).

Note also that equation (20.2) is in general non-linear because the matrix elements, $A_{ij}(\delta T, \delta \log X_i, \delta\xi)$, are a function of unknowns, $(\delta T, \delta \log X_i, \delta\xi)$, as will be shown later. Therefore equation (20.2) must be solved through iteration.

If we focus only on seismic wave velocities and density, then using the relations $V_S = \sqrt{\mu/\rho}$, $V_P = \sqrt{(K_S + \frac{4}{3}\mu)/\rho}$ and $V_\phi \equiv \sqrt{V_P^2 - \frac{4}{3}V_S^2} = \sqrt{K_S/\rho}$ (V_S, shear wave velocity; V_P, compressional wave velocity; V_ϕ, bulk sound wave velocity), the lateral variation in seismic wave velocities can be written in terms of the lateral variation in elastic constants and density, namely,

$$\delta \log V_S = \frac{1}{2}(\delta \log \mu - \delta \log \rho) \tag{20.3a}$$

$$\delta \log V_P = \frac{1}{2}[B \cdot \delta \log \mu + (1 - B) \cdot \delta \log K_S - \delta \log \rho] \tag{20.3b}$$

and

$$\delta \log V_\phi = \frac{1}{2}(\delta \log K_S - \delta \log \rho) \tag{20.3c}$$

where $B \equiv 2(1 - 2\nu)/3(1 - \nu) = \frac{4}{3}\,\mu/(K + \frac{4}{3}\mu) = \frac{4}{3}V_S^2/V_P^2$ and ν is the Poisson ratio. Among these three velocities, two are independent, and one can choose two out of three (V_S, V_P and V_ϕ).[4]

[4] Although V_P contains both K and μ, V_P and V_S are independent. In an (isotropic) elastic material, there are two independent elastic moduli and therefore there are two independent elastic wave velocities. Either one chooses (V_P, V_S) or (V_ϕ, V_S) does not matter.

In inferring the origin of velocity heterogeneity, it is useful to consider the ratios of velocity anomalies. If one takes a ratio of anomalies of two different velocities, then the factor that caused the anomalies (e.g., temperature anomalies) will be cancelled, and the ratio depends only on the partial derivatives that are the material properties.[5] Consequently, the inference of the cause of velocity anomalies is less non-unique when a ratio of two velocity anomalies is used.

The ratios of various anomalies can be written as

$$R_{\rho/S} \equiv \frac{\delta \log \rho}{\delta \log V_S} = \frac{2}{\tilde{\delta}_\mu - 1} \tag{20.4a}$$

$$R_{\rho/P} \equiv \frac{\delta \log \rho}{\delta \log V_P} = \frac{2}{B(\tilde{\delta}_\mu - 1) + (1 - B)(\tilde{\delta}_K - 1)} \tag{20.4b}$$

$$R_{S/P} \equiv \frac{\delta \log V_S}{\delta \log V_P} = \frac{1}{(1 - B)((\tilde{\delta}_K - 1)/(\tilde{\delta}_\mu - 1)) + B} \tag{20.4c}$$

and

$$R_{\phi/S} \equiv \frac{\delta \log V_\phi}{\delta \log V_S} = \frac{\tilde{\delta}_K - 1}{\tilde{\delta}_\mu - 1} \tag{20.4d}$$

where

$$\tilde{\delta}_K \equiv \frac{\delta \log K_S}{\delta \log \rho} \tag{20.5a}$$

and

$$\tilde{\delta}_\mu \equiv \frac{\delta \log \mu}{\delta \log \rho}. \tag{20.5b}$$

It immediately follows from (20.4c) and (20.4d) that

$$R_{S/P} = \frac{1}{(1 - B)R_{\phi/S} + B}. \tag{20.6}$$

Note that $R_{\phi/S} < 0$ corresponds to $R_{S/P} > 1/B = 3V_P^2/4V_S^2$.

These relations (i.e., equations (20.4) and (20.5)) are useful because elastic constants are in many cases essentially dependent on density (i.e., Birch's law; see Chapter 4). If the lateral variation of properties is due to temperature variation then $\tilde{\delta}_{K,\mu}$ is related to the

[5] This is true only when anomalies depend on temperature (and other variables) in a linear fashion. When anelasticity is important, for example, anomalies depend on temperature in a non-linear fashion, then the ratio itself depends on temperature.

Grüneisen parameters as $\tilde{\delta}_{K,\mu} = 2\gamma_{K,\mu} + \frac{1}{3}$ (see equation (4.34)). The behavior of the Grüneisen parameters has been investigated in theoretical and experimental studies (see equation (4.43)), and therefore one can place some constraints on the ratios defined by equation (20.4). A comparison of mineral physics predictions with seismological observations was made by KARATO and KARKI (2001), TRAMPERT *et al.* (2001) and DESCHAMPS and TRAMPERT (2003) and will be discussed later in this chapter.

Thermal origin

Let us first consider the simplest case where chemical composition is kept constant and partial melting or other phase transformations do not have any effect. In this ideal case, equations (20.1a), (20.1b) and (20.1c) are simplified to give

$$\delta \log V_{P(S,\phi)} = A_{V_{P(S,\phi)}T}\, \delta T \tag{20.7a}$$

$$\delta \log Q^{-1}_{P(S)} = A_{Q^{-1}_{P(S)}T}\, \delta T. \tag{20.7b}$$

and

$$\delta \log \rho = A_{\rho T}\, \delta T. \tag{20.7c}$$

In this case, we have only one unknown quantity, the temperature anomaly, δT, so if we have lateral variations of P- and S-wave (or bulk sound wave) velocities (and attenuation), they are not independent but there must be relationships among them. For instance,

$$R_{S/P(\phi)} \equiv \frac{\delta \log V_S}{\delta \log V_{P(\phi)}} = \frac{A_{V_S T}}{A_{V_{P(S,\phi)}T}} \tag{20.8a}$$

or

$$R_{\rho/P(S,\phi)} \equiv \frac{\delta \log \rho}{\delta \log V_{P(S,\phi)}} = \frac{A_{\rho T}}{A_{V_{P(S,\phi)}T}} \tag{20.8b}$$

or

$$R_{P(S,\phi)/Q^{-1}_{P(S)}} \equiv \frac{\delta \log V_{P(S,\phi)}}{\delta \log Q^{-1}_{P(S)}} = \frac{A_{V_{P(S,\phi)}T}}{A_{Q^{-1}_{P(S)}T}}. \tag{20.8c}$$

These ratios are (nearly) independent of δT and are determined only by material properties.

The detailed physics of the temperature dependence of seismic wave velocities is discussed in Chapter 4. Let us first consider the simplest case where the temperature dependence of seismic wave velocities arises only from *anharmonicity*. In this case, the temperature sensitivity of the seismic wave velocity is given approximately by a linear function of temperature as (see Chapter 4)

$$V^{\infty}_{P(S,\phi)}(T, P, X_i, \xi) = V^{\infty}_{P(S,\phi)}(T_0, P, X_i, \xi)\cdot \left[1 - \theta_{P(S,\phi)}(P, X_i)\cdot (T - T_0)\right] \tag{20.9}$$

where I write the seismic wave velocity as $V^{\infty}_{P(S,\phi)}(T, P, X_i, \xi)$ where the suffix ∞ is used to indicate that this velocity corresponds to the velocity at an infinite frequency, and $\theta_{P(S,\phi)}(P, X_i)(\equiv -\left(\partial \log V^{\infty}_{P(S,\phi)}/\partial T\right)_{P,X_i,\xi} = -(A_{V_{P(S,\phi)}T})_{ah}$, the suffix "ah" means anharmonicity) is a constant that is nearly independent of temperature but depends strongly on pressure (and chemical composition). This constant can be determined by high-frequency elasticity measurements or by first-principle calculations (see Chapter 4). The experimentally determined values of $\theta_{P(S,\phi)}(P, X_i)$ for representative minerals are summarized in Table 20.1. In the simplest case, this effect is through the change in density due to thermal expansion and the resultant change in the elastic constant (i.e., *Birch's law*; see Chapter 4). Consequently, $\theta_{P(S,\phi)}(P, X_i)$ is related to the coefficient of thermal expansion and the Grüneisen parameter as (Problem 20.1)

$$\theta_{P(S,\phi)} = -A_{V_{P(S,\phi)}T} = \alpha_{th}\left(\gamma_{P(S,\phi)} - \frac{1}{3}\right) \tag{20.10}$$

with $\gamma_P \equiv B\gamma_K + (1 - B)\gamma_\mu$, $\gamma_S = \gamma_\mu$ and $\gamma_\phi = \gamma_K$.

Therefore if the lateral variation in seismic wave velocities is due to the lateral variation in temperature and the temperature effect is only through the anharmonic effect (through thermal expansion), then,

$$R_{S/P(\phi)} = \frac{\gamma_S - \frac{1}{3}}{\gamma_{P(\phi)} - \frac{1}{3}} \tag{20.11a}$$

and

$$R_{\rho/P(S,\phi)} = \frac{1}{\gamma_{P(\phi)} - \frac{1}{3}}. \tag{20.11b}$$

Equation (20.11b) implies that the conditions for $R_{\rho/P(S,\phi)}>0$ are equivalent to the conditions for $\gamma_{P(S,\phi)}>\frac{1}{3}$ (or $\tilde{\delta}_{K,\mu} = 2\gamma_{K,\mu} + \frac{1}{3} > 1$) and $\partial \log V_{P(S,\phi)}/\partial \log \rho > 0$.

Problem 20.1

Derive equation (20.10).

Solution

By taking the derivative of equation (20.9) with respect to T, one has $\partial \log V^{\infty}_{P(S,\phi)}/\partial T = -\theta_{P(S,\phi)}$. Now if one writes an equation for seismic wave velocities as $V_{P(S,\phi)} = \sqrt{C_{P(S,\phi)}/\rho}$ with $C_P = K_S + \frac{4}{3}\mu$, $C_S = \mu$, $C_\phi = K_S$, then $\partial \log V_{P(S,\phi)}/\partial T = \frac{1}{2}(\partial \log C_{P(S,\phi)}/\partial T - \partial \log \rho/\partial T) = \frac{1}{2}(\partial \log \rho/\partial T)(\partial \log C_{P(S,\phi)}/\partial \log \rho - 1)$. By definition, $\partial \log \rho/\partial T = -\alpha_{th}$ and from the definition of the Grüneisen parameter, $\partial \log C_{P(S,\phi)}/\partial \log \rho = 2\gamma_{P(S,\phi)} + \frac{1}{3}$ (Chapter 4) where $\gamma_P = B\gamma_\mu + (1-B)\gamma_K$ $(B \equiv 2(1-2\nu)/3(1-\nu) = \frac{4}{3}(\mu/(K+\frac{4}{3}\mu)) = \frac{4}{3}(V_S^2/V_P^2))$, $\gamma_S = \gamma_\mu$ and $\gamma_\phi = \gamma_K$. Therefore $\theta_{P(S,\phi)} = -\frac{1}{2}(\partial \log \rho/\partial T)(\partial \log C_{P(S,\phi)}/\partial \log \rho - 1) = \alpha_{th}(\gamma_{P(S,\phi)} - \frac{1}{3})$.

In many cases, experimental studies are conducted only under limited conditions. Consequently some extrapolation is often needed. The most important extrapolation is extrapolation to different (higher) pressures. In order to extrapolate laboratory results to higher pressures, we use equation (20.10) and two empirical relationships for anharmonic properties, namely, $\alpha_{th}/(\alpha_{th})_0 = (\rho/\rho_0)^{-\delta_T}$ (equation (4.42)) and $\gamma/\gamma_0 = (\rho/\rho_0)^{-1}$ (equation (4.44)) (the density dependence of thermal expansion is more important than that of the Grüneisen parameter). This simple approach with δ_T=4–5 yields results that are closely in agreement with those from a more sophisticated approach by Stixrude and Lithgow-Bertelloni (2005b). The values of $\theta_{P(S,\phi)}(P)$ (i.e., $-\partial \log V^{\infty}_{P(S,\phi)}/\partial T$) for Earth's mantle are given in Table 20.2 and are plotted in Fig. 20.8 (thick solid line). I used the experimental results on $\theta_{P(S,\phi)}(P)$ under limited pressure conditions (Table 20.1), and used the above-mentioned semi-empirical relations to extrapolate in terms of pressure. $R_{S/P}$ and $R_{\rho/S}$ are calculated from $\theta_{P(S,\phi)}(P)$ and thermal expansion and the results are shown in Fig. 20.9a,b (thick solid lines).

The results are subject to large uncertainties when a phase transformation occurs (deep upper mantle and transition zones). However, the following gross features are robust. The absolute magnitude of this coefficient decreases significantly with depth. The absolute values of $\theta_S(P)$ are larger than those of $\theta_P(P)$, implying that the lateral variation in V_S is larger than that of V_P.

TABLE 20.1 Temperature derivatives ($\partial \log V^{\infty}_{P(S)}/\partial T = -\theta_{P(S)}$) of seismic wave velocities in typical mantle minerals. $\partial \log V^{\infty}_{P(S)}/\partial T = -\theta_{P(S)}$ is almost independent of temperature but decreases strongly with pressure (see Chapter 4). Units are P (GPa), T (K), θ_P, θ_S (10^{-5} K^{-1}).

Mineral	P	T	θ_P	θ_S	Remarks	Ref.
MgO	RP[1]	300–1800	6.8	8.8	Single crystal[2]	(1)
Al_2O_3	RP	300–1800	4.9	6.9	Single crystal	(1)
Olivine (Fo)	RP	300–1700	6.8	9.0	Single crystal	(1)
Olivine (Fa)	RP	300–700	7.0	7.5	Single crystal	(1)
Olivine ($Fo_{90}Fa_{10}$)	RP	300–1500	6.7	7.9	Single crystal	(1)
Pyrope	RP	300–1000	4.5	4.4	Single crystal	(1)
Majorite	RP	280–1073	3.0	4.0	Polycrystal	(2)
Wadsleyite	RP to 7	300–873	4.1	5.8	Polycrystal	(3)
Wadsleyite	RP	278–318	6.2	6.7	Polycrystal	(4)
Ringwoodite	RP	295–923	4.8	4.9	Single crystal	(5)
$MgSiO_3$ perovskite	RP	257–318	7.9	8.4	Polycrystal	(6)
$MgSiO_3$ perovskite	1.5–8.0	300–800	–	9.8	Polycrystal	(7)

(1): Data compiled by Anderson and Isaak (1995).
(2): Calculated from Sinogeikin and Bass (2002).
(3): Calculated from Li *et al.* (1998).
(4): Calculated from Katsura *et al.* (2001).
(5): Calculated from Sinogeikin *et al.* (2003).
(6): Calculated from Aizawa *et al.* (2004).
(7): Calculated from Sinogeikin *et al.* (1998).
[1] Room pressure (0.0001 GPa).
[2] Elastic wave velocities are average velocities using the Voigt–Reuss–Hill averaging scheme.

TABLE 20.2 Temperature derivatives of seismic wave velocities, $A_{V_{S,P}T}$, and thermal expansion, α_{th}, in Earth's mantle. These values can be used to convert velocity anomalies to temperature anomalies, namely, $\delta T = \delta \log V_{P,S}/(A_{V_{P,S}T})_{total}$, if the velocity anomalies are inferred to be due to temperature anomalies. The errors are ~ 10% for the upper mantle and ~ 20% for the transition zone and the lower mantle. Note that the correction for the anelastic effects depends on *Q*. The average one-dimensional *Q* model of QLM9 (Lawrence and Wysession, 2005) is used but the actual *Q* likely varies from one region to another. The unit is $10^{-5}\,K^{-1}$.

Depth (km)	$-(A_{V_S T})_{ah}$	$-(A_{V_P T})_{ah}$	$-(A_{V_S T})_{an}$	$-(A_{V_P T})_{an}$	$-(A_{V_S T})_{total}$	$-(A_{V_P T})_{total}$	α_{th}
24.4	7.25	5.90	–	–	–	–	3.00
40	7.71	6.08	–	–	–	–	3.00
60	8.09	6.24	–	–	–	–	3.00
80	8.60	6.48	8.24	3.36	16.84	9.85	3.04
115	8.87	6.59	8.30	3.38	17.17	9.97	3.04
185	8.97	6.67	8.35	3.40	17.32	10.08	3.09
220	8.97	6.55	8.35	3.44	17.32	10.03	3.09
220^+	8.65	6.23	4.67	1.83	13.32	8.06	2.75
265	8.36	6.15	4.66	1.82	13.02	7.97	2.83
310	7.71	5.66	4.63	1.80	12.34	7.46	2.56
355	7.45	5.47	4.58	1.77	12.04	7.24	2.45
400	7.28	5.33	4.53	1.73	11.8	7.06	2.38
400^+	6.48	4.01	2.44	0.95	8.92	4.96	2.50
450	6.01	3.72	2.49	0.97	8.50	4.70	2.28
500	5.64	3.50	2.52	0.98	8.16	4.48	2.11
550	5.31	3.29	2.54	1.00	7.85	4.29	1.95
550^+	5.31	4.84	2.54	1.00	7.85	5.83	1.80
600	4.94	4.50	2.55	1.00	7.49	5.50	1.65
635	4.94	4.50	2.55	1.00	7.49	5.50	1.65
670	4.89	4.46	2.55	1.00	7.44	5.46	1.63
670	7.07	3.81	1.38	0.56	8.46	4.37	2.00
721	6.88	3.74	1.37	0.57	8.26	4.31	1.93
771	6.70	3.68	1.37	0.58	8.07	4.26	1.87
871	6.35	3.48	1.35	0.57	7.70	4.04	1.75
971	6.02	3.28	1.33	0.56	7.35	3.84	1.64
1071	5.71	3.09	1.46	0.60	7.18	3.70	1.53
1171	5.43	2.92	1.44	0.59	6.87	3.51	1.44
1271	5.20	2.79	1.42	0.58	6.62	3.37	1.36
1371	4.95	2.65	1.58	0.64	6.53	3.29	1.28
1471	4.74	2.53	1.55	0.62	6.30	3.15	1.21
1571	4.52	2.40	1.53	0.61	6.04	3.00	1.14
1671	4.34	2.29	1.50	0.59	5.84	2.89	1.08
1771	4.17	2.20	1.38	0.54	5.54	2.74	1.03
1871	4.00	2.10	1.35	0.53	5.36	2.63	0.98
1971	3.82	2.00	1.33	0.52	5.14	2.52	0.93
2071	3.67	1.92	1.04	0.41	4.71	2.33	0.88
2171	3.53	1.84	1.02	0.39	4.56	2.24	0.84
2271	3.40	1.77	0.99	0.38	4.39	2.15	0.80
2371	3.27	1.70	0.97	0.37	4.25	2.07	0.76
2471	3.15	1.63	0.80	0.30	3.96	1.93	0.73
2571	3.04	1.57	0.78	0.30	3.82	1.86	0.70
2671	2.93	1.50	0.77	0.29	3.70	1.79	0.66
2771	2.82	1.45	0.76	0.29	3.59	1.74	0.63
2871	2.72	1.39	1.21	0.45	3.94	1.85	0.61
2891	2.70	1.38	1.20	0.45	3.91	1.83	0.60

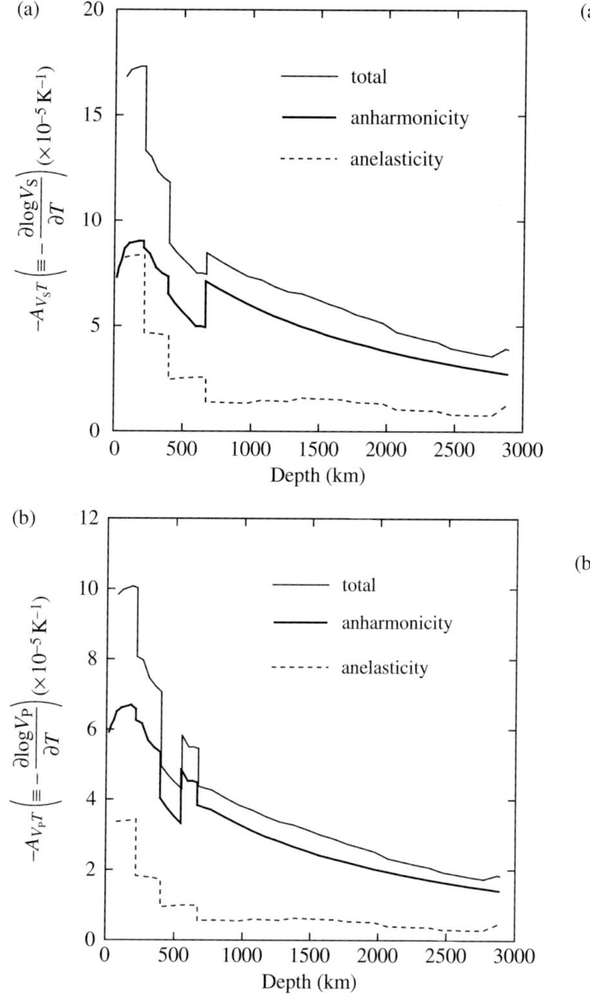

FIGURE 20.8 Plots of $A_{V_{S,P}T}$ $(\equiv \partial \log V_{P,S}/\partial T)$ as a function of depth. The data summarized in Table 20.2 are used together with a theoretical model of the pressure dependence of the anharmonicity. The representative error bar is 10–20%. (a) For S-wave velocities, (b) for P-wave velocities. Thick lines correspond to $-\partial \log V^{\infty}_{P,S}/\partial T$, i.e., the anharmonic effects. Contributions from anelasticity are shown by thin broken lines. This effect is dependent on Q, which also depends on the local temperature and other factors. Shown here are the anelastic effects corresponding to average Q. This effect is large in low-Q regions and small in high-Q regions. The net $-\partial \log V_{P,S}/\partial T$ is the sum of these two effects (shown by thin lines).

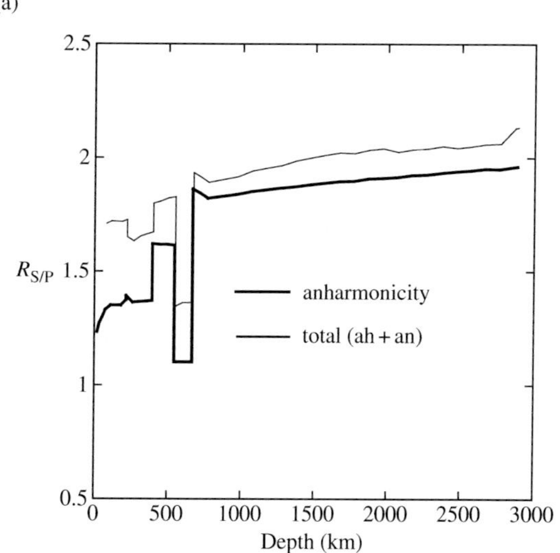

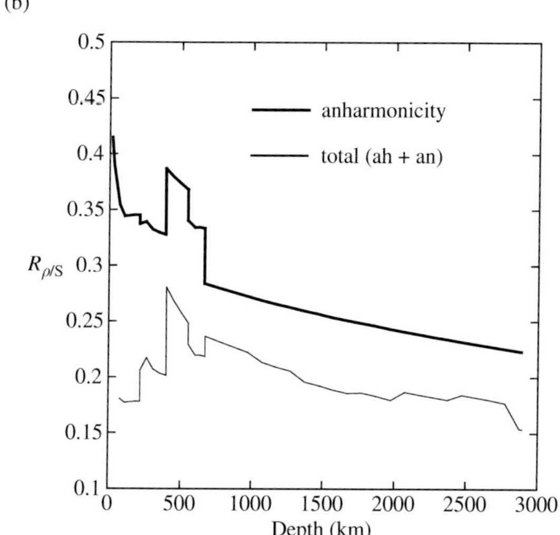

FIGURE 20.9 A plot of (a) $R_{S/P}$ and (b) $R_{\rho/S}$ as a function of depth predicted from mineral physics models. Thick lines are the results calculated from $-\partial \log V^{\infty}/\partial T$ and thermal expansion. Thin lines show the results including the influence of anelasticity.

These predictions are consistent with the observations noted before (items (7) and (8)). However, this simple model has three major problems:

(1) The magnitude of temperature anomalies corresponding to the observed velocity anomalies calculated from this model is too high. A typical shear wave velocity anomaly in the shallow regions near mid-ocean ridges is $\sim$4–6% (e.g., Masters *et al.*, 2000). Using the high-frequency values of $A_{V_S T} = \partial \log V_S/\partial T$, one would find a temperature anomaly of $\sim$ 500–750 K. This would cause a much higher degree of partial melting and a much thicker oceanic crust than is observed.

(2) The S-wave velocity anomalies to P-wave velocity anomaly ratio, $R_{S/P}$, would be $\sim$1.0–1.5 for this mechanism throughout the mantle whereas the

observed values significantly exceed these values in the deep lower mantle (see Fig. 20.5).[6]

(3) The density to S-wave velocity anomaly ratio $R_{\rho/S}$ is ~0.3–0.5 if this model is applied. The density to S-wave velocity anomaly ratio inferred from geodynamic modeling is much less than these values, ~0.1–0.3 or even negative (e.g., FORTE *et al.*, 1994).

Consequently, I conclude that the model considering only the anharmonic effect is not consistent with geophysical observations. However, at seismic frequencies, the temperature dependence of seismic wave velocities comes not only from anharmonicity but also from *anelasticity*, and if one includes the influence of anelasticity these three problems can be solved simultaneously, at least to a large extent as firstly shown by KARATO (1993a). For weak anelasticity, the functional form of the temperature dependence is approximately given by (see Chapter 3)

$$V_{P(S,\phi)}(\omega,T,P,X_i,\xi) = V^{\infty}_{P(S,\phi)}(T,P,X_i,\xi)\cdot \left[1-\frac{1}{2}\cot\frac{\alpha\pi}{2}Q^{-1}_{P(S,\phi)}(\omega,T,P,X_i,\xi)\right] \tag{20.12}$$

where $Q^{-1} \propto \omega^{-\alpha}$, $0<\alpha<1$.

The term in square brackets in equation (20.12) represents the anelastic effects and the temperature dependence is non-linear (see Chapter 11),

$$\frac{Q^{-1}_{P(S,\phi)}(\omega,T,P,X_i,\xi)}{Q^{-1}_{P(S,\phi)}(\omega,T_0,P,X_i,\xi)} = \exp\left[-\frac{\alpha H^*_{P(S,\phi)}}{R}\left(\frac{1}{T}-\frac{1}{T_0}\right)\right] \tag{20.13}$$

where $H^*_{P(S,\phi)}$ is the activation enthalpy for attenuation. From equations (20.9), (20.10), (20.12) and (20.13), one has (Problem 20.2)

$$A_{V_{P(S,\phi)}T} = (A_{V_{P(S,\phi)}T})_{ah} + (A_{V_{P(S,\phi)}T})_{an} \tag{20.14}$$

with

$$(A_{V_{P(S,\phi)}T})_{ah} = -\theta_{P(S,\phi)} \tag{20.15a}$$

and

$$(A_{V_{P(S,\phi)}T})_{an} \approx -\frac{H^*_{P(S,\phi)}}{\pi RT^2}Q^{-1}_{P(S,\phi)} \tag{20.15b}$$

where I used an approximation, $(\alpha\pi/2)\cot(\alpha\pi/2) \approx 1$. Here the parameter $H^*_{P(S,\phi)}$ must be determined by low-frequency attenuation measurements (see Chapter 11). Given this parameter and the values of $Q_{P(S,\phi)}$ determined from seismology, one can make corrections for the effects of anelasticity, $-(H^*_{P(S,\phi)}/\pi RT^2)Q^{-1}_{P(S,\phi)}$. As discussed in Chapter 11, experimental studies of $Q_{P(S,\phi)}$ are far more complicated than those of high-frequency elasticity and currently available data are limited to a small number of minerals (olivine, MgO, Al_2O_3, calcite, titanate perovskites) at low pressures. Theoretical calculations of this term, i.e., the activation enthalpy for attenuation, are also difficult. Consequently the uncertainties of parameters related to anelastic effects are larger than those of anharmonic effects. However, several general points can be noted.

(1) Except for a case where a material is highly heterogeneous, attenuation for volumetric deformation, i.e., Q_{ϕ}^{-1}, is significantly smaller than attenuation for shear deformation, $Q_S^{-1}(=Q_{\mu}^{-1})$. In such a case, $Q_P/Q_S \approx 3V_P^2/4V_S^2 = 1/B$. The exception is attenuation in a partially molten material in which bulk attenuation in not negligible (see Chapter 11; KARATO, 1977; BUDIANSKI and O'CONNELL, 1980; HEINZ *et al.*, 1982).

(2) Attenuation is in general highly sensitive to temperature and water content (Chapter 11). Consequently, lateral variation in attenuation is usually larger than that of seismic wave velocities.

Problem 20.2

Derive equation (20.14).

Solution

Taking the logarithm of equation (20.12), one has $\log V_{P(S,\phi)} = \log V^{\infty}_{P(S,\phi)} + \log\left[1-\frac{1}{2}\cot(\alpha\pi/2)Q^{-1}_{P(S,\phi)}\right]$. Now taking the derivative of this equation with respect to temperature, one finds

$$\frac{\partial \log V_{P(S,\phi)}}{\partial T} = \frac{\partial \log V^{\infty}_{P(S,\phi)}}{\partial T} - \frac{\frac{1}{2}\cot(\alpha\pi/2)}{1-\frac{1}{2}\cot(\alpha\pi/2)Q^{-1}_{P(S,\phi)}}\frac{\partial Q^{-1}_{P(S,\phi)}}{\partial T}.$$

Using equation (20.13), $\partial Q^{-1}_{P(S,\phi)}/\partial T = Q^{-1}_{P(S,\phi)}\alpha H^*_{P(S,\phi)}/RT^2$. Assuming that the effect of anelasticity is

[6] A comparison of mineral physics models with seismological observation of $R_{S/P}$ needs to be made cautiously. The results shown in Fig. 20.4 are average $R_{S/P}$ for a given depth. If chemical heterogeneity exists or if anelasticity plays an important role $R_{S/P}$ should change laterally.

small, i.e., $\frac{1}{2}\cot(\alpha\pi/2)Q^{-1}_{P(S,\phi)} \ll 1$, and using $(\alpha\pi/2)\cdot(\alpha\pi/2) \approx 1$, one obtains $\partial \log V_{P(S,\phi)}/\partial T \approx -\theta_{P(S,\phi)} - (H^*_{P(S,\phi)}/\pi RT^2)Q^{-1}_{P(S,\phi)}$.

Combining the values for anharmonic effects $(\theta_{P(S,\phi)}(P))$ with the experimental data on activation enthalpy of attenuation and seismologically determined $Q_{P(S,\phi)}$, one can calculate $A_{V_{P(S)}T}(\omega, T, P)$. One major uncertainty in doing this is the estimation of the influence of anelasticity. In fact, there are no experimental studies on anelasticity under deep Earth conditions at this time. However, the formulation (20.14) indicates that one does not have to predict the magnitude of attenuation from laboratory (or theoretical) studies. Instead seismologically observed values of attenuation, i.e., Q, can be used to estimate the effect of anelasticity. In this formulation, the only parameter that we need to calculate $A_{V_{P(S)}T}$ is the activation enthalpy $H^*_{P(S,\phi)}$. Again, the activation enthalpy changes with pressure, although there are no data on the influence of pressure on anelasticity. A common practice in such a case is to use a first-order approximation that the activation enthalpy is proportional to the melting temperature, $H^*_{P(S,\phi)} = \beta_{P(S,\phi)} RT_m$. The value of β and $T_m(P)$ (melting temperature) are obtained from the experimental studies. Equation (20.15b) is translated to $(A_{V_{P(S,\phi)}T})_{an} \approx -(\beta_{P(S,\phi)} T_m/\pi T^2)Q^{-1}_{P(S,\phi)}$. Therefore the ratio of these two terms is given by $\Pi \equiv (A_{V_{P(S,\phi)}T})_{an}/(A_{V_{P(S,\phi)}T})_{ah} = (1/\theta_{P(S,\phi)}T)\cdot(T_m/T)\,(\beta_{P(S,\phi)}/\pi)\cdot Q^{-1}_{P(S,\phi)}$. If one inserts typical values of parameters, one will obtain $\Pi \approx 1.0$ for the upper mantle and $\Pi \approx 0.2$–0.5 for the lower mantle.

The results are shown in Figures 20.8 and 20.9. It is noted that:

(1) The absolute values of $A_{V_{P(S)}T}$ decrease significantly with depth due to the pressure effects on thermal expansion and attenuation. This means that the amplitude of heterogeneity of seismic wave velocities should decrease with depth if thermal effects are dominant. This is consistent with the observation (7). However, the values of $A_{V_{P(S)}T}$ are mineral-dependent and as a consequence, there are jumps in this value at seismic discontinuities.

(2) The influence of anelasticity on $A_{V_{P(S)}T}$ is very large in the shallow mantle where attenuation is large ($Q_S \sim 100$): the effect of attenuation is to change the absolute values of $A_{V_{P(S)}T}$ by a factor of ~ 2. In the lower mantle where the average attenuation is small ($Q_S \sim 300$), the influence of anelasticity is less important in the shallow lower mantle but its effects are not negligible, particularly in regions of high attenuation (e.g., plumes) and in the deep lower mantle.

(3) In most cases, attenuation is larger for S-waves than for P-waves. Consequently, the incorporation of anelasticity effects increases $R_{S/P(\phi)} \equiv \delta \log V_S/\delta \log V_{P(\phi)} = A_{V_S T}/A_{V_{P(S,\phi)}T}$. This effect is strong in regions of high attenuation. Consequently, there should be a systematic variation in $R_{S/P(\phi)}$ at a given depth such that it is larger in hot regions than in cold regions. Oki (2006) reported a regional variation in $R_{S/P(\phi)}$ in the lower mantle that is consistent with this model. Such a systematic variation in $R_{S/P(\phi)}$ is a natural consequence of the temperature sensitivity of anelasticity, and this observation suggests that anelasticity is important in the lower mantle. In the extreme case where the influence of anelasticity dominates, we have $R_{S/P(\phi)} \approx 3V_P^2/4V_S^2 = 1/B$. (However, the regional variation in $R_{S/P}$ may also be due to the regional variation in chemical composition.)

(4) Similarly, the incorporation of anelasticity reduces $R_{\rho/P(S,\phi)} \equiv \delta \log \rho/\delta \log V_{P(S,\phi)} = A_{\rho T}/A_{V_{P(S,\phi)}T}$. This effect is strong in regions of high attenuation. However, this value is always positive if lateral heterogeneity is caused by the heterogeneity in temperature.

Problem 20.3

Assuming that the lateral heterogeneity in seismic wave velocities is caused by the heterogeneity in temperature, discuss the lateral variation in $R_{S/P}$.

Solution

When a thermal origin is assumed, then $R_{S/P} = A_{V_S T}/A_{V_P T}$. If only an anharmonic effect is considered, then, $R_{S/P} = (A_{V_S T})_{ah}/(A_{V_P T})_{ah} = \theta_S/\theta_P$. Now, as we learned in Chapter 4, $\theta_{S,P}$ is nearly independent of temperature, and consequently, $R_{S/P}$ should not vary regionally if only an anharmonic effect is important. However, the anelastic effect is sensitive to temperature and hence $R_{S/P}$ will vary laterally if the anelastic effect is important. Let us consider the lower mantle. In the lower mantle $\left|(A_{V_{S,P}T})_{ah}\right|$ decreases strongly with depth due to the increase in pressure, but $\left|(A_{V_{S,P}T})_{an}\right|$ is only weakly dependent on depth. Consequently, the relative contribution of the anelastic effects increases with depth. At 1000 km, the contribution from anelasticity is only ~20% so the change in anelasticity has little effect and $R_{S/P} \sim 1.8$ (Fig. 20.9). In contrast, near the bottom of the lower mantle the contribution of anelasticity is larger (~40%). In a hot region, the influence of the anelastic effect is large and $R_{S/P} \sim 2.3$, whereas in cold regions $R_{S/P} \sim 1.9$ (Fig. 20.9).

The last two points can be illustrated by an $R_{S/P} - R_{\rho/S}$ plot (Fig. 20.10). If the observed lateral variation in seismic wave velocities and density had a thermal origin, the data must lie in the right bottom region in this figure. In particular, $R_{S/P} > 3V_P^2/4V_S^2 = 1/B$ and $R_{\rho/S} < 0$ are conditions that cannot be attained by any thermally induced anomalies (for a material with $\frac{\theta_s}{\theta_p} < \frac{(A_{V_S T})_{an}}{(A_{V_P T})_{an}}$). When observed data are plotted on this diagram, it is seen that most of the data from Earth's mantle do indeed fall within this region, although the data from very deep regions (deeper than ~2300 km) and some data from the upper mantle are outside these regions. I conclude that when the influence of anelasticity is included, the model assuming a thermal origin for the lateral heterogeneity in seismic wave velocities and density can explain the majority of large-scale seismological and geodynamic observations except for limited regions such as the ocean–continent contrast in the upper mantle and the very deep lower mantle (deeper than ~2300 km). In the very deep lower mantle, current seismological and geodynamic observations showing very large values of $R_{S/P}$ (and equivalently a negative $R_{S/\phi}$) and negative values of $R_{\rho/S}$ are difficult to explain using a thermal model.

I must also note that the one-to-one correlation between velocity and attenuation assumed in equation (20.14) is valid only within the absorption band (i.e., in the frequency region where $Q^{-1} \propto \omega^{-\alpha}$) and when there are no other mechanisms of anelasticity at higher frequencies. A particularly important case is anelastic relaxation due to partial melting which may have peak frequencies that are higher than the seismic frequency band (see Chapters 3 and 11). In these cases, the method of inversion using the above formula has to be

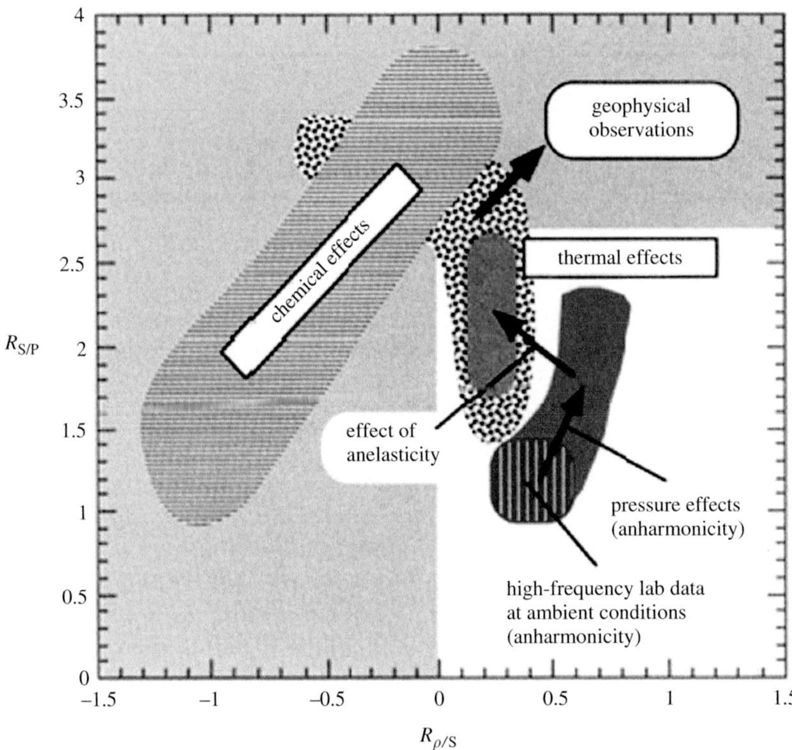

FIGURE 20.10 A plot of $R_{S/P}$ against $R_{\rho/S}$ predicted from the thermal and chemical origin of heterogeneity as compared with actual observations (after KARATO and KARKI, 2001). Plate 11.

modified to include the influence of relaxation caused by processes whose characteristic frequencies are higher than seismic frequencies.

Effects of phase transformations

A lateral variation in temperature can cause the lateral variation in dominant phases if the temperature and pressure conditions are favorable for phase transformations. If the discontinuity is sharp enough, then this effect is seen as topography on discontinuities (e.g., Shearer, 2000). If the discontinuity is not very sharp, then this effect will be seen as a modification to the lateral variation in seismic wave velocities. The influence of phase transformation is particularly important when the absolute value of the Clapeyron slope, $(dP/dT)_{eq}$, is large. For example, for $(dP/dT)_{eq} \approx 8$ MPa/K (2 MPa/K), the topography of the discontinuity corresponding to a ~ 500 K variation in temperature would be ~ 130 km (~ 30 km). Phase transformations usually cause a large variation in seismic wave velocities and densities and consequently it will have important effects on the lateral heterogeneity in seismic wave velocities. Anderson (1987b) discussed possible roles of phase transformations in the interpretation of lateral heterogeneity, see Fig. 20.11.

Let us consider the variation of seismic wave velocity across a finite temperature interval in which a phase transformation from phase 1 to phase 2 occurs. The velocity change for this temperature interval is given by

$$\delta \log V^1_{P(S)} = \frac{\partial \log V^1_{P(S)}}{\partial T} \delta T_1 + \frac{V^2_{P(S)} - V^1_{P(S)}}{V^1_{P(S)}} + \zeta \frac{\partial \log V^2_{P(S)}}{\partial T} (\delta T - \delta T_1) \tag{20.16}$$

where $V^{1,2}_{P(S)}$ is the P- (S-) wave velocity of phase 1 and 2 and $\zeta \equiv V^2_{P(S)}/V^1_{P(S)}$, where I have assumed that a phase transformation from phase 1 to phase 2 occurs at $T = T_0 + \delta T_1$. Similar equations can be written for density. Because of the presence of a number of phase

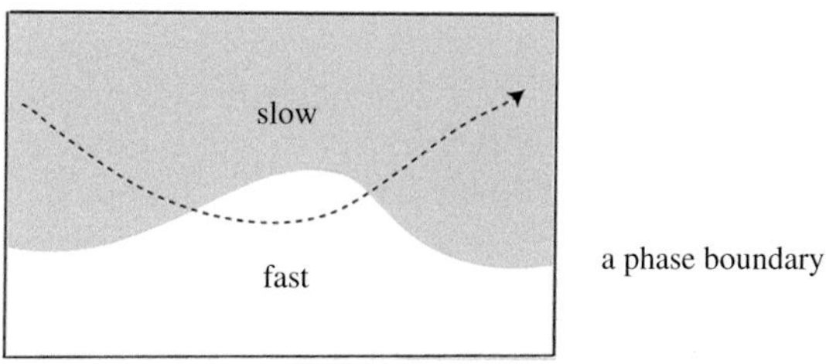

FIGURE 20.11 Lateral variation in phases due to the lateral variation in temperature and/or chemical composition leads to large lateral variation in seismic wave velocities

transformations, the interpretation of tomographic images of the mantle transition zone is complicated (Anderson, 1987b; see also Stixrude and Lithgow-Bertelloni, 2005a). Oganov and Ono (2004) and Tsuchiya *et al.* (2004a) discussed the possible role of a phase transformation of (Mg, Fe)SiO_3 perovskite to a post-perovskite phase for the lateral variation in seismic wave velocities in the D″ layer. Their calculations show that this phase transformation has a large Clapeyron slope ($\sim$ 8–10 MPa/K) and the transformation from perovskite to a post-perovskite phase increases the shear wave velocity but not the bulk sound wave velocity. Consequently, the perovskite to post-perovskite transformation can provide an explanation for large values of $R_{S/P}$ and negative values of $R_{S/\phi}$. However, the negative values of $R_{\rho/S}$ cannot be explained by this model and require the chemical heterogeneity.

Chemical origin

Now let us consider the effects of variation in chemical composition. In a chemically evolving planet like Earth, there must be some degree of chemical heterogeneity. Chemical heterogeneity is usually investigated through a petrological or geochemical approach based on the study of rock samples (e.g., Hofmann, 1997). However, the sampling range of these *direct* approaches is limited. In this section, I will briefly review the issues in identifying chemical heterogeneity through a *geophysical* approach in which geophysical (mostly seismological) observations are interpreted in terms of chemical composition. The most important points to recognize are (i) velocity variations due to a plausible chemical heterogeneity in the Earth are of the same order of magnitude as those due to temperature variation, but (ii) there are several distinct differences between thermal and chemical anomalies that allow us to distinguish a thermal origin from a chemical origin for the anomalies. To show these points, let us consider the influence of a variation of Fe/(Fe + Mg). A 5% increase in Fe/(Fe + Mg) will result in a reduction of the shear wave velocity of ~ 2% (for olivine), which would correspond to the effect of a $\sim$ 150–200 K increase in temperature (in the upper mantle). Also note that when a change in Fe/(Fe + Mg) is the cause of a velocity variation, the velocity variation has a negative correlation with the density variation. Table 4.4 summarizes some of the experimental data on the influence of chemical composition on seismic wave velocities. It is clearly seen that when one atom is substituted by another with a large difference in atomic weight (e.g., the substitution of Fe with Mg), then many partial derivatives are negative, whereas if the substitution occurs with a similar mass (e.g., substitution of Al with Mg + Si), then the partial derivatives are positive.

Two issues must be addressed: (i) how to identify the influence of chemical heterogeneity as opposed to that of thermal anomalies and (ii) how to reduce the number of unknowns in order to uniquely infer the nature of chemical heterogeneity. The first issue has already been discussed. A ratio–ratio plot such as Fig. 20.10 provides good guidance for identifying non-thermal effects. As discussed above, the negative value of $R_{\rho/S}$ and the very large values of $R_{S/P}$ in the deep lower mantle, and very small positive or negative values of $R_{\rho/S}$ in the upper mantle, strongly suggest either the effects of phase transformations or the influence of the chemical heterogeneity.

How many parameters (unknowns) do we need to make such an inversion? Unlike petrological or geochemical observations, geophysical observations provide only a small number of data values from a single position ($\delta \log V_{P(S)}, \delta \log Q^{-1}_{P(S)}, \delta \log \rho$). Therefore one needs to have a good model to reduce the number of unknowns to infer the anomalies in chemical composition from seismological observations. If the concentrations of all the possible major elements are treated independently, then there would be too large a number of unknowns and one could not uniquely determine the chemical heterogeneity from seismological observations. In many previous studies, the number of unknowns was chosen to be arbitrarily small to allow inversion (e.g., Forte and Mitrovica, 2001; Deschamps *et al.*, 2002), but there have not been many studies to justify the validity of such an assumption. In doing so, it is important to distinguish the influence of major element composition such as Mg, Fe, Ca, Al and Si from that of trace elements such as hydrogen. Matsukage *et al.*, (2005) addressed this question in an attempt to investigate the heterogeneity in the major element composition of the upper mantle (see also Schutt and Lesher, 2006). They found that the chemical heterogeneity of rocks in the oceanic upper mantle can be characterized by a single parameter such as Mg# (Mg# is the molar ratio of Mg to (Mg + Fe)). This is due to the fact that in the oceanic upper mantle, the compositional variation is controlled mainly by the variation in the degree of partial melting of a single common rock (i.e., pyrolite) at more or less the same conditions (beneath mid-ocean ridges). However, for rocks in the continental upper mantle, another parameter, such as opx# (the molar fraction of orthopyroxene) is also needed to characterize the chemical composition. This reflects the fact that the composition of rocks in the continental regions is controlled not only by partial melting under shallow mantle conditions but also by a mixing of several components with different Si/(Si + Mg + Fe) ratios. Therefore the number of unknowns in chemical composition in equation (20.2) becomes $n = 2$ (in addition to Mg#, the hydrogen content is also important as we will learn later) or $n = 3$ (hydrogen content + Mg# + opx#). Consequently, it is possible to determine the lateral heterogeneity in the chemical composition from seismic tomography if both the velocity and the attenuation tomography data are available with similar resolution (Shito *et al.*, 2006).

It should be noted that even in the simplest case of oceanic upper mantle where a single parameter (Mg#) is enough to characterize the composition, one should not calculate the dependence of the seismic wave velocity on the chemical composition from $[\partial \log V_{P(S,\phi)}/\partial X_{Mg}]$ ($X_{Mg} \equiv Mg/(Fe + Mg) \equiv Mg\#$) and the abundance of each mineral as is often done (e.g., Forte and Mitrovica, 2001; Deschamps *et al.*, 2002). When Mg# changes, all other compositions such as the abundance of Ca, Al, Si, etc. will also change following the rule of chemical reaction associated with partial melting. The influence of all of these factors must be included when the dependence of the seismic wave velocity on the chemical composition is parameterized (Matsukage *et al.*, 2005). This can be shown mathematically as $d \log V_{P(S,\phi)}/dX_{Mg} = \partial \log V_{P(S,\phi)}/\partial X_{Mg} + \sum(\partial \log V_{P(S,\phi)}/\partial X_i)\ (\partial X_i/\partial X_{Mg})$. When one uses Mg# as a parameter to represent the major element composition, one should use $d \log V_{P(S,\phi)}/dX_{Mg}$ and not $\partial \log V_{P(S,\phi)}/\partial X_{Mg}$. In some cases, these two can even have a different sign (see Matsukage *et al.*, 2005).

The effects of chemical composition on seismic wave propagation are three-fold. First, when one element replaces another in a solid solution, then usually, the effect on the elastic constants is mostly through the change in density. This is because the electrostatic charge and stiffness of ions that can be replaced are similar except for their atomic weight. In fact, for most minerals, the Fe/(Fe + Mg) ratio does not change the bulk modulus so much (see Chapter 4). In such a case, the term $-\delta \log \rho$ dominates in equation (20.3), and hence $R_{\rho/P(S)} \equiv \delta \log \rho / \delta \log V_{P,S} < 0$. Second, when the amount of one element is modified, the volume fraction of minerals could change. For instance, when the amount of Al (and/or Ca) is reduced in the upper mantle, the volume fraction of garnet will be reduced, which has an important effect on seismic wave velocities. Third is the influence on anelasticity. To see how the change in chemical composition may affect attenuation, let us consider the general relationship

$$Q^{-1}_{P(S)}(\omega, T, P, X_H, \xi) \propto (\omega\tau)^{-\alpha} \tag{20.17}$$

where $\alpha = 0.2$–0.4 and τ is the characteristic time of the relation process, which is inversely proportional to the mobility of defects (see Chapter 11). Experimental studies on the influence of major element composition (such as Fe/(Fe + Mg)) on the defect mobility indicate that a variation in Fe/(Fe + Mg) by 3% (usually Fe/(Fe + Mg) = 10 ± 3%, see Chapter 17) would result in a change in Q of only ~ 5%, which is less than the error bars of anelasticity measurements in seismology and the resultant effects on the seismic wave velocity are also too small to be detected (less than ~ 10^{-2} %).

However, the variation in water (hydrogen) content could have an important effect. Dissolution of hydrogen changes the concentration of point defects and hence has a large influence on the characteristic relaxation time, τ. In most cases, $\tau \propto X_H^{-r}$, where X_H is the concentration of hydrogen ($r \sim 1$–2, Chapters 10 and 11) and $A_{Q^{-1}_{P(S)}X_H} \equiv \partial \log Q^{-1}_{P(S)} / \partial \log X_H = \alpha r$. One expects a broad range of water content in Earth's mantle (~ 10^{-5}–10^{-1} wt%, see Chapter 17). This will lead to a variation of Q by a factor of ~ 10, and hence the variation of velocity of a few percent. The presence of water (hydrogen) can also affect the unrelaxed seismic wave velocities, $V^{\infty}_{P(S,\phi)}$ (e.g., JACOBSEN, 2006). However a large effect is limited to a relatively large water content (~ 1 wt%). Consequently, the most important influence of water (hydrogen) on the seismological signature, in most cases (less than 1 wt% water), is its influence on attenuation. Electrical conductivity can also provide a robust estimate of water (hydrogen) content (for a more complete discussion of these issues see KARATO (2006c)).

One important point to be noted about the chemical effect is that unlike the thermal effect, which is reduced significantly by pressure (due to the large pressure effect of thermal expansion, Chapter 4), there are no obvious reasons for the chemical effect to change with pressure, and in fact, for many minerals, the pressure effect on $A_{V_{P(S,\phi)}X_i}$ is generally less than that on $A_{V_{P(S,\phi)}T}$. Consequently, the chemical effect will be more visible in the deeper mantle (Problem 20.4).

Problem 20.4*

Assuming that Fe content does not affect the bulk modulus (Chapter 4), but does affect the density, show that $\partial \log V_\phi / \partial X_{Fe}$ changes with pressure much less than $\partial \log V_\phi / \partial T$ (Fig. 20.12).

Solution

When the bulk modulus is insensitive to the Fe content, then equation (20.3c) leads to

$$\frac{\partial \log V_\phi}{\partial X_{Fe}} = -\frac{1}{2}\left[\frac{\partial \log\{X_{Fe}M_{Fe} + (1 - X_{Fe})M_{Mg}\}}{\partial X_{Fe}} - \frac{\partial \log\{X_{Fe}V_{Fe} + (1 - X_{Fe})V_{Mg}\}}{\partial X_{Fe}}\right]$$

where $M_{Fe,Mg}$ is the molecular weight of the Fe (Mg) end-member, and $V_{Fe,Mg}$ is the molar volume of the Fe (Mg) end-member. Therefore $\partial \log V_\phi / \partial X_{Fe} = -\frac{1}{2}(\Delta M/M - \Delta V/V) \approx -\frac{1}{2}(\Delta M/M)$ (<0) ($\Delta M \equiv M_{Fe} - M_{Mg}$, $\Delta V \equiv V_{Fe} - V_{Mg}$), where I have used an approximation that the ionic radii of Fe and Mg are similar but the atomic weights are largely different (i.e., $|\Delta M/M| \gg |\Delta V/V|$). Since the atomic weight does not change with pressure (and temperature), $\partial \log V_\phi / \partial X_{Fe}$ is nearly independent of pressure (and temperature). In contrast, $\partial \log V_\phi / \partial T$ changes

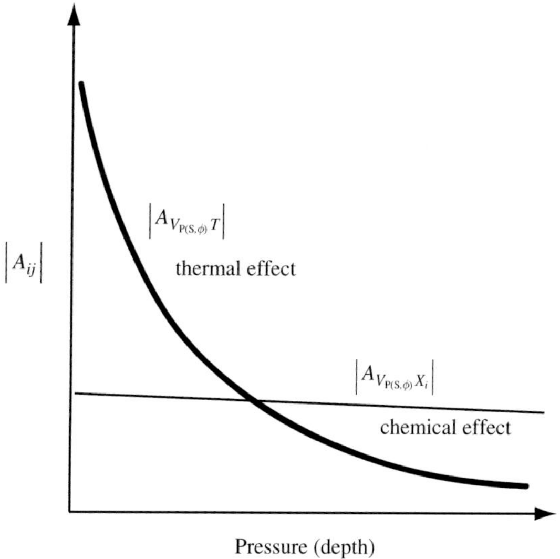

FIGURE 20.12 A schematic diagram showing the depth (pressure) variation of partial derivatives indicating that the chemical effect becomes more visible than the thermal effect in the deeper portions of the Earth.

significantly with pressure as can be seen from $\partial \log V_\phi / \partial T \approx -\theta_\phi(P) \approx -\theta_\phi(0)(\rho/\rho_0)^{-\delta_T}(\gamma_\phi(0)(\rho_0/\rho) - \frac{1}{3})$, where I have used the relation $\alpha_{\text{th}}(P)/\alpha_{\text{th}}(0) = (\rho/\rho_0)^{-\delta_T}$ and $\gamma\rho \approx$ constant (Chapter 4). From this equation it is obvious that $|\partial \log V_\phi / \partial T|$ is reduced significantly with depth, whereas $|\partial \log V_\phi / \partial X_{\text{Fe}}|$ is nearly independent of depth.

Direct effects of partial melting

In addition to its effects through the change in major element chemistry and water content, partial melting has direct effects on seismic properties, i.e., the effect of the presence of melts on mechanical properties. Some details of the effects of partial melting on seismic properties are discussed in Chapter 11. The presence of melt affects the seismic wave propagation in several ways. First, melt has different elastic moduli (zero shear modulus and a much smaller bulk modulus) than those of solid minerals. Consequently, a mixture of melt and solid has different elastic properties than a pure solid. Second, deformation of melt occurs by viscous flow as well as by elastic compression. Consequently, deformation becomes time-dependent and there will be attenuation and seismic wave velocities become frequency-dependent.

O'Connell and Budianski (1977) provided a detailed analysis of both elastic and anelastic properties of partially molten materials (liquid and solid mixtures) (Takei (1998) obtained similar results for relaxed moduli through the analysis of deformation due to grain–grain contact, see also Takei (2000, 2002)). Briefly, the elastic and anelastic properties of a partial melt depend on the contrast in bulk modulus, the viscosity of the melt and the geometry of the melt pocket. When the contrast in bulk modulus is large (this is usually the case for silicates), then the presence of a melt phase reduces the compressional wave velocity and increases bulk attenuation (Chapter 11). This effect is also sensitive to the melt geometry. Fig. 20.13 shows the $R_{S/P}$ (for low-frequency, relaxed velocities) as a function of the aspect ratio of the melt pocket and the bulk modulus ratio. For equilibrium geometry, partial melting in typical mantle conditions will lead to $R_{S/P} \sim 1-1.2$, whereas a highly flattened melt pocket could yield $R_{S/P} \sim 2-2.5$. The range of values of $R_S/_P$ for partially molten materials is within the range of thermally induced anomalies (see Fig. 20.9) and therefore the values of $R_{S/P}$ cannot be used to uniquely identify the presence of flattened melt pockets.

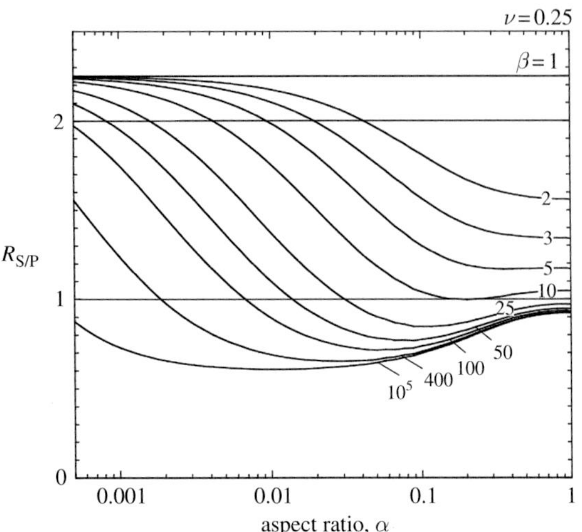

FIGURE 20.13 A plot showing the influence of the bulk modulus ratio, $\beta \equiv K_{\text{solid}}/K_{\text{melt}}$, and the aspect ratio, α, of melt pocket on $R_{S/P}$ for (relaxed) seismic wave velocities of partially molten rocks (after Takei, 2002).

The recent experimental study showed that the frequency dependence of attenuation is weak for a partially molten peridotite due to the presence of a broad peak in the seismic frequency range (Jackson *et al.*, 2004a) as compared to that of melt-free olivine aggregates for which attenuation depends on frequency through a power law, $Q \propto \omega^\alpha$ with $\alpha = 0.2$–0.4 (see Chapter 11). The seismological observations on the frequency dependence of attenuation in the Philippine Sea upper mantle are, however, consistent with melt-free olivine and not with a partially molten peridotite (Shito *et al.*, 2004). This indicates that either partial melt is not present in this region or that even if there is partial melting the characteristic frequency of attenuation is away from seismic frequencies. Furthermore, rather surprisingly, the extensive studies on the structure of the mantle beneath a mid-ocean ridge failed to find evidence for melt-induced anisotropy (Wolfe and Solomon, 1998), suggesting that regions of significant melt (>1 %) are highly localized. Alternatively, partial melt may exist in these regions but the shape of the melt pocket is not modified by stress. Similarly, high electrical conductivity is often considered as evidence for partial melting (Shankland *et al.*, 1981). However, Karato (1990) suggested that high electrical conductivity can be attributed to high water (hydrogen) content without invoking partial melting (this was confirmed later by Wang *et al.* (2006)). Geochemical/geodynamic observations indicate that at a given time and point beneath mid-ocean ridges, the volume fraction of melt is less than 0.1% (Spiegelman and Kenyon, 1992; Spiegelman and Elliott, 1993). Consequently, I conclude that there is little seismological observation to support the presence of a significant fraction of partial melting in most of the Earth's mantle. Melting must occur at least beneath volcanoes, but these observations suggest that the fraction of melt in a broad region beneath volcanoes (mid-ocean ridges) is small, presumably due to efficient melt transport. Also petrological studies have shown that the degree of melting in the deep upper mantle (> 70 km) is controlled by the amount of water (and carbon dioxide) and is ~0.1% (Langmuir *et al.*, 1992). The ultra-low velocity zone near the bottom of the mantle might be a region of partial melting (Williams and Garnero, 1996; Lay *et al.*, 2004), but an alternative explanation in terms of high Fe content is also possible (Garnero and Jeanloz, 2000).

The above summary suggests that regions with a significant amount of melt are highly localized. Identifying these regions is important for Earth science. There are several ways by which the presence of melt can be inferred. For example, the presence of melt often leads to bulk attenuation (see Chapter 11). Bulk attenuation is expected to be very small for solid-state mechanisms of anelasticity. Consequently, the detection of bulk attenuation is a way to identify the presence of partial melt. Also, if a large lateral variation in seismic wave velocities is observed without a large variation in attenuation, the presence of partial melt is inferred when the influence of major element chemistry is small.

Grain size

Grain size should not have any effect on high-frequency seismic wave velocity, $V^{\infty}_{P(S,\phi)}$. However, grain size may affect anelasticity. Faul and Jackson (2005) proposed that in addition to temperature and pressure, grain size has some effects on seismic wave attenuation and hence indirectly seismic wave velocities in the upper mantle. Note, however, that the influence of grain size on seismic wave attenuation is probably not very large for a plausible range of grain size in Earth's upper mantle. The grain size for most of the upper mantle xenoliths is in the range of ~3–10 mm (e.g., Mercier, 1980; Karato, 1984), and using the experimentally obtained grain-size sensitivity of attenuation, $Q \propto L^q$ with $q \sim 0.25$, one expects a change in Q of ~30%, which results in a change in velocity of only 0.3% (0.1%) for $Q = 100$ ($Q = 300$). For thermoelasticity (see Chapter 11), grain size sensitivity is stronger, $q \sim 0.5$ ($Q \propto (\omega\tau)^\alpha$, $\alpha \approx 0.25$, $\tau \propto L^2$). But the absolute value of attenuation due to thermoelasticity is small ($Q > 1000$, Chapter 11) and the influence of grain size on attenuation through this mechanism is not large. I conclude that the grain-size effect is modest compared to the effect of water (hydrogen) content and temperature.

20.3.2. Topography on discontinuities and its geodynamic significance

I have already discussed the lateral variation in seismic wave velocity due to the lateral variation in mineral phases. When the radial variation in mineral phases occurs as a seismic discontinuity, this lateral variation will appear as topography on the discontinuity. Fig. 20.6 shows some of the examples of topography on mantle discontinuities (for reviews on this topic see Gu *et al.* (1998), Shearer (2000) and Romanowicz (2003)). If we know what controls the depth of the discontinuity, then from the topography of the discontinuity we can learn the lateral variation of parameters that control the depth of the discontinuity. Seismic discontinuities may be caused by: (1) phase transformations, (2) changes in chemical composition and/or (3) changes in microstructures (such as the lattice-preferred orientation). Consequently, when the parameters (physical or chemical) that control the depth of discontinuities change laterally, then there will be lateral variation in the depth of the discontinuity. Generally, one can write

$$\delta z = \left(\frac{\partial z}{\partial T}\right)_{X_i,M} \delta T + \sum_i \left(\frac{\partial z}{\partial X_i}\right)_{T,M} \delta X_i + \left(\frac{\partial z}{\partial M}\right)_{T,X_i} \delta M \tag{20.18a}$$

or

$$\delta z = \frac{(\partial P/\partial T)_{X_i,M}\, \delta T + \sum_i (\partial P/\partial X_i)_{T,M} \delta X_i + (\partial P/\partial M)_{T,X_i} \delta M}{\rho g} \tag{20.18b}$$

where δz is the topography on a boundary and δT, δX_i and δM are anomalies in temperature, chemical composition and microstructure respectively. $(\partial P/\partial T)_{X_i,M}$, $(\partial P/\partial X_i)_{T,M}$ and $(\partial P/\partial M)_{T,X_i}$ are the parameters that are determined by material properties that can be investigated by laboratory studies (or by theoretical models).

The nature of these parameters depends on the physical origin of the discontinuity. Let us consider two examples. One is a discontinuity caused by a phase transformation. In this case, the effects of microstructure are negligible. We need to consider only $(\partial P/\partial T)_{X_i,M}$ and $(\partial P/\partial X_i)_{T,M}$. $(\partial P/\partial T)_{X_i,M}$ is the Clapeyron slope that is known for most phase transformations (Table 2.3, Chapter 2). $(\partial P/\partial T)_{X_i,M} > 0$ for olivine to wadsleyite transition (410 km) and $(\partial P/\partial T)_{X_i,M} < 0$ for the ringwoodite to perovskite + magnesiowüstite transition (660 km). Consequently, the topography on the 410- and 660-km discontinuity would be anti-correlated if the topography were caused by the lateral variation in temperature.

The influence of the chemical composition on the phase boundary, $(\partial P/\partial X_i)_{T,M}$, is well known only for the variation in Fe/(Fe + Mg). For the olivine to wadsleyite transformation (the 410-km discontinuity), $(\partial P/\partial X_{Fe})_{T,M} < 0$, whereas for the ringwoodite to perovskite + magnesiowüstite transformation, $(\partial P/\partial X_{Fe})_{T,M} \approx 0$. In addition to Fe/(Fe + Mg), water content may also modify the depth of discontinuity. Briefly, the presence of water (hydrogen) increases the stability fields of transition zone minerals relative to upper and lower mantle minerals, and consequently, the 410-km discontinuity will be shallower and the 660-km discontinuity will be deeper when a large amount of water is present (Wood, 1995; Higo *et al.*, 2001; Chen *et al.*, 2002a; Smyth and Frost, 2002; see also Karato, 2006b).

Another is the case of a discontinuity caused by microstructural variations. They include the Lehmann discontinuity at around 220 km (Karato, 1992; Gaherty and Jordan, 1995) and the discontinuity at the top of the D″ layer (at around 2700 km). The regional variation in the depths of these boundaries may reflect the regional variation in microstructures such as the lattice-preferred orientation. The nature of lattice-preferred orientation is sensitive to a number of parameters including stress, temperature and water content (Chapter 14), the details of which remain to be determined (e.g., Jung *et al.*, 2006). The above analysis shows that in order to infer δT, δX_i and δM from the topography of discontinuity, one needs to have more than one observation, a situation much the same as the situation in inferring thermal and chemical anomalies from tomographic images. Such studies include Gu *et al.* (2003) and Blum and Shen (2004), but separating various effects is not straightforward. An alternative way is to investigate the correlation of topography of various discontinuities. For example, if the topography of "410-" and "660-km" discontinuities in a certain region is caused by thermal anomalies associated with a vertical flow of materials, then temperature anomalies near these boundaries are likely to be similar (except for the effect due to latent heat). Consequently, $\delta z^{410}/\delta z^{660} = (dP/dT)^{410} \cdot \delta T^{410}/((dP/dT)^{660}\delta T^{660}) \propto (dP/dT)^{410}/(dP/dT)^{660}$. The value of $(dP/dT)^{410}/(dP/dT)^{660}$ is known from laboratory studies (the Clapeyron slope, see Chapter 2), so that one can determine if the thermal origin of the topography of the boundaries is consistent with observations or not. We know $(dP/dT)^{410}/(dP/dT)^{660} < 0$ from mineral physics, so if the anomaly is thermal, then $\delta z^{410}/\delta z^{660} < 0$. Any deviation from this would imply that at least one of the assumptions is not met in this region.

As has been discussed above, several factors could cause the topography on the discontinuity. Therefore when the topography alone is investigated, its interpretation is non-unique. A combination with other observations could narrow the plausible causes for topography. For example, Blum and Shen (2004) combined the observations of topography of the 410-km discontinuity and velocity anomalies in the transition zone to infer the water content in the transition zone.

20.3.3. Summary of inversion scheme

A key issue in inverting tomographic images in terms of the physical and chemical state of the Earth's interior is to identify the cause for lateral variation (in velocities and in attenuation). If only one data value, say the anomaly in V_S, is available for a given point in the Earth, then the interpretation of anomalies in terms of physical or chemical conditions is non-unique. The same velocity anomaly could be due to an anomaly in temperature or in chemical composition or due to partial melting. For this reason it is critical to have multiple data sets for a given region. These multiple data sets must be for those quantities whose dependence on thermal or chemical anomalies are distinctly different. In this sense a joint inversion of velocity and attenuation anomalies is very useful because the sensitivity of velocity and attenuation to thermal and chemical anomalies are quite distinct as summarized above. Similarly, a joint inversion of V_P and V_S (or V_ϕ and V_S) is useful because the sensitivity of these two velocities, particularly V_ϕ and V_S, to thermal and chemical anomalies is different. In doing a joint inversion of multiple data sets, one must make sure that the data sets that one uses have a similar resolution. Technical issues for obtaining joint V_P and V_S (or V_ϕ and V_S) tomographic images are discussed by Masters *et al.*

(2000). In many cases, the quality of data from which anomalies of V_P and V_S are obtained is quite different and consequently, there can be systematic difference in errors for these anomalies. In addition, the wavelengths are different between the two waves for the same frequency, which could also cause a systematic difference in errors. Consequently, it is important to check the consistency of the data when tomographic images of two different velocities are compared. One way is to see if the relation (20.6) is satisfied by the data on V_P and V_S (or V_ϕ and V_S) (KARATO and KARKI, 2001). Deviation from the relation (20.6) means that the joint inversion contains some inconsistency. When tomographic images are analyzed in regions where seismic discontinuities occur (e.g., a transition zone), then the topography on the discontinuities can also be used to infer the origin of lateral heterogeneity.

Some issues in identifying the cause for heterogeneity have already been discussed using a ratio–ratio plot (Fig. 20.10). Here I will summarize some additional points using the velocity and attenuation tomography together. For simplicity, let us first treat a case in which the influence of partial melting and grain size is negligible. In this case, the influence of temperature, water content and major element chemistry can be summarized as follows.

(1) If the major cause for lateral variation of properties is the lateral variation in major element chemistry, then attenuation will be nearly constant whereas seismic wave velocities will have a large variation.
(2) If the major cause for lateral variation of properties is the lateral variation in water content, then attenuation will have a large lateral variation, whereas seismic wave velocities will have a relatively small lateral variation.
(3) If the major cause for lateral variation of properties is the lateral variation in temperature, then both seismic wave velocities and attenuation will have a large lateral variation.

Consequently, a plot such as that shown in Fig. 20.14 provides a guide for inferring the cause of lateral variation. If the variation in seismic wave velocities is large but the variation in attenuation is small, then we can infer that there is a large variation in major element chemistry. If, in contrast, there is a large variation in attenuation but a relatively small variation in seismic wave velocities, the most likely cause of lateral variation is the lateral variation in water content. More quantitatively, equation (20.2) can be solved to determine the lateral variation of temperature, major element composition and water content, if more than three observations are available and a simplified parameterization for the influence of major element chemistry is used. SHITO *et al.* (2006) performed such an inversion for a region where high-resolution velocity and attenuation tomography are available.

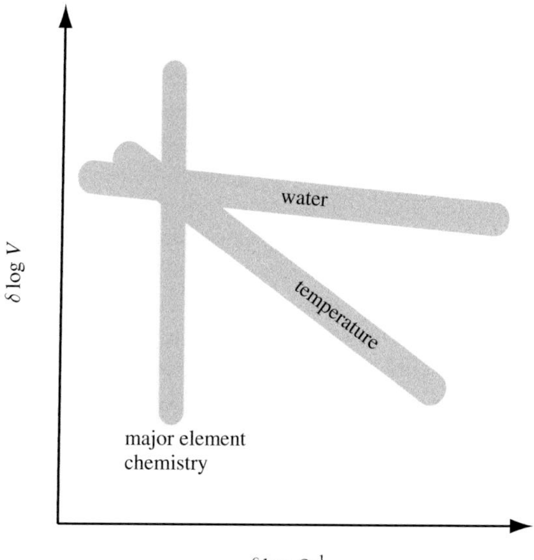

FIGURE 20.14 A schematic diagram of the attenuation anomaly–velocity anomaly relationship showing distinct trends for temperature-control, major element chemistry-control and water-control cases (after SHITO and SHIBUTANI, 2003b).

I note that although such an inversion is possible in certain regions, the ability to determine thermal and chemical anomalies is limited at this time. One of the main reasons for this is the resolution of seismic tomography. In order to separately determine these parameters, one needs to have high-resolution tomography of two seismic wave velocities (V_P and V_S (or V_ϕ and V_S)) and attenuation from the same region with a similar level of resolution. Another obvious limitation is the incomplete knowledge of mineral physics parameters, i.e., A_{ij}. Progress in these two areas is critical to make this approach useful. In the next section, I will review some of the results of such an inversion to infer thermal and chemical anomalies in Earth's interior.

20.3.4. Origin and geodynamic significance of lateral heterogeneity: Some examples

Upper mantle

A large lateral variation in seismic waves is observed in the upper mantle. Key mineral physics parameters, $A_{V_{P(S,\phi)}T}$, are reasonably well constrained for the upper mantle. The application of these mineral physics results to tomographic images of the oceanic upper mantle yields plausible temperature and density anomalies if the effects of anelasticity are included.

One of the most significant conclusions on the oceanic upper mantle is that the low-velocity anomalies associated with mid-ocean ridges are shallow and sometimes asymmetric (e.g., ZHANG and TANIMOTO, 1992, 1993). These observations provide strong support for the notion that upwelling at mid-ocean ridges is in most cases passive: upwelling occurs because of the separation of the overlying plates but not by the buoyancy force of hot materials. In contrast, low velocity regions associated with ocean islands such as Hawaii and Island extends deeper to the transition zone or deeper (e.g., MONTELLI *et al.*, 2004). Using the model presented here, amplitude of temperature anomalies in plumes can be estimated. The results yield $\sim$200–300 K for plumes in most cases that is much less than the temperature anomalies expected from a classic model of plume originating from the bottom thermal boundary layer.

Continental roots are deep but the exact depth of continents is somewhat controversial. Influence of anisotropy has an important effect on the estimated depth (e.g., ANDERSON, 1979; GUNG *et al.*, 2003). Velocity and density anomalies associated with continents cannot be solely attributed to thermal anomalies (e.g., JORDAN, 1981; FORTE *et al.*, 1994). Continental upper mantle has different major element chemical composition than that of oceanic upper mantle. However, mapping the anomalies in chemical compositions is challenging because the seismic wave velocities are less sensitive to major element chemistry than temperature (e.g., MATSUKAGE *et al.*, 2005; SCHUTT and LESHER, 2006). The contrast between oceanic and continental upper mantle is mostly in density but the anomalies in seismic wave velocities are subtle.

In addition to the heterogeneity in major element chemistry, evidence for heterogeneity in water (hydrogen) content has been suggested. Based on the combination of high-resolution velocity and attenuation tomography, SHITO *et al.* (2006) and SHITO and SHIBUTANI (2003b) found that in addition to the high lateral variation in major element chemistry in the shallow upper mantle ($<$ 150 km), there is a water-rich region in the deep upper mantle ($\sim$ 300–400 km) beneath the Philippine Sea (Fig. 20.15). Subduction of an old and cold lithosphere in this region is likely to have transported water deep into the upper mantle. However, inverting the velocity and attenuation anomalies is challenging both from seismological and mineral physics points of view. One needs reliable high-resolution velocity and attenuation tomography from the same region, and also the water content dependence of seismic wave velocities and attenuation must be characterized in detail.

Transition zone

One of the most prominent features of tomographic images on the transition zone is the presence of high-velocity anomalies in the broad regions of the transition zone particularly in the western Pacific. This implies that subducted slabs are deformed and stagnant in some of the transition zone (e.g., FUKAO *et al.*, 1992, 2001; ROMANOWICZ, 2003; ZHAO, 2004). The transition zone or the 660-km discontinuity provides a strong resistance to subduction, presumably due to the negative Clapeyron slope at the 660-km discontinuity or high viscosity of the lower mantle. Also note that this means that subducted slabs are relatively soft (KARATO *et al.*, 2001). However, there is a distinct ring of high-velocity anomalies in the bottom of the lower mantle corresponding to surface trenches, suggesting that stagnation of materials at the transition zone is only a temporal feature and these cold materials ultimately sink to the bottom of the mantle. This type of hybrid convection was demonstrated by numerical modeling incorporating the negative Clapeyron slope corresponding to the phase transformation at the 660-km discontinuity (e.g., HONDA *et al.*, 1993; TACKLEY *et al.*, 1993).

Observations on the topography on discontinuities can be used to infer the presence of chemical heterogeneity. If the lateral variation in temperature is the cause for lateral variation in topography, the topography on the two discontinuities ("410-" and "660-km") would be anti-correlated because the Clapeyron slope has different signs for the phase transformation corresponding to these two discontinuities. In general, topography on the 410-km discontinuity is less pronounced than that on the 660-km discontinuity and the correlation (or anti-correlation) between the topography on the 410-km and the 660-km discontinuity is poor (e.g., FLANAGAN and SHEARER, 1998; GU *et al.*, 1998, 2003). These results suggest the presence of chemical heterogeneity in the transition zone, although the nature of chemical heterogeneity is not clearly understood.

An important subject about the transition zone is its water content. The mineral physics studies clearly show that the transition zone minerals can dissolve up to $\sim$3 wt% of water (see Chapter 17). If indeed

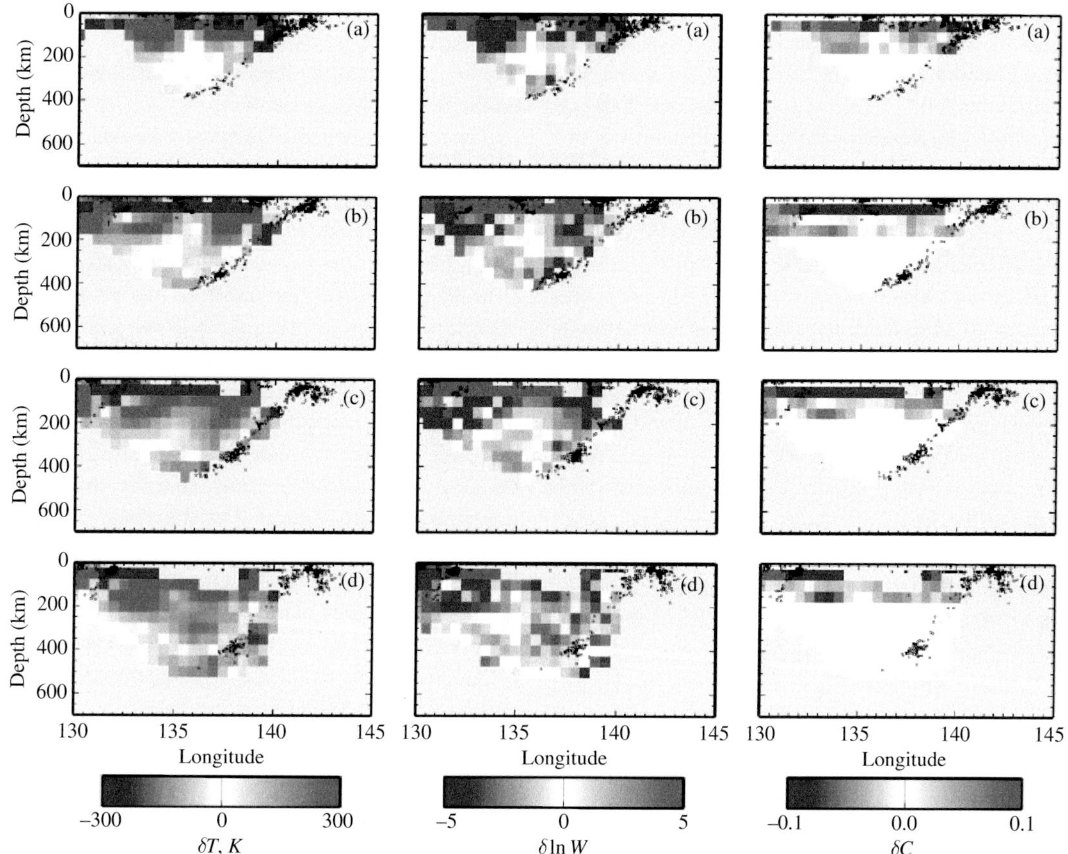

FIGURE 20.15 Inversion of velocity and attenuation tomography in terms of temperature (*T*), major element composition (*C*; Mg#) and water content (*W*) anomalies in the upper mantle beneath the Philippine Sea (Shito *et al.*, 2006). High-resolution tomography is available both for seismic wave velocities and attenuation from this region. A region of unusually high water content is identified in the deep upper mantle where attenuation is high but velocities are not so low. Plate 12.

the water content is this much, it will have an important influence on the dynamics and chemical evolution of Earth. There have been some attempts at inferring water content of the transition zone based on seismological observations (e.g., van der Meijde *et al.*, 2003; Blum and Shen, 2004; Suetsugu *et al.*, 2006). van der Meijde *et al.* (2003) used the sharpness of the 410-km discontinuity to infer the water content, and Blum and Shen (2004) and Suetsugu *et al.* (2006) used the topography of the 410-km discontinuity and the velocity anomalies in the transition zone to infer the water content. Again, inferring the water content remains a challenging subject.

Lower mantle

The most prominent feature of the lower mantle is the high-velocity ring corresponding roughly to the surface trenches, and the presence of two broad regions of low velocities (beneath south Pacific and Africa). Also the continuation of high-velocity anomalies associated with subducting slabs can be seen down to the deep lower mantle at least in the eastern Pacific and beneath Tibet to western Asia (e.g., Grand, 1994; van der Hilst *et al.*, 1997). These observations, together with the observations on the transition zone, suggest that mantle convection is, at least at present and in the recent past, whole mantle wide although evidence of significant resistance for convection currents is clear in the transition zone. The two broad low-velocity regions correspond to the broad swell in the Pacific (e.g., McNutt, 1998) as well as in Africa (Lithgow-Bertelloni and Silver, 1998). The location of these broad low-velocity regions corresponds roughly to the location of surface hotspots (e.g., Woodhouse and Dziewonski, 1989; Garnero, 2000). Consequently, it is considered that these low velocity regions represent upwelling. However, these low velocities are not all due to high temperatures. The anti-correlation of the S-wave velocity anomaly and the bulk-sound velocity anomaly ($R_{S/\phi} < 0$), and the very large values of $R_{S/P}$ indicate that anomalies in this region cannot be due solely to thermal anomalies (Karato and Karki, 2001). A sharp boundary extending from the core–mantle boundary (CMB) to ~ 500 km above the CMB suggests that chemical heterogeneity is likely to be present in the deep lower mantle (Wen, 2001). However, the extent to which chemical heterogeneity is present in the lower mantle is not clear. Kellogg *et al.* (1999) proposed, based largely on the interpretation of tomographic images of van der Hilst and Kárason (1999), that the bottom ~ 1000 km of the lower mantle is chemically distinct from the regions above. I should emphasize that the physical basis for such an argument is weak. Because $|A_{V_{P(S,\phi)T}}|$ decreases significantly with depth while $|A_{V_{P(S,\phi)X_i}}|$ does not, chemical effect becomes more visible than thermal effect with depth (Problem 20.2). Consequently, the observations suggesting a more important role for chemical effect with depth do not necessarily mean that the degree of chemical heterogeneity increases with depth. Karato and Karki (2001) discussed that chemical anomalies are required to explain available seismological data only in the bottom ~ 500 km (or less). Forte and Mitrovica (2001) attempted to map the chemical heterogeneity in the deep lower mantle. Presence of partial melting was proposed by Williams and Garnero (1996).

The recent discovery of the post-perovskite phase of $(Mg, Fe)SiO_3$ has introduced another complexity in interpreting seismological data on the bottom of the mantle (e.g., Murakami *et al.*, 2004; Oganov and Ono, 2004; Tsuchiya *et al.*, 2004a). Oganov and Ono (2004) argued that most of the seismological observations such as the large $R_{S/P}$ and negative $R_{\phi}/_{S}$ can be attributed to the phase transformation from perovskite to post-perovskite. However, the currently available data on seismic wave velocities such as $R_{S/P}$ and $R_{\phi}/_{S}$ are not enough to distinguish thermally-induced phase transformation from the influence of this phase transformation. The sharpness of velocity contrast (e.g., Wen, 2001) and/or some data on density distribution (e.g., Ishii and Tromp, 1999; Garnero and Jeanloz, 2000; Trampert *et al.*, 2004) suggest the presence of chemical heterogeneity in the D″ layer. Detecting density anomalies from geophysical observations remains challenging, however (e.g., Masters and Gubbins, 2003).

21 Seismic anisotropy and its geodynamic implications

Earth materials are anisotropic to some extent and the nature of anisotropic structures can be inferred from a range of seismological observations. Anisotropic structures carry valuable information as to the anisotropic fields that reflect the dynamic processes in Earth's interior including the geometry of flow. Therefore seismic anisotropy provides an important insight into the dynamics of Earth's interior. However, both observations and the interpretation of seismic anisotropy are not straightforward. Following the description of some of the fundamentals of seismic wave propagation in an anisotropic material, I will review the basic seismological observations on anisotropy. This is followed by a brief summary of the essence of the proccsses that may cause anisotropic structure formation. They include lattice-preferred orientation of anisotropic crystals and shape-preferred orientation of (isotropic or anisotropic) materials with distinct elastic moduli (or other types of layered structure). In most cases, seismic anisotropy is caused by lattice-preferred orientation of elastically anisotropic minerals. The relation between seismic anisotropy and deformation in such a case is mineral specific, and also depends on the physical and chemical conditions that change the elastic and plastic anisotropy of minerals. Following detailed discussions on the mechanisms of lattice-preferred orientation and other microscopic processes of anisotropic structure formation, some geodynamic implications of seismic anisotropy are discussed in the final section.

Key words azimuthal anisotropy, polarization anisotropy, shear wave splitting, Christoffel equation, lattice-preferred orientation, shape-preferred orientation.

21.1. Introduction

In the previous chapter, we learned how to investigate the dynamics of Earth's interior through the study of small deviations of real Earth structure from an ideal model assuming that Earth materials are isotropic. However, most rocks in Earth's interior show some degree of anisotropy and the magnitude of velocity variation with direction is in many cases comparable to that of heterogeneity. Therefore the results of seismic tomography summarized in the previous chapter are valid only when they represent anomalies of velocities averaged over the directions.

However, by taking the average over the directions, one will lose some important information on Earth's structure. In this chapter, we will focus on the directional dependence of seismic wave velocities, i.e., *seismic anisotropy*. The anisotropic structure of a rock reflects the *anisotropic field* in which the rock was formed or deformed. Anisotropic field includes stress or strain field as well as a temperature gradient or a magnetic field, all of which have an important influence on the way in which Earth works. Therefore studies of anisotropic structure can provide valuable information for understanding the dynamics of Earth's interior.

Although its potential importance is obvious, there are two important complications in investigating the dynamics of Earth's interior from seismic anisotropy. First, the measurement of anisotropy from seismology is not straightforward. In most cases, complete characterization of anisotropic structure is impossible and consequently, some simplifying schemes are used in the seismological measurements of anisotropy. Furthermore, depending on the technique used, the spatial resolution for the determination of anisotropic structure can be quite different. In order to interpret seismological observations on anisotropic structures, it is important to understand the merits and limitations of each technique. Second, geodynamic interpretation of seismic anisotropy requires the understanding of processes of anisotropic structure formation. There has been major progress in both the experimental and the theoretical sides of this issue, but there remain some major issues particularly concerning the physical processes of anisotropic structure development for materials in Earth's deep interior.

In this chapter I will first provide a brief summary of the essence of seismic wave propagation in an anisotropic material (section 21.2). Key concepts such as azimuthal and polarization anisotropy will be defined in section 21.3 and the nature of seismological observations on anisotropy is reviewed. Then some important observations are summarized in section 21.4 including anisotropy in the crust, the upper mantle, the transition zone, the lower mantle and the inner core. In section 21.5, some key mineral physics models are summarized that provide the bases for the geodynamic interpretation of seismic anisotropy. A brief summary of the processes of lattice-preferred orientation and the formation of a layer structure will be presented. Finally, in section 21.6, some geodynamic interpretations of seismic anisotropy are discussed.

21.2. Some fundamentals of elastic wave propagation in anisotropic media

Wave propagation in an anisotropic media is complicated. Recall that in the most general case, there are 21 independent elastic constants (Chapter 4) and consequently, in order to fully characterize anisotropic structures, one would need an impractically large number of data. Consequently, in almost all seismological studies on anisotropy, only some aspects of anisotropy are studied. In interpreting these results, it is important to understand the nature of the approximations or assumptions behind these studies. This section provides a brief summary of the fundamentals of seismic wave propagation in anisotropic media to help the geodynamical interpretation of seismological data. A more detailed account of elastic wave propagation in an anisotropic material is given in Smith and Dahlen (1973), Crampin (1981) and Montagner and Nataf (1986).

When the body-force such as gravity can be ignored (this is the case for short wavelength waves), the equation of motion of an anisotropic elastic material is given by (e.g., Landau and Lifshitz, 1959),

$$\rho \frac{\partial^2 u_i}{\partial t^2} = \sum_{j,k,l} C_{ijkl} \frac{\partial^2 u_k}{\partial x_j \partial x_l} \tag{21.1}$$

where ρ is density, u_i is displacement, t is time, x is the spatial coordinate and C_{ijkl} is the elastic constant tensor.

Let us consider a plane wave,

$$u_i = a_i \exp\left[i\omega \left(t - \sum_k q_k x_k \right) \right] \tag{21.2}$$

where a_i is the polarization vector that defines the direction of displacement, q_k is the slowness vector. Substituting equation (21.2) into (21.1), one has,

$$\rho a_i = \sum_{j,k,l} C_{ijkl} q_j q_l a_k. \tag{21.3}$$

Now let us define

$$q_j = \frac{1}{V} n_j \tag{21.4}$$

where V is the (phase) velocity[1] of seismic wave, and $\boldsymbol{n}$ is the unit vector parallel to the direction of propagation of waves. Then

$$\sum_k (T_{ik} - \rho V^2 \delta_{ik}) a_k = 0 \tag{21.5}$$

where

$$T_{ik} = \sum_{j,l} C_{ijkl} n_j n_l. \tag{21.6}$$

[1] The velocity defined here is called "phase velocity" because it is a velocity with which the phase of a plane wave, $-\omega\,(t - \sum_k q_k x_k)$, propagates. There is another definition of velocity of elastic waves, called "group velocity" that is the velocity at which the energy propagates with a wave. The distinction between them is important when a seismic wave contains various frequency components.

Problem 21.1

Derive equation (21.5).

Solution

Inserting (21.4) into (21.3), one has $\rho V^2 a_i = \sum_{j,k,l} C_{ijkl} q_j q_l a_k$. Therefore $\sum_{j,k,l} C_{ijkl} n_j n_l a_k - \rho V^2 a_i = \sum_k \left(\sum_{j,l} C_{ijkl} n_j n_l - \rho V^2 \delta_{ik} \right) a_k = \sum_k (T_{ik} - \rho V^2 \delta_{ik}) a_k = 0.$

Equation (21.5) is called the *Christoffel equation.* T_{ik} is a 3×3 symmetric matrix and therefore, for a given propagation direction n_i, (21.5) has three solutions. For an isotropic material, it can be shown from (21.6) that T_{ik} has a form (we choose $i = 1$ as a direction of propagation) (note that $K = (C_{11} + 2C_{12})/3$ and $\mu = C_{44} = (C_{11} - C_{12})/2$),[2]

$$T = \begin{bmatrix} K + \frac{4}{3}\mu & 0 & 0 \\ 0 & \mu & 0 \\ 0 & 0 & \mu \end{bmatrix}. \quad (21.7)$$

Problem 21.2

Derive equation (21.7).

Solution

By definition $\boldsymbol{n} = (1,0,0)$. So $T_{11} = \sum_{j,l} C_{1j1l} n_j n_l = C_{1111} = C_{11}$, $T_{22} = \sum_{j,l} C_{2j2l} n_j n_l = C_{2121} = C_{44} = T_{33}$ and all other $T_{ij} = 0$. Now for an isotropic material, $K = (C_{11} + 2C_{12})/3$ and $\mu = C_{44} = (C_{11} - C_{12})/2$, so $C_{11} = K + \frac{4}{3}\mu$, hence $T = \begin{bmatrix} K + \frac{4}{3}\mu & 0 & 0 \\ 0 & \mu & 0 \\ 0 & 0 & \mu \end{bmatrix}$.

For an isotropic material, the Christoffel equation becomes

$$\begin{bmatrix} (K + \frac{4}{3}\mu) - \rho V^2 & 0 & 0 \\ 0 & \mu - \rho V^2 & 0 \\ 0 & 0 & \mu - \rho V^2 \end{bmatrix} \cdot \begin{bmatrix} a_1 \\ a_2 \\ a_3 \end{bmatrix} = 0. \quad (21.8)$$

[2] For the relationship between the Voigt notation of elastic constants, C_{ij}, and the full tensor notation of elastic constants, C_{ijkl}, see Chapter 4. Typical values of elastic constants are given in Table 21.1 (for olivine).

This equation has three eigenvalues ($\rho V^2 = K + \frac{4}{3}\mu$, and two degenerated solutions $\rho V^2 = \mu$) and corresponding eigenvectors, $\boldsymbol{a} = (100)$, (010) and (001). Thus the first solution corresponds to the P-wave for which the polarization direction is parallel to the direction of propagation, whereas the other two solutions correspond to S-waves for which the polarization directions are perpendicular to the propagation direction (transverse waves). Such a clear distinction between P- and S-waves is no longer present in a general anisotropic material. However, in a weakly anisotropic material, such a relation holds approximately and terms quasi-P (qP) wave and quasi-S (qS) waves are sometimes used to refer to these waves. Note that the directions of polarization of transverse waves are controlled by the symmetry of the anisotropic structure of the material (this is how an optical polarizer works). In the above case, two orthogonal directions of polarization are (010) and (001) directions.

From the above analysis, one should note that there are two different ways of observing anisotropy in an elastic medium. First is the variation of elastic wave velocity with the orientation of wave *propagation*. This is often referred to as *azimuthal anisotropy*. Second is the difference in two shear wave velocities (of the same propagation direction) with different *polarization* directions. This is referred to as *polarization anisotropy*. Obviously, polarization anisotropy occurs only for shear waves, and the nature of polarization anisotropy, in general, changes with the direction of wave propagation. These two concepts are schematically shown in Fig. 21.1.

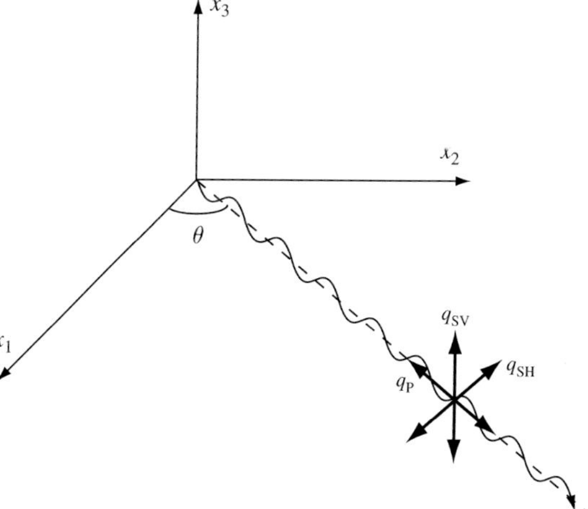

FIGURE 21.1 A schematic diagram showing the wave propagation in an anisotropic medium. Velocity can be dependent on the direction of wave propagation (azimuthal anisotropy). For shear waves (transverse wave), velocities of two waves with different polarization can be different (polarization anisotropy).

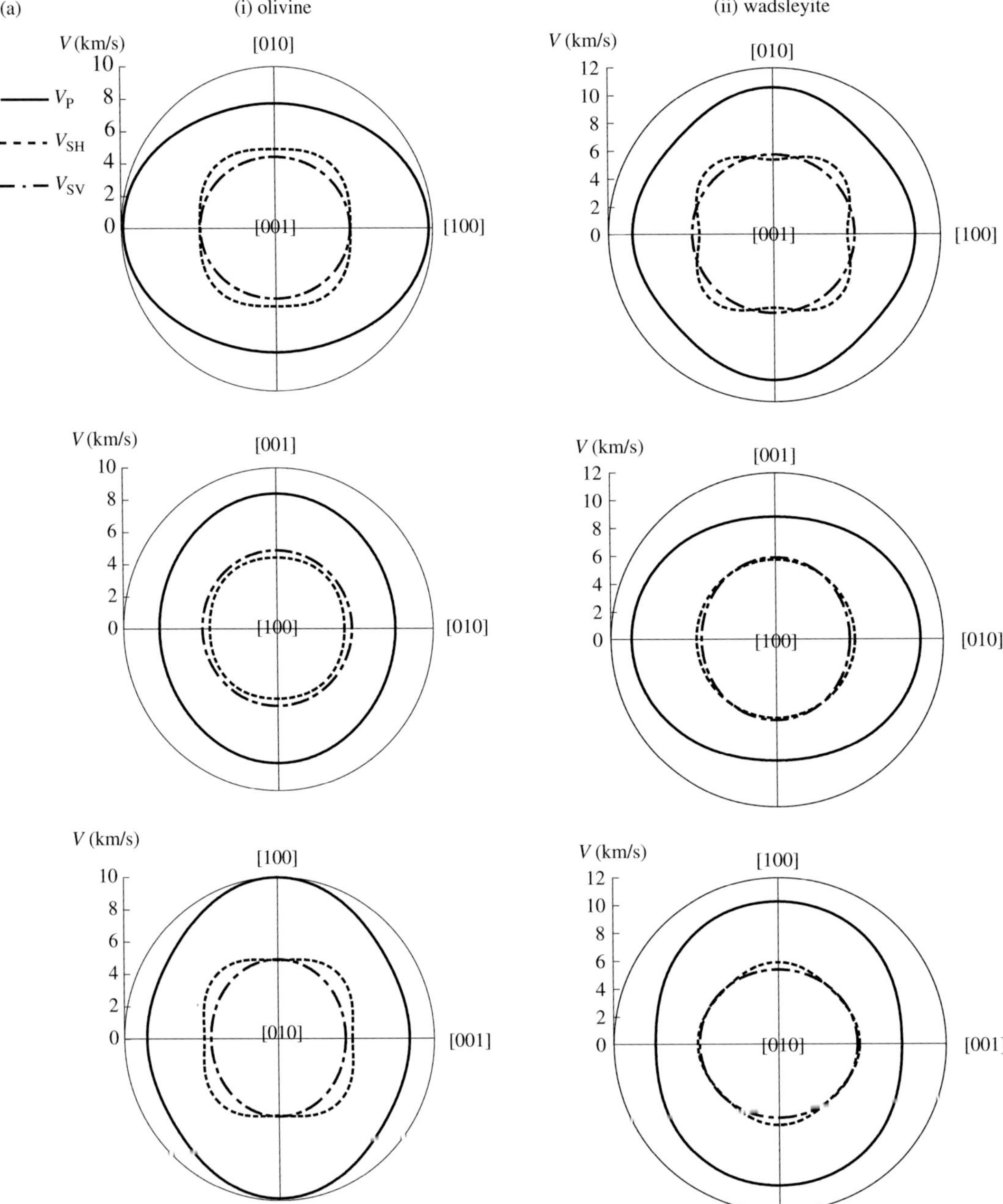

FIGURE 21.2 (a) Three elastic wave velocities as a function of the propagation direction in typical mantle minerals, (b) directions (shown by bars) of polarization of the faster S-wave as a function of the direction of the wave propagation. Both results are under ambient conditions (note that anisotropy can change with pressure, see Fig. 21.14). In (a), V_{SH} indicates the phase velocity of shear waves whose polarization direction is on the plane of wave propagation and V_{SV} is the phase velocity of shear waves whose polarization direction is normal to the plane of wave propagation. For pyrope these two velocities are nearly identical.

Velocity anisotropy can be illustrated by several diagrams. Because seismic wave velocities are the function of direction of propagation and (in the case of shear waves) also the function of polarization, one can represent the values of three velocities as a function of direction of propagation using the coordinate fixed to a crystal. Figure 21.2a is an example in which two-dimensional cross sections of such velocity surfaces are

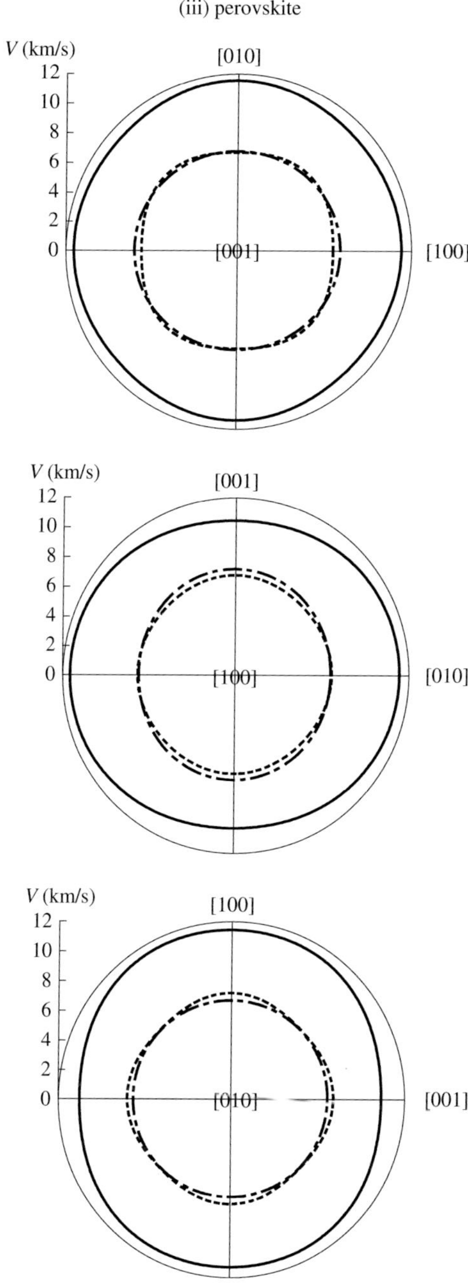

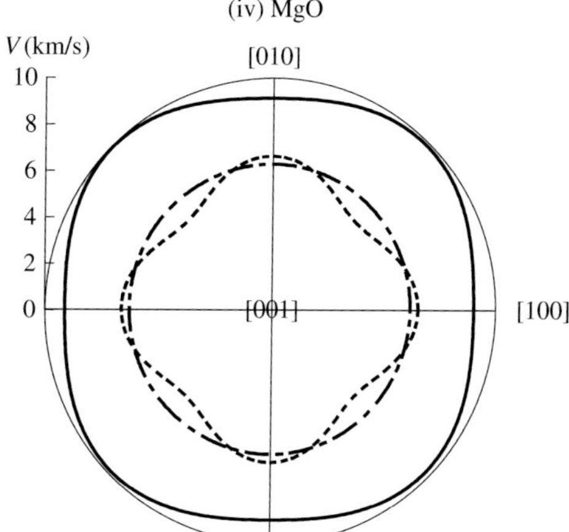

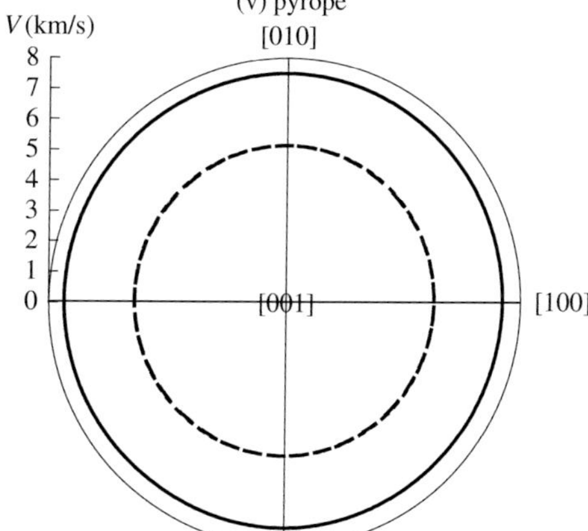

FIGURE 21.2 (cont.)

shown for three different waves as a function of direction of propagation. For a given direction of wave propagation, there are three velocities. Polarization of each wave can be determined by solving the Christoffel equation (21.5). For a certain symmetry orientation, two shear wave velocities are the same. For example, in a cubic system (spinel or MgO), the two shear waves propagating along the ⟨100⟩ and ⟨111⟩ orientations have the same velocity. Waves propagating along such directions will show no shear wave splitting. This type of diagram does not show the direction of the polarization of the faster S-waves. Figure 21.2b is another type of diagram in which the orientations of the polarization of the faster S-waves are plotted (using a stereographic projection) as a function of the direction of wave propagation.

The nature of azimuthal anisotropy can be shown from the extension of the above discussions. In the following we shall make an assumption of weak anisotropy. In this case we shall investigate a small deviation from isotropic case, so that the directions of quasi-P and quasi-S waves' polarization are the same as those in an isotropic material. We define the coordinate system as follows: x_1 is the direction of propagation

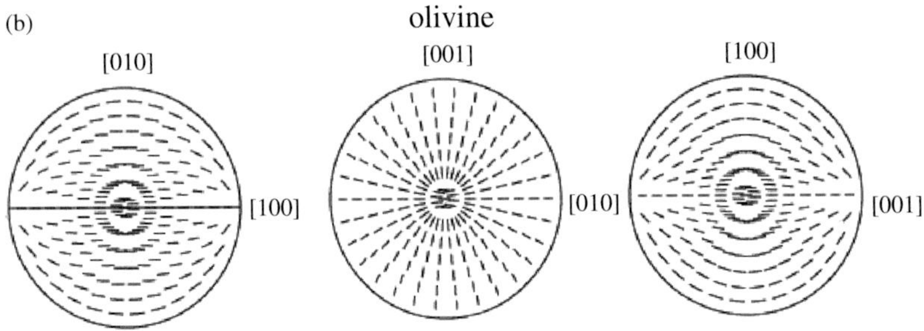

FIGURE 21.2 (cont.)

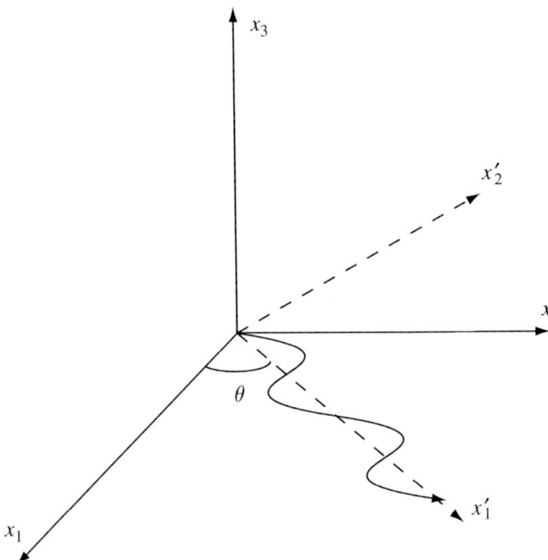

FIGURE 21.3 Rotation of the coordinate system.

of waves, x_3 is the vertical plane normal and x_2 is the direction normal to both x_1 and x_3. Azimuthal anisotropy can be investigated either by changing the orientation of n_i in the $x_3 = 0$ plane or by rotating the elastic constant matrix but fixing the propagation direction of waves.

Let us use the second approach (which is equivalent to the first approach). In this case, the nature of azimuthal anisotropy is determined by the nature of change in elastic constant matrix, C_{ijkl}, upon rotation. The change in matrix due to rotation across an axis can be determined by the operation of a 3×3 matrix, a_{ij}, (see Fig. 21.3)

$$a_{ij} = \begin{bmatrix} \cos\theta & \sin\theta & 0 \\ -\sin\theta & \cos\theta & 0 \\ 0 & 0 & 1 \end{bmatrix}. \tag{21.9}$$

The elastic constant matrix is transformed into a new form by this rotation,

$$C'_{ijkl} = \sum_{r,s,t,u=1}^{3} a_{ir}a_{js}a_{kt}a_{lu}C_{rstu}. \tag{21.10}$$

The seismic wave velocities have azimuthal dependence only up to 4θ terms ($\cos 4\theta, \sin 4\theta$). This is not a result of truncation but the exact result. The reason for this is the fact that the azimuthal anisotropy comes from the rotation of C_{ijkl} matrix (a fourth rank tensor), so that the rotation involves multiplying four matrices containing $\sin\theta$ and $\cos\theta$ (see (21.10)).

The matrix T_{ik} in this case is given by,

$$T_{ik} = \begin{bmatrix} C_{11} & C_{16} & C_{15} \\ C_{16} & C_{66} & C_{56} \\ C_{15} & C_{56} & C_{55} \end{bmatrix}. \tag{21.11}$$

If the $x_3 = 0$ plane is a symmetry plane, then all the elastic constants which contain an odd number of suffices equal to 3 will vanish so that C_{15} $(= C_{1113}) = C_{56}$ $(= C_{1312}) = 0$, then,

$$T_{ik} = \begin{bmatrix} C_{11} & C_{16} & 0 \\ C_{16} & C_{66} & 0 \\ 0 & 0 & C_{55} \end{bmatrix} \tag{21.12}$$

with $C_{11}, C_{66}, C_{55} \gg C_{16}$. Solving the Christoffel equation one obtains,

$$\begin{aligned} \rho V_P^2 &= C_{11} - \frac{C_{16}^2}{C_{11} - C_{66}} \\ \rho V_{SH}^2 &= C_{66} + \frac{C_{16}^2}{C_{11} - C_{66}} \\ \rho V_{SV}^2 &= C_{55}. \end{aligned} \tag{21.13}$$

Here V_{SH} is the velocity of quasi-shear wave with the polarization direction in the x_2 direction (i.e., the horizontal polarization), and V_{SV} is the velocity of quasi-shear waves with the polarization direction in the x_3 (the vertical polarization) (for the definition of SH and SV waves see Box 21.1).

Box 21.1 Seismological nomenclatures

Seismology is a science that is more than a century old, and consequently some technical jargon has been introduced in seismology. In order to read and understand the seismological literature, an understanding of the basic terminology is necessary. For an introduction to seismology, Lay and Wallace (1995) is a good start, and for a higher level presentation, see Dahlen and Tromp (1998) and Aki and Richards (2002).

The displacement of materials associated with elastic waves is called *polarization*. The polarization vector for the P-wave is always parallel to the propagation direction. For S-waves, the polarization vectors are on a plane that is perpendicular to the propagation direction (for simplicity, we consider only weak anisotropy). A combination of the polarization direction and the direction of wave propagation defines the plane on which material displacement occurs. Seismologists often classify S-waves into SH and SV waves based on the orientation of these planes. As discussed in the text, the directions of polarization of S-waves are controlled by the symmetry of anisotropy. When one of the symmetry axes of the elastic constants lies on the horizontal plane, then one of the polarization directions of the S-waves will be in the horizontal plane. When one considers a wave that propagates nearly horizontally (i.e., surface waves), then the other direction of polarization will be the vertical direction. In these cases, one can classify S-waves into SH (S-waves in which material motion is in the horizontal plane) and SV waves (material motion in the vertical plane). Similarly, PH (PV) wave is a P-wave in which material motion (and propagation direction) is horizontal (vertical). Other technical terms include:

Love wave a type of surface wave in which displacement is in the horizontal plane;

Rayleigh wave a type of surface wave in which displacement is mostly in the vertical plane (both A. E. H. Love (1863–1940) and Lord Rayleigh (J. W. Strutt; 1842–1919) were physicists who did their major work in late nineteenth century);

S (*P*) an S- (P-) wave directly coming from a source to a receiver;

ScS a body wave that originates as an S-wave and is reflected at the core–mantle boundary and comes back as an S-wave;

SKS a body wave that originates as an S-wave and is converted to a P-wave in the outer core and converted back to the mantle as an S-wave. An advantage of using this wave is that any information on the polarization in the initial S-wave phase is eliminated as the wave goes into the liquid outer core, so any information on anisotropy must be from the place after the last P- to S-wave conversion. This helps to locate the region of anisotropy below the receiver.

Problem 21.3

Derive equation (21.13).

Solution

Inserting equation (21.12) into equation (21.5), one obtains, $\begin{bmatrix} C_{11}-\rho V^2 & C_{16} & 0 \\ C_{16} & C_{66}-\rho V^2 & 0 \\ 0 & 0 & C_{55}-\rho V^2 \end{bmatrix} \cdot \begin{bmatrix} a_1 \\ a_2 \\ a_3 \end{bmatrix} = \begin{bmatrix} 0 \\ 0 \\ 0 \end{bmatrix}$.

For this equation to have non-trivial solution, one needs to have $\begin{vmatrix} C_{11}-\rho V^2 & C_{16} & 0 \\ C_{16} & C_{66}-\rho V^2 & 0 \\ 0 & 0 & C_{55}-\rho V^2 \end{vmatrix} = 0$. Therefore $(C_{55}-\rho V^2)\{(C_{11}-\rho V^2)(C_{66}-\rho V^2)-C_{16}^2\}=0$, and the solutions (the eigen-values) are: $\rho V_1^2 = C_{55}$, $\rho V_2^2 \approx C_{11} - C_{16}^2/(C_{11}-C_{66})$ and $\rho V_3^2 \approx C_{66} + C_{16}^2/(C_{11}-C_{66})$ where I used an approximation $C_{16} \ll C_{11}, C_{66}$. The corresponding eigenvectors can be calculated from (21.5) to obtain the relation (21.13).

Now in order to investigate the azimuthal anisotropy, we will rotate the crystal around the x_3 axis. Using equations (21.9) and (21.10), one gets a new elastic constant matrix. Then using this new elastic constant matrix, one calculates T_{ik} and solves equation (21.5) to obtain velocities and the polarizations of three elastic waves. In the case where $x_3=0$ is a symmetry plane, the velocities of seismic waves propagating on the $x_3=0$ plane are given by (Crampin, 1981),

$$\rho V_P^2 = A + B_c \cos 2\theta + B_s \sin 2\theta + C_c \cos 4\theta + C_s \sin 4\theta \quad (21.14a)$$

$$\rho V_{SH}^2 = D + E_c \cos 4\theta + E_s \sin 4\theta \quad (21.14b)$$

$$\rho V_{\mathrm{SV}}^2 = F + G_c \cos 2\theta + G_s \sin 2\theta \tag{21.14c}$$

with

$$A = \frac{3(C_{11}+C_{22})+2(C_{12}+2C_{66})}{8}$$
$$B_c = \frac{(C_{11}-C_{22})}{2}$$
$$B_s = C_{16}+C_{26}$$
$$C_c = \frac{C_{11}+C_{22}-2(C_{12}+2C_{66})}{8}$$
$$C_s = \frac{C_{16}-C_{26}}{2} \tag{21.15}$$
$$D = \frac{C_{11}+C_{22}-2(C_{12}-2C_{66})}{8}$$
$$E_c = -C_c$$
$$E_s = -C_s$$
$$F = \frac{C_{44}+C_{55}}{2}$$
$$G_c = \frac{C_{55}-C_{44}}{2}$$
$$G_s = C_{45}$$

where the elastic constants are defined with respect to a coordinate system in which x_3 is the vertical axis, and the seismic wave propagates in the horizontal plane and the angle θ is measured from the x_1 axis.

If in addition, the $x_2 = 0$ plane is the symmetry plane then out of C_{ijkl}, those containing an odd number of suffices equal to 2 vanish (e.g., C_{16} $(= C_{1112}) = C_{26}$ $(= C_{2212}) = 0$). Then equation (21.14) is reduced to

$$\rho V_{\mathrm{P}}^2 = A + B_c \cos 2\theta + C_c \cos 4\theta \tag{21.16a}$$

$$\rho V_{\mathrm{SH}}^2 = D + E_c \cos 4\theta \tag{21.16b}$$

$$\rho V_{\mathrm{SV}}^2 = F + G_c \cos 2\theta. \tag{21.16c}$$

Note that in both cases an SH-wave has only 4θ terms but an SV-wave has only 2θ terms.

Problem 21.4*

Explain why an SH-wave has the four-fold symmetry (only 4θ terms) whereas an SV-wave has the two-fold symmetry (only 2θ terms) in the plane of wave propagation if the plane is a symmetry plane.

Solution

Without the loss of generality, we can choose a coordinate system in which x_3 is the vertical axis, and the seismic wave propagates in the horizontal plane normal to x_3 (the angle θ is measured from the x_1 axis). Then remembering that the particle motion occurs on the x_1–x_2 plane for an SH-wave, we note that the velocity of an SH-wave is determined to the first order by an elastic constant corresponding to the shear strain ε_{12} associated with the shear stress σ_{12}, i.e., $C_{1212} (= C_{66})$. If one rotates the coordinate (around the x_3 axis) by $\pi/4$, then by exchanging $x_1 \leftrightarrow x_2$ one finds $C_{1212}(= C_{66}) \leftrightarrow C_{2121}(= C_{66})$. Therefore the velocity of an SH-wave is invariant for the rotation of $\pi/4$ so that it has four-fold symmetry. The velocity of an SV-wave propagating along the x_1 axis is controlled by the shear strain of ε_{13} associated with the shear stress σ_{13}, i.e., C_{1313} $(= C_{55})$. Upon exchanging $x_1 \leftrightarrow x_2$, it will change to $C_{2323}(= C_{44} \neq C_{55})$. Consequently, it has the two-fold symmetry but not the four-fold symmetry (obviously by exchanging $x_1 \leftrightarrow -x_1$, the elastic constant does not change).

21.3. Seismological methods for detecting anisotropic structures

21.3.1. Azimuthal anisotropy

The case for azimuthal anisotropy is straightforward from the above analysis. The azimuthal anisotropy of P-waves propagating in the horizontal plane can be analyzed using equation (21.14a) (or (21.16a)). In many cases, the cos 2θ term dominates and

$$\frac{\Delta V_{\mathrm{P}}}{V_{\mathrm{P}}} \approx \frac{B_c}{A} = \frac{4}{3}\frac{C_{11}-C_{22}}{(C_{11}+C_{22})+2(C_{12}+2C_{66})} \quad (\cos 2\theta \text{ term}). \tag{21.17}$$

The azimuthal anisotropy was first proposed by Hess (1964) and found in the uppermost mantle by Raitt *et al.* (1969) in the eastern Pacific by an explosion seismological experiment (they only determined P-wave anisotropy, i.e., A and B_c). Similar azimuthal anisotropy was also found in some portions of continents (Bamford, 1977) and in the inner core (Song, 1997; Tromp, 2001).

Azimuthal anisotropy can also be observed for shear waves. For an SH-wave,

$$\frac{\Delta V_{\mathrm{SH}}}{V_{\mathrm{SH}}} \approx \frac{E_c}{D} = -\frac{(C_{11}+C_{22})-2(C_{12}+2C_{66})}{(C_{11}+C_{22})-2(C_{12}-2C_{66})} \quad (\cos 4\theta \text{ term}) \tag{21.18}$$

and for an SV-wave,

$$\frac{\Delta V_{SV}}{V_{SV}} \approx \frac{C_{55} - C_{44}}{C_{55} + C_{44}} \quad (\cos 2\theta \text{ term}). \tag{21.19}$$

Note that the anisotropy of a Rayleigh wave is controlled by that of P- and SV-waves, so that the $\cos 2\theta$ term usually dominates. Azimuthal anisotropy of surface waves has been studied by FORSYTH (1975), TANIMOTO and ANDERSON (1984, 1985), MONTAGNER and TANIMOTO (1990, 1991) and MONTAGNER (2002). Most of these studies are on the upper mantle because surface waves are sensitive to the structures of shallow portions. However, by using the overtones (higher modes), one can infer anisotropy in the deeper portions. TRAMPERT and VAN HEIJST (2002) reported azimuthal anisotropy of the mantle transition zone using the overtones of surface waves (signals contained in the overtones are weak and the robustness of the results of this paper is not yet well established).

Azimuthal anisotropy is determined by measuring the wave velocities in a place in Earth as a function of the direction of wave propagation. To do this one needs a large number of source–station combinations. One of the important limitations of this method is that different waves passing through a place with different directions must pass different regions outside the study region. Consequently, the influence of lateral heterogeneity needs to be corrected (Fig. 21.4). This is not trivial because the amplitude of lateral heterogeneity is similar to that of anisotropy in most cases.

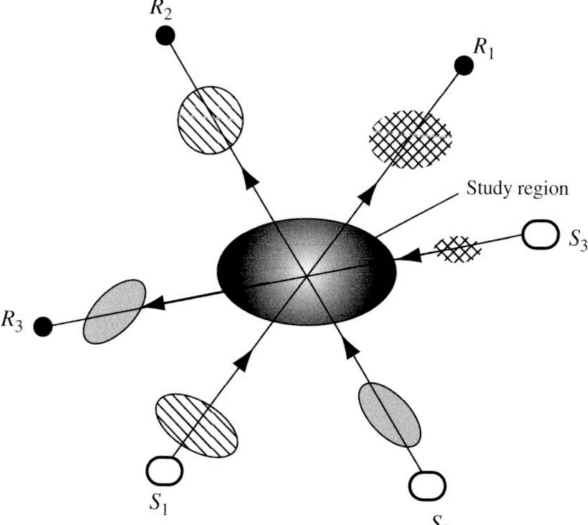

FIGURE 21.4 A schematic diagram showing a way in which azimuthal anisotropy is measured. S_i indicates the source and R_i denotes the receiver. Note that waves used to determine azimuthal anisotropy in a study region must pass through different regions between a source and a receiver.

Problem 21.5

Derive equation (21.19).

Solution

Following (21.14c), we obtain $\rho(V_{S_1}^2 - V_{S_1}^2) = 2c_{7c}$ where V_{S_1}, V_{S_2} are the fast and slow velocities of SV waves. Therefore

$$\frac{V_{S1}^2 - V_{S2}^2}{V_{S1}^2 + V_{S2}^2} = \frac{C_{55} - C_{44}}{C_{55} + C_{44}} \approx \frac{\Delta V_S}{V_S}.$$

21.3.2. Polarization anisotropy

The difficulty of correcting the influence of lateral heterogeneity can be eliminated if one focuses on the difference in seismic wave velocities between two shear waves with different polarizations. In this case, the polarization anisotropy can be determined from one seismic record and consequently there is no influence of lateral heterogeneity. In practice, this polarization anisotropy is analyzed in two different ways. First, when a seismic wave propagates through an anisotropic medium, then two shear waves with different polarization vectors will have different velocities and therefore they arrive at different times. This phenomenon is called *shear wave splitting*. By looking at seismic records one can determine the travel time difference of two shear waves, and if one has records of shear waves with two different directions of displacement, then one can also determine the directions of polarization (Fig. 21.5). In the case where one looks at a vertically traveling wave (e.g., SKS wave), then one can use the analysis from (21.11) through (21.13) to find the velocities and the polarization vectors of two shear waves. The directions of the two polarization vectors are controlled by the symmetry of the elastic constants, and the magnitude of splitting is given by

$$\frac{\Delta V_S}{V_S} \approx \frac{C_{55} - C_{44}}{C_{55} + C_{44}} \tag{21.20}$$

where I use the coordinate system such that x_3 is the vertical axis. Note that this result is identical to (21.19) because we are looking at the same type of waves (SV-waves). The relation between the body- and the surface-wave anisotropy was investigated by MONTAGNER *et al.* (2000).

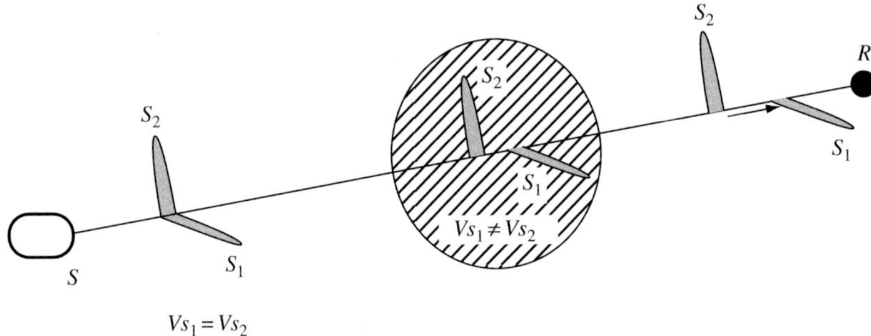

FIGURE 21.5 A schematic diagram showing how shear wave splitting occurs.

Shear wave splitting is a powerful tool to investigate anisotropic structure. One advantage of this method is that because anisotropic structure can be inferred from a single record, the result is free from the influence of lateral heterogeneity. However, a major limitation is that it is hard to identify where shear wave splitting occurs. For example, when one uses an SKS wave (shear wave that goes to the outer core and comes back from the core–mantle boundary, see Box 21.1), any place from the core–mantle boundary to the station can cause shear wave splitting. There is some progress to determine three-dimensional anisotropic structures from shear wave splitting measurements. When one uses a wave converted from P- to S- at a discontinuity, then shear wave splitting must occur in the materials above the discontinuity (e.g., PARK and LEVIN, 2002). Also using the back-azimuth and/or frequency dependence of splitting one can obtain some constraints on the depth variation of anisotropy (e.g., CHEVROT, 2000; SALTZER *et al.*, 2000; CHEVROT and VAN DER HILST, 2003), but the uncertainties in locating the anisotropic structures are still large.

Second, and another commonly used approach, is to analyze seismological data assuming *transverse isotropy* and determine the ratio of V_{SH}/V_{SV}. Transverse isotropy is a particular type of anisotropic structure in which the elastic properties do not depend on the orientation in the horizontal plane, but they are different between properties in the horizontal and vertical directions. One example is a horizontally layered structure. In such a case, the P-wave has anisotropy with respect to the direction of propagation (PH, PV), and the horizontally propagating S-wave has anisotropy with respect to the direction of polarization (SH, SV). Since polarization anisotropy is better constrained, the ratio of velocities of SH and SV waves, V_{SH}/V_{SV}, is investigated in most of the literature. The amplitude of SH–SV anisotropy can be calculated from the previously given relation, (21.14), as

$$\left(\frac{V_{SH}}{V_{SV}}\right)^2 = \frac{D}{F} = \frac{C_{11} + C_{22} - 2(C_{12} - 2C_{66})}{4(C_{44} + C_{55})} \qquad (21.21)$$

where as before the elastic constants are defined using a coordinate system in which x_3 is the vertical axis.

In contrast to the azimuthal anisotropy, the measurements of the polarization anisotropy of S-waves can be made with less problems of heterogeneity. In a measurement of polarization anisotropy, one investigates a seismogram at a single station and determines the difference in velocities of two shear waves that have different polarizations. One may compare the phase velocities of Love and Rayleigh waves that depend on SH- and SV-wave velocities respectively. Indeed, this SH–SV *polarization anisotropy* was recognized in the early 1960s in one of the first reports on the evidence of seismic anisotropy (AKI and KAMINUMA, 1963). By using the fact that surface waves with different frequencies sample different ranges of depth, one can determine the depth variation of anisotropic structures. Such studies include FORSYTH (1975), NATAF *et al.* (1986), MONTAGNER and TANIMOTO (1990, 1991) and BEGHEIN and TRAMPERT (2003a).

Similarly, the analysis of free oscillation provides information as to the anisotropic structure of Earth. The principle is similar to the surface waves, because both surface waves and free oscillation belong to the same class of elastic vibration in which the displacement of materials is constrained by the boundary conditions. Anisotropic structures can be determined by investigating the *splitting* of peaks of free oscillation. Free oscillation of Earth has many peaks corresponding to various "modes." If Earth were a perfectly homogeneous sphere, then these peaks would be sharp and well defined. Anisotropy (as well as rotation and heterogeneity) will cause the splitting of these peaks. Consequently, by investigating the nature of splitting of free oscillation peaks with various frequencies, one can investigate the depth variation of anisotropic structures (e.g., WOODHOUSE *et al.*, 1986; MONTAGNER and KENNETT, 1996).

21.4. Major seismological observations

Important seismological observations on anisotropy can be summarized as follows.

(i) Significant anisotropy is observed in the crust, particularly in the areas with intense deformation (e.g., CRAMPIN, 1984).

(ii) Significant azimuthal anisotropy of P-waves (Fig. 21.6, RAITT *et al.*, 1969; SHIMAMURA *et al.*, 1983) and Rayleigh waves (Fig. 21.7, TANIMOTO and ANDERSON, 1984, 1985; MONTAGNER and TANIMOTO, 1990, 1991; DEBAYLE *et al.*, 2005) is observed in the shallow oceanic upper mantle and in some of the continental upper mantle (BAMFORD, 1977; FUCHS, 1983; SAVAGE, 1999).

(iii) In most of the shallow upper mantle $V_{SH} > V_{SV}$ (AKI and KAMINUMA, 1963; NATAF *et al.*, 1984, 1986; GAHERTY and JORDAN, 1995; BEGHEIN and TRAMPERT, 2003a). However, in the old oceanic upper mantle and also beneath mid-ocean ridges $V_{SV} > V_{SH}$ (NATAF *et al.*, 1986; MONTAGNER and GUILLOT, 2000, 2002).

(iv) Seismic anisotropy beneath hotspots is somewhat different. In the upper mantle beneath Hawaii, $V_{SH} > V_{SV}$ anisotropy is stronger than other regions (MONTAGNER and GUILLOT, 2000) (Fig. 21.8). GAHERTY (2001) reported that beneath Iceland, $V_{SH} > V_{SV}$ is observed in the deep upper mantle (> 100 km) whereas $V_{SV} > V_{SH}$ anisotropy is observed in the shallow upper mantle (< 100 km). The modifications to the azimuthal anisotropy (shear wave splitting pattern) are found around hotspots (e.g., WALKER *et al.*, 2001).

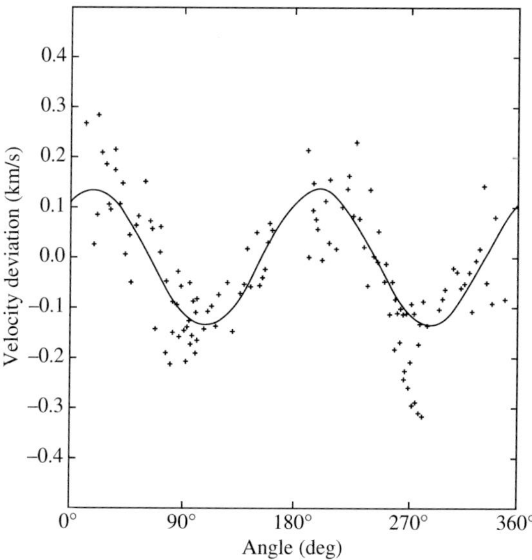

FIGURE 21.6 Azimuthal anisotropy in the uppermost mantle in the Pacific (after RAITT *et al.*, 1969). The measured P-wave velocities are plotted as a function of the azimuth.

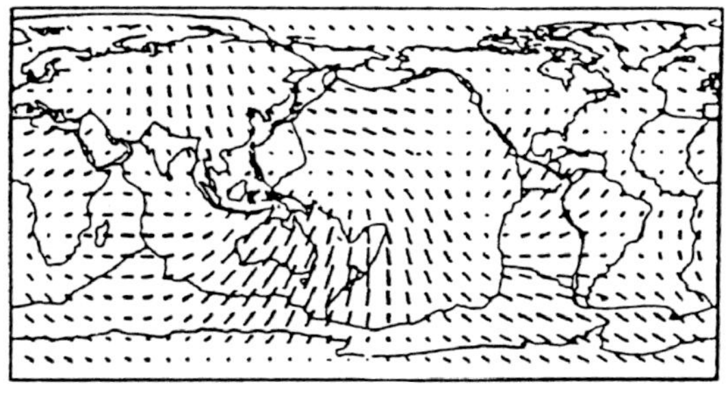

FIGURE 21.7 Azimuthal anisotropy of Rayleigh waves of 200 s period. Bars indicate the propagation direction of fast Rayleigh waves. The results represent the anisotropic structure of the asthenosphere, ~100–200 km depth (after TANIMOTO and ANDERSON, 1984).

(v) In most regions, strong shear wave splitting, up to ~2 s (typically ~1 s) is observed. Usually such a splitting result shows good correlation with surface geology and is interpreted to be due to the anisotropic structure in the shallow upper mantle. The polarization of the fast shear wave is nearly parallel to the direction of geological structure (e.g., Fig. 21.9, SILVER, 1996; see also SAVAGE, 1999). Similarly, the direction of polarization of the faster S-wave in a convergent boundary is nearly parallel to the strike of a trench near the trench whereas it becomes orthogonal to the trench away from the trench (e.g., SMITH *et al.*, 2001; NAKAJIMA and HASEGAWA, 2004; LONG and VAN DER HILST, 2005; Fig. 21.10).

(vi) The amplitude of anisotropy in the upper mantle decreases with depth (MONTAGNER and TANIMOTO, 1990, 1991) (there are some exceptions, see DEBAYLE *et al.*, 2005).

(vii) However, there are some hints as to the presence of anisotropy in the *transition zone* around 400–800 km depth (MONTAGNER and KENNETT, 1996; VINNIK and MONTAGNER, 1996; WOOKEY *et al.*, 2002; BEGHEIN and TRAMPERT, 2003a). TRAMPERT and VAN HEIJST (2002) reported azimuthal anisotropy in the transition zone using the higher mode data of surface waves (Fig. 21.11).

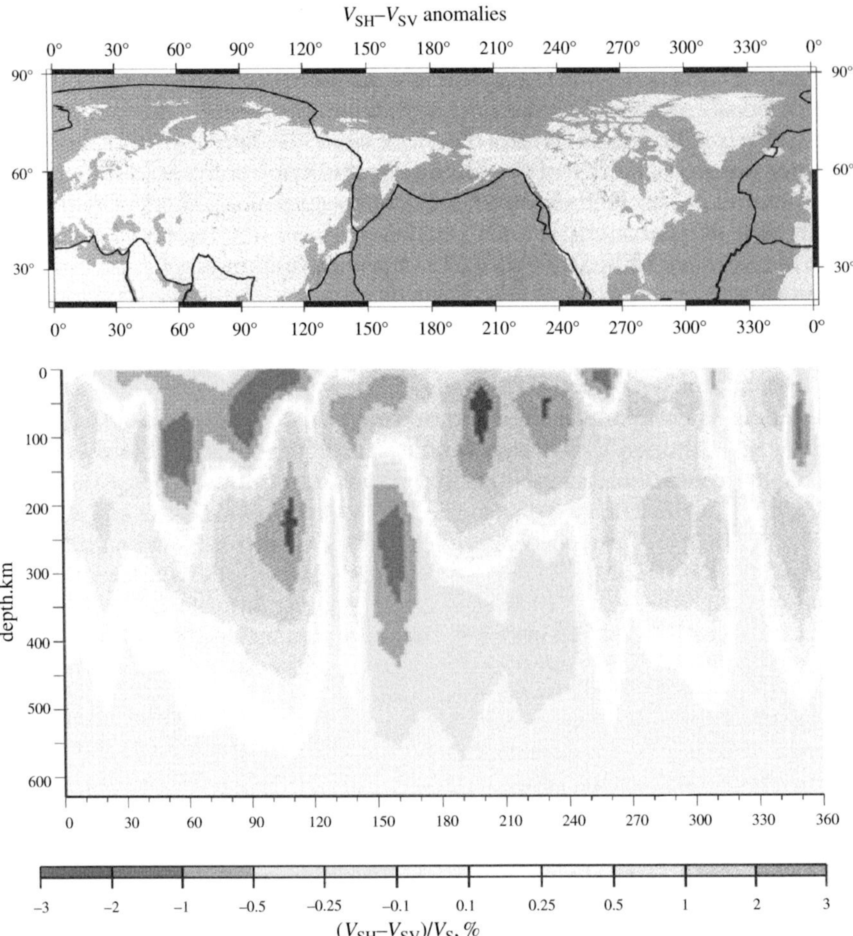

FIGURE 21.8 The V_{SH}/V_{SV} polarization anisotropy of the upper mantle (after MONTAGNER and GUILLOT, 2000).
The deviation of V_{SH}/V_{SV} polarization anisotropy in the upper mantle from the PREM model is shown along a cross section shown in the top map (the unit is %). Plate 13.

(viii) Most of the *lower mantle* is devoid of any significant anisotropy (MEADE *et al.*, 1995; MONTAGNER and KENNETT, 1996). However, significant shear wave anisotropy is documented in some portions of the D″ layer (the bottom of the mantle) (KENDALL and SILVER, 1996; VINNIK *et al.*, 1998; GARNERO *et al.*, 2004; PANNING and ROMANOWICZ, 2004). Usually $V_{SH} > V_{SV}$ in the D″ layer, but the magnitude of anisotropy changes from one place to another. Anisotropy is strong in regions where average velocity is higher than normal. Evidence for azimuthal anisotropy in the D″ layer is controversial.

(ix) There is a strong anisotropy in the *inner core*. The anisotropy has nearly axial symmetry and the fast P-waves are along the rotation axis (Fig. 21.12; MORELLI *et al.*, 1986; WOODHOUSE *et al.*, 1986; DUREK and ROMANOWICZ, 1999; TROMP, 2001). However, anisotropy is depth-dependent (e.g., SONG, 1997) and there is some suggestion for hemispheric asymmetry (e.g., TANAKA and HAMAGUCHI, 1997).

21.5. Mineral physics bases of geodynamic interpretation of seismic anisotropy

In order to translate observed seismic anisotropy in the geodynamic context, one needs to understand the physical mechanisms by which anisotropic structures are formed. There are two distinct structures that can cause seismic anisotropy. One is a layered structure including the shape-preferred orientation (SPO) of isotropic or anisotropic materials and another is lattice-preferred orientation (LPO) of anisotropic minerals.

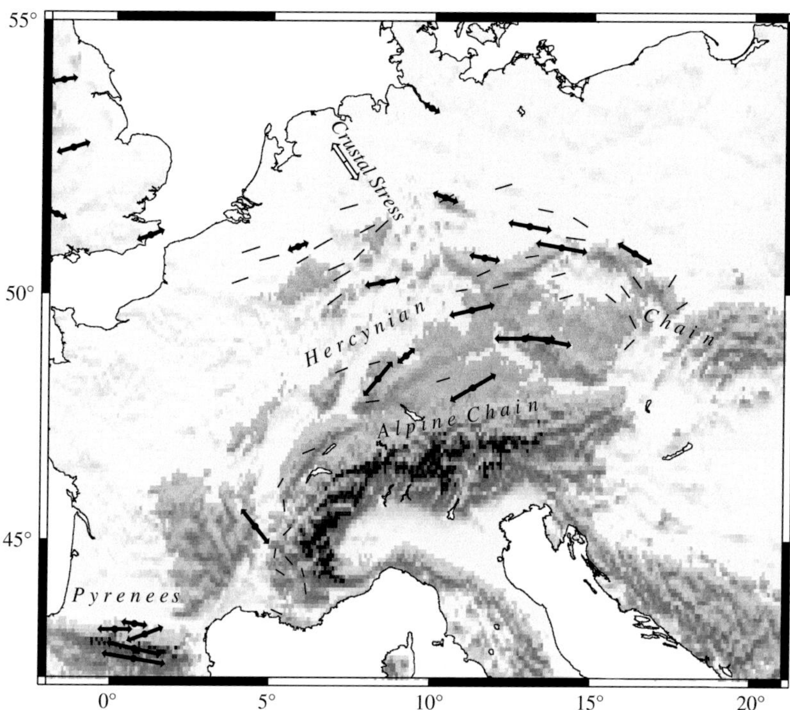

FIGURE 21.9 The direction of polarization of faster shear waves (SKS) in the continent (after SILVER, 1996). The direction of polarization of a faster S-wave is nearly parallel to the trend of the orogenic belt.

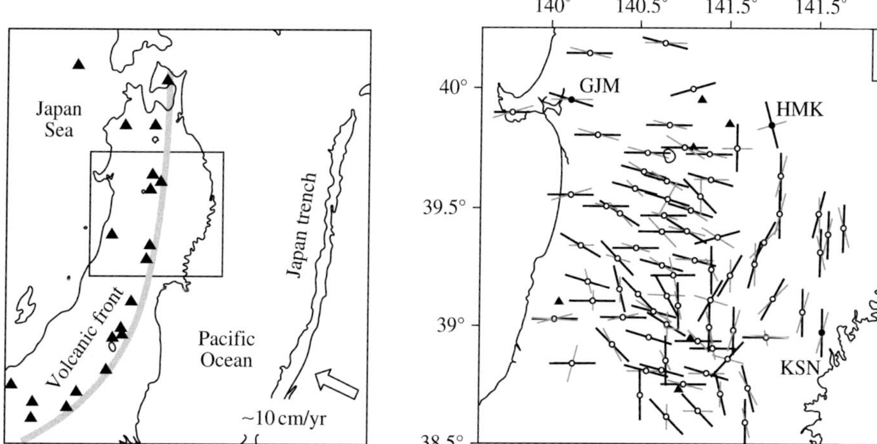

FIGURE 21.10 Shear wave splitting in the Tohoku, Japan (after NAKAJIMA and HASEGAWA, 2004). In regions near the trench, the direction of polarization of the faster S-wave is nearly parallel to the trench, whereas in regions far from the trench its direction becomes normal to the trench. Such a trend is common to many subduction zones.

21.5.1. Anisotropy due to a layered structure

When a material contains two components with different properties (chemical composition, density or viscosity), upon deformation the two regions will change their shape. After large deformation, one could obtain a highly layered structure (e.g., CHRISTENSEN and HOFMANN, 1994). The scale of layering, i.e., the mean thickness of layers depends on the mechanisms of deformation (stirring), but in many cases it can reach a small size compared to the wavelength of seismic waves (~ 10–100 km for body waves, ~ 100–1000 km for surface waves). If the layer thickness is much smaller than the wavelength of seismic waves, then

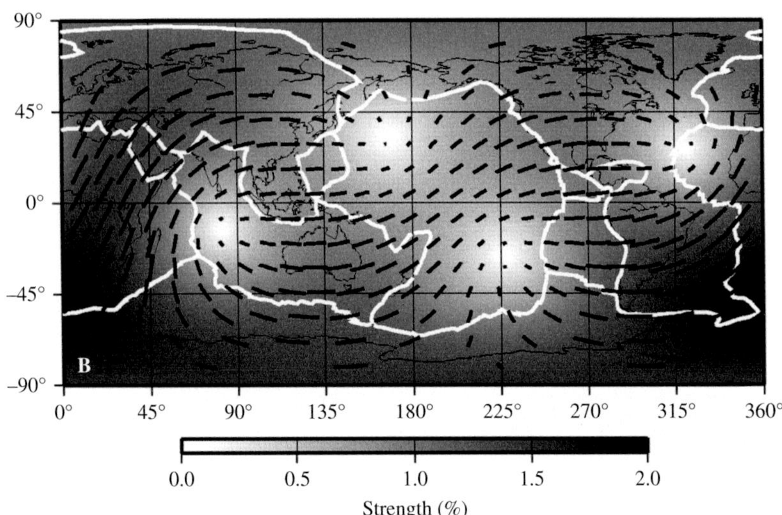

FIGURE 21.11 Azimuthal anisotropy of the transition zone as determined by the overtones of surface waves (after TRAMPERT and VAN HEIJST, 2002).

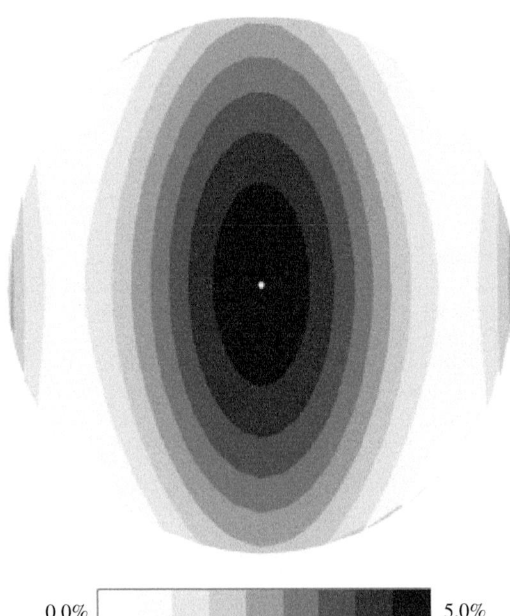

FIGURE 21.12 Anisotropy in the inner core (after DUREK and ROMANOWICZ, 1999; TROMP, 2001). The velocities of P-waves propagating nearly parallel to the rotation axis are significantly faster than those propagating normal to the rotation axis. The details of the anisotropic structure such as the depth variation of anisotropy are not well constrained.

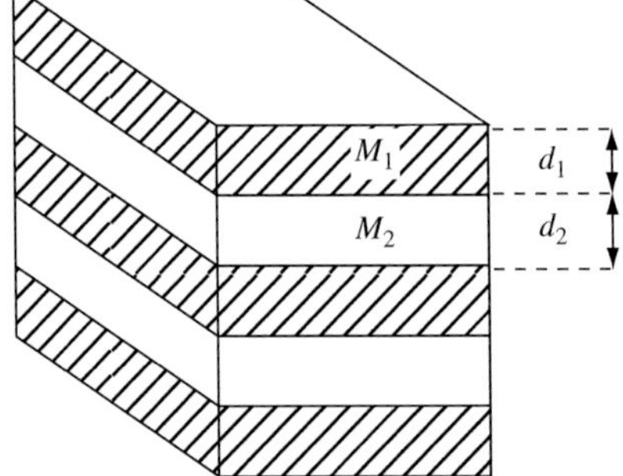

FIGURE 21.13 A finely layered structure. The elastic constants for deformation in the layering plane and normal to the layering plane are different, and are characterized by five independent elastic constants.

such a layered structure will appear to be a homogeneous but anisotropic structure.

Consider a layered structure made of two isotropic materials (Fig. 21.13). By definition, there is a symmetry axis normal to the plane. The elastic constants corresponding to such a structure have the same symmetry as the elastic constants for a hexagonal structure and have five independent components (see Chapter 4).[3] This type of anisotropy is referred to as *transverse isotropy* or *radial anisotropy*. In order to fully characterize the nature of a layered structure, one would need to determine five independent seismic wave velocities (for a more complete analysis, see BACKUS, 1962).

Consider the deformation of a layered material. For compression normal to the layer (corresponding to deformation associated with a PV-wave), stress in each layer must be identical, whereas for compression within the plane of a layer (deformation associated with a PH-wave), strain must be the same for each layer (because of the assumption that the wavelength of the wave is much larger than that of the thickness of each layer). A similar argument applies to the

[3] Using the elastic constants of a material with a hexagonal symmetry (see Chapter 4), it can easily be shown that all the coefficients of azimuthal anisotropy in equation (21.14) vanish ($B_c = B_s = C_c = C_s = E_c = G_c = G_s = 0$).

deformation associated with SH- and SV-motion. For SH-wave propagation, the particle motion is in the layering plane. By assumption the wavelength of the seismic wave is much larger than that of the thickness of each layer, and therefore the strain in each layer must be the same for SH-motion. For SV-motion, materials move vertically in each layer. Strain in each layer can be different, but stress must be the same. Therefore the elastic moduli for the two types of deformation correspond to homogeneous stress for the horizontal motion and homogeneous stress for vertical motion and hence,

$$M_H = M_1 d_1 + M_2 d_2 \tag{21.22}$$

and

$$M_V = \frac{1}{d_1/M_1 + d_2/M_2} \tag{21.23}$$

where $M_{1(2)}$ is the elastic modulus of phase 1 (2) and $d_1 + d_2 = 1$. It follows that for a horizontally layered structure with any elastic constants, we *always* have

$$V_{SH}(V_{PH}) > V_{SV}(V_{PV}) \tag{21.24}$$

where $V_{SH(SV, PH, PV)}$ are the velocities of SH(SV, PH, PV)-waves.[4]

The contrast in elastic modulus is given by

$$\frac{M_H - M_V}{\langle M \rangle} \approx \phi \left(\frac{\Delta M}{\langle M \rangle} \right)^2 \tag{21.25}$$

and hence

$$\frac{V_{PH(SH)} - V_{PV(SV)}}{\langle V_{P(S)} \rangle} \approx \frac{\phi}{2} \left(\frac{\Delta M}{\langle M \rangle} \right)^2 \tag{21.26}$$

where ϕ is the fraction of a secondary phase, $\Delta M \equiv M_1 - M_2$ and $\langle M \rangle \equiv \sqrt{M_1 M_2}$. Note that the elastic modulus ratio is proportional to the square of elastic modulus contrast. This makes sense because the elastic constant ratio does not depend on which material is softer (always $M_H > M_V$). But this also means that in order for this mechanism to cause significant seismic anisotropy, one needs a large contrast in elastic moduli. For example, for $\phi \sim 1\%$, $\Delta M/M$ must be on the order of 1. The only plausible mechanism for this to happen is partial melting. If $\Delta M/M \sim 10\%$, then the only way to obtain substantial anisotropy is when the fraction of secondary material is large, $\phi \sim 50\%$. The latter combination works only for special cases. A plausible case is a layering due to the stirring in a two-phase mixture such as the deformed mixture of a mixture of peridotite and eclogite (e.g., Allègre and Turcotte, 1986). Similar layering may occur in the transition zone (a mixture of garnetite and peridotite (with wadsleyite or ringwoodite)) and in the D″ layer (a mixture of paleo-oceanic crust and peridotitic components (made of perovskite and magnesiowüstite)).

When a two-phase mixture is deformed to large strains, a layered structure will be formed. However, the plane of layering is parallel to the shear plane only at infinite strain. At a finite strain there will be a finite tilt (Gay (1968) investigated the geometry of an inclusion in an infinite matrix and concluded that the tilt depends on the viscosity contrast between the inclusion and the matrix). In this case, a layered structure will result in azimuthal anisotropy with azimuthal dependence through a 2θ term (Karato, 1998e). In such a case, the sense of shear can be inferred if the sign of 2θ is determined from seismology.

21.5.2. Lattice-preferred orientation (see Chapter 14)

Most rock-forming minerals are elastically anisotropic. Therefore when the crystallographic orientations are non-random, an aggregate of minerals will show macroscopic anisotropy (when viewed at a scale significantly larger than grain size).[5] The plastic deformation of a polycrystalline material often leads to a non-random distribution of crystallographic axes called *lattice-preferred orientation* (*LPO*). LPO can also be made by anisotropic nucleation and/or growth of crystals during crystallization from melt, phase transformations and recrystallization.

Elastic anisotropy caused by LPO depends on the elastic anisotropy of constituent minerals and the orientations of each mineral (i.e., LPO). Elastic constants of most minerals are anisotropic (Chapter 4), except for garnet (see Fig. 21.2). Consequently, any non-random distribution of crystallographic axes will cause elastic anisotropy of an aggregate. A typical case is summarized in Table 21.1 where the elastic constants of deformed olivine aggregates under typical upper mantle conditions are shown based on the experimental

[4] The second subscript indicates the direction of the particle motion. Therefore PH (PV) is a P-wave that *propagates* along the horizontal (vertical) direction.

[5] The assumption that the grain size is significantly smaller than the wavelength of seismic waves is valid, except perhaps for the inner core where grain size is probably not much smaller than the wavelength of body waves (this causes significant scattering of body waves, see Chapter 17).

TABLE 21.1 Elastic constants, C_{ij} (GPa), of deformed olivine aggregates evaluated at pressure of 5 GPa and temperature of 1573 K (these are typical values and the actual values can vary from these values dependent on the strength of lattice-preferred orientation).

$i\backslash j$	1	2	3	4	5	6
A-type fabric						
1	236.3	84.5	81.5	0.4	3.4	0.3
2		218.5	82.9	−1.8	1.2	0.3
3			208.0	−1.3	6.1	0.2
4				64.9	−0.1	−1.9
5					68.7	0.3
6						66.6
B-type fabric						
1	221.3	84.3	83.3	−0.3	0.8	1.6
2		223.5	81.7	−1.4	1.0	1.6
3			215.5	−1.3	1.1	−0.4
4				68.9	0.7	−0.4
5					67.4	−0.4
6						69.4
C-type fabric						
1	223.2	83.6	83.3	0.3	−3.7	0.3
2		209.8	81.9	0.8	1.5	0.3
3			228.5	0.4	−5.9	0.2
4				67.9	0.2	−1.9
5					71.1	0.3
6						66.6
E-type fabric						
1	236.8	82.3	84.1	−0.6	0.4	0.1
2		207.7	82.7	−2.6	−0.3	−1.0
3			217.4	−2.1	−1.9	−0.7
4				65.0	0.1	0.4
5					71.1	1.4
6						68.5

Reference axes are defined as 1: shear direction, 3: shear plane normal and 2: perpendicular to both 1 and 3 directions.

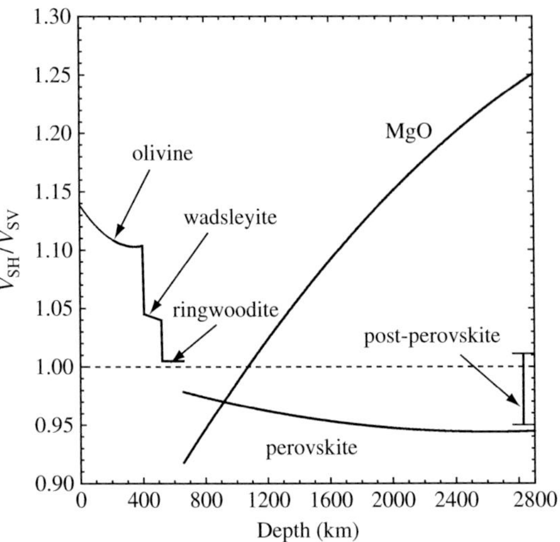

FIGURE 21.14 The depth variation of V_{SH}/V_{SV} due to the perfect alignment of major constituent minerals. The horizontal plane is assumed to be (010) for all minerals. Note the large effects of pressure on elastic anisotropy. The seismic anisotropy may change with depth due to the depth variation in LPO (the horizontal plane may change from (010) to another plane or to a more complicated LPO).

data. Also note that the nature of elastic anisotropy for a given mineral can change with pressure (and temperature) (Fig. 21.14). A particularly notable case is MgO for which even the qualitative feature of anisotropy (faster or slower direction) changes with pressure (see Chapter 4). This unique feature is common to many minerals with cubic structure (e.g., ringwoodite (silicate spinel)) that have an intrinsic instability caused by the softening of C_{44} upon compression (see Chapter 4).

The elastic constants of an aggregate can be calculated from the elastic constants of its constituent minerals and their orientations (see Chapter 12).

Two points need to be emphasized concerning the nature of LPO. First, LPO will be developed by plastic deformation only when deformation occurs by some specific mechanisms (dislocation creep and/or twinning). Among them, deformation by dislocation creep is the most important mechanism to cause LPO in Earth's interior. Deformation by diffusional creep (or grain-boundary sliding) does not usually produce LPO. Deformation by dislocation creep (or twinning) occurs at relatively high stress and/or large grain size. Consequently, anisotropic structures tend to develop in certain regions in Earth where stress is high (i.e., boundary layers, see Karato, 1998d; Montagner, 1998).

Second, when deformation is due to dislocation creep, the resultant LPO depends on (i) the physical and chemical conditions of deformation and (ii) the geometry of deformation (and in some cases the geometry of stress and other fields such as the magnetic field). The physical conditions of deformation control the relative ease of deformation by different slip systems. The relative contributions from different slip systems determine the nature of LPO for a given deformation geometry. Consequently, in order to infer the geometry of deformation from seismic anisotropy, one needs to know the dominant slip system(s) under a given physical and chemical condition.

One obvious, but important conclusion is that the relation between seismic anisotropy and the geometry of flow is mineral specific and also depends on the physical and chemical conditions. The $V_{SH}/V_{SV} > 1$ corresponds to the horizontal flow for a case where anisotropy is caused by LPO of olivine under limited conditions (A-, B-, E-type, but not for C-type; for olivine LPO see Chapter 14, also see Problem 12.6), but is not necessarily true for other conditions and for other minerals. Similarly, the notion that the direction of polarization of the faster S-wave is parallel to the flow direction is true only for the olivine-dominated case under limited physical and chemical conditions (A-, C- and E-type, but not for B-type). It is also important to recall that the nature of elastic anisotropy can change with pressure (and temperature). In short, both elastic and plastic anisotropies are different among different minerals, and both of them can change with physical and chemical conditions, and therefore the relation between seismic anisotropy and the nature of flow (e.g., flow geometry) can change from one region of Earth to another. *In order to interpret seismic anisotropy in terms of geodynamics in one region (say the lower mantle), one needs to know the relation between seismic anisotropy and the nature of flow appropriate for materials in that region. A relation appropriate for a certain region of Earth (e.g., the lithosphere) cannot be applied to other regions in general.*

Important progress in recent years includes: (i) experimental studies on the variation of LPO (in olivine) under various physical and chemical conditions that have revealed a rich variety of LPO (e.g., JUNG and KARATO, 2001b; KATAYAMA *et al.*, 2004; JUNG *et al.*, 2006; KATAYAMA and KARATO, 2006), and (ii) the new development of a high-pressure large-strain deformation technique (e.g., XU *et al.*, 2005) that allows us to investigate LPOs in deep Earth materials. (iii) Also there has been much progress in the numerical modeling of LPO (e.g., KAMINSKI and RIBE, 2001; KAMINSKI, 2002; WENK, 2002). One advantage of this approach is the ability to simulate complicated deformation history. However, a major limitation of this approach is the fact that most of the conclusions from this approach, such as the geometry of LPO and the rate at which LPO changes with deformation, hinge on the material properties that are not always well constrained.

Problem 21.6

Using the results given in Table 21.1, calculate V_{SH}/V_{SV} for A-, B-, C- and E type olivine fabric corresponding to the horizontal shear. Also calculate the direction of the polarization of the faster shear waves.

Solution

Simply use the numbers from Table 21.1 into equation (21.21) to obtain: $V_{SH}/V_{SV} = 1.02$ (A-type), 1.01 (B-type), 0.98 (C-type) and 1.01 (E-type). The direction of polarization can be inferred from the comparison of C_{44} and C_{55}. For A-, C- and E-type fabrics, $C_{55} > C_{44}$ and therefore the polarization of the fast S-wave is along the x_1 direction (flow direction), whereas for B-type fabric $C_{55} < C_{44}$, so the flow direction is normal to the polarization direction of the fast S-wave.

21.6. Geodynamic interpretation of seismic anisotropy

21.6.1. Generalities

The first issue in the geodynamic interpretation of anisotropy is to identify the type of anisotropic structure that causes seismic anisotropy, namely either it is due to a layered structure or to lattice-preferred orientation (LPO). This can be made through several methods. For example, anisotropy due to horizontal layering would result in transverse isotropy, namely a type of anisotropy in which wave propagation is isotropic in the horizontal plane. Therefore if azimuthal anisotropy is significant, one can conclude that horizontal layering is unlikely to be the cause of anisotropy. One caveat in this inference, however, is that even in strictly horizontal shear flow, the resulting layered structure may have some tilt (KARATO, 1998e). In this case, the structure will have azimuthal anisotropy where the seismic wave velocities depend on 2θ. If anisotropy is found to be due to LPO, then one needs to find out the relation between LPO and flow field for the materials in the relevant region of Earth.

21.6.2. Seismic anisotropy and its geodynamic implications

Gross distribution of anisotropy in Earth's interior

Figure 21.15 shows a gross depth distribution of anisotropy in Earth's interior. Distribution of anisotropy is three dimensional and this diagram is certainly an over-simplification of the actual distribution, but does show some essential features. The average amplitude of

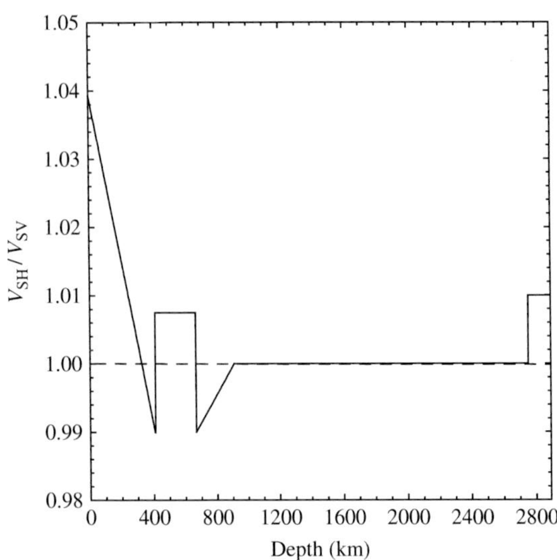

FIGURE 21.15 The depth variation of the polarization anisotropy, V_{SH}/V_{SV}, in the Earth's mantle (after MONTAGNER and KENNETT, 1996).

seismic anisotropy is large in the shallow Earth (upper mantle) and at the very bottom (D″ layer). The amplitude of anisotropy in the inner core (not shown) is again large. The depth variation of seismic anisotropy partly reflects the depth variation of elastic anisotropy of constituent materials, but it also reflects the depth variation of dynamic processes that cause LPO. For example, the elastic anisotropy of minerals in the upper mantle does not change with depth so much whereas the amplitude of anisotropy decreases significantly with depth. Similarly, the elastic anisotropy of lower mantle minerals is comparable to that of the upper mantle, yet the amplitude of anisotropy of the bulk of the lower mantle is much less than that of the upper mantle. These observations suggest that the depth variation of the amplitude of seismic anisotropy mainly reflects the depth variation in the dynamic processes of Earth's interior.

Crust

The deformation of the crust occurs mostly by brittle fracture in its shallow part (e.g., continental upper crust) but by ductile deformation in the deeper part (e.g., continental lower crust). Consequently, the most important anisotropic structure in the shallow crust is aligned cracks (i.e., SPO). In this case, the pattern of anisotropy reflects the stress field as opposed to the strain field[6] (e.g., CRAMPIN, 1978; KANESHIMA, 1990). In the lower crust, however, deformation occurs by plastic flow. Most of the plastically deformed crustal rocks show significant lattice-preferred orientation and have some degree of elastic anisotropy (e.g., SIEGESMUND *et al.*, 1989). However, the anisotropic layer in the crust is thin (compared to the mantle) and the anisotropic structure is not always persistent at a large scale, and consequently, there is a relatively small signal from the crustal anisotropy in the seismic record compared to those from the upper mantle (e.g., SILVER, 1996).

Upper mantle

Seismic anisotropy in the upper mantle is well documented both by body- and surface-wave studies. The classic work by RAITT *et al.* (1969) demonstrated the azimuthal anisotropy in the uppermost oceanic mantle. They found significant azimuthal anisotropy of P-waves ($\sim\pm3\%$), the fast direction nearly parallel to the (paleo) spreading direction. Global studies include TANIMOTO and ANDERSON (1984, 1985), MONTAGNER and TANIMOTO (1990, 1991), MONTAGNER and RITSEMA (2001), BEGHEIN and TRAMPERT (2003a). Seismic anisotropy in the upper mantle is in most cases attributed to the lattice-preferred orientation of minerals particularly that of olivine (e.g., NICOLAS and CHRISTENSEN, 1987). The role of layering such as layering of partial melt was also suggested (e.g., AKI, 1968; BLACKMAN and KENDALL, 1997), but evidence for melt-induced anisotropy has not been convincingly demonstrated yet. In fact, one of the most striking observations on the oceanic upper mantle is the failure to observe evidence for partial melting from shear wave splitting measurements near a mid-ocean ridge (e.g., WOLFE and SOLOMON, 1998). Melt appears to be concentrated in very narrow regions beneath a mid-ocean ridge.

When anisotropy is attributed to LPO, then the relation between LPO and the tectonic field must be understood to interpret seismic anisotropy in terms of tectonic field. The classic work by Nicolas (reviewed in NICOLAS and CHRISTENSEN, 1987; BEN ISMAIL and MAINPRICE, 1998) is used in most geodynamic interpretations of upper mantle seismic anisotropy (e.g., VAUCHEZ and NICOLAS, 1991; NICOLAS, 1993; SILVER, 1996; SAVAGE, 1999; MONTAGNER and GUILLOT, 2000; TOMMASI *et al.*, 2000). In this model, anisotropy is due mainly to the A-type olivine fabric. Consequently, the direction of polarization of the fast shear wave is parallel to the flow direction, the direction of the fast Rayleigh wave is parallel to the flow direction and $V_{SH}/V_{SV} > 1$ (< 1) corresponds to the

[6] The distinction between stress and strain is critical for finite strain (see Chapter 1).

horizontal (vertical) flow. Such a relation was used to infer the convection pattern in the upper mantle (e.g., TANIMOTO and ANDERSON, 1984; NATAF *et al.*, 1986). The observed convergent boundary-parallel shear wave splitting is interpreted, in this model, by the shear motion *along* the convergent boundary called *transpression* (e.g., NICOLAS, 1993). When this model is applied to interpret subduction zone anisotropy, then one needs to invoke a complicated flow pattern where flow near the trench is dominated by trench parallel flow as opposed to trench normal flow (e.g., RUSSO and SILVER, 1994; SMITH *et al.*, 2001). A major challenge to this paradigm came from the experimental finding that olivine LPO changes with physical and chemical conditions particularly with water content as reviewed in Chapter 14 (JUNG and KARATO, 2001b). When deformation in olivine occurs (with some water) at relatively low temperatures (above a certain stress), then the direction of polarization of the faster shear wave will be perpendicular to the flow direction (B-type fabric). Similarly, when olivine is deformed at high temperature with a large amount of water, then the horizontal flow will result in $V_{SH}/V_{SV} < 1$ (C-type fabric). KARATO (2003b) proposed an alternative model to explain trench-parallel (or convergent boundary-parallel) anisotropy by the B-type olivine fabric. KNELLER *et al.* (2005) made a detailed numerical modeling to investigate the spatial variation of olivine LPO in the subduction zones. However, in most of the lithosphere or in the asthenosphere, olivine fabric is either A-, E- or C-type, and consequently, the correlation between the fast shear waves (Rayleigh waves) and flow pattern is similar to what has been assumed based on the A-type olivine fabric (see Problem 12.6). There is evidence of convergent boundary-parallel flow in the structures of deformed rocks from the shallow upper mantle (e.g., NICOLAS, 1993; MEHL *et al.*, 2003). However, the degree to which transpression occurs at a greater depth is not known. To distinguish these models, it is critical to map out the depth variation of anisotropy in the subduction zone.

In addition to olivine, another volumetrically important mineral that may affect seismic anisotropy is orthopyroxene. However, the LPO of orthopyroxene is less extensively studied than that of olivine (e.g., CHRISTENSEN and LUNDQUIST, 1982).

Transition zone

The transition zone is the key to a number of geodynamic problems. In Chapter 20, we learned that there is evidence for a significant modification to the convection pattern in the transition zone. Therefore one expects a signature in the seismic anisotropy of a change in flow pattern in the transition zone. However, very little is known about the nature of anisotropy in the transition zone at the present time. Earlier works using shear wave splitting suggested only weak anisotropy in the transition zone (e.g., FISCHER and WIENS, 1996), but the resolution of these studies is not high. Some later studies using overtones of surface waves or free oscillations have revealed some degree of anisotropy in the transition zone. MONTAGNER and KENNETT (1996) and BEGHEIN and TRAMPERT (2003a) showed evidence for SH–SV polarization anisotropy in the transition zone. TRAMPERT and VAN HEIJST (2002) showed the presence of azimuthal anisotropy (see also VINNIK and MONTAGNER, 1996). The pattern of transition zone azimuthal anisotropy, from their study, is quite distinct from that of the upper mantle. However, the geodynamic significance of this observation is unknown because we do not know the nature of LPO of minerals in the transition zone (at the time of writing, 2007). In the transition zone, a volumetrically important mineral that has a large elastic anisotropy is wadsleyite (in the upper transition zone). However the LPO of wadsleyite is not well known. TOMMASI *et al.* (2004) presented the results of numerical modeling of LPO in wadsleyite and discussed the geodynamic implications of transition zone anisotropy. However, the results of such a model depend entirely on the assumed plastic anisotropy of wadsleyite that is unconstrained at the present time. In the lower transition zone, volumetrically important minerals, i.e., ringwoodite and majorite, are all nearly elastically isotropic and they cannot contribute to seismic anisotropy. If there is any seismic anisotropy present it must be due to a layered structure or LPO of a volumetrically small but highly anisotropic mineral such as akimotoite (silicate ilmenite).

Lower mantle

The majority of the lower mantle is devoid of anisotropy. This was first noted by the study of shear wave splitting by MEADE *et al.* (1995), who showed that the magnitude of shear wave splitting from most of the lower mantle is less than ~ 0.2 s. This is surprising because the lower mantle minerals are elastically as anisotropic as upper mantle minerals (Figs. 21.2 and 21.15). KARATO *et al.* (1995b) compared this value with the expected shear wave splitting resulting from LPO in perovskite (and magnesiowüstite) (~ 10 to

~ 20 s) and concluded that the most likely cause for the lack of anisotropy is that lower mantle minerals deform by diffusional creep (or superplasticity). The absence of anisotropy is also suggested by the analysis of free oscillation (Montagner and Kennett, 1996). However, anisotropy does exist in some regions of the lowermost mantle (e.g., Vinnik *et al.*, 1995; Kendall and Silver, 1996, 1998; Garnero *et al.*, 2004).

The absence of strong anisotropy in the bulk of the lower mantle, and the presence of significant anisotropy in the D″ layer can be naturally explained by the dynamics of mantle convection and the general nature of deformation mechanisms in polycrystalline aggregates. The D″ layer is likely to be a boundary layer of convecting mantle and therefore deformation there probably occurs at higher stress and strain than deformation elsewhere. If this view is combined with another well-established mineral physics concept that LPO develops only at high stress, then the localized distribution of anisotropy can be attributed to the stress–strain distribution and the change in deformation mechanisms in minerals (Karato, 1998d). In particular, Karato (1998e) emphasized the role of highly anisotropic (Mg, Fe)O (see also Yamazaki and Karato, 2002). This hypothesis has been explored by numerical modeling (McNamara *et al.*, 2001, 2002, 2003) who showed that anisotropy in the lower mantle is confined to regions where subducted slabs collide at the core–mantle boundary in which deformation occurs at high stress to large strains. A serious limitation to current knowledge is the lack of definitive experimental data on the LPO of silicate perovskite. Karato *et al.* (1995b) determined the LPO of $CaTiO_3$ perovskite (an orthorhombic perovskite) under the high-temperature, power-law dislocation creep regime, and inferred that the [100](010) slip system dominates. But a similar study on $(Mg, Fe)SiO_3$ perovskite is missing at this time (2007).

Panning and Romanowicz (2004) reported highly variable anisotropy in the D″ layer beneath the central Pacific and suggested that variable anisotropy might reflect a complicated flow pattern in that region from which plumes are considered to rise. However, the mineral physics basis for such an inference is missing because the relation between the anisotropic structure of minerals in the D″ layer and the flow pattern is not yet well known. Similarly, a detailed study on anisotropy was conducted in the D″ layer beneath Cocos plate (Rokosky *et al.*, 2004). A major complication to the interpretation of D″ layer anisotropy is the discovery of a post-perovskite phase. According to theoretical calculations, this phase is likely to have modest elastic anisotropy (Oganov and Ono, 2004; Tsuchiya *et al.*, 2004a), and therefore its role in seismic anisotropy needs to be included in the analysis, but the nature of LPO of this mineral is not known. Oganov *et al.* (2005) estimated the slip system of a post-perovskite phase based on the first-principle calculation in which the concept of crystal dislocation is not included. They predicted slip planes of (110). However, the recent experimental results on an analog material of post-perovskite ($CaIrO_3$) by Yamazaki *et al.* (2006) show a slip plane of (010) (with a slip direction of [100]). I note that the validity of the use of analog materials to infer the dominant slip system is unknown. For materials with a B-1 structure (NaCl-type crystals), NaCl and MgO show quite different LPOs. The slip systems and LPO of perovskite and post-perovskite phases of $(Mg, Fe)SiO_3$ are poorly constrained at this time.

In addition to the anisotropy in the D″ layer, there are reports on anisotropy in the top lower mantle (Wookey *et al.*, 2002; Wookey and Kendall, 2004). These authors reported evidence for anisotropy by shear wave splitting and the nature of anisotropy there is $V_{SH} > V_{SV}$. Anisotropy caused by (Mg, Fe)O will be very small there because the elastic properties of (Mg, Fe)O at this depth are nearly isotropic. Consequently, anisotropy in the shallow lower mantle is likely to be caused by LPO of perovskite, but the nature of LPO in perovskite is poorly known at this time.

Inner core

The inner core, the central part of Earth, has strong anisotropy (e.g., Song, 1997; Tromp, 2001). To a good approximation, anisotropy has an axial symmetry and P-waves propagate faster along the rotation axis than perpendicular to it (by ~ 3%). The presence of strong anisotropy in the inner core is remarkable. The inner core is very likely to be in a calm environment because the lateral variation in temperature should be low because of the efficient heat transfer both in the outer and the inner core. Melt-induced anisotropy (aligned melt pocket) was suggested by Singh *et al.* (2000), but the presence of a significant melt fraction is unlikely because of efficient compaction due to gravity (Sumita *et al.*, 1996). Therefore anisotropy due to LPO is the most likely cause. Indeed, the material composing the inner core of Earth is most likely to be hcp iron, which has a highly anisotropic crystal structure and hence anisotropic elastic constants. This anisotropy also implies that either plastic deformation in the inner core occurs by dislocation creep or the crystallization of hcp iron occurs anisotropically or both. If the latter is the dominant mechanism, then the influence of deformation must be small.

Since the stress levels in the inner core are likely to be low, it is not obvious if dislocation creep operates or not. Consequently, KARATO (1993b) and BERGMAN (1997) proposed that LPO in the inner core may develop by anisotropic crystallization. Similarly, YOSHIDA *et al.* (1996) proposed that LPO in the inner core may develop by stress-induced grain-boundary migration. However, as I discussed in Chapter 17, the grain size of the inner core is likely to be large, of the order of 100–1000 m. In such a case, dislocation creep can operate if stress is modest (e.g., above $\sim 10^2$ Pa). After the evaluation of various sources of stress in the inner core (KARATO, 1999, 2000) concluded that the stress caused by the magnetic field is the most important stress in the inner core. In this case, there is a connection between the magnetic field and the anisotropy of the inner core, and seismological observations of anisotropy may provide a clue as to the generation of a geomagnetic field. However, the details of flow due to the magnetic force are unknown (also see BUFFETT and BLOXHAM, 2000).

References

Aaronson H. I. (1990) Atomic mechanisms of diffusional nucleation and growth and comparisons with their counterparts in shear transformations. *Metallurgical Transactions A* **24**, 241–276.

Abe Y. (1997) Thermal and chemical evolution of terrestrial magma ocean. *Physics of Earth and Planetary Interiors* **100**, 27–39.

Adams B. L., Wright S. I., and Kunze K. (1993) Orientation imaging: the emergence of a new microscopy. *Metallurgical Transactions A* **24**, 819–831.

Agee C. B. (1993) Petrology of the mantle transition zone. *Annual Review of Earth and Planetary Sciences* **21**, 19–42.

Aines R. D. and Rossman G. R. (1984) The hydrous component in garnets: pyralsites. *American Mineralogist* **69**, 1116–1126.

Aizawa Y., Yoneda A., Katsura T., Ito E., Saito T., and Suzuki I. (2004) Temperature derivatives of elastic moduli of $MgSiO_3$ perovskite. *Geophysical Research Letters* **31**, 10.1029/2003GL018762.

Akaogi M., Ito E., and Navrotsky A. (1989) Olivine-modified spinel–spinel transitions in the system Mg_2SiO_4–Fe_2SiO_4: calorimetric measurements, thermochemical calculation, and geophysical application. *Journal of Geophysical Research* **94**, 15 671–15 685.

Aki K. (1968) Seismological evidence for the existence of soft thin layers in the upper mantle under Japan. *Journal of Geophysical Research* **73**, 585–594

Aki K., Christoffersson A., and Husebye F. S. (1977) Determination of three-dimensional seismic structure of the lithosphere. *Journal of Geophysical Research* **82**, 277–296.

Aki K. and Kaminuma K. (1963) Phase velocity of Love waves in Japan (part 1): Love waves from the Aleutian shock of March 1957. *Bulletin of Earthquake Research Institute* **41**, 243–259.

Aki K. and Richards P. G. (2002) *Quantitative Seismology*. University Science Books.

Akimoto S., Akaogi M., Kawada K., and Nishizawa O. (1976) Mineralogic distribution of iron in the upper half of the transition zone in the Earth's mantle. In *The Geophysics of the Pacific Ocean Basin and Its Margin*, pp. 399–405. American Geophysical Union.

Akulov N. S. (1964) On dislocation kinetics. *Acta Metallurgica* **12**, 1195–1196.

Alfé D., Gillan M. J., and Price G. D. (2002a) Composition and temperature of Earth's core constrained by combining *ab-initio* calculations and seismic data. *Earth and Planetary Science Letters* **95**, 91–98.

Alfé D., Price G. D., and Gillan M. J. (2000) Constraints on the composition of the Earth's core from *ab-initio* calculations. *Nature* **405**, 172–175.

Alfé D., Price G. D., and Gillan M. J. (2002b) *Ab initio* chemical potentials of solid and liquid alloys and the chemistry of the Earth's core. *Journal of Chemical Physics* **116**, 7127–7136.

Allègre C. J. and Turcotte D. L. (1986) Implications of a two-component marble-cake mantle. *Nature* **323**, 123–127.

Allen F. M., Smith B. K., and Buseck P. R. (1987) Direct observation of dissociated dislocations in garnet. *Science* **238**, 1695–1697.

Amadeo R. J. and Ghoniem N. M. (1988) A review of experimental observations and theoretical models of dislocation cells and subgrains. *Res Mechanica* **23**, 137–160.

Amin K. E., Mukherjee A. K., and Dorn J. E. (1970) A universal law for high temperature diffusion controlled transient creep. *Journal of Mechanics and Physics of Solids* **18**, 413–426.

Anand L., Kim K. H., and Shawki T. G. (1987) Onset of shear localization in viscoplastic solids. *Journal of Mechanics and Physics of Solids* **35**, 407–429.

Anderson D. L. (1979) The deep structure of continents. *Journal of Geophysical Research* **84**, 7555–7560.

Anderson D. L. (1987a) A seismic equation of state II. Shear properties and thermodynamics of the lower mantle. *Physics of Earth and Planetary Interiors* **45**, 307–323.

Anderson D. L. (1987b) Thermally induced phase changes, lateral heterogeneity of the mantle, continental roots, and deep slab anomalies. *Journal of Geophysical Research* **92**, 13 968–13 980.

Anderson D. L. and Bass J. D. (1986) Transition region of the Earth's upper mantle. *Nature* **320**, 321–328.

Anderson D. L. and Given J. W. (1982) Absorption band Q model for the Earth. *Journal of Geophysical Research* **87**, 3893–3904.

Anderson D. L. and Minster J. B. (1979) The frequency dependence of Q in the Earth and implications for mantle rheology and Chandler wobble. *Geophysical Journal of Royal Astronomical Society* **58**, 431–440.

Anderson D. L., Sammis C. G., and Phinney R. A. (1969) Brillouin scattering – A new geophysical tool. In *The Application of Modern Physics to the Earth and Planetary Interiors* (ed. S. K. Runcorn), pp. 465–477. Wiley-Interscience.

Anderson D. L., Tanimoto T., and Zhang Y. (1992) Plate tectonics and hotspots – the third dimension. *Science* **256**, 1645–1651.

Anderson O. L. (1968) Comments on the negative pressure dependence of the shear modulus found in some oxides. *Journal of Geophysical Research* **73**, 7707–7712.

Anderson O. L. (1996) *Equation of State of Solids for Geophysics and Ceramic Sciences*. Oxford University Press.

Anderson O. L. (2002) The three-dimensional phase diagram of iron. In *Earth's Core: Dynamics, Structure, Rotation* (ed. V. Dehant, K. C. Creager, S. Karato, and S. Zatman), pp. 83–103. American Geophysical Union.

Anderson O. L. and Isaak D. G. (1995) Elastic constants of mantle minerals at high temperature. In *Mineral Physics & Crystallography* (ed. T. J. Ahrens), pp. 64–97. American Geophysical Union.

Anderson O. L. and Liebermann R. C. (1970) Equations for elastic constants and their pressure derivatives for three cubic lattices and some geophysical applications. *Physics of Earth and Planetary Interiors* **3**, 61–85.

Ando J., Shibata Y., Okajima Y., Kanagawa K., and Furusho M. (2001) Striped iron zoning of olivine induced by dislocation creep in deformed olivine. *Nature* **414**, 893–895.

Ando K. (1989) Self-diffusion in oxides. In *Rheology of Solids and of the Earth* (ed. S. Karato and M. Toriumi), pp. 57–82. Oxford University Press.

Andrade E. N. d. C. (1910) On the viscous flow in metals and allies phenomena. *Proceedings of the Royal Society of London A* **84**, 1–12.

Andrault D., Fiquet G., Guyot F., and Hanfland M. (1998) Pressure-induced Landau-type transition in stishovite. *Science* **282**, 720–724.

Ardell A. J. (1997) Harper–Dorn creep – Prediction of the dislocation network theory of high temperature deformation. *Acta Materialia* **45**, 2971–2981.

Argon A. S. (1973) Stability of plastic deformation. In *The Inhomogeneity of Plastic Deformation* (ed. R. E. Reed-Hill), pp. 161–189. American Society of Metals.

Artemieva I. M. (2006) Global 1° × 1° thermal model TC1 for the continental lithosphere: implications for lithosphere secular evolution. *Tectonophysics* **416**, 245–277.

Arzi A. A. (1978) Critical phenomena in the rheology of partially melted rocks. *Tectonophysics* **44**, 173–184.

Arzt E., Ashby M. F., and Verrall R. A. (1983) Interface controlled diffusional creep. *Acta Metallurgica* **31**, 1977–1989.

Ashby M. F. (1969) On interface reaction-control of Nabarro–Herring creep and sintering. *Scripta Metallurgica* **3**, 837–842.

Ashby M. F. (1970) The deformation of plastically non-homogeneous crystals. *Philosophical Magazine* **21**, 399–424.

Ashby M. F. (1972) A first report on deformation-mechanism maps. *Acta Metallurgica* **20**, 887–897.

Ashby M. F. and Brown A. M. (1982) Flow in polycrystals and the scaling of mechanical properties. In *Deformation of Polycrystals: Mechanisms and Microstructures* (ed. N. Hansen, A. Horsewell, T. Leffers, and H. Lilholt), pp. 1–13. RISØ National Laboratory.

Ashby M. F., Edward G. H., Davenport J., and Verrall R. A. (1978) Application of bound theorems for creeping solids and their application to large strain diffusional flow. *Acta Metallurgica* **26**, 1379–1388.

Ashby M. F. and Verrall R. A. (1973) Diffusion accommodated flow and superplasticity. *Acta Metallurgica* **21**, 149–163.

Atkinson H. V. (1988) Theories of normal grain growth in pure single phase systems. *Acta Metallurgica* **36**, 469–491.

Auten T. A., Davis L. A., and Gordon R. B. (1973) Hydrostatic pressure and the mechanical properties of NaCl polycrystals. *Philosophical Magazine* **28**, 335–341.

Auten T. A., Radcliffe S. V., and Gordon R. B. (1976) Flow stress of MgO single crystals compressed along [100] at high hydrostatic pressure. *Journal of the American Ceramic Society* **59**, 40–42.

Avé Lallemant H. G. (1978) Experimental deformation of diopside and websterite. *Tectonophysics* **48**, 1–27.

Avé Lallemant H. G. and Carter N. L. (1970) Syntectonic recrystallization of olivine and modes of flow in the upper mantle. *Geological Society of America Bulletin* **81**, 2203–2220.

Avé Lallemant H. G., Mercier J.-C. C., and Carter N. L. (1980) Rheology of the upper mantle: inference from peridotite xenoliths. *Tectonophysics* **70**, 85–114.

Backus G. E. (1962) Long wave elastic anisotropy produced by horizontal layering. *Journal of Geophysical Research* **67**, 4427–4440.

Bai Q., Mackwell S. J., and Kohlstedt D. L. (1991) High temperature creep of olivine single crystals 1. Mechanical results for buffered samples. *Journal of Geophysical Research* **96**, 2441–2463.

Bai Y. L. (1982) Thermo-plastic instability in simple shear. *Journal of Mechanics and Physics of Solids* **30**, 195–207.

Bamford D. (1977) P_n velocity anisotropy in a continental upper mantle. *Geophysical Journal of Royal Astronomical Society* **57**, 397–429.

Barrat J.-L. and Hansen J.-P. (2003) *Basic Concepts for Simple and Complex Liquids*. Cambridge University Press.

Bass J. D. (1995) Elasticity of minerals, glasses, and melts. In *Mineral Physics and Crystallography: a Handbook of Physical Constants* (ed. T. J. Ahrens), pp. 46–63. American Geophysical Union.

Bass J. D., Weidner D. J., Hamaya N., Ozima M., and Akimoto S. (1984) Elasticity of the olivine and spinel polymorphs of Ni_2SiO_4. *Physics and Chemistry of Minerals* **10**, 261–272.

Beauchesne S. and Poirier J.-P. (1989) Creep of barium titanate perovskite: a contribution to a systematic approach to the viscosity of the lower mantle. *Physics of Earth and Planetary Interiors* **55**, 187–199.

Beauchesne S. and Poirier J.-P. (1990) In search of systematics for the viscosity of perovskite: creep of potassium tantalate and niobate. *Physics of Earth and Planetary Interiors* **61**, 182–198.

Becker T. W., Kellogg J. B., and O'Connell R. J. (1999) Thermal constraints on the survival of primitive blobs in the lower mantle. *Earth and Planetary Science Letters* **171**, 351–365.

Beghein C. and Trampert J. (2003a) Probability density functions for radial anisotropy: implications for the upper 1200 km of the mantle. *Earth and Planetary Science Letters* **217**, 151–162.

Beghein C. and Trampert J. (2003b) Robust normal mode constraints on inner-core anisotropy from model space search. *Science* **299**, 552–555.

Behrmann J. H. and Mainprice D. (1987) Deformation mechanisms in a high temperature quartz–feldspar mylonite: evidence for superplastic flow in the lower continental crust. *Tectonophysics* **140**, 297–305.

Béjina F., Jaoul O., and Liebermann R. C. (1999) Activation volume of Si diffusion in San Carlos olivine: implications for upper mantle rheology. *Journal of Geophysical Research* **104**, 25 529–25 542.

Bell D. R. and Rossman G. R. (1992) Water in Earth's mantle: the role of nominally anhydrous minerals. *Science* **255**, 1391–1397.

Bell D. R., Rossman G. R., Maldener J., Endisch D., and Rauch F. (2003) Hydroxide in olivine: a quantitative determination of the absolute amount and calibration of the IR spectrum. *Journal of Geophysical Research* **108**, 10.1029/2001JB000679.

Bell T. H. and Etheridge M. A. (1973) Microstructures of mylonites and their descriptive terminology. *Lithos* **6**, 337–348.

Ben Ismail W. and Mainprice D. (1998) An olivine fabric database: an overview of upper mantle fabrics and seismic anisotropy. *Tectonophysics* **296**, 145–157.

Benn K. and Allard B. (1989) Preferred mineral orientations related to magmatic flow in ophiolite layered gabbros. *Journal of Petrology* **30**, 925–946.

Bennington K. O. (1963) Some crystal growth features of sea ice. *Journal of Glaciology* **4**, 669–688.

Benz H. and Vidale J. E. (1993) Sharpness of upper-mantle discontinuities determined from high-frequency reflections. *Nature* **365**, 147–150.

Beran A. and Libowitzky E. (2006) Water in natural mantle minerals II: olivine, garnet and accessary minerals. In *Water in Nominally Anhydrous Minerals* (ed. H. Keppler and J. R. Smyth), pp. 169–191. Mineralogical Society of America.

Berckhemer H., Kampfmann W., Aulbach E., and Schmeling H. (1982) Shear modulus and Q of forsterite and dunite near partial melting from forced oscillation experiments. *Physics of Earth and Planetary Interiors* **29**, 30–41.

Bercovici D. (2003) The generation of plate tectonics from mantle convection. *Earth and Planetary Science Letters* **205**, 107–121.

Bercovici D. and Karato S. (2002) Some theoretical concepts of shear localization in the lithosphere. In *Plastic Deformation of Minerals and Rocks* (ed. S. Karato and H.-R. Wenk), pp. 387–420. American Mineralogical Society.

Bercovici D. and Karato S. (2003) Whole mantle convection and transition-zone water filter. *Nature* **425**, 39–44.

Bercovici D. and Ricard Y. (2003) Energetics of two-phase model of lithospheric damage, shear localization and plate-boundary formation. *Geophysical Journal International* **152**, 1–16.

Bercovici D. and Ricard Y. (2005) Tectonic plate generation and two-phase damage: void growth versus grain-size reduction. *Journal of Geophysical Research* **110**, 10.1029/2004JB003181.

Bercovici D., Ricard Y., and Schubert G. (2001a) A two-phase model for compaction and damage 1. General theory. *Journal of Geophysical Research* **106**, 8887–8906.

Bercovici D., Ricard Y., and Schubert G. (2001b) A two-phase model for compaction and damage 3. Applications to shear localization and plate boundary formation. *Journal of Geophysical Research* **106**, 8925–8939.

Bergman M. I. (1997) Measurements of elastic anisotropy due to solidification texturing and the implications for the Earth's inner core. *Nature* **389**, 60–63.

Bergman M. I. (1998) Estimates of the Earth's inner core grain size. *Geophysical Research Letters* **25**, 1593–1596.

Berthe D., Choukrouse P., and Jegouzo P. (1979) Orthogneiss, mylonite and non coaxial deformation of granite: the example of the South American shear zone. *Journal of Structural Geology* **1**, 31–42.

Bethe H. A. (1935) Statistical theory of superlattice. *Proceedings of the Royal Society of London A* **150**, 552–575.

Bhattacharya J., Shearer P. M., and Masters G. (1993) Inner core attenuation for short-period PKP (BC) versus PKP (DF) waveforms. *Geophysical Journal International* **114**, 1–11.

Bilde-Sorenson J. B. and Smith D. A. (1994) Comment on 'Refutation of the relationship between denuded zones and diffusional creep'. *Scripta Metallurgica et Material* **30**, 383–386.

Billen M. I. and Gurnis M. (2001) A low viscosity wedge in subduction zones. *Earth and Planetary Science Letters* **193**, 227–236.

Billen M. I., Gurnis M., and Simons M. (2003) Multiscale dynamics of the Tonga–Kermadec subduction zone. *Geophysical Journal International* **153**, 359–388.

Billien M., Lébeque J.-J., and Trampert J. (2000) Global maps of Rayleigh wave attenuation for periods between 40 and 150 seconds. *Geophysical Research Letters* **27**, 3619–3622.

Bina C. B. and Helffrich G. (1994) Phase transition Clapeyron slopes and transition zone seismic discontinuity topography. *Journal of Geophysical Research* **99**, 15 853–15 860.

Birch F. (1952) Elasticity and constitution of the Earth's interior. *Journal of Geophysical Research* **57**, 227–286.

Birch F. (1961) The velocity of compressional waves in rocks to 10 kilobars, Part 2. *Journal of Geophysical Research* **66**, 2199–2224.

Birch F. (1964) Density and composition of mantle and core. *Journal of Geophysical Research* **69**, 4377–4388.

Blacic J. D. (1972) Effects of water in the experimental deformation of olivine. In *Flow and Fracture of Rocks* (ed. H. C. Heard, I. Y. Borg, N. L. Carter, and C. B. Raleigh), pp. 109–115. American Geophysical Union.

Blacic J. D. (1975) Plastic-deformation mechanisms of quartz: the effect of water. *Tectonophysics* **27**, 271–294.

Blackman D. K. and Kendall J.-M. (1997) Sensitivity of teleseismic body waves to mineral texture and melt in the mantle beneath a mid-ocean ridge. *Philosophical Transactions of Royal Society of London A* **355**, 217–231.

Bloomfield J. P. and Covey-Crump S. J. (1993) Correlating mechanical data with microstructural observations in deformation experiments on synthetic two-phase aggregates. *Journal of Structural Geology* **15**, 1007–1019.

Bloss F. D. (1971) *Crystallography and Crystal Chemistry*. Holt, Reinhart and Winston, Inc.

Blum J. and Shen Y. (2004) Thermal, hydrous, and mechanical states of the mantle transition zone beneath southern Africa. *Earth and Planetary Science Letters* **217**, 367–378.

Blum W., Eisenlohr P. and Breutinger F. (2002) Understanding creep – A review. *Metallurgical and Materials Transactions A* **33**, 291–303.

Boland J. N. and Tullis T. E. (1986) Deformation behaviour of wet and dry clinopyroxenite in the brittle to ductile transition region. In *Mineral and Rock Deformation: Laboratory Studies* (ed. B. E. Hobbs and H. C. Heard), pp. 35–50. American Geophysical Union.

Bolfan-Casanova N. (2005) Water in the Earth's mantle. *Mineralogical Magazine* **69**, 229–257.

Bolfan-Casanova N., Keppler H., and Rubie D. C. (2000) Water partitioning between nominally anhydrous minerals in the $MgO-SiO_2-H_2O$ system up to 24 GPa: implications for the distribution of water in the Earth's mantle. *Earth and Planetary Science Letters* **182**, 209–221.

Bolfan-Casanova N., Keppler H., and Rubie D. C. (2003) Water partitioning at 660 km depth and evidence for very low water solubility in magnesium silicate perovskite. *Geophysical Research Letters* **30**, 10.1029/2003GL017182.

Bolfan-Casanova N., Mackwell S. J., Keppler H., McCammon C., and Rubie D. C. (2002) Pressure dependence of H solubility in magnesiowüstite up to 25 GPa: implications for the storage of water in the Earth's lower mantle. *Geophysical Research Letters* **29**, 89–1/89–4.

Bons P. D. and Cox, S. J. D. (1994) Analogue experiments and numerical modelling on the relation between microgeometry and flow properties of polyphase materials. *Materials Science and Engineering A* **175**, 237–245.

Bons P. D. and Den Brok B. (2000) Crystallographic preferred orientation development by dissolution–precipitation creep. *Journal of Structural Geology* **22**, 1713–1722.

Bons P. D. and Urai J. L. (1994) Experimental deformation of two-phase rock analogues. *Materials Science and Engineering A* **175**, 221–229.

Borch R. S. and Green H. W., II. (1987) Dependence of creep in olivine on homologous temperature and its implication for flow in the mantle. *Nature* **330**, 345–348.

Borch R. S. and Green H. W., II. (1989) Deformation of peridotite at high pressure in a new molten cell: comparison of traditional and homologous temperature treatments. *Physics of Earth and Planetary Interiors* **55**, 269–276.

Born M. (1940) On the stability of crystal lattice, 1. *Proceedings of Cambridge Philosophical Society* **36**, 160–165.

Born M. and Huang K. (1954) *Dynamical Theory of Crystal Lattice*. Clarendon Press.

Bouchez J. L., Lister G. S., and Nicolas A. (1983) Fabric asymmetry and shear sense in movement zones. *Geologische Rundschau* **72**, 401–419.

Boullier A. M. and Gueguen Y. (1975) SP-mylonites: origin of some mylonites by superplastic flow. *Contributions to Mineralogy and Petrology* **50**, 93–104.

Boysen H., Dorner B., Frey F. A., and Grimm H. (1980) Dynamic structure determination of two interacting modes at the M-point in α- and β-quartz by inelastic neutron scattering. *Journal of Physics C: Solid State Physics* **13**, 6127–6146.

Braithwaite J. S., Wright K., and Catlow C. R. A. (2003) A theoretical study of the energetics and IR frequencies of hydroxyl defects in forsterite. *Journal of Geophysical Research* **108**, 10.1029/2002JB002126.

Braun J., Cherny J., Poliakov A. N. B., *et al.* (1999) A simple parameterization of strain localization in the ductile regime due to grain size reduction: a case study for olivine. *Journal of Geophysical Research* **104**, 25 167–25 181.

Brennan B. J. and Stacey F. D. (1977) Frequency dependence of elasticity of rock – test of seismic velocity dispersion. *Nature* **268**, 220–222.

Breuer D. and Spohn T. (1995) Possible flushing instability in mantle convection at the Archean–Proterozoic transition. *Nature* **378**, 608–610.

Brodholt J. P. and Refson K. (2000) An *ab initio* study of hydrogen in forsterite and a possible mechanism for hydrolytic weakening. *Journal of Geophysical Research* **105**, 18 977–18 982.

Bromiley G. D., Keppler H., McCammon C., Bromiley F. A., and Jacobsen S. B. (2004) Hydrogen solubility and speciation in natural, gem quality chromian diopside. *American Mineralogist* **89**, 941–949.

Brook R. J. (1969) Pore-grain boundary interactions and grain growth. *Journal of the American Ceramic Society* **52**, 65–67.

Brookes C. A., O'Neill J. B., and Redfern B. A. W. (1971) Anisotropy in the hardness of single crystals. *Proceedings of the Royal Society of London A* **322**, 73–88.

Brown A. M. and Ashby M. F. (1980) Correlations for diffusion constants. *Acta Metallurgica* **28**, 1085–1101.

Brown J. M. and McQueen R. G. (1980) Melting of iron under shock conditions. *Geophysical Research Letters* **7**, 533–536.

Brown J. M. and McQueen R. G. (1982) The equation of state for iron and the Earth's core. In *High-Pressure Research in Geophysics* (ed. S. Akimoto and M. H. Manghnani), pp. 611–623. Reidel.

Brown J. M. and Shankland T. J. (1981) Thermodynamic parameters in the Earth as determined from seismic profiles. *Geophysical Journal of Royal Astronomical Society* **66**, 579–596.

Brown L. M. (1961) Mobile charged dislocations in ionic crystals. *Physica Status Solidi* **1**, 585–599.

Bruhn D. F. and Casey M. (1997) Texture development in experimentally deformed two-phase aggregates of calcite and anhydrite. *Journal of Structural Geology* **19**, 909–925.

Bruhn D. F., Olgaard D. L., and Dell'Angelo L. N. (1999) Evidence for enhanced deformation in two-phase rocks: experiments on the rheology of calcite–anhydrite aggregates. *Journal of Geophysical Research* **104**, 707–724.

Budianski B. and O'Connell R. J. (1980) Bulk dissipation in heterogeneous media. In *Solid Earth Geophysics and Geotechnology* (ed. S. N. Nasser), pp. 1–10. American Society of Mechanical Engineering.

Buffett B. A. (1997) Geodynamic estimates of the viscosity of the Earth's inner core. *Nature* **388**, 571–573.

Buffett B. A. and Bloxham J. (2000) Deformation of Earth's inner core by electromagnetic forces. *Geophysical Research Letters* **27**, 4001–4004.

Buffett B. A., Garnero E. J., and Jeanloz R. (2000) Sediments at the top of Earth's core. *Science* **290**, 1338–1342.

Bunge H.-J. (1982) *Texture Analysis in Materials Science – Mathematical Methods*. Butterworth.

Bunge H.-P., Ricard Y., and Matas J. (2001) Non-adiabaticity in mantle convection. *Geophysical Research Letters* **28**, 879–882.

Burns R. G. (1970) *Mineralogical Applications of Crystal Field Theory*. Cambridge University Press.

Burton B. and Reynolds G. L. (1994) In defense of diffusional creep. *Materials Science and Engineering A* **191**, 135–141.

Bussod G. Y. and Christie J. C. (1991) Textural development and melt topology in spinel lherzolite experimentally deformed at hypersolidus conditions. In *Orogenic Lherzolites and Mantle Processes* (ed. M. A. Menzies, C. Dupuy, and A. Nicolas), pp. 17–39. Oxford University Press.

Bussod G. Y., Katsura T., and Rubie D. C. (1993) The large volume multi-anvil press as a high *P–T* deformation apparatus. *Pure and Applied Geophysics* **141**, 579–599.

Byerlee J. D. (1978) Friction of rocks. *Pure and Applied Geophysics* **116**, 615–626.

Bystricky M. and Mackwell S. J. (2001) Creep of dry clinopyroxene aggregates. *Journal of Geophysical Research* **106**, 13 443–13 454.

Cadek O. and Fleitout L. (2003) Effect of lateral viscosity variation in the top 300 km on the geoid and dynamic topography. *Geophysical Journal International* **152**, 566–580.

Cadek O. and van den Berg A. (1998) Radial profile of temperature and viscosity in the Earth's mantle inferred from the geoid and lateral seismic structure. *Earth and Planetary Science Letters* **164**, 607–615.

Cahn J. W. and Balluffi R. W. (1979) On diffusional mass tranport in polycrystals containing stationary or migrating boundaries. *Scripta Metallurgica* **13**, 499–502.

Cahn J. W. and Fullman R. L. (1956) On the use of lineal analysis for obtaining particle size distribution functions in opaque samples. *Transaction of AIME* **206**, 610–612.

Callen H. B. (1960) *Thermodynamics*. John Wiley and Sons.

Campbell I. H. and Taylor S. R. (1983) No water, no granites – no oceans, no continents. *Geophysical Research Letters* **10**, 1061–1064.

Cannon R. M. and Coble R. L. (1975) Review of diffusional creep of Al_2O_3. In *Deformation of Ceramic Materials* (ed. R. C. Bradt and R. E. Tressler), pp. 61–100. Plenum Press.

Cannon R. M. and Langdon T. G. (1988) Creep of ceramics, Part 2. An examination of flow mechanisms. *Journal of Materials Science* **23**, 1–20.

Cannon W. R. and Sherby O. D. (1973) Third-power stress dependence in creep of polycrystalline nonmetals. *Journal of the American Ceramic Society* **56**, 157–160.

Caristan Y. (1982) The transition from high-temperature creep to fracture in Maryland diabase. *Journal of Geophysical Research* **887**, 6781–6790.

Carlson R. W., Shirey S. B., Pearson D. G., and Boyd F. R. (1994) The mantle beneath continents. *Carnegie Institution of Washington, Yearbook* **93**, 109–117.

Carmichael I. S. E., Turner F. J., and Verhoogen J. (1974) *Igneous Petrology*. McGraw-Hill.

Carpenter M. A. (2006) Elastic properties of minerals and the influence of phase transitions. *American Mineralogist* **91**, 229–246.

Carter C. H., Jr., Stone C. A., and Davis R. F. (1980) High-temperature, multi-atmosphere, constant stress compression creep apparatus. *Review of Scientific Instruments* **51**, 1352–1357.

Carter N. L. and Avé Lallemant H. G. (1970) High temperature deformation of dunite and peridotite. *Geological Society of America Bulletin* **81**, 2181–2202.

Cathles L. M. (1975) *The Viscosity of the Earth's Mantle*. Princeton University Press.

Chaix C. and Lasalmonie A. (1981) Transformation induced plasticity in titanium. *Res Mechanica* **2**, 241–249.

Chaklader A. C. D. (1963) Deformation of quartz crystals at the transformation temperature. *Nature* **197**, 791–792.

Chakraborty S. and Ganguly J. (1992) Cation diffusion in aluminosilicate garnets: experimental determination in spessartine–almandine diffusion couples, evaluation of effective binary diffusion coefficients, and applications. *Contributions to Mineralogy and Petrology* **111**, 74–86.

Chang R. (1961) Dislocation relaxation phenomena in oxide crystals. *Journal of Applied Physics* **32**, 1127–1132.

Chen G., Miletich R., Mueller K., and Spetzler H. A. (1997) Shear and compressional mode measurements with GHz interferometry and velocity-composition systematics for the purope–almandine solid solution series. *Physics of the Earth and Planetary Interiors* **99**, 273–287.

Chen I.-W. (1982) Diffusional creep of two-phase materials. *Acta Metallurgica* **30**, 1655–1664.

Chen I.-W. and Argon A. S. (1979) Steady state power-law creep in heterogeneous alloys with coarse microstructures. *Acta Metallurgica* **27**, 785–791.

Chen I.-W. and Argon A. S. (1981) Creep cavitation in 304 steel. *Acta Metallurgica* **29**, 1321–1333.

Chen J., Inoue T., Weidner D. J., Wu Y., and Vaughan M. T. (1998) Strength and water weakening of mantle minerals, olivine, wadsleyite and ringwoodite. *Geophysical Research Letters* **25**, 575–578.

Chen J., Inoue T., Yurimoto H., and Weidner D. J. (2002a) Effect of water on olivine–wadsleyite phase boundary in the $(Mg, Fe)_2SiO_4$ system. *Geophysical Research Letters* **29**, 10.1029/2001GRL014429.

Chen J., Li L., Weidner D. J., and Vaughan M. T. (2004) Deformation experiments using synchrotron X-rays: *in situ* stress and strain measurements at high pressure and temperature. *Physics of Earth and Planetary Interiors* **143–144**, 347–356.

Chen J., Weidner D. J., and Vaughan M. T. (2002b) Strength of $Mg_{0.9}Fe_{0.1}SiO_3$ perovskite at high pressure and temperature. *Nature* **419**, 824–826.

Chen W. K. and Peterson N. L. (1973) Cation diffusion, semiconductivity and nonstoichiometry in (Co, Ni)O crystals. *Journal of Physics and Chemistry of Solids* **34**, 1093–1108.

Chen W.-P. and Molnar P. (1983) Focal depths of intracontinental and intraplate earthquakes and their implications for the thermal structure and mechanical properties of the lithosphere. *Journal of Geophysical Research* **88**, 4183–4214.

Chester F. M. (1988) The brittle–ductile transition in a deformation-mechanism map for halite. *Tectonophysics* **154**, 125–136.

Chevrot S. (2000) Multichannel analysis of shear wave splitting. *Journal of Geophysical Research* **105**, 21 579–21 590.

Chevrot S. and van der Hilst R. D. (2003) On the effects of a dipping axis of symmetry on shear wave splitting measurements in a transversely isotropic medium. *Geophysical Journal International* **152**, 497–505.

Chiang Y.-M. and Takagi T. (1990) Grain-boundary chemistry of barium titanate and strontium titanate: I. High-temperature equilibrium space charge. *Journal of American Ceramic Society* **73**, 3278–3285.

Chinh N. Q., Horvath G., Horita Z., and Langdon T. G. (2004) A new constitutive relationship for the homogeneous deformation of metals over a wide range of strain. *Acta Materialia* **52**, 3555–3563.

Chopra P. N. and Paterson M. S. (1981) The experimental deformation of dunite. *Tectonophysics* **78**, 453–573.

Chopra P. N. and Paterson M. S. (1984) The role of water in the deformation of dunite. *Journal of Geophysical Research* **89**, 7861–7876.

Christensen N. I. and Lundquist S. M. (1982) Pyroxene orientations within the upper mantle. *Geological Society of America Bulletin* **93**, 279–288.

Christensen U. R. (1989) Mantle rheology, constitution and convection. In *Mantle Convection* (ed. W. R. Peltier), pp. 595–655. Gordon and Breach.

Christensen U. R. and Hofmann A. W. (1994) Segregation of subducted oceanic crust in the convecting mantle. *Journal of Geophysical Research* **99**, 19 867–19 884.

Chung D. H. (1972) Birch's law: why is it so good? *Science* **177**, 261–263.

Clinard F. W. and Sherby O. D. (1964) Strength of iron during allotropic transformation. *Acta Metallurgica* **12**, 911–919.

Coble R. L. (1963) A model for boundary-diffusion controlled creep in polycrystalline materials. *Journal of Applied Physics* **34**, 1679–1682.

Cocks A. C. F. and Gill S. P. A. (1996) A variational approach to two dimensional grain growth – I. Theory. *Acta Materialia* **44**, 4765–4775.

Colombo L., Kataoka T., and Li J. C. M. (1982) Movement of edge dislocations in KCl by large electric fields. *Philosophical Magazine A* **46**, 211–215.

Cooper R. F. (2002) Seismic wave attenuation: energy dissipation in viscoelastic crystalline solids. In *Plastic Deformation of Minerals and Rocks* (ed. S. Karato and H.-R. Wenk), pp. 253–290. Mineralogical Society of America.

Cooper R. F. and Kohlstedt D. L. (1982) Interfacial energies in the olivine–basalt system. In *High Pressure Research in Geophysics* (ed. S. Akimoto and M. H. Manghnani), pp. 217–228. Center for Academic Publication.

Cooper R. F. and Kohlstedt D. L. (1986) Rheology and structure of olivine–basalt partial melts. *Journal of Geophysical Research* **91**, 9315–9323.

Cooper R. F., Kohlstedt D. L., and Chyung K. (1989) Solution-precipitation enhanced creep in solid–liquid aggregates which displays a non-zero dihedral angle. *Acta Metallurgica* **37**, 1759–1771.

Cordier P. (2002) Dislocations and slip systems of mantle minerals. In *Plastic Deformation of Minerals and Rocks* (ed. S. Karato and H.-R. Wenk), pp. 137–179. American Mineralogical Society.

Cordier P. and Doukhan J.-C. (1989) Water solubility in quartz and its influence on ductility. *European Journal of Mineralogy* **1**, 221–237.

Cordier P. and Doukhan J. C. (1995) Plasticity and dissociation of dislocations in water-poor quartz. *Philosophical Magazine A* **72**, 497–514.

Cordier P., Ungar T., Zsoldos L., and Tichy G. (2004) Dislocation creep in $MgSiO_3$ perovskite at conditions of the Earth's uppermost lower mantle. *Nature* **428**, 837–840.

Cormier V. F. and Li X.-D. (2002) Frequency-dependent seismic attenuation in the inner core, 2. A scattering and fabric interpretation. *Journal of Geophysical Research* **107**, 10.1029/2002JB001796.

Cormier V. F., Xu L., and Choy G. L. (1998) Seismic attenuation of the inner core: viscoelastic or stratigraphic? *Geophysical Research Letters* **25**, 4019–4022.

Cottrell A. H. (1953) *Dislocations and Plastic Flow in Crystals*. Clarendon Press.

Cottrell A. H. (1964) *The Mechanical Properties of Matter*. Wiley.

Covey-Crump S. J. and Rutter E. H. (1989) Thermally-induced grain growth of calcite marbles on Naxos Island, Greece. *Contributions to Mineralogy and Petrology* **101**, 69–86.

Crampin S. (1978) Seismic waves propagating through a cracked solid: polarization as a possible dilatancy diagnostic. *Geophysical Journal of Royal Astronomical Society* **53**, 467–496.

Crampin S. (1981) A review of wave motion in anisotropic and cracked elastic-media. *Wave Motion* **3**, 343–391.

Crampin S. (1984) An introduction to wave propagation in anisotropic media. *Geophysical Journal of Royal Astronomical Society* **76**, 17–28.

Creager K. C. (2000) Inner core anisotropy and rotation. In *Earth's Deep Interior: Mineral Physics and Tomography* (ed. S. Karato, A. M. Forte, R. C. Liebermann, G. Masters, and L. Stixrude), pp. 89–114. American Geophysical Union.

Dahlen F. A. and Tromp J. (1998) *Theoretical Global Seismology*. Princeton University Press.

Davies G. F. (1974) Elasticity, crystal structure and phase transitions. *Earth and Planetary Science Letters* **22**, 339–346.

Davis L. A. and Gordon R. B. (1968) Pressure dependence of the plastic flow stress of alkali halide single crystals. *Journal of Applied Physics* **39**, 3885–3897.

De Bresser J. H. P., Peach C. J., Reijs J. P. J., and Spiers C. J. (1998) On dynamic recrystallization during solid state flow: effects of stress and temperature. *Geophysical Research Letters* **25**, 3457–3460.

De Bresser J. H. P., ter Heege J. H., and Spiers C. J. (2001) Grain size reduction by dynamic recrystallization: can it result in major rheological weakening? *International Journal of Earth Sciences* **90**, 28–45.

de Groot S. R. and Mazur P. (1962) *Non-Equilibrium Thermodynamics*. North-Holland.

de Jong M. and Rathenau G. W. (1959) Mechanical properties of iron and some iron alloys while undergoing allotropic transformations. *Acta Metallurgica* **7**, 246–253.

de Jong M. and Rathenau G. W. (1961) Mechanical properties of an iron carbon alloy during allotropic transformation. *Acta Metallurgica* **11**, 714–720.

Debayle E., Kennett B. L. N., and Priestley K. (2005) Global azimuthal seismic anisotropy and the unique plate-motion deformation of Australia. *Nature* **433**, 509–512.

Demouchy S., Deloule E., Frost D. J., and Keppler H. (2005) Pressure and temperature-dependence of water solubility in iron-free wadsleyite. *American Mineralogist* **90**, 1084–1091.

Dennis P. F. (1984) Oxygen self-diffusion in quartz under hydrothermal conditions. *Journal of Geophysical Research* **89**, 4047–4057.

Derby B. and Ashby M. F. (1987) On dynamic recrystallization. *Scripta Metallurgica* **21**, 879–884.

Deschamps F., Snieder R., and Trampert J. (2001) The relative density-to-shear velocity scaling in the uppermost mantle. *Physics of Earth and Planetary Interiors* **124**, 193–211.

Deschamps F. and Trampert J. (2003) Mantle tomography and its relation to temperature and composition. *Physics of the Earth and Planetary Interiors* **140**, 277–291.

Deschamps F., Trampert J., and Snieder R. (2002) Anomalies of temperature and iron in the uppermost mantle inferred from gravity data and tomographic models. *Physics of Earth and Planetary Interiors* **129**, 245–264.

Deuss A. and Woodhouse J. H. (2002) A systematic search for mantle discontinuities using SS-precursors. *Geophysical Research Letters* **29**, 10.1029/2002GL014768.

Deuss A. and Woodhouse J. H. (2004) The nature of the Lehmann discontinuity from its seismological Clapeyron slope. *Earth and Planetary Science Letters* **225**, 295–304.

Dhalenne G., Dechamps M., and Revcolevschi A. (1982) Relative energies of <011> tilt boundaries in NiO. *Journal of the American Ceramic Society* **65**, 611–612.

Dick H. J. B. and Sinton J. M. (1979) Compositional layering in alpine peridotites: evidence for pressure solution creep in the mantle. *Journal of Geology* **87**, 403–416.

Dieterich J. H. (1978) Time-dependent friction and mechanism of stick-slip. *Pure and Applied Geophysics* **116**, 790–806.

Dimos D., Wolfensteine J., and Kohlstedt D. L. (1988) Kinetic demixing and decomposition of multicomponent oxides due to a nonhydrostatic stress. *Acta Metallurgica* **36**, 1543–1552.

Dingley D. J. and Randle V. (1992) Microtexture determination by electron back-scatter diffraction. *Journal of Materials Sciences* **27**, 4545–4566.

Doherty R. D., Hughes D. A., Humphreys F. J., *et al.* (1997) Current issues in recrystallization: a review. *Materials Science and Engineering A* **238**, 219–274.

Doin M., Fleitout L., and Christensen U. R. (1997) Mantle convection and stability of depleted and undepleted continental lithosphere. *Journal of Geophysical Research* **102**, 2771–2787.

Doukhan J.-C. and Paterson M. S. (1986) Solubility of water in quartz. *Bulletin Mineralogie* **109**, 193–198.

Doukhan J.-C. and Trépied L. (1985) Plastic deformation of quartz single crystals. *Bulletin Mineralogie* **108**, 97–123.

Dresen G., Wang Z., and Bai Q. (1996) Kinetics of grain growth in anorthite. *Tectonophysics* **258**, 251–262.

Drury M. R. and Fitz Gerald J. D. (1996) Grain boundary melt films in an experimentally deformed olivine–pyroxene rock: implications for melt distribution in upper mantle rocks. *Geophysical Research Letters* **23**, 701–704.

Drury M. R. and Fitz Gerald J. D. (1998) Mantle rheology: insights from laboratory studies of deformation and phase transition. In *The Earth's Mantle: Composition, Structure and Evolution* (ed. I. Jackson), pp. 503–559. Cambridge University Press.

Drury M. R., Humphreys F. J. and White S. H. (1985) Large strain deformation studies using polycrystalline magnesium as a rock analogue, part II: dynamic recrystallization mechanisms at high temperature. *Physics of the Earth and Planetary Interiors* **40**, 208–222.

Drury M. R. and Urai J. (1990) Deformation-related recrystallization processes. *Tectonophysics* **172**, 235–253.

Drury M. R. and van Roermund H. L. M. (1989) Fluid assisted recrystallization in upper mantle peridotite xenoliths from kimberlite. *Journal of Petrology* **30**, 133–152.

Drury M. R., Vissers R. L. M., van der Wal D., and Hoogerduin Strating E. H. (1991) Shear localization in upper mantle peridotites. *Pure and Applied Geophysics* **137**, 439–460.

Duclos R., Doukhan N., and Escaig B. (1978) High-temperature creep behaviour of nearly stoichiometric alumina spinel. *Journal of Materials Science* **13**, 1740–1748.

Duffy T. S. and Ahrens T. H. (1992) Lateral variation in lower mantle seismic velocity. In *High-Pressure Research: Application to Earth and Planetary Sciences* (ed. Y. Syono and M. H. Manghnani), pp. 197–205. Terra Scientific Publishers.

Duffy T. S., Shen G., Heinz D. L., Shu J., Ma Y., Mao H.-K., Hemley R. J., and Singh A. K. (1999) Lattice strain in gold and rhenium under nonhydrostatic compression to 37 GPa. *Physical Review B* **60**, 15 063–15 073.

Duffy T. S., Zha C.-S., Downs R. T., Mao H.-K., and Hemley R. J. (1995) Elasticity of forsterite to 16 GPa and the composition of the upper mantle. *Nature* **378**, 170–173.

Dupas-Bruzek C., Sharp T. G., Rubie D. C., and Durham W. B. (1998) Mechanisms of transformation and deformation in $Mg_{1.8}Fe_{0.2}SiO_4$ olivine and wadsleyite under nonhydrostatic stress. *Physics of Earth and Planetary Interiors* **108**, 33–48.

Durek J. J. and Ekström G. (1995) Evidence of bulk attenuation in the asthenosphere from recordings of the Bolivia earthquake. *Geophysical Research Letters* **22**, 2309–2312.

Durek J. J. and Romanowicz B. (1999) Inner core anisotropy inferred by direct inversion of normal mode spectra. *Geophysical Journal International* **139**, 599–622.

Durham W. B., Goetze C., and Blake B. (1977) Plastic flow of oriented single crystals of olivine, 2. Observations and interpretations of the dislocation structure. *Journal of Geophysical Research* **82**, 5755–5770.

Durham W. B. and Stern L. A. (2001) Rheological properties of water ice – Applications to satellites of the outer planets. *Annual Review of Earth and Planetary Sciences* **29**, 295–330.

Durham W. B., Weidner D. J., Karato S., and Wang Y. (2002) New developments in deformation experiments at high pressure. In *Plastic Deformation of Minerals and Rocks* (ed. S. Karato and H.-R. Wenk), pp. 21–49. Mineralogical Society of America.

Durinck J., Legris A., and Cordier P. (2005a) Influence of crystal chemistry on ideal plastic shear anisotropy in forsterite: first principles calculations. *American Mineralogist* **90**, 1072–1077.

Durinck J., Legris A., and Cordier P. (2005b) Pressure sensitivity of olivine slip systems: first-principle calculations of generalised stacking faults. *Physics and Chemistry of Minerals* **32**, 646–654.

Duval P., Ashby, M. F., and Anderman, I. (1978) Anelastic behaviour of polycrystalline ice. *Journal of Glaciology* **21**, 621–628.

Duval P., Ashby M. F., and Anderman I. (1983) Rate-controlling processes in the creep of polycrystalline ice. *Journal of Physical Chemistry* **87**, 4066–4074.

Duyster J. and Stöckhert B. (2001) Grain boundary energies in olivine derived from natural microstructures. *Contributions to Mineralogy and Petrology* **140**, 567–576.

Dziewonski A. M. (1984) Mapping the lower mantle: determination of lateral heterogeneity in P velocity up to degree and order 6. *Journal of Geophysical Research* **89**, 5929–5952.

Dziewonski A. M. and Anderson D. L. (1981) Preliminary reference Earth model. *Physics of Earth and Planetary Interiors* **25**, 297–356.

Dziewonski A. M., Hager B. H., and O'Connell R. J. (1977) Large-scale heterogeneities in the lower mantle. *Journal of Geophysical Research* **82**, 239–255.

Edington J. W., Melton K. N., and Cutler C. P. (1976) Superplasticity. *Progress in Materials Sciences* **21**, 63–170.

Edmond J. M. and Paterson M. S. (1972) Volume changes during the deformation of rocks at high pressures. *International Journal of Rock Mechanics and Mining Sciences* **9**, 161–182.

Ehrenfest P. (1933) Phase conversions in a general and enhanced sense, classified according to the specific singularities of the thermodynamic potential. *Proceedings of the Koninklijke Akademie van Wetenschappen te Amsterdam* **36**, 153–157.

Einstein A. (1906) A new determination of molecular dimensions. *Annalen der Physik* **19**, 289–306.

Elliott D. (1973) Diffusion flow laws in metamorphic rocks. *Geological Society of America Bulletin* **84**, 2645–2664.

Engdahl E. R., van der Hilst R. D., and Buland R. P. (1998) Global teleseismic earthquake relocation with improved travel times and procedures for depth determination. *Bulletin of Seismological Society of America* **88**, 722–743.

Escaig B. (1968) Sur le glissement dévie des dislocations dans la structure cubique à faces centrées. *Journal de Physique* **29**, 225–239.

Eshelby J. D. (1956) The continuum theory of lattice defects. In *Solid State Physics*, Vol. 3 (ed. F. Seitz and D. Turnbull), pp. 79–144. Academic Press.

Eshelby J. D. (1957) The determination of the elastic strain field of an ellipsoidal inclusion and related problems. *Proceedings of the Royal Society of London, Ser. A* **241**, 376–396.

Eshelby J. D., Newey C. W. A., and Pratt P. L. (1958) Charged dislocations and the strength of ionic crystals. *Philosophical Magazine* **3**, 75–89.

Estrin Y. and Kubin L. (1986) Local strain hardening and nonuniformity of plastic deformation. *Acta Metallurgica* **34**, 2455–2464.

Estrin Y. and Kubin L. P. (1988) Plastic instabilities: classification and physical mechanisms. *Res Mechanica* **23**, 197–221.

Estrin Y. and Kubin L. P. (1991) Plastic instabilities: phenomenology and theory. *Materials Science and Engineering A* **137**, 125–134.

Etchecopar A. and Visseur G. (1987) A 3-D kinematic model of fabric development in polycrystalline aggregates: comparisons with experimental and natural examples. *Journal of Structural Geology* **9**, 705–717.

Etheridge M. A., Wall V. J., and Vernon R. H. (1983) The role of fluid phase during regional metamorphism and deformation. *Journal of Metamorphic Geology* **1**, 205–226.

Evans B., Fredrich J. T., and Wong T.-F. (1990) The brittle–ductile transition in rocks: recent experimental and theoretical progress. In *The Brittle–Ductile Transition in Rocks: the Heard Volume* (ed. A. G. Duba, W. B. Durham, J. W. Handin, and H. F. Wang), pp. 1–20. American Geophysical Union.

Evans B. and Goetze C. (1979) Temperature variation of hardness of olivine and its implication for polycrystalline yield stress. *Journal of Geophysical Research* **84**, 5505–5524.

Evans B., Renner J. and Hirth G. (2001) A few remarks on the kinetics of static grain growth in rocks. *International Journal of Earth Sciences* **90**, 88–103.

Evans B. and Wong T.-F. (1985) Shear localization in rocks induced by tectonic deformation. In *Mechanics of Geomaterials* (ed. Z. Bazant), pp. 189–210. John Wiley & Sons.

Eyring H. (1935) The activated complex and the absolute rate of chemical reaction. *Chemical Review* **17**, 65–82.

Fan D., Chen L.-Q., and Chen S.-P. P. (1998) Numerical simulation of Zener pinning with growing second-phase particle. *Journal of the American Ceramic Society* **81**, 526–531.

Fantozzi G., Esnouf C., Benoit W., and Ritchie I. G. (1982) Internal friction and microdeformation due to the intrinsic properties of dislocations: the Bordoni relaxation. *Progress in Materials Sciences* **27**, 311–451.

Farber D. L., Williams Q., and Ryerson F. J. (2000) Divalent cation diffusion in Mg_2SiO_4 spinel (ringwoodite), β phase (wadsleyite), and olivine: implications for the electrical conductivity of the mantle. *Journal of Geophysical Research* **105**, 513–529.

Farver J. R. and Yund R. A. (1991) Oxygen diffusion in quartz: dependence on temperature and water fugacity. *Chemical Geology* **90**, 55–70.

Faul U. H. (2001) Melt retention and segregation beneath mid-ocean ridges. *Nature* **410**, 920–923.

Faul U. H., Fitz Gerald J. D., and Jackson I. (2004) Shear-wave attenuation and dispersion in melt-bearing olivine polycrystals II. Microstructural interpretation and seismological implications. *Journal of Geophysical Research* **109**, 10.1029/2003JB002407.

Faul U. H. and Jackson I. (2005) The seismological signature of temperature and grain size variations in the upper mantle. *Earth and Planetary Science Letters* **234**, 119–134.

Faul U. H., Toomey D. R., and Waff H. S. (1994) Intergranular basaltic melt is distributed in thin, elongated inclusions. *Geophysical Research Letters* **21**, 29–32.

Fearn D. R., Loper D. E., and Roberts P. H. (1981) Structure of the Earth's inner core. *Nature* **292**, 232–233.

Fischer K. M. and Wiens D. G. (1996) The depth of mantle anisotropy beneath the Tonga subduction zone. *Earth and Planetary Science Letters* **142**, 253–260.

Flanagan M. P. and Shearer P. M. (1998) Global mapping of topography on transition zone velocity discontinuities by stacking SS precursors. *Journal of Geophysical Research* **103**, 2673–2692.

Flanagan M. P. and Wiens D. A. (1994) Radial upper mantle attenuation structure of inactive back arc basins from differential shear wave attenuation measurements. *Journal of Geophysical Research* **99**, 15 469–15 485.

Flesh L. M., Li B., and Liebermann R. C. (1998) Sound velocities of polycrystalline $MgSiO_3$-orthopyroxene to 10 GPa at room temperature. *American Mineralogist* **83**, 444–450.

Flinn D. (1962) On folding during three-dimensional progressive deformation. *Geological Society of London, Quaternary Journal* **118**, 385–433.

Flynn C. P. (1968) Atomic migration in monoatomic crystals. *Physical Review* **171**, 682–698.

Flynn C. P. (1972) *Point Defects and Diffusion*. Oxford University Press.

Foreman A. J., Jawson M. A., and Wood J. K. (1951) Factors controlling dislocation width. *Proceedings of Physical Society A* **64**, 156–163.

Forsyth D. W. (1975) The early structural evolution and anisotropy of the oceanic upper mantle. *Geophysical Journal of Royal Astronomical Society* **43**, 103–162.

Forsyth D. W. (1985) Subsurface loading and estimates of the flexural rigidity of continental lithosphere. *Journal of Geophysical Research* **90**, 12 623–12 632.

Forsyth D. W. (1992) Geophysical constraints on mantle flow and melt generation beneath mid-ocean ridge. In *Mantle Flow and Melt Generation at Mid-Ocean Ridges* (ed. J. P. Morgan, D. K. Blackman, and J. M. Sinton), pp. 1–66. American Geophysical Union.

Forte A. M. (2000) Seismic–geodynamic constraints on mantle flow: implications for layered convection, mantle viscosity, and seismic anisotropy in the deep mantle. In *Earth's Deep Interior: Mineral Physics and Seismic Tomography* (ed. S. Karato, A. M. Forte, R. C. Liebermann, G. Masters, and L. Stixrude), pp. 3–36. American Geophysical Union.

Forte A. M., Dziewonski A. M., and O'Connell R. J. (1994) Continent–ocean chemical heterogeneity in the mantle based on seismic tomography. *Science* **268**, 386–388.

Forte A. M. and Mitrovica J. X. (1996) New inferences of mantle viscosity from joint inversion of long-wavelength mantle convection and post-glacial rebound data. *Geophysical Research Letters* **23**, 1147–1150.

Forte A. M. and Mitrovica J. X. (2001) Deep-mantle high-viscosity flow and thermochemical structure inferred from seismic and geodynamic data. *Nature* **410**, 1049–1056.

Forte A. M., Woodward, R. L., and Dziewonski, A. M. (1994) Joint inversion of seismic and geodynamic data for models of three-dimensional mantle heterogeneity. *Journal of Geophysical Research* **99**, 21 857–21 877.

French J. D., Zhao J., Harmer M. P., Chan H. M., and Miller G. A. (1994) Creep of duplex microstructures. *Journal of the American Ceramic Society* **77**, 2857–2865.

Fressengas C. and Molinari A. (1987) Instability and localization of plastic flow in shear at high strain rate. *Journal of Mechanics and Physics of Solids* **35**, 185–211.

Freund F. and Wengeler H. (1982) The infrared spectrum of OH-compensated defect sites in C-doped MgO and CaO single crystals. *Journal of Physics and Chemistry of Solids* **43**, 129–145.

Frisillo A. L. and Barsch G. R. (1972) Measurement of single-crystal elastic constants of bronzite as a function of pressure and temperature. *Journal of Geophysical Research* **77**, 6360–6384.

Frost D. J., Liebske C., Langenhorst F., *et al.* (2004) Experimental evidence for the existence of iron-rich metal in the Earth's lower mantle. *Nature* **428**, 409–412.

Frost D. J. and Wood B. J. (1997a) Experimental measurements of the fugacity of CO_2 and graphite/diamond stability from 35 to 77 kbar at 925 to 1650 °C. *Geochimica et Cosmochimica Acta* **61**, 1565–1574.

Frost D. J. and Wood B. J. (1997b) Experimental measurements of the properties of H_2O–CO_2 mixtures at high pressures and temperatures. *Geochimica et Cosmochimica Acta* **61**, 3301–3309.

Frost H. J. and Ashby M. F. (1982) *Deformation Mechanism Maps*. Pergamon Press.

Fuchs K. (1983) Recently formed elastic anisotropy and petrological models for the continental subcrustal lithosphere in southern Germany. *Physics of Earth and Planetary Interiors* **31**, 93–118.

Fujimura A., Endo S., Kato M., and Kumazawa M. (1981) Preferred orientation of β-Mn_2GeO_4. *Programme and Abstracts, The Japan Seismological Society*, **185**.

Fujino K., Nakazaki H., Momoi H., Karato S., and Kohlstedt D. L. (1992) TEM observation of dissociated dislocations with b = [010] in naturally deformed olivine. *Physics of Earth and Planetary Interiors* **78**, 131–137.

Fukao Y., Obayashi M., Inoue H., and Nenbai M. (1992) Subducting slabs stagnant in the mantle transition zone. *Journal of Geophysical Research* **97**, 4809–4822.

Fukao Y., To A., and Obayashi M. (2003) Whole mantle P wave tomography using P and PP-P data. *Journal of Geophysical Research* **108**, 10.1029/2001JB000989.

Fukao Y., Widiyantoro R. D. S., and Obayashi M. (2001) Stagnant slabs in the upper and lower mantle transition zone. *Review of Geophysics* **39**, 291–323.

Funamori N., Yagi T., and Uchida T. (1994) Deviatoric stress measurement under uniaxial compression by a powder X-ray diffraction method. *Journal of Applied Physics* **75**, 4327–4331.

Furusho M. and Kanagawa K. (1999) Reaction-induced strain localization in a lherzolite mylonite from the Hidaka metamorphic belt of central Hokkaido, Japan. *Tectonophysics* **313**, 411–432.

Gaherty J. B. (2001) Seismic evidence for hotspot-induced buoyant flow beneath the Reykjanes ridge. *Science* **293**, 1645–1647.

Gaherty J. B. and Jordan T. H. (1995) Lehmann discontinuity as the base of an anisotropic layer beneath continent. *Science* **268**, 1468–1471.

Gaherty J. B., Jordan T. H., and Gee L. S. (1996) Seismic structure of the upper mantle in a central Pacific corridor. *Journal of Geophysical Research* **101**, 22 291–22 309.

Gaherty J. B., Kato M., and Jordan T. H. (1999) Seismological structure of the upper mantle: a regional comparison of seismic layering. *Physics of Earth and Planetary Interiors* **110**, 21–41.

Gandin C.-A., Rappaz M., West D., and Adams B. L. (1995) Grain texture evolution during the columnar growth of dendritic alloys. *Metallurgical Materials Transaction A* **26**, 1543–1551.

Gannarelli C. M. S., Alfé D., and Gillan M. J. (2005) The axial ratio of hcp iron at the conditions of the Earth's inner core. *Physics of Earth and Planetary Interiors* **152**, 67–77.

Garnero E. J. (2000) Heterogeneity of the lowermost mantle. *Annual Review of Earth and Planetary Sciences* **28**, 509–537.

Garnero E. J. (2004) A new paradigm for Earth's core–mantle boundary. *Science* **304**, 834–836.

Garnero E. J. and Jeanloz R. (2000) Fuzzy patches on the Earth's core–mantle boundary? *Geophysical Research Letters* **27**, 2777–2780.

Garnero E. J., Maupin V., Lay T., and Fouch M. J. (2004) Variable azimuthal anisotropy in Earth's lowermost mantle. *Science* **306**, 259–261.

Garnero E. J., Revenaugh J., Williams Q., Lay T., and Kellogg L. H. (1998) Ultralow velocity zone at the core–mantle boundary. In *The Core–Mantle Boundary Regions* (ed. M. E. W. M. Gurnis, E. Knittle and B. A. Fuffett), pp. 319–334. American Geophysical Union.

Garofalo F. (1965) *Fundamentals of Creep and Creep-Rupture in Metals*. MacMillan.

Gay N. C. (1968) Pure shear and simple shear deformation of inhomogeneous viscous fluids. 1. Theory. *Tectonophysics* **5**, 211–234.

Getting I. C., Dutton S. J., Burnley P. C., Karato S., and Spetzler H. A. (1997) Shear attenuation and dispersion in MgO. *Physics of Earth and Planetary Interiors* **99**, 249–257.

Getting I. C. and Kennedy G. C. (1970) Effect of pressure on the EMF of chromel–alumel and platinum–platinum 10% rhodium thermocouples. *Journal of Applied Physics* **41**, 4552–4562.

Ghose S. (1985) Lattice dynamics, phase transitions and soft modes. In *Microscopic to Macroscopic* (ed. S. W. Kiefer and A. Navrotsky), pp. 127–163. Mineralogical Society of America.

Gill S. P. A. and Cocks A. C. F. (1996) A variational approach to two dimensional grain growth – II. numerical results. *Acta mater* **44**, 4777–4789.

Gilman J. J. (1985) Hardness test: a mechanical microprobe. In *Science of Hardness Testing* (ed. J. H. Westbrook and H. Conrad), pp. 51–74. American Society for Metals.

Gilman M. J. (1981) The volume of formation of defects in ionic crystals. *Philosophical Magazine A* **43**, 301–312.

Glansdorff P. and Prigogine I. (1971) *Thermodynamic Theory of Stability, Structure and Fluctuation*. John Wiley & Sons.

Gleason G. C. and Tullis J. (1995) A flow law for dislocation creep of quartz aggregates determined with the molten slat cell. *Tectonophysics* **247**, 1–23.

Goetze C. and Evans B. (1979) Stress and temperature in the bending lithosphere as constrained by experimental rock mechanics. *Geophysical Journal of Royal Astronomical Society* **59**, 463–478.

Goldsby D. L. and Kohlstedt D. L. (2001) Superplastic deformation of ice: experimental observations. *Journal of Geophysical Research* **106**, 11 017–11 030.

Gordon R. B. (1965) Diffusion creep in the Earth's mantle. *Journal of Geophysical Research* **70**, 2413–2418.

Gordon R. B. (1971) Observation of crystal plasticity under high pressure with application to the Earth's mantle. *Journal of Geophysical Research* **76**, 1248–1254.

Gordon R. S. (1973) Mass transport in the diffusional creep of ionic solids. *Journal of the American Ceramic Society* **65**, 147–152.

Gordon R. S. and Terwillinger G. R. (1972) Transient creep in Fe-doped polycrystalline MgO. *Journal of the American Ceramic Society* **55**, 450–455.

Gottstein G. and Mecking H. (1985) Recrystallization. In *Preferred Orientation in Deformed Metals and Rocks* (ed. H.-R. Wenk), pp. 183–232. Academic Press.

Grand S. (1994) Mantle shear structure beneath Americas and surrounding oceans. *Journal of Geophysical Research* **99**, 11 591–11 621.

Green D. H. and Falloon, T. J. (1998) Pyrolite: a Ringwood concept and its current expression. In *The Earth's Mantle* (ed. I. Jackson), pp. 311–378. Cambridge University Press.

Green H. W., II. (1984) "Pressure solution" creep: some causes and mechanisms. *Journal of Geophysical Research* **89**, 4313–4318.

Green H. W., II. (1970) Diffusional flow in polycrystalline materials. *Journal of Applied Physics* **41**, 3899–3902.

Green H. W., II. and Borch R. S. (1987) The pressure dependence of creep. *Acta Metallurgica* **35**, 1301–1305.

Green H. W., II. and Houston H. (1995) The mechanics of deep earthquakes. *Annual Review of Earth and Planetary Sciences* **23**, 169–213.

Greenwood G. W. (1994) Denuded zones and diffusional creep. *Scripta Metallurgica et Material* **30**, 1527–1530.

Greenwood G. W. and Johnson R. H. (1965) The deformation of metals under small stresses during phase transformations. *Proceedings of the Royal Society of London A* **238**, 403–422.

Gribb T. T. and Cooper R. F. (1998) Low-frequency shear attenuation in polycrystalline olivine: grain boundary diffusion and the physical significance of the Andrade model for viscoelastic rheology. *Journal of Geophysical Research* **103**, 27 267–27 279.

Gribb T. T. and Cooper R. F. (2000) The effect of an equilibrated melt phase on the shear creep and attenuation behavior of polycrystalline olivine. *Geophysical Research Letters* **27**, 2341–2344.

Griggs D. T. (1967) Hydrolytic weakening of quartz and other silicates. *Geophysical Journal of Royal Astronomical Society* **14**, 19–31.

Griggs D. T. (1974) A model of hydrolytic weakening in quartz. *Journal of Geophysical Research* **79**, 1653–1661.

Griggs D. T. and Baker D. W. (1969) The origin of deep-focus earthquakes. In *Properties of Matter Under Unusual Conditions* (ed. H. Marks and S. Feshbach), pp. 23–42. Interscience.

Griggs D. T. and Blacic J. D. (1965) Quartz: anomalous weakness of synthetic crystals. *Science* **147**, 292–295.

Grimmer H. (1979) The distribution of disorientation angles if all relative orientations of neighbouring grains are equally probable. *Scripta Metallurgica* **13**, 161–164.

Grüneisen E. (1912) Theories des festen Zustands einatomiger elemente. *Annalen der Physik, Berlin* **39**, 257–306.

Gu Y. J., Dziewonski A. M., and Agee C. B. (1998) Global de-correlation of the topography of transition zone discontinuities. *Earth and Planetary Science Letters* **157**, 57–67.

Gu Y. J., Dziewonski A. M., and Ekström G. (2001) Preferential detection of the Lehmann discontinuity beneath continents. *Geophysical Research Letters* **28**, 4655–4658.

Gu Y. J., Dziewonski A. M., and Ekström G. (2003) Simultaneous inversion for mantle velocity and topography of transition zone discontinuities. *Geophysical Journal International* **154**, 559–583.

Gubbins D., Alfé D., Masters G., Price G. D., and Gillan M. J. (2003) Can the Earth's dynamo run on heat alone? *Geophysical Journal International* **155**, 609–622.

Gueguen Y., Darot M., Mazot P., and Woirgard J. (1989) Q^{-1} of forsterite single crystals. *Physics of Earth and Planetary Interiors* **55**, 254–258.

Gueguen Y. and Mercier J. M. (1973) High attenuation and low velocity zone. *Physics of Earth and Planetary Interiors* **7**, 39–46.

Gueguen Y. and Palciauskas V. (1994) *Introduction to the Physics of Rocks*. Princeton University Press.

Guillopé M. and Poirier J.-P. (1979) Dynamic recrystallization during creep of single-crystalline halite: an experimental study. *Journal of Geophysical Research* **84**, 5557–5567.

Gung Y. and Romanowicz B. (2004) Q tomography of the upper mantle using three-component long-period waveforms. *Geophysical Journal International* **157**, 813–830.

Gung Y., Romanowicz B., and Panning M. (2003) Global anisotropy and the thickness of continents. *Nature* **422**, 707–711.

Gutenberg B. (1926) Untersuchen zur Frage, bis zu welcher Tiefe die Erde kristallin ist. *Zeitschrift für Geophisik* **2**, 24–29.

Gutenberg B. (1948) On the layer of relatively low wave velocity at a depth of about 80 kilometers. *Bulletin of Seismological Society of America* **35**, 117–130.

Gutenberg B. (1954) Low-velocity layers in the Earth's mantle. *Bulletin of Seismological Society of America* **65**, 337–348.

Haasen P. (1979) Kink formation and migration as dependent on the Fermi level. *Journal de Physique C* **6**, 111–116.

Hacker B. R., Gnos E., Ratschbacher L., *et al.* (2000) Hot and dry lower crustal xenoliths from Tibet. *Science* **287**, 2463–2466.

Hager B. H. (1984) Subducted slabs and the geoid: constraints on mantle rheology and flow. *Journal of Geophysical Research* **89**, 6003–6015.

Hager B. H. and Clayton R. W. (1989) Constraints on the structure of mantle convection using seismic observations, flow models and the geoid. In *Mantle Convection* (ed. W. R. Peltier), pp. 657–763. Gordon and Breach.

Haggerty S. E. and Sautter V. (1990) Ultra deep (> 300 km) ultramafic, upper mantle xenoliths. *Science* **248**, 993–996.

Hall C. E. and Parmentier E. M. (2002) The influence of grain size evolution on a composite dislocation–diffusion creep rheology. *Journal of Geophysical Research.*

Hammond W. C. and Humphreys E. D. (2000a) Upper mantle seismic wave velocity: effects of realistic partial melt distribution. *Journal of Geophysical Research* **105**, 10 987–10 999.

Hammond W. C. and Humphreys E. D. (2000b) Upper mantle seismic wave velocity: effects of realistic partial melt geometries. *Journal of Geophysical Research* **105**, 10 975–10 986.

Handy M. R. (1989) Deformation regimes and the rheological evolution of fault zones in the lithosphere: the effects of pressure, temperature, grain size and time. *Tectonophysics* **163**, 119–152.

Handy M. R. (1994) Flow laws for rocks containing two nonlinear viscous phases: a phenomenological approach. *Journal of Structural Geology* **16**, 287–301.

Handy M. R. and Brun J. P. (2004) Seismicity, structure and strength of the continental lithosphere. *Earth and Planetary Science Letters* **223**, 427–441.

Harper J. and Dorn J. E. (1957) Viscous creep of aluminium near its melting temperature. *Acta Metallurgica* **5**, 654–665.

Harren S. V., Dève H. E., and Asaro R. J. (1988) Shear band formation in plane strain compression. *Acta Metallurgica* **36**, 2435–2480

Harrison R. J. and Redfern S. A. T. (2002) The influence of transformation twins on the seismic-frequency elastic and anelastic properties of perovskite: dynamical mechanical analysis of single crystal $LaAlO_3$. *Physics of the Earth and Planetary Interiors* **134**, 253–272.

Hart E. W. (1967) Theory of tensile test. *Acta Metallurgica* **15**, 351–355.

Hart E. W. (1970) A phenomenological theory for plastic deformation of polycrystalline metals. *Acta Metallurgica* **18**, 599–610.

Hartmann W. K. (1999) *Moons and Planets.* Wadsworth Publishers.

Hashida T. (1989) Three-dimensional seismic attenuation structure beneath the Japanese islands and its tectonic and thermal implications. *Tectonophysics* **159**, 163–180.

Hashin Z. and Shtrikman S. (1963) A variational approach to the theory of the elastic behavior of multiphase materials. *Journal of Mechanics and Physics of Solids* **11**, 127–140.

Haskell N. A. (1935a) The motion of a viscous fluid under a surface load. *Physics* **6**, 265–269.

Haskell N. A. (1935b) The motion of a viscous fluid under a surface load. Part II. *Physics* **7**, 56–61.

Haskell N. A. (1937) The viscosity of the asthenosphere. *American Journal of Science* **33**, 22–28.

Hazen R. M. and Finger L. W. (1979) Bulk modulus–volume relationship for cation–anion polyhedra. *Journal of Geophysical Research* **84**, 6723–6728.

Heard H. C., Borg I. Y., Carter N. L., and Raleigh C. B. (1972) *Flow and Fracture of Rocks.* American Geophysical Union.

Heard H. C. and Kirby S. H. (1981) Activation volume for steady-state creep in polycrystalline CsCl: cesium chloride structure. In *Mechanical Behavior of Crustal Rocks* (ed. N. L. Carter, M. Friedman, J. M. Logan, and O. W. Stearns), pp. 83–91. American Geophysical Union.

Heggie M. and Jones R. (1986) Models of hydrolytic weakening in quartz. *Philosophical Magazine, A* **53**, L65–L70.

Heinz D. L., Jeanloz R., and O'Connell R. J. (1982) Bulk attenuation in a polycrystalline Earth. *Journal of Geophysical Research* **87**, 7772–7778.

Helffrich G. (2000) Topography of the transition zone discontinuities. *Review of Geophysics* **38**, 141–158.

Herring C. (1950) Diffusional viscosity of a polycrystalline solid. *Journal of Applied Physics* **21**, 437–445.

Hess H. H. (1964) Seismic anisotropy of the uppermost mantle under oceans. *Nature* **203**, 629–631.

Hier-Majumder S., Anderson I. M., and Kohlstedt D. L. (2005a) Influence of protons on Fe–Mg interdiffusion in olivine. *Journal of Geophysical Research* **110**, 10.1029/2004JB003292.

Hier-Majumder S., Leo P. H., and Kohlstedt D. L. (2004) On grain boundary wetting during deformation. *Acta Materialia* **52**, 3425–3433.

Hier-Majumder S., Mei S., and Kohlstedt D. L. (2005b) Water weakening of clinopyroxene in diffusion creep regime. *Journal of Geophysical Research* **110**, 10.1029/2004JB003414.

Higo Y., Inoue T., Irifune T., and Yurimoto H. (2001) Effect of water on the spinel–postspinel transformation in Mg_2SiO_4. *Geophysical Research Letters* **28**, 3505–3508.

Hill R. (1952) The elastic behaviour of a crystalline aggregate. *Proceedings of the Physical Society, London A* **65**, 349–354.

Hill R. (1958) A general theory of uniqueness and stability of elastic plastic models. *Journal of Mechanics and Physics of Solids* **6**, 236–249.

Hill R. (1965) A self consistent mechanics of composite materials. *Journal of the Mechanics and Physics of Solids* **13**, 213–222.

Hill R. I., Campbell I. H., Davies G. F., and Griffiths R. W. (1992) Mantle plumes and continental tectonics. *Science* **256**, 186–193.

Hillert M. (1965) On the theory of normal and abnormal grain growth. *Acta Metallurgica* **13**, 227–238.

Hippertt J. F. and Hongn F. D. (1998) Deformation mechanisms in the mylonite/ultramylonite transition. *Journal of Structural Geology* **20**, 1435–1448.

Hiraga T., Anderson I. M., and Kohlstedt D. L. (2004) Grain-boundaries as reservoirs of incompatible elements in the Earth's mantle. *Nature* **427**, 699–703.

Hiraga T., Anderson I. M., Zimmerman M. E., Mei S., and Kohlstedt D. L. (2002) Structure and chemistry of grain boundaries in deformed, olivine + basalt and partially molten lherzolite aggregates: evidence of melt-free grain boundaries. *Contributions to Mineralogy and Petrology* **144**, 163–175.

Hirsch P. B. (1979) A mechanism for the effect of doping on dislocation mobility. *Journal de Physique C* **6**, 117–121.

Hirth G. (2002) Laboratory constraints on the rheology of the upper mantle. In *Plastic Deformation of Minerals and Rocks*, Vol. 51 (ed. S. Karato and H.-R. Wenk), pp. 97–120. Mineralogical Society of America.

Hirth G. and Kohlstedt D. L. (1995a) Experimental constraints on the dynamics of partially molten upper mantle: deformation in the diffusion creep regime. *Journal of Geophysical Research* **100**, 1981–2001.

Hirth G. and Kohlstedt D. L. (1995b) Experimental constraints on the dynamics of partially molten upper mantle: deformation in the dislocation creep regime. *Journal of Geophysical Research* **100**, 15 441–15 450.

Hirth G. and Kohlstedt D. L. (1996) Water in the oceanic upper mantle – implications for rheology, melt extraction and the evolution of the lithosphere. *Earth and Planetary Science Letters* **144**, 93–108.

Hirth G. and Kohlstedt D. L. (2003) Rheology of the upper mantle and the mantle wedge: a view from the experimentalists. In *Inside the Subduction Factory* (ed. J. E. Eiler), pp. 83–105. American Geophysical Union.

Hirth G., Teyssier C., and Dunlop D. J. (2001) An evaluation of quartzite flow law based on comparisons between experimentally and naturally deformed rocks. *International Journal of Earth Sciences* **90**, 77–87.

Hirth G. and Tullis J. (1992) Dislocation creep regimes in quartz aggregates. *Journal of Structural Geology* **14**, 145–159.

Hirth J. P. and Lothe J. (1982) *Theory of Dislocations*. Krieger Publishing Company.

Hitchings R. S., Paterson M. S., and Bitmead J. (1989) Effects of iron and magnetite additions in olivine–pyroxene rheology. *Physics of Earth and Planetary Interiors* **55**, 277–291.

Hobbs B. E. (1968) Recrystallization of single crystal of quartz. *Tectonophysics* **6**, 353–401.

Hobbs B. E. (1981) The influence of metamorphic environment upon the deformation of minerals. *Tectonophysics* **78**, 335–383.

Hobbs B. E. (1983) Constraints on the mechanism of deformation of olivine imposed by defect chemistry. *Tectonophysics* **92**, 35–69.

Hobbs B. E. (1984) Point defect chemistry of minerals under hydrothermal environment. *Journal of Geophysical Research* **89**, 4026–4038.

Hobbs B. E. (1985) The geological significance of microfabric analysis. In *Preferred Orientation in Deformed Metals and Rocks: an Introduction to Modern Texture Analysis* (ed. H.-R. Wenk), pp. 463–484. Academic Press.

Hobbs B. E., McLaren A. C., and Paterson M. S. (1972) Plasticity of single crystals of quartz. In *Flow and Fracture of Rocks* (ed. H. C. Heard, I. Y. Borg, N. L. Carter, and C. B. Raleigh), pp. 29–53. American Geophysical Union.

Hobbs B. E., Means W. D., and Williams P. F. (1976) *The Outline of Structural Geology*. Addison & Wiley.

Hobbs B. E., Mulhaus H.-B., and Ord A. (1990) Instability, softening and localization of deformation. In *Deformation Mechanisms, Rheology and Tectonics* (ed. R. J. Knipe and E. H. Rutter), pp. 143–165. The Geological Society.

Hobbs B. E. and Ord A. (1988) Plastic instabilities: implications for the origin of intermediate and deep focus earthquakes. *Journal of Geophysical Research* **89**, 10 521–10 540.

Hobbs B. E., Ord A., and Teyssier C. (1986) Earthquakes in the ductile regime. *Pure and Applied Geophysics* **124**, 310–336.

Hoff N. J. (1954) Approximate analysis of structures in the presence of moderately large creep deformations. *Quarterly Journal of Applied Mathematics* **12**, 49–55.

Hofmann A. W. (1997) Mantle geochemistry: the message from oceanic volcanism. *Nature* **385**, 219–228.

Hollomon J. H. (1947) The mechanical equation of state. *Trans. AIME* **171**, 535–545.

Holness M. B. (1993) Temperature and pressure dependence of quartz-aqueous fluid dihedral angles: the control of absorbed H_2O on the permeability of quartzites. *Earth and Planetary Science Letters* **117**, 363–377.

Holt D. L. (1970) Dislocation cell formation in metals. *Journal of Applied Physics* **41**, 3197–3201.

Holtzman B. K., Groebner N. J., Zimmerman M. E., Ginsberg S. B., and Kohlstedt D. L. (2003a) Stress-driven melt segregation in partially molten rocks. *Geochemistry, Geophysics, Geosystems* **4**, 10.1029/2001GC000258.

Holtzman B. K., Kohlstedt D. L., and Phipps Morgan J. (2005) Viscous energy dissipation and strain partitioning in partially molten rocks. *Journal of Petrology* **46**, 2569–2592.

Holtzman B. K., Kohlstedt D. L., Zimmerman M. E., *et al.* (2003b) Melt segregation and strain partitioning: implications for seismic anisotropy and mantle flow. *Science* **301**, 1227–1230.

Honda S., Yuen D. A., Balachandar S., and Reuteler D. (1993) Three-dimensional instabilities of mantle convection with multiple phase transitions. *Science* **259**, 1308–1311.

Horn R. G., Smith D. T., and Haller W. (1989) Surface forces and viscosity of water measured between silica sheets. *Chemical Physics Letters* **162**, 404–408.

Houlier B., Cheraghmakni M., and Jaoul O. (1990) Silicon diffusion in San Carlos olivine. *Physics of Earth and Planetary Interiors* **62**, 329–340.

Houlier B., Jaoul O., Abel F., and Liebermann R. C. (1988) Oxygen and silicon diffusion in natural olivine at $T = 1300\,°C$. *Physics of Earth and Planetary Interiors* **50**, 240–250.

Huang X., Xu Y., and Karato S. (2005) Water content of the mantle transition zone from the electrical conductivity of wadsleyite and ringwoodite. *Nature* **434**, 746–749.

Hutchinson J. W. (1976) Bounds and self-consistent estimates for creep of polycrystalline materials. *Proceedings of the Royal Society of London A* **348**, 101–127.

Hutchinson J. W. (1977) Creep and plasticity of hexagonal polycrystals as related to single crystal slip. *Metallurgical Transactions A* **8**, 1465–1469.

Ingrin J. and Skogby H. (2000) Hydrogen in nominally anhydrous upper-mantle minerals: concentration levels and implications. *European Journal of Mineralogy* **12**, 543–570.

Inoue T., Yurimoto H., and Kudoh Y. (1995) Hydrous modified spinel, $Mg_{1.75}SiH_{0.5}O_4$: a new water reservoir in the mantle transition zone. *Geophysical Research Letters* **22**, 117–120.

Irifune T. and Ringwood A. E. (1987) Phase transformations in primitive MORB and pyrolyte composition to 25 GPa and some geophysical implications. In *High-Pressure Research in Mineral Physics* (ed. M. H. Manghnani and Y. Syono), pp. 231–242. American Geophysical Union.

Isaak D. G. (1992) High-temperature elasticity of iron-bearing olivines. *Journal of Geophysical Research* **97**, 1871–1885.

Ishii M. and Dziewonski A. M. (2002) The innermost inner core of the earth: evidence for a change in anisotropic behavior at the radius of about 300 km. *Proceedings of American Academy of Arts and Sciences* **99**, 14 026–14 030.

Ishii M. and Tromp J. (1999) Normal mode and free-air gravity constraints on lateral variation in density of Earth's mantle. *Science* **285**, 1231–1236.

Ita J. and Cohen R. E. (1997) Effects of pressure on diffusion and vacancy formation in MgO from nonempirical free-energy integration. *Physical Review Letters* **79**, 3198–3201.

Ito E. and Katsura T. (1989) A temperature profile of the mantle transition zone. *Geophysical Research Letters* **16**, 425–428.

Ito Y. and Toriumi M. (2007) Pressure effect on self-diffusion in periclase (MgO) by molecular dynamics. *Journal of Geophysical Research*, in press.

Iyer H. M. and Hirahara K. (1993) *Seismic Tomography: Theory and Practice*. Chapman and Hall.

Jackson I. (1998) Elasticity, composition and temperature of the Earth's lower mantle: a reappraisal. *Geophysical Journal International* **134**, 291–311.

Jackson I. (2000) Laboratory measurements of seismic wave dispersion and attenuation: recent progress. In *Earth's Deep Interior: Mineral Physics and Tomography from the Atomic to the Global Scale* (ed. S. Karato, A. M. Forte, R. C. Liebermann, G. Masters, and L. Stixrude), pp. 265–289. American Geophysical Union.

Jackson I., Faul U. H., Fitz Gerald J. D., and Tan B. (2004a) Shear wave attenuation and dispersion in melt-bearing olivine polycrystals: 1. Specimen fabrication and mechanical testing. *Journal of Geophysical Research* **109**, 10.1029/2003JB002406.

Jackson I., Fitz Gerald J. D., Faul U. H., and Tan B. H. (2002) Grain-size sensitive seismic-wave attenuation in polycrystalline olivine. *Journal of Geophysical Research* **107**, 10.1029/2002JB001225.

Jackson I., Fitz Gerald J. D., and Kokkonen H. (2000a) High-temperature viscoelastic relaxation in iron and its implications for the shear modulus and attenuation of the Earth's inner core. *Journal of Geophysical Research* **105**, 23 605–23 634.

Jackson I. and Niesler H. (1982) The elasticity of periclase to 3 GPa and some geophysical implications. In *High-Pressure Research in Geophysics* (ed. S. Akimoto and M. H. Manghnani), pp. 93–113. Center for Academic Publications, Japan.

Jackson I. and Paterson M. S. (1987) Shear modulus and internal friction of calcite rocks at seismic frequencies. *Physics of Earth and Planetary Interiors* **45**, 349–367.

Jackson I. and Paterson M. S. (1993) A high-pressure, high-temperature apparatus for studies of seismic wave dispersion and attenuation. *Pure and Applied Geophysics* **141**, 445–466.

Jackson I., Paterson M. S., and Fitz Gerald J. D. (1992) Seismic wave dispersion and attenuation in Åheim dunite. *Geophysical Journal International* **108**, 517–534.

Jackson J. A. (2002a) Faulting, flow, and strength of the continental lithosphere. *International Geological Review* **11**, 39–61.

Jackson J. A. (2002b) Strength of the continental lithosphere: time to abandon the jelly sandwich? *GSA Today* **12**, 4–10.

Jackson J. M., Sinogeikin S. V., and Bass J. D. (2000b) Sound velocities and elastic properties of γ-Mg_2SiO_4 to 873 K by Brillouin spectroscopy. *American Mineralogist* **85**, 296–303.

Jackson J. M., Zhang J., and Bass J. D. (2004b) Sound velocities and elasticity of aluminous $MgSiO_3$ perovskite: implications for aluminium heterogeneity in Earth's lower mantle. *Geophysical Research Letters* **31**, 10.1029/2004GL019918.

Jacobsen S. D. (2006) Effect of water on the equation of state of nominally anhydrous minerals. In *Water in Nominally Anhydrous Minerals* (ed. H. Keppler and J. R. Smyth), pp. 321–342. Mineralogical Society of America.

Jagoutz E., Palme H., Baddenhausen H., *et al.* (1979) The abundances of major, minor, and trace elements in the Earth's mantle as derived from primitive ultramafic nodules. *Proceedings of 10th Lunar and Planetary Science Conference*, 2031–2050.

Jaoul O. (1990) Multicomponent diffusion and creep in olivine. *Journal of Geophysical Research* **95**, 17631–17642.

Jaoul O. and Houlier B. (1983) Study of ^{18}O diffusion in magnesium orthosilicate by nuclear micro analysis. *Journal of Geophysical Research* **88**, 613–624.

Jaoul O., Poumellec M., Froidevaux C., and Havette A. (1981) Silicon diffusion in forsterite: a new constraint for understanding mantle deformation. In *Anelasticity in the Earth* (ed. F. D. Stacey, M. S. Paterson, and A. Nicolas). American Geophysical Union.

Jaoul O., Sautter V., and Abel F. (1991) Nuclear microanalysis: a powerful tool for measuring low atomic diffusivity with mineralogical applications. In *Diffusion, Atomic Ordering, and Mass Transport* (ed. J. Ganguly), pp. 198–220. Springer-Verlag.

Jaroslow G. E., Hirth G., and Dick H. J. B. (1996) Abyssal peridotite mylonites: implications for grain-size sensitive flow and strain localization in the oceanic lithosphere. *Tectonophysics* **256**, 17–37.

Jessel M. W. (1988a) Simulation of fabric development in recrystallizing aggregates – I. Description of the model. *Journal of Structural Geology* **10**, 771–778.

Jessel M. W. (1988b) Simulation of fabric development in recrystallizing aggregates – II. Example model runs. *Journal of Structural Geology* **10**, 779–793.

Jesser W. A. and Kuhlmann-Wilsdorf D. (1972) The flow stress and dislocation structure of nickel deformed at very high pressure. *Materials Science and Engineering* **9**, 111–117.

Ji S., Wang Z., and Wirth R. (2001) Bulk flow strength of forsterite–enstatite composites as a function of forsterite content. *Tectonophysics* **341**, 69–93.

Ji S. and Zhao P. (1993) Flow laws of multiphase rocks calculated from experimental data on the constituent phases. *Earth and Planetary Science Letters* **117**, 181–187.

Jin D., Karato S., and Obata M. (1998) Mechanisms of shear localization in the continental lithosphere: inference from the deformation microstructures of peridotites from the Ivrea zone, northern Italy. *Journal of Structural Geology* **20**, 195–209.

Jin Z. M., Green H. W., II., and Zhou Y. (1994) Melt topology in partially molten mantle peridotite during ductile deformation. *Nature* **372**, 164–167.

Jin Z. M., Zhang J., Green H. W., II., and Jin S. (2001) Eclogite rheology: implications for subducting lithosphere. *Geology* **29**, 667–670.

Joesten R. (1983) Grain growth and grain-boundary diffusion in quartz from the Christmas Mountains (Texas) contact aurole. *American Journal of Science* **283**, 233–254.

Johnson W. C. (1984) On the elastic stabilization of precipitates against coarsening under applied load. *Acta Metallurgica* **32**, 465–475.

Johnson W. C. and Cahn J. W. (1984) Elastically induced shape bifurcations of inclusions. *Acta Metallurgica* **32**, 1925–1933.

Johnson W. C. and Schmalzried H. (1992) Gibbs–Duhem and Clausius–Clapeyron type equations for elastically stressed crystals. *Acta Metallurgica et Materialia* **40**, 2337–2342.

Johnston W. G. (1962) Yield points and delay times in single crystals. *Journal of Applied Physics* **33**, 2716–2730.

Johnston W. G. and Gilman J. J. (1959) Dislocation velocities, dislocation densities, and plastic flow in lithium fluoride crystals. *Journal of Applied Physics* **30**, 129–144.

Jones R. (1980) The structure of kinks on the 90° partial in silicon and a 'strain-bond model' for dislocation motion. *Philosophical Magazine B* **42**, 213–219.

Jordan P. G. (1987) The deformation behaviour of bimineralic limestone–halite aggregates. *Tectonophysics* **135**, 185–197.

Jordan P. G. (1988) The rheology of polymineralic rocks: an approach. *Geologiches Rundschau* **77**, 285–294.

Jordan T. H. (1975) The continental tectosphere. *Review of Geophysics and Space Physics* **13**, 1–12.

Jordan T. H. (1981) Continents as a chemical boundary layer. *Philosophical Transactions of the Royal Society of London A* **301**, 359–373.

Jung H. and Karato S. (2001a) Effect of water on the size of dynamically recrystallized grains in olivine. *Journal of Structural Geology* **23**, 1337–1344.

Jung H. and Karato S. (2001b) Water-induced fabric transitions in olivine. *Science* **293**, 1460–1463.

Jung H., Katayama I., Jiang Z., Hiraga T., and Karato S. (2006) Effects of water and stress on the lattice preferred orientation in olivine. *Tectonophysics* **421**, 1–22.

Kamb W. B. (1961) The thermodynamic theory of non-hydrostatically stressed solids. *Journal of Geophysical Research* **66**, 259–271.

Kamb W. B. (1972) Experimental recrystallization of ice under stress. In *Flow and Fracture of Rocks* (ed. H. C. Heard,

I. Y. Borg, N. L. Carter, and C. B. Raleigh), pp. 211–241. American Geophysical Union.

Kaminski E. (2002) The influence of water on the development of lattice preferred orientation in olivine aggregates. *Geophysical Research Letters* **29**, 17-1/17-4.

Kaminski E. and Ribe N. M. (2001) A kinematic model for dynamic recrystallization and texture development in olivine polycrystals. *Earth and Planetary Science Letters* **189**, 253–267.

Kaminski E., Ribe N. M., and Browaeys J. T. (2004) D-rex, a program for calculation of seismic anisotropy due to crystal lattice preferred orientation in the convective upper mantle. *Geophysical Journal International* **158**, 744–752.

Kamiya S. and Kobayashi Y. (2000) Seismological evidence for the presence of serpentinized wedge mantle. *Geophysical Research Letters* **27**, 819–822.

Kampfmann W. and Berckhemer H. (1985) High temperature experiments on the elastic and anelastic behaviour of magmatic rocks. *Physics of Earth and Planetary Interiors* **40**, 223–247.

Kanamori H. and Anderson D. L. (1977) Importance of physical dispersion in surface wave and free oscillation problems: review. *Review of Geophysics and Space Physics* **15**, 105–112.

Kanamori H. and Press F. (1970) How thick is the lithosphere? *Nature* **226**, 330–331.

Kaneshima S. (1990) Origin of crustal anisotropy: shear wave splitting studies in Japan. *Journal of Geophysical Research* **95**, 11 121–11 133.

Kaneshima S. and Helffrich G. (1999) Dipping low-velocity layer in the mid-mantle: evidence for geochemical heterogeneity. *Science* **283**, 1888–1891.

Kanzaki H. (1957) Point defects in face-centred cubic lattice – I. Distortion around defects. *Journal of Physics and Chemistry of Solids* **2**, 24–36.

Karato S. (1977) *Rheological Properties of Materials Composing the Earth's Mantle*. Ph.D., University of Tokyo.

Karato S. (1978) The concentration minimum of point defects under high pressures and the viscosity of the lower mantle. *Programme and Abstracts, The Seismological Society of Japan* **1**, D31.

Karato S. (1981a) Pressure dependence of diffusion in ionic solids. *Physics of Earth and Planetary Interiors* **25**, 38–51.

Karato S. (1981b) Rheology of the lower mantle. *Physics of Earth and Planetary Interiors* **24**, 1–14.

Karato S. (1984) Grain-size distribution and rheology of the upper mantle. *Tectonophysics* **104**, 155–176.

Karato S. (1986) Does partial melting reduce the creep strength of the upper mantle? *Nature* **319**, 309–310.

Karato S. (1987a) Scanning electron microscope observation of dislocations in olivine. *Physics and Chemistry of Minerals* **14**, 245–248.

Karato S. (1987b) Seismic anisotropy due to lattice preferred orientation of minerals: kinematic or dynamic? In *High-Pressure Research in Geophysics* (ed. M. H. Manghnani and Y. Syono), pp. 455–471. American Geophysical Union.

Karato S. (1988) The role of recrystallization in the preferred orientation in olivine. *Physics of Earth and Planetary Interiors* **51**, 107–122.

Karato S. (1989a) Defects and plastic deformation in olivine. In *Rheology of Solids and of the Earth* (ed. S. Karato and M. Toriumi), pp. 176–208. Oxford University Press.

Karato S. (1989b) Grain growth kinetics in olivine aggregates. *Tectonophysics* **155**, 255–273.

Karato S. (1989c) Plasticity-crystal structure systematics in dense oxides and its implications for creep strength of the Earth's deep interior: a preliminary result. *Physics of Earth and Planetary Interiors* **55**, 234–240.

Karato S. (1990) The role of hydrogen in the electrical conductivity of the upper mantle. *Nature* **347**, 272–273.

Karato S. (1992) On the Lehmann discontinuity. *Geophysical Research Letters* **19**, 2255–2258.

Karato S. (1993a) Importance of anelasticity in the interpretation of seismic tomography. *Geophysical Research Letters* **20**, 1623–1626.

Karato S. (1993b) Inner core anisotropy due to the magnetic field-induced preferred orientation of iron. *Science* **262**, 1708–1711.

Karato S. (1995) Effects of water on seismic wave velocities in the upper mantle. *Proceedings of the Japan Academy* **71**, 61–66.

Karato S. (1997a) On the separation of crustal component from subducted oceanic lithosphere near the 660 km discontinuity. *Physics of Earth and Planetary Interiors* **99**, 103–111.

Karato S. (1997b) Phase transformations and rheological properties of mantle minerals. In *Earth's Deep Interior* (ed. D. Crossley), pp. 223–272. Gordon and Breach.

Karato S. (1998a) A dislocation model of seismic wave attenuation and velocity dispersion and microcreep of the solid Earth: Harold Jeffreys and the rheology of the solid Earth. *Pure and Applied Geophysics* **153**, 239–256.

Karato S. (1998b) Effects of pressure on plastic deformation of polycrystalline solids: some geological applications. In *High Pressure Research in Materials Sciences* (ed. R. M. Wentzcovitch, R. J. Hemley, W. J. Neillis, and P. Y. Yu), pp. 3–14. Materials Research Society.

Karato S. (1998c) Micro-physics of post glacial rebound (In *Dynamics of the Ice Age Earth* ed. P. Wu), pp. 351–364. *Trans. Tech.*

Karato S. (1998d) Seismic anisotropy in the deep mantle, boundary layers and geometry of mantle convection. *Pure and Applied Geophysics* **151**, 565–587.

Karato S. (1998e) Some remarks on seismic anisotropy in the D″ layer. *Earth, Planets, Space* **50**, 1019–1028.

Karato S. (1999) Seismic anisotropy of the Earth's inner core resulting from flow induced by Maxwell stress. *Nature* **402**, 871–873.

Karato S. (2000) Dynamics and anisotropy of the Earth's inner core. *Proceedings of Japan Academy B* **76**, 1–6.

Karato S. (2003a) *Dynamic Structure of the Deep Earth: an Interdisciplinary Approach*. Princeton University Press.

Karato S. (2003b) Mapping water content in Earth's upper mantle. In *Inside the Subduction Factory* (ed. J. E. Eiler), pp. 135–152. American Geophysical Union.

Karato S. (2006a) Influence of hydrogen-related defects on the electrical conductivity and plastic deformation of mantle minerals: a critical review. In *Earth's Deep Water Cycle* (ed. S. D. Jacobsen and S. van der Lee), pp. 113–129. American Geophysical Union.

Karato S. (2006b) Remote sensing of hydrogen in Earth's mantle. In *Water in Nominally Anhydrous Minerals* (ed. H. Keppler and J. R. Smyth), pp. 343–375. Mineralogical Society of America.

Karato S. (2007) Microscopic models for the influence of hydrogen on physical and chemical properties of minerals. In *Superplume: Beyond Plate Tectonics* (ed. D. A. Yuen, S. Maruyama, S. Karato, and B. F. Windley), 321–355. Springer-Verlag.

Karato S., Bercovici D., Leahy G., Richard G., and Jing Z. (2006) Transition zone water filter model for global material circulation: where do we stand? In *Earth's Deep Water Cycle* (ed. S. D. Jacobsen and S. van der Lee), pp. 289–313. American Geophysical Union.

Karato S., Dupas-Bruzek C., and Rubie D. C. (1998) Plastic deformation of silicate spinel under the transition zone conditions of the Earth. *Nature* **395**, 266–269.

Karato S., Ito E., and Fujino K. (1990) Plasticity of $MgSiO_3$ perovskite: the results of microhardness tests on single crystals. *Geophysical Research Letters* **17**, 13–16.

Karato S. and Jung H. (1998) Water, partial melting and the origin of seismic low velocity and high attenuation zone in the upper mantle. *Earth and Planetary Science Letters* **157**, 193–207.

Karato S. and Jung H. (2003) Effects of pressure on high-temperature dislocation creep in olivine polycrystals. *Philosophical Magazine A* **83**, 401–414.

Karato S. and Karki B. B. (2001) Origin of lateral heterogeneity of seismic wave velocities and density in Earth's deep mantle. *Journal of Geophysical Research* **106**, 21 771–21 783.

Karato S. and Lee K.-H. (1999) Stress–strain distribution in deformed olivine aggregates: inference from microstructural observations and implications for texture development. *12th International Conference on Textures of Materials*, 1546–1555.

Karato S. and Li P. (1992) Diffusion creep in the perovskite: implications for the rheology of the lower mantle. *Science* **255**, 1238–1240.

Karato S. and Murthy V. R. (1997) Core formation and chemical equilibrium in the Earth I. Physical considerations. *Physics of Earth and Planetary Interiors* **100**, 61–79.

Karato S., Paterson M. S., and Fitz Gerald J. D. (1986) Rheology of synthetic olivine aggregates: influence of grain-size and water. *Journal of Geophysical Research* **91**, 8151–8176.

Karato S., Riedel M. R., and Yuen D. A. (2001) Rheological structure and deformation of subducted slabs in the mantle transition zone: implications for mantle circulation and deep earthquakes. *Physics of Earth and Planetary Interiors* **127**, 83–108.

Karato S. and Rubie D. C. (1997) Toward experimental study of plastic deformation under deep mantle conditions: a new multianvil sample assembly for deformation experiments under high pressures and temperatures. *Journal of Geophysical Research* **102**, 20 111–20 122.

Karato S. and Sato H. (1982) The effect of oxygen partial pressure on the dislocation recovery in olivine: a new constraint on creep mechanisms. *Physics of Earth and Planetary Interiors* **28**, 312–319.

Karato S. and Spetzler H. A. (1990) Defect microdynamics in minerals and solid state mechanisms of seismic wave attenuation and velocity dispersion in the mantle. *Review of Geophysics* **28**, 399–421.

Karato S., Toriumi M., and Fujii T. (1980) Dynamic recrystallization of olivine single crystals during high temperature creep. *Geophysical Research Letters* **7**, 649–652.

Karato S., Wang Z., Liu B., and Fujino K. (1995a) Plastic deformation of garnets: systematics and implications for the rheology of the mantle transition zone. *Earth and Planetary Science Letters* **130**, 13–30.

Karato S. and Wu P. (1993) Rheology of the upper mantle: a synthesis. *Science* **260**, 771–778.

Karato S., Zhang S., and Wenk H.-R. (1995b) Superplasticity in Earth's lower mantle: evidence from seismic anisotropy and rock physics. *Science* **270**, 458–461.

Karki B. B., Stixrude L., Clark S. J., *et al.* (1997) Structure and elasticity of MgO at high pressure. *American Mineralogist* **82**, 635–639.

Karki B. B., Stixrude L., and Wentzcovitch R. M. (2001) High-pressure elastic properties of major materials of Earth's mantle from first principles. *Review of Geophysics* **39**, 507–534.

Kataoka T., Colombo L., and Li J. C. M. (1983) Dislocation charges in pure and Ca^{2+}-doped KCl in the temperature range from 82 to 294 K. *Radiation Effects* **75**, 227–234.

Kataoka T., Colombo L., and Li J. C. M. (1984a) Direct measurements of dislocation charges in Ca^{2+}-doped KCl by using large electric fields. *Philosophical Magazine A* **49**, 395–407.

Kataoka T., Colombo L., and Li J. C. M. (1984b) Dislocation charges in Ca^{2+}-doped KCl. Effects of impurity concentration and temperature. *Philosophical Magazine A* **49**, 409–423.

Katayama I., Hirose K., Yurimoto H., and Nakashima S. (2003) Water solubility in majorite garnet in subducting oceanic crust. *Geophysical Research Letters* **30**, 10.1029/2003GL018127.

Katayama I., Jung H., and Karato S. (2004) New type of olivine fabric at modest water content and low stress. *Geology* **32**, 1045–1048.

Katayama I. and Karato S. (2006) Effects of temperature on the B- to C-type fabric transition in olivine. *Physics of the Earth and Planetary Interiors* **157**, 33–45.

Katayama I. and Karato S. (2007) The role of water and iron content on the rheological contrast between garnet and olivine. *Physics of the Earth and Planetary Interiors*, submitted.

Katayama I. and Nakashima S. (2003) Hydroxyl in clinopyroxene from the deep subducted crust: evidence for H_2O transport into the mantle. *American Mineralogist* **88**, 229–234.

Kato T. and Kumazawa M. (1985) Garnet phase of $MgSiO_3$ filling the pyroxene–ilmenite gap at very high temperature. *Nature* **316**, 803–805.

Katsura T., Mayama N., Shouno K., *et al.* (2001) Temperature derivatives of elastic moduli of $(Mg_{0.91}, Fe_{0.09})_2SiO_4$ modified spinel. *Physics of Earth and Planetary Interiors* **124**, 163–166.

Katz R. F., Spiegelman M., and Holtzman B. (2006) The dynamics of melt and shear localization in partially molten aggregates. *Nature* **442**, 676–679.

Kaula W. M. (1964) Tidal dissipation by solid friction and the resulting orbital evolution. *Review of Geophysics* **2**, 661–685.

Kawakatsu H. and Niu F. (1995) Seismic evidence for the 920-km discontinuity in the mantle. *Nature* **371**, 301–305.

Kawamoto T., Hertig R. J., and Holloway J. R. (1996) Experimental evidence for a hydrous transition zone in Earth's early mantle. *Earth and Planetary Science Letters* **142**, 587–592.

Kê T. S. (1947) Experimental evidence of the viscous behavior of grain boundaries in metals. *Physical Review* **71**, 533–546.

Kekulawala K. R. S. S., Paterson M. S., and Boland J. N. (1978) Hydrolytic weakening in quartz. *Tectonophysics* **46**, T1–T6.

Kekulawala K. R. S. S., Paterson M. S., and Boland J. N. (1981) An experimental study of the role of water in quartz deformation. In *Mechanical Behavior of Crustal Rocks: the Handin Volume* (ed. N. L. Carter, M. Friedman, J. M. Logan, and D. W. Stearns), pp. 49–60. American Geophysical Union.

Kelemen P. B., Hirth G., Shimizu N., Spiegelman M., and Dick H. J. B. (1997) A review of melt migration processes in the adiabatically upwelling mantle beneath oceanic spreading ridges. *Philosophical Transactions of the Royal Society of London* **355**, 283–318.

Kellogg L. H., Hager B. H., and van der Hilst R. D. (1999) Compositional stratification in the deep mantle. *Science* **283**, 1881–1884.

Kelly A., Tyson W. R., and Cottrell A. H. (1967) Ductile and brittle crystals. *Philosophical Magazine* **15**, 567–586.

Kendall J.-M. and Silver P. G. (1996) Constraints from seismic anisotropy on the nature of the lowermost mantle. *Nature* **381**, 409–412.

Kendall J.-M. and Silver P. G. (1998) Investigating causes of D″ anisotropy. In *The Core–Mantle Boundary Region* (ed. M. E. W. M. Gurnis, E. Knittle and B. A. Buffett), pp. 97–118. American Geophysical Union.

Kennett B. L. N., Engdahl E. R., and Buland R. P. (1995) Constraints on seismic wave velocities in the Earth from travel times. *Geophysical Journal International* **122**, 108–124.

Keppler H., Wiedenbeck M., and Shcheka S. S. (2003) Carbon solubility in olivine and the mode of carbon storage in the Earth's mantle. *Nature* **424**, 414–416.

Keyes R. W. (1963) Continuum models of the effect of pressure on activated processes. In *Solids Under Pressure* (ed. W. Paul and D. M. Warschauer), pp. 71–91. McGraw-Hill.

Khisina N. R., Wirth R., Andrut M., and Ukhanov A. V. (2001) Extrinsic and intrinsic mode of hydrogen occurrence in natural olivines: FTIR and TEM investigation. *Physics and Chemistry of Minerals* **28**, 291–301.

Kido M. and Cadek O. (1997) Inferences of viscosity from the oceanic geoid: indication of a low viscosity zone below the 660-km discontinuity. *Earth and Planetary Science Letters* **151**, 125–137.

King S. D. (1995a) Radial models of mantle viscosity: results from a genetic algorithm. *Geophysical Journal International* **122**, 725–734.

King S. D. (1995b) The viscosity structure of the mantle. *Review of Geophysics* **33**, 11–17.

Kingery W. D. (1974a) Plausible concepts necessary and sufficient for interpretation of ceramic grain-boundary phenomena: I. Grain boundary characteristics, structure, and electrostatic potential. *Journal of the American Ceramic Society* **57**, 1–8.

Kingery W. D. (1974b) Plausible concepts necessary and sufficient for interpretation of ceramic grain-boundary phenomena: II. Solute segregation, grain-boundary diffusion, and general discussion. *Journal of the American Ceramic Society* **57**, 74–83.

Kingery W. D., Bowen H. K., and Uhlmann D. R. (1976) *Introduction to Ceramics*. John Wiley & Sons.

Kinsland G. L. and Bassett W. A. (1977) Strength of MgO and NaCl polycrystals to confining pressures of 250 kbar at 25 °C. *Journal of Applied Physics* **48**, 978–985.

Kirby S. H. (1977) The effects of the α–β phase transformation on the creep properties of hydrolytically-weakened synthetic quartz. *Geophysical Research Letters* **4**, 97–100.

Kitamura M., Kondoh S., Morimoto N., *et al.* (1987) Planar OH-bearing defects in mantle olivine. *Nature* **328**, 143–145.

Kitamura M., Matsuda H., and Morimoto N. (1986) Direct observation of the Cottrell atmosphere in olivine. *Proceedings of Japan Academy* **62**, 149–152.

Kittel C. (1986) *Introduction to Solid State Physics*. John Wiley & Sons.

Kliewer K. L. and Koehler J. S. (1965) Space charge in ionic crystals. I. General approach with application to NaCl. *Physical Review* **140**, A1226–A1240.

Kneller E. A., van Keken P. E., Karato S., and Park J. (2005) B-type olivine fabric in the mantle wedge: insights from high-resolution non-Newtonian subduction zone models. *Earth and Planetary Science Letters* **237**, 781–797.

Kocks U. F. (1970) The relation between polycrystal deformation and single crystal deformation. *Metallurgical Transactions* **1**, 1121–1143.

Kocks U. F., Argon A. S., and Ashby M. F. (1975) Thermodynamics and kinetics of slip. *Progress in Materials Sciences* **19**, 1–288.

Kocks U. F., Jonas J. J., and Mecking H. (1979) The development of strain-rate gradients. *Acta Metallurgica* **27**, 419–432.

Kocks U. F., Tomé C. N., and Wenk H.-R. (1998) *Texture and Anisotropy*. Cambridge University Press.

Kogiso T., Hirose K., and Takahashi E. (1998) Melting experiments on homogeneous mixtures of peridotites and basalts: application to the genesis of ocean island basalts. *Earth and Planetary Science Letters* **162**, 45–61.

Kohlstedt D. L. (2002) Partial melting and deformation. In *Plastic Deformation of Minerals and Rocks*, Vol. 51 (ed. S. Karato and H.-R. Wenk), pp. 121–135. Mineralogical Society of America.

Kohlstedt D. L. (2006) The role of water in high-temperature rock deformation. In *Water in Nominally Anhydrous Minerals* (ed. H. Keppler and J. R. Smyth), pp. 377–396. Mineralogical Society of America.

Kohlstedt D. L., Evans B., and Mackwell S. J. (1995) Strength of the lithosphere: constraints imposed by laboratory measurements. *Journal of Geophysical Research* **100**, 17 587–17 602.

Kohlstedt D. L. and Goetze C. (1974) Low-stress, high-temperature creep in olivine single crystals. *Journal of Geophysical Research* **79**, 2045–2051.

Kohlstedt D. L., Goetze C., and Durham W. B. (1976) A new technique for decorating dislocations in olivine. *Science* **191**, 1945–1046.

Kohlstedt D. L., Keppler H., and Rubie D. C. (1996) Solubility of water in the α, β and γ phases of $(Mg, Fe)_2SiO_4$. *Contributions to Mineralogy and Petrology* **123**, 345–357.

Kohlstedt D. L. and Mackwell S. J. (1998) Diffusion of hydrogen and intrinsic point defects in olivine. *Zeitschrift für Phisikalische Chemie* **207**, 147–162.

Kohlstedt D. L. and Mackwell S. J. (1999) Solubility and diffusion of 'water' in silicates. In *Microscopic Properties and Processes in Minerals* (ed. K. Wright and R. Catlow), pp. 539–559. Kluwer Academic Publishers.

Kohlstedt D. L. and Weathers M. S. (1980) Deformation-induced microstructures, paleopiezometers, and differential stresses in deeply eroded fault zones. *Journal of Geophysical Research* **85**, 6269–6285.

Kolsky H. (1956) The propagation of stress pulses in viscoelastic solids. *Philosophical Magazine* **1**, 693–710.

Korenaga J. (2005) Firm mantle plumes and the nature of the core–mantle region. *Earth and Planetary Science Letters* **232**, 29–37.

Korenaga J. (2007) Thermal cracking and the deep hydration of oceanic lithosphere: a key to the generation of plate tectonics? *Journal of Geophysical Research* **112**, 10.1029/2006JB004502.

Krajewski P. E., Jones J. W., and Allison J. E. (1995) The effect of particle reinforcement on the creep behavior of single-phase aluminum. *Metallurgical Materials Transactions A* **26**, 3107–3118.

Kronenberg A. K., Kirby S. H., and Aines R. D. (1986) Solubility and diffusional uptake of hydrogen in quartz at high water pressures: implications for hydrolytic weakening. *Journal of Geophysical Research* **91**, 12 723–12 744.

Kronenberg A. K. and Tullis J. (1984) Flow strength of quartz aggregates: grain size and pressure effects due to hydrolytic weakening. *Journal of Geophysical Research* **89**, 4281–4297.

Kubin L. P. (1993) Dislocation patterning. In *Materials Science and Technology*, Vol. 6 (ed. R. W. Cahn, P. Haasen, and E. J. Kramer), pp. 137–190. VCH.

Kubo T., Ohtani E., Kato T., Shinmei T., and Fujino K. (1998) Effects of water on the α–β transformation kinetics in San Carlos olivine. *Science* **281**, 85–87.

Kubo T., Ohtani E., Kato T., *et al.* (2000) Formation of metastable assemblages and mechanisms of the grain-size reduction in the postspinel transformation of Mg_2SiO_4. *Geophysical Research Letters* **27**, 807–810.

Kumazawa M. (1974) On the relation between plastic flow properties and elastic wave velocities. In *Flow of Solids: From Earth to Crystals* (ed. S. Uyeda), pp. 246–262. Tokai University Press.

Kurishita H., Yoshinaga H., and Nakashima H. (1989) The high temperature deformation mechanism in pure metals. *Acta Metallurgica* **37**, 499–505.

Kurz W. and Fischer D. J. (1998) *Fundamentals of Solidification*. Trans Tech.

Kushiro I. (1975) On the nature of silicate melt and its significance in magma genesis: regularities in the shift of the liquidus boundaries involving olivine. *American Journal of Science* **275**, 411–431.

Kushiro I. (1976) Viscosities of basalt and andesite melts at high pressures. *Journal of Geophysical Research* **81**, 6351–6356.

Labrosse S., Poirier J.-P., and Le Mouel J.-L. (2001) The age of the inner core. *Physics of Earth and Planetary Interiors* **190**, 111–123.

Lager G. A., Armbruster T., Rotella F. J., and Rossman G. R. (1989) The OH substitution in garnets: X-ray and neutron diffraction, infrared and geometric-modelling studies. *American Mineralogist* **74**, 840–851.

Lakki A., Schaller R., Carry C., and Benoit W. (1998) High temperature anelastic and viscoelastic deformation of fine-grained MgO-doped Al_2O_3. *Acta Materialia* **46**, 689–700.

Lambeck K. and Johnston P. (1998) The viscosity of the mantle: evidence from analyses of glacial-rebound phenomena. In *The Earth's Mantle* (ed. I. Jackson), pp. 461–502. Cambridge University Press.

Landau L. D. and Lifshitz E. M. (1959) *Theory of Elasticity*. Pergamon Press.

Landau L. D. and Lifshitz E. M. (1964) *Statistical Physics*. Pergamon Press.

Landau L. D. and Lifshitz E. M. (1987) *Fluid Dynamics*. Pergamon Press.

Langdon T. G., Dehghan A., and Sammis C. G. (1982) Deformation of olivine, and the application to lunar and planetary interiors. *Strength of Metals and Alloys, Proceedings of the 6th International Conference*, pp. 757–762 Pergamon Press.

Langdon T. G. and Yavari P. (1982) An investigation of Harper–Dorn creep – II. The flow process. *Acta Metallurgica* **30**, 881–887.

Langmuir J. W., Klein E. M., and Plank T. (1992) Petrological systematics of mid-ocean ridge basalt: constraints on melt generation beneath ocean ridges. In *Mantle Flow and Melt Generation at Mid-Ocean Ridges* (ed. J. P. Morgan, D. K. Blackman, and J. M. Sinton), pp. 183–280. American Geophysical Union.

Lasaga A. C. (1997) *Kinetic Theory in Earth Sciences*. Princeton University Press.

Lawlis J. D. (1998) *High Temperature Creep of Synthetic Olivine-Enstatite Aggregates*. Ph.D., The Pennsylvania State University.

Lawrence J. F. and Wysession M. E. (2005) QLM9: a new radial quality factor (*Q*) model for the mantle. *Earth and Planetary Science Letters* **241**, 962–971.

Lawrence J. F. and Wysession M. E. (2006) Seismic evidence for subduction-transported water in the lower mantle. In *Earth's Deep Water Cycle* (ed. S. D. Jacobsen and S. v. d. Lee), pp. 251–261. American Geophysical Union.

Lay T., Garnero E. J., and Williams Q. (2004) Partial melting in a thermo-chemical boundary layer at the base of the mantle. *Physics of Earth and Planetary Interiors* **146**, 441–467.

Lay T., Heinz D. L., Ishii M., *et al.* (2005) Multidisciplinary impact of the deep mantle phase transition in perovskite structure. *EOS, Transactions of American Geophysical Union* **86**, 1–5.

Lay T. and Wallace T. C. (1995) *Modern Global Seismology*. Academic Press.

Lay T., Williams Q., and Garnero E. J. (1998) The core–mantle boundary layer and deep Earth dynamics. *Nature* **392**, 461–468.

Lebensohn R. A. and Tomé C. N. (1993) A self-consistent anisotropic approach for the simulation of plastic deformation and texture development of polycrystals: application to zirconium alloys. *Acta Metallurgica et Materials* **41**, 2611–2624.

Lee D.-C. and Halliday A. N. (1995) Hafnium–tungsten chronometry and the timing of terrestrial core formation. *Nature* **392**, 771–774.

Lee K.-H., Jiang Z., and Karato S. (2002) A scanning electron microscope study of effects of dynamic recrystallization on the lattice preferred orientation in olivine. *Tectonophysics* **351**, 331–341.

Lemaire C., Kohn S. C., and Brooker R. A. (2004) The effect of silica activity on the incorporation mechanisms of water in synthetic forsterite: a polarized infrared spectroscopic study. *Contributions to Mineralogy and Petrology* **147**, 48–57.

Lenardic A. and Moresi L. N. (1999) Some thoughts on the stability of cratonic lithosphere: effects of buoyancy and viscosity. *Journal of Geophysical Research* **104**, 12 747–12 759.

Levien L. and Prewitt C. T. (1981) High-pressure structural study of diopside. *American Mineralogist* **66**, 315–323.

Lewis J. S. (1974) Chemical composition of the solar system. *Scientific American* **230**, 50–65.

Li B., Liebermann R. C., and Weidner D. J. (1998) Elastic moduli of wadsleyite (β-Mg_2SiO_4) to 7 GPa and 873 K. *Science* **281**, 675–677.

Li J., Hadidiacos C., Mao H.-K., Fei Y., and Hemley R. J. (2003a) Behavior of thermocouples under high pressure in a multi-anvil apparatus. *High Pressure Research* **23**, 389–401.

Li J. C. M. (1963) A dislocation mechanism of transient creep. *Acta Metallurgica* **11**, 1269–1270.

Li L., Raterron P., Weidner D. J., and Long H. (2006) Plastic flow of pyrope at mantle pressure and temperature. *American Mineralogist* **91**, 517–525.

Li L., Ratteron P., Weidner D. J., and Chen J. (2003b) Olivine flow mechanisms at 8 GPa. *Physics of Earth and Planetary Interiors* **138**, 113–129.

Li L., Weidner D. J., Chen J., *et al.* (2004a) X-ray strain analysis at high pressure: effect of plastic deformation in MgO. *Journal of Applied Physics* **95**, 8357–8365.

Li L., Weidner D. J., Ratteron P., Chen J., and Vaughan M. T. (2004b) Stress measurements of deforming olivine at high pressure. *Physics of Earth and Planetary Interiors* **143/144**, 357–367.

Li P., Karato S., and Wang Z. (1996) High-temperature creep in fine-grained polycrystalline $CaTiO_3$, an analogue material of (Mg, Fe)SiO_3 perovskite. *Physics of Earth and Planetary Interiors* **95**, 19–36.

Li X. and Cormier V. F. (2002) Frequency-dependent seismic attenuation in the inner core 1. A viscoelastic interpretation. *Journal of Geophysical Research* **107**, 10.1029/2002JB001795.

Li Y. and Langdon T. G. (1998) High strain rate superplasticity in metal matrix composites: the role of load transfer. *Acta Materialia* **46**, 3937–3948.

Lidiard A. B. (1981) The volume of formation of Schottky defects in ionic solids. *Philosophical Magazine A* **43**, 292–300.

Liebermann R. C. (1982) Elasticity of minerals at high pressure and temperature. In *High Pressure Research in Geosciences* (ed. W. Schreyer), pp. 1–14. Schweizerbartsche.

Liebermann R. C. (2000) Elasticity of mantle minerals (experimental studies). In *Earth's Deep Interior: Mineral Physics and Tomography* (ed. S. Karato, A. M. Forte, R. C. Liebermann, G. Masters, and L. Stixrude), pp. 181–199. American Geophysical Union.

Liebermann R. C. and Ringwood A. E. (1973) Birch's law and polymorphic phase transformations. *Journal of Geophysical Research* **78**, 6926–6932.

Lifshitz I. M. (1963) On the theory of diffusion–viscous flow of polycrystalline bodies. *Soviet Physics JETP* **17**, 909–920.

Lifshitz I. M. and Shikin V. B. (1965) The theory of diffusional viscous flow of polycrystalline solids. *Soviet Physics, Solid State* **6**, 2211–2218.

Lin J.-F., Sturhahn W., Zhao J., *et al.* (2005) Sound velocities of hot dense iron: Birch's law revisited. *Science* **308**, 1892–1894.

Lindemann F. A. (1910) Über die Berechnung Molecular Eigenfrequnzen. *Physikalische Zeitschrift* **11**, 609–612.

Linker M. F. and Kirby S. H. (1981) Anisotropy in the rheology of hydrolytically weakened quartz crystals. In *Mechanical Behavior of Crustal Rocks* (ed. N. L. Carter, M. Friedman, J. M. Logan, and D. W. Stearns), pp. 29–48. American Geophysical Union.

Linker M. F., Kirby S. H., Ord A., and Christie J. M. (1984) Effects of compression direction on the plasticity and rheology of hydrolytically weakened synthetic quartz crystals at atmospheric pressure. *Journal of Geophysical Research* **89**, 4241–4255.

Lister G. S. (1979) Fabric transitions in plastically deformed quartzites: competition between basal, prism and rhomb systems. *Bulletin Mineralogie* **102**, 232–241.

Lister G. S. and Hobbs B. E. (1980) The simulation of fabric development during plastic deformation and its application to quartzite: the influence of deformation history. *Journal of Structural Geology* **2**, 355–370.

Lister G. S. and Paterson M. S. (1979) The simulation of fabric development during plastic deformation and its application to quartzite: fabric transition. *Journal of Structural Geology* **1**, 99–115.

Lister G. S., Paterson M. S., and Hobbs B. E. (1978) The simulation of fabric development during plastic deformation and its application to quartzite: the model. *Tectonophysics* **45**, 107–158.

Lister G. S. and Snoke A. W. (1984) S-C mylonite. *Journal of Structural Geology* **6**, 617–638.

Litasov K., Ohtani E., Langenhorst F., *et al.* (2003) Water solubility in Mg-perovskite and water storage capacity in the lower mantle. *Earth and Planetary Science Letters* **211**, 189–203.

Lithgow-Bertelloni C. and Silver P. G. (1998) Dynamic topography, plate driving forces and the African superswell. *Nature* **395**, 269–272.

Long M., Xiao X., Jiang Z., Evans B., and Karato S. (2006) Lattice preferred orientation in deformed polycrystalline (Mg, Fe)O and implications for seismic anisotropy in D″. *Physics of Earth and Planetary Interiors* **156**, 75–88.

Long M. D. and van der Hilst R. D. (2005) Upper mantle anisotropy beneath Japan from shear wave splitting. *Physics of Earth and Planetary Interiors* **151**, 206–222.

Loper D. E. and Fearn D. R. (1983) A seismic model of a partially molten inner core. *Journal of Geophysical Research* **88**, 1235–1242.

Louat N. P. and Duesbery M. S. (1994) On the theory of normal grain growth. *Philosophical Magazine A* **69**, 841–854.

Louchet F. and George A. (1983) Dislocation mobility measurements: an essential tool for understanding the atomic and electronic core structures of dislocations in semiconductors. *Journal de Physique C* **4**, 51–58.

Lu R. and Keppler H. (1997) Water solubility in pyrope to 100 kbar. *Contributions to Mineralogy and Petrology* **129**, 35–42.

Luan F. C. and Paterson M. S. (1992) Preparation and deformation of synthetic aggregates of quartz. *Journal of Geophysical Research* **97**, 301–320.

Luton M. J. and Sellars C. M. (1969) Dynamic recrystallization in nickel and nickel–iron alloys during high temperature deformation. *Acta Metallurgica* **17**, 1033–1043.

Mackwell S. J. (1991) High-temperature rheology of enstatite: implications for creep in the upper mantle. *Geophysical Research Letters* **18**, 2027–2030.

Mackwell S. J. and Kohlstedt D. L. (1990) Diffusion of hydrogen in olivine: implications for water in the mantle. *Journal of Geophysical Research* **95**, 5079–5088.

Mackwell S. J., Kohlstedt D. L., and Paterson M. S. (1985) The role of water in the deformation of olivine single crystals. *Journal of Geophysical Research* **90**, 11 319–11 333.

Mackwell S. J. and Paterson M. S. (1985) Water-related diffusion and deformation effects in quartz at pressure of 1500 and 300 MPa. In *Point Defects in Minerals* (ed. R. N. Schock), pp. 141–150. American Geophysical Union.

Mackwell S. J., Zimmerman M. E., and Kohlstedt D. L. (1998) High-temperature deformation of dry diabase with application to tectonics on Venus. *Journal of Geophysical Research* **103**, 975–984.

Mainprice D. H. and Nicolas A. (1989) Development of shape and lattice preferred orientations: application to the seismic anisotropy of the lower crust. *Journal of Structural Geology* **11**, 175–189.

Mainprice D. H. and Paterson M. S. (1984) Experimental studies on the role of water in the plasticity of quartzite. *Journal of Geophysical Research* **89**, 4257–4269.

Mainprice D. H. and Silver P. G. (1993) Interpretation of SKS-waves using samples from the subcontinental lithosphere. *Physics of Earth and Planetary Interiors* **78**, 257–280.

Malvern L. E. (1969) *Introduction to the Mechanics of a Continuous Medium*. Prentice-Hall.

Manga M. (1996) Mixing of heterogeneities in the mantle: effect of viscosity differences. *Geophysical Research Letters* **23**, 403–406.

Mao H.-K., Shu J., Shen G., *et al.* (1998) Elasticity and rheology of iron above 200 GPa and the nature of the Earth's inner core. *Nature* **396**, 741–743.

Maradudin A. A., Montroll E. W., Weiss G. H., and Ipanova I. P. (1971) *Theory of Lattice Dynamics in the Harmonic Approximation*. Academic Press.

March N. H. and Tosi M. P. (2002) *Introduction to Liquid State Physics*. World Scientific.

Marone C. (1998) Laboratory-derived friction laws and their application to seismic faulting. *Annual Review of Earth and Planetary Sciences* **26**, 643–696.

Martin R. F. and Donnay G. (1972) Hydroxyl in the mantle. *American Mineralogist* **57**, 554–570.

Mase G. E. (1970) *Continuum Mechanics*. McGraw-Hill.

Masters G. and Gubbins D. (2003) On the resolution of density within the Earth. *Physics of the Earth and Planetary Interior* **139**, 159–167.

Masters G., Laske G., Bolton H., and Dziewonski A. M. (2000) The relative behavior of shear velocity, bulk sound speed, and compressional velocity in the mantle: implications for chemical and thermal structure. In *Earth's Deep Interior* (ed. S. Karato, A. M. Forte, R. C. Liebermann, G. Masters, and L. Stixrude), pp. 63–87. American Geophysical Union.

Matsukage K. N., Nishihara Y., and Karato S. (2005) Seismological signature of chemical evolution of Earth's upper mantle. *Journal of Geophysical Research* **110**, 10.1029/2004JB003504.

Matthies S. and Wagner F. (1996) On a $1/n$ law in texture related single orientation analysis. *Physica Status Solidi* **196**, K11–K15.

Mavko G. M. (1980) Velocity and attenuation in partially molten rocks. *Journal of Geophysical Research* **85**, 5173–5189.

Mavko G. M. and Nur A. (1975) Melt squirt in the asthenosphere. *Journal of Geophysical Research* **80**, 1444–1448.

Mayama N., Suzuki I., and Saito T. (2004) Temperature dependence of elastic moduli of β-(Mg, Fe)$_2$SiO$_4$. *Geophysical Research Letters* **31**, 10.1029/2003GL019247.

McBirney A. R. and Murase T. (1984) Rheological properties of magmas. *Annual Review of Earth and Planetary Sciences* **12**, 337–357.

McCartney L. N. (1976) No time-gentlemen please! *Philosophical Magazine* **33**, 689–695.

McDonnell R. D., Peach C. J., van Roemund H. L. M., and Spiers C. J. (2000) Effect of varying enstatite content on the deformation behavior of fine-grained synthetic peridotite under wet conditions. *Journal of Geophysical Research* **105**, 13 535–13 553.

McDonough W. F. and Sun S.-S. (1995) The composition of the Earth. *Chemical Geology* **120**, 223–253.

McGovern P. J. and Schubert G. (1989) Thermal evolution of the Earth: effects of volatile exchange between atmosphere and interior. *Earth and Planetary Science Letters* **96**, 27–37.

McKenzie D. P. (1969) Speculations on the consequences and cause of plate motion. *Geophysical Journal of Royal Astronomical Society* **18**, 1–32.

McKenzie D. P. (1984) The generation and compaction of partially molten rocks. *Journal of Petrology* **25**, 713–765.

McKenzie D. P. (2003) Estimating T_e in the presence of internal loads. *Journal of Geophysical Research* **108**, 10.1029/JB001766.

McLaren A. C., Cook R. F., Hyde S. T., and Tobin R. C. (1983) The mechanisms of the formation and growth of water bubbles and associated dislocation loops in synthetic quartz. *Physics and Chemistry of Minerals* **9**, 79–94.

McLaren A. C., Fitz Gerald J. D., and Gerretsen J. (1989) Dislocation nucleation and multiplication in synthetic quartz: relevance to water weakening. *Physics and Chemistry of Minerals* **16**, 465–482.

McNamara A., Karato S., and van Keken P. E. (2001) Localization of dislocation creep in Earth's lower mantle: implications for seismic anisotropy. *Earth and Planetary Science Letters* **191**, 85–99.

McNamara A., van Keken P. E., and Karato S. (2002) Development of anisotropic structure by solid-state convection in the Earth's lower mantle. *Nature* **416**, 310–314.

McNamara A., van Keken P. E., and Karato S. (2003) Development of finite strain in the convecting lower mantle and its implications for seismic anisotropy. *Journal of Geophysical Research* **108**, 10.1029/2002JB001970, 2003.

McNutt M. K. (1998) Superswells. *Review of Geophysics* **36**, 211–244.

Meade C. and Jeanloz R. (1988) Yield strength of the B1 and B2 phases of NaCl. *Journal of Geophysical Research* **93**, 3270–3274.

Meade C. and Jeanloz R. (1990) The strength of mantle silicates at high pressures and room temperature: implications for the viscosity of the mantle. *Nature* **348**, 533–535.

Meade C., Reffner J. A., and Ito E. (1993) Synchrotron infrared absorbance measurements of hydrogen in $MgSiO_3$ perovskite. *Science* **264**, 1558–1560.

Meade C., Silver P. G., and Kaneshima S. (1995) Laboratory and seismological observations of lower mantle isotropy. *Geophysical Research Letters* **22**, 1293–1296.

Means W. D. (1976) *Stress and Strain*. Springer-Verlag.

Mecklenburgh J. and Rutter E. H. (2003) On the rheology of partially molten synthetic granite. *Journal of Structural Geology* **25**, 1575–1585.

Mehl L., Hacker B. R., and Hirth G. (2003) Arc-parallel flow within the mantle wedge: evidence from the accreted Talkeetna arc, south central Alaska. *Journal of Geophysical Research* **108**, 10.1029/2002JB002233.

Mei S., Bai W., Hiraga T., and Kohlstedt D. L. (2002) Influence of melt on the creep behavior of olivine–basalt aggregates under hydrous conditions. *Earth and Planetary Science Letters* **201**, 491–507.

Mei S. and Kohlstedt D. L. (2000a) Influence of water on plastic deformation of olivine aggregates, 1. Diffusion creep regime. *Journal of Geophysical Research* **105**, 21 457–21 469.

Mei S. and Kohlstedt D. L. (2000b) Influence of water on plastic deformation of olivine aggregates, 2. Dislocation creep regime. *Journal of Geophysical Research* **105**, 21 471–21 481.

Meike A. (1993) A critical review of investigation into transformational plasticity. In *Defects and Processes in the Solid States* (ed. J. N. Boland and J. D. Fitz Gerald), pp. 5–25. Elsevier.

Meissner R. and Mooney W. D. (1998) Weakness of lower continental crust: a condition for delamination, uplift, and escape. *Tectonophysics* **296**, 47–60.

Mendelson M. I. (1969) Average grain size in polycrystalline ceramics. *Journal of the American Ceramic Society* **55**, 19–24.

Mercier J.-C. C. (1980) Magnitude of the continental lithospheric stresses inferred from rheomorphic petrology. *Journal of Geophysical Research* **85**, 6293–6303.

Merkel S., Wenk H.-R., Badro J., *et al.* (2003) Deformation of $(Mg_{0.9}, Fe_{0.1})SiO_3$ perovskite aggregates up to 32 GPa. *Earth and Planetary Science Letters* **209**, 351–360.

Merkel S., Wenk H.-R., Gillet P., Mao H.-K., and Hemley R. J. (2004) Deformation of polycrystalline iron up to 30 GPa and 1000 K. *Physics of Earth and Planetary Interiors* **145**, 239–251.

Merkel S., Wenk H.-R., Shu J., *et al.* (2002) Deformation of polycrystalline MgO at pressures of the lower mantle. *Journal of Geophysical Research* **107**, 10.1029/2001JB000920.

Merrill R. T., McElhinny M. W., and McFaddon P. L. (1998) *The Magnetic Field of the Earth*. Academic Press.

Mibe K., Fujii T., and Yasuda A. (1998) Connectivity of aqueous fluid in the Earth's upper mantle. *Geophysical Research Letters* **25**, 1233–1236.

Minster J. B. and Anderson D. L. (1980) Dislocations and nonelastic processes in the mantle. *Journal of Geophysical Research* **85**, 6347–6352.

Minster J. B. and Anderson D. L. (1981) A model of dislocation-controlled rheology for the mantle. *Philosophical Transaction of Royal Society of London A* **299**, 319–356.

Misener D. J. (1974) Cationic diffusion in olivine to 1400 °C and 35 kbar. In *Geochemistry and Reaction Kinetics* (ed. A. W. Hofmann, B. J. Giletti, J. H. S. Yorder, and R. A. Yund), pp. 117–129. Carnegie Institution of Washington.

Mistler R. E. and Coble R. L. (1974) Grain-boundary diffusion and boundary widths in metals and ceramics. *Journal of Applied Physics* **45**, 1507–1509.

Mitrovica J. X. (1996) Haskell [1935] revisited. *Journal of Geophysical Research* **101**, 555–569.

Mitrovica J. X. and Peltier W. R. (1991a) A complete formalism for the inversion of postglacial rebound data: resolving power analysis. *Geophysical Journal International* **104**, 267–288.

Mitrovica J. X. and Peltier W. R. (1991b) Radial resolution in the inference of mantle viscosity from observations of glacial isostatic adjustment. In *Glacial Isostasy, Sea-Level and Mantle Rheology* (ed. R. Sabadini, K. Lambeck, and E. Boschi), pp. 63–78. Kluwer Academic Publisher.

Mizutani H. and Kanamori H. (1964) Variation in elastic wave velocity and attenuative property near the melting temperature. *Journal of Physics of the Earth* **12**, 43–49.

Molinari A., Canova G. R., and Ahzi S. (1987) A self-consistent approach of the large deformation polycrystal viscoplasticity. *Acta Metallurgica* **35**, 2983–2994.

Möller H.-J. (1978) The movement of dissociated dislocations in the diamond-cubic structure. *Acta Metallurgica* **26**, 963–973.

Montagner J.-P. (1998) Where can seismic anisotropy be detected in the Earth's mantle? In boundary layers …. *Pure and Applied Geophysics* **151**, 223–256.

Montagner J.-P. (2002) Upper mantle low anisotropy channels below the Pacific Plate. *Earth and Planetary Science Letters* **202**, 263–274.

Montagner J.-P., Griot-Pommera D.-A., and Lavé J. (2000) How to relate body wave and surface wave anisotropy? *Journal of Geophysical Research* **105**, 19 015–19 028.

Montagner J.-P. and Guillot L. (2000) Seismic anisotropy in the Earth's mantle. In *Problems in Geophysics for the*

New Millennium (ed. E. Boschi, G. Ekström, and A. Morelli), pp. 217–253. Editrice Compositori.

Montagner J.-P. and Guillot L. (2002) Seismic anisotropy and global geodynamics. In *Plastic Deformation of Minerals and Rocks*, Vol. 51 (ed. S. Karato and H.-R. Wenk), pp. 353–385. Mineralogical Society of America.

Montagner J.-P. and Kennett B. L. N. (1996) How to reconcile body-wave and normal-mode reference Earth models. *Geophysical Journal International* **125**, 229–248.

Montagner J.-P. and Nataf H.-C. (1986) A simple method for inverting the azimuthal anisotropy of surface waves. *Journal of Geophysical Research* **91**, 511–520.

Montagner J.-P. and Ritsema J. (2001) Interactions between ridges and plumes. *Science* **294**, 1472–1473.

Montagner J.-P. and Tanimoto T. (1990) Global anisotropy in the upper mantle inferred from the regionalization of phase velocities. *Journal of Geophysical Research* **95**, 4797–4819.

Montagner J.-P. and Tanimoto T. (1991) Global upper mantle tomography of seismic wave velocities and anisotropies. *Journal of Geophysical Research* **96**, 20337–20351.

Montelli R., Nolet G., Dahlen F. A., *et al.* (2004) Finite-frequency tomography reveals a variety of plumes in the mantle. *Science* **303**, 338–343.

Montési L. and Hirth G. (2003) Grain size evolution and the rheology of ductile shear zone: from laboratory experiments to postseismic creep. *Earth and Planetary Science Letters* **211**, 97–110.

Montési L. and Zuber M. T. (2002) A unified description of localization for application to large-scale tectonics. *Journal of Geophysical Research* **107**, 1/1–1/21.

Morelli A., Dziewonski A. M., and Woodhouse J. H. (1986) Anisotropy of the inner core inferred from PKIKP travel times. *Geophysical Research Letters* **13**, 1545–1548.

Mosenfelder J. L., Connelly J. A. D., Rubie D. C., and Liu M. (2000) Strength of $(Mg,Fe)_2SiO_4$ wadsleyite determined by relaxation of transformational stress. *Physics of Earth and Planetary Interiors* **120**, 63–78.

Mott N. F. and Littleton M. J. (1938) Conduction in polar crystals: I. Electrolytic conduction in solid salts. *Transactions of Faraday Society* **34**, 485–491.

Mukherjee A. K. (1971) The rate controlling mechanism in superplasticity. *Materials Science and Engineering* **8**, 83–89.

Murakami M., Hirose K., Kawamura K., Sata N., and Ohnishi Y. (2004) Post-perovskite phase transition in $MgSiO_3$. *Science* **304**, 855–858.

Murakami M., Hirose K., Yurimoto H., Nakashima S., and Takafuji N. (2002) Water in Earth's lower mantle. *Science* **295**, 1885–1887.

Nabarro F. R. N. (1948) Deformation of crystals by the motion of single ions. *Report of a Conference on Strength of Solids*, 75–90.

Nabarro F. R. N. (1967a) Steady state diffusional creep. *Philosophical Magazine* **16**, 231–237.

Nabarro F. R. N. (1967b) *Theory of Crystal Dislocations*. Oxford University Press.

Nabarro F. R. N. (1989) The mechanism of Harper-Dorn creep. *Acta Metallurgica* **37**, 2217–2222.

Nakada M. (1986) Holocene sea levels in oceanic islands: implications for the rheological structure of the Earth's mantle. *Tectonophysics* **121**, 263–276.

Nakada M. and Lambeck K. (1987) Glacial rebound and relative sea level variations: a new appraisal. *Geophysical Journal of Royal Astronomical Society* **90**, 171–224.

Nakada M. and Lambeck K. (1989) Late Pleistocene and Holocene sea-level change in the Australian region and mantle rheology. *Geophysical Journal International* **96**, 497–517.

Nakada M. and Lambeck K. (1991) Late Pleistocene and Holocene sea-level change: evidence for lateral mantle viscosity variation? In *Glacial Isostasy, Sea Level and Mantle Rheology* (ed. R. Sabadini, K. Lambeck, and E. Boschi), pp. 79–94. Kluwer Academic.

Nakajima J. and Hasegawa A. (2004) Shear-wave polarization anisotropy and subduction-induced flow in the mantle wedge of northern Japan. *Earth and Planetary Science Letters* **225**, 365–377.

Nakashima S. (1995) Diffusivity of ions in pore water as a quantitative basis for rock deformation rate estimate. *Tectonophysics* **245**, 185–203.

Nakashima S., De Meer S., and Spiers C. J. (2004) Distribution of thin film water in grain boundaries of crustal rocks and implications for crustal strength. In *Physicochemistry of Water in Geological and Biological Systems* (ed. S. Nakashima, C. J. Spiers, L. Mercury, P. A. Fenter, and M. F. M. F. Hochella, Jr.), pp. 159–178. Universal Academy Press, Inc.

Nakatani M. (2001) Conceptual and physical clarification of rate and state friction: frictional sliding and thermally activated rheology. *Journal of Geophysical Research* **106**, 13 347–13 380.

Nataf H.-C., Nakanishi I., and Anderson D. L. (1984) Anisotropy and shear wave heterogeneities in the upper mantle. *Geophysical Research Letters* **11**, 109–112.

Nataf H.-C., Nakanishi I., and Anderson D. L. (1986) Measurement of mantle wave velocities and inversion for lateral heterogeneities and anisotropy, 3. Inversion. *Journal of Geophysical Research* **91**, 7261–7307.

Navrotsky A. (1994) *Physics and Chemistry of Earth Materials*. Cambridge University Press.

Nes E., Hirsch J., and Lücke K. (1984) On the origin of the cube recrystallization texture in directionally solidified aluminium. *Seventh International Conference on Texture of Materials*, 663–674.

Newman J., Lamb W. M., Drury M. R., and Vissers R. L. M. (1999) Deformation processes in a peridotite shear zone: reaction-softening by a H_2O-deficit, continuous net transfer reaction. *Tectonophysics* **303**, 193–222.

Nicolas A. (1978) Stress estimates from structural studies in some mantle peridotites. *Philosophical Transactions of the Royal Society of London A* **288**, 49–57.

Nicolas A. (1993) Why fast polarization direction of SKS seismic waves are parallel to mountain belts. *Physics of Earth and Planetary Interiors* **78**, 337–342.

Nicolas A. and Christensen N. I. (1987) Formation of anisotropy in upper mantle peridotite: a review. In *Composition, Structure and Dynamics of the Lithosphere–Asthenosphere System* (ed. K. Fuchs and C. Foridevaux), pp. 111–123. American Geophysical Union.

Nieh T. G., Wadsworth J., and Sherby O. D. (1997) *Superplasticity in Metals and Ceramics.* Cambridge University Press.

Nimmo F., Price G. D., Brodholt J. P., and Gubbins D. (2004) The influence of potassium on core and geodynamo evolution. *Geophysical Journal International* **156**, 363–376.

Nishihara Y., Shinmei T., and Karato S. (2006) Grain-growth kinetics in wadsleyite: effects of chemical environment. *Physics of Earth and Planetary Interiors* **154**, 30–43.

Nishihara Y., Shinmei T., and Karato S. (2007a) Effects of chemical environments on the hydrogen-defects in wadsleyite. *American Mineralogist* in press.

Nishihara Y., Tinker D., Xu Y., *et al.* (2007b) Plastic deformation of wadsleyite and olivine at high-pressures and high-temperatures using a rotational Drickamer apparatus (RDA). *Physics of the Earth and Planetary Interiors* submitted.

Nitsan U. (1974) Stability field of olivine with respect to oxidation and reduction. *Journal of Geophysical Research* **79**, 706–711.

Nixon P. H. and Boyd F. R. (1973) Petrogenesis of the granular and sheared ultrabasic nodule site in kimberlites. In *Lesotho Kimberlites* (ed. P. H. Nixon), pp. 48–56. Lesotho National Development.

Nolet G. (1987a) *Seismic Tomography*. Reidel Publishing Company.

Nolet G. (1987b) Waveform tomography. In *Seismic Tomography* (ed. G. Nolet), pp. 301–322. Reidel Publishing Company.

Nolet G. (2000) Interpreting seismic waveforms: forward and inverse problems for heterogeneous media. In *Problems in Geophysics for the New Millennium* (ed. E. Boschi, G. Ekström, and A. Morelli), pp. 373–401. Edrice Compositori.

Nolet G. and Dahlen F. A. (2000) Wave front healing and the evolution of seismic delay times. *Journal of Geophysical Research* **105**, 19 043–19 054.

Nolet G. and Zielhuis A. (1994) Low S velocities under the Tornquist–Teisseyre zone: evidence for water injection into the transition zone by subduction. *Journal of Geophysical Research* **99**, 15 813–15 820.

Nowick A. S. and Berry B. S. (1972) *Anelastic Relaxation in Crystalline Solids*. Academic Press.

Nye J. F. and Mae S. (1972) The effect of non-hydrostatic stress on intergranular water veins and lenses in ice. *Journal of Glaciology* **11**, 81–101.

O'Connell R. J. (1977) On the scale of mantle convection. *Tectonophysics* **38**, 119–136.

O'Connell R. J. and Budianski B. (1974) Seismic velocities in dry and saturated cracked solids. *Journal of Geophysical Research* **79**, 5412–5426.

O'Connell R. J. and Budianski B. (1977) Viscoelastic properties of fluid-saturated cracked solids. *Journal of Geophysical Research* **82**, 5719–5735.

O'Neill H. S. C., McCammon C. A., Canil D., *et al.* (1993) Mössbauer spectroscopy of mantle transition zone phases and determination of minimum Fe^{3+} content. *American Mineralogist* **78**, 456–460.

Obata M. and Karato S. (1995) Ultramafic pseudotachylyte from Balmuccia peridotite, Ivrea–Verbana zone, northern Italy. *Tectonophysics* **242**, 313–328.

Oganov A. R., Martonak R., Laio A., Raiteri P., and Parrinello M. (2005) Anisotropy of Earth's D″ layer and stacking faults in the $MgSiO_3$ post-perovskite phase. *Nature* **438**, 1142–1144.

Oganov A. R. and Ono S. (2004) Theoretical and experimental evidence for a post-perovskite phase of $MgSiO_3$ in Earth's D″ layer. *Nature* **430**, 445–448.

Ogawa M. (1987) Shear instability in a viscoelastic material as the cause for deep earthquakes. *Journal of Geophysical Research* **92**, 13 801–13 810.

Ohtani E. (1988) Chemical stratification of the mantle formed by melting in the early stage of the terrestrial evolution. *Tectonophysics* **154**, 201–210.

Ohtani E., Mizobata H., and Yurimoto H. (2000) Stability of dense hydrous magnesium silicate phases in the system Mg_2SiO_4–H_2O and $MgSiO_3$–H_2O at pressures up to 27 GPa. *Physics and Chemistry of Minerals* **27**, 533–544.

Ohuchi T. and Nakamura M. (2006) Grain growth in the forsterite–diopside system. *Physics of the Earth and Planetary Interiors* **160**, 1–21.

Oki S. (2006) *Whole mantle Vp/Vs tomography*, University of Tokyo.

Omori S., Kamiya S., Maruyama S., and Zhao D. (2002) Morphology of the intraslab seismic zone and devolatilization phase equilibria of the subducting slab peridotite. *Bulletin of Earthquake Research Institute* **76**, 455–478.

Orowan E. (1934) Zur Kristallplastizität. *Zeitschrift für Phisik* **89**, 605–659.

Ozawa K. (1989) Stress induced Al-Cr zoning of spinel in deformed peridotite. *Nature* **338**, 141–144.

Panasyuk S. V. and Hager B. H. (1998) A model of transformational superplasticity in the upper mantle. *Geophysical Journal International* **133**, 741–755.

Panning M. and Romanowicz B. (2004) Inferences on flow at the base of the Earth's mantle based on seismic anisotropy. *Science* **303**, 351–353.

Park J. and Levin V. (2002) Seismic anisotropy: tracing plate dynamics in the mantle. *Science* **296**, 485–489.

Park K. T. and Mohamed F. A. (1995) Creep strengthening in a discontinuous SiC–Al composite. *Metallurgical Materials Transaction A* **26**, 3119–3129.

Parmentier E. M. (1981) A possible mantle instability due to superplastic deformation associated with phase transitions. *Geophysical Research Letters* **8**, 143–146.

Pasteris J. D. (1984) Kimberlites: complex mantle melts. *Annual Review of Earth and Planetary Sciences* **12**, 133–153.

Paterson M. S. (1970) A high temperature high pressure apparatus for rock deformation. *International Journal of Rock Mechanics and Mining Sciences* **7**, 517–526.

Paterson M. S. (1973) Non-hydrostatic thermodynamics and its geologic applications. *Review of Geophysics and Space Physics* **11**, 355–389.

Paterson M. S. (1982) The determination of hydroxyl by infrared absorption in quartz, silicate glass and similar materials. *Bulletin Mineralogie* **105**, 20–29.

Paterson M. S. (1983) Creep of transforming materials. *Mechanics of Materials* **2**, 103–109.

Paterson M. S. (1989) The interaction of water with quartz and its influence in dislocation flow – an overview. In *Rheology of Solids and of the Earth* (ed. S. Karato and M. Toriumi), pp. 107–142. Oxford University Press.

Paterson M. S. (1990) Rock deformation experimentation. In *The Brittle–Ductile Transition in Rocks: the Heard Volume* (ed. A. G. Duba, W. B. Durham, J. W. Handin, and H. F. Wang), pp. 187–194. American Geophysical Union.

Paterson M. S. and Kekulawala K. R. S. S. (1979) The role of water in quartz deformation. *Bulletin Mineralogie* **102**, 92–98.

Paterson M. S. and Olgaard D. L. (2000) Rock deformation tests to large shear strains in torsion. *Journal of Structural Geology* **22**, 1341–1358.

Paterson M. S. and Weiss L. E. (1961) Symmetry concepts in the structural analysis of deformed rocks. *Geological Society of America Bulletin* **72**, 841–882.

Paterson M. S. and Wong T.-F. (2005) *Experimental Rock Deformation – The Brittle Field.* Springer-Verlag.

Pauling L. (1960) *The Nature of the Chemical Bonds.* Cornell University Press.

Pearson D. G. (1999) Evolution of cratonic lithospheric mantle: an isotopic perspective. In *Mantle petrology: Field Observations and High Pressure Experimentation* (ed. Y. Fei, C. M. Bertka, and B. O. Mysen), pp. 57–78. The Geochemical Society.

Peltier W. R. (1984) The thickness of the continental lithosphere. *Journal of Geophysical Research* **89**, 11303–11316.

Peltier W. R. (1985a) New constraints on transient lower mantle rheology and internal mantle buoyancy from glacial rebound data. *Nature* **318**, 614–617.

Peltier W. R. (1985b) The LAGEOS constraint on deep mantle viscosity: results from a new normal mode method for the inversion of viscoelastic spectra. *Journal of Geophysical Research* **90**, 9411–9421.

Peltier W. R. (1989) Mantle viscosity. In *Mantle Convection* (ed. W. R. Peltier), pp. 389–478. Gordon & Breach.

Peltier W. R. (1998) Postglacial variation in the level of the sea: implications for climate dynamics and solid-Earth geophysics. *Review of Geophysics* **36**, 603–689.

Pharr G. M. and Ashby M. F. (1983) On creep enhanced by a liquid phase. *Acta Metallurgica* **31**, 129–138.

Phipps Morgan J. and Shearer P. M. (1993) Seismic constraints on mantle flow and topography of the 660-km discontinuity. *Nature* **365**, 506–511.

Pieri M., Kunze K., Burlini L., Stretton I., Olgaard D. L., Burg J.-P., and Wenk H.-R. (2001) Texture development of calcite by deformation and dynamic recrystallization at 1000 K during torsion experiments of marble to large strains. *Tectonophysics* **330**, 119–142.

Pitzer K. S. and Sterner S. M. (1994) Equations of state valid continuously from zero to extreme pressures for H_2O and CO_2. *Journal of Chemical Physics* **101**, 3111–3116.

Plank T. and Langmuir A. H. (1992) Effects of melting regime on the composition of the oceanic crust. *Journal of Geophysical Research* **97**, 19 749–19 770.

Poirier J.-P. (1976a) On the symmetrical role of cross-slip of screw dislocations and climb of edge dislocations as recovery processes controlling high-temperature creep. *Revue de Physique Appliquée* **11**, 731–738.

Poirier J.-P. (1976b) *Plasticité a Haute Température des Solides Cristallins.* Editions Eyrolles.

Poirier J.-P. (1980) Shear localization and shear instability in materials in the ductile field. *Journal of Structural Geology* **2**, 135–142.

Poirier J.-P. (1982) On transformation plasticity. *Journal of Geophysical Research* **87**, 6791–6797.

Poirier J.-P. (1985) *Creep of Crystals.* Cambridge University Press.

Poirier J.-P. (1988) Transport properties of liquid metals and viscosity of the Earth's core. *Geophysical Journal of Royal Astronomical Society* **92**, 99–105.

Poirier J.-P. (1994) Light elements in the Earth's outer core: a critical review. *Physics of Earth and Planetary Interiors* **85**, 319–337.

Poirier J.-P. (2000) *Introduction to the Physics of Earth's Interior.* Cambridge University Press.

Poirier J.-P. and Guillopé M. (1979) Deformation induced recrystallization of minerals. *Bulletin Mineralogie* **102**, 67–74.

Poirier J.-P. and Liebermann R. C. (1984) On the activation volume for creep and its variation with depth in the Earth's lower mantle. *Physics of Earth and Planetary Interiors* **35**, 283–293.

Poirier J.-P., Peyronneau J., Gesland J. Y., and Brebec G. (1983) Viscosity and conductivity of the lower mantle: and experimental study on a $MgSiO_3$ analogue, $KZnF_3$. *Physics of Earth and Planetary Interiors* **32**, 273–287.

Poirier J.-P., Peyronneau J., Madon M., Guyot F., and Revcoleshi A. (1986) Eutectoid phase transformation of olivine and spinel into perovskite and rock salt structures. *Nature* **321**, 603–605.

Poirier J.-P., Sotin C., and Peyronneau J. (1981) Viscosity of high-pressure ice VI and evolution and dynamics of Ganymede. *Nature* **292**, 225–227.

Poirier J.-P. and Vergobbi B. (1978) Splitting of dislocations in olivine, cross-slip controlled creep and mantle rheology. *Physics of Earth and Planetary Interiors* **16**, 370–378.

Polanyi M. (1934) Über eine Art Gitterstörung, die einen Kristall plastisch machen könnte. *Zeitschrift für Physik* **89**, 660–664.

Pollack H. K. (1986) Cratonization and the thermal evolution of the mantle. *Earth and Planetary Science Letters* **80**, 175–182.

Pollack H. K., Hurter S. J., and Johnson J. R. (1993) Heat flow from the Earth's interior: analysis of the global data. *Review of Geophysics* **31**, 267–280.

Ponte Castañeda P. and Willis J. R. (1988) On the overall properties of nonlinearly viscous composites. *Proceedings of the Royal Society of London A* **416**, 217–244.

Post A. and Tullis J. A. (1999) A recrystallized grain size piezometer for experimentally deformed feldspar. *Tectonophysics* **303**, 159–173.

Post A. D., Tullis J., and Yund R. A. (1996) Effects of chemical environment on dislocation creep of quartzite. *Journal of Geophysical Research* **101**, 22 143–22 155.

Post R. L. (1977) High-temperature creep of Mt. Burnett dunite. *Tectonophysics* **42**, 75–110.

Prigogine I. and Defay R. (1950) *Thermodynamique Chemique*. Editions Desoer.

Prior D. and Wheeler J. (1999) Feldspar fabrics in a greenshist facies albite-rich mylonite from electron backscatter diffraction. *Tectonophysics* **303**, 29–49.

Przystupa M. A. and Ardell A. J. (2002) Predictive capabilities of the dislocation-network theory of Harper–Dorn creep. *Metallurgical and Materials Transactions A* **33**, 231–239.

Raitt R. W., Shor G. G., Francis T. J. G., and Morris G. B. (1969) Anisotropy of the Pacific upper mantle. *Journal of Geophysical Research* **74**, 3095–3109.

Raj R. (1982) Separation of cavitation-strain and creep strain during deformation. *Journal of the American Ceramic Society* **65**, 46–48.

Raj R. (1986) Unstable spreading of a film inclusion in a grain boundary under normal stress. *Journal of the American Ceramic Society* **69**, 708–712.

Raj R. and Ashby M. F. (1971) On grain boundary sliding and diffusional creep. *Metallurgical Transactions* **2**, 1113–1127.

Raj R. and Chung C. K. (1981) Solution-precipitation creep in glass ceramics. *Acta Metallurgica* **29**, 159–166.

Raleigh C. B., Kirby S. H., Carter N. L., and Avé Lallemant H. G. (1971) Slip and the clinoenstatite transformation as competing processes in enstatite. *Journal of Geophysical Research* **76**, 4011–4022.

Raleigh C. B. and Paterson M. S. (1965) Experimental deformation of serpentine and its tectonic implications. *Journal of Geophysical Research* **70**, 3965–3985.

Randle V. (2003) *Microtexture Determination and its Applications*. The Institute of Materials, Minerals and Mining.

Rauch M. and Keppler H. (2002) Water solubility in orthopyroxene. *Contributions to Mineralogy and Petrology* **143**, 525–536.

Regan J. and Anderson D. L. (1984) Anisotropic models of the upper mantle. *Physics of Earth and Planetary Interiors* **35**, 227–263.

Regenauer-Lieb K. and Yuen D. A. (2003) Modeling shear zones in geological and planetary sciences: solid- and fluid-mechanical approaches. *Review of Earth Sciences* **63**, 295–349.

Renner J., Stöckhert B., Zerbian A., Roller K., and Rummel F. (2001) An experimental study into the rheology of synthetic polycrystalline coesite aggregates. *Journal of Geophysical Research* **106**, 19 411–19 429.

Reppich B., Haasen P., and Ilschner B. (1964) Kriechen von Silizium-Einkristallen. *Acta Metallurgica* **12**, 1283–1288.

Revenaugh J. and Jordan T. H. (1991) Mantle layering from ScS reverberations, 3. Upper mantle. *Journal of Geophysical Research* **96**, 19 781–19 810.

Revenaugh J. and Sipkin S. A. (1994) Seismic evidence for silicate melt atop the 410-km mantle discontinuity. *Nature* **369**, 474–476.

Ribe N. M. (1989a) A continuum theory for lattice preferred orientation. *Geophysical Journal* **97**, 199–207.

Ribe N. M. (1989b) Seismic anisotropy and mantle flow. *Journal of Geophysical Research* **94**, 4213–4223.

Ribe N. M. and Yu Y. (1991) A theory of plastic deformation and textural evolution of olivine polycrystals. *Journal of Geophysical Research* **96**, 8325–8335.

Ricard Y., Bercovici D., and Schubert G. (2001) A two-phase model for compaction and damage 2. Applications to compaction, deformation, and the role of interfacial tension. *Journal of Geophysical Research* **106**, 8907–8924.

Rice J.R. (1976) The localization of plastic deformation. In *Theoretical and Applied Mechanics* (ed. W.T. Koitier), pp. 207–220. North-Holland.

Richard G., Monnereau M., and Ingrin J. (2002) Is the transition zone an empty water reservoir? Inference from numerical model of mantle dynamics. *Earth and Planetary Science Letters* **205**, 37–51.

Richards M.A. and Hager B.H. (1984) Geoid anomalies in the dynamic Earth. *Journal of Geophysical Research* **89**, 5987–6002.

Richards M.A., Yang W.S., Baumgardnner J.R., and Bunge H.-P. (2001) Role of a low-viscosity zone in stabilizing plate tectonics: implications for comparative planetology. *Geochemistry, Geophysics, Geosystems* **2**, 2000GC000115.

Ricoult D.L. and Kohlstedt D.L. (1983) Structural width of low-angle grain boundaries in olivine. *Physics and Chemistry of Minerals* **9**, 133–138.

Riedel M.R. and Karato S. (1996) Microstructural development during nucleation and growth. *Geophysical Journal International* **125**, 397–414.

Riedel M.R. and Karato S. (1997) Grain-size evolution in subducted oceanic lithosphere associated with the olivine-spinel transformation and its effects on rheology. *Earth and Planetary Science Letters* **148**, 27–43.

Ringwood A.E. (1975) *Composition and Structure of the Earth's Mantle*. McGraw-Hill.

Ritsema J., van Heijst H.J., and Woodhouse J.H. (1999) Complex shear wave velocity structure imaged beneath Africa and Iceland. *Science* **286**, 1925–1928.

Rogers H.C. (1979) Adiabatic plastic deformation. *Annual Review of Materials Science* **9**, 283–311.

Rokosky J.M., Lay T., Garnero E.J., and Russell S.A. (2004) High-resolution investigation of shear wave anisotropy in D″ beneath the Cocos Plate. *Geophysical Research Letters* **31**, 10.1029/2003GL018902.

Romanowicz B. (1994) Anelastic tomography: a new perspective on upper mantle thermal structure. *Earth and Planetary Science Letters* **128**, 113–121.

Romanowicz B. (1995) A global tomographic model of shear attenuation in the upper mantle. *Journal of Geophysical Research* **100**, 12 375–12 394.

Romanowicz B. (2003) Global mantle tomography: progress status in the past 10 years. *Annual Review of Earth and Planetary Sciences* **31**, 303–328.

Romanowicz B. and Durek J.J. (2000) Seismological constraints on attenuation in the Earth: a review. In *Earth's Deep Interior* (ed. S. Karato, A.M. Forte, R.C. Liebermann, G. Masters, and L. Stixrude), pp. 161–179. American Geophysical Union.

Roscoe R. (1952) The viscosity of suspensions of rigid spheres. *British Journal of Applied Physics* **3**, 267–269.

Ross J.V., Avé Lallemant H.G., and Carter N.L. (1979) Activation volume for creep in the upper mantle. *Science* **203**, 261–263.

Ross J.V., Avé Lallemant H.G., and Carter N.L. (1980) Stress dependence of recrystallized grain and subgrain size in olivine. *Tectonophysics* **70**, 39–61.

Ross J.V., Bauer S.J., and Hansen F.D. (1987) Textural evolution of synthetic anhydrite–halite mylonites. *Tectonophysics* **140**, 307–326.

Ross J.V. and Nielsen K.C. (1978) High-temperature flow of wet polycrystalline enstatite. *Tectonophysics* **44**, 233–261.

Rossman G.R. and Aines R.D. (1991) The hydrous components in garnets: grossular-hydrogrossular. *American Mineralogist* **76**, 1153–1164.

Rossman G.R., Beran A., and Lange M.A. (1989) The hydrous component of pyrope from the Dora Maira Massif, western Alps. *European Journal of Mineralogy* **1**, 151–154.

Roth E.G., Wiens D.A., Dorman L.M., Hildebrand J., and Webb S.C. (1999) Seismic attenuation tomography of the Toga-Fiji region using phase pair methods. *Journal of Geophysical Research* **104**, 4795–4809.

Royden L.H., Burchfiel B.C., King R.W., Chen Z., Shen F., and Liu Y. (1997) Surface deformation and lower crust flow in eastern Tibet. *Science* **276**, 788–790.

Ruano O.A., Wadsworth J., Wolfensteine J., and Sherby O.D. (1993) Evidence for Nabarro–Herring creep in metals: fiction or reality? *Materials Science and Engineering A* **165**, 133–141.

Rubie D.C. (1983) Reaction-enhanced ductility: the role of solid–solid univariant reactions in deformation of the crust and mantle. *Tectonophysics* **96**, 331–352.

Rubie D.C. (1984) The olivine -> spinel transformation and the rheology of subducting lithosphere. *Nature* **308**, 505–508.

Rubie D.C., Karato S., Yan H., and O'Neill H.S.C. (1993) Low differential stress and controlled chemical environment in multianvil high-pressure experiments. *Physics and Chemistry of Minerals* **20**, 315–322.

Rubie D.C. and Ross C.R., II. (1994) Kinetics of the olivine–spinel transformation in subducting lithosphere: experimental constraints and implications for deep slab processes. *Physics of Earth and Planetary Interiors* **86**, 223–241.

Rudnick R.L. and Fountain D.M. (1995) Nature and composition of the continental lower crust: the lower crustal perspective. *Review of Geophysics and Space Physics* **33**, 267–309.

Rudnick R.L., McDonough W.F., and O'Connell R.J. (1998) Thermal structure, thickness and comopsition of continental lithosphere. *Chemical Geology* **145**, 395–411.

Rudnicki J.W. and Rice J.R. (1975) Conditions for localization of deformation in pressure-sensitive dilatant materials. *Journal of Mechanics and Physics of Solids* **23**, 371–394.

Ruina A. (1983) Slip instability and state variable friction laws. *Journal of Geophysical Research* **88**, 10 359–10 370.

Ruoff A. L. (1965) Mass transfer problems in ionic crystals with charge neutrality. *Journal of Applied Physics* **36**, 2903–2907.

Rüpke L. H., Phipps Morgan J., Hort M., and Connolly J. A. D. (2004) Serpentine and the subduction zone water cycle. *Earth and Planetary Science Letters* **223**, 17–34.

Russo R. and Silver P. G. (1994) Trench-parallel flow beneath the Nazca plate from seismic anisotropy. *Science* **263**, 1105–1111.

Rutter E. H. (1972) The influence of interstitial water on the rheological behaviour of calcite rocks. *Tectonophysics* **14**, 13–33.

Rutter E. H. (1976) The kinetics of rock deformation by pressure solution. *Philosophical Transactions of the Royal Society of London A* **283**, 203–219.

Rutter E. H. (1983) Pressure solution in nature, theory and experiment. *Journal of the Geological Society of London* **140**, 725–740.

Rutter E. H. (1986) On the nomenclature of failure transitions in rocks. *Tectonophysics* **122**, 381–387.

Rutter E. H. (1995) Experimental study of the influence of stress, temperature, and strain on the dynamic recrystallization of Carrara marble. *Journal of Geophysical Research* **100**, 24 651–24 663.

Rutter E. H. (1998) Use of extension testing to investigate the influence of finite strain on the rheological behaviour of marble. *Journal of Structural Geology* **20**, 243–254.

Rutter E. H. and Brodie K. (1988) The role of tectonic grainsize reduction in the rheological stratification of the lithosphere. *Geologische Rundschau* **77**, 295–308.

Rutter E. H. and Brodie K. H. (1992) Rheology of the lower crust. In *Continental Lower Crust* (ed. D. M. Fountain, R. Arculus, and R. W. Key), pp. 201–267. Elsevier.

Rutter E. H. and Brodie K. H. (2004) Experimental grain size-sensitive flow of hot-pressed Brasilian quartz aggregates. *Journal of Structural Geology* **26**, 2011–2023.

Rutter E. H., Casey M., and Burlini L. (1994) Preferred crystallographic orientation development during the plastic and superplastic flow of calcite rocks. *Journal of Structural Geology* **16**, 1431–1446.

Rybacki E. and Dresen G. (2000) Dislocation and diffusion creep of synthetic anorthite aggregates. *Journal of Geophysical Research* **105**, 26 017–26 036.

Rybacki E. and Dresen G. (2004) Deformation mechanism maps for feldspar rocks. *Tectonophysics* **382**, 173–187.

Rybacki E., Gootschalk M., Wirth R., and Dresen G. (2006) Influence of water fugacity and activation volume on the flow properties of fine-grained anorthite aggregates. *Journal of Geophysical Research* **111**, 10.1029/2005JB003663.

Rybacki E., Paterson M. S., Wirth R., and Dreibus G. (2003) Rheology of calcite–quartz aggregates deformed to large strain in torsion. *Journal of Geophysical Research* **108**, 10.1029/2002JB001833.

Rychert C. A., Fischer K. M., and Rodenay S. (2005) A sharp lithosphere–asthenosphere boundary imaged beneath eastern North America. *Nature* **434**, 542–545.

Ryerson F. J., Weed H. C., and Piwinskii A. J. (1988) Rheology of subliquidus magmas 1. Picritic compositions. *Journal of Geophysical Research* **93**, 3421–3436.

Sabadini R., Smith B. K., and Yuen D. A. (1987) Consequences of experimental transient rheology. *Geophysical Research Letters* **14**, 816–819.

Sakai T. and Jonas J. J. (1984) Dynamic recrystallization: mechanical and microstructural considerations. *Acta Metallurgica* **32**, 189–209.

Saltzer R. L., Gaherty J. B., and Jordan T. H. (2000) How are vertical shear wave splitting measurements affected by variations in the orientation of azimuthal anisotropy with depth? *Geophysical Journal International* **141**, 374–390.

Saltzer R. L., van der Hilst R. D., and Karason H. (2001) Comparing P and S wave heterogeneity in the mantle. *Geophysical Research Letters* **28**, 1335–1338.

Sammis C. G. and Dein J. L. (1974) On the possibility of transformational superplasticity in the Earth's mantle. *Journal of Geophysical Research* **79**, 2961–2965.

Sammis C. G., Smith J. C., and Schubert G. (1981) A critical assessment of estimation methods for activation volume. *Journal of Geophysical Research* **86**, 10 707–10 718.

Sammis C. G., Smith J. C., Schubert G., and Yuen D. A. (1977) Viscosity depth profile of the Earth's mantle: effect of polymorphic transitions. *Journal of Geophysical Research* **82**, 3747–3761.

Sandström R. (1977) Subgrain growth occurring by boundary migration. *Acta Metallurgica* **25**, 905–911.

Sato H., Sacks I. S., Murase T., Munchill G., and Fukuyama H. (1989) Q_p-melting temperature relation in peridotite at high pressure and temperature: attenuation mechanism and implications for the mechanical properties of the upper mantle. *Journal of Geophysical Research* **94**, 10 647–10 661.

Sato M. (1971) Electrochemical measurements and control of oxygen fugacity and other gaseous fugacities with solid electrolyte sensors. In *Research Techniques for High Pressure and High Temperature* (ed. G. C. Ulmer), pp. 43–99. Springer-Verlag.

Savage M. K. (1999) Seismic anisotropy and mantle deformation: what have we learned from shear wave splitting? *Review of Geophysics* **37**, 65–106.

Saxena S. K., Dubrovinski L. S., and Lazor P. (1996) Stability of perovskite ($MgSiO_3$) in the Earth's lower mantle. *Science* **274**, 1357–1359.

Schmalzried H. (1995) *Chemical Kinetics of Solids*. VCH.

Schmeling H. (1985) Numerical models on the influence of partial melt on elastic, anelastic and electric properties of rocks. Part I: elasticity and anelasticity. *Physics of Earth and Planetary Interiors* **41**, 34–57.

Schmeling H. (1987) On the interaction between small- and large-scale convection and postglacial rebound flow in a power-law mantle. *Earth and Planetary Science Letters* **84**, 254–262.

Schmid S. M. and Casey M. (1986) Complete fabric analysis of some commonly observed quartz c-axis patterns. In *Mineral and Rock Deformation: Laboratory Studies, The Paterson Volume* (ed. B. E. Hobbs and H. C. Heard), pp. 263–286. American Geophysical Union.

Schmidt C., Bruhn D., and Wirth R. (2003) Experimental evidence of transformation plasticity in silicates: mimimum of creep strength in quartz. *Earth and Planetary Science Letters* **205**, 273–280.

Scholz C. H. (2002) *The Mechanics of Earthquake and Faulting*. Cambridge University Press.

Schubert G., Turcotte D. L., and Olson P. (2001) *Mantle Convection in the Earth and Planets*. Cambridge University Press.

Schutt D. L. and Lesher C. E. (2006) Effects of melt depletion on the density and seismic velocity of garnet and spinel lherzolite. *Journal of Geophysical Research* **111**, 10.1029/2003JB002950.

Sclater J. C., Parsons B., and Jaupart C. (1981) Oceans and continents: similarities and differences in the mechanisms of heat loss. *Journal of Geophysical Research* **86**, 11535–11552.

Sclater J. G., Jaupart C., and Galson D. (1980) The heat flow through the oceanic and continental crust and the heat loss of the earth. *Review of Geophysics and Space Physics* **18**, 269–312.

Scott D. R. and Stevenson D. J. (1984) Magma solitons. *Geophysical Research Letters* **11**, 1161–1164.

Scott T. and Kohlstedt D. L. (2006) The effect of large melt fraction on the deformation behavior of peridotite. *Earth and Planetary Science Letters* **246**, 177–187.

Secco R. A. (1995) Viscosity of the outer core. In *Mineral Physics & Crystallography* (ed. T. H. Ahrens), pp. 218–226. American Geophysical Union.

Seeger A. and Schiller P. (1966) Kinks in dislocation lines and their effects on the internal friction in crystals. In *Physical Acoustics*, Vol. III – Part A (ed. W. P. Mason), pp. 361–495. Academic Press.

Selitser S. I. and Morris J. W., Jr. (1994) Substructure formation during plastic deformation. *Acta Metallurgica et Materials* **42**, 3985–3991.

Shankland T. J. (1977) Elastic properties, chemical composition, and crystal structures of minerals. *Geophysical Survey* **3**, 69–100.

Shankland T. J., O'Connell R. J., and Waff H. S. (1981) Geophysical constraints on partial melt in the upper mantle. *Review of Geophysics and Space Physics* **19**, 394–406.

Sharp T. G., Bussod G. Y., and Katsura T. (1994) Misrostructures in beta-$Mg_{1.8}Fe_{0.2}SiO_4$ experimentally deformed at transition-zone conditions. *Physics of Earth and Planetary Interiors* **86**, 69–83.

Shearer P. M. (2000) Upper mantle discontinuities. In *Earth's Deep Interior: Mineral Physics and Tomography from the Atomic to the Global Scales* (ed. S. Karato, A. M. Forte, R. C. Liebermann, G. Masters, and L. Stixrude), pp. 115–131. American Geophysical Union.

Shearer P. M. and Masters G. (1992) Global mapping of topography on the 660-km discontinuity. *Nature* **355**, 791–796.

Shen G., Mao H.-K., Hemley R. J., and Duffy T. S. (1998) Melting and crystal structure of iron at high pressures and temperatures. *Geophysical Research Letters* **25**, 373–376.

Shen Y. and Blum J. (2003) Seismic evidence for accumulated oceanic crust above the 660-km discontinuity beneath southern Africa. *Geophysical Research Letters* **30**, 10.1029/2003GL017991.

Sherby O. D., Klundt R. H., and Miller A. L. (1977) Flow-stress, subgrain size and subgrain stability at elevated temperatures. *Metallurgical Transactions A* **8**, 843–850.

Sherby O. D., Robbins J. L., and Goldberg A. (1970) Calculation of activation volumes for self-diffusion and creep at high temperatures. *Journal of Applied Physics* **41**, 3961–3968.

Shewmon P. G. (1989) *Diffusion in Solids*. The Minerals, Metals & Materials Society.

Shieh S. R., Duffy T. S., and Li B. (2002) Strength and elasticity of SiO_2 across the stishovite–$CaCl_2$-type phase boundary. *Physical Review Letters* **89**, 10.1103/PhysRevLett. 89.255507.

Shim S.-H., Duffy T. S., and Shen G. (2001) Stability and structure of $MgSiO_3$ perovskite to 2300-kilometer depth in Earth's mantle. *Science* **293**, 2437–2440.

Shimamoto T. and Logan J. M. (1981) Effects of simulated gouges on the sliding behavior of Tennessee sandstone. *Tectonophysics* **75**, 243–255.

Shimamura H., Asada T., Suyehiro K., Yamada T., and Inatani H. (1983) Longshot experiments to study velocity anisotropy in the oceanic lithosphere of the northwestern Pacific. *Physics of Earth and Planetary Interiors* **31**, 348–362.

Shimazu Y. (1954) Equation of state of materials composing the Earth's interior. *Journal of Earth Science, Nagoya University* **2**, 15–172.

Shimizu I. (1992) Nonhydrostatic and nonequilibrium thermodynamics of deformable materials. *Journal of Geophysical Research* **97**, 4587–4597.

Shimizu I. (1994) Rock deformation by pressure solution and its implications to the rheology of lithosphere: a review. *Structural Geology* **39**, 153–164.

Shimizu I. (1998) Stress and temperature dependence of recrystallized grain size: a subgrain misorientation model. *Geophysical Research Letters* **25**, 4237–4240.

Shimozuru D. (1963) On the possibility of the existence of the molten portion in the upper mantle of the earth. *Journal of Physics of the Earth* **11**, 49–55.

Shinmei T., Tomioka N., Fujino K., Kuroda K., and Irifune T. (1999) *In situ* X-ray diffraction of enstatite up to 12 GPa and 1473 K and equation of state. *American Mineralogist* **84**, 1588–1594.

Shito A., Karato S., Matsukage K. N., and Nishihara Y. (2006) Toward mapping water content, temperature and major element chemistry in Earth's upper mantle from seismic tomography. In *Earth's Deep Water Cycle* (ed. S. D. Jacobsen and S. van der Lee), pp. 225–236. American Geophysical Union.

Shito A., Karato S., and Park J. (2004) Frequency dependence of Q in Earth's upper mantle inferred from continuous spectra of body wave. *Geophysical Research Letters* **31**, 10.1029/2004GL019582.

Shito A. and Shibutani T. (2003a) Anelastic structure of the upper mantle beneath the northern Philippine Sea. *Physics of Earth and Planetary Interiors* **140**, 319–329.

Shito A. and Shibutani T. (2003b) Nature of heterogeneity of the upper mantle beneath the northern Philippine Sea as inferred from attenuation and velocity tomography. *Physics of Earth and Planetary Interiors* **140**, 331–341.

Sibson R. H. (1975) Generation of pseudotachylyte by ancient seismic faulting. *Geophysical Journal of Royal Astronomical Society* **43**, 775–794.

Sibson R. H. (1977) Fault rocks and fault mechanics. *Journal of Geological Society of London* **133**, 191–213.

Siegesmund S., Takeshita T., and Kern H. (1989) Anisotropy of V_p and V_s in an amphibolite of the deeper crust and its relationship to the mineralogical, microstructural and textural characteristics of the rock. *Tectonophysics* **157**, 25–38.

Silver P. G. (1996) Seismic anisotropy and mantle deformation: probing the depths of geology. *Annual Review of Earth and Planetary Sciences* **24**, 385–432.

Silver P. G., Mainprice D., Ben Ismail W., Tommasi A., and Barroul G. (1999) Mantle structural geology from seismic anisotropy. In *Mantle Petrology: Field Observations and High Pressure Experimentation* (ed. Y. Fei, C. M. Bertka, and B. O. Mysen), pp. 79–103. The Geochemical Society.

Simpson C. and Schmid S. (1983) An evaluation of criteria to deduce the sense of movement in sheared rocks. *Geological Society of America Bulletin* **94**, 1281–1288.

Singh A. K. (1993) The lattice strain in a specimen (cubic system) compressed nonhydrostatically in an opposed anvil device. *Journal of Applied Physics* **73**, 4278–4286.

Singh A. K., Mao H.-K., Shu J., and Hemley R. J. (1998) Estimation of single-crystal elastic moduli from polycrystalline X-ray diffraction at high pressure: application to FeO and iron. *Physical Review Letters* **80**, 2157–2160.

Singh S. C., Taylor M. A. J., and Montagner J.-P. (2000) On the presence of liquid in Earth's inner core. *Science* **287**, 2471–2474.

Sinogeikin S. V. and Bass J. D. (1999) Single-crystal elasticity of MgO at high pressure. *Physical Review B* **59**, R14 141–R14 144.

Sinogeikin S. V. and Bass J. D. (2002) Elasticity of majorite and a majorite–pyrope solid solution to high pressure: implications for the transition zone. *Geophysical Research Letters* **29**, 10.1029/2001GL013937.

Sinogeikin S. V., Bass J. D., and Katsura T. (2003) Single-crystal elasticity of ringwoodite to high pressures and high temperatures: implications for 520 km seismic discontinuity. *Physics of Earth and Planetary Interiors* **136**, 41–66.

Sinogeikin S. V., Chen G., Neuville D. R., Vaughan M. T., and Lierbermann R. C. (1998) Ultrasonic shear wave velocities of $MgSiO_3$ perovskite at 8 GPa and 800 K and lower mantle composition. *Science* **281**, 677–679.

Sipkin S. and Jordan T. H. (1979) Frequency dependence of Q_{ScS}. *Bulletin of Seismological Society of America* **69**, 1055–1079.

Skemer P. A., Katayama I., Jiang Z., and Karato S. (2005) The misorientation index: development of a new method for calculating the strength of lattice-preferred orientation. *Tectonophysics* **411**, 157–167.

Skemer P. A., Katayama I., and Karato S. (2006) Deformation fabrics of a peridotite from Cima di Gagnone, central Alps, Switzerland: evidence of deformation under water-rich condition at low temperatures. *Contributions to Mineralogy and Petrology* **152**, 43–51.

Skinner B. J., Porter S. C., and Park J. (2004) *Dynamic Earth: an Introduction to Physical Geology*. John Wiley & Sons.

Skogby H. (1994) OH incorporation in synthetic clinopyroxene. *American Mineralogist* **79**, 240–249.

Skogby H., Bell D. R., and Rossman G. R. (1990) Hydroxide in pyroxene: variations in the natural environment. *American Mineralogist* **75**, 764–774.

Skogby H. and Rossman G. R. (1989) OH^- in pyroxene: an experimental study of incorporation mechanisms and stability. *American Mineralogist* **74**, 1059–1069.

Smith B. K. (1985) The influence of defect crystallography on some properties of orthosilicates. In *Metamorphic Reactions, Kinetics, Textures and Deformation* (ed. T. A. B. and D. C. Rubie), pp. 98–117. Springer-Verlag.

Smith B. K. and Carpenter F. O. (1987) Transient creep in orthosilicates. *Physics of Earth and Planetary Interiors* **49**, 314–324.

Smith G. P., Wiens D. A., Fischer K. M., Dorman L. M., and Hildebrand J. A. (2001) A complex pattern of mantle flow in the Lau back-arc. *Science* **292**, 713–716.

Smith M. F. and Dahlen F. A. (1981) The period and Q of the Chandler wobble. *Geophysical Journal of Royal Astronomical Society* **64**, 223–281.

Smith M. L. and Dahlen F. A. (1973) The azimuthal dependence of Love and Rayleigh wave propagation in a slightly anisotropic medium. *Journal of Geophysical Research* **78**, 3321–3333.

Smyth J. R., Bell D. R., and Rossman G. R. (1991) Incorporation of hydroxyl in upper-mantle clinopyroxenes. *Nature* **351**, 732–735.

Smyth J. R. and Frost D. J. (2002) The effect of water on the 410-km discontinuity: an experimental study. *Geophysical Research Letters* **29**, 10.129/2001GL014418.

Solomatov V. S. (1996) Can hot mantle be stronger than cold mantle? *Geophysical Research Letters* **23**, 937–940.

Solomatov V. S. (2001) Grain size-dependent viscosity convection and the thermal evolution of the Earth. *Earth and Planetary Science Letters* **191**, 203–212.

Solomatov V. S., El-Khozondar R., and Tikare V. (2002) Grain size in the lower mantle: constraints from numerical modeling of grain growth in two-phase systems. *Physics of Earth and Planetary Interiors* **129**, 265–282.

Solomatov V. S. and Moresi L. N. (1996) Stagnant lid convection on Venus. *Journal of Geophysical Research* **101**, 4737–4753.

Solomon S. C. (1972) Seismic wave attenuation and partial melting in the upper mantle of North America. *Journal of Geophysical Research* **77**, 1483–1502.

Solomon S. C., Head J. W., Kaula W. M., *et al.* (1991) Venus tectonics: initial analysis from Magellan. *Science* **252**, 297–312.

Song T.-R. A., Helmberger D. V., and Grand S. P. (2004) Low-velocity zone atop the 410-km seismic discontinuity in the northwestern United States. *Nature* **427**, 530–533.

Song X. (1997) Anisotropy of the Earth's inner core. *Review of Geophysics* **35**, 297–313.

Song X. and Helmberger D. V. (1998) Seismic evidence for an inner core transition zone. *Science* **282**, 924–927.

Song X. and Richards P. G. (1996) Seismic evidence for the rotation of the inner core. *Nature* **382**, 221–224.

Souriau A. (1998) Earth's inner core: is the rotation real? *Science* **281**, 55–56.

Souriau A. and Poupinet G. (2002) Inner core rotation: a critical appraisal. In *Earth's Core: Dynamics, Structure, Rotation* (ed. V. Dehant, K. C. Creager, S. Karato, and S. Zatman), pp. 65–82.

Souriau A. and Roudil P. (1995) Attenuation in the uppermost inner core from broadband Geoscope PKP data. *Geophysical Journal International* **123**, 572–587.

Spetzler H. A. and Anderson D. L. (1968) The effect of temperature and partial melting on velocity and attenuation in a simple binary system. *Journal of Geophysical Research* **73**, 6051–6060.

Speziale S., Jiang F., and Duffy T. S. (2005) Compositional dependence of the elastic wave velocities of mantle minerals: implications for seismic properties of mantle rocks. In *Earth's Deep Mantle* (ed. R. D. v. d. Hilst, J. D. Bass, J. Matas, and J. Trampert), pp. 301–320. American Geophysical Union.

Spiegelman M. (2003) Linear analysis of melt band formation by simple shear. *Geochemical Geophysical Geosystems* **4**, 10.1029/2002GC000499.

Spiegelman M. and Elliott T. (1993) Consequences of melt transport for uranium series disequilibrium in young lavas. *Earth and Planetary Science Letters* **118**, 1–20.

Spiegelman M. and Kenyon P. M. (1992) The requirement of chemical disequilibrium during magma migration. *Earth and Planetary Science Letters* **109**, 611–620.

Spiers C. J., De Meer S., Niemeijer A. R., and Zhang X. (2004) Kinetics of rock deformation by pressure solution and the role of thin aqueous films. In *Physicochemistry of Water in Geological and Biological Systems* (ed. S. Nakashima, C. J. Spiers, L. Mercury, P. A. Fenter, and J. M. F. Hochella), pp. 129–158. Universal Academy Press.

Spiers C. J., Schutjens P. M. T. M., Brezesowsky P. H., *et al.* (1990) Experimental determination of constitutive parameters governing creep of rocksalt by pressure solution. In *Deformation Mechanisms, Rheology and Tectonics* (ed. R. J. Knipe and E. H. Rutter), pp. 215–227. The Geological Society.

Spingarn J. R., Barnett D. M., and Nix W. D. (1979) Theoretical description of climb controlled steady state creep at high and intermediate temperatures. *Acta Metallurgica* **27**, 1549–1562.

Spingarn J. R. and Nix W. D. (1978) Diffusional creep and diffusionally accommodated grain rearrangement. *Acta Metallurgica* **26**, 1388–1398.

Spray J. G. (1987) Artificial generation of pseudotachylyte using friction welding apparatus: simulation of melting on a fault plane. *Journal of Structural Geology* **9**, 49–60.

Srolovitz D. J. and Davis S. H. (2001) Do stresses modify wetting angles? *Acta Materialia* **49**, 1005–1007.

Stacey F. D. (1992) *Physics of the Earth*. Brookfield Press.

Stauffer D. and Aharony A. (1992) *Introduction to Percolation Theory*. Taylor and Francis.

Steinle-Neumann G., Stixrude L., Cohen R. E., and Gülseren O. (2001) Elasticity of iron at the temperature of the Earth's inner core. *Nature* **413**, 57–60.

Stevenson D. J. (1989) Spontaneous small-scale melt segregation in partial melts undergoing deformation. *Geophysical Research Letters* **16**, 1067–1070.

Stipp M. and Tullis J. (2003) The recrystallized grain size piezometer for quartz. *Geophysical Research Letters* **30**, 10.1029/2003GL018444.

Stipp M., Tullis J., and Behrens H. (2006) Dislocation creep of quartz: the effect of water on flow stress and microstructure. *Journal of Geophysical Research* **111**, 10.1029/2005JB003852.

Stixrude L. and Lithgow-Bertelloni C. (2005a) Mineralogy and elasticity of the oceanic upper mantle: origin of the low-velocity zone. *Journal of Geophysical Research* **110**, 10.1029/2004JB002965.

Stixrude L. and Lithgow-Bertelloni C. (2005b) Thermodynamics of mantle minerals – I. Physical properties. *Geophysical Journal International* **162**, 610–632.

Stocker R. L. and Ashby M. F. (1973) Rheology of the upper mantle. *Review of Geophysics and Space Physics* **11**, 391–426.

Stroh A. N. (1954) The formation of cracks as a result of plastic flow. *Proceedings of the Royal Society of London A* **223**, 404–414.

Stroh A. N. (1955) The formation of cracks in plastic flow II. *Proceedings of the Royal Society of London A* **232**, 548–560.

Stünitz H., Fitz Gerald J. D., and Tullis J. (2003) Dislocation generation, slip systems, and dynamic recrystallization in experimentally deformed plagioclase single crystals. *Tectonophysics* **372**, 215–233.

Stünitz H. and Tullis J. (2001) Weakening and strain localization produced by syn-deformational reaction of plagioclase. *International Journal of Earth Sciences* **90**, 136–148.

Sturhahn W., Toellner T. S., Alp E. E., *et al.* (1995) Phonon density of states measured by inelastic nuclear resonant scattering. *Physical Review Letters* **74**, 3832–3835.

Su W.-J., Dziewonski A. M., and Jeanloz R. (1996) Planet within a planet – rotation of the inner core of the Earth. *Science* **274**, 1883–1887.

Suetsugu D., Inoue T., Yamada A., Zhao D., and Obayashi M. (2006) Towards mapping three-dimensional distribution of water in the transition zone from P-wave velocity tomography and 660-km discontinuity depths. In *Earth's Deep Water Cycle* (ed. S. D. Jacobsen and S. van der Lee), pp. 237–249. American Geophysical Union.

Sumino K. (1974) A model for the dynamical state of dislocations in crystals. *Materials Science and Engineering* **13**, 269–275.

Sumita I., Yoshida H., Hamano Y., and Kumazawa M. (1996) A model for sedimentary compaction in a viscous medium and its application to inner-core growth. *Geophysical Journal International* **124**, 502–524.

Sung C.-M., Goetze C., and Mao H.-K. (1977) Pressure distribution in the diamond anvil press and shear strength of fayalite. *Review of Scientific Instruments* **48**, 1386–1391.

Suzuki H. (1962) Segregation of solute atoms to stacking faults. *Journal of Physical Society of Japan* **17**, 322–325.

Tackley P. J. (2000a) Self-consistent generation of tectonic plates in time-dependent, three dimensional mantle convection simulations, 1. Pseudoplastic yielding. *Geochemistry, Geophysics, Geosystems* **1**, 2000GC000, 036.

Tackley P. J. (2000b) Self-consistent generation of tectonic plates in time-dependent, three dimensional mantle convection simulations, 2. Strain weakening and asthenosphere. *Geochemistry, Geophysics, Geosystems* **1**, 2000GC000, 043.

Tackley P. J., Stevenson D. J., Glatzmaier G. A., and Schubert G. (1993) Effects of endothermic phase transition at 670 km depth in a spherical model of mantle convection in the Earth's mantle. *Nature* **361**, 699–704.

Tada R. and Siever R. (1986) Experimental knife-edge pressure solution of halite. *Geochemica et Cosmochemica Acta* **50**, 29–36.

Tada R. and Siever R. (1987) A new mechanism for pressure solution in porous quartzose sandstone. *Geochemica et Cosmochemica Acta* **51**, 2295–2301.

Takanami T., Sacks I. S., and Hasegawa A. (2000) Attenuation structure beneath the volcanic front in northeastern Japan from broad-band seismograms. *Physics of Earth and Planetary Interiors* **121**, 339–357.

Takei Y. (1998) Constitutive mechanical relations of solid–liquid composites in terms of grain-boundary contiguity. *Journal of Geophysical Research* **103**, 18 183–18 203.

Takei Y. (2000) Acoustic properties of partially molten media studied on a simple binary system with a controllable dihedral angle. *Journal of Geophysical Research* **105**, 16 665–16 682.

Takei Y. (2002) Effect of pore geometry on Vp/Vs: from equilibrium geometry to crack. *Journal of Geophysical Research* **107**, 10.1029/2001JB000522.

Takeshita T. (1989) Plastic anisotropy in textured mineral aggregates: theories and geological applications. In *Rheology of Solids and of the Earth* (ed. S. Karato and M. Toriumi), pp. 237–262. Oxford University Press.

Takeshita T. and Wenk H.-R. (1988) Plastic anisotropy and geometric hardening in quartzites. *Tectonophysics* **149**, 345–361.

Takeshita T., Wenk H.-R., Molinari A., and Canova G. (1990) Simulation of dislocation assisted plastic deformation in olivine polycrystals. In *Deformation Processes in Minerals, Ceramics and Rocks* (ed. D. J. Barber and P. G. Meredith), pp. 365–377. Unwin Hyman.

Takeuchi S. and Argon A. S. (1976) Steady-state creep of alloys due to viscous motion of dislocations. *Acta Metallurgica* **24**, 883–889.

Takeuchi S. and Suzuki T. (1988) Deformation of crystals controlled by the Peierls mechanism. *Strength of Metals and Alloys (ICSMA 8)*, 161–166.

Tan B., Jackson I., and Fitz Gerald J. D. (1997) Shear wave dispersion and attenuation in fine-grained synthetic olivine aggregates: preliminary results. *Geophysical Research Letters* **24**, 1055–1058.

Tan B., Jackson I., and Fitz Gerald J. D. (2001) High-temperature viscoelasticity of fine-grained polycrystalline olivine. *Physics and Chemistry of Minerals* **28**, 641–664.

Tanaka S. and Hamaguchi H. (1997) Degree one heterogeneity and hemispherical variation of anisotropy in the inner core from PKP(BC)-PKP(DF) times. *Journal of Geophysical Research* **102**, 2925–2938.

Tanimoto T. and Anderson D. L. (1984) Mapping mantle convection. *Geophysical Research Letters* **11**, 287–290.

Tanimoto T. and Anderson D. L. (1985) Lateral heterogeneities and azimuthal anisotropy of the upper mantle: Love and Rayleigh waves 100–250 s. *Journal of Geophysical Research* **90**, 1842–1858.

Tanimoto T. and Anderson D. L. (1990) Long-wavelength S-wave velocity structure throughout the mantle. *Geophysical Journal International* **100**, 327–336.

Tapponnier P. and Francheteau J. (1978) Necking of the lithosphere and the mechanics of accreting plate boundaries. *Journal of Geophysical Research* **83**, 3955–3970.

Tarits P., Hautot S., and Perrier F. (2004) Water in the mantle: results from electrical conductivity beneath the French Alps. *Geophysical Research Letters* **31**, 10.1029/2003GL019277.

Taylor G. I. (1934) The mechanism of plastic deformation of crystals. *Proceedings of the Royal Society A* **145**, 362–415.

Tharp T. M. (1983) Analogies between the high-temperature deformation of polyphase rocks and the mechanical behavior of porous powder metal. *Tectonophysics* **96**, T1–T11.

Thompson A. B. (1992) Water in the Earth's upper mantle. *Nature* **358**, 295–302.

Thoraval C. and Richards M. A. (1997) The geoid constraint in global geodynamics: viscosity structure, mantle heterogeneity models and boundary conditions. *Geophysical Journal International* **131**, 1–8.

Thurel E. and Cordier P. (2003) Plastic deformation of wadsleyite: I. High-pressure deformation in compression. *Physics and Chemistry of Minerals* **30**, 256–266.

Thurel E., Cordier P., Frost D. J., and Karato S. (2003a) Plastic deformation of wadsleyite: II. High-pressure deformation in shear. *Physics and Chemistry of Minerals* **30**, 267–270.

Thurel E., Douin J., and Cordier P. (2003b) Plastic deformation of wadsleyite: III. Interpretation of dislocation slip systems. *Physics and Chemistry of Minerals* **30**, 271–279.

Tingle T. N., Green H. W., II., Young T. E., and Koczynski T. A. (1993) Improvements to Griggs-type apparatus for mechanical testing at high pressures and temperatures. *Pure and Applied Geophysics* **141**, 523–543.

Tommasi A., Mainprice D., Canova G., and Chastel Y. (2000) Viscoelastic self-consistent and equilibrium-based modeling of olivine preferred orientations: implications for the upper mantle seismic anisotropy. *Journal of Geophysical Research* **105**, 7893–7908.

Tommasi A., Mainprice D., Cordier P., Thoraval C., and Couvy H. (2004) Strain-induced seismic anisotropy of wadsleyite polycrystals and flow patterns in the mantle transition zone. *Journal of Geophysical Research* **109**, 10.1029/2004JB003158.

Toomey D. R., Wilcock W. S. D., Solomon S. C., Hammond W. C., and Orcott J. A. (1998) Mantle structure beneath the MELT region of the East Pacific Rise from P and S wave tomography. *Science* **280**, 1224–1227.

Toramaru A. and Fujii N. (1986) Connectivity of melt phase in a partially molten peridotite. *Journal of Geophysical Research* **91**, 9239–9252.

Toriumi M. (1982) Grain boundary migration in olivine at atmospheric pressure. *Physics of Earth and Planetary Interiors* **30**, 26–35.

Toriumi M. and Karato S. (1985) Preferred orientation development of dynamically recrystallized olivine during high temperature creep. *Journal of Geology* **93**, 407–417.

Tosi M. P. (1964) Cohesion of ionic solids in the Born model. *Solid State Physics* **16**, 1–120.

Trampert J., Deschamps F., Resovsky J. S., and Yuen D. A. (2004) Probabilistic tomography maps chemical heterogeneities throughout the lower mantle. *Science* **306**, 853–856.

Trampert J., Vacher P., and Vlaar N. J. (2001) Sensitivities of seismic velocities to temperature, pressure and composition in the lower-mantle. *Physics of Earth and Planetary Interiors* **124**, 255–267.

Trampert J. and van Heijst H. J. (2002) Global azimuthal anisotropy in the transition zone. *Science* **296**, 1297–1299.

Treagus S. H. (2002) Modelling the bulk viscosity of two-phase mixtures in terms of clast shape. *Journal of Structural Geology* **24**, 57–76.

Treagus S. H. (2003) Viscous anisotropy of two-phase composites, and applications to rocks and structures. *Tectonophysics* **372**, 121–133.

Tromp J. (2001) Inner-core anisotropy and rotation. *Review of Earth and Planetary Sciences* **29**, 47–69.

Tsenn M. C. and Carter N. L. (1987) Upper limits of power law creep of rocks. *Tectonophysics* **136**, 1–26.

Tsuchiya T., Tsuchiya J., Umemoto K., and Wentzcovitch R. M. (2004a) Elasticity of post-perovskite $MgSiO_3$. *Geophysical Research Letters* **31**, 10.1029/2004GL020278.

Tsuchiya T., Tsuchiya J., Umemoto K., and Wentzcovitch R. M. (2004b) Phase transition in $MgSiO_3$ in the Earth's lower mantle. *Earth and Planetary Science Letters* **224**, 241–248.

Tsumura N., Matsumoto S., Horiuchi S., and Hasegawa A. (2000) Three-dimensional attenuation structure beneath the northeastern Japan arc estimated from spectra of small earthquakes. *Tectonophysics* **319**, 241–260.

Tsutsumi A. and Shimamoto T. (1997) High-velocity frictional properties of gabbro. *Geophysical Research Letters* **24**, 699–702.

Tullis J. (2002) Deformation of granitic rocks: Experimental studies and natural examples. In *Plastic Deformation of Minerals and Rocks*, Vol. 51 (ed. S. Karato and H.-R. Wenk), pp. 51–95. Mineralogical Society of America.

Tullis J., Christie J. M., and Griggs D. T. (1973) Microstructure and preferred orientations of experimentally deformed quartzites. *Geological Society of America Bulletin* **84**, 297–314.

Tullis J., Shelton G. L., and Yund R. A. (1979) Pressure dependence of rock strength: implications for hydrolytic weakening. *Bulletin Mineralogie* **102**, 110–114.

Tullis J. and Yund R. A. (1980) Hydrolytic weakening of experimentally deformed Westerly granite and Hale albite rock. *Journal of Structural Geology* **2**, 439–451.

Tullis J. and Yund R. A. (1982) Grain growth kinetics of quartz and calcite aggregates. *Journal of Geology* **90**, 301–318.

Tullis J. and Yund R. A. (1985) Dynamic recrystallization of feldspar: a mechanism of shear zone formation. *Geology* **13**, 238–241.

Tullis T. E., Horowitz F. G., and Tullis J. (1991) Flow laws of polyphase aggregates from end-member flow laws. *Journal of Geophysical Research* **96**, 8081–8096.

Tullis T. E. and Tullis J. (1986) Experimental rock deformation. In *Mineral and Rock Deformation* (ed. B. E. Hobbs and H. C. Heard), pp. 297–324. American Geophysical Union.

Tungatt P. D. and Humphreys F. J. (1984) The plastic deformation and dynamic recrystallization of polycrystalline sodium nitrate. *Acta Metallurgica* **32**, 1625–1635.

Turcotte D. L. and Schubert G. (1982) *Geodynamics: Applications of Continuum Physics to Geological Problems*. John Wiley & Sons.

Twiss R. J. (1977) Theory and applicability of a recrystallized grain size paleopiezometer. *Pure and Applied Geophysics* **115**, 227–244.

Uchida T., Funamori N., and Yagi T. (1996) Lattice strains in crystals under uniaxial stress field. *Journal of Applied Physics* **80**, 739–746.

Underwood E. E. (1969) *Quantitative Stereology*. Addison-Wesley.

Urai J. L. (1983) Water-assisted dynamic recrystallization and weakening in polycrystalline bischofite. *Tectonophysics* **96**, 125–157.

Urai J. L. (1987) Development of microstructure during deformation of carnalite and bischofite in transmitted light. *Tectonophysics* **135**, 251–263.

Urai J. L., Means W. D., and Lister G. S. (1986a) Dynamic recrystallization in minerals. In *Mineral and Rock Deformation: Laboratory Studies* (ed. B. E. Hobbs and H. C. Heard), pp. 166–199. American Geophysical Union.

Urai J. L., Spiers C. J., Zwart H. J., and Lister G. S. (1986b) Water weakening in rock salt during long-term creep. *Nature* **324**, 554–557.

van der Hilst R. D. and Kárason H. (1999) Compositional heterogeneity in the bottom 1000 kilometers of Earth's mantle: toward a hybrid convection model. *Science* **283**, 1885–1888.

van der Hilst R. D., Widiyantoro R. D. S., and Engdahl E. R. (1997) Evidence for deep mantle circulation from global tomography. *Nature* **386**, 578–584.

van der Meijde M., Marone F., Giardini D., and van der Lee S. (2003) Seismic evidence for water deep in Earth's upper mantle. *Science* **300**, 1556–1558.

van der Molen I. and Paterson M. S. (1979) Experimental deformation of partially-melted granite. *Contributions to Mineralogy and Petrology* **70**, 299–318.

van der Wal D., Chopra P. N., Drury M., and Fitz Gerald J. D. (1993) Relationships between dynamically recrystallized grain size and deformation conditions in experimentally deformed olivine rocks. *Geophysical Research Letters* **20**, 1479–1482.

van Houtte P. and Wagner F. (1985) Development of texture by slip and twinning. In *Preferred Orientation in Deformed Metals and Rocks* (ed. H.-R. Wenk), pp. 233–258. Academic Press.

Van Orman J. A. (2004) On the viscosity and creep mechanism of Earth's inner core. *Geophysical Research Letters* **31**, 10.1029/2004GL021209.

Van Orman J. A., Fei Y., Hauri E. H., and Wang J. (2003) Diffusion in MgO at high pressure: constraints on deformation mechanisms and chemical transport at the core–mantle boundary. *Geophysical Research Letters* **30**, 10.1029/2002GL016343.

Vauchez A. and Nicolas A. (1991) Mountain building: strike-parallel motion and mantle anisotropy. *Tectonophysics* **185**, 183–191.

Vaughan P. J. and Coe R. S. (1978) Geometric flow properties of the germanate analog of forsterite. *Tectonophysics* **46**, 187–196.

Vaughan P. J. and Coe R. S. (1981) Creep mechanisms in Mg_2GeO_4: effects of a phase transition. *Journal of Geophysical Research* **86**, 389–404

Vidale J. E., Dodge D. A., and Earle P. S. (2000) Slow differential rotation of the Earth's inner core indicated by temporal change in scattering. *Nature* **405**, 445–448.

Vidale J. E. and Earle P. S. (2000) Fine-scale heterogeneity in the Earth's inner core. *Nature* **404**, 273–275.

Vineyard G. H. (1957) Frequency factors and isotope effects in solid state rate processes. *Journal of Physics and Chemistry of Solids* **3**, 121–127.

Vinnik L., Breger L., and Romanowicz. (1998) Anisotropic structures at the base of the Earth's mantle. *Nature* **393**, 564–567.

Vinnik L. and Montagner J.-P. (1996) Shear wave splitting in the mantle from Ps phases. *Geophysical Research Letters* **23**, 2449–2452.

Vinnik L., Romanowicz B., Le Stunff Y., and Makayeva L. I. (1995) Seismic anisotropy in D″-layer. *Geophysical Research Letters* **22**, 1657–1660.

Voce E. (1948) The relationship between stress and strain for homogeneous deformation. *Journal of Institute of Metals* **74**, 537–562.

von Mises R. (1928) Mechanik der plastischen Formändern von Kristallen. *Zeitschrift für Angewandte Mathematik und Mechanik* **8**, 161–185.

Voorhees P. W. (1985) The theory of Ostwald ripening. *Journal of Statistical Physics* **38**, 231–252.

Voorhees P. W. (1992) Ostwald ripening of 2-phase mixtures. *Annual Review of Materials Science* **22**, 197–215.

Waff H. S. and Blau J. R. (1979) Equilibrium fluid distribution in an ultramafic partial melt under hydrostatic stress conditions. *Journal of Geophysical Research* **84**, 6109–6114.

Waff H. S. and Blau J. R. (1982) Experimental determination of near equilibrium textures in partially molten silicates at high pressures. In *High-pressure Research in Geophysics* (ed. S. Akimoto and M. H. Manghnani), pp. 229–236. Center for Academic Publication.

Walcott R. I. (1970) Flexural rigidity, thickness, and viscosity of the lithosphere. *Journal of Geophysical Research* **75**, 3941–3954.

Walker K. T., Bokelmann G. H., and Klemperer S. L. (2001) Shear-wave splitting to test mantle deformation models around Hawaii. *Geophysical Research Letters* **28**, 4319–4322.

Wall A. and Price G. D. (1989) Electrical conductivity of the lower mantle: a molecular dynamics simulation of $MgSiO_3$. *Physics of Earth and Planetary Interiors* **58**, 192–204.

Wallace D. C. (1972) *Thermodynamics of Crystals*. Wiley.

Walsh J. B. (1968) Attenuation in partially molten material. *Journal of Geophysical Research* **73**, 2209–2216.

Walsh J. B. (1969) New analysis of attenuation in partially molten rock. *Journal of Geophysical Research* **74**, 4333–4337.

Wang D., Mookherjee M., Xu Y., and Karato S. (2006) The effect of water on the electrical conductivity in olivine. *Nature* **443**, 977–980.

Wang J. N. (1994) Harper–Dorn creep in olivine. *Materials Science and Engineering A* **183**, 267–272.

Wang J. N., Hobbs B. E., Ord A., Shimamoto T., and Toriumi M. (1994) Newtonian dislocation creep in quartzites: implications for the rheology of the lower crust. *Science* **265**, 1203–1205.

Wang J. N. and Nieh T. G. (1995) Effects of the Peierls stress on the transition from power-law creep to Harper–Dorn creep. *Acta Metallurgica et Materialia* **43**, 1415–1419.

Wang Y., Durham W. B., Getting I. C., and Weidner D. J. (2003) The deformation-DIA: a new apparatus for high temperature triaxial deformation to pressures up to 15 GPa. *Review of Scientific Instruments* **74**, 3002–3011.

Wang Y., Guyot F., and Liebermann R. C. (1992) Electron microscopy of (Mg, Fe)SiO_3 perovskite: evidence for structural phase transitions and implications for the lower mantle. *Journal of Geophysical Research* **97**, 12 327–12 347.

Wang Z. and Ji S. (2000) Diffusion creep of fine-grained garnetite: implications for the flow strength of subducting slabs. *Geophysical Research Letters* **27**, 2333–2336.

Wang Z., Karato S., and Fujino K. (1993) High temperature creep of single crystal strontium titanate: a contribution to creep systematics in perovskites. *Physics of Earth and Planetary Interiors* **79**, 299–312.

Wang Z., Karato S., and Fujino K. (1996) High temperature creep of single crystal gadolinium gallium garnet. *Physics and Chemistry of Minerals* **23**, 73–80.

Wang Z., Mei S., Karato S., and Wirth R. (1999) Grain growth in $CaTiO_3$–perovskite + FeO–wüstite aggregates. *Physics and Chemistry of Minerals* **27**, 11–19.

Watson E. B. and Brenan J. M. (1987) Fluids in the lithosphere 1. Experimentally-determined wetting characteristics of CO_2–H_2O fluids and their implications for fluid transport, host-rock physical-properties, and fluid inclusion formation. *Earth and Planetary Science Letters* **85**, 497–515.

Watt J. P., Davies G. F., and O'Connell R. J. (1976) The elastic properties of composite materials. *Review of Geophysics and Space Physics* **14**, 541–563.

Webb S. and Jackson I. (1990) Polyhedral rationalization of variation among the single crystal elastic moduli for upper-mantle silicates: garnets, olivine and orthopyroxene. *American Mineralogist* **75**, 731–738.

Webb S. and Jackson I. (2003) Anelasticity and microcreep in polycrystalline MgO at high temperature: an exploratory study. *Physics and Chemistry of Minerals* **30**, 157–166.

Webb S., Jackson I., and Fitz Gerald J. (1999) Viscoelasticity of the titanate perovskite $CaTiO_3$ and $SrTiO_3$ at high temperature. *Physics of Earth and Planetary Interiors* **115**, 259–291.

Webster G. A. (1966a) A widely applicable dislocation model of creep. *Philosophical Magazine* **14**, 775–783.

Webster G. A. (1966b) In support of a model of creep based on dislocation dynamics. *Philosophical Magazine* **14**, 1303–1307.

Weertman J. (1957) Steady state creep of crystals. *Journal of Applied Physics* **28**, 1185–1191.

Weertman J. (1968) Dislocation climb theory of steady state creep. *Transactions of the American Society of Metals* **61**, 681–694.

Weertman J. (1970) The creep strength of the Earth's mantle. *Review of Geophysics and Space Physics* **8**, 145–168.

Weertman J. (1978) Creep laws for the mantle of the Earth. *Philosophical Transactions of the Royal Society of London A* **228**, 9–26.

Weertman J. and Blacic J. D. (1984) Harper–Dorn creep; an artifact of low-amplitude temperature cycling? *Geophysical Research Letters* **11**, 117–120.

Weertman J. and Weertman J. R. (1975) High temperature creep of rock and mantle viscosity. *Annual Review of Earth and Planetary Sciences* **3**, 293–315.

Weidner D. J. (1987) Elastic properties of rocks and minerals. In *Methods of Experimental Physics*, Vol. 1–30 (ed. C. G. Sammis and T. L. Henyey). Academic Press.

Weidner D. J. (1998) Rheological studies at high pressure. In *Ultrahigh-Pressure Mineralogy* (ed. R. J. Hemley), pp. 492–524. The Mineralogical Society of America.

Weidner D. J., Li L., Davis M., and Chen J. (2004) Effect of plasticity on elastic modulus measurements. *Geophysical Research Letters* **31**, 10.1029/2003GL019090.

Weidner D. J. and Wang Y. (2000) Phase transformations: implications for mantle structure. In *Earth's Deep Interior: Mineral Physics and Tomography* (ed. S. Karato, A. M. Forte, R. C. Liebermann, G. Masters, and L. Stixrude), pp. 215–235. American Geophysical Union.

Weidner D. J., Wang Y., Chen G., Ando J., and Vaughan M. T. (1998) Rheology measurements at high pressure and temperature. In *Properties of Earth and Planetary Materials at High Pressure and Temperature* (ed. M. H. Manghnani and T. Yagi), pp. 473–480. American Geophysical Union.

Weiss L. E. and Wenk H.-R. (1985) Symmetry of pole figures and textures. In *Preferred Orientation in Deformed Metals and Rocks: an Introduction to Modern Texture Analysis* (ed. H.-R. Wenk), pp. 49–72. Academic Press.

Wen L. (2001) Seismic evidence for a rapidly varying compositional anomaly at the base of the Earth's mantle beneath the Indian Ocean. *Earth and Planetary Science Letters* **194**, 83–95.

Wen L. and Anderson D. L. (1997) Layered mantle convection: a model for geoid and topography. *Earth and Planetary Science Letters* **146**, 367–377.

Wen L. and Niu F. (2002) Seismic velocity and attenuation structures in the top of the Earth's inner core. *Journal of Geophysical Research* **107**, 10.1029/2001JB000170.

Wen L., Silver P. G., James D. E., and Kuehnel R. (2001) Seismic evidence of a thermo-chemical boundary at the base of the Earth's mantle. *Earth and Planetary Science Letters* **189**, 141–153.

Wenk H.-R. (1985) *Preferred Orientation in Deformed Metals and Rocks: an Introduction to Modern Texture Analysis*. Academic Press.

Wenk H.-R. (2002) Texture and anisotropy. In *Plastic Deformation of Minerals and Rocks*, Vol. 51 (ed. S. Karato and H.-R. Wenk), pp. 291–329. The Mineralogical Society of America.

Wenk H.-R., Bennett K., Canova G., and Molinari A. (1991) Modelling plastic deformation of peridotite with the self-consistent theory. *Journal of Geophysical Research* **96**, 8337–8349.

Wenk H.-R., Canova G., Molinari A., and Mecking H. (1989) Texture development in halite: comparison of Taylor model and self-consistent theory. *Acta Metallurgica* **37**, 2017–2029.

Wenk H.-R., Canova G. C., Brechet Y., and Flandin L. (1997) A deformation-based model for recrystallization of anisotropic materials. *Acta Mater* **45**, 3283–3296.

Wenk H.-R. and Christie J. M. (1991) Comments on the interpretation of deformation textures in rocks. *Journal of Structural Geology* **13**, 1091–1110.

Wenk H.-R., Lonardelli I., Pehl J., Devine J. D., Prakapenka V., Shen G., and Mao H.-K. (2004) *In situ* observation of texture development in olivine, ringwoodite, magnesiowüstite and silicate perovskite at high pressure. *Earth and Planetary Science Letters* **226**, 507–519.

Wenk H.-R., Matthius S., Hemley R. J., Mao H.-K., and Shu J. (2000) The plastic deformation of iron at pressures of the Earth's inner core. *Nature* **405**, 1044–1047.

Wettlaufer J. S., Worster M. G., and Huppert H. E. (1997) Natural convection during solidification of an alloy from above with application to the evolution of sea ice. *Journal of Fluid Dynamics* **344**, 291–316.

Wheeler J. (1992) Importance of pressure solution and Coble creep in the deformation of polymineralic rocks. *Journal of Geophysical Research* **97**, 4579–4586.

Wheeler J., Prior D. J., Jiang Z., Spiess R., and Trimbly P. W. (2001) The petrological significance of misorientations between grains. *Contributions to Mineralogy and Petrology* **141**, 109–124.

White S. H. (1979) Grain and sub-grain size variations across a mylonite shear zone. *Contributions to Mineralogy and Petrology* **70**, 193–202.

White S. H., Burrows S. E., Carreras J., Shaw N. D., and Humphreys F. J. (1980) On mylonites in ductile shear zones. *Journal of Structural Geology* **2**, 175–187.

White S. H. and Knipe R. J. (1978) Transformation- and reaction-induced ductility in rocks. *Journal of the Geological Society of London* **135**, 513–516.

Whitehead J. A., Jr. and Luther P. S. (1975) Dynamics of laboratory diapir and plume models. *Journal of Geophysical Research* **80**, 705–717.

Williams D. B. and Carter C. B. (1996) *Transmission Electron Microscopy*. Plenum Press.

Williams Q. and Garnero E. J. (1996) Seismic evidence for partial melt at the base of Earth's mantle. *Science* **273**, 1528–1530.

Williams Q. and Hemley R. J. (2001) Hydrogen in the deep Earth. *Annual Review of Earth and Planetary Sciences* **29**, 365–418.

Winger L. A., Bradt R. C., and Hoke J. H. (1980) Transformational superplasticity of Bi_2WO_6 and Bi_2MoO_6. *Journal of the American Ceramic Society* **63**, 291–294.

Withers A. C., Wood B. J., and Carroll M. R. (1998) The OH content of pyrope at high pressure. *Chemical Geology* **147**, 161–171.

Woirgard J., Rivière A., and De Fouquet J. (1981) Experimental and theoretical aspect of the high temperature damping of pure metals. *Journal de Physique Colloque C* **5**, 407–419.

Wolf G. H. and Jeanloz R. (1984) Lindemann melting law: anharmonic correction and test of its validity for minerals. *Journal of Geophysical Research* **89**, 7821–7835.

Wolfe C. J. and Solomon S. C. (1998) Shear-wave splitting and implications for mantle flow beneath the MELT region of the East Pacific. *Science* **280**, 1230–1232.

Wolfenstein J., Ruano O. A., Wadsworth J., and Sherby O. D. (1993) Refutation of the relationship between denuded zones and diffusional creep. *Scripta Metallurgica et Material* **29**, 515–520.

Wood B. J. (1995) The effect of H_2O on the 410-kilometer seismic discontinuity. *Science* **268**, 74–76.

Wood B. J. and Fraser D. G. (1976) *Elementary Thermodynamics for Geologists*. Oxford University Press.

Wood B. J. and Haliday A. N. (2005) Cooling of the Earth and core formation after the giant impact. *Nature* **437**, 1345–1348.

Wood B. J., Pawley A. R., and Frost D. R. (1996) Water and carbon in the Earth's mantle. *Philosophical Transactions of the Royal Society of London* **354**, 1495–1511.

Wood B. J., Walter M. J., and Wade J. (2006) Accretion of the Earth and segregation of its core. *Nature* **441**, 825–833.

Woodhouse J. H. and Dziewonski A. M. (1984) Mapping the upper mantle: three-dimensional modeling of Earth structure by inversion of seismic waveforms. *Journal of Geophysical Research* **89**, 5953–5986.

Woodhouse J. H. and Dziewonski A. M. (1989) Seismic modelling of the Earth's large-scale three-dimensional structure. *Philosophical Transactions of the Royal Society of London A* **328**, 291–308.

Woodhouse J. H., Giardini D., and Li X. D. (1986) Evidence for inner core anisotropy from free oscillations. *Geophysical Research Letters* **13**, 1549–1552.

Wookey J. and Kendall J.-M. (2004) Evidence of midmantle anisotropy from shear wave splitting and the influence of shear-coupled P waves. *Journal of Geophysical Research* **109**, 10.1029/2003JB002871.

Wookey J., Kendall J. M., and Barruol G. (2002) Mid-mantle deformation inferred from seismic anisotropy. *Nature* **415**, 777–780.

Wright K. and Price G. D. (1993) Computer simulation of defects and diffusion in perovskites. *Journal of Geophysical Research* **98**, 22 245–22 253.

Wyllie P. J. and Huang W. L. (1976) Carbonation and melting reactions in the system CaO-MgO-SiO_2-CO_2 at mantle pressure with geophysical and petrological applications. *Contributions to Mineralogy and Petrology* **54**, 79–107.

Xiao X., Wirth R., and Dresen G. (2002) Diffusion creep of anorthite–quartz aggregates. *Journal of Geophysical Research* **107**, 10.1029/2001JB000789.

Xie Y., Wenk H.-R., and Matthies S. (2003) Plagioclase preferred orientation by TOF neutron diffraction and SEM-EBSD. *Tectonophysics* **370**, 269–286.

Xu Y., Nishihara Y., and Karato S. (2005) Development of a rotational Drickamer apparatus for large-strain deformation experiments under deep Earth conditions. In *Frontiers in High-pressure Research: Applications to Geophysics* (ed. J. Chen, Y. Wang, T. S. Duffy, G. Shen, and L. F. Dobrzhinetskaya), pp. 167–182. Elsevier.

Xu Y., Weidner D. J., Chen J., Vaughan M. T., Wang Y., and Uchida T. (2003) Flow-law for ringwoodite at subduction zone conditions. *Physics of Earth and Planetary Interiors* **136**, 3–9.

Yamazaki D., Inoue T., Okamoto M., and Irifune T. (2005) Grain growth kinetics of ringwoodite and its implication for rheology of the subducting slab. *Earth and Planetary Science Letters* **236**, 871–881.

Yamazaki D. and Irifune T. (2003) Fe–Mg interdiffusion in magnesiowüstite up to 35 GPa. *Earth and Planetary Science Letters* **216**, 301–311.

Yamazaki D. and Karato S. (2001a) High pressure rotational deformation apparatus to 15 GPa. *Review of Scientific Instruments* **72**, 4207–4211.

Yamazaki D. and Karato S. (2001b) Some mineral physics constraints on the rheology and geothermal structure of Earth's lower mantle. *American Mineralogist* **86**, 385–391.

Yamazaki D. and Karato S. (2002) Fabric development in (Mg, Fe)O during large strain, shear deformation: implications for seismic anisotropy in Earth's lower mantle. *Physics of Earth and Planetary Interiors* **131**, 251–267.

Yamazaki D., Kato T., Ohtani E., and Toriumi M. (1996) Grain growth rates of $MgSiO_3$ perovskite and periclase under lower mantle conditions. *Science* **274**, 2052–2054.

Yamazaki D., Kato T., Toriumi M., and Ohtani E. (2001) Silicon self-diffusion in $MgSiO_3$ perovskite at 25 GPa. *Physics of Earth and Planetary Interiors* **119**, 299–309.

Yamazaki D., Yishino T., Ohfuji H., Ando J., and Yoneda A. (2006) Origin of seismic anisotropy in the D″ layer inferred from shear deformation experiments on post-perovskite phase. *Earth and Planetary Science Letters*.

Yan H. (1992) Dislocation Recovery in Olivine. Master of Science, University of Minnesota.

Yan M. F., Cannon R. F., and Bowen H. K. (1977) Grain boundary migration in ceramics. In *Ceramic Microstructures '76* (ed. R. M. Fulrath and J. A. Pask), pp. 276–307. Westview Press.

Yan M. F., Cannon R. M., and Bowen H. K. (1983) Space charge, elastic field and dipole contributions to equilibrium solute segregation at interfaces. *Journal of Applied Physics* **54**, 764–777.

Yokobori T. (1968) Criteria for nearly brittle fracture. *The International Journal of Fracture Mechanics* **4**, 179–205.

Yoon C. K. and Chen I.-W. (1990) Superplastic flow of two-phase ceramics containing rigid inclusions: zirconia/mullite composites. *Journal of the American Ceramic Society* **73**, 1555–1565.

Yoon D. N. and Lazarus D. (1972) Pressure dependence of ionic conductivity in KCl, NaCl, KBr and NaBr. *Physical Review B* **5**, 4935–4945.

Yoshida H., Ikuhara Y., and Sakuma T. (2002) Grain boundary electronic structure related to the high-temperature creep resistance in polycrystalline Al_2O_3. *Acta Materialia* **50**, 2955–2966.

Yoshida S., Sumita I., and Kumazawa M. (1996) Growth model of the inner core coupled with the outer core dynamics and the resulting elastic anisotropy. *Journal of Geophysical Research* **101**, 28 085–28 103.

Yoshii T. (1973) Upper mantle structure beneath the north Pacific and marginal seas. *Journal of Physics of the Earth* **21**, 313–328.

Yoshii T., Kono Y., and Ito K. (1976) Thickening of the oceanic lithosphere. In *The Geophysics of the Pacific Ocean Basin and Its Margin* (ed. G. H. Sutton, M. H. Manghnani, and R. Moberly), pp. 423–430. American Geophysical Union.

Yoshino T., Nishihara Y., and Karato S. (2007) Complete wetting of olivine grain-boundaries by a hydrous melt near the mantle transition zone. *Earth and Planetary Science Letters* **256**, 466–472.

Yoshinobu A. S. and Hirth G. (2002) Microstructural and experimental constraints on the rheology of partially molten gabbro beneath oceanic spreading centers. *Journal of Structural Geology* **24**, 1101–1107.

Yuen D. A., Sabadini R., and Boschi E. V. (1982) Viscosity of the lower mantle as inferred from rotational data. *Journal of Geophysical Research* **87**, 10 745–10 762.

Zamora M. and Poirier J.-P. (1983) Experiments in anisothermal transformation plasticity: the case of cobalt. Geophysical implications. *Mechanics of Materials* **2**, 193–202.

Zener C. (1942) Theory of lattice expansion introduced by cold-work. *Transactions of the Metallurgical Society of AIME* **147**, 104–110.

Zener C. (1948a) *Elasticity and Anelasticiy of Metals.* University of Chicago Press.

Zener C. (1948b) The micro-mechanism of fracture. In *Fracturing of Metals* (ed. F. Johnson, W. P. Roop, and R. T. Bayles), pp. 3–31. ASM.

Zener C. and Hollomon J. H. (1946) Problems in non-elastic deformation of metals. *Journal of Applied Physics* **17**, 69–82.

Zeuch D. H. (1982) Ductile faulting, dynamic recrystallization and grain-size-sensitive flow in olivine. *Tectonophysics* **83**, 293–308.

Zeuch D. H. (1983) On the inter-relationship between grain-size sensitive creep and dynamic recrystallization of olivine. *Tectonophysics* **93**, 151–168.

Zha C.-S., Duffy T. S., Downs R. T., *et al.* (1998) Brillouin scattering and X-ray diffraction of San Carlos olivine: direct pressure determination to 32 GPa. *Earth and Planetary Science Letters* **159**, 25–33.

Zhang S. and Christensen U. R. (1993) Some effects of lateral viscosity variations on geoid and surface velocities induced by density anomalies in the mantle. *Geophysical Journal International* **114**, 531–547.

Zhang S. and Karato S. (1995) Lattice preferred orientation of olivine aggregates deformed in simple shear. *Nature* **375**, 774–777

Zhang S., Karato S., Fitz Gerald J., Faul U. H., and Zhou Y. (2000) Simple shear deformation of olivine aggregates. *Tectonophysics* **316**, 133–152.

Zhang Y. and Xu Z. (1995) Atomic radii of noble gas elements in condensed phases. *American Mineralogist* **80**, 670–675.

Zhang Y. S. and Tanimoto T. (1992) Ridges, hotspots and their interactions as observed in seismic velocity maps. *Nature* **355**, 45–49.

Zhang Y. S. and Tanimoto T. (1993) High-resolution global upper mantle structure and plate tectonics. *Journal of Geophysical Research* **98**, 9793–9823.

Zhao D. (2004) Global tomographic images of mantle plumes and subducting slabs: insights into deep mantle dynamics. *Physics of the Earth and Planetary Interiors* **146**, 3–34.

Zhao Y.-H., Ginsberg S. B., and Kohlstedt D. L. (2004) Solubility of hydrogen in olivine: dependence on temperature and iron content. *Contributions to Mineralogy and Petrology* **147**, 155–161.

Zimmerman M. E. and Kohlstedt D. L. (2004) Rheological properties of partially molten lherzolite. *Journal of Petrology* **45**, 275–298.

Zimmerman M. R., Zhang S., Kohlstedt D. L., and Karato S. (1999) Melt distribution in mantle rocks deformed in shear. *Geophysical Research Letters* **26**, 1505–1508.

Materials index

Subject index

Printed in the United States
By Bookmasters